29.95

Biochemistry
for Medical Sciences

Biochemistry
for Medical Sciences

Isidore Danishefsky, Ph.D.
Professor and Chairman
Department of Biochemistry
New York Medical College

LITTLE, BROWN AND COMPANY   BOSTON

Copyright © 1980 by Little, Brown and Company (Inc.)

First Edition

All rights reserved. No part of this book may be reproduced in any form or by any electronic or mechanical means, including information storage and retrieval systems, without permission in writing from the publisher, except by a reviewer who may quote brief passages in a review.

Library of Congress Catalog Card No. 77-81498

ISBN 0-316-17198-0 (C)

Printed in the United States of America

HAL

*To my wife, Madeleine,
our son, Kenneth,
and daughter, Avis*

Preface

The primary goal of this book is to present the biochemical foundation necessary to students of medicine and related fields. More specifically, the aim is to introduce the reader to the basic aspects of normal metabolism that are required for an understanding of the causes and consequences of various disorders. As such, this text should also be useful to biologists and clinicians who wish to keep abreast of ongoing developments and current hypotheses.

A major portion of the book deals with intermediary metabolism and its relationship to the overall functioning of mammalian organisms. In order to maintain the emphasis on these subjects, digressions into historical developments are avoided, and descriptions of experimental methodologies are kept to a minimum. In conformance with the primary objective, attention is centered on the reactions or pathways that regulate metabolism and the disorders that result from disturbances or deficiencies in those processes.

The basic principles and the structural chemistry required for the description of intermediary metabolism are discussed in the first two chapters. More specific details of molecular structures and enzyme characteristics are presented in conjunction with their application to metabolism. For example, only the general aspects of lipid structure are described in Chapter 1, whereas the individual structures are given when lipid metabolism is discussed in Chapters 3 and 5. Similarly, the principles of enzyme catalysis are described in Chapter 2, but the features of allosteric enzymes are defined in connection with isocitrate dehydrogenase (Chapter 3) and aspartate transcarbamoylase (Chapter 9).

Chapters 3, 4, and 5 deal with the metabolism of lipids and carbohydrates. The interrelationships of the metabolism of these materials, their transport in the circulatory system, and their disposition in different tissues are discussed in Chapter 6. Various control mechanisms and disorders of lipid and carbohydrate metabolism are also outlined in this chapter.

Chapter 7 discusses the reactions that are common to most amino acids and their relationship to carbohydrate and lipid metabolism. This discussion is followed by an outline of the metabolism of each amino acid (Chapter 8). The biosynthesis and degradation of nucleotides and porphyrins are then described in Chapters 9 and 10, respectively. Chapter 11 focuses on the general field of molecular biology, i.e., synthesis of nucleic acids and proteins. If one desires, the material covered in Chapter 11 may be inserted in the teaching sequence before the discussions of intermediary metabolism.

Blood, respiration, and electrolyte balance are the subjects of Chapters 12 and 13. Although some material introduced in those chapters overlaps topics that are treated in physiology courses, the chemical aspects are extremely important to medical students and are, therefore, covered in this text.

Chapter 14 is concerned with the structure and function of specialized tissues, and Chapter 15 deals with various aspects of nutrition. In addition to the basic importance of nutrition, discussions of this subject also serve to integrate various aspects of metabolism described in earlier chapters.

The Suggested Reading lists include articles that expand on the subject matter in each chapter, as well as papers involving clinical applications.

As a research science, the field of biochemistry has become so vast and diverse that its specific boundaries cannot always be defined. Therefore, in writing a book on this subject, one is faced with the prospect of preparing an encyclopedic compendium but still not including all the important topics. In this textbook I have attempted to restrict myself to material that can be covered within the amount of time assigned to biochemistry in most medical schools. It includes the core subject matter required for clinical medicine and for comprehension of the current literature. Conceivably, there will be some objections regarding subjects that were omitted. It is hoped there will not be too many.

Acknowledgments for review of various sections of the book are due to professional colleagues—biochemists and clinicians—as well as medical and graduate students. I am especially indebted, for detailed and thoughtful reviews of specific chapters, to Professors Harold Appleton, Martin I. Horowitz, Frank S. Parker, Milton Tabachnick, and Joseph M. H. Wu. Most of the manuscript as well as the galley proofs were checked by Kenneth Danishefsky.

I am grateful to Mrs. Alice Cross and Mrs. Ruth Prunty for typing some of the chapters of the manuscript and to Mrs. Evelyn Roberts for spending many hours of her own time to help in the organization of the work.

Finally, I express my appreciation to Mrs. Lin Richter, Medical Editor of Little, Brown and Company, for her cooperation, advice, and patience, and to Mrs. Marcia Mirski, Copyediting Supervisor, for expert assistance in coordinating the production of this book.

I. D.

Contents

Preface   vii

1. **Fundamentals**   1
   1.1. Scope of Biochemistry   1
   1.2. Chemistry of Cellular Components   1
      A. Water and Electrolytes   1
      B. Amino Acids   13
      C. Proteins   21
      D. Carbohydrates   42
      E. Purines, Pyrimidines, Nucleosides, and Nucleotides   49
      F. Nucleic Acids   52
      G. Lipids   55
   1.3. Membranes and Transport   58
      A. Membrane Structure   58
      B. Transport Through Membranes   61
      C. Osmotic Pressure, Membrane Equilibria, and the Gibbs-Donnan Equilibrium   61
   1.4. Biochemistry of Metabolism   64
      A. Overview and Aims   64
      B. Experimental Approaches   65
      C. Isotopes   66
   1.5. Intracellular Compartments   69

2. **Characteristics and Functions of Enzymes**   73
   2.1. Thermodynamic Principles   73
   2.2. Free Energy and Reversible Reactions   75
   2.3. Adenosine Triphosphate   78
   2.4. Rate and Progress of Enzymatic Reactions   82
   2.5. Effect of Substrate Concentration   85
   2.6. Enzyme Concentration and Reaction Rate   92
   2.7. Influence of pH on Enzyme Activity   93
   2.8. Effect of Temperature   94
   2.9. Coenzymes and Cofactors   95
   2.10. Enzyme Nomenclature   96
   2.11. Oxidoreductases   97
   2.12. The Active Site of Enzymes   101
   2.13. Enzyme Inhibition   106

3. Catabolism of Triglycerides and the Citric Acid Cycle   111
    3.1.   Structural Chemistry   111
    3.2.   Digestion and Absorption   114
    3.3.   Degradation of Fatty Acids to Acetyl-Coenzyme A   116
    3.4.   Citric Acid Cycle   122
    3.5.   Oxidation, Reduction, and Free Energy   128
    3.6.   Biologic Oxidations   132
    3.7.   Energy from Biologic Oxidations   141
    3.8.   Energy from Fatty-Acid Catabolism   144
    3.9.   Special Features of the Citric Acid Cycle   147
          A. Asymmetry of Citrate Reactions   147
          B. Randomization After Succinate   149
          C. Fate of the Carbons from Acetate   150
    3.10. Isocitrate Dehydrogenase and Allosteric Enzymes   150
    3.11. Processes of Oxidative Phosphorylation   154
          A. Mitochondrial Structure and Enzymes   154
          B. Sites of ATP Generation   155
          C. Uncouplers of Oxidative Phosphorylation   156
          D. Mechanisms of Oxidative Phosphorylation   157

4. Metabolism of Carbohydrates   163
    4.1.   Digestion and Absorption   163
    4.2.   Glucose 6-Phosphate: Synthesis and Disposition   167
    4.3.   Glycolysis   169
    4.4.   Conversion of Pyruvate to Acetyl-Coenzyme A   175
    4.5.   Aerobic Oxidation of Glucose   179
          A. Shuttle Systems   179
          B. Energy from Glucose Metabolism   181
          C. Pasteur Effect   181
    4.6.   Pentose Pathway   184
    4.7.   Glycogen Metabolism   190
          A. Synthetic Pathway   192
          B. Glycogenolysis   195
          C. Hormonal Effects on Glycogen Metabolism   196
    4.8.   Gluconeogenesis   201
    4.9.   Nucleotide Sugar Interconversions   210
    4.10. Metabolism of Galactose and Fructose   211
    4.11. Hexosamines   213

5. Biosynthesis of Lipids   219
    5.1.   Intracellular Transport of Acetyl-Coenzyme A   219

5.2. Synthesis of Fatty Acids 220
   A. Malonyl-Coenzyme A 220
   B. Fatty-Acid Synthetase Multienzyme System 220
   C. Elongation of Fatty Acids 223
   D. NADPH Requirements and Sources 223
   E. Unsaturated Fatty Acids 225
5.3. Synthesis of Triglycerides 226
5.4. Synthesis of Phospholipids 228
5.5. Cholesterol Metabolism 232
   A. Biosynthesis of Cholesterol 232
   B. Nomenclature of Steroids 237
   C. Esterification of Cholesterol 238
   D. Turnover and Elimination of Cholesterol and Bile-Salt Formation 239
5.6. Steroid Hormones 240
   A. Pregnenolone 240
   B. Progesterone 241
   C. Cortisol and Aldosterone 242
   D. Androgens 243
   E. Estrogens 244
5.7. Sphingolipids 244

6. Tissue Disposition and Transport of Carbohydrates and Lipids 253

6.1. Ketone Bodies: Formation and Disposition 253
6.2. Lactate Metabolism 257
   A. Cori Cycle 257
   B. Isozymes of Lactate Dehydrogenase 259
6.3. Blood Glucose and Its Regulation 262
6.4. Circulation and Mobilization of Lipids 264
   A. Lipoproteins 267
   B. Physiologic Significance of Plasma Lipid Levels 273
6.5. Regulation of Glycolysis and Gluconeogenesis 274
6.6. Hormonal Regulation of Carbohydrate and Lipid Metabolism 277
   A. Insulin 277
   B. Epinephrine 279
   C. Glucagon 281
   D. Glucocorticoids 281
   E. Thyroid Hormone 281
6.7. Metabolism of Lipids and Carbohydrates in Individual Tissues and Organs 281
   A. Liver 282
   B. Kidney 284

   C. Muscle   284
   D. Adipose Tissue   284
   E. Brain   286
   F. Erythrocytes   286
 6.8. Glycogen Storage Diseases   288
 6.9. Abnormalities in Lipid Disposition   291

7. Metabolism of Proteins and Amino Acids   297
 7.1. Digestion and Absorption   297
 7.2. Transamination   301
 7.3. Deamination Reactions   306
   A. Oxidative Deamination   306
   B. Non-oxidative Deamination   307
 7.4. Ammonia Metabolism   308
   A. Synthesis of Glutamine   309
   B. Synthesis of Carbamoyl Phosphate   309
 7.5. Formation of Urea   310
 7.6. Catabolism of the Carbon Skeleton of Amino Acids   313
 7.7. Glucogenic and Ketogenic Amino Acids   314
 7.8. Transmethylation   316
 7.9. Creatine Metabolism   319
 7.10. Single-Carbon Transfers Other than Methylation   320
 7.11. Essential and Non-essential Amino Acids   322
   A. Definition   322
   B. Synthesis of Non-essential Amino Acids   324

8. Metabolism of Specific Amino Acids   331
 8.1. Alanine   331
 8.2. Serine   332
 8.3. Glycine   333
 8.4. Threonine   337
 8.5. Cysteine   340
 8.6. Aspartate and Asparagine   342
 8.7. Glutamate and Glutamine   344
 8.8. Arginine   347
 8.9. Histidine   348
 8.10. Proline   351
 8.11. Methionine   352
 8.12. Valine   355
 8.13. Isoleucine   356
 8.14. Leucine   358
 8.15. Lysine   359

8.16. Tyrosine   361
8.17. Phenylalanine   365
8.18. Tryptophan   367

9. **Metabolism of Nucleotides**   373
    9.1. Biosynthesis of Purine Ribonucleotides   373
    9.2. Control Mechanisms in Purine Nucleotide Biosynthesis   378
    9.3. Biosynthesis of Pyrimidine Ribonucleotides   379
    9.4. Control Mechanisms in Pyrimidine Nucleotide Biosynthesis   383
    9.5. Biosynthesis of Deoxyribonucleotides   385
    9.6. Salvage Pathways   387
    9.7. Inhibitors of Nucleotide Synthesis   389
    9.8. Digestion and Absorption of Nucleotides   393
    9.9. Catabolism of Pyrimidines   393
    9.10. Catabolism of Purines   395
    9.11. Uric Acid and Hyperuricemic Disorders   397

10. **Metabolism of Porphyrins**   405
    10.1. Structure and Functions   405
    10.2. Biosynthesis of Porphyrins   406
    10.3. Methemoglobin and Other Derivatives of Hemoglobin   408
    10.4. Porphyrias   410
    10.5. Heme Catabolism   410

11. **Biosynthesis of Nucleic Acids and Proteins**   413
    11.1. Structure of DNA and Its Replication   413
    11.2. Mechanisms of DNA Replication   418
        A. Reactions Involved in Replication and Their Probable Sequence   418
        B. Enzymes Involved in DNA Replication   423
    11.3. Transcription   425
    11.4. Differentiation of RNA   428
        A. Messenger RNA   428
        B. Transfer RNA   430
        C. Ribosomal RNA   434
    11.5. Translation and the Genetic Code   435
    11.6. Protein Synthesis   438
        A. Synthesis and Properties of Aminoacyl-tRNA   438
        B. Base-Pairing Between tRNA and mRNA   440
        C. Initiation of Protein Synthesis   441
        D. Elongation of the Polypeptide Chain   445

　　　　　E. Termination in Protein Synthesis   445
　　　　　F. Posttranslational Modification of Proteins   448
　　11.7.　Inhibitors of Transcription and Translation   450
　　11.8.　DNA and Nuclear Proteins   450
　　11.9.　Control of Gene Expression   451
　　11.10.　Viruses   455
　　11.11.　Tumor Viruses   458
　　11.12.　Mutations   458

12. **Blood**   465
　　12.1.　Composition and Function   465
　　12.2.　Cellular Components of Blood   466
　　12.3.　Plasma Proteins   469
　　　　　A. Protein Fractions   469
　　　　　B. Albumin   470
　　　　　C. Globulins   471
　　12.4.　Blood Coagulation and Its Controls   472
　　　　　A. The Coagulation Process   472
　　　　　B. Vitamin K and Dicumarol   478
　　　　　C. Heparin   479
　　　　　D. Reactions of the Blood to Tissue Injury   480
　　12.5.　Immunoglobulins   481
　　　　　A. Antibodies   481
　　　　　B. Immunoglobulin Structure   482
　　　　　C. Myeloma Proteins   487
　　　　　D. Immunoglobulin Synthesis   487
　　12.6.　Complement   488
　　12.7.　Kinins   490
　　12.8.　Diagnostic Enzymes   491

13. **Respiration and Electrolyte Balance: Functions of the Lungs and Kidneys**   495
　　13.1.　Transport of Oxygen and Carbon Dioxide   495
　　13.2.　Hemoglobin and Oxyhemoglobin   497
　　13.3.　Transport of Carbon Dioxide   504
　　13.4.　Variations in Hemoglobin Structure   507
　　13.5.　Blood Buffers   509
　　13.6.　Renal Regulation of Body pH   515
　　13.7.　Compensatory Control of Blood pH   519
　　13.8.　Compartmentalization of Fluids and Electrolytes   522
　　　　　A. Water   522
　　　　　B. Electrolytes   522
　　　　　C. Excretion of Urine   523

14. Specialized Tissues: Their Structures and Functions   527
   14.1.   Muscle Contraction   527
           A. The Muscle Cell and Contractile Proteins   527
           B. Structural Aspects of Contractile Proteins   529
           C. Reaction Sequence in Muscle Contraction   531
           D. Myoglobin   532
           E. Role of Phosphocreatine   534
   14.2.   Neural Transmission   535
           A. The Nerve Cell   535
           B. Neuronal Stimulation   535
           C. Synaptic Transmission   536
           D. Acetylcholinesterase Inhibitors   538
   14.3.   Connective Tissue   539
           A. General Composition   539
           B. Collagen   540
           C. Elastin   542
           D. Glycosaminoglycan Structure   542
           E. Metabolism of Glycosaminoglycans   547
           F. Bone and Teeth   549
   14.4.   Endocrine System   551
           A. Epinephrine   551
           B. Insulin   551
           C. Glucagon   553
           D. Thyroid Hormones and Thyroid-Stimulating Hormone   553
           E. Parathyroid Hormone and Calcitonin   555
           F. Growth Hormone   556
           G. Sex Hormones   556
           H. Adrenocortical Hormones   557
           I. Prostaglandins   559

15. Nutrition   561
   15.1.   Food Energy and Calories   561
           A. Caloric Contribution of Nutrients   561
           B. Physiologic Energy Expenditure   563
           C. Basal Metabolism   568
           D. Specific Dynamic Action   569
           E. Energy Requirements   569
   15.2.   Protein Requirements   571
   15.3.   Carbohydrates   572
   15.4.   Lipids   573
   15.5.   Water and Minerals   574

15.6. Vitamins 580
  A. Vitamin A 580
  B. Vitamin D 583
  C. Vitamin E 586
  D. Vitamin K 588
  E. B-Complex Vitamins 589
  F. Vitamin C 599
  G. Other Vitamin-Like Nutrients 600

Index 605

Biochemistry
for Medical Sciences

1. Fundamentals

1.1. Scope of Biochemistry

The aim of biochemistry is to define the chemical or molecular properties of living entities. This involves characterization of the structural components of cells as well as elucidation of the chemical reactions that occur within the organism. Living systems are unique in several respects. Cells can utilize materials from the environment to maintain their functional capabilities and structural integrity. They also contain the machinery for self-repair and reproduction. Additionally, higher organisms undergo regulated growth and develop tissues with specialized functions. Biochemistry deals with the molecular structures and transformations that give rise to these distinctive manifestations of life.

The maintenance of life involves numerous dynamic systems of sequential and concurrent reactions. Most of the subject matter of biochemistry is concerned with these processes. Before this material can be considered, however, it is necessary to describe some of the structural features and chemical properties of the principal cellular and tissue components. This chapter deals with the basic concepts and structures required for understanding biochemistry. Details of some of the more complex molecular structures will be described in conjunction with discussions of their metabolism.

1.2. Chemistry of Cellular Components

A. WATER AND ELECTROLYTES

Water constitutes about 80% of the weight of animal tissues. It also serves as the medium in which tissue components are dissolved or suspended. Substances that dissociate in water to yield ions are called *electrolytes*. These include acids, bases, and salts. *Acids* are defined as electrolytes that release hydrogen ions, or protons. *Bases* are substances that combine with hydrogen ions. Acids are classified further according to their degree of ionization or dissociation: those that are completely dissociated (or ionized) are termed *strong acids*, and those that are partially dissociated (or ionized) are called *weak acids*.

When a weak acid is dissolved in water, the solution contains a mixture of undissociated acid, hydrogen ions, and anions. Thus, for such an acid

(HA), the dissociation can be expressed as a reversible reaction:*

$$HA \rightleftharpoons H^+ + A^-$$

The concentrations of these components when the process reaches equilibrium are governed by the equation:

$$K_a = \frac{[H^+][A^-]}{[HA]} \tag{1-1}$$

In this relationship, the square brackets indicate that the concentrations of H^+, A^-, and HA are in moles per liter. The term K_a is the *ionization constant*, or the dissociation constant, of the acid, and it has a specific value for a given temperature. The ionization constant for acetic acid (CH_3COOH) at 25°C, for example, is 1.86×10^{-5}. From the relatively low value of this ionization constant, we know that only a very small fraction of acetic acid is dissociated.

Water itself dissociates to a minute but significant extent, and this process can also be defined by an equilibrium constant:

$$H_2O \rightleftharpoons H^+ + OH^-$$

$$K_a = \frac{[H^+][OH^-]}{[H_2O]} = 1.8 \times 10^{-16} \quad \text{at} \quad 25°C$$

Since the molar concentration of undissociated water (1000/18) is not affected significantly by the amount that is ionized, this equation may be written as:

$$K_a = \frac{[H^+][OH^-]}{55.5} = 1.8 \times 10^{-16}$$

A new constant, K_w, the *ion-product of water*, is thus obtained:

$$K_w = [H^+][OH^-] = 1 \times 10^{-14} \tag{1-2}$$

In pure water, $[H^+]$ is equal to $[OH^-]$, and the concentrations of the hydrogen ions and of the hydroxyl ions are equal to 10^{-7} molar.

Upon addition of acid to water, the hydrogen-ion concentration is in-

---

* The dissociation of acids in aqueous solutions yields hydronium ions rather than protons:

$$HA + H_2O \rightleftharpoons H_3O^+ + A^-$$

However, since this does not affect the ensuing discussions or calculations, H^+ will be employed for simplification.

Table 1-1. Relationship Between [H$^+$], [OH$^-$], and pH

Molarity of H$^+$	Molarity of OH$^-$	pH
1.0	10^{-14}	0
0.01 (10^{-2})	10^{-12}	2
10^{-4}	10^{-10}	4
10^{-6}	10^{-8}	6
10^{-7}	10^{-7}	7
10^{-8}	10^{-6}	8
10^{-10}	10^{-4}	10
10^{-12}	10^{-2}	12

creased. However, since the ion-product of water is a constant, the hydroxyl-ion concentration must be decreased to a value less than 10^{-7} M. The hydrogen-ion concentration of an aqueous solution can be expressed in terms of pH. The *pH* is defined as the negative logarithm of the hydrogen-ion concentration:

$$\text{pH} = -\log[\text{H}^+] = \log\frac{1}{[\text{H}^+]} \qquad (1\text{-}3)$$

Thus, for neutral solutions, when [H$^+$] is 10^{-7}, the pH is 7. Acidic solutions have a hydrogen-ion concentration greater than 10^{-7}; hence, their pH is below 7. The relationship between pH and hydrogen-ion concentration is shown in Table 1-1.

The ionization constants of acids can also be expressed in a manner analogous to the pH scale. Thus, the pK_a of an acid is the negative logarithm of the ionization constant of the acid:

$$\text{p}K_a = -\log K_a = \log\frac{1}{K_a} \qquad (1\text{-}4)$$

In this system, an acid with an ionization constant of 10^{-6} has a pK_a of 6, one with K_a of 10^{-3} has a pK_a of 3, and so on. From this consideration, it should be obvious that the magnitude of the pK_a is related inversely to the strength of the acid, i.e., a strong acid will have a low pK_a value and a weak acid a high one.

Problem 1   If the hydrogen-ion concentration of a solution is 3.2×10^{-5} moles per liter, what is the pH?

Answer
$$\text{pH} = -\log[\text{H}^+] = -\log(3.2 \times 10^{-5})$$
$$= -\log 3.2 + 5$$
$$= -0.505 + 5 = 4.49$$

Problem 2   If $[H^+]$ is 6.4×10^{-5} M, what is the pH?

Answer $pH = -\log(6.4 \times 10^{-5}) = -0.806 + 5.0 = 4.19$

(Note in Problem 2 that the $[H^+]$ is double that in Problem 1, but the difference in pH is only 0.3. This is because a change of one in the pH scale represents a 10-fold change in the hydrogen-ion concentration.)

Problem 3   What is the hydrogen-ion concentration of a solution with pH 4.23?

Answer $-\log[H^+] = 4.23$
$\log[H^+] = -4.23$
$[H^+] = 10^{-4.23} = 10^{0.77} \times 10^{-5}$
$= 5.9 \times 10^{-5}$ M

Problem 4   What is the hydroxyl-ion concentration in the solution discussed in Problem 3?

Answer $[H^+][OH^-] = K_w = 1 \times 10^{-14}$
$(5.9 \times 10^{-5})[OH^-] = 1 \times 10^{-14}$
$[OH^-] = \dfrac{1 \times 10^{-14}}{5.9 \times 10^{-5}} = 0.169 \times 10^{-9}$
$= 1.69 \times 10^{-10}$ M

Problem 5   An acid has a K_a of 3.4×10^{-4}. What is the pK_a?

Answer $pK_a = -\log K_a$
$pK_a = -\log(3.4 \times 10^{-5}) = -0.5315 + 5 = 4.47$

A strong acid is completely dissociated when dissolved in water. The hydrogen-ion concentration of such an acid, therefore, is essentially the same as the normality of the acid. Thus, for example, the hydrogen-ion concentration of 0.01 M HCl is 10^{-2} and that of 0.01 M H_2SO_4 is 2×10^{-2}. This is not the case, however, with weak acids, which dissociate to a very limited extent. With solutions of these acids, the concentration of hydrogen ions depends on the K_a of the acid and on its concentration.

When protons are released during the dissociation of a monoprotic acid, an equal number of anions (A^-) are generated. Thus, equation 1-1 for K_a may be written as:

$$K_a = \dfrac{[H^+]^2}{[HA]}$$

The hydrogen-ion concentration for a solution of a weak acid is thus:

$$[H^+] = \sqrt{[HA]K_a}$$

Hence, the information required for ascertaining the hydrogen-ion concentration of a solution of weak acid is the K_a and the molar concentration of the undissociated acid. Since a fraction of the acid is dissociated, the amount of undissociated acid in solution is actually less than the total that was added. However, the degree of dissociation of weak acids is extremely small, so only a minimal error is introduced if the value for the total acid is used for the concentration of undissociated acid, [HA], in the calculation of hydrogen-ion concentration. The equation for the hydrogen-ion concentration of solutions of weak acids can therefore be expressed as:

$$[H^+] = \sqrt{K_a C_a} \tag{1-5}$$

where C_a = normality of acid. (It should be noted that when the degree of dissociation of an acid in solution is greater than 5%, the effect of dissociation on the concentration of undissociated acid becomes significant and $[H^+]$ must be determined by alternative procedures.)

Problem 6   The K_a for acetic acid is 1.86×10^{-5}. What is the hydrogen-ion concentration of a 0.015 M solution of acetic acid? What is its pH?

Answer $[H^+] = \sqrt{0.015 \times 1.86 \times 10^{-5}}$
$= 5.28 \times 10^{-4}$
pH $= 3.28$

Certain acids contain more than one ionizable hydrogen. These are called *polyprotic* or *polybasic* acids. The ionization of these acids proceeds in discrete steps; for example, carbonic acid, a *dibasic* (or diprotic) acid, dissociates in the following sequence:

(1) $H_2CO_3 \rightleftharpoons H^+ + HCO_3^-$

(2) $HCO_3^- \rightleftharpoons H^+ + CO_3^{2-}$

The degree of each dissociation is governed by a specific ionization constant. The constant for the release of a proton and bicarbonate ion from carbonic acid is termed the first ionization constant, K_1. The second step, i.e., the formation of carbonate ions and the concomitant release of the second proton, is governed by the second ionization constant, K_2. Both the constants and the molar ratios that they define are expressed as follows:

$$K_1 = \frac{[H^+][HCO_3^-]}{[H_2CO_3]} = 1.70 \times 10^{-4}$$

$$K_2 = \frac{[H^+][CO_3^{2-}]}{[HCO_3^-]} = 6.31 \times 10^{-11}$$

It may be noted that HCO_3^- functions both as an acid and as a base:

$$HCO_3^- \rightleftharpoons CO_3^{2-} + H^+$$

$$HCO_3^- + H^+ \rightleftharpoons H_2CO_3$$

Such substances, which can act as both proton donors and proton acceptors, are called *amphoteric* materials.

Another important example of a polyprotic acid is phosphoric acid, H_3PO_4. The sequence for its ionization is represented by the following equations:

(1) $H_3PO_4 \rightleftharpoons H^+ + H_2PO_4^-$; $K_1 = 10^{-2}$

(2) $H_2PO_4^- \rightleftharpoons H^+ + HPO_4^{2-}$; $K_2 = 10^{-7}$

(3) $HPO_4^{2-} \rightleftharpoons H^+ + PO_4^{3-}$; $K_3 = 10^{-13}$

It can be seen from both of these examples that the first ionization occurs to a considerably greater extent than the subsequent release of additional hydrogens.

The interrelationship between acids and bases can be expressed in terms of the ionization reaction of the acid. The dissociation of acetic acid yields protons and acetate ions:

$$CH_3COOH \rightleftharpoons CH_3COO^- + H^+$$

Since the acetate ion can accept a proton to form acetic acid, it can be categorized as a base. Acetate ion is therefore called the *conjugate base* of acetic acid. Similar relationships can be defined for all acid-base pairs, as shown in Table 1-2.

Table 1-2. Acids and Their Conjugate Bases

Acid	Conjugate Base	
CH_3COOH	$\rightleftharpoons CH_3COO^-$	$+ H^+$
H_2CO_3	$\rightleftharpoons HCO_3^-$	$+ H^+$
HCO_3^-	$\rightleftharpoons CO_3^{2-}$	$+ H^+$
H_3PO_4	$\rightleftharpoons H_2PO_4^-$	$+ H^+$
$H_2PO_4^-$	$\rightleftharpoons HPO_4^{2-}$	$+ H^+$
HPO_4^{2-}	$\rightleftharpoons PO_4^{3-}$	$+ H^+$
H_2O	$\rightleftharpoons OH^-$	$+ H^+$
H_3O^+	$\rightleftharpoons H_2O$	$+ H^+$
NH_4^+	$\rightleftharpoons NH_3$	$+ H^+$
RNH_3^+	$\rightleftharpoons RNH_2$	$+ H^+$

It may be noted that an amphoteric substance is both an acid and a conjugate base of an acid. For example, bicarbonate ion (HCO_3^-) is the conjugate base of carbonic acid (H_2CO_3), and it is also an acid that ionizes to yield the conjugate base, carbonate ion (CO_3^{2-}). Noteworthy, too, is the fact that substances we generally call bases, such as NH_3 or amines (RNH_2), can be considered conjugate bases of their acidic protonated forms, i.e., NH_4^+ or RNH_3^+. This concept will be utilized further in the discussions of amino acids.

From a practical point of view, solutions of conjugate bases of specific acids can be prepared by dissolving the salts of the respective acids. For example, when sodium acetate is dissolved in water, the solution contains acetate ion (conjugate base of acetic acid) and sodium ion. The hydrogen-ion concentration of a solution containing a mixture of an acid and its salt (conjugate base) can be derived from the equation that defines the equilibrium constant of the acid:

$$K_a = \frac{[H^+][A^-]}{[HA]}$$

where $[A^-]$ is the molar concentration of the salt. Then,

$$[H^+] = K_a \frac{[HA]}{[A^-]} = K_a \frac{[Acid]}{[Conjugate\ base]} = K_a \frac{[Proton\ donor]}{[Proton\ acceptor]}$$

$$[H^+] = K_a \frac{[Acid]}{[Salt]} \tag{1-6}$$

In order to derive an expression for the pH of a solution of an acid and its salt, we take the negative logarithm of both sides of equation 1-6:

$$-\log[H^+] = -\log K_a - \log \frac{[Acid]}{[Salt]}$$

Since $-\log[H]$ is equal to pH and $-\log K_a$ is the pK_a, the equation may be rewritten as follows:

$$pH = pK_a - \log \frac{[Acid]}{[Salt]}$$

or

$$pH = pK_a + \log \frac{[Salt]}{[Acid]} \tag{1-7}$$

Equation 1-7 is known as the *Henderson-Hasselbalch equation*. As will be shown subsequently, this is a general equation for calculating the pH of a buffer.

In the specific case in which the molar concentration of acid is equal to that of the salt, the numerator and denominator in the second term on the right side of the Henderson-Hasselbalch equation are equal. Since the logarithm of 1 is zero, the equation becomes:

$$pH = pK_a$$

This situation occurs when a weak acid is titrated with a base to the point at which half the acid is neutralized. The pH at this point is equal to the pK_a because the solution contains equimolar amounts of the acid and its salt.

The Henderson-Hasselbalch equation can be used to determine the pH during the course of an acid-base titration. A plot of these data, as exemplified in Figure 1-1, yields a *titration curve*, or a graph showing the pH of the solution during the entire process. If we consider the change in pH as base is added, we see that the increase in pH is least in the vicinity of the midpoint in the titration. In other words, when the concentration of acid and salt are equimolar and the pH is equal to the pK_a, the addition of acid or base has a minimal effect on the pH.

Solutions of weak acids and their salts are called *buffers* because they maintain a relatively constant pH upon the addition of moderate amounts of acid or alkali. The maximum buffering action of such a mixture is at the pH value that is equal to the pK_a of the acid. Actually, the effect of a buffer is consider-

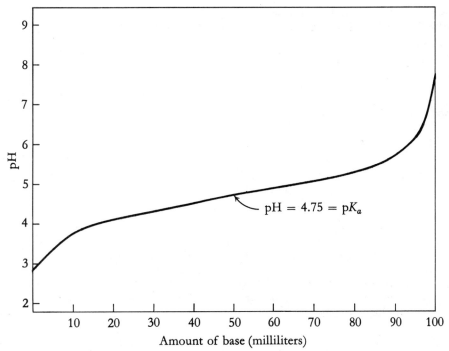

Fig. 1-1. Titration curve for the neutralization of a weak acid by a strong base. The graph shows the pH of the solution when 100 ml of 0.1 N acetic acid is titrated with 0.1 N sodium hydroxide. Note that the change in pH is relatively small when alkali is added to a solution whose pH is close to the pK_a of the weak acid.

able, even within a range of one pH unit above or below the pK_a. Beyond this range, its buffering action is minimal. Thus, a mixture of acetic acid and sodium acetate (pK_a, 4.75) is a good buffer between pH 3.7 and pH 5.7. The combination NaH_2PO_4-Na_2HPO_4 is an effective buffer around pH 7, since the pK_a of $H_2PO_4^-$ is 7.2.

The pH of a solution composed of a mixture of a weak acid and its salt is a function of the pK_a of the acid and the molar ratio of the two components (Henderson-Hasselbalch equation). The buffering efficiency of the mixture depends on the closeness of the pH to the pK_a of the acid. Additionally, since buffers are ultimately consumed by acid or alkali, their buffering capacity depends on the actual concentrations of acid and salt.

Problem 7   What is the pH of a solution composed of 10 milliliters of 0.1 M acetic acid and 10 milliliters of 0.1 M sodium acetate? (The pK_a of acetic acid is 4.75.)

Answer
$$pH = pK_a + \log \frac{[Sodium\ acetate]}{[Acetic\ acid]}$$
$$= 4.75 + \log \frac{0.05}{0.05}$$
$$= 4.75$$

Problem 8 What is the pH of the following solutions?
(a) 10 milliliters of 0.1 M acetic acid and 2 milliliters of 0.1 M sodium acetate.
(b) 2 milliliters of 0.1 M acetic acid and 10 milliliters of 0.1 M sodium acetate.

Answer (a) $pH = 4.75 + \log\dfrac{[0.017]}{[0.083]}$

$= 4.75 + \log 0.2 = 4.75 + 9.3010 - 10$

$= 4.05$

(b) $pH = 4.75 + \log\dfrac{[0.083]}{[0.017]}$

$= 4.75 + \log 5 = 4.75 + 0.6990$

$= 5.45$

(Note that the critical factor in the logarithmic term is the ratio of the two concentrations, rather than the absolute concentrations.)

Problem 9 Buffer A is a 10-milliliter solution that is 0.1 M with respect to both acetic acid and sodium acetate. Buffer B is 0.02 M for these two components. What is the pH of each buffer when 1 milliliter of 0.1 M HCl is added?

Answer Since both buffers have equimolar concentrations of acid and salt, they have the same pH, 4.75.

Buffer A originally contains 1 millimole of acetic acid and 1 millimole of sodium acetate. The addition of 1 milliliter of 0.1 M HCl introduces 0.1 millimole of H^+ into the solution. This reacts with sodium acetate to produce acetic acid according to the equation:

$$CH_3COO^- + H^+ \longrightarrow CH_3COOH$$

Hence the final amount of sodium acetate is $1.0 - 0.1 = 0.9$ millimole. In the process, the amount of acetic acid has increased by 0.1 millimole. The ratio of sodium acetate to acetic acid is then 0.9/1.1 and the pH is:

$$4.75 + \log\dfrac{0.9}{1.1} = 4.66$$

Buffer B originally contains 0.2 millimole acetic acid and 0.2 millimole acetate. After adding 1 milliliter of 0.1 M HCl, the final solution contains 0.1 millimole acetate and 0.3 millimole acetic acid. Then,

$$pH = 4.75 + \log\frac{0.1}{0.3} = 4.27$$

Problem 10 The effective concentration of H_2CO_3 in blood plasma is 1.25 millimolar. For the specific conditions in plasma, the pK_a for the first ionization of carbonic acid ($H_2CO_3 \rightarrow H^+ + HCO_3^-$) is 6.1 rather than 3.8 (Chap. 13). If the blood pH is 7.4, what is the molar concentration of bicarbonate ion? What is the molar ratio of HCO_3^- to H_2CO_3?

Answer Substituting the given information in the Henderson-Hasselbalch equation, we obtain:

$$7.4 = 6.1 + \log\frac{[HCO_3^-]}{0.00125}$$

$$1.3 = \log[HCO_3^-] - \log 0.00125$$

$$\log[HCO_3^-] = 1.3 + 7.0969 - 10 = 8.3969 - 10$$

$$[HCO_3^-] = 0.025 \text{ M}$$

$$\frac{[HCO_3^-]}{[H_2CO_3]} = \frac{0.025}{0.00125} = \frac{20}{1}$$

An important acid-conjugate base, or acid-salt, system that deserves special emphasis because of its physiologic functions is that involving the different dissociated forms of phosphoric acid (p. 7). In addition to the activity of inorganic phosphate anions in the buffering of blood, organic phosphate derivatives are components of the structures of nucleic acids, coenzymes, energy-storage systems, and metabolic intermediates. An appreciation of the dissociation properties of these substances is fundamental for the understanding of some of their functions within the cell.

The structures of phosphoric acid and its anions are depicted as follows:

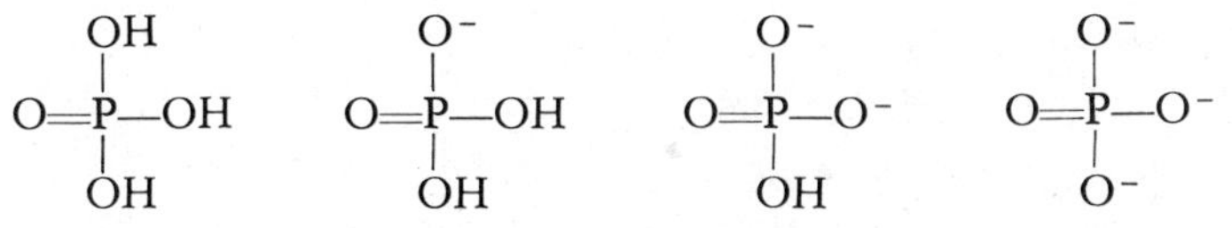

H_3PO_4 $H_2PO_4^-$ HPO_4^{2-} PO_4^{3-}

The dissociation of hydrogen ions from phosphoric acid proceeds in a sequential manner, since a proton is released more easily from H_3PO_4 than

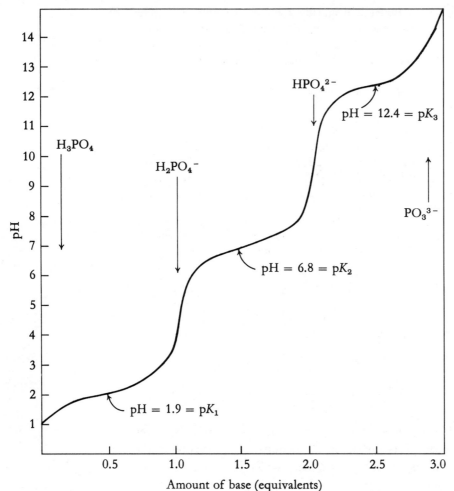

Fig. 1-2. Titration curve for phosphoric acid with three equivalents of alkali.

from $H_2PO_4^-$, and so on. Consequently, upon addition of alkali to a solution of phosphoric acid, the first equivalent of base added produces $H_2PO_4^-$, the second yields primarily HPO_4^{2-}, and the third yields PO_4^{3-}. Thus, although the neutralization of H_3PO_4 can be accomplished with three equivalents of alkali, the titration curve of the process shows three discrete steps (Fig. 1-2). The pK_a values, as obtained from the pH of solutions at the midpoint in titrations of H_3PO_4, $H_2PO_4^-$, and HPO_4^{2-} are 1.9, 6.8, and 12.4, respectively. The relative amounts of each species of phosphate at a given pH can be estimated from the graph or calculated from the Henderson-Hasselbalch equation.

Problem 11   The pH of plasma is normally 7.4. In what form will most of the phosphate in plasma be found?

Answer   Inspection of Figure 1-2 reveals that at pH 7.4, phosphate is in the form of $H_2PO_4^-$ and HPO_4^{2-}.

Problem 12   What is the relative concentration of these two forms of phosphate in plasma at pH 7.4?

Answer   According to the Henderson-Hasselbalch equation,

$$pH = pK_a + \log\frac{[\text{Conjugate base}]}{[\text{Acid}]}$$

In the present case, the acid is $H_2PO_4^-$ and the conjugate base is HPO_4^{2-}. Substitution of the given information into the equation yields:

$$7.4 = 6.8 + \log\frac{[HPO_4^{2-}]}{[H_2PO_4^-]}$$

Then,

$$\log\frac{[HPO_4^{2-}]}{[H_2PO_4^-]} = 0.6$$

$$\frac{[HPO_4^{2-}]}{[H_2PO_4^-]} = 4$$

The molar ratio of HPO_4^{2-} to $H_2PO_4^-$ in plasma is 4 to 1.

B. AMINO ACIDS

The distinguishing characteristic of the *amino acids* contained in proteins is that they are carboxylic acids that contain an amino group on the alpha carbon, i.e., the carbon atom that is immediately adjacent to the one of the carboxyl group:

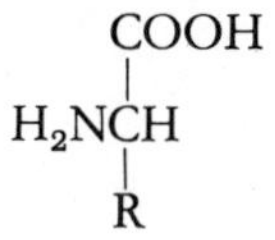

$$\begin{array}{c} \text{COOH} \\ | \\ H_2N\text{CH} \\ | \\ R \end{array}$$

The primary function of α-amino acids is to serve as the building blocks or fundamental components of proteins. Animal proteins are composed of twenty different amino acids. These differ from each other in the nature of the R groups. Except for glycine, the alpha carbon of amino acids is asymmetric. Consequently, amino acids can exist as optical isomers. Those occurring in proteins are L-amino acids and are represented as shown above.

Amino acids are electrolytes since the carboxyl group can dissociate to release a proton. Moreover, if we consider the amino group in the form of

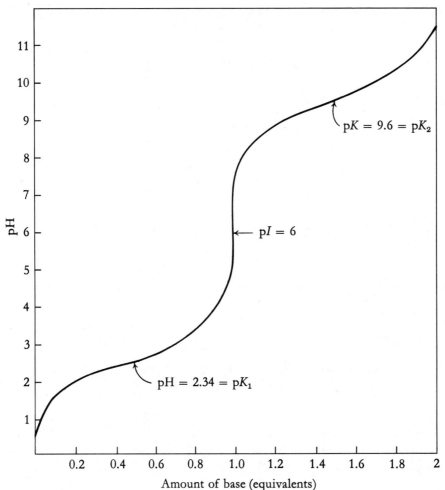

Fig. 1-3. *Titration curve for glycine hydrochloride.*

the ammonium ion ($RNH_2 + H^+ \rightarrow RNH_3^+$), a fully protonated amino acid can be visualized as a diprotic acid with two ionization constants. Thus, for glycine the following discrete dissociations will occur:

(1) $\underset{COOH}{CH_2NH_3^+} \rightleftharpoons \underset{COO^-}{CH_2NH_3^+} + H^+$

(2) $\underset{COO^-}{CH_2NH_3^+} \rightleftharpoons \underset{COO^-}{CH_2NH_2} + H^+$

The pK_a for the ionization shown in step 1 is 2.34 (pK_1) and that for step 2 is 9.60 (pK_2). As expected, two equivalents of alkali are required for the complete conversion of $^+H_3NCH_2COOH$ to $H_2NCH_2COO^-$.

The titration curve for glycine is shown in Figure 1-3. The principal species at pH 2.34 are $^+H_3NCH_2COOH$ and $^+H_3NCH_2COO^-$. At this pH, the solution contains equimolar amounts of the two forms. Similarly, at pH 9.60, there are equimolar concentrations of $^+H_3NCH_2COO^-$ and $H_2NCH_2COO^-$. At a pH of about 6, the amino acid contains both a positive and a negative charge, and these charges compensate each other so that the *net* charge is zero. Such a dipolar ion is called a *zwitterion*. The pH at which this form predominates is known as the *isoelectric point*, or pI. For amino acids that do not have charged groups in the side chain,

$$pI = \frac{pK_1 + pK_2}{2} \tag{1-8}$$

Although amino acids are generally represented as uncharged structures, it should be remembered that practically all the molecules are in a charged state in solution. The specific form depends on the pH of the solution, i.e., at low pH amino acids are positively charged, at high pH they carry a negative charge, and at the isoelectric pH the number of negative charges equals the number of positive charges. It should also be noted that the α-carboxyl group is a stronger acid than the α-ammonium unit.

Amino acids that do not contain carboxyl or basic groups other than the amino and carboxyl groups on the alpha carbon are called *neutral amino acids*. The structures for these amino acids are depicted as follows:

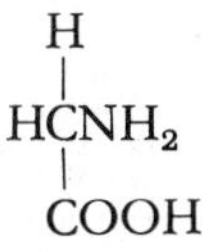

Glycine, 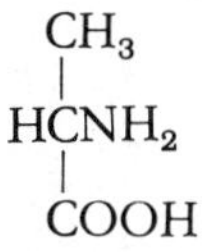 Alanine, 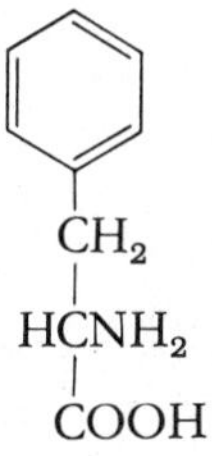 Phenylalanine, 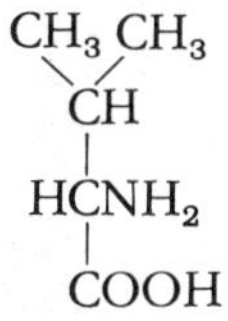 Valine

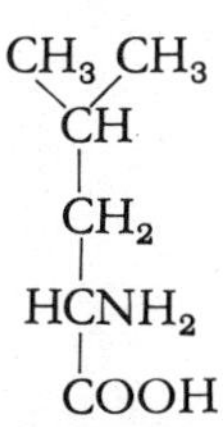

Leucine, 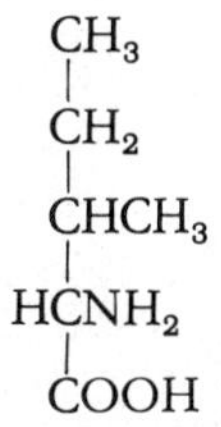 Isoleucine, 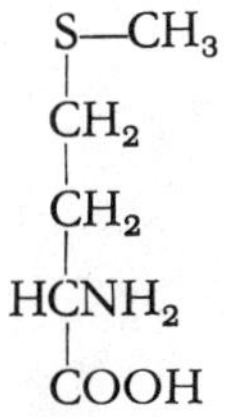 Methionine, 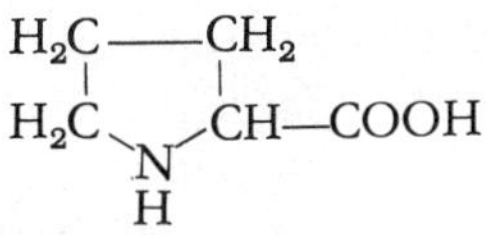 Proline

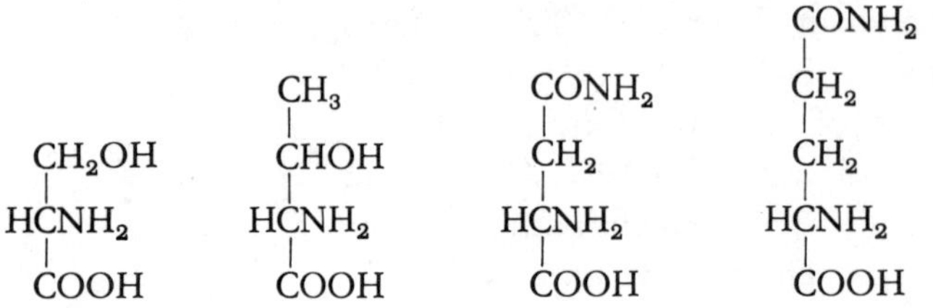

CH₂OH	CH₃ CHOH	CONH₂ CH₂	CONH₂ CH₂ CH₂
HCNH₂	HCNH₂	HCNH₂	HCNH₂
COOH	COOH	COOH	COOH
Serine	Threonine	Asparagine	Glutamine

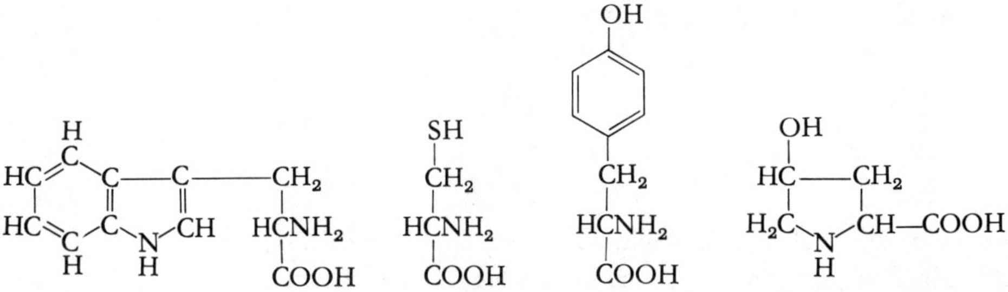

Tryptophan, Cysteine*, Tyrosine*, Hydroxyproline

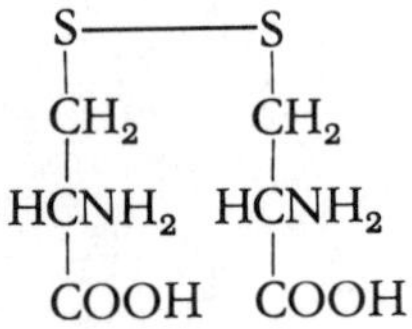

Cystine

Some amino acids have acidic groups on their side chains. These are classified as *acidic amino acids*. Included in this category are:

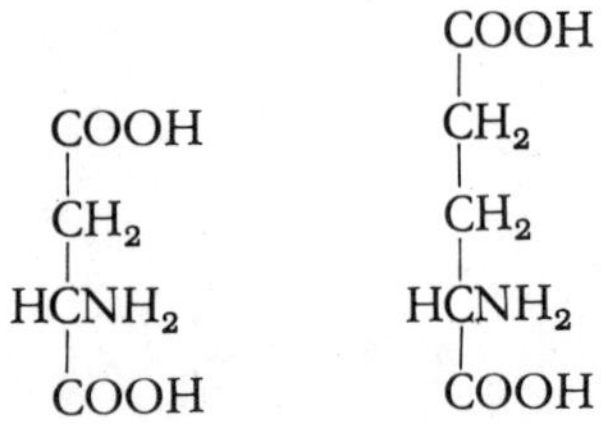

Aspartic acid, Glutamic acid

Amino acids that contain additional basic units on the side chains or R groups are called *basic amino acids*. These include the following:

---

* The hydroxyl group of tyrosine and the sulfhydryl group of cysteine do not dissociate appreciably and are therefore considered "neutral" substituents.

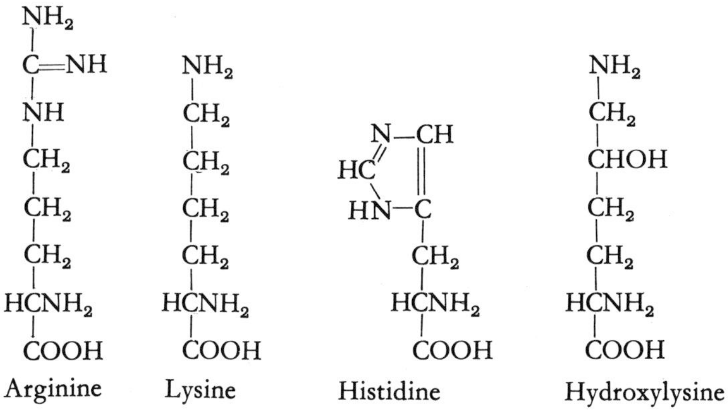

Arginine, Lysine, Histidine, Hydroxylysine (structural formulas)

Table 1-3. Some Properties of the Common Amino Acids

Amino Acid	Abbreviations		pK			pI
	Common	Single Letter	α-Carboxyl	α-Amino	Other (R)	
Alanine	Ala	A	2.34	9.69		6.02
Arginine	Arg	R	2.17	9.04	12.48 (guanidino)	10.76
Asparagine	Asn	N	2.02	8.80		5.41
Aspartic acid	Asp	D	2.09	9.82	3.86 (carboxyl)	2.97
Asparagine or aspartic acid	Asx*	B	—	—		—
Cysteine	Cys	C	1.96	10.28	8.18 (sulfhydryl)	5.07
Cystine	—	—	1.65 and 2.26	7.85, 9.85		5.06
Glutamic acid	Glu	E	2.19	9.67	4.25 (carboxyl)	3.22
Glutamine	Gln	Q	2.17	9.13		5.65
Glutamic acid or glutamine	Glx*	Z	—	—		—
Glycine	Gly	G	2.34	9.60		5.97
Histidine	His	H	1.82	9.17	6.0 (imidazole)	7.59
Hydroxylysine	Hyl	—	2.13	8.62	9.67 (ε-amino)	9.15
Hydroxyproline	Hyp	—	1.92	9.73		5.83
Isoleucine	Ile	I	2.36	9.68		6.02
Leucine	Leu	L	2.36	9.60		5.98
Lysine	Lys	K	2.18	8.95	10.53 (ε-amino)	9.74
Methionine	Met	M	2.28	9.21		5.75
Phenylalanine	Phe	F	1.83	9.13		5.48
Proline	Pro	P	1.99	10.60		6.30
Serine	Ser	S	2.21	9.15		5.68
Threonine	Thr	T	2.63	10.43		6.53
Tryptophan	Trp	W	2.38	9.39		5.89
Tyrosine	Tyr	Y	2.20	9.11	10.07 (phenol)	5.66
Valine	Val	V	2.32	9.62		5.97

* The abbreviations Asx and Glx are used when the proportions of free carboxyl to amide are not defined.

18   1. Fundamentals

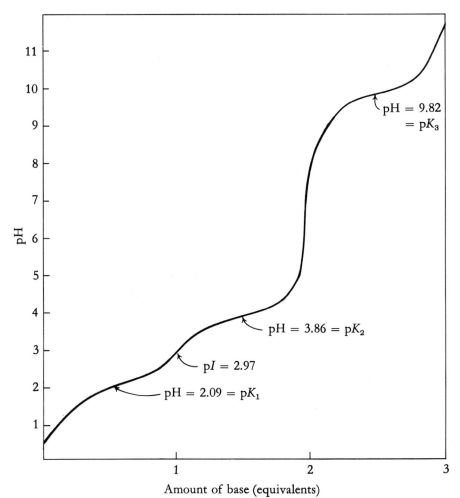

Fig. 1-4. *The pH profile during neutralization of aspartic acid with alkali (titration curve).*

Table 1-3 lists the common amino acids, together with their abbreviations and pK and pI values.

It can be seen from Table 1-3 that the α-carboxyl groups in acidic amino acids (aspartic and glutamic) are stronger acids than the side-chain carboxyl groups. In turn, the latter are more acidic than the α-ammonium groups. Thus, the dissociation of aspartic acid may be depicted as follows:

Since the β-carboxyl is the second proton-releasing group, its pK is designated as pK_2. The dissociation of the α-ammonium group is characterized by pK_3. The isoelectric pH for dicarboxylic amino acids is the mean of pK_1 and pK_2. Hence for aspartic acid,

$$pI = \frac{2.09 + 3.86}{2} = 2.97$$

The titration curve for this acid is shown in Figure 1-4. Analysis of the graph reveals that at pH 2.97, only one of the carboxyl groups is ionized. This structure therefore has one negative and one positive charge, so the net charge is zero.

The basic amino acid, lysine, contains two amino groups: an α-amino group and an ε-amino group on the side chain. This amino acid can also exist in four forms, depending on the pH of the solution:

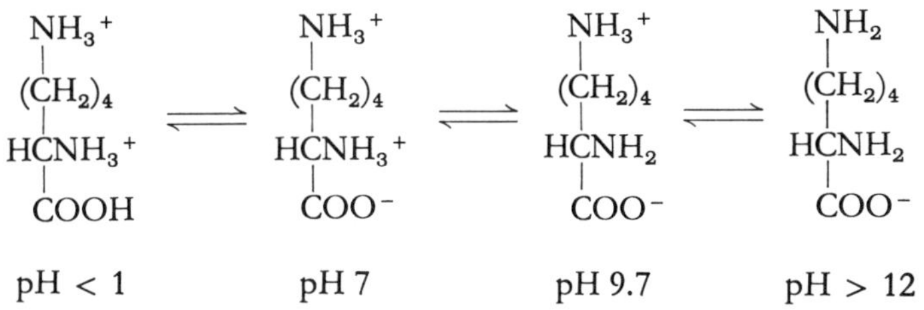

```
    NH₃⁺              NH₃⁺              NH₃⁺              NH₂
     |                 |                 |                 |
   (CH₂)₄           (CH₂)₄           (CH₂)₄           (CH₂)₄
     |       ⇌        |       ⇌        |       ⇌        |
   HCNH₃⁺           HCNH₃⁺           HCNH₂            HCNH₂
     |                 |                 |                 |
    COOH             COO⁻              COO⁻              COO⁻

   pH < 1            pH 7             pH 9.7           pH > 12
```

It should be noted that the α-ammonium group is a stronger acid than the ε-ammonium group. Conversely, the ε-amino group is a stronger base than the α-amino group. Conversion of the totally protonated amino acid to the negatively charged form requires three equivalents of alkali (Fig. 1-5). The first equivalent neutralizes the carboxyl group ($pK_1 = 2.2$). Addition of another equivalent neutralizes the α-ammonium group ($pK_2 = 8.9$). At this stage, the lysine molecule has a net charge of zero. The third equivalent of base removes the proton from the ε-ammonium group ($pK_3 = 10.5$). The pI for basic amino acids can be obtained by applying the formula:

$$pI = \frac{pK_2 + pK_3}{2} \tag{1-9}$$

The dissociation characteristics of histidine are of special interest because of the effects of its side-chain imidazole group. The interactions of histidine with protons and its dissociations provide the following structures:

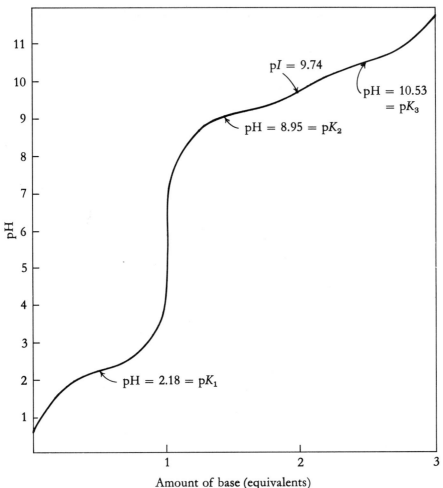

Fig. 1-5. Titration curve for lysine.

The titration curve for this amino acid is shown in Figure 1-6. Unlike the other amino acids, histidine has an appreciable buffering effect between pH 6 and pH 8. This is due to the dissociation of the imidazolium group. As a result of this ionization, histidine moieties in proteins contribute to the main-

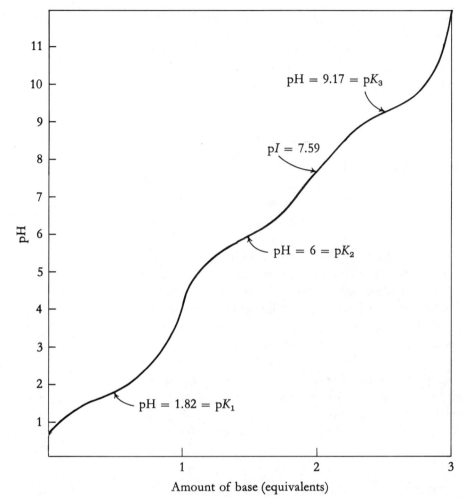

Fig. 1-6. Titration curve for histidine.

tenance of *physiologic pH*, i.e., pH 7.4. This will be discussed further in subsequent sections.

C. PROTEINS
1. Structures of Peptides and Polypeptides

Proteins constitute over 50% of the dry weight of animal cells. They have both structural and functional roles in cells and tissues. Proteins are integral components of cell membranes, skin, bone, tendons, and muscle. All the enzymes and a number of hormones are proteins. Blood proteins function in the transport of oxygen and numerous compounds that are vital for the maintenance and protection of the organism. Various proteins also serve as mediators in the action of nerves and as regulators in the expression of genetic information. The pivotal role of proteins is appreciated when it is realized that the principal activities of the genetic processes are the specification and control of protein synthesis.

Proteins are polymers of amino acids that are linked to each other by peptide bonds. *Peptide bonds* are the amide bonds formed between the carboxyl carbon atom of one amino acid and the amino nitrogen atom of another amino acid:

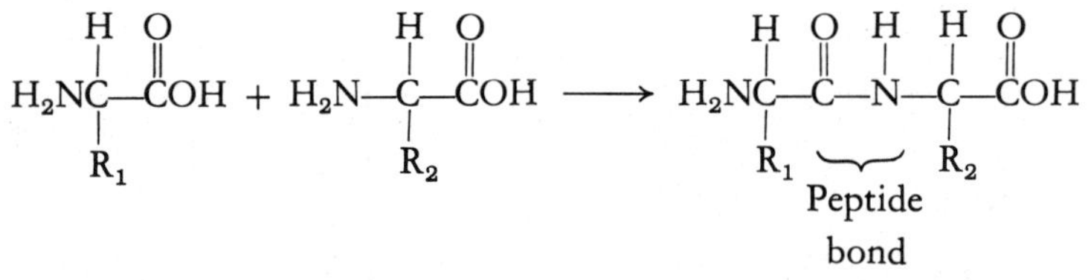

$$\underset{R_1}{H_2NCH-COH} + \underset{R_2}{H_2N-CH-COH} \longrightarrow \underset{R_1 \quad R_2}{H_2NCH-C-N-CH-COH}$$
$$\text{Peptide bond}$$

Proteins are composed of hundreds or thousands of amino-acid residues linked covalently by such bonds. Although the same twenty amino acids are present in most proteins, individual amino acids may recur many times in the macromolecule. The fundamental features that differentiate one protein from another are the relative proportion of each amino acid, the specific sequence of amino acids, and the total number of amino-acid units in the chain, or the molecular weight. Molecular weights of the proteins found in nature vary from about six thousand to several million. Other factors involved in the specificity of protein structure are discussed next.

A compound formed from two amino-acid units is called a *dipeptide*. Alanine and serine, for example, can provide the dipeptide serylalanine:

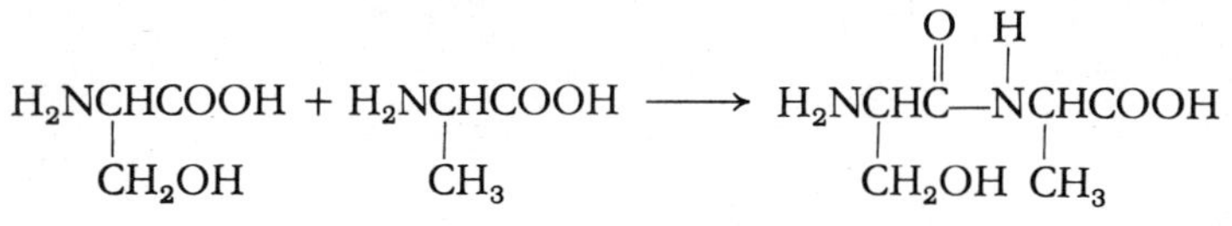

$$\underset{CH_2OH}{H_2NCHCOOH} + \underset{CH_3}{H_2NCHCOOH} \longrightarrow \underset{CH_2OH \; CH_3}{H_2NCHC-NCHCOOH}$$

Serine Alanine Serylalanine

In this case, alanine is the carboxyl-terminal, or *C-terminal*, amino acid and serine constitutes the amino-terminal, or *N-terminal*, amino acid. Peptides are named as derivatives of the C-terminal amino acid. The peptide shown is therefore a derivative of alanine, i.e., serylalanine. The alternative dipeptide that may be formed from serine and alanine is *alanylserine*:

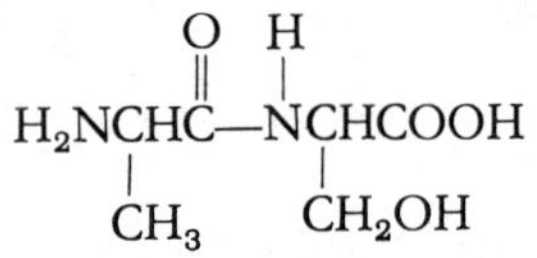

$$\underset{CH_3 \quad CH_2OH}{H_2NCHC-NCHCOOH}$$

If three amino-acid residues are involved, the compound is a *tripeptide*. An example is tyrosylvalylthreonine:

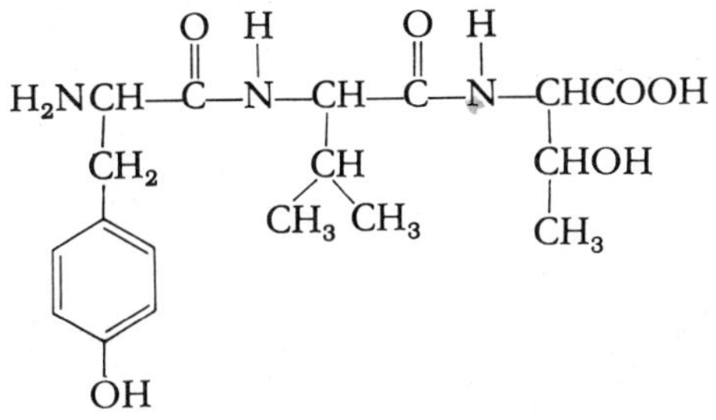

$$\text{H}_2\text{NCH}-\overset{\overset{\text{O}}{\|}}{\text{C}}-\overset{\overset{\text{H}}{|}}{\text{N}}-\text{CH}-\overset{\overset{\text{O}}{\|}}{\text{C}}-\overset{\overset{\text{H}}{|}}{\text{N}}-\text{CHCOOH}$$

with side chains: CH₂–(C₆H₄)–OH (tyrosine); CH(CH₃)(CH₃) (valine); CHOH–CH₃ (threonine).

Since threonine is the C-terminal amino acid, this tripeptide is named as a derivative of threonine. The other amino acids are indicated according to their specific sequence from the N-terminal amino acid. One more example, a *tetrapeptide*, will serve to clarify the nomenclature system:

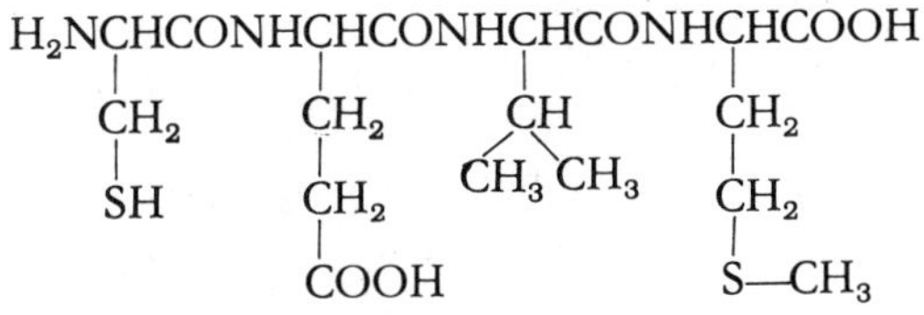

$$\text{H}_2\text{NCHCONHCHCONHCHCONHCHCOOH}$$

with side chains: CH₂–SH (cysteine); CH₂–CH₂–COOH (glutamic acid); CH(CH₃)(CH₃) (valine); CH₂–CH₂–S–CH₃ (methionine).

Cysteinylglutamylvalylmethionine

A protein macromolecule may consist of a single polypeptide chain, or it may involve several chains that are held together by either covalent bonds or other types of attractive forces. An example of a comparatively low-molecular-weight protein is bovine insulin. This is composed of two polypeptide chains linked to each other by disulfide bonds. The two polypeptides are termed the *A chain* and the *B chain*, as shown in Figure 1-7.

Consideration of Figure 1-7 will help in the understanding of some aspects of protein structure and terminology. (The diagram is not meant to define any conformational structure; the polypeptide chains were bent to fit on the page.) The N-terminal amino acid of the A chain is glycine and the C-terminal is asparagine. Cysteine-6 of this chain is linked to cysteine-11 by a disulfide linkage. The A chain and B chain are also linked to each other by disulfide bonds. Thus, there is a disulfide bridge between cysteine-7 of each chain, and another disulfide link occurs between cysteine-20 of the A chain and cysteine-19 of the B chain. The complete molecule has fifty-one amino-acid residues. Some amino acids appear several times in a polypeptide chain, e.g., leucine or asparagine. The sequence of amino acids in each polypeptide is fixed and

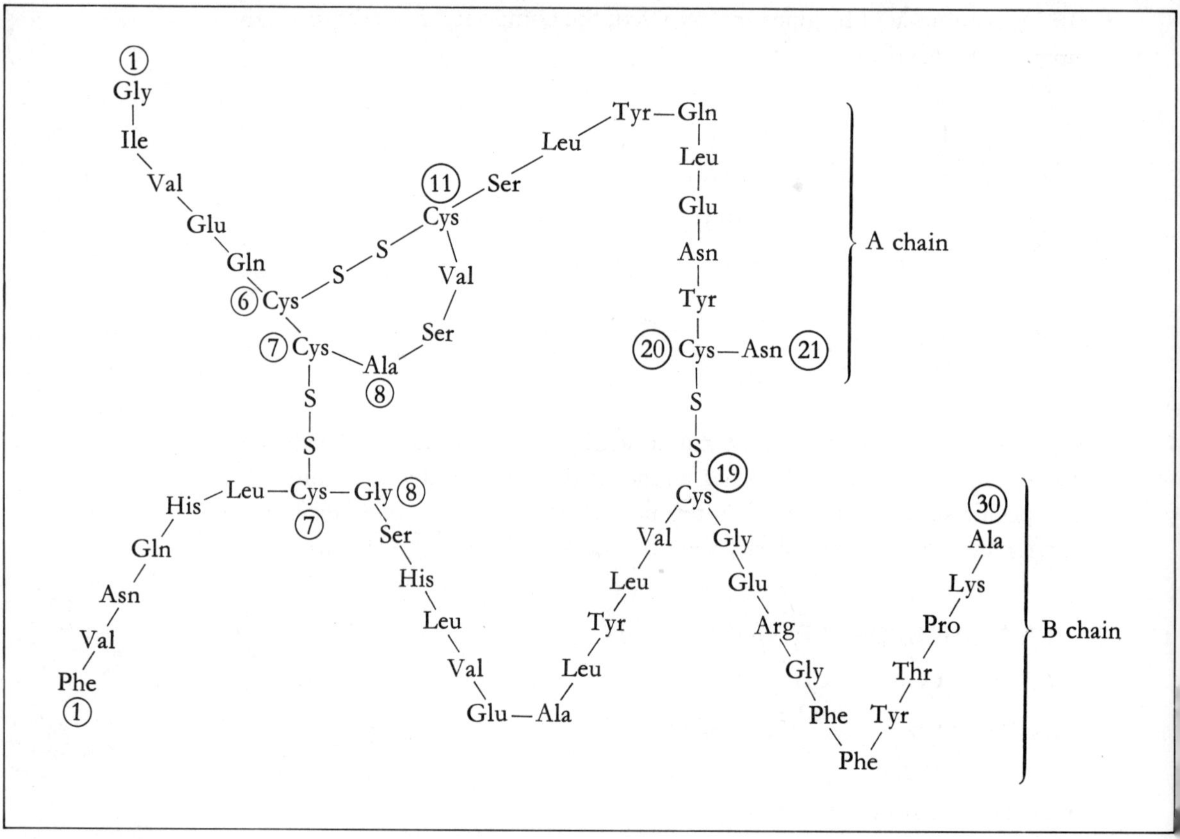

Fig. 1-7. *Primary structure of bovine insulin.*

specific for this protein. The abbreviations as written from left to right indicate the sequence of the amino acids, i.e., the N-terminal is on the left side and the C-terminal is on the right.

2. *Spatial Characteristics of Proteins*

The structure of proteins is described in terms of primary, secondary, tertiary, and quaternary levels. The basic characteristic of any protein is the sequence of the amino-acid residues in the chain; this is termed the *primary structure* of the protein. Various aspects of the chemistry of amino-acid residues in a polypeptide chain determine whether it will form a coil (helix) or zigzag chain (pleated sheet), or some mixture of the two; this physical structure of the polypeptide is called the *secondary structure*. Additionally, the total chain may bend or coil over itself to produce various three-dimensional conformations (e.g., a globular form); this is the *tertiary structure*. Another

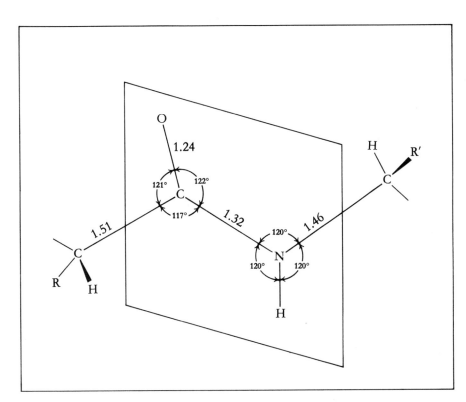

Fig. 1-8. Dimensions and steric aspects of a peptide bond. Bond distances are given in angstrom units (Å). An angstrom unit is equal to 0.1 nanometers (nm) or 10^{-8} cm.

level in the structural order of many proteins is due to the interaction of individual chains with each other. Certain proteins are known to be composed of a specific number of distinct polypeptide chains or units. The overall dimensions and spatial relationships of the complete multi-unit system make up the *quaternary structure* of the protein. Each of these aspects of protein structure will be considered separately.

The peptide bond confers certain conformational or spatial characteristics on a protein chain. Although the carbon-to-nitrogen linkage in the peptide bond is written as a single bond, there is essentially no free rotation at this site. The linkage can be visualized as a partial double bond or as a resonating system:

As a result, the peptide bond assumes a rigid planar structure with the bond angles and distances shown in Figure 1-8. Free rotation is possible around

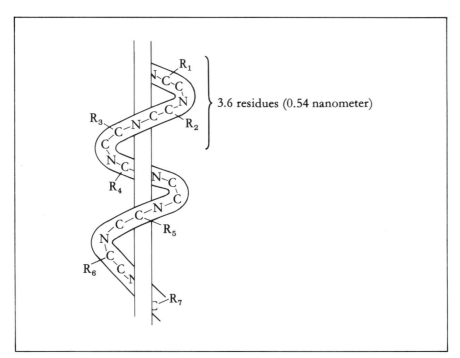

Fig. 1-9. *Helical structure of a polypeptide chain.*

the carbon-to-carbon bond and the nitrogen-to-carbon (non-carbonyl) bond, allowing variations in conformation.

Two basic types of secondary structures of polypeptides are conceivable, the *alpha* and *beta* structures. The alpha structure is a helical system, also called the α helix. It may be visualized as a spring-like coil formed from the sequence of atoms shown below in bold type:

$$
\begin{array}{c}
\text{H}\text{H}\text{O}R_2\text{H}\text{H}\text{O}R_4\text{H}\text{H}\text{O}R_6 \\
-\mathbf{N}-\mathbf{C}-\mathbf{C}-\mathbf{N}-\mathbf{C}-\mathbf{C}-\mathbf{N}-\mathbf{C}-\mathbf{C}-\mathbf{N}-\mathbf{C}-\mathbf{C}-\mathbf{N}-\mathbf{C}-\mathbf{C}-\mathbf{N}-\mathbf{C}-\mathbf{C}- \\
R_1\text{H}\text{H}\text{O}R_3\text{H}\text{H}\text{O}R_5\text{H}\text{H}\text{O}
\end{array}
$$

The R groups in this system are directed outward, somewhat perpendicular to the rodlike backbone of the helix (Fig. 1-9). The helical structure is maintained by hydrogen bonds between the nitrogen atoms and the carbonyl oxygen atoms in the same peptide chain. Since each turn in the helix corresponds to 3.6 amino-acid residues, the stabilizing hydrogen bonds can be visualized to occur between the —C═O and —NH of amino acids that are separated from each other by three intervening residues (Fig. 1-10).

Instead of forming a coiled, or helical, structure, the polypeptide system can exist as an extended macromolecule with a zigzag backbone:

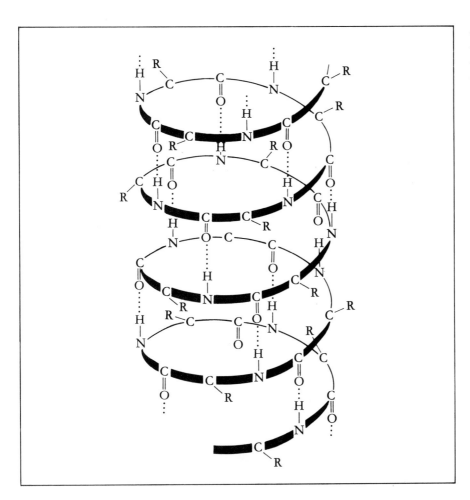

Fig. 1-10. *Alpha-helical structure of a polypeptide segment showing the sites of intrachain hydrogen bonds.*

This representation depicts a two-dimensional view of the beta structure. It should be realized, however, that if the sequence ····–C–C–N–C–C–N–C–···· is in the plane of the paper, then R_1, R_2, R_3, and so on, are perpendicular to this plane. This structure is stabilized by *interchain* hydrogen bonding between individual polypeptide molecules, as shown in Figure 1-11. When viewed from the edge of the arrangement of parallel strands, the overall conformation is that of a pleated sheet, hence the term *beta(β) pleated sheet*. The side chains, or R groups, in this system project above and below the plane of the sheet as shown in the diagram.

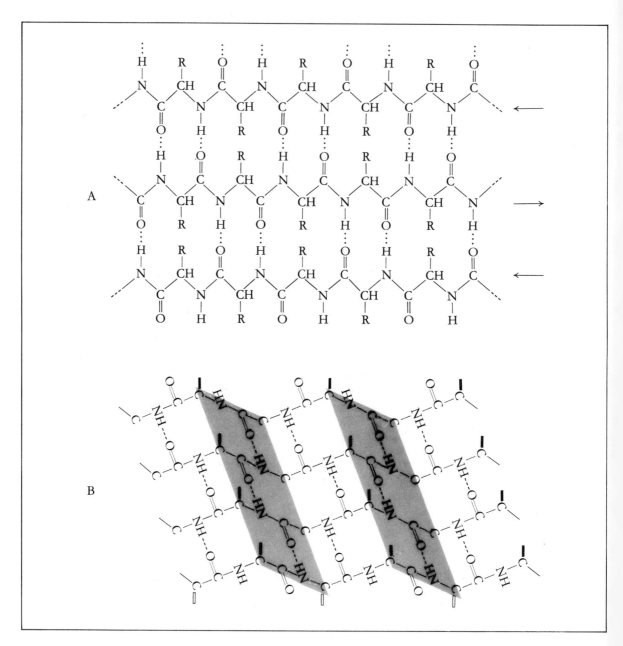

Fig. 1-11. Beta pleated sheet. Interchain hydrogen bonds occur between polypeptide chains. Note that the polypeptide chains run in opposite directions. A. View from above; the side chains (R groups) of the amino acid residues project above and below the plane of the paper. B. Edge-view of the sheet gives a pleated appearance. The dark and light perpendicular bars represent the R groups above and below the plane, respectively.

Proteins are classified, in a very general way, as either *fibrous* or *globular* proteins. Fibrous proteins are categorized further as α-keratins, β-keratins, and collagen. Included among α-keratins are the constituent proteins of hair, skin, wool, and nails. An example of β-keratin is found in silk. Collagen is the primary protein component of connective tissues, such as bone and tendon. A common feature of fibrous proteins is that they are insoluble, and thus they compose the structural components of the tissues of higher organisms.

The principal conformation of α-keratins is the helical structure. As already indicated, this type of secondary structure is stabilized by hydrogen bonds between —C=O and —NH units of the same polypeptide chain, i.e., intrachain bonding. In some instances, several chains may coil around each other to produce a rope-like structure. This is the case with the keratin of hair, in which the individual peptide chains are also linked to each other by disulfide bonds.

In beta structures, the chains are extended and the system is stabilized by *interchain* hydrogen bonding. As is the case with the helical conformation, the bonds are between the —C=O and —NH of the peptide linkages, but the peptide bonds belong to different chains. Certain proteins (e.g., those of hair) can change from a helical conformation to a beta structure as the result of heating or stretching. Such treatment causes the disruption of the intrachain hydrogen bonds that stabilize the helix and it yields protein strands with extended beta conformations.

The structures of the amino acids in collagen do not allow formation of the same type of helical system as exists in α-keratins. The presence of comparatively high amounts of proline and hydroxyproline creates rigid segments in the polypeptide chain that interfere with the development of α-helical conformation. Collagen also contains a high percentage of glycine, which destabilizes the α helix. The structure of collagen consists of three protein chains wound around each other to form an extended triple helix (Fig. 1-12). Details of the structure of collagen are described in Chapter 14.

Globular proteins do not have a uniform secondary structure throughout the entire chain. A single protein may be helical in some sections and have beta-type extended structures at other sites. Some segments of these proteins are bent or folded in different directions. Because of their irregularity, such regions are termed *random coils*. The macromolecular chains of globular proteins, which represent different types of secondary structure, are coiled and folded around each other so their overall spatial form, or tertiary structure, is considerably more compact than that of the extended fibrous proteins (Fig. 1-13). In contrast with fibrous proteins, which have rod-like structures, globular proteins have ellipsoid or spherical shapes. As a result of various physical interactions, the water-insoluble, or hydrophobic, groups of the component amino acids (e.g., the side chains of phenylalanine, leucine, iso-

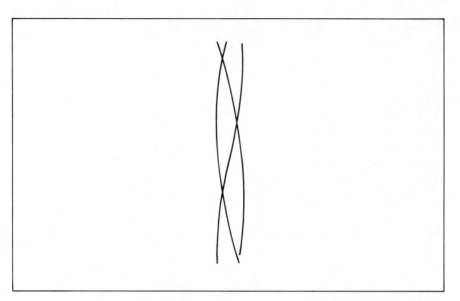

Fig. 1-12. Conformation of a section of collagen.

leucine, or valine) are oriented toward the interior of the protein structure, whereas the hydrophilic polar groups (e.g., the acidic or basic R groups) are positioned toward the exterior surface. Globular proteins are therefore soluble in aqueous media. These proteins include most of the intracellular enzymes, blood proteins, and hormones.

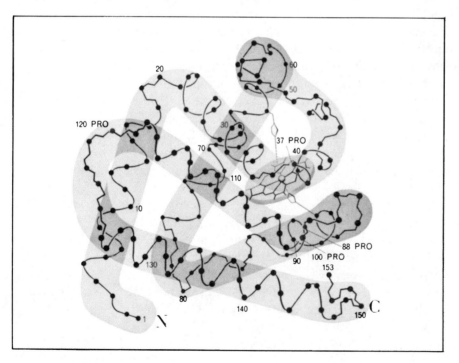

Fig. 1-13. Representation of the folding in a globular protein. The structure shown is myoglobin. (Reprinted with permission from "The Hemoglobin Molecule" by M. F. Perutz. Copyright © November 1964 by Scientific American, Inc. All rights reserved.)

The specific array of folds and coils in each globular protein confers upon it a distinctive shape or three-dimensional structure. This aspect of the structure of the protein, i.e., its tertiary structure, is uniquely determined by the component amino acids and their sequence. In other words, a specific protein invariably assumes a characteristic tertiary structure that depends on the sequence of its amino acids.

Certain amino-acid sequences usually confer a helical secondary structure. Amino acids that have this effect are called *helix-stabilizing* amino acids. These include alanine, asparagine, cysteine, glutamine, histidine, leucine, methionine, phenylalanine, tryptophan, tyrosine, and valine. The presence of proline or hydroxyproline in which the geometry of the nitrogen atom attached to the alpha carbon and most of the other atoms is fixed by the ring (p. 15–16), causes an interruption in the helical strand and a characteristic bend in the chain. These two amino acids, as well as glycine, favor formation of the extended beta-type conformation. Charged amino acids—such as aspartic acid, glutamic acid, arginine, or lysine—destabilize the helical structure through the effects of their side chains. Segments of proteins containing these residues will generally assume random-coil conformations. Another factor that contributes to helix destabilization is the steric nature or reactivity of the amino-acid side chains. Thus, isoleucine, with its bulky side chain, cannot serve as a component in a helical structure. Also, serine, which can form hydrogen bonds via its side-chain hydroxyl group, effectively inhibits helix formation in a polypeptide sequence.

The coiling and folding of the protein chain into a compact system has a notable effect on the distances between specific amino-acid residues. In globular proteins, certain amino acids may be close to each other spatially, even though there is a considerable distance between them in the primary sequence. For example, in the protein ribonuclease (Fig. 1-14), cysteine-40 appears next to cysteine-95. The two amino acids are connected by a disulfide bridge, even though they are separated by fifty-four other amino acids in the primary sequence. Thus, the coils in the chain form a loop that allows the cysteine residues to be close enough to each other so they can be linked by a covalent bond. The grouping of certain amino acids within a specific area is also critical for the catalytic functions of enzymes and the biologic activities of numerous proteins.

The tertiary structure of proteins is maintained and stabilized by several types of forces. These include hydrophobic interactions, hydrogen bonds, and ionic attractions. It is generally accepted that hydrophobic interactions make the greatest contribution to stabilizing a specific structure. This interaction is the result of the tendency for the non-polar side chains—such as those of alanine, isoleucine, leucine, phenylalanine, and valine—to be drawn toward each other and to cluster in specific, defined areas. The intimate

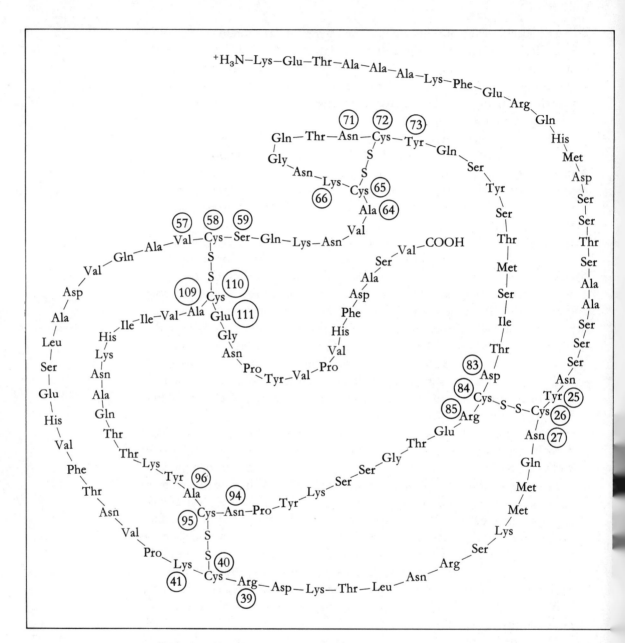

Fig. 1-14. Structure of bovine ribonuclease.

association of water-insoluble, or hydrophobic, units is induced by the high affinity of water molecules for each other (hydrogen bonding). When surrounded by molecules of water, any non-polar, water-insoluble groups are caused to adhere to each other in order to occupy the smallest possible

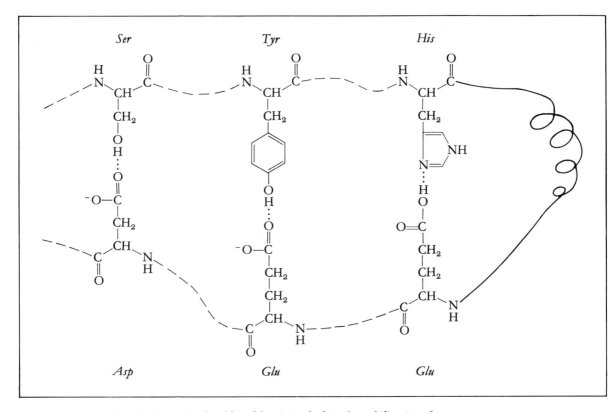

Fig. 1-15. Examples of hydrogen bonding (dotted lines) involved in the stabilization of the tertiary structure of proteins.

volume. The overall effect of the affinity of hydrophobic groups toward each other is to stabilize specific arrangements and patterns in the three-dimensional structure of a protein. These interactions also lead to the positioning of hydrophobic groups in the interior of globular proteins.

The tertiary structure of proteins is also stabilized by hydrogen bonds between the side chains of various amino-acid residues. (These are not to be confused with the hydrogen bonds involved in maintaining the α helix or β pleated sheets.) For instance, hydrogen bonding can occur between the side-chain carboxyl groups of aspartate or glutamate and the hydroxyl groups of serine, threonine, or tyrosine. Other units that participate in such bonds are the imidazole rings of histidine and the amide groups of glutamine or asparagine. Examples of some hydrogen-bond interactions are shown in Figure 1-15.

The three-dimensional structure of proteins is also influenced by ionic interactions, i.e., those between residues bearing opposite charges. Thus, terminal or side-chain carboxylate anions tend to attract the positively

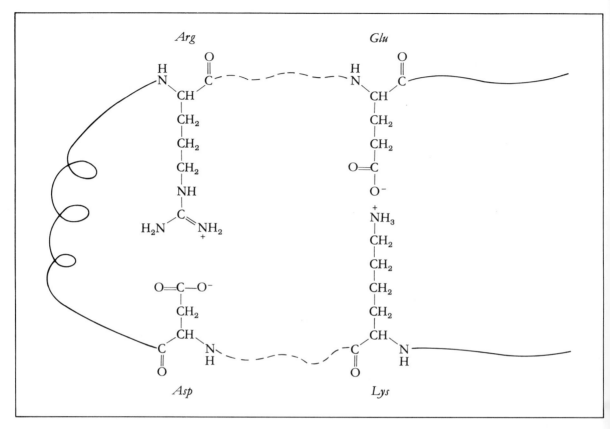

Fig. 1-16. Examples of ionic interactions (between + and −) involved in the stabilization of the tertiary structure of proteins.

charged ε-amino groups of lysine or guanidino groups of arginine (Fig. 1-16).

In addition to the above-mentioned forces, van der Waals interactions occur at various sites. These can exist between phenylalanine and tyrosine units that are close to each other or between proximal serine residues. Van der Waals attractive forces are also prominent between side chains that are involved in hydrophobic interactions. Although van der Waals attractions are comparatively weak, the cumulative effect of numerous interacting sites has a considerable influence on protein structure.

The three-dimensional structure of proteins may be modified or disrupted by various reagents. For instance, concentrated solutions of urea (8 M) or guanidine hydrochloride (6 M) cause the dissociation of units held together by hydrogen bonds. A large increase or decrease in pH will change the state of ionization of various residues and modify ionic interactions. Moderate elevation of temperature can also disrupt a number of weak non-covalent attractive forces. Although these conditions usually do not break covalent linkages, they may effect a change in the conformation of a protein. The result of such structural alterations is termed *denaturation*. Denaturation may or may not be reversible. In many instances, the protein regains its original structure (or biologic activity) when the denaturing agent is removed or when the temperature is returned to normal; in this case, the denaturation is reversible. When subjected to high temperatures (e.g., the boiling point of water), the protein, when cooled, will not regain its original structure, and the denaturation is irreversible.

The tertiary structure of a number of proteins is stabilized by disulfide bonds between oxidized cysteine residues (cystine dimers) that are distant from each other in amino-acid sequence (see Fig. 1-14). Disruption of these bonds by conversion of the cystine to cysteine units can produce a significant alteration in the three-dimensional structure. This disruption can be effected with β-mercaptoethanol (Fig. 1-17). If the latter is subsequently removed from the protein solution, the sulfhydryl groups are oxidized by atmospheric oxygen, and the disulfide bridges are re-formed. Thus, the original tertiary structure may be restored.

The finding that denaturation can often be reversed indicates that the normal three-dimensional structure of a protein is specified by its primary structure, i.e., by its component amino acids and their sequence. During denaturation, the bonds that stabilize the secondary and tertiary structure of the protein are disrupted, and portions of the macromolecule assume new conformations. Yet, when the denaturing condition is eliminated, the protein returns to its original, unique system of folds and coils. This implies that the specific amino-acid sequence programs the secondary and tertiary structure of the protein. The structure that results is the one that is thermodynamically most stable; it therefore arises spontaneously and is maintained by all available stabilizing forces. The concepts that (1) secondary and tertiary structures are inherent in the primary structure and (2) that their production does not require any additional mechanism are fundamental to protein biochemistry and have important implications with respect to protein synthesis.

All the structural aspects discussed in this section have dealt with the spatial relationships of a single macromolecular chain. A number of proteins are composed of multiple polypeptide chains (oligomers); these chains are linked to one another by non-covalent bonds. Hemoglobin, for example, consists of four individual globular polypeptides. Various enzymes are also composed of multiple units. The biologic properties of these proteins depend on the cooperative effect of all the units. For their concerted action, these polypeptide units must be juxtaposed in a precise manner. The interaction among these subunits results in a complex spatial arrangement, termed the *quaternary structure of the protein*, which is defined by a variety of parameters.

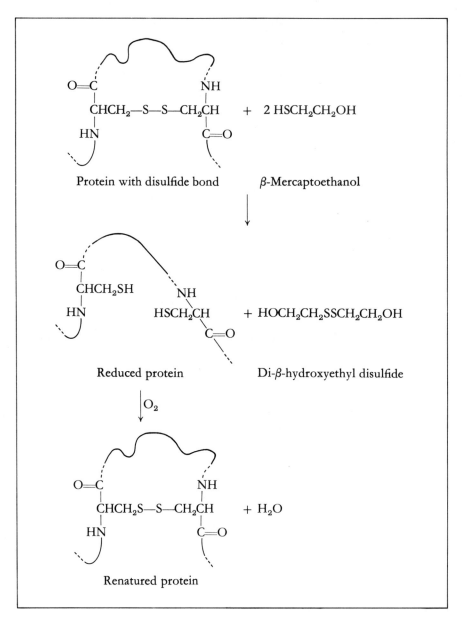

Fig. 1-17. Reduction of disulfide bonds by β-mercaptoethanol. Renaturation may occur upon reoxidation of the sulfhydryl units.

3. Conjugated Proteins

Many proteins are linked covalently to non-polypeptide components; these are called *conjugated* proteins. For instance, hemoglobin contains non-protein heme units, and the flavoproteins are composed of polypeptide chains linked to flavin groups. Numerous examples of such proteins will be encountered throughout the ensuing discussions.

Many proteins are also associated with lipids or carbohydrates; these are called *lipoproteins* and *glycoproteins*, respectively.

4. Electrolyte Characteristics of Proteins

With the exception of the N-terminal and C-terminal amino acids, the α-amino and α-carboxyl groups of the amino-acid residues in proteins are involved in the peptide linkages. The acid-base properties of proteins are therefore due primarily to the ionizable units on the side chains, or R groups. The pK values for these groups in proteins differ to some extent from those for the free amino acids, and they depend on the nature of the other residues in their vicinity. The ranges of values for different units are shown in Table 1-4.

The net charge on a protein depends on the number of ionizable side chains, their pK value, and the pH of the medium. The pH at which the sum of the negative charges is equal to that of the positive charges is called the *isoelectric point* of the protein. When placed in an electrical field, as in the procedure of electrophoresis, a protein at its isoelectric pH will not migrate in either direction. In solutions with a pH below its isoelectric point, a protein will have a net positive charge; above the isoelectric point, it will have a net negative charge. When it has a net positive charge, the protein will migrate toward the cathode on electrophoresis; when its net charge is negative, its migration will be toward the anode. Each protein has its characteristic isoelectric point, and this property is utilized in many procedures involving the separation of proteins.

The degree of dissociation of various charged groups in a protein at a given pH can be calculated from their pK values by the Henderson-Hasselbalch equation (Eq. 1-7). However, some qualitative conclusions may be drawn from inspection of Table 1-4. It can be seen, for example, that at pH 7, the terminal and side-chain carboxyl groups are almost entirely in the ionized (negatively charged) form. Similarly, the ε-amino groups of lysine ($pK_a \approx 10$)

Table 1-4. pK Values for Ionizable Groups in Proteins

Group	pK Range
C-terminal α-carboxyl	3.8–4.7
β- or γ-Carboxyl of aspartate and glutamate	4.0–5.6
N-terminal α-amino	7.4–7.9
ε-Amino of lysine	10.0–10.6
Imidazole of histidine	6.0–7.0
Hydroxyl of tyrosine	9.5–10.8
Sulfhydryl of cysteine	9.1–10.8
Guanidinyl of arginine	11.9–13.3

will be positively charged. The imidazole groups of the histidine residues are about 50% ionized at pH 7, since this is very close to the value of the pK.

The isoelectric point of a protein depends mainly on the relative number of acidic or basic residues. Proteins that contain a large number of aspartate or glutamate residues carry a negative charge at pH 7. Neutralization of the side-chain carboxyl group requires a decrease in pH of the medium. Even at pH 5, half of the carboxyl groups are dissociated. For such proteins to be uncharged, a solution with a pH of less than 3 may be required. Proteins of this type are called *acidic proteins*. An example of such a protein is the enzyme pepsin, which has an isoelectric point of less than 1.

Proteins that contain large amounts of lysine and arginine (e.g., lysozyme or various histones) are positively charged even at pH 8 or pH 9. The isoelectric point of these proteins is about 11 or 12. These are termed *basic proteins*.

Most proteins contain acidic and basic amino acids, and the isoelectric points depend on their relative proportions. The isoelectric points are essentially at the pH where the charge of the negative groups is equal to that of the positively charged units, generally between pH 5 and pH 8.

5. *Separation and Physical Characterization of Proteins*

Two fundamental features that serve to distinguish one protein from another are the net charge and the molecular weight. A number of procedures are available for the separation and quantitative assay of proteins on the basis of these parameters. The ones that have been adapted to the greatest extent for clinical purposes involve either electrophoresis or ultracentrifugation. Various modifications of electrophoresis are employed routinely for the diagnosis of protein or enzyme abnormalities. Ultracentrifugation involves highly complex and more expensive systems and is therefore utilized only for special studies.

In *ultracentrifugation*, when a solution of a high-molecular-weight substance is centrifuged at speeds that yield a centrifugal force of about $10^5 \times$ gravity, the particles of the solute sediment at a rate S defined by:

$$S = \frac{dr}{dt} \bigg/ r\omega^2 \qquad (1\text{-}10)$$

where r is the distance (meters) of the particle from the axis of rotation and ω is the angular velocity in radians per second (360 degrees $= 2\pi$ radians). The *sedimentation coefficient* S is thus proportional to the rate of "downward movement" of the particle. The sedimentation coefficient is expressed in svedberg units (S). Experimentally, S is affected by a number of variables. However, conditions can be standardized so that it will depend primarily on the mass of the molecule (molecular weight) and its shape. Macromolecules, such as proteins or nucleic acids, can therefore be characterized by the value of the sedimentation coefficient.

Electrophoresis is a technique in which molecules are separated on the basis of their charge. The underlying principle of the method is that when proteins or other charged substances are subjected to an electric field, they migrate toward the anode (positive pole) or cathode (negative pole), depending on the sign of their charge. If several compounds with about the same mass and charges of the same sign are studied by this procedure, the one with the highest numerical value of its net charge will move the greatest distance in a given amount of time. Since the charge on a protein depends on the hydrogen-ion concentration of the solution, the pH must be specified and kept constant during the electrophoretic procedure. Most clinical determinations on plasma or serum proteins are performed at pH 8.6. This is above the isoelectric point of the proteins involved, so they are all negatively charged. Under these conditions, albumin, which has a higher net charge than the globulins, has a greater electrophoretic mobility.

The original electrophoretic procedures were performed in solution ("free" electrophoresis) and required special arrangements for introducing samples and minimizing disturbing factors, such as convection. In free electrophoresis, the determination of the position of the migrating protein required complex optical systems. Most methods now in use, however, employ a matrix, such as a gel, to support the migrating proteins and allow for staining processes with which the protein can be visualized. Supporting matrices or media include filter paper and cellulose acetate, starch, or polyacrylamide gels. In paper electrophoresis, the protein is applied as a narrow band to a strip of filter paper that is wetted in a buffered solution and connected at both ends to a voltage source (Fig. 1-18). When a given time has elapsed, the applied voltage is turned off, and the paper is dried and stained by soaking in a dye that is specific for proteins. After final processing, each protein appears as a discrete zone or band that can be identified and quantitated; hence the procedure is termed *zone electrophoresis*.

6. *Hemoglobin*

Some of the principles of protein chemistry may be more clearly understood if we consider their application to the structure of a specific protein, hemoglobin.

Hemoglobin is found in erythrocytes, and its primary function is the transport of oxygen from the lungs to all the cells. It is a conjugated protein, that is, it is linked to a non-protein, iron-porphyrin component. Hemoglobin is composed of four polypeptide chains. In the most common type found in adults, there are two identical chains termed *alpha* (α) and two others called

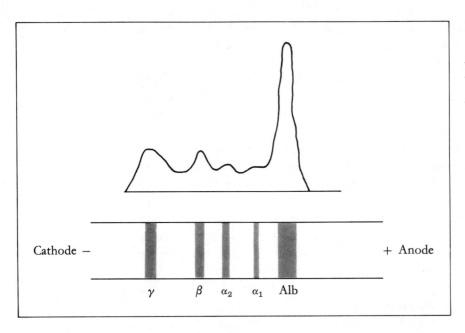

Fig. 1-18. Paper electrophoresis of serum proteins. Top: Concentration profile. Bottom: Visualization of zones on staining. Alb = albumin; α_1, α_2, β, and γ correspond to globulin classes.

beta (β). The total protein may thus be designated as $\alpha_2\beta_2$. Since several hemoglobin variants have been described, the one under consideration is identified as hemoglobin A_1, or HbA_1.

Each α chain has a molecular weight of about 15,750, and the molecular weight of each β chain is about 16,500. Hemoglobin HbA_1 thus has a molecular weight of 64,500.

The primary structure of a protein is the amino-acid sequence of each of its individual chains. The α chains of hemoglobin are composed of 141 amino-acid residues. The amino-terminal (N-terminal) amino acid is valine, and the carboxyl-terminal (C-terminal) residue is arginine. The β chains consist of 146 amino acids, of which valine is the N-terminal and histidine is the C-terminal residue.

With respect to the secondary structure, each chain of hemoglobin consists of several helical regions separated from each other by non-helical segments. The tertiary structure involves a number of bends or coils, so each individual chain resembles a globular protein with an approximately spherical shape. Each α chain contains a Fe^{2+}-porphyrin (heme) group sandwiched between histidine-58 and histidine-87. The heme unit in each β chain is complexed with histidines 63 and 92 (Fig. 1-19).

The quaternary structure defines the mode of interaction between the four chains and the three-dimensional appearance, or shape, of the complete macromolecule. The results of x-ray crystallography indicate that there is a close fit between the chains and that the overall arrangement approximates a

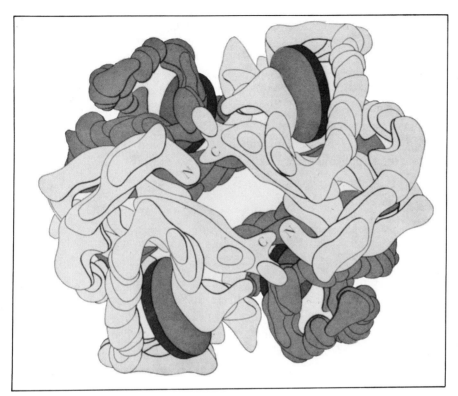

Fig. 1-19. Structural characteristics of hemoglobin. (Reprinted with permission from "The Hemoglobin Molecule" by M. F. Perutz. Copyright © November 1964 by Scientific American, Inc. All rights reserved.)

tetrahedron, with each subunit occupying a corner. The α chains interact more strongly with a β chain than with another α chain. Conditions that dissociate hemoglobin into dimers therefore yield αβ units rather than αα or ββ units.

The effect of specific amino acids on electrophoretic mobility can also be seen with hemoglobin. At pH 8.6, hemoglobin A_1 is negatively charged and migrates toward the anode (Fig. 1-20). In the hemoglobin of individuals

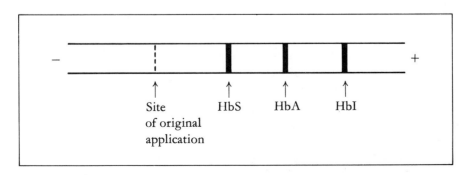

Fig. 1-20. Electrophoresis of different types of hemoglobin. HbA, normal, adult hemoglobin; HbI, HbS, genetic variants (see text).

with sickle-cell anemia, a glutamic acid residue at position 6 of the β chains is replaced by a valine. As a result, the net charge of this abnormal hemoglobin (termed HbS) is decreased by two. This protein therefore does not migrate as rapidly on electrophoresis as does HbA (Fig. 1-20). In another genetic disease, a lysine at position 16 on the α chain of the hemoglobin (HbI) is replaced by glutamic acid. This effectively increases the net negative charge on the protein. As may be expected, the electrophoretic mobility of HbI is greater than that of HbA (Fig. 1-20). The diagnosis of hemoglobinopathies is only one of numerous applications of electrophoresis in clinical practice.

D. CARBOHYDRATES

The fundamental units of carbohydrate structure are the monosaccharides, which are either polyhydroxy aldehydes or ketones, called *aldoses* and *ketoses*, respectively. The most common sugars in animal tissues are the pentoses and hexoses and their derivatives. Only these will be described in this section; the structures of other sugars and of polysaccharides will be discussed in relation to their metabolism.

A *pentose* is composed of a chain of five carbons, and a *hexose* contains six carbons. Glucose is an example of an *aldohexose*, whereas fructose is a *ketohexose*.

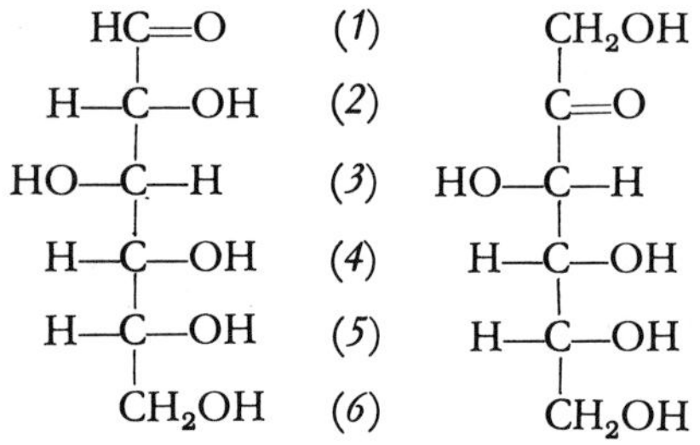

$$
\begin{array}{cc}
\text{HC=O} & (1) \quad \text{CH}_2\text{OH} \\
\text{H–C–OH} & (2) \quad \text{C=O} \\
\text{HO–C–H} & (3) \quad \text{HO–C–H} \\
\text{H–C–OH} & (4) \quad \text{H–C–OH} \\
\text{H–C–OH} & (5) \quad \text{H–C–OH} \\
\text{CH}_2\text{OH} & (6) \quad \text{CH}_2\text{OH} \\
\text{Glucose} & \quad \text{Fructose}
\end{array}
$$

The individual carbon atoms are numbered as shown. Note that carbons 2, 3, 4, and 5 of the aldohexoses are *asymmetric* since each of these carbons is linked to four different substituents. In these structures, if the positions of the hydrogen atoms and the hydroxyl groups on any of the asymmetric carbons are interchanged, the resultant compound cannot be superimposed on the original compound, i.e., the compounds will not be the same. For example, if the substituents on carbon-4 of glucose are transposed, the new compound is galactose:

```
    HC=O      (1)
     |
  H—C—OH     (2)
     |
  HO—C—H     (3)
     |
  HO—C—H     (4)
     |
  H—C—OH     (5)
     |
    CH₂OH    (6)
```
Galactose

The structures of monosaccharides, as written, are planar projections of a three-dimensional system. The bonds from each carbon (except for carbon-1) extend above and below the plane of the paper as shown:

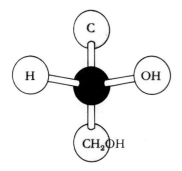

The planar projection of monosaccharides should be visualized in terms of this model. Hence, the horizontal bonds to H and OH are, in reality, in front of the plane of the paper. Similarly, the vertical lines above and below the asymmetric carbons extend to the back of the plane. Sugars in which the asymmetric carbon farthest from the carbonyl group has the configuration shown in the structures illustrated (and in the model) are classified as D-sugars. Those having the opposite configuration are L-sugars. Hence, for hexoses, the D series differs from the L series with respect to the configuration on carbon-5:

```
     |                    |
  H—C—OH              HO—C—H
     |                    |
    CH₂OH               CH₂OH
    D                    L
```

The structures of D and L glucose are mirror images of one another:

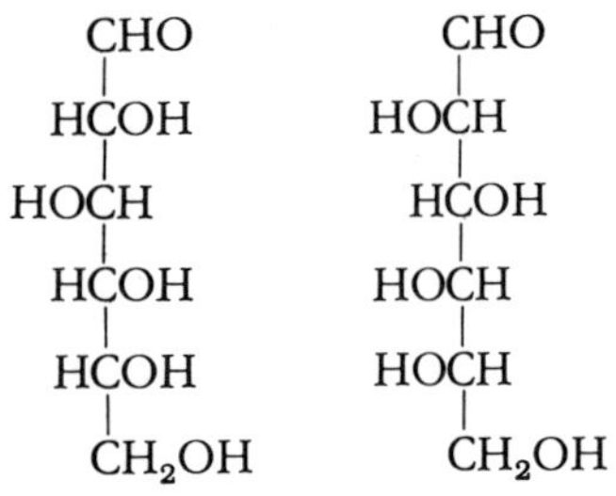

```
    CHO              CHO
     |                |
   HCOH             HOCH
     |                |
   HOCH             HCOH
     |                |
   HCOH             HOCH
     |                |
   HCOH             HOCH
     |                |
   CH₂OH            CH₂OH
  D-Glucose        L-Glucose
```

The structure of glucose allows yet another variation, since it can form an oxygen bridge between carbon-5 and carbon-1. The latter carbon, termed the *anomeric carbon*, constitutes an additional center of asymmetry. The two forms of glucose that result are called α and β. If the configuration of the anomeric carbon is the same as that of carbon-5, it is designated as α; otherwise, it is β. A simple way to visualize this is that in the α anomer, the OH on carbon-1 projects in the same direction from the plane of the ring as the H on carbon-5, shown in the following equation. In solution, there is an equilibrium between the straight chain of D-glucose and the two ring forms; however, the predominant form is that of the rings:

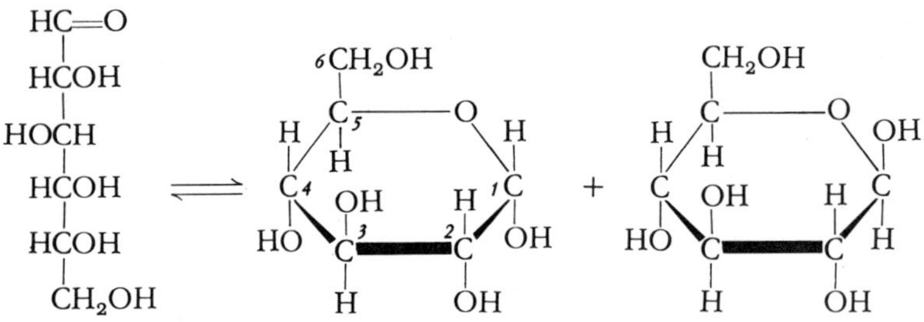

[Structural diagram: D-Glucose ⇌ α-D-Glucose + β-D-Glucose]

If the ring contains six members, it is designated as a *pyranose* ring. Thus, the complete names for these sugars are α-D-glucopyranose and β-D-glucopyranose.

In a number of sugars, the ring form is five-membered; this is called a *furanose* ring. The most important hexose (from a metabolic point of view) that forms a furanose structure is fructose:

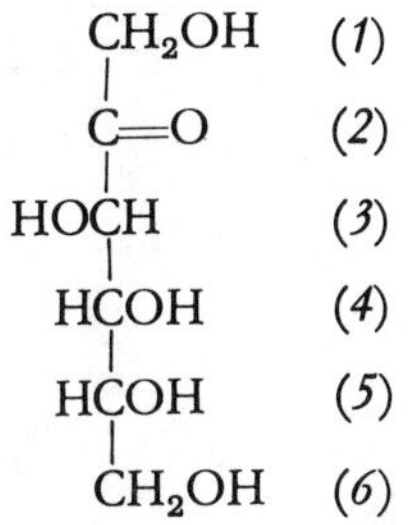

$\begin{array}{ll} CH_2OH & (1) \\ C\!\!=\!\!O & (2) \\ HOCH & (3) \\ HCOH & (4) \\ HCOH & (5) \\ CH_2OH & (6) \end{array}$

D-Fructose 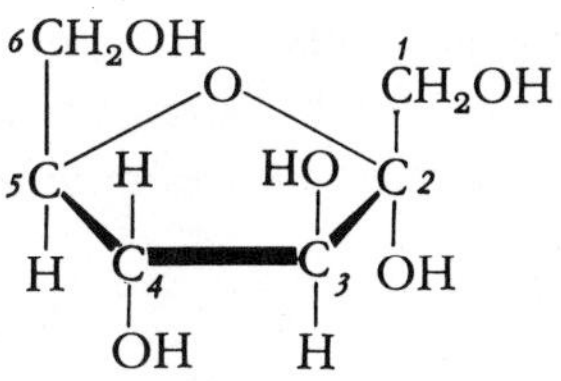   α-D-Fructose 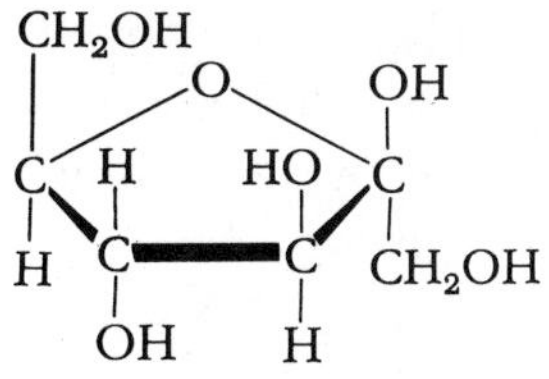   β-D-Fructose

The most abundant *pentose* in animal tissue is D-ribose. Its straight-chain and ring structures are:

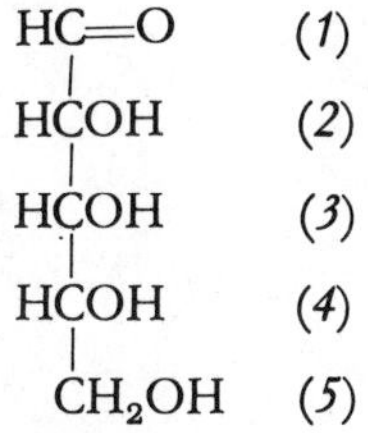

$\begin{array}{ll} HC\!\!=\!\!O & (1) \\ HCOH & (2) \\ HCOH & (3) \\ HCOH & (4) \\ CH_2OH & (5) \end{array}$

D-Ribose 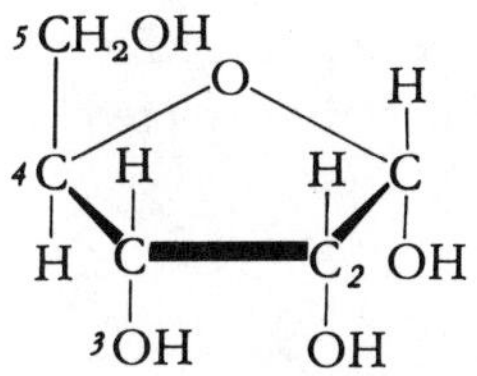   α-D-Ribofuranose 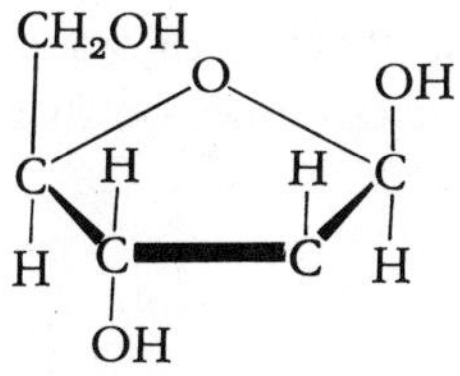   β-D-Ribofuranose

A derivative of ribose that is present in certain types of nucleic acids (i.e., DNA) is 2-deoxyribose:

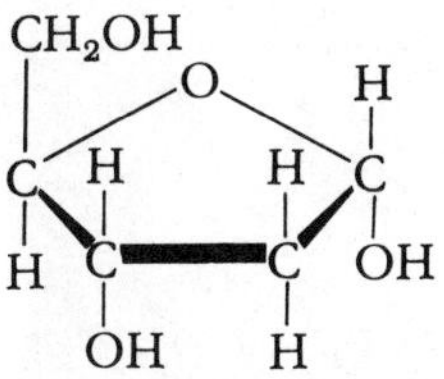

2-Deoxyribose

Although monosaccharides occur primarily as ring forms, the potential carbonyl-carbons (carbon-1 of aldoses and carbon-2 of ketoses) remain susceptible to oxidation to sugar acids by reagents that contain divalent copper, such as Fehling's or Benedict's solution. The dynamic equilibrium that exists between the straight-chain and the ring form permits formation of the former structure, which can be oxidized. Since the oxidation by Cu^{2+} results in its reduction to Cu^+, the sugars that undergo this reaction are designated as *reducing sugars*. When the hydroxyl group on carbon-1 of the ring form of an aldose is replaced with a methyl group or similar unit, the reducing activity of the sugar is lost, because it cannot revert to the straight-chain form. Acetal compounds of this type are called *glycosides*. An example of a glycoside is methyl α-D-glucopyranoside:

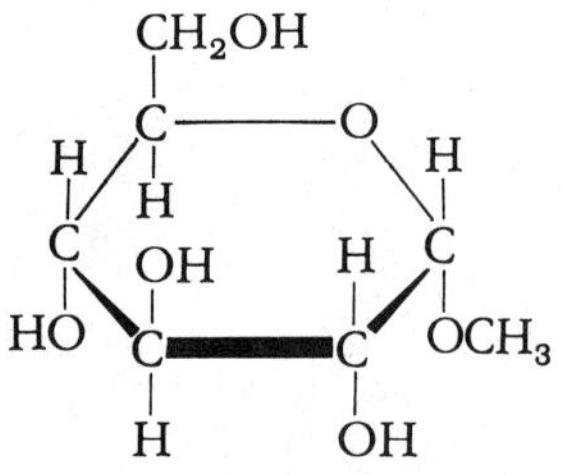

Disaccharides are sugars composed of two monosaccharides linked together as glycosides. The structures of several common disaccharides are depicted as follows:

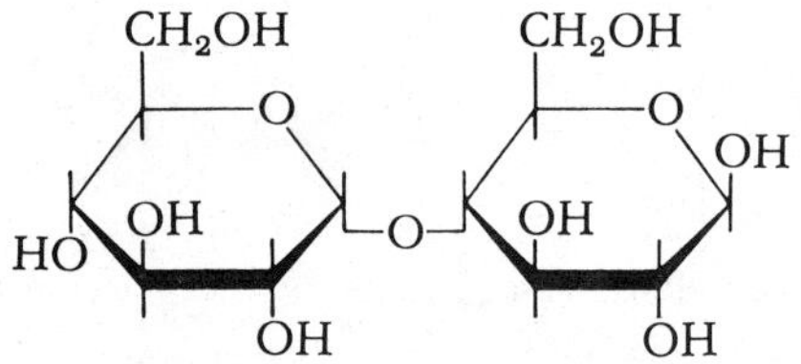

β-Maltose
4-O-α-D-glucopyranosyl-D-glucopyranose

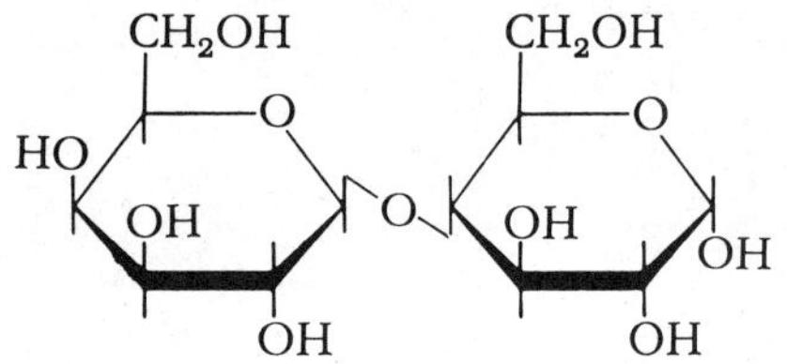

α-Lactose
4-O-β-D-galactopyranosyl-D-glucopyranose

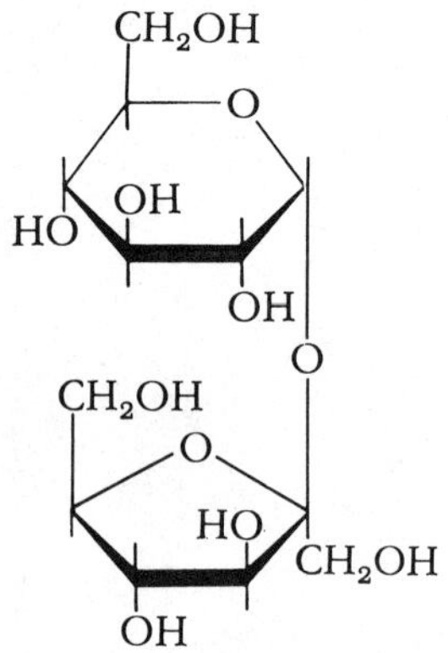

Sucrose
β-D-fructofuranosyl α-D-glucopyranoside

Maltose is composed of two glucose units connected to each other by a glycosidic bond between carbon-1 of one glucose component and carbon-4 of the other. It should also be noted that the glycosidic linkage has an α configuration. Lactose is composed of galactose and glucose. The linkage is between carbon-1 (β) of galactose and carbon-4 of glucose. Both maltose and lactose are reducing sugars, since they contain one unit (the glucose unit) in which the anomeric carbon is a potential aldehyde group, i.e., the glucopyranose can open to a straight-chain form. In sucrose, a glucose and a fructose are linked to each other through their anomeric carbons. Sucrose does not have a free anomeric carbon, and it therefore does not have reducing activity.

Polysaccharides consist of a large number of sugar residues linked to each other by glycosidic linkages. Although numerous types occur in nature, only a few will be described in this section. Certain complex polysaccharides (hyaluronic acid, chondroitin sulfates, dermatan sulfate, and heparin) will be discussed in later chapters.

The major dietary polysaccharide is starch. Starch is actually a mixture of two polysaccharides: *amylose* and *amylopectin*. Amylose is composed of glucose units linked to each other by α-1,4-linkages:

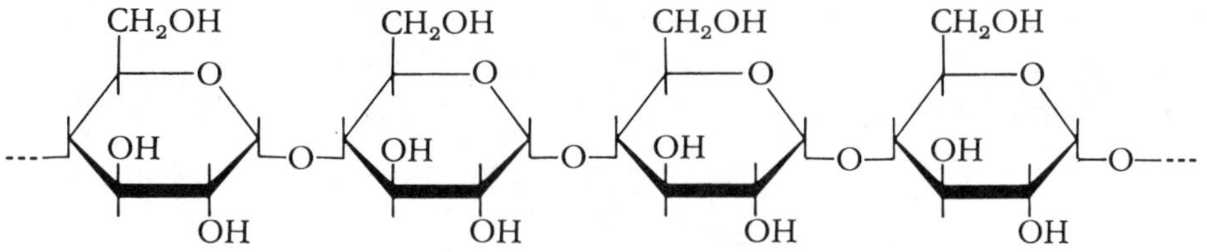

Amylopectin differs from amylose in that it has branches through carbon-6 at various sites:

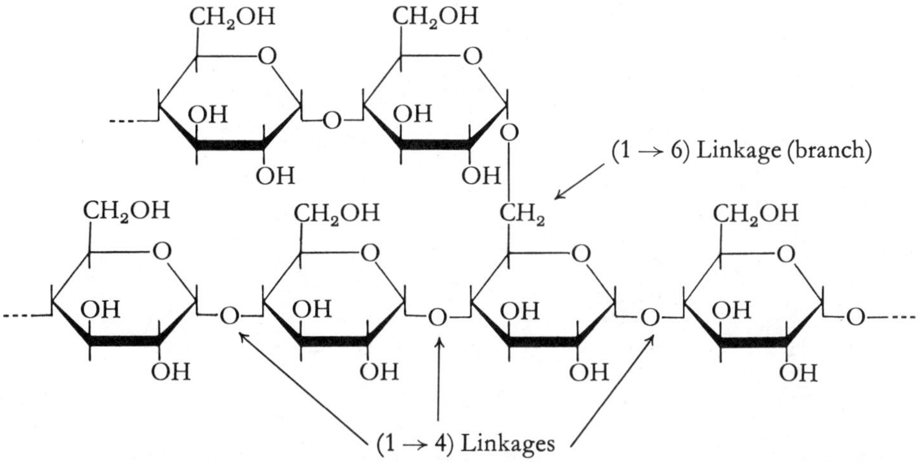

Starch is the carbohydrate storage material of higher plants. In animals, this function is served by *glycogen*. The latter is similar to amylopectin except that it has a greater degree of branching, i.e., branch points occur at intervals of 8 to 10 glucose units. Glycogen is found in highest concentration in liver (8% to 10% of wet weight). Cells of many other tissues contain glycogen; for example, the amount of glycogen in skeletal muscle is about 1% to 2% by weight.

Another abundant polysaccharide in plants is *cellulose*, the major structural component of cell walls. It is a polymer of glucose linked by β-1,4-linkages:

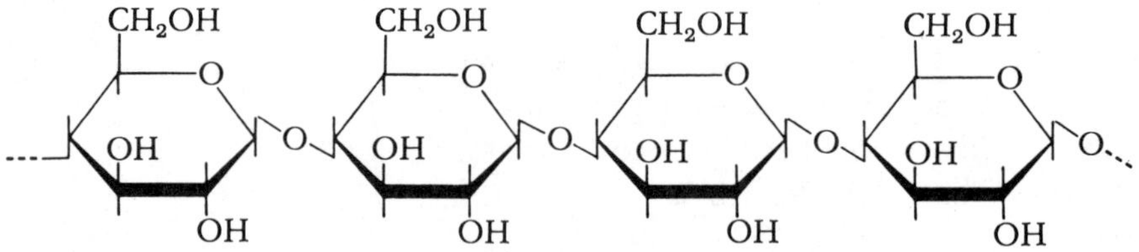

Unlike starch, this polysaccharide does not have dietary value, because human digestive enzymes do not catalyze the hydrolysis of β-1,4-glycosidic linkages (Chap. 4).

Various derivatives of the hexoses occur as components of complex polysaccharides, glycoproteins, and glycolipids. These include D-glucosamine, D-galactosamine, D-glucuronic acid, L-iduronic acid, sialic acid, and their derivatives:

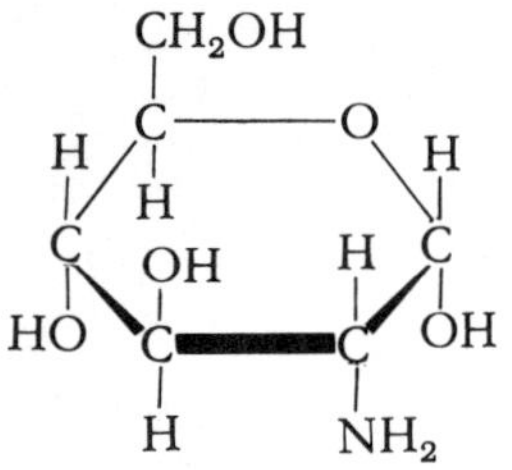

D-Glucosamine
(2-amino-2-deoxy-α-D-glucopyranose)

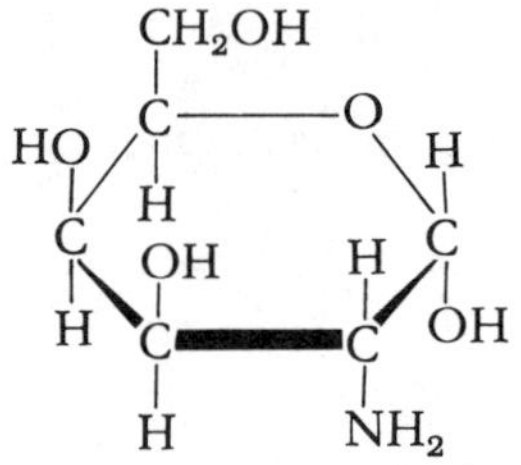

D-Galactosamine
(2-amino-2-deoxy-α-D-galactopyranose)

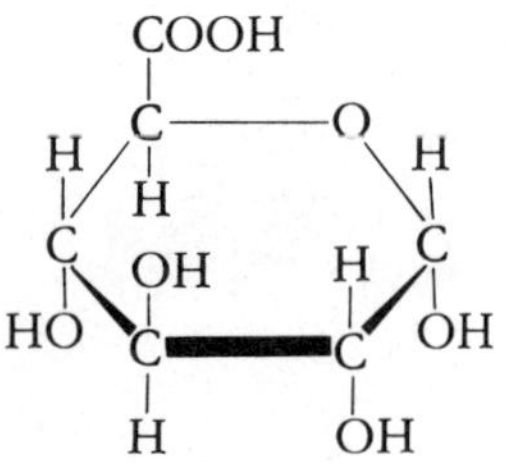

α-D-Glucuronic acid

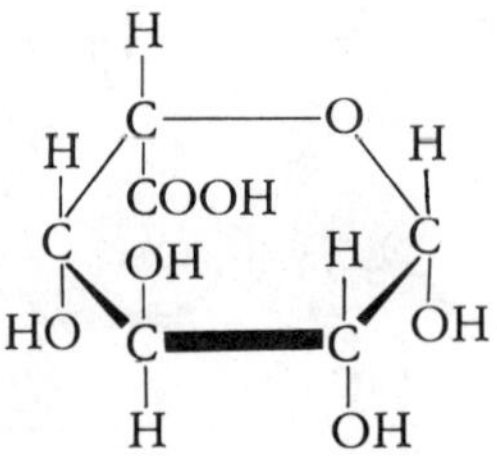

β-L-Iduronic acid

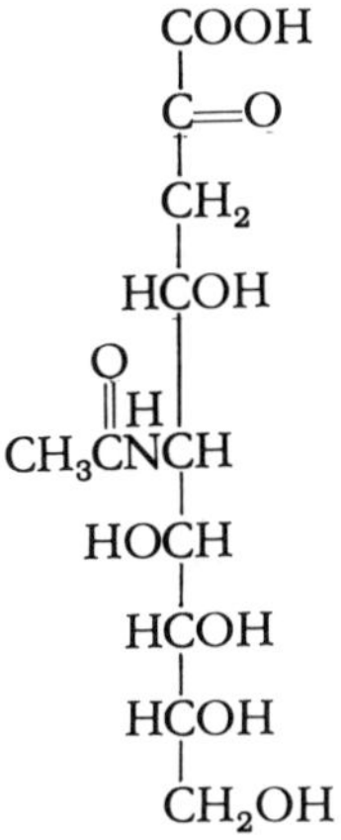

```
      COOH
       |
       C=O
       |
       CH₂
       |
       HCOH
    O  |
    ‖H |
CH₃CNCH
       |
       HOCH
       |
       HCOH
       |
       HCOH
       |
       CH₂OH
```
A sialic acid
(*N*-acetylneuraminic acid)

E. PURINES, PYRIMIDINES, NUCLEOSIDES, AND NUCLEOTIDES

In addition to their being components of the nucleic acids, various nucleotides function as cofactors for enzymes and as sources of energy for endergonic reactions. This section deals with the basic structural aspects of the nucleotides. *Nucleotides* are composed of three building blocks: an organic nitrogenous base, ribose or deoxyribose, and phosphate. The nitrogenous bases are either purines or pyrimidines.

The *purines* of animal cells are *adenine* and *guanine*:

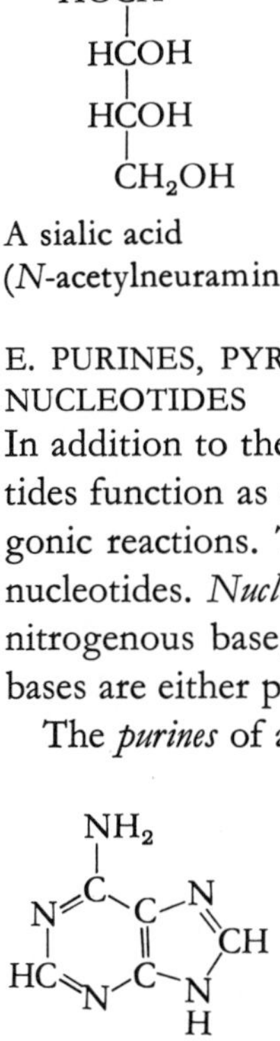

Adenine 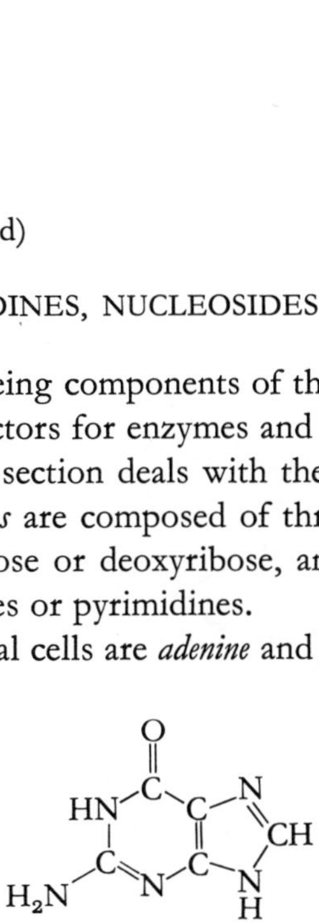   Guanine

The most common *pyrimidines* are *cytosine*, *uracil*, and *thymine*:

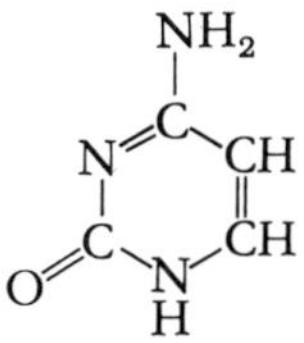

Cytosine 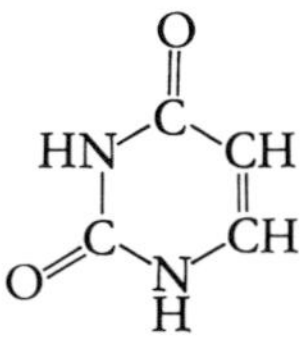   Uracil 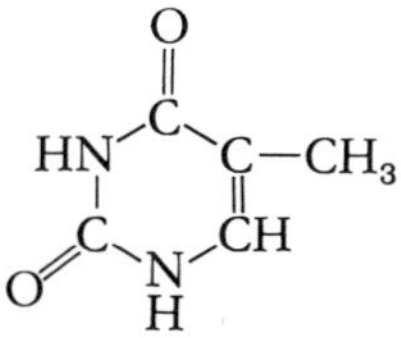   Thymine

Minute amounts of methyl derivatives and other modifications of these purines and pyrimidines are found in specific types of nucleic acids.

Compounds of nitrogenous bases linked to D-ribose (or deoxyribose) are called *nucleosides*; they differ from nucleotides in that they lack the phosphate group (see page 51). The nucleosides of the five nitrogenous bases are:

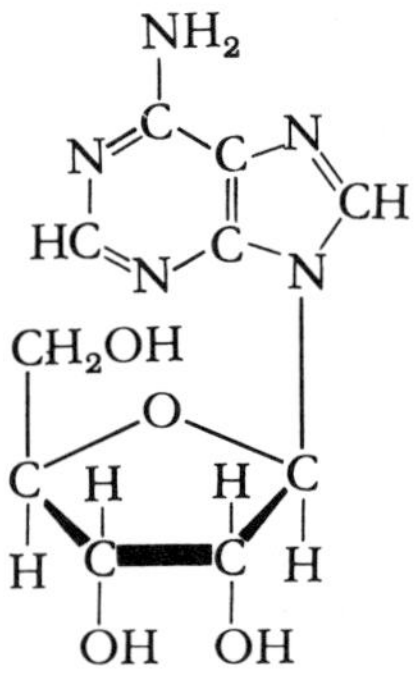

Adenosine 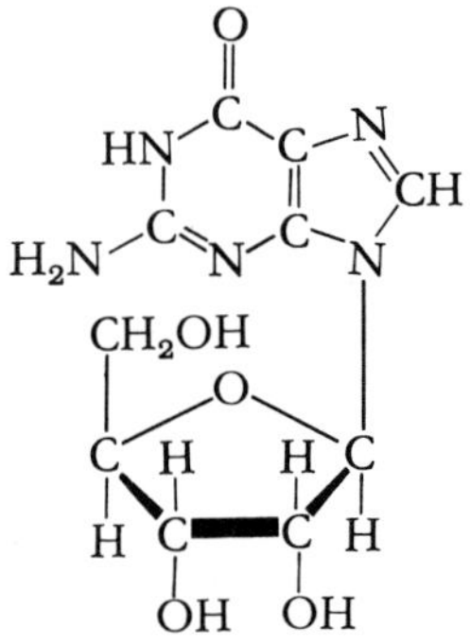   Guanosine

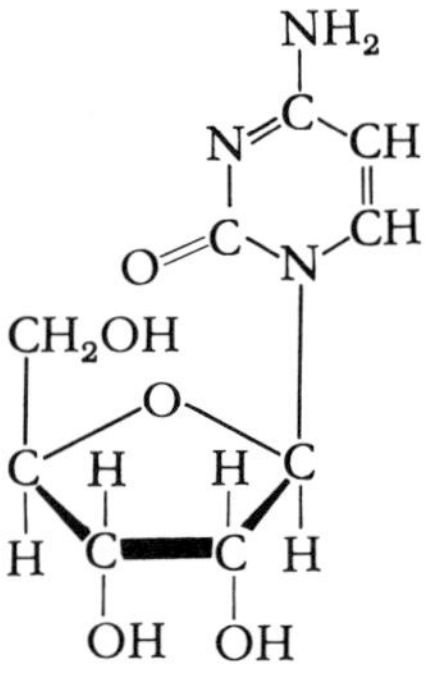

Cytidine 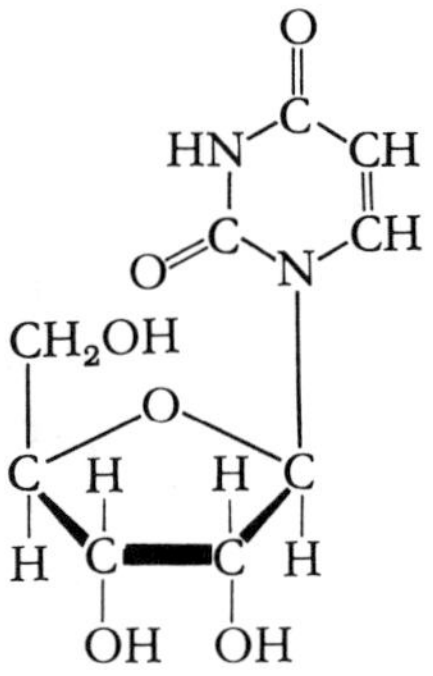   Uridine 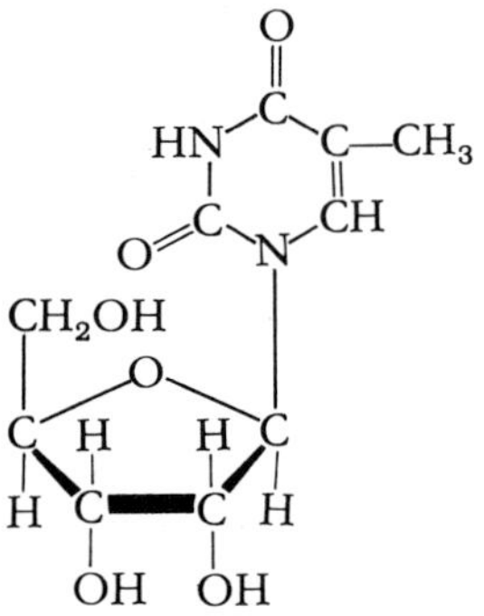   Thymidine

For purposes of nomenclature, the atoms of the rings in purine and pyrimidine nucleosides are numbered as follows:

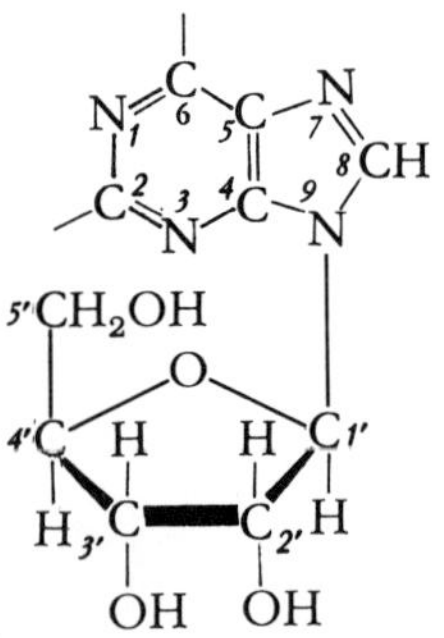

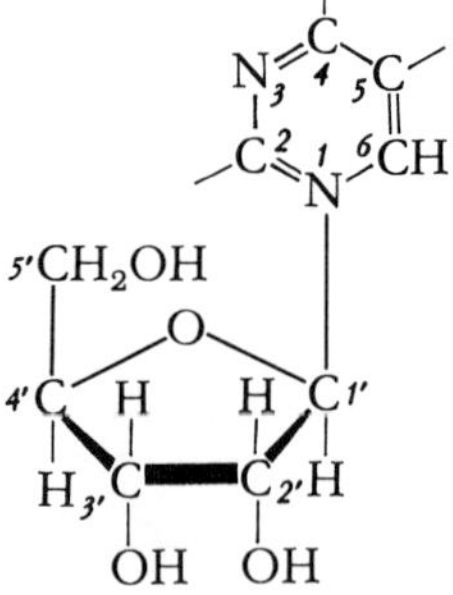

Note that the atoms in the ribose units are designated by primes, i.e., 1′, 2′, 3′, etc.

Nucleotides are phosphate esters of nucleosides. These are shown in the following structures (abbreviations and alternative names are given in parentheses):

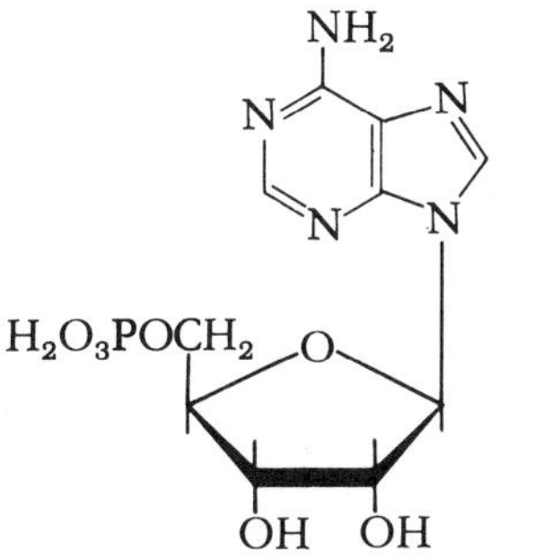

Adenosine 5'-phosphate
(adenosine monophosphate; AMP; adenylic acid)

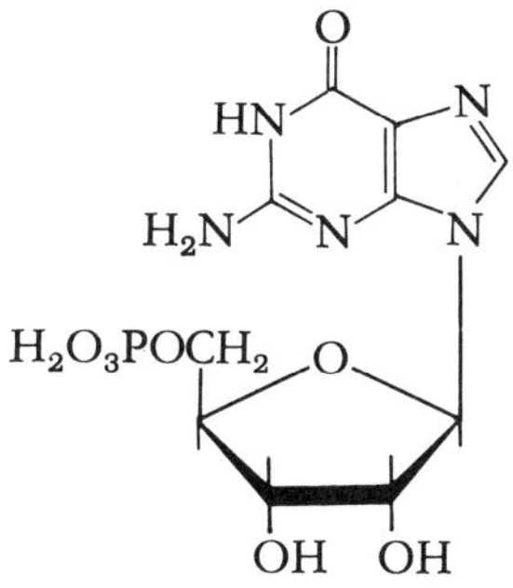

Guanosine 5'-phosphate
(guanosine monophosphate; GMP; guanylic acid)

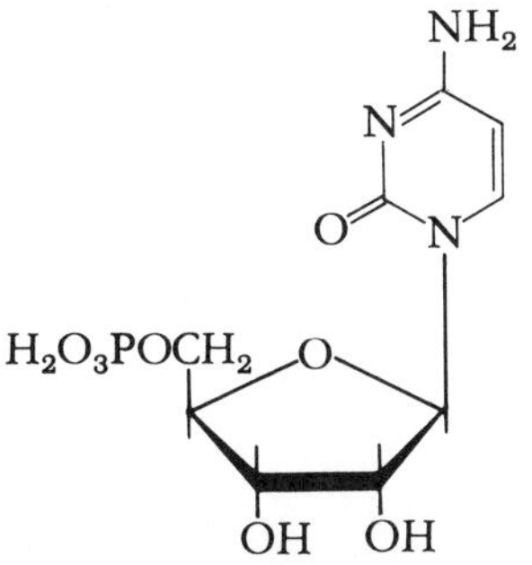

Cytidine 5'-phosphate
(cytidine monophosphate; CMP; cytidylic acid)

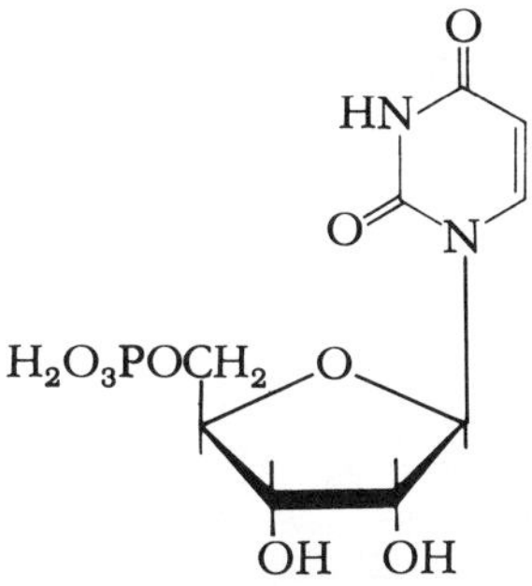

Uridine 5'-phosphate
(uridine monophosphate; UMP; uridylic acid)

Derivatives of the nucleotides in which additional phosphates are linked as anhydrides with the phosphate groups are called *nucleoside diphosphates* and *nucleoside triphosphates*. Two of the most common of these are *adenosine diphosphate* (ADP) and *adenosine triphosphate* (ATP):

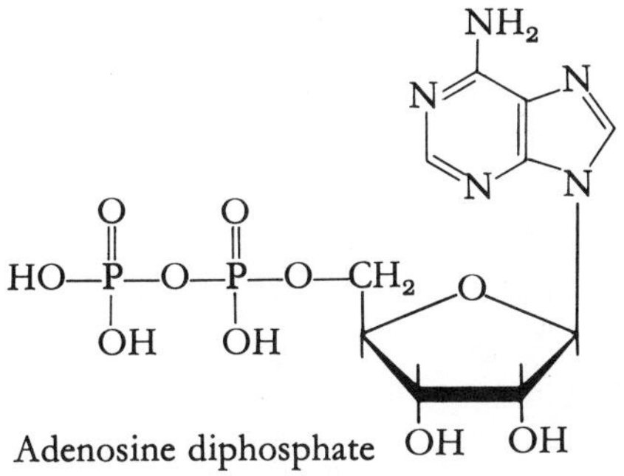

Adenosine diphosphate

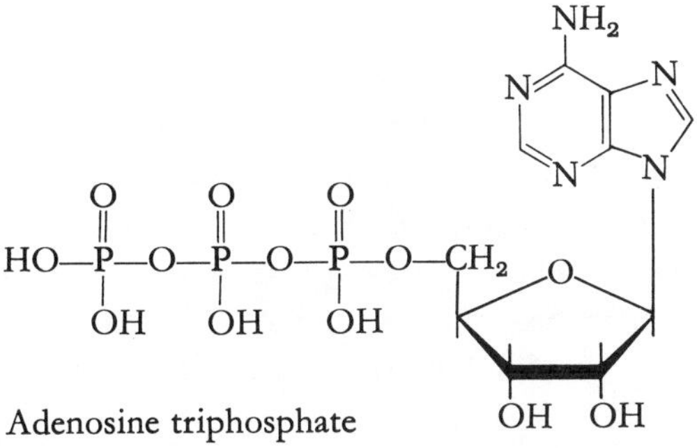

Adenosine triphosphate

The diphosphates and triphosphates of the other nucleosides are named in an analogous manner.

F. NUCLEIC ACIDS

Nucleic acids are macromolecules composed of long chains of nucleotides. The latter are linked covalently by phosphodiester bridges between the 5' position and the 3' position of adjacent nucleosides:

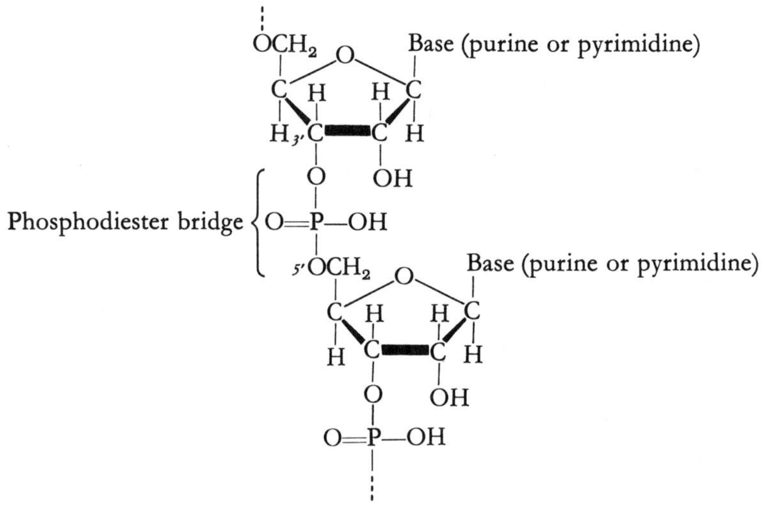

$$\text{Phosphodiester bridge} \begin{cases} \\ \\ \end{cases}$$

Nucleic acids are classified into two general types: *ribonucleic acid* (RNA) and *deoxyribonucleic acid* (DNA). The pentose component in the former is ribose, and in the latter it is 2-deoxyribose. The nitrogenous bases in RNA are adenine, guanine, cytosine, and uracil; DNA contains thymine instead of uracil. (More detailed characteristics of DNA and RNA will be described in Chapter 11.)

Three classes of RNA occur in cells. These are categorized as *ribosomal RNA* (rRNA), *messenger RNA* (mRNA), and *transfer RNA* (tRNA). Each RNA class comprises a number of different, individual molecular entities. For example, there are at least as many tRNAs within the cell as there are amino acids; in fact, there appear to be over fifty different forms. The number of forms of mRNA is not known; however, in animal cells, it is probably in the thousands. As far as rRNA is concerned, eukaryotic cells have four types. These are categorized according to their sedimentation constants, that is, 5S, 7S, 18S, and 28S. Ribosomes consist of about 60% RNA and 40% protein. The ribosomes are also characterized by their sedimentation constants. Thus, in eukaryotic ribosomes, 18S RNA is associated with the 40S ribosomal subunit and 28S RNA is part of the 60S ribosomal subunit; together, these subunits constitute the total ribosome.

The base sequence in nucleic acids or oligonucleotides and the attachment sites of the phosphates are described according to several systems of notation. The ribonucleosides adenosine, guanosine, cytidine, uridine, and thymidine are denoted by A, G, C, U, and T, respectively. The deoxyribonucleosides are indicated as dA, dG, dC, or dT. The position of the phosphate on an individual nucleoside is shown by the order in which the symbols for the phosphate (p) and the nucleoside are written. Thus, pA indicates that the phosphate is on the 5' position of adenosine, whereas Ap means that it is on the 3' position. Accordingly, pApU means that adenosine 5'-phosphate is linked through another phosphate (on carbon-3' of its ribose) to the 5' position of uridine. The same dinucleotide is also denoted as pA—U, that is, the phosphodiester link may be indicated by a dash:

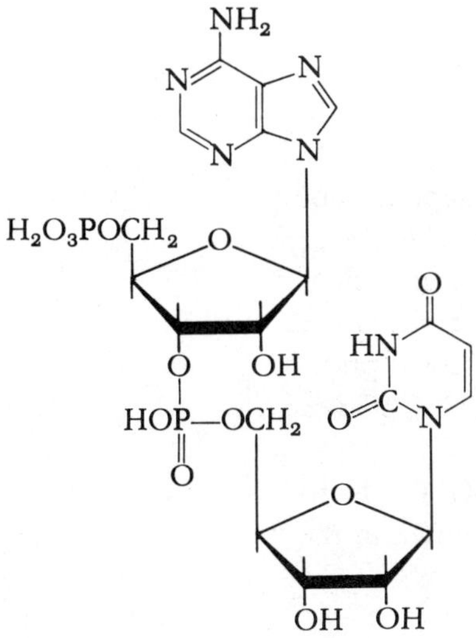

pApU or pA—U

Similarly, pApUpGpC indicates that there are phosphate bridges between the 3' and 5' positions of adenine and uridine, uridine and guanosine, and guanosine and cytidine; in addition, adenosine contains phosphate on position 5':

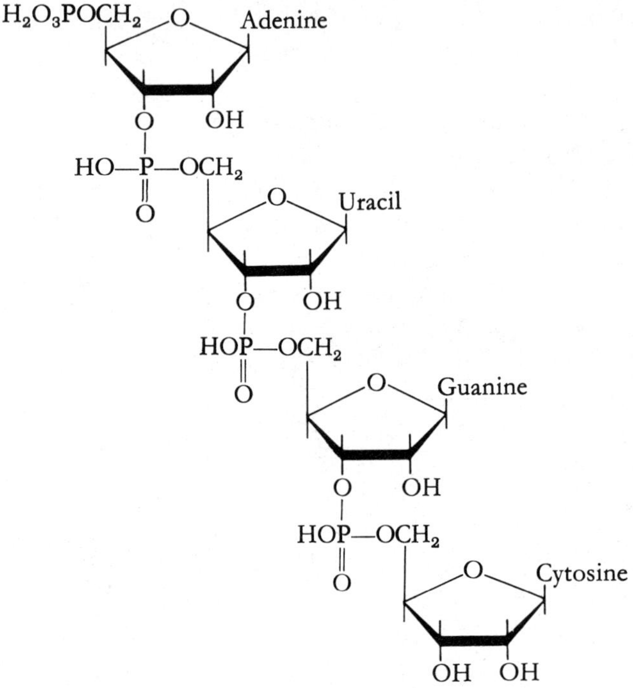

pApUpGpC or pA—U—G—C

In this structure, position 5' of adenosine and position 3' of cytidine are not involved in phosphodiester bridges. Adenosine is thus termed the *5' terminus*; and cytidine, the *3' terminus*. According to this notation system, the nucleoside on the left is the 5' terminus, and the one on the extreme right is the 3' terminus. If, in the structure just shown, there were an additional phosphate on position 3' of the cytidine, it would be denoted as pApUpGpCp or pA—U—G—Cp.

Another way of representing the oligonucleotide pApUpGpCp is to indicate the pentose by a vertical line. The 3' position is then shown in the center of the line and the 5' position at the lower end:

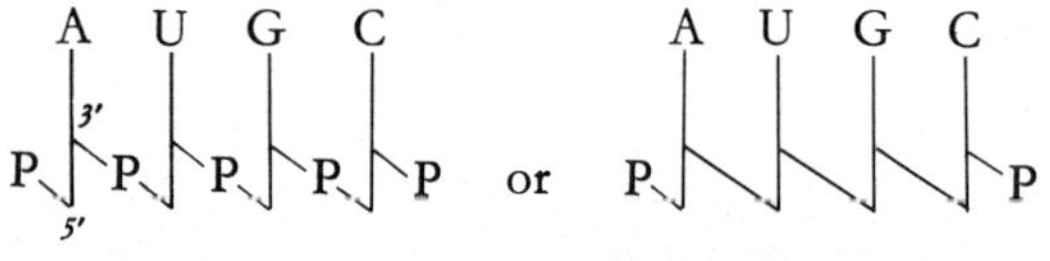

Nucleic acids are characterized by the numbers of each of the bases and their sequence. Hence, although the macromolecule is composed of comparatively few types of units, variations in the sequences of these components allow for large numbers of different structures. Moreover, as the molecular weight of the nucleic acid increases, the number of nucleotide units is increased and thus the possibilities for differences in the sequence are increased. The uniqueness of the sequences confers characteristic physical, chemical, and biologic properties on the nucleic acids. More specific structural details of nucleic acids will be described in conjunction with discussions of their biologic function (see Chap. 11).

G. LIPIDS

Compounds categorized as *lipids* include a wide variety of substances with different types of structures; the basis of their classification in a single group is their solubility in organic solvents (such as carbon tetrachloride, chloroform, or methanol) and their relative insolubility in water. The structures and chemistry of specific lipids will be described in detail in conjunction with discussions of their metabolism. This section is limited to a general classification and definition of the different types of lipids.

1. *Free Fatty Acids*

Free fatty acids are composed of carboxylic acid with long hydrocarbon chains, or RCOOH, where the R group may be a saturated, monounsaturated, or polyunsaturated chain (Chap. 3).

2. *Glycerol Esters of Fatty Acids*

The compounds in this class are esters of glycerol with either one, two, or three fatty acids; these are termed *monoglycerides*, *diglycerides*, and *triglycerides*, respectively:

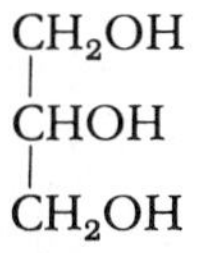

$$
\begin{array}{llll}
\text{CH}_2\text{OH} & \text{CH}_2\text{OCR} & \text{CH}_2\text{OCR}_1 & \text{CH}_2\text{OCR}_1 \\
| & | & | & | \\
\text{CHOH} & \text{HOCH} & \text{R}_2\text{COCH} & \text{R}_2\text{COCH} \\
| & | & | & | \\
\text{CH}_2\text{OH} & \text{CH}_2\text{OH} & \text{CH}_2\text{OH} & \text{CH}_2\text{OCR}_3 \\
\text{Glycerol} & \text{Monoglyceride} & \text{Diglyceride} & \text{Triglyceride}
\end{array}
$$

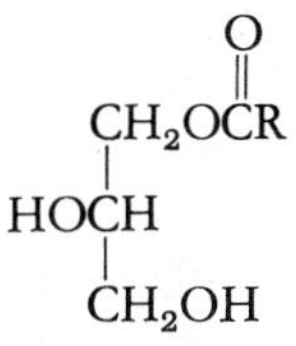

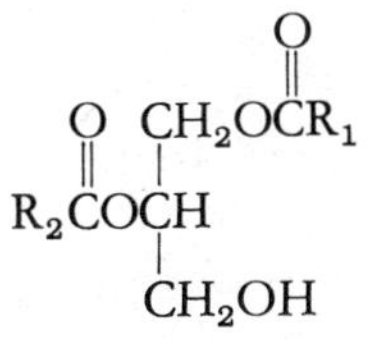

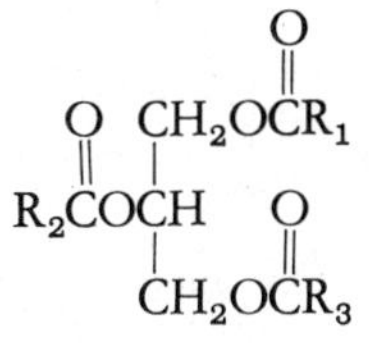

(carbonyl C=O groups on the acyl substituents)

Specific aspects of glyceride structure are discussed in Chapter 3. The primary function of the glycerides in the body is in the storage of fatty acids.

3. *Glycerol Phospholipids*

This class of lipids, also known as *phosphoglycerides*, comprises various derivatives of phosphatidic acid:

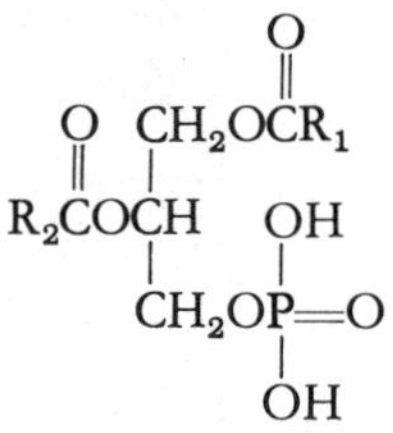

$$\begin{array}{l}\quad\quad\quad\; O\\ \quad\quad O\;\; CH_2OCR_1\\ \;\;\|\quad\;\;|\\ R_2COCH\;\; OH\\ \quad\quad\;|\quad\;|\\ \quad\quad CH_2OP=O\\ \quad\quad\quad\;|\\ \quad\quad\quad\;OH\end{array}$$

Phosphatidic acid

In phosphatidic acid, two hydroxyls of the parent glycerol are esterified with fatty acids (saturated or unsaturated) and the third hydroxyl is linked to a phosphate. If the phosphate is bound covalently to choline, the resultant phospholipid is classified as a phosphatidylcholine, or *lecithin*.

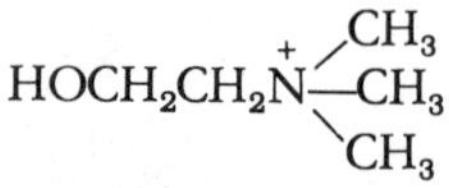

$$HOCH_2CH_2\overset{+}{N}\!\!\diagdown\!\!\!\!\diagup\!\!\begin{array}{l}CH_3\\CH_3\\CH_3\end{array}$$
Choline

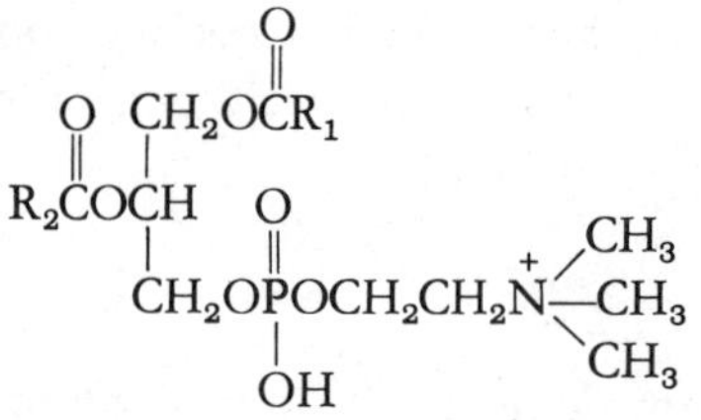

$$\begin{array}{l}\quad\quad\quad\; O\\ \quad\quad O\;\; CH_2OCR_1\\ \;\;\|\quad\;\;|\\ R_2COCH\;\; O\\ \quad\quad\;|\quad\;\|\quad\quad\quad\quad\quad CH_3\\ \quad\quad CH_2OPOCH_2CH_2\overset{+}{N}\!\!\diagdown\!\!\!\!\diagup CH_3\\ \quad\quad\quad\;|\quad\quad\quad\quad\quad\quad\quad CH_3\\ \quad\quad\quad\;OH\end{array}$$
Lecithin

Alternatively, the phosphatidic acid may be esterified with serine or ethanolamine. The latter compounds are termed *cephalins*:

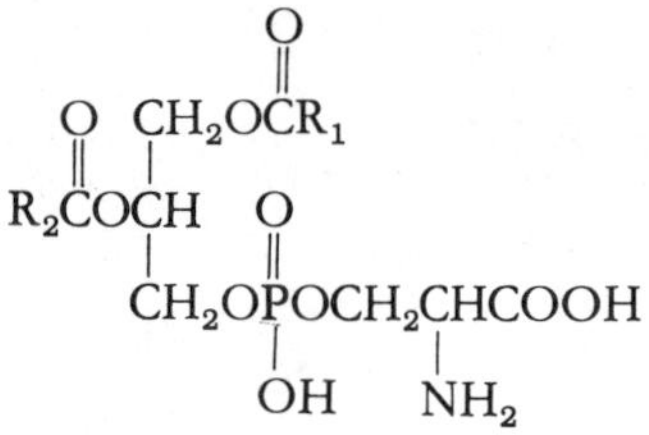

$$\begin{array}{l}\quad\quad\quad\; O\\ \quad\quad O\;\; CH_2OCR_1\\ \;\;\|\quad\;\;|\\ R_2COCH\;\; O\\ \quad\quad\;|\quad\;\|\\ \quad\quad CH_2OPOCH_2CHCOOH\\ \quad\quad\quad\;|\quad\quad\quad\;|\\ \quad\quad\quad\;OH\quad\;\;NH_2\end{array}$$
Phosphatidylserine

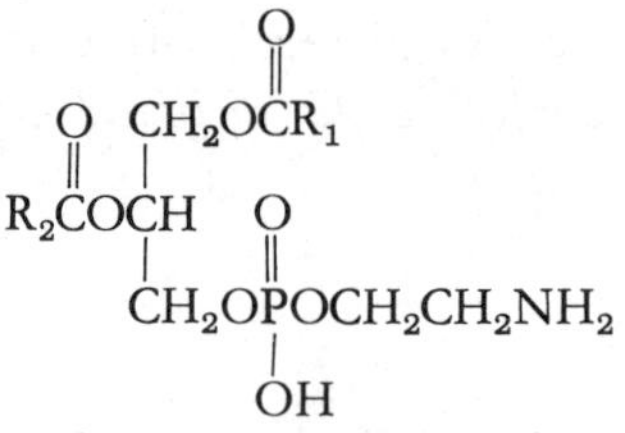

$$\begin{array}{l}\quad\quad\quad\; O\\ \quad\quad O\;\; CH_2OCR_1\\ \;\;\|\quad\;\;|\\ R_2COCH\;\; O\\ \quad\quad\;|\quad\;\|\\ \quad\quad CH_2OPOCH_2CH_2NH_2\\ \quad\quad\quad\;|\\ \quad\quad\quad\;OH\end{array}$$
Phosphatidylethanolamine

In another type of glycerol phospholipid, the phosphate is linked to inositol instead of one of the nitrogenous bases shown in these structures:

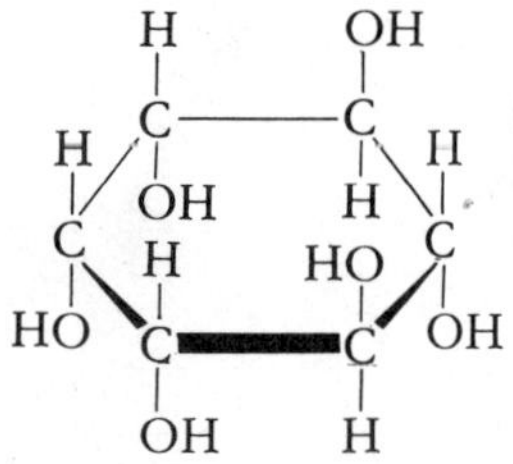

Inositol

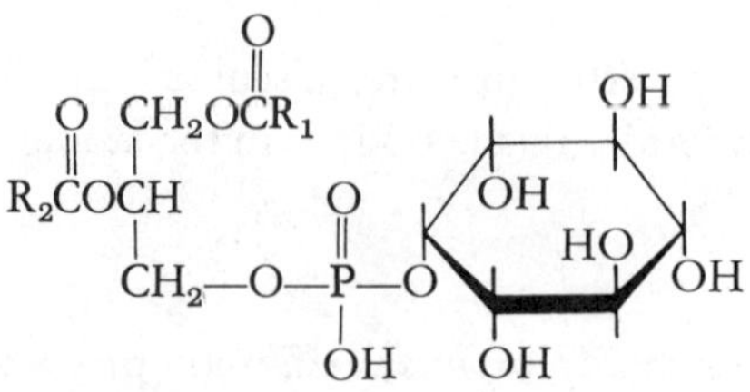

Phosphatidylinositol

4. Sphingolipids

Sphingolipids are derivatives of a long-chain, unsaturated, amino alcohol called *sphingosine*:

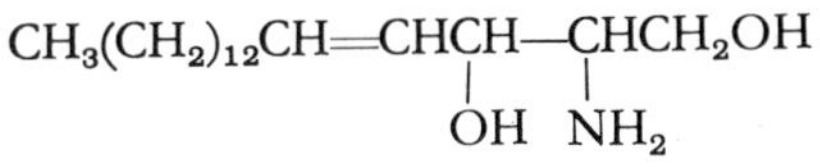

$$CH_3(CH_2)_{12}CH=CHCH(OH)-CH(NH_2)CH_2OH$$

Sphingosine

The sphingolipids are grouped into sphingomyelins, cerebrosides, and gangliosides.

The *sphingomyelins* may be considered phospholipids, similar to lecithins and cephalins, in which glycerol is replaced by sphingosine. In the representative structure that follows, note that the amino group is linked as an amide to a long-chain fatty acid. Such amides of sphingosine are termed *ceramides* (N-acylsphingosine); thus, a sphingomyelin is a phosphorylcholine (or phosphorylethanolamine) derivative of a ceramide:

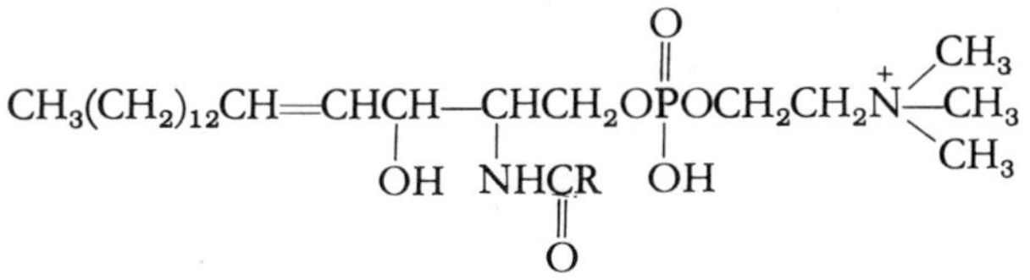

$$CH_3(CH_2)_{12}CH=CHCH(OH)-CH(NHCR{=}O)CH_2OP(=O)(OH)OCH_2CH_2\overset{+}{N}(CH_3)_3$$

Sphingomyelin

The sphingomyelins are found in relatively large amounts in brain, liver, and kidney.

In *cerebrosides*, a ceramide is linked to a sugar, e.g., galactose or glucose:

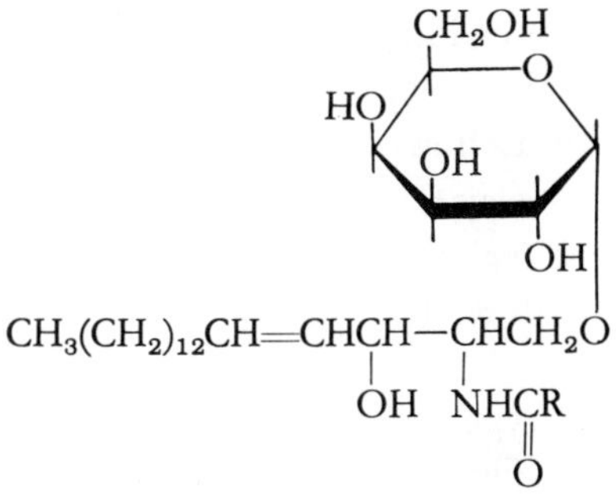

$$CH_3(CH_2)_{12}CH=CHCH(OH)-CH(NHCR{=}O)CH_2O-\text{galactose}$$

Galactocerebroside

As the name implies, cerebrosides are found in brain. However, like sphingomyelins, they are also components of numerous other tissues.

Gangliosides are structurally the most complex of the sphingolipids. They are composed of amides of sphingosine (ceramide), hexose units (glucose or galactose), hexosamines (N-acetylglucosamine or N-acetylgalactosamine), and sialic acid (e.g., N-acetylneuraminic acid). Gangliosides are found in nerve tissue, spleen, and erythrocytes. Various gangliosides are involved in genetic diseases (e.g., Tay-Sachs disease) and are discussed in the description of these disorders (Chap. 5).

5. *Sterols and Steroids*
Sterols and steroids are fundamentally composed of three six-membered rings and one five-membered carbon ring:

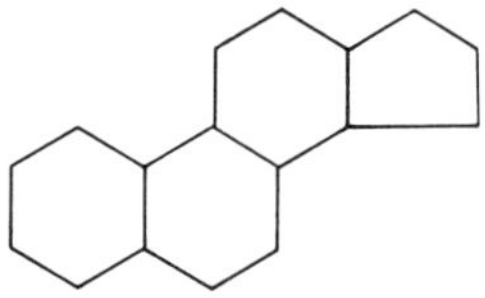

Steroid nucleus

Various combinations of unsaturated bonds and substituent groups on different carbon atoms of the rings provide a large number of compounds. These include cholesterol, a wide variety of hormones, and bile acids. The details of individual structures will be discussed in later chapters dealing with the functions and metabolism of these compounds (Chaps. 5, 6, and 14).

1.3. Membranes and Transport
A. MEMBRANE STRUCTURE
Membranes of cells and intracellular organelles are highly organized systems composed mainly of proteins and lipids. In addition to serving as structures that enclose the cells, membranes function to control the movement of ions and compounds into and out of cells. Membranes also act as barriers that discriminate and select the materials that can pass into cells or intracellular structures. Another vital activity of membranes is to function as a relay for transmitting information from the extracellular environment to intracellular systems. In this way, certain hormones can produce their effect within the cell without passing through the cell membrane.

Although the structures of membranes are not the same in all types of cells, they are similar with respect to a number of features. They are films of double layers (bilayers) of lipids with areas in which proteins are enmeshed. The molecules of lipids and proteins are held together by non-covalent interactions. The total arrangement of a membrane is highly specific both in structure and in function.

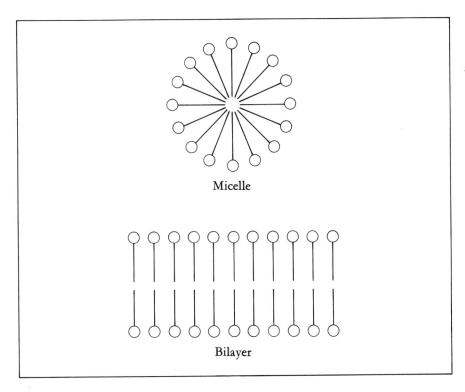

Fig. 1-21. *Arrangement of lipid molecules into micelles and bilayers. The circles represent hydrophilic groups; the straight lines, hydrophobic ones.*

The principal lipid components of membranes are phospholipids, sphingolipids, and cholesterol. The structures of the first two types were discussed in the previous section; that of cholesterol is depicted as follows:

Cholesterol

An important feature of these lipids is that they are composed of hydrophobic (water-insoluble) hydrocarbon sections and hydrophilic (water-soluble) units. The latter include charged units (e.g., phosphate or amino groups) and uncharged hexose and hydroxyl groups. In water, such compounds orient themselves so that only the hydrophilic section is exposed to the water. The hydrophobic portions of individual molecules tend to contact each other; this is accomplished by either arrangement into micelles or the formation of bilayers (Fig. 1-21). The prevalent arrangement of lipids in membranes is

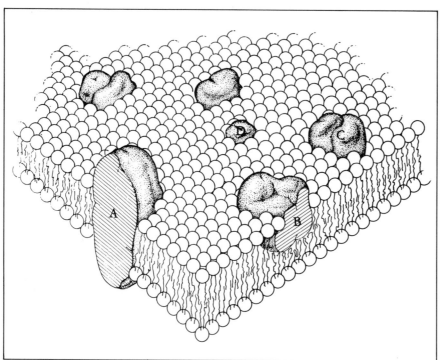

Fig. 1-22. Position of proteins in the structure of membranes. Protein A extends through entire thickness. Proteins B and C occur in only a portion of the bilayer. Protein D is on the periphery of the membrane. (Reprinted with permission from S. J. Singer and G. L. Nicolson, Science 175: 720–731, 1972. Copyright © 1972 by the American Association for the Advancement of Science.)

the bilayer system, in which hydrophilic units are exposed to the aqueous media both outside and within the cell.

Proteins are interspersed in various ways in different sections of membranes. For example, the protein may extend through the entire thickness of the membrane, or it may be located only in part of the total film. In addition, certain proteins are bound only to the periphery of the membrane. These different types of arrangements are shown in Figure 1-22.

The general concept of the membranes described is termed the *fluid-mosaic model*. The overall picture is one of a lipid-bilayer fluid or matrix that is interspersed with a mosaic of numerous globular proteins. The proteins in the lipid matrix can undergo a certain degree of lateral movement, but penetration in the direction perpendicular to the surface of the membrane is highly restricted.

Differences in the structure of membranes and the reactions that occur on them can be ascribed to the relative amounts of membrane proteins and their specific binding or catalytic properties. The inner and outer surfaces of membranes may differ with respect to the nature and amount of proteins, which accounts for membrane asymmetry. Another factor that contributes to membrane asymmetry is the presence of glycoprotein oligosaccharides on the outer surface of many membranes.

B. TRANSPORT THROUGH MEMBRANES

The lipid sections of membranes are essentially impermeable to water-soluble materials, i.e., ions or polar substances. The movement of such substances across cell membranes depends on the protein components. The processes by which compounds flow across membranes are classified as *passive transport* and *active transport*. In the former case, transport requires a difference between the concentrations of the mobile substance on the two sides of the membrane. Thus, the substance will move only toward a compartment in which its concentration is lower than in the one in which it existed originally. Expressed in another way, such movement only occurs down a concentration gradient. A passive transport system does not require any input of energy; that is, from a thermodynamic point of view, the reaction is spontaneous.

In active transport, the material involved is transferred into an environment in which its concentration is higher than that of its original medium. As would be expected, this type of transport requires energy and therefore must be linked to an energy-yielding system. Unlike passive transport, which resembles a diffusion system and does not discriminate according to the direction of movement, active transport is a one-directional process. In other words, the compound or ion that is being transported is transferred across the membrane only in one direction. Thus, the active transport system that transfers potassium and sodium ions across erythrocyte membranes, for example, drives the former into the cell and pumps the latter out of the same cell.

The protein component of the membrane binds the substances that are being transferred, and it is involved in the mechanisms of their transport. Even in cases of passive transport, most of the diffusion activities are mediated by membrane proteins. These protein-mediated processes are termed *facilitated diffusion* (as opposed to purely passive diffusion). Although certain types of passive transport may be visualized as movement that occurs through pores in the membrane, most mechanisms involve some type of facilitated diffusion.

C. OSMOTIC PRESSURE, MEMBRANE EQUILIBRIA, AND THE GIBBS-DONNAN EQUILIBRIUM

Many processes involving the movement of water and ions through biologic membranes can be interpreted in terms of the principles of the physical chemistry of solutions. For this purpose, the cell may be visualized as an enclosed compartment, separated from its environment by a selective or semipermeable membrane. Various ions, water, and low-molecular-weight solutes can pass through the membrane, but macromolecules, such as proteins and polysaccharides, are confined within the cell. If a chamber containing a protein is separated by a semipermeable membrane from another chamber containing only water, the water tends to flow into the compartment with the protein. The amount of pressure required to arrest the influx of water into the chamber with the protein is termed the *osmotic pressure*. This movement of water is not only a property of protein solutions, but it is also an aspect of a more general phenomenon, namely, that water is transferred spontaneously from a solution of lower concentration of solute to one with a higher concentration. Thus, if cells are placed in a medium that is more dilute than that within the cell, water from the surroundings will flow into the cell; when the medium has a higher solute concentration than that which is present intracellularly, the water will flow out of the cell.

Another characteristic of solutions concerns the movement of solutes. When a concentrated solution is placed in contact with a dilute solution, solute will diffuse from the former to the latter. This will also occur when a semipermeable membrane is inserted between the two solutions; the solute will diffuse across the membrane until equilibrium is attained. If the solute is sodium chloride, for example, the concentrations of each ion will be the same on both sides of the membrane when equilibrium is reached.

A more complex situation applies if, in addition to the diffusible salt, another charged, non-diffusible component is present in one of the compartments. For example, if a solution of NaCl is separated by a semipermeable membrane (Fig. 1-23) from a solution of the sodium salt of a protein, the concentration of NaCl on each side of the membrane, at equilibrium, will not be the same. This condition arises because Na^+ and Cl^- must diffuse in pairs in order to preserve electrical neutrality. However, since Na^+ is also the counterion for protein, the compartment with the protein will contain a greater amount of Na^+ than the one in which protein is absent. Similarly, the latter compartment will have a relatively higher concentration of Cl^-. The actual concentrations of diffusible ions are related as shown in Figure 1-23. The molar concentrations of Na^+ and Cl^- are related by the equation:

$$[Na^+]_1[Cl^-]_1 = [Na^+]_2[Cl^-]_2$$

Therefore,

$$\frac{[Na^+]_1}{[Na^+]_2} = \frac{[Cl^-]_2}{[Cl^-]_1}$$

It can also be shown that in the presence of several diffusible electrolytes (e.g., NaCl, KCl, $NaHCO_3$), the molar ratios of the ions will be:

$$\frac{[Na^+]_1}{[Na^+]_2} = \frac{[K^+]_1}{[K^+]_2} = \frac{[Cl^-]_2}{[Cl^-]_1} = \frac{[HCO_3^-]_2}{[HCO_3^-]_1}$$

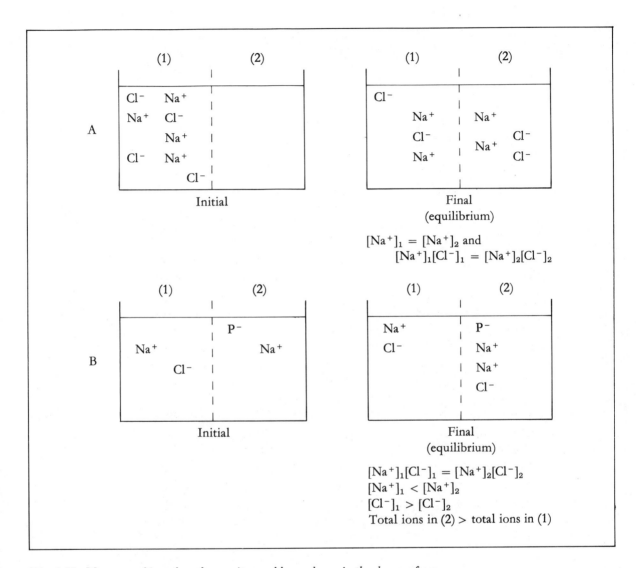

Fig. 1-23. *Movement of ions through a semipermeable membrane in the absence of nondiffusible protein (A) and in the presence of protein P (B), when initially chamber 1 contains only NaCl solution and chamber 2 contains only a solution of the sodium salt of the protein.*

In the condition in which protein is present, shown in Figure 1-23, analysis will demonstrate that at equilibrium (if there is no change in volume), the total amount of diffusible ions in the solution containing the protein is greater than the amount in the compartment without the protein. As a result, the protein-containing solution will have a higher osmotic pressure than the one that does not. This phenomenon is of special importance in salt and fluid exchange between cells and their environment. It is of particular physiologic significance and clinical interest in understanding the exchange of electrolytes and water between erythrocytes and plasma and between blood plasma and the interstitial fluids.

1.4. Biochemistry of Metabolism
A. OVERVIEW AND AIMS

An individual cell or even an entire higher organism may be viewed as an extremely intricate, yet precisely integrated, system of degradative and synthetic biochemical reactions. A cell is a dynamic entity in which numerous substances are constantly being utilized or broken down and new molecules are being synthesized to compensate for the losses or to satisfy specific requirements. *Metabolism* is the combination or sum total of these activities. Although metabolic processes are interrelated and interdependent, it is heuristically helpful to consider certain aspects individually. One of these aspects concerns the degradation of nutrients and various components of cells and tissues, e.g., carbohydrates, proteins, lipids, and nucleic acids. Another facet of metabolism involves the biosynthesis of the constituents of the organism. The synthetic processes are categorized as *anabolism*; and the degradative, as *catabolism*. A third consideration, which involves both anabolism and catabolism, is the derivation and utilization of energy. As a rule, the synthesis of large molecules from smaller ones requires an input of energy, whereas degradative and oxidative reactions release energy. An integral aspect of metabolism is thus the mechanism whereby energy is delivered and channeled to serve the requirements of the organism.

Both the degradation and the synthesis of cellular components take place by a series of steps or pathways. These sequential reactions and the properties of the enzymes that catalyze each transformation are dealt with in the study of *intermediary metabolism*. An auxiliary concern of intermediary metabolism is the mode of regulation of various reactions and pathways. This involves the characteristics of individual enzymes, the availability or concentration of cofactors and substrates, the effects of compartmentalization within the cell, the effect of hormones, and the contribution of genetic factors.

A knowledge of intermediary metabolism is important in a rational approach to medicine, since many diseases are the results of aberrations of specific metabolic reactions. A number of disorders can actually be diagnosed on the basis of metabolic or enzymatic findings. In many situations, although the exact cause of the disease has not been clarified, certain of its effects can be predicted on the basis of the knowledge of intermediary metabolism. The mode of treatment in such situations must take into account various metabolic interactions and the problems that may arise. The elucidation of normal metabolism and of the abnormalities caused by metabolic defects is the ultimate goal of investigations into intermediary metabolism. In many instances, this aim has been realized, but a large amount of work remains to be done.

B. EXPERIMENTAL APPROACHES

In order to follow the metabolic pathways that have been proposed for various nutrients and cellular components (as well as to understand the reservations regarding their conclusiveness), it is useful to be familiar with some of the general procedures utilized for their study.

Experiments on metabolism with intact animals—i.e., where various substances are administered either by feeding or by injection—provide data that are important for defining nutritional requirements. Such studies yield limited information on intermediary metabolism because, in most situations, only the final degradation products can be identified, and they are usually simple compounds that shed little light on the intermediate steps in the pathway. Furthermore, a number of different types of substances yield the same terminal products; for example, both carbohydrates and triglycerides are converted to carbon dioxide and water. Another limitation in the use of intact animals is that the interpretation of results in terms of cellular reactions may be complicated by unknown effects of hormones or other regulatory factors, differences in permeability of various tissues to the administered material, and variations due to the different roles played by specific organs. Despite these considerations, however, a considerable amount of information has been obtained through experiments with whole animals. Such procedures are especially important when the aim is to elucidate the disposition of various substances by the total organism and the effect of various conditions on specific modes of metabolism. Thus, one can learn comparatively little by feeding carbohydrates to an animal and then analyzing the urine for breakdown products. However, if it is desired to compare the enzyme levels in various tissues or the lipids in blood under normal conditions and during starvation, it is necessary to start with an intact organism. The use of intact animals also yields a great deal of information when experiments are conducted with radioactively labeled isotopes.

A somewhat less complex mode of investigation involves procedures that deal with metabolism in a specific organ. These are generally carried out by perfusion experiments in which specific substances are introduced through a circulating medium (blood, plasma, or an artificial mixture). The fluid that emerges can then be studied with respect to the products formed. Investigations of this type are employed especially for studies of the liver, heart, and kidney, although other organs may also be studied by perfusion methods. The principal limitations in interpreting the results are due to the artificial or nonphysiologic conditions of the specific experiments and the barriers to permeability that may minimize cellular metabolism of the substance under study.

More direct investigations on the cellular disposition of various substances and on metabolic intermediates are performed by incubating thin slices or homogenates of tissues. The assumption is that the enzymatic transformations that operate in the intact tissue or whole animal will also occur under the conditions of the experiment. This assumption generally requires additional experimental substantiation.

Since many fundamental reactions of intermediary metabolism are common to many organisms, including unicellular microorganisms, a great deal of information about intermediary metabolism in higher animals is obtained by studies with bacteria or other simple cellular systems. Naturally, such results cannot be extrapolated directly to higher animals; however, investigations of this type serve to suggest possible pathways.

A further refinement involves investigations of metabolic reactions by individual cellular components. The most general method for separating cell fractions is *differential centrifugation*. In a typical experiment, the tissue is homogenized at low temperature with 0.25 M sucrose. The mixture, which contains the disrupted cells, is strained to separate insoluble tissue components (connective tissue, blood vessels, and so on), and the filtrate is centrifuged at a comparatively low speed (600–1000 × g for 10 minutes). The fraction that settles to the bottom of the tube contains *nuclei* and possibly some unbroken cells. Subsequent centrifugation of the supernatant fluid at 15,000 × g for about 5 minutes yields another precipitate, which is composed of *mitochondria*, *lysosomes*, and *microbodies*. The fluid fraction is separated and centrifuged at 100,000 × g for an hour. The resultant precipitate is termed the *microsomal fraction*; this is composed of *ribosomes, endoplasmic reticulum,* and *Golgi bodies*. The remaining supernatant fluid is called the *soluble fraction* (Fig. 1-24).

Experiments with individual fractions obtained by differential centrifugation revealed that the enzymes for certain reactions or for complete metabolic pathways may be localized in specific cellular fractions or organelles. Thus, it has been found that the enzymes are not distributed in a homogeneous manner; rather, they are compartmentalized in specific areas or organelles. This physical separation of the enzymes from each other is important in that it allows cellular control of the mechanisms in which the enzymes participate. Furthermore, since cellular organelles are not equally permeable to all metabolic intermediates, the reactions that may occur with any given substance are subject to restrictions that are set by such physical isolation.

C. ISOTOPES

Molecules labeled with specific isotopes are employed in biologic research for essentially two reasons: the assay procedures for isotopes are extremely sensitive and allow for detection and quantitative analysis of minute quanti-

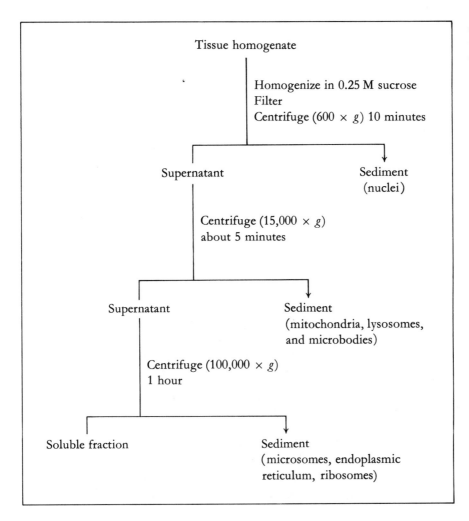

Fig. 1-24. *Fractionation of cellular organelles by differential centrifugation.*

ties of material, and second, metabolic studies with labeled substances can yield information on their fate when administered, even though such substances are already present in the organism. Thus, the fate of a specific molecule can be followed during various intermediary states, and similar components, which were originally in the tissues, will not interfere with the analysis.

Isotopes are atoms of the same element (i.e., they have the same atomic number) that differ in atomic mass. This difference in the atomic mass of isotopes is due to a difference in the number of neutrons in the nucleus. Since isotopes of the same element have identical numbers of protons and extranuclear electrons, their chemical properties are the same. Among the stable isotopes utilized in biochemical investigations are ^2H, ^{13}C, ^{18}O, and ^{15}N.

Table 1-5. Properties of Radioactive Isotopes

Isotope	Radiation	Half-life	Energy*
^{14}C	Beta	5700 years	0.156
^{3}H	Beta	12.1 years	0.0185
^{35}S	Beta	87.1 days	0.169
^{32}P	Beta	14.3 days	1.712
^{131}I	Beta; gamma	8 days	0.68; 0.72
^{60}Co	Beta; gamma	5.3 days	0.31; 1.33
^{24}Na	Beta; gamma	15 hours	1.39; 2.89
^{59}Fe	Beta; gamma	45 days	0.46; 1.30

* Energy of the emitted radiation measured in million electron volts (mev).

In addition to mass differences, a number of isotopes are radioactive. In such atoms, the nucleus has an unstable configuration and it consequently undergoes decomposition and rearrangement. This is accompanied by emission of particles from the nucleus. The unstable isotopes disintegrate at a specific rate, which is such that only a fraction of the total number of atoms will disintegrate in a specific amount of time. This decay rate may be expressed as the *half-life*, or the amount of time that it takes for half of the atoms of the radioactive isotope to disintegrate. The decay rate is characteristic of the specific isotope and is not affected by experimental conditions such as concentration, temperature, and pressure. Some of the radioactive isotopes employed in biochemistry or medicine are shown in Table 1-5.

A common unit for the quantitation of stable isotopes is *atoms percent*, which is the number of atoms of a particular isotope present per one hundred atoms of all isotopes of that atomic species, i.e., of all isotopes with the same atomic number. *Atoms percent excess* is the excess of the isotope in question relative to its concentration in nature or in a standard sample. The mass spectrometer is employed for measuring the relative numbers of atoms of isotopes.

The unit for measuring radioactive isotopes is the *curie*; this is the quantity of isotope that yields 3.7×10^{10} disintegrations per second (dps). More workable units for ordinary biochemical research are the *millicurie* (mc), i.e., 0.001 curie, and the *microcurie* (μc), 0.001 millicurie. Measurements of radioactivity in terms of disintegrations per unit of time are made with Geiger or scintillation counters. Very often in biochemical studies, radioactivities are measured with respect to a standard or starting material. Results can then be given in counts per minute as measured with a specific counting instrument, such as a scintillation counter. Specific activities in such studies could be given, say, as counts per minute (cpm) per milligram. When the conditions are controlled, the number of counts registered by a given instrument is related to the number of disintegrations within the same time.

1.5. Intracellular Compartments

Elementary microscopic examination of all types of cells reveals that they have a high amount of internal structure. However, there are considerable qualitative and quantitative differences in these structures among different types of organisms. Moreover, in higher organisms, the cells vary in their structures from one tissue to another. Nonetheless, some generalizations at this point will serve as a basis for subsequent discussion.

Cells may be categorized as either *prokaryotic* or *eukaryotic*. The prokaryotic organisms include bacteria, blue-green algae, and other lower organisms with minimal internal organization. The prokaryotes are surrounded by a cell membrane and a cell wall composed of polysaccharides. However, they do not have intracellular, membrane-bound structures, such as nuclei or mitochondria; they do, however, have nuclear areas that contain double-stranded DNA in a helical conformation. The chromosome consists of DNA in a circular structure. In addition, prokaryotic cells contain ribosomes and storage granules suspended in a viscous matrix solution. Most of the metabolic enzymes are found in this matrix. A number of enzymes and the processes that they catalyze are associated with the cell membrane.

The eukaryotes include all the higher organisms. Their cells have characteristic internal structures or *organelles* (e.g., a nucleus, mitochondria, and so on) bounded by membranous material (Fig. 1-25).

An integral aim of modern biochemistry is to elucidate the metabolic functions of the cellular components or, conversely, to determine which chemical transformations are specific for a given organelle. Indeed, it is known that certain processes occur within confined sections of the cell. The individual metabolic processes are separated from each other by compartmentalization of the necessary enzymes and by restrictions on the movement of key substrates through the membranes of organelles. This confinement of certain reactions and metabolic pathways to specific sections in the cell serves as an important control mechanism for efficient utilization of various cellular materials.

The individual cellular organelles that are visualized by light or electron microscopy can be separated from each other by various fractionation methods. One common procedure utilizes differential centrifugation (see Fig. 1-24). Analysis of each such fraction obtained reveals that it contains the components and enzymes to carry out specific cellular reactions. For example, mitochondria have been shown to be the site of various aerobic oxidation processes and energy transformations. Glycolytic reactions are

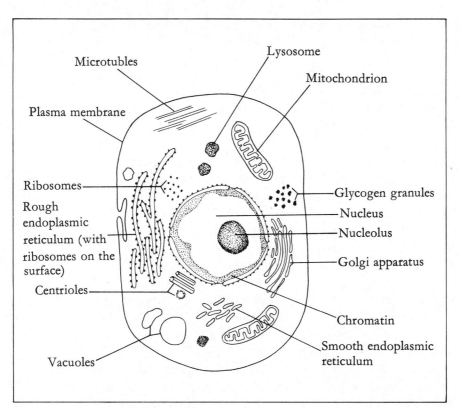

Fig. 1-25. Organelles of eukaryotic cells.

catalyzed by enzymes in the soluble fraction (cytosol). Protein synthesis occurs on components of the endoplasmic reticulum called the *ribosomes*. More specific aspects of the sites of intracellular reactions and the effects of compartmentalization will be discussed in conjunction with individual metabolic transformations and processes in the chapters to follow.

Suggested Reading

ACID-BASE CHEMISTRY
Davenport, H. W. *ABC of Acid-Base Chemistry* (5th ed.). Chicago: University of Chicago Press, 1969.

Dawes, E. A. *Quantitative Problems in Biochemistry* (5th ed.). Baltimore: Williams & Wilkins, 1969.

Montgomery, R., and Swenson, C. A. *Quantitative Problems in Biochemical Sciences*. San Francisco: Freeman, 1969.

AMINO ACIDS AND PROTEINS
Dickerson, R. E. The Structure and History of an Ancient Protein. *Scientific American* 226:58–72, 1968.

Meister, A. *Biochemistry of the Amino Acids* (2nd ed.), vols. 1 and 2. New York: Academic Press, 1965.

Neurath, H. *The Proteins* (2nd ed.), vols. 1–4. New York: Academic Press, 1966.

Perutz, M. F. The Hemoglobin Molecule. *Scientific American* 211:2–14, 1964.

Perutz, M. F. Stereospecificity of the Cooperative Effects of Hemoglobin. *Nature* 228:726–729, 1970.

CARBOHYDRATES

Pigman, W., Horton, D., and Herp, A. *The Carbohydrates*, vols. 1 and 2. New York: Academic Press, 1970.

NUCLEOTIDES AND NUCLEIC ACIDS

Davidson, J. N. *The Biochemistry of Nucleic Acids*. New York: Academic Press, 1972.

LIPIDS

Gurr, M. I., and James, A. T. *Lipid Biochemistry*. Ithaca, N.Y.: Cornell University Press, 1971.

MEMBRANES

Bretscher, M. S. Membrane Structure: Some General Principles. *Science* 181:622–629, 1973.

Fox, C. F. The Structure of Cell Membranes. *Scientific American* 226:30–38, 1972.

Singer, S. J., and Nicolson, G. L. The Fluid Mosaic Model of the Structure of Membranes. *Science* 175:720–731, 1972.

2. Characteristics and Functions of Enzymes*

2.1. Thermodynamic Principles

Although living systems and biologic phenomena are generally visualized as constituting a unique field of study, the major premise of biochemistry is that animate materials obey the same laws of physics and chemistry as inanimate matter. It is therefore expected that the chemical and energy transformations that occur within cells and higher organisms should be explicable by the laws and hypotheses of chemistry and thermodynamics. Conversely, any explanation of a process or event that conflicts with thermodynamic principles is not acceptable to the biologist. Before embarking on a discussion of enzymes and intermediary metabolism, it will therefore be useful to summarize some of the physical principles that apply to the energy changes involved in chemical transformations.

Energy is a highly abstract concept, which is best understood and measured by its effects. *Energy* is defined as the capacity to do work, and it is denoted by the symbol E. It can manifest itself in different forms, such as mechanical, thermal, chemical, electrical, or radiant energy. Thus, "mechanical energy" means the capacity to move an object from one place to another, or to bring about the contraction of muscle fibers. Thermal energy can be sensed as heat, and heat in turn can drive machines. Chemical energy is the energy inherent in the structure, bonds, and configuration of molecules; it may be released or absorbed as a result of structural rearrangements and chemical transformations.

In most situations, we do not deal with absolute energy values but with changes or transfers of energy that occur in specific processes. For example, chemical reactions may either release or take up energy. If the total energy of the products of a reaction is lower than that of the reactants, the difference in energy, ΔE, is released to the environment. Conversely, when it is desired

---

* This chapter deals with some general principles of enzyme properties and kinetics. A number of topics are not included, because they can be discussed more coherently when related to specific reactions. Among these subjects are *isozymes*, which will be covered together with lactate dehydrogenase activity (Chap. 6), and *allosteric enzymes*, which will be described in detail in conjunction with isocitrate dehydrogenase and aspartate transcarbamoylase (Chaps. 3 and 9).

to generate products with higher energy than those of the reactants, additional energy, ΔE, has to be imparted into the system.

When energy is brought into a system from the surroundings, the energy of the system increases, but the *total* energy of the system plus its surroundings does not change. This follows from the *first law of thermodynamics*, which states that the sum total of the energy of the universe remains constant, or that energy is neither created nor destroyed. It should be noted, however, that energy can be converted from one form to another; heat, for example, can be transformed to mechanical energy, or chemical energy can be released as heat.

The energy absorbed or released in different processes (ΔE) may consist of several components, as defined by the equation:

$$\Delta E = \Delta H - P\Delta V \tag{2-1}$$

The term $P\Delta V$ accounts for the energy involved in the changes of volume (ΔV) at constant pressure (P) that occur in certain processes. The term ΔH, called the change in *enthalpy*, is the energy obtainable as heat when a reaction proceeds at constant pressure. In biologic reactions, the volume change is minimal and can be neglected; thus, the value for ΔH equals that of ΔE. The enthalpy component of the energy can be visualized as composed of two constituents, i.e., the change in *free energy*, ΔG, and the product of the *entropy* change, ΔS, and the absolute temperature, T:

$$\Delta H = \Delta G + T\Delta S \tag{2-2}$$

Entropy is the component of energy that is expended or dissipated in transformations from an organized to a more randomized system. Thus, reactions that yield smaller molecules from larger, internally organized precursors generally result in an increase in entropy. This energy is not destroyed but is spread out; it cannot be channeled for specific purposes, such as performing work. The free-energy change ΔG is the energy that can be utilized for work in reactions at constant temperature and pressure. Biologic systems, unlike most machines, release, utilize, and channel energy within very narrow ranges of temperature or pressure. The change in free energy, ΔG, is therefore the primary energy factor to be considered in the understanding of cellular or tissue functions.

By convention, ΔG is negative for reactions that proceed with a release of free energy. Such processes are also termed *exergonic* and *spontaneous*. Conversely, reactions that require an input of energy—i.e., in which the products have a higher energy than the reactants—have a positive ΔG and are termed *endergonic*. The unit for ΔG in biochemical reactions is the *calorie*, which is defined as the amount of energy necessary to raise the temperature of one gram of water from 14.5 to 15.5 degrees centigrade.

The free-energy change for a reaction can be calculated by a number of procedures. The relationships generally employed in biochemistry are those among ΔG, the equilibrium constant, and the redox potential. The former relationship is the subject of the next section, and the relationship between free energy and the redox potential is discussed in Chapter 3.

2.2. Free Energy and Reversible Reactions

Most chemical reactions do not go to completion. Instead, the process reaches a point at which the system contains both products and reactants. Thus, for the reaction

$$A + B \rightleftharpoons C + D$$

A reacts with B until specific amounts of products C and D are formed. When this stage, termed *equilibrium*, is reached, the concentrations of A, B, C, and D do not change; this is because the products C and D also react with each other to form A and B. At equilibrium, the forward and reverse reactions are operating simultaneously, and the concentrations of the materials are functions of the rates for the reactions in each direction. Similarly, when the reaction is initiated between C and D, the concentrations of the four materials at equilibrium are the same as those obtained when the reaction is initiated between A and B. Reactions of this type are called *reversible reactions*.

The equilibrium concentrations of products and reactants in reversible reactions are related to a constant, termed the *equilibrium constant*, K, which is defined as the ratio of the molar concentrations of the products to those of the reactants when the reaction attains equilibrium. For the hypothetical reaction just described,

$$K = \frac{[C][D]}{[A][B]} \qquad (2\text{-}3)$$

The value of this constant depends on temperature.

When the equilibrium constant for this reaction is greater than 1, the reaction proceeds from left to right, and at equilibrium the concentration of products—i.e., the value of [C] times [D]—is greater than that of the reactants. If the product of the equilibrium concentrations of the products of the reaction is the same as that of the reactants, the equilibrium constant will be 1. Finally, if there is more reactant than product when equilibrium is reached, K will be less than 1.

If, for example, the reaction is initiated with 1 M solutions of A and B and if 10% of the reactants remain unchanged when equilibrium is reached, then:

$$K = \frac{(0.9)(0.9)}{(0.1)(0.1)} = 81$$

However, in a process in which only 5% of the reactants are utilized when equilibrium is attained,

$$K = \frac{(0.05)(0.05)}{(0.95)(0.95)} = 0.0028 = 2.8 \times 10^{-3}$$

The equilibrium constant is related to the *standard free-energy change* ΔG^0 by the equation:

$$\Delta G^0 = -RT \ln K \tag{2-4}$$

where R is the gas constant (1.987 calories per mole per degree), *ln* is the natural logarithm, which can be converted to *log* by multiplying by 2.303, and T is the absolute temperature (degrees Kelvin, or degrees Celsius plus 273). The standard free-energy change is that obtained if initially all the products and reactants are present in one-molar concentrations.

The standard free-energy change indicates whether a reaction is exergonic or endergonic under standard conditions. Thus, if all the components A, B, C, and D are at one-molar concentrations and the reaction is allowed to proceed to equilibrium at 25°C,

$$\Delta G^0 = -(1.987)(298)(2.303) \log K$$
$$= -1363 \log K$$

For a reaction in which K is equal to 81,

$$\Delta G^0 = -1363 \log 81$$
$$= -1363(1.91) = -2603 \text{ calories per mole}$$

Table 2-1. Relationship Between K, ΔG^0, and Temperature

K^*	$\log K$	ΔG^0 (25°C) cal/mole	ΔG^0 (37°C) cal/mole
1000	+3	−4089	−4260
100	+2	−2726	−2840
10	+1	−1363	−1420
1	0	0	0
0.1	−1	+1363	+1420
0.01	−2	+2726	+2840

* The temperature at which K was obtained must be specified, e.g., 25°C or 37°C.

Since in this case ΔG^0 has a large negative value, the reaction is exergonic, i.e., it is spontaneous and proceeds with a release of free energy. If the equilibrium constant for the reaction has a value less than 1, e.g., 2.8×10^{-3}, then:

$$\Delta G^0 = -1363 \log 2.8 \times 10^{-3}$$
$$= -1363(-2.553) = +3480 \text{ calories per mole}$$

This reaction would be endergonic. In this case, if one-molar amounts of the four components were present initially, the spontaneous reaction would be from right to left, i.e., more A and B would accumulate. In order for the process to go in the opposite direction, energy would have to be imparted into the system.

The relationship between the equilibrium constant, the standard free-energy change, and temperature is shown in Table 2-1. It can be seen that for a reaction in which K is 1000, if the initial mixture is composed of one-molar concentrations of reactants and products, the change in free energy as the reaction proceeds to equilibrium is −4089 calories at 25°C or −4260 calories at 37°C. Since ΔG^0 is negative, the reaction proceeds with a release of energy and is spontaneous. If K is equal to 0.1, however, ΔG^0 is positive. This means that no reaction will occur from left to right unless energy is imparted to the system. Another way to look at it is that if the components in the initial mixture were at standard concentrations, the reaction would proceed in the opposite direction from the way it is written (i.e., from right to left) until equilibrium is reached. In order to bring the concentrations from equilibrium back to standard conditions, 1363 calories of energy input would be required. It is also seen from Table 2-1 that if $K = 1$, $\Delta G^0 = $ zero. Thus, for systems in which the equilibrium state occurs at standard concentrations, there is no change in free energy.

Generally, reactions are not initiated with equimolar concentrations of products and reactants. Moreover, the substances in cells and tissues are not at standard conditions and not in equimolar amounts. In the case when the materials are not at standard concentration, the observed free-energy change for the reaction depends on the actual molar concentrations of the components:

$$\Delta G = -2.303 RT \log K + 2.303 RT \log \frac{[\text{Products}]}{[\text{Reactants}]}$$

or

$$\Delta G = \Delta G^0 + 2.303 RT \log \frac{[\text{Products}]}{[\text{Reactants}]} \tag{2-5}$$

At 37°C

$$\Delta G = \Delta G^0{}_{310°K} + 1420 \log \frac{[\text{Products}]}{[\text{Reactants}]}$$

Products and *reactants* in these equations refer to the actual initial concentrations of the substances and should not be confused with those at equilibrium (i.e., those used in computing the value of K). Note that the observed free-energy change is zero at equilibrium.

Consideration of these equations reveals that under appropriate conditions, a reaction may be spontaneous—i.e., have a negative ΔG—even though the standard free-energy change (ΔG^0) is positive. For example, if K for the reaction $S \rightleftharpoons P$ is 0.1, then ΔG^0 at 37°C is $+1420$ calories per mole. However, the reaction will have a negative ΔG if the initial concentrations of S and P are 0.1 M and 0.001 M, respectively:

$$\Delta G = +1420 + 2.303 RT \log \frac{0.001}{0.1}$$

$$= +1420 + (1420)(-2) = -1420 \text{ calories per mole}$$

On the basis of the foregoing discussion, ΔG can be defined as the free-energy change for a reaction as it proceeds to equilibrium. The standard free-energy change, ΔG^0, is thus the theoretical maximum energy that can be utilized when all the components of a reaction are converted from one-molar concentrations to equilibrium concentrations.

In biochemical systems, "standard" conditions are often specified for pH 7. The standard free energy at this pH is denoted as $\Delta G^{0\prime}$.

2.3. Adenosine Triphosphate

The mechanisms by which animals derive energy involve the degradation of various nutrients. When glucose, for example, is oxidized to yield carbon dioxide and water, there is a release of energy:

$$C_6H_{12}O_6 + 6O_2 \longrightarrow 6CO_2 + 6H_2O$$
$\Delta G^{0'} = -686$ kilocalories per mole

Oxidation of palmitic acid is also an exergonic reaction:

$$C_{15}H_{31}COOH + 23O_2 \longrightarrow 16CO_2 + 16H_2O$$
$\Delta G^{0'} = -2338$ kilocalories per mole

Although part of this energy appears as heat, a large fraction is utilized for driving various endergonic processes and effecting mechanical work. In order to use such energy for specific reactions or as fuel for the biologic processes, it must be "collected," stored, and channeled. Essentially, the energy released by the degradation of nutrients is converted to a specific intermediate form of chemical energy, which the cells can channel for the metabolic and mechanical activities of life. This central energy intermediate is *adenosine triphosphate* (ATP). It is unique in that the cellular machinery is geared to its formation when energy is released and to its utilization when energy is required.

The energy-releasing reaction of adenosine triphosphate is its hydrolysis, which yields adenosine diphosphate (ADP) and inorganic phosphate (P_i):

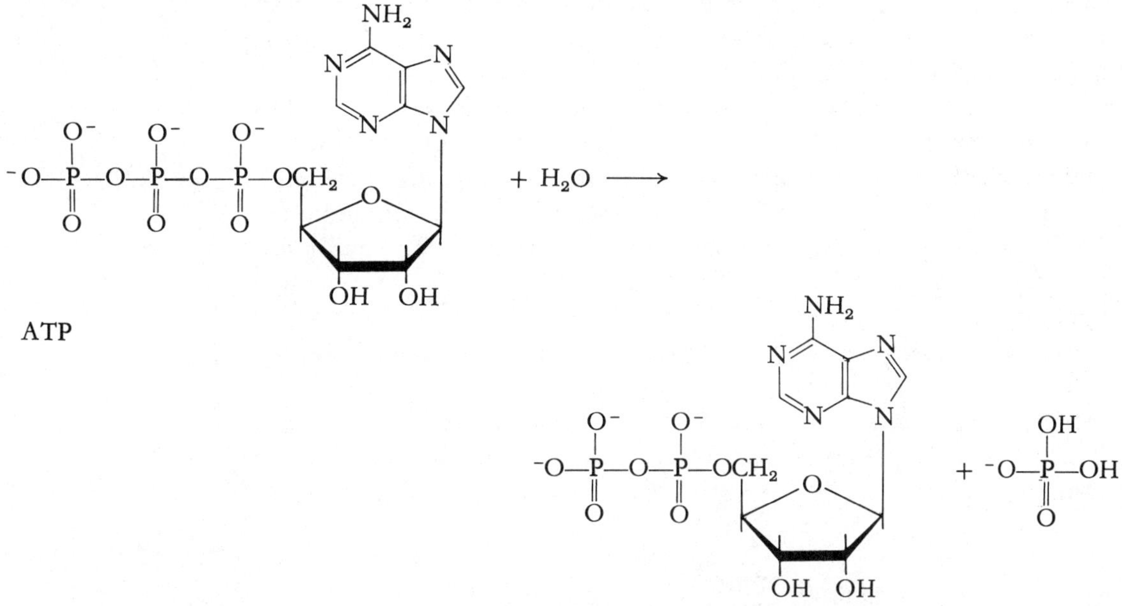

ATP + $H_2O \longrightarrow$ ADP + P_i

$\Delta G^{0'} = -7.3$ kilocalories per mole

The bonds between the phosphates are anhydride bonds, whereas that between phosphate and carbon-5 of ribose is an ester bond. Hydrolysis of any anhydride (e.g., acetic anhydride) is a highly exergonic process. Consequently, hydrolytic cleavage of either the terminal or the second phosphate of ATP has a high, negative ΔG. In contrast, hydrolysis of the remaining phosphate from adenosine monophosphate (AMP) does not result in the release of exceedingly high amounts of energy ($\Delta G^{0\prime} = -3.4$ kilocalories per mole).

The conversion of ATP to ADP and P_i is especially exergonic for several other reasons. ATP has four ionizable protons. Three of these have ionization constants of about 10^{-2} to 10^{-3} (pK_a 2 to 3), and the fourth has a pK_a of approximately 6.5. At pH 7, therefore, three of the acidic groups are completely dissociated and the fourth is about 75% ionized. ADP has three ionizable protons; two of these are dissociated at pH 7 and the third is about 40% dissociated (pK_a 7.2). Thus, one of the important factors that provide the large free-energy release upon hydrolysis of ATP results from the difference between ATP and ADP plus inorganic phosphate with respect to the number and proximity of charges. In ATP, there are four neighboring negative charges that repel each other; this creates an electrical tension or high-energy state in the molecule. ADP has three such charges, and the inorganic phosphate is cleaved as a separate entity, which is also negatively charged. The equilibrium constant for the hydrolysis of ATP thus is extremely high, since the likelihood for the two negatively charged products to approach each other and recombine is very small. Another factor operating to push the reaction in the direction of ATP hydrolysis is the greater resonance stabilization of HPO_4^{2-} compared to that of the phosphates in ATP. The conversion of ATP to ADP and P_i thus provides products that are more stable and of lower internal energy; that is, the products contain less free energy. ATP has a high affinity for magnesium ions, and within the cell, it exists primarily as a magnesium complex. This interaction with magnesium and other intracellular ions also has a critical effect on the free energy of hydrolysis.

It should be noted that 7.3 kilocalories per mole is the *standard* free energy of hydrolysis of ATP. However, the concentrations of ATP, ADP, and P_i within the cell are not one-molar. The actual amounts are of the order of 0.001 M. The free-energy change for the hydrolysis of ATP at 37°C can be approximated as:

$$\Delta G = \Delta G^{0\prime} + 2.303 RT \log \frac{[ADP][P_i]}{[ATP]}$$

$$= -7.3 - 4.3 = -11.6 \text{ kilocalories per mole}$$

It is clear that the actual free-energy change is considerably greater than 7.3 kilocalories and it will depend on the momentary concentrations of ATP, ADP, and P_i. In order to simplify calculations, a ΔG value of -7.3 kcal will be employed for the hydrolysis of ATP; however, it should be realized that this is a minimal value.

Since the hydrolysis of ATP provides at least 7.3 kilocalories per mole, in the reverse reaction,

$$ADP + P_i \rightleftharpoons ATP$$

an input of an equivalent amount of energy is required. The energy for this synthesis is provided by the controlled exergonic processes in the metabolic degradation of nutrients. ATP is thus the central energy carrier in metabolic processes. The energy released in these reactions is channeled to the formation of ATP from ADP and P_i. When energy is required by the cell, it is reobtained from ATP through the hydrolysis of the latter to ADP and P_i. The system is analogous to a battery: the formation of ATP from ADP and P_i can be considered to be the charging reaction, whereas the hydrolysis of ATP to ADP and P_i constitutes the discharging of the battery.

Many reactions that utilize the energy of ATP involve a hydrolysis that yields adenosine monophosphate (AMP) and inorganic pyrophosphate (PP_i):

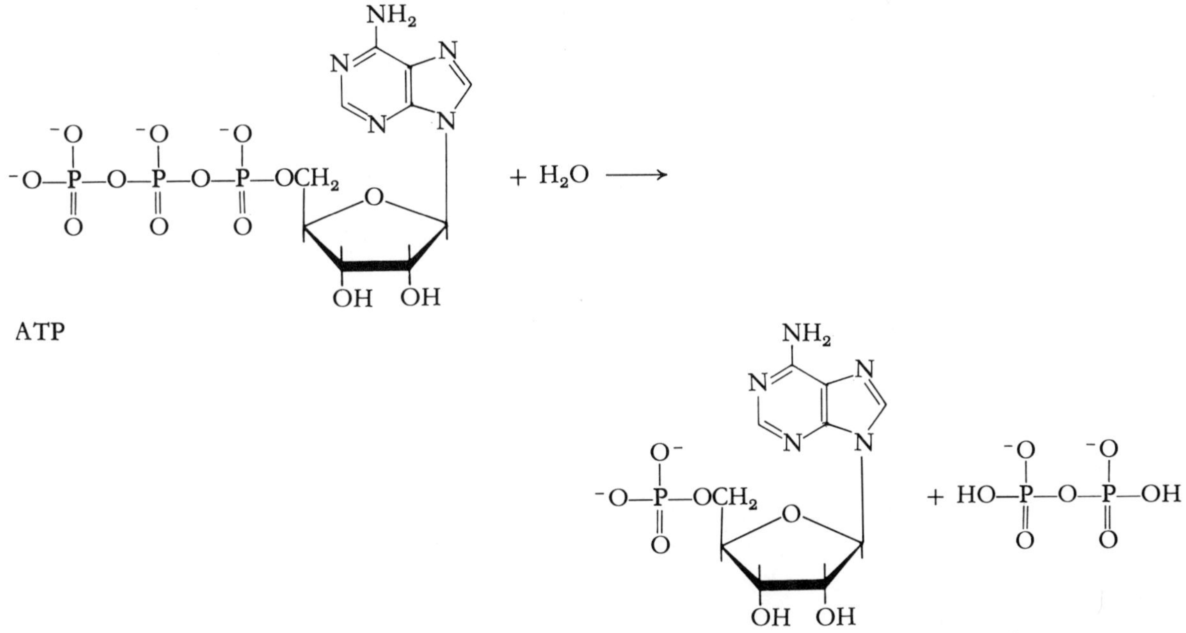

This type of reaction is termed a *pyrophosphate cleavage*, and it is distinguished from the *orthophosphate* cleavage that occurs when ADP is formed. The energy released as a result of pyrophosphate cleavage—$\Delta G^{0\prime} = -10$ kilocalories per mole—is even greater than that released by orthophosphate cleavage.

The two terminal phosphates of ATP are sometimes termed "high-energy phosphates" and their bonds are written with a wavy line to indicate this:

A—R—P∼P∼P

It is more accurate, however, to visualize the energy as residing in the total molecule rather than in the specific bond. ATP is thus a high-energy *molecule* that releases large amounts of energy on hydrolysis. By the same token, inorganic pyrophosphate is also a high-energy compound that releases large amounts of energy upon hydrolysis.

2.4. Rate and Progress of Enzymatic Reactions

The fact that a specific reaction is exergonic indicates that it can proceed spontaneously on the basis of thermodynamic considerations. However, a negative value of ΔG does not reflect the *rate* at which the process will occur. For example, the oxidation of glucose to carbon dioxide and water results in a release of free energy, yet glucose is quite stable and does not decompose on standing in air. In order for a substance to react, the molecules must be activated or energized. The most common method for producing this state is by heating the reactants. The energy required for this purpose is called the *energy of activation*, $\Delta G^\ddagger$. The effect of $\Delta G^\ddagger$ can be shown by a graph of the energy versus the progress of the reaction (Fig. 2-1). In Figure 2-1, it can be seen that although the product has a lower energy than the reactant and ΔG is negative, the energy of activation must be overcome before any transformation can take place.

The velocity of a reaction depends on the reaction's energy of activation. It may be increased either by imparting the required activation energy into the system or by lowering the energy requirement. Catalysts operate by the latter mode, that is, they increase the rate by providing a "progress pathway" involving a lower energy of activation.

Enzymes are proteins that function as catalysts to increase reaction rates. They do not alter the equilibrium constant or the free-energy change of a reaction. In reversible reactions, enzymes decrease the time required for attaining equilibrium. Consequently, they increase the rate of the forward reaction as well as the reverse reaction. Hence, processes catalyzed by

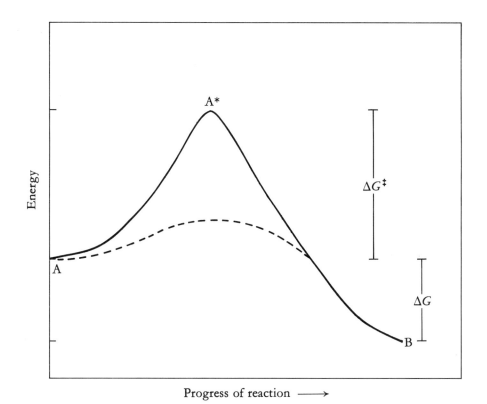

Fig. 2-1. Hypothetical sequence of energy changes as reactant A is converted to product B. Since B has a lower energy than A, the reaction is exergonic (ΔG has a negative value). However, the reaction velocity is low because energy of activation ($\Delta G^{\ddagger}$) is required to raise the energy level of the molecules to their active state (A^*). Catalysts or enzymes provide a pathway with lower activation-energy requirements (dashed line).

enzymes must be thermodynamically feasible; the action of the enzymes is merely to provide a reaction mechanism with a lower $\Delta G^{\ddagger}$. As is the case with inorganic catalysts, enzymes are not consumed in the conversion of reactants to products; although they are involved as intermediates in the reaction mechanism, they are released or regenerated as the products are formed.

Two distinctive properties of enzymes are their high specificity with respect to their substrates and the narrow range of conditions under which they will be effective. As a rule, each cellular reaction is catalyzed by a specific enzyme. Although we categorize enzymes according to the type of reaction that they catalyze (i.e., oxidation, hydrolysis, isomerization, and so forth), an enzyme usually influences only one substrate and will not affect another one in a similar manner. In addition, the activities of enzymes are

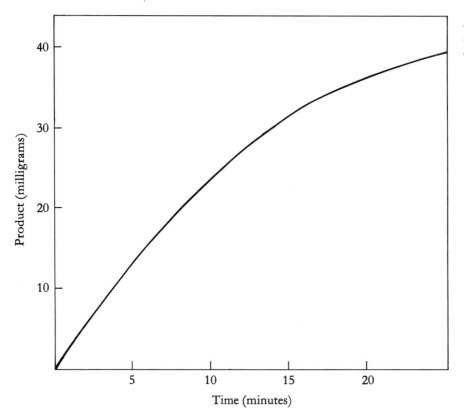

Fig. 2-2. Progress of a typical enzymatically catalyzed reaction.

highly dependent on pH, the presence of various cofactors (e.g., nucleotide phosphates, vitamin derivatives, or metal ions), and the concentrations of substrates and products. These unique characteristics of enzymes allow them to play a central role in promoting and regulating cellular processes and make them vital components of living systems.

The effect of an enzyme on a specific process can be followed by the progressive formation of products. The *rate of the reaction* is then obtained by determining the amount of product generated in a specified time interval. In performing such measurements with enzymatically catalyzed reactions, special consideration must be given to the time interval employed in the computations. This is necessitated by the fact that the progress of reactions generally decreases with time (Fig. 2-2). In the initial period, the concentration of product increases dramatically, and the amount generated is practically linear with time. As the reaction progresses, the rate of formation of product decreases even when a considerable amount of substrate is still available.

Several factors may contribute to this variation of the reaction rate. As a reaction proceeds, the concentration of substrate decreases. Since the rate is related to the concentration of reactant, it will decrease as substrate is consumed. If there is a change in hydrogen-ion concentration as a result of the reaction, the catalytic capability of the enzyme will be affected. Also, the enzyme may become deactivated or denatured under the conditions of the specific experiment. This last factor is especially prominent with unstable enzymes. Another factor that is important in many reactions is the inhibition of the enzyme by the product. In such cases, as the product accumulates, the activity of the enzyme decreases. Finally, in reversible processes, the rate of the reverse reaction becomes significant as the concentration of product increases.

Whatever the reason for the decrease in reaction rate, the fact that it does occur requires that the measurement of rate be made under specified conditions. If the result shown in Figure 2-2 is taken as an example, it can be seen that when the activity is measured for the first 2 minutes, the rate is 2.5 milligrams of product per minute. A similar value for the rate will be obtained if it is determined after 4 minutes. However, if the rate is measured by the amount of product formed in 16 minutes, a value of 2 milligrams per minute will be obtained. It will also be noted that the shorter the time interval following the initiation of the reaction, the higher the value of the calculated rate will be and the closer it will approximate linearity with time. It is thus reasonable to assume that this is the characteristic rate for the reaction. During this interval, the diverse factors that modify the rate will make a minimal contribution. The rate of a reaction is thus best defined by its *initial velocity*, designated as v_0. The critical factor in measuring v_0 is that it must be determined within the period when the reaction velocity or rate is linear with time.

2.5. Effect of Substrate Concentration

The study of the kinetics of enzymes involves elucidation of the parameters that influence the rate of reactions and the quantitative determination of their effects. The major factors of interest in physiologic processes are the effect of substrate concentration, the influences of hydrogen-ion concentration and of various cofactors, and the effect of substances that increase or decrease the rate of reactions.

The effect of substrate concentration can be analyzed by considering the simplest possible reaction, that is, a case in which one reactant (S) is converted to a single product (P):

$$S \rightleftharpoons P$$

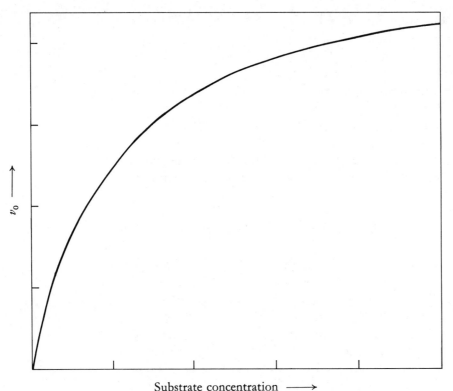

Fig. 2-3. *Effect of substrate concentration on initial velocity* (v_0) *in enzymatically catalyzed reactions. The concentration of substrate is generally expressed in molarity.*

If this process is catalyzed by enzyme E, the concentration of S usually affects the rate or initial velocity (v_0) of the reaction in a manner described by the graph in Figure 2-3. As is expected from the law of mass action, the velocity increases as the concentration of substrate (reactant) is increased. When the concentration of substrate is relatively low, the reaction velocity is directly proportional to the substrate concentration. When such a relationship occurs in an ordinary chemical reaction, the process is characterized as exhibiting *first-order kinetics*, i.e., the reaction rate is directly proportional to the concentration of one reactant. At intermediate substrate concentrations, the influence of substrate concentration on initial velocity diminishes. Finally, at high substrate concentrations, the velocity reaches a maximum and cannot be increased by increasing the concentration of substrate; the velocity for an enzymatic reaction at this stage is called the *maximum velocity* and is denoted as V_{max}.

The observed relationship between the initial reaction velocity (v_0) and substrate concentration can be explained by assuming that one step in the mechanism involves the combination of substrate with enzyme to form an enzyme-substrate complex, ES. The latter undergoes an intramolecular transformation, which results in the release of product (P) and regenerated enzyme:

$$E + S \underset{k_2}{\overset{k_1}{\rightleftharpoons}} ES \underset{k_4}{\overset{k_3}{\rightleftharpoons}} P + E$$

The constants k_1, k_2, k_3, and k_4 are the *rate constants* for the designated steps. The velocity of reaction (v_0) as measured by the rate of appearance of product P is affected by the concentration of substrate as long as the functional amount of enzyme exceeds that of the substrate. When the concentration of substrate relative to that of enzyme is sufficient to maintain all the enzyme as the ES complex, addition of more substrate does not bring about any enhancement in the reaction rate. At this point, the enzyme is saturated and has reached its maximal catalytic effect. The kinetics are then termed *zero order kinetics*, i.e., the reaction rate is not influenced by substrate concentration.

The mathematical formulation of these concepts yields an equation giving the relationship between substrate concentration and initial velocity. As the reaction is initiated, a portion of the enzyme becomes bound to the substrate to form the enzyme-substrate complex ES, and the remainder of the enzyme, E, is free. The total of free and bound enzyme, E_t, is equal to E plus ES; thus, the concentration of free enzyme [E] is equal to $[E_t]$ minus [ES]. The *rate of formation* of the complex from free enzyme and substrate can be defined by the equation:

$$\frac{d[ES]}{dt} = k_1([E_t] - [ES])[S] \tag{2-6}$$

This means that the rate of production of ES—which is expressed by the derivative $d[ES]/dt$— is proportional to the concentration of substrate and free enzyme. The *rate of breakdown* of the complex can be given by the relationship:

$$-\frac{d[ES]}{dt} = k_2[ES] + k_3[ES] \tag{2-7}$$

The complex ES breaks down to both E + P and E + S, so the rate constants for both reactions are included as terms in this equation. The rate of formation of ES from E + P can be neglected in the initial reaction stage, however, since the concentration of product at this stage is minimal.

By combining these expressions, an equation for the net change in concentration of ES with time can be obtained:

$$\frac{d[ES]}{dt} = k_1([E_t] - [ES])[S] - k_2[ES] - k_3[ES] \qquad (2\text{-}8)$$

As the reaction proceeds, a *steady state* is reached in which the concentration of ES remains essentially constant, i.e., the formation of ES from S occurs essentially at the same rate as that of its decomposition. During this period, $d[ES]/dt$ equals zero. At this steady state, equation 2-8 becomes:

$$k_1([E_t] - [ES])[S] = k_2[ES] + k_3[ES] \qquad (2\text{-}9)$$

Solving equation 2-9 for [ES] yields:

$$[ES] = \frac{[E_t][S]}{[S] + \dfrac{k_2 + k_3}{k_1}} \qquad (2\text{-}10)$$

The ratio $(k_2 + k_3)/k_1$ defines a new constant K_M, termed the *Michaelis constant*. Substituting K_M in equation 2-10 yields:

$$[ES] = \frac{[E_t][S]}{[S] + K_M} \qquad (2\text{-}11)$$

The velocity of formation of product—i.e., the overall rate of the reaction—is proportional to the concentration of ES:

$$v_0 = k_3[ES] \qquad (2\text{-}12)$$

Substituting from equation 2-11, we obtain:

$$v_0 = \frac{k_3[E_t][S]}{[S] + K_M} \qquad (2\text{-}13)$$

When the substrate concentration is high enough to convert *all* the enzyme to ES, the reaction velocity will be at its maximum. The numerical value of this velocity, denoted by V_{max}, will then be proportional to the total concentration of enzyme, E_t:

$$V_{max} = k_3[E_t] \qquad (2\text{-}14)$$

Substitution of V_{max} for $k_3[E_t]$ in equation 2-13 yields:

$$v_0 = \frac{V_{max}[S]}{K_M + [S]} \tag{2-15}$$

This expression is termed the *Michaelis-Menten equation*; it defines the relationship between substrate concentration and initial reaction velocity (v_0). The two constants, K_M and V_{max}, are specific for each enzyme and are discussed further in the ensuing paragraphs.

The Michaelis-Menten equation shows that when the substrate concentration is comparatively small, the numerical contribution of its value to the denominator is negligible. The initial reaction velocity is then directly proportional to the substrate concentration. In the opposite situation—i.e., when the value of [S] is relatively large compared with K_M—the latter may be neglected, and the equation then approaches the relationship $v_0 = V_{max}$. This is essentially what occurs at high substrate concentrations; i.e., at this point, the velocity reaches a maximal value and is independent of substrate concentration.

Another relationship implicit in the Michaelis-Menten equation is that between K_M and substrate concentration. If we consider the condition in which the molar concentration of the substrate is equal to the numerical value of K_M, i.e., $[S] = K_M$, then K_M in equation 2-15 can be replaced by [S]. This yields:

$$v_0 = \frac{V_{max}[S]}{[S] + [S]} = \frac{V_{max}}{2} \tag{2-16}$$

Hence, K_M may be defined as the substrate concentration when the reaction velocity is half the maximum velocity (Fig. 2-4). In effect, the dimensions for K_M are concentration units, e.g., moles per liter.

In determining K_M for a particular reaction, its value may conceivably be obtained by measuring the initial velocities for different concentrations of substrate and determining the concentration when the velocity is half the maximum. However, this is generally not practical since it requires numerous individual experiments in which large amounts of enzyme and substrate are expended. Moreover, the numerical value for V_{max} obtained in this way is usually not precise. The error is compounded further by the necessity of finding the K_M from a line whose exact curvature must be estimated.

In practice, K_M and V_{max} are determined by one of several other methods that obviate the above-mentioned difficulties. The most common procedure,

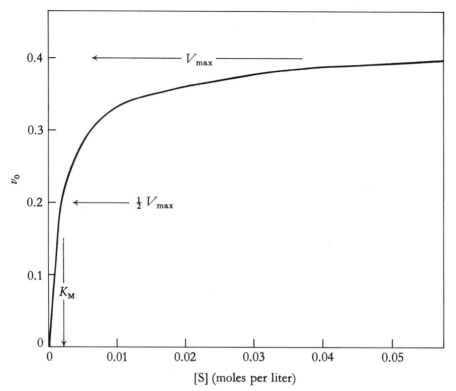

Fig. 2-4. *Relation of* K_M, V_{max}, *and* $[S]$. *In this example,* V_{max} *is equal to 0.4 and* K_M *is equal to 0.002 moles per liter or* 2×10^{-3} *M.*

that of Lineweaver and Burk, is based on a modification of the Michaelis-Menten equation. If equation 2-15 is rewritten in reciprocal form, it becomes:

$$\frac{1}{v_0} = \frac{K_M + [S]}{V_{max}[S]} = \frac{K_M}{V_{max}}\left(\frac{1}{[S]}\right) + \frac{1}{V_{max}} \tag{2-17}$$

Since K_M and V_{max} are constants for the specific reaction, a plot of $1/v_0$ versus $1/[S]$ should give a straight line, whose slope is equal to K_M/V_{max}. The point at which this line intercepts the ordinate is equal to $1/V_{max}$, and the interception point on the abscissa is $-1/K_M$, as shown in Figure 2-5.

Considering the definition of K_M as the substrate concentration at which the reaction reaches half maximal velocity, it should be realized that this constant is a characteristic of the *substrate* in a particular enzymatically catalyzed reaction. If a process involves the interaction of two or more components, each substrate has its specific K_M. The experimental procedures for determining the K_M for each reactant are cumbersome, and more than

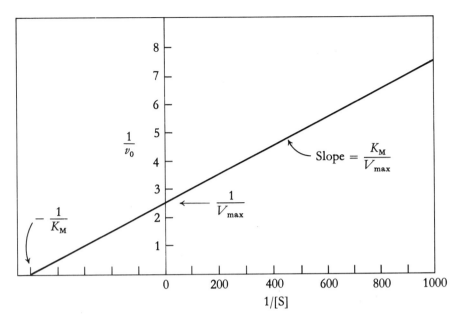

Fig. 2-5. Plot of $1/v_0$ versus $1/[S]$ derived from initial velocity data with several different concentrations of substrate. The straight line obtained by connecting the individual points is extended to intercept the ordinate and abscissa. The data plotted on this graph are the same as those in Figure 2-4. The values of K_M and V_{max} are determined as shown on the graph.

one approach is required to obtain reasonably accurate values. The simplest method for a reaction such as

$$A + B \longrightarrow P + Q$$

is to determine the effect of the concentration of A on the reaction velocity when the concentration of B is in excess. Similar studies are then performed on the effect of different concentrations of B in the presence of excess A. When the data for each set of experiments are plotted and analyzed, the K_M values for A and B can be obtained. Although this procedure is subject to various pitfalls, it can serve as a first approach.

The significance of the relative magnitude of the Michaelis constant is not the same for all enzymes, since the K_M is a combination of the rate constants for the intermediate steps in the reaction mechanism. In the simple case of one substrate, one product, and a single enzyme-substrate complex, K_M is the ratio $(k_2 + k_3)/k_1$. When several transformations occur in the enzyme-substrate complex before the product is released, the rate constants for these intermediate steps also may contribute to the value for K_M. For reactions involving more than one substrate, the Michaelis constant involves numerous additional rate constants. Whatever the factors entering into the composition of K_M, a low value for the constant indicates that the enzyme becomes saturated by the low concentrations of substrate and that maximum reaction velocity is attained with relatively small concentrations of substrate. Conversely, when the value of K_M is large, the implication is that high substrate concentrations are required for maximum velocity.

Although it might be inferred from these generalizations that a low K_M implies a high affinity of enzyme for substrate, such a conclusion is not justified in the absence of data from additional experiments. For example, if it is known from other studies that k_2 is much greater than k_3, then the latter factor can be neglected and K_M becomes equal to k_2/k_1. For such a reaction, the Michaelis constant is actually the dissociation constant of the enzyme-substrate complex, and its numerical value is inversely proportional to the affinity of the enzyme for the substrate. (When this is known to be the case, K_M is written as K_S, or substrate dissociation constant.) Otherwise, K_M is to be considered only a reflection of the concentration of substrate necessary to saturate the enzyme or a constant that specifies the relationship between the rate of reaction and the substrate concentration.

The K_M value is a characteristic constant for each enzymatic reaction. If an enzyme catalyzes a reaction involving several substrates, each substrate has its specific K_M. Some of the actual values of K_M will be indicated in subsequent discussions of individual enzymes.

2.6. Enzyme Concentration and Reaction Rate

The foregoing discussions dealt with the effect of substrate concentration on the initial velocity of a reaction when the amount of enzyme is kept constant. It was seen that when a certain concentration of substrate is exceeded, the reaction velocity reaches its maximum, and any additional increase in substrate does not enhance the reaction rate. The reason for this phenomenon is that the enzyme is saturated and therefore becomes the limiting factor. Since the maximal reaction velocity (V_{max}) is a function of the amount of enzyme available, it will be increased by introducing more enzyme into the system. In other words, V_{max} is a value that is obtained for a given amount of enzyme when the reaction mixture is saturated with substrate.

On the basis of these considerations, it should be realized that if constant amounts of *excess* substrate (relative to enzyme) are employed, the velocity of the reaction will increase with increasing concentrations of enzyme. A typical result is shown in Figure 2-6. The proportionality between enzyme concentration and the initial velocity of a reaction is the basis of most experimental and clinical enzyme assay systems.

The results of such assays are expressed in enzyme-activity units. According to the International Commission on Enzymes, one *enzyme-activity unit* is

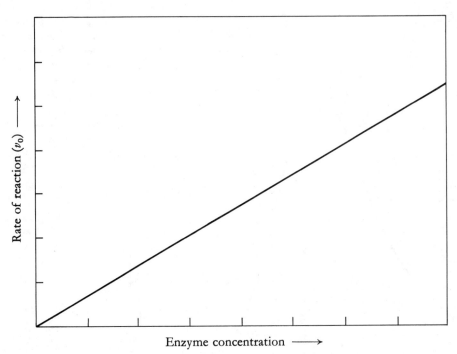

Fig. 2-6. *Effect of enzyme concentration on initial velocity* (v_0). *The concentration of substrate is such that it is above that required for maximal velocity. Hence, the above diagram expresses the effect of enzyme concentration on V_{max} for the reaction.*

defined as the amount of enzyme that catalyzes the transformation of 1 micromole of substrate per minute. Assays are generally performed at 25°C; nonetheless, the temperature should be specified. The *specific activity* of an enzyme is generally expressed as the number of enzyme-activity units per milligram of protein. Another unit, the *katal* (kat), is the amount of enzyme that converts one mole of substrate to product per second.

2.7. Influence of pH on Enzyme Activity

Most enzymes are effective only within a relatively narrow pH range. Generally, for a given enzyme, there is a specific pH at which it is most active. This is termed the *pH optimum* for the enzyme. The relationship between pH and activity of a typical enzyme is shown in Figure 2-7. Although a large number of enzymes have pH optima between pH 6 and pH 8, a considerable number exhibit their greatest activities below or above these pH values. For example, pepsin, which catalyzes the hydrolysis of peptide bonds of various proteins, has a pH optimum of about 2. Furthermore, the pH-activity curve does not necessarily have to be bell-shaped; with some enzymes, the decrease in activity at lower or higher hydrogen-ion concentrations may be comparatively small.

The influence of pH on enzyme activity may be the result of one or more factors. Since enzymes are proteins, the hydrogen-ion concentration of the

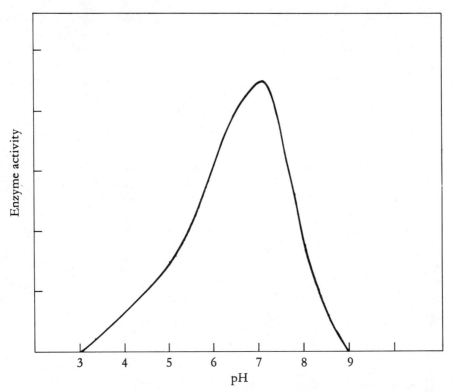

Fig. 2-7. Enzyme activity versus pH. Measurements are generally made when the substrate is in excess with respect to the enzyme, so the reaction velocity may also be expressed as enzyme activity. All other factors are kept constant.

medium determines whether certain groups are ionized or remain in an undissociated form. For example, certain enzymes may be active only if a specific functional group on a side chain is in the ionized state. These enzymes will be active only in the pH range in which ionization can occur. Another factor involves the effect of pH on the substrate and the specific limitations on activity set by the degree of ionization of the substrate. Both the effect on the enzyme and that on the substrate may be important either for the binding of the substrate to the enzyme or for the catalytic effect of the enzyme. The observed relationship between pH and activity expresses the combined effect of these factors.

Enzymes are generally unstable in solutions of very low or very high pH. Since they are proteins, enzymes may be denatured and precipitated at certain hydrogen-ion concentrations. Although the pH for optimal stability is not necessarily related to the pH for maximal catalytic effect, both factors must be considered in assaying for enzyme activity.

2.8. Effect of Temperature

The rate of most chemical reactions increases as the temperature is elevated. In many enzymatic reactions, if the temperature is increased by 10 degrees, the rate increases as much as two-fold. However, since proteins are rapidly denatured at higher temperature, especially above 40°C, an excessive increase in temperature has a deactivating effect on the enzyme. The optimal temperature for the activity of a specific enzyme is thus the resultant of two factors: the effect on reaction rate and the enzyme's stability. Generally, enzyme activity determinations are carried out at 25°C. For precise comparison of activities, the temperature must be kept constant so the above-mentioned factors do not affect the experimental results.

2.9. Coenzymes and Cofactors

A large number of enzymes are bound to non-protein components that are essential for their activity. In many instances, these substances are linked covalently to the protein; such units are called *prosthetic groups*. In other cases, the bond either is undefined or may involve electrostatic attractions. Such non-protein components, whether they are bound covalently or otherwise, are termed *coenzymes* or *cofactors*. Although the terminology is not rigid, when the non-protein material is a complex organic molecule, it is generally called a coenzyme, and when it is a metal ion or simple organic compound, it is called a cofactor. The complete enzyme with its coenzyme is termed the *holoenzyme*. The protein moiety by itself is called the *apoenzyme*. Thus, the apoenzyme and the coenzyme together form the holoenzyme.

A number of the coenzymes together with their structures are given in Table 2-2. With some enzymes, the coenzymes may be bound so weakly that they can be separated by physical means. For example, nicotinamide adenine dinucleotide (NAD^+) can be removed from a number of NAD-dependent enzymes by dialysis. Since the coenzyme has a comparatively low molecular

Table 2-2. Some Important Coenzymes

Name	Abbreviation	Structure Shown on Page
Nicotinamide adenine dinucleotide	NAD^+	99
Nicotinamide adenine dinucleotide phosphate	$NADP^+$	99
Flavin mononucleotide	FMN	133
Flavin adenine dinucleotide	FAD	119
Thiamine pyrophosphate	TPP	124
Pyridoxal phosphate	—	302
Lipoic acid	—	124
Biotin	—	220
Iron-porphyrin complex	—	139
Cobalamine	—	597

weight, it passes through the dialysis membrane, whereas the larger protein is not dialyzable (i.e., it remains in the dialysis chamber). Flavin mononucleotide (FMN) and flavin adenine dinucleotide (FAD) generally cannot be separated from their respective apoenzymes by dialysis. When more drastic methods are employed for the dissociation, the enzyme is often destroyed irreversibly. It should thus be realized that although all the substances listed in Table 2-2 are classified as coenzymes, their molecular relationships with their enzymes vary considerably.

Many enzymes require specific metal ions for their activity. Such metals include Zn^{2+}, Mg^{2+}, Mn^{2+}, Fe^{2+}, Ca^{2+}, Cu^{2+}, Na^+, and K^+. Details of specific enzymes and their cofactors will be discussed in subsequent sections where the actions of the enzymes are described.

2.10. Enzyme Nomenclature

Enzymes are usually named according to the reaction that they catalyze with the suffix "ase" added to the name. For example, an enzyme that catalyzes an oxidation with the concomitant removal of hydrogen from the substrate is called a *dehydrogenase*. Enzymes that effect the transfer of amino groups from one substrate to another are termed *transaminases*. Those that effect the hydrolysis of phosphate esters are *phosphatases*, and those that hydrolyze proteins are *proteases*. There are also numerous exceptions to this mode of nomenclature. Examples of names that give no indication as to the specific action of the enzymes include catalase, pepsin, papain, and trypsin.

As is the case with many nomenclature systems, enzymes were named at the time they were discovered without consideration for any precise rules. As the number of characterized enzymes increased, a uniform classification and nomenclature had to be devised for communication purposes and indexing systems. Although such a system has been constructed, the shorter and more familiar trivial names are still used widely. It is therefore necessary for biologic scientists to be familiar with the common names for various enzymes as well as with the system devised by international convention. The method of classification by the International Enzyme Commission is given here, although its full significance will not be appreciated until individual enzymes are described. Additional references to this classification and nomenclature will be made in subsequent chapters.

Enzymes are categorized into six groups, as shown in Table 2-3. Each group is divided further into subclasses according to more specific aspects of the individual reactions. In the international classification, the subclasses are denoted by a numbering system, e.g., the numeral 3.1 denotes a hydrolase that hydrolyzes an ester. Specific examples of the classification system will be indicated when the activities of individual enzymes are described.

Table 2-3. Enzyme Classification

Class	Action
1. Oxidoreductase	Removal or addition of electrons (oxidation-reduction reactions)
2. Transferase	Transfer of a functional group (e.g., phosphate, amino, methyl) from one substrate to another
3. Hydrolase	Bond splitting by addition of water (hydrolysis reactions)
4. Lyase	Reactions involving additions to double bonds or cleavages with formation of double bonds
5. Isomerase	Isomerization reactions (internal rearrangements; e.g., ketose to aldose)
6. Ligase	Formation of new bonds with concomitant cleavage of ATP (energy-requiring syntheses)

2.11. Oxidoreductases

Processes involving the transfer of electrons from one substance to another are called *redox reactions*. Such reactions may involve the concomitant transfer of hydrogen, but this is not a universal requirement. Enzymes that catalyze redox reactions are classified as *oxidoreductases*. These enzymes operate in conjunction with specific coenzymes, which function either to accept or to donate the required electrons. These coenzymes are not merely prosthetic groups or cofactors of the protein catalyst; rather, they can be visualized as substrates that undergo changes during the transformation. This is most evident in the pyridine nucleotide coenzymes (NAD^+ and $NADP^+$), which are not bound covalently to the apoenzyme. Since a large number of the oxidoreductases that catalyze the direct oxidation steps of metabolic intermediates operate in conjunction with pyridine nucleotides, the latter are discussed here.

Enzymes that catalyze the oxidation of a substrate with the concomitant reduction of NAD^+ have also been called *dehydrogenases*. Some enzymes of this type are specific for NAD^+ (NAD-linked dehydrogenases) and others are specific for $NADP^+$ ((NADP-linked dehydrogenases). The specific pyridine nucleotide is generally not interchangeable. Examples of NAD-linked dehydrogenases are alcohol dehydrogenase, lactate dehydrogenase, and malate dehydrogenase. Some NADP-linked dehydrogenases are glucose 6-phosphate dehydrogenase and gluconic-acid 6-phosphate dehydrogenase. The reactions catalyzed by these enzymes are shown below:

Alcohol dehydrogenase

$$CH_3CH_2OH + NAD^+ \rightleftharpoons CH_3\overset{O}{\overset{\|}{C}}H + NADH + H^+$$

Ethanol   Acetaldehyde

Lactate dehydrogenase

$$CH_3CHOHCOOH + NAD^+ \rightleftharpoons CH_3\overset{O}{\underset{\|}{C}}COOH + NADH + H^+$$

Lactic acid Pyruvic acid

Malate dehydrogenase

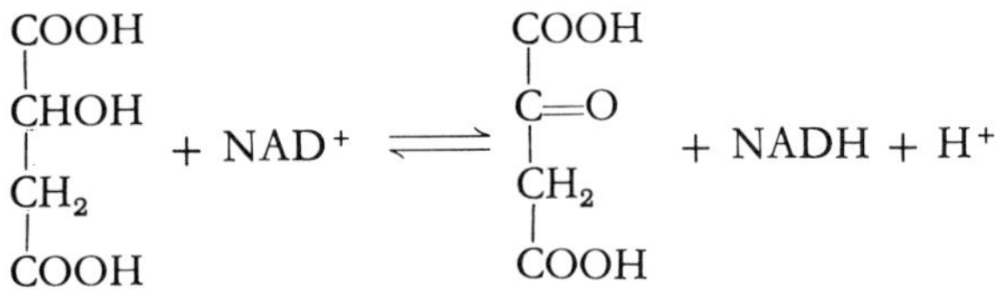

$$\begin{array}{c} COOH \\ | \\ CHOH \\ | \\ CH_2 \\ | \\ COOH \end{array} + NAD^+ \rightleftharpoons \begin{array}{c} COOH \\ | \\ C=O \\ | \\ CH_2 \\ | \\ COOH \end{array} + NADH + H^+$$

Malic acid Oxaloacetic acid

Glucose 6-phosphate dehydrogenase

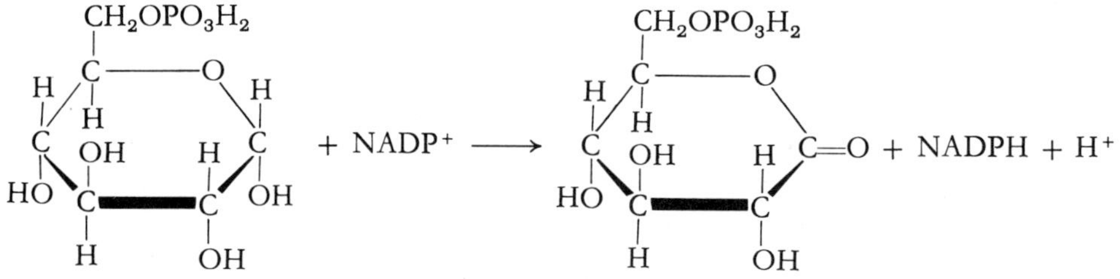

[Glucose 6-phosphate] + NADP$^+$ ⟶ [Gluconolactone 6-phosphate] + NADPH + H$^+$

The systematic name for alcohol dehydrogenase is *alcohol:NAD oxidoreductase*, since it catalyzes an oxidation-reduction reaction between alcohol and NAD$^+$. It is also designated by the Enzyme Commission number 1.1.1.1. The first of the four numbers indicates that the enzyme is an oxidoreductase. The second number refers to the group that is oxidized; when it is a hydroxyl group, the number is 1. The third number designates that the electron acceptor is either NAD$^+$ or NADP$^+$, and the fourth number is specific for the given enzyme. Similarly, the systematic name for lactate dehydrogenase is L-*lactate:NAD oxidoreductase* and is designated by the Enzyme Commission number EC 1.1.1.27. The other two enzymes shown are, respectively, L-*malate:NAD oxidoreductase* (EC 1.1.1.37) and D-*glucose 6-phosphate:NADP oxidoreductase* (EC 1.1.1.49). It may be noted that the first three numerals in the Enzyme Commission numbers for all the enzymes shown are the same— i.e., 1.1.1— since they are all oxidoreductases involving hydroxyl groups of the substrate and pyridine nucleotide coenzymes.

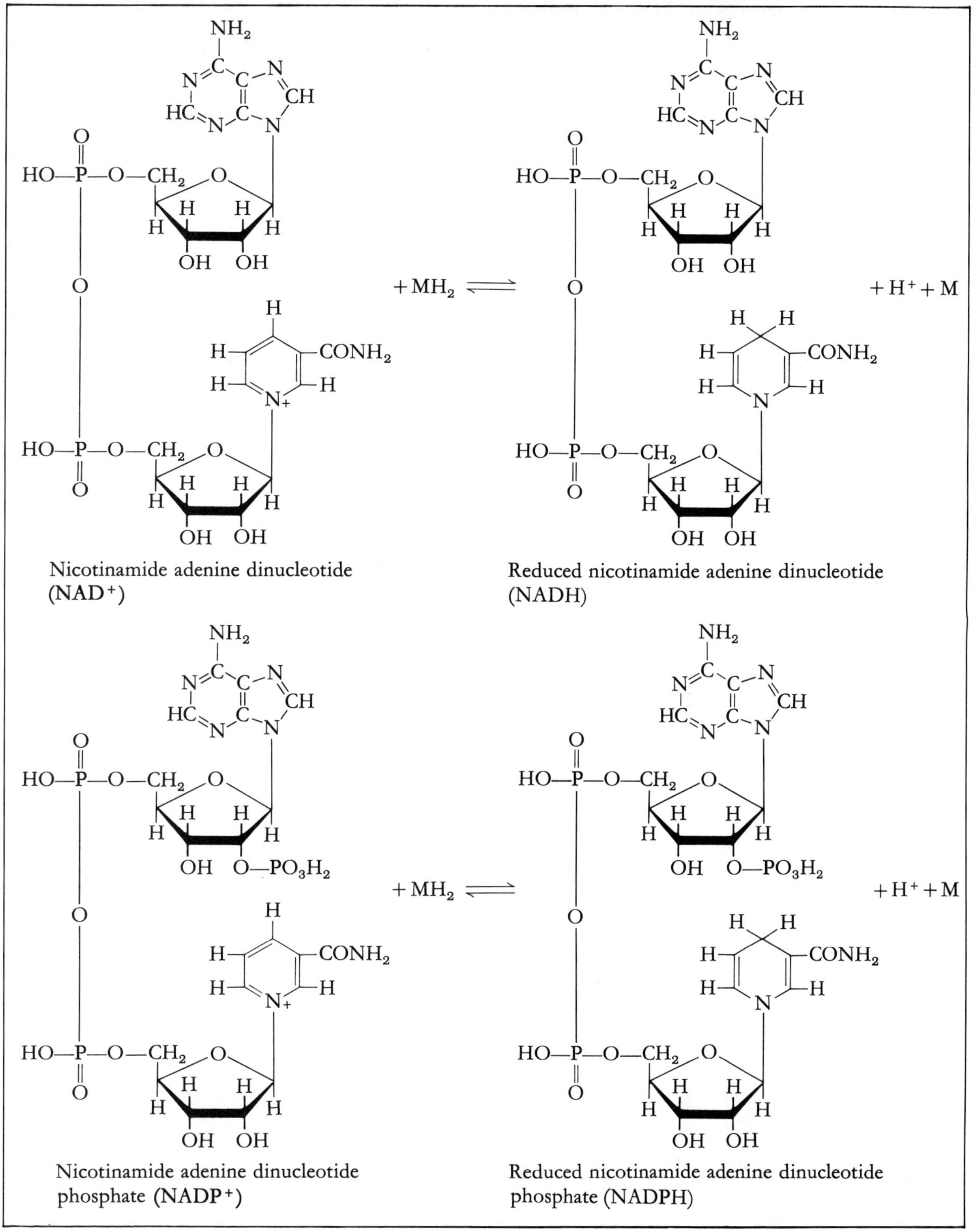

Fig. 2-8. Oxidation of substrate, MH_2, by NAD^+ and $NADP^+$, where M is the portion of the substrate molecule bearing the electrons and protons.

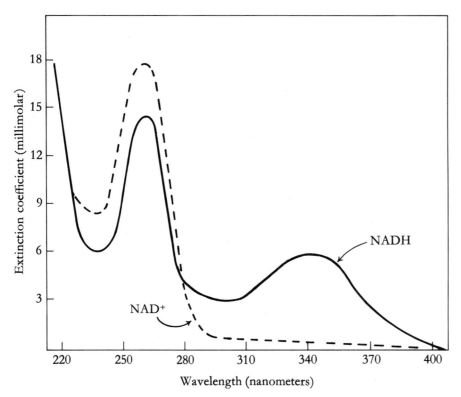

Fig. 2-9. Absorption spectra of NAD^+ and $NADH$.

In each of the reactions shown, two electrons and two protons are removed from the substrate. (The loss of the electrons from the substrate constitutes the oxidation.) One proton is released to the medium, and the other proton plus two electrons are incorporated into the pyridine nucleotides, as shown in Figure 2-8. Since protons are released in the process, the pH of the medium decreases if the reaction is performed in an unbuffered system.

The principal method for assaying the activity of enzymes involving pyridine nucleotides utilizes the difference in the absorption spectra for the oxidized and the reduced coenzymes (Fig. 2-9). As can be seen in the diagram, the absorbance at 340 nanometers for NAD^+ is minimal, but in any reaction in which NAD^+ is converted to NADH, there will be an increase in absorbance at 340 nanometers, which can be determined spectrophotometrically. This change serves as an index of the amount of NAD^+ reduced and consequently of the amount of substrate oxidized (Fig. 2-10).

It should be noted that one of the components of both NAD^+ and $NADP^+$ is nicotinamide, which is related to the vitamin niacin (nicotinic acid):

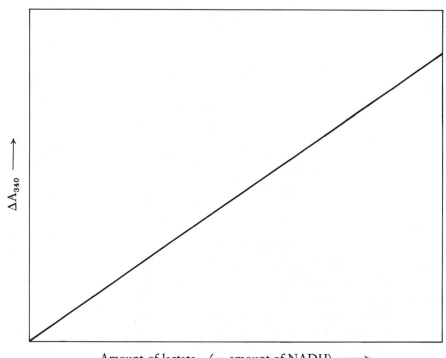

Fig. 2-10. *Assay for lactate by means of lactate dehydrogenase and NAD^+. The amount of NADH formed is equivalent to the amount of lactate present. The NADH, in turn, is assayed spectrophotometrically by its absorption at 340 nanometers.*

It will be seen in subsequent sections that many of the other vitamins also function as coenzymes in a number of metabolic processes.

2.12. The Active Site of Enzymes

The numerous and complicated intermediate steps in an enzymatically catalyzed reaction can be divided into two general processes: the enzyme binds to the substrate to form an enzyme-substrate complex, and the enzyme promotes a chemical change in the substrate. Both these processes involve interactions of the substrate with a particular section of the enzyme or, more precisely, with specific amino-acid components of the enzyme. The area in the enzyme structure that is involved in its catalytic functions is called the *active site*.

Most substrates have considerably smaller dimensions or lower molecular weights than the protein molecules of their respective enzymes. Even when

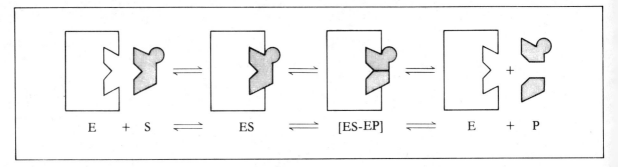

Fig. 2-11. "Lock-and-key" concept of the interaction between an enzyme and its substrate.

the substrate is a polysaccharide or a protein, only the section in which the eventual transformation occurs appears to be bound to the enzyme. The active site thus constitutes a relatively small fraction of the enzyme structure. This does not imply, however, that the amino acids in the active site are near each other in their sequence or in the primary structure of the protein. As a result of the folding of the polypeptide chain (secondary structure) and the overall conformation of the enzyme (tertiary structure), certain amino-acid residues may be distant from each other in the primary sequence yet near one another on the active site of the completed protein. This is known to be the case for a number of enzymes.

One explanation for the high specificity of enzymes is that their tertiary and quaternary structures are such as to permit a precise physical interlocking with the substrate. In other words, the shape of the protein and its surface are uniquely adapted to accommodate its particular substrate. Many enzymes can be visualized as containing grooves with fixed dimensions that permit only compounds with a specific shape to be inserted. Substances that cannot fit into the groove to form the enzyme-substrate complex will not react, even though they have the same functional groups as the true substrate. This is called the "lock-and-key" concept since it presumes that the substrate must fit the enzyme like a key in a lock (Fig. 2-11).

Investigations into the structural and mechanistic details of a number of enzymes demonstrated that they undergo certain conformational changes as they become bound to their substrates. Thus, although the enzyme-substrate complex comprises the two components in a spatially fixed relationship, the unattached enzyme does not necessarily contain a receptacle with the *exact* dimensions of the substrate. The original concept was therefore extended by the formulation of the *induced-fit theory* (Fig. 2-12). According to this view, the topography of the original enzyme may allow for an initial, superficial attachment of the substrate. However, this preliminary interaction effects a conformational change that provides a more perfect apposition between the

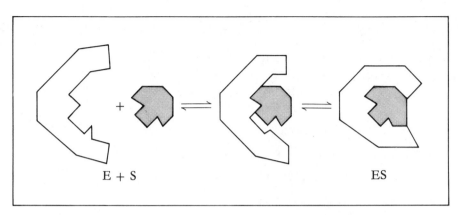

Fig. 2-12. Induced-fit hypothesis. The initial, superficial binding of the substrate (*middle*) induces conformational changes in the enzyme to provide a better fit to its critical sites (*right*).

critical sites (or amino-acid residues) of the enzyme and the specific functional groups of the substrate.

The underlying factors that confer the unique catalytic properties on a given enzyme are its three-dimensional structure and the specific architecture of the active site. Rather than being a mere physical receptacle with the correct dimensions for uniting with the substrate, the active site is also the area in the enzyme structure that contains the amino-acid residues involved in positioning the substrate and in catalyzing its transformation. For example, certain substrates interact with the side-chain carboxyl groups of aspartic or glutamic acid residues in the respective enzymes. Other reactions involve the ϵ-amino groups of lysine, the imidazole units of histidine, or the hydroxyl groups of serine. Generally, several amino-acid residues in the enzyme are critical for its action. An important aspect in elucidating the mechanism for a specific process is therefore the identification of the functional amino acids in the active site.

Some of the methods employed to determine the active-site amino acids will be described in conjunction with discussions of the function of specific enzymes. However, a few examples may be indicated at this point. The catalytic effect of a number of enzymes (e.g., chymotrypsin, trypsin, cholinesterase, and thrombin) involves interaction with a specific serine residue. Such enzymes are inhibited by diisopropylphosphofluoridate (DFP). This compound forms an ester with the hydroxyl group of an active serine, so that the enzymes that contain serine in their active site cannot interact with their substrates.

$$E\begin{bmatrix} C=O \\ | \\ CHCH_2OH \\ | \\ -NH \end{bmatrix} + FP\begin{matrix} OCH(CH_3)_2 \\ \\ \| \\ O \end{matrix}OCH(CH_3)_2 \longrightarrow E\begin{bmatrix} C=O \\ | \\ CHCH_2OP \\ | \\ -NH \end{bmatrix}\begin{matrix} OCH(CH_3)_2 \\ \\ \| \\ O \end{matrix}OCH(CH_3)_2 + HF$$

Enzyme　　　　　　DFP　　　　　　　　Deactivated enzyme

Reaction with DFP may therefore be used to determine whether a serine unit is involved in the active site of an enzyme.

Another group that appears as a key factor in many enzymes (e.g., glyceraldehyde 3-phosphate dehydrogenase) is the sulfhydryl group of a cysteine residue. The activity of such enzymes can be blocked by iodoacetate:

$$\left[\begin{array}{c}-C=O \\ | \\ E\ CHCH_2SH \\ | \\ -NH\end{array}\right] + ICH_2COO^- \longrightarrow \left[\begin{array}{c}-C=O \\ | \\ E\ CHCH_2SCH_2COO^- \\ | \\ -NH\end{array}\right] + HI$$

Enzyme   Iodoacetate   Deactivated enzyme

Similarly, certain enzymes that have a critical requirement for magnesium ions are inhibited by fluoride, which forms a complex with magnesium and thus renders it unavailable for action with the enzyme.

Although the specific mechanism for the action of each enzyme is unique to the process that it catalyzes, certain sequences of steps are common to many enzymes. The initial step is the non-covalent binding of the substrate to the enzyme. This may involve hydrophobic interactions, hydrogen bonding, ionic attractions, or other weak, reversible interactions. Specific groups or residues of the enzyme then react further with the substrate at the site where the latter is transformed. As a result, various bonds in the substrate are weakened, its electronic structure may be rearranged, or new bonds may be introduced. In this step (or sequence of steps), the enzyme and substrate may form various types of transitory covalent structures. Ultimately, the products of these interactions are released from the enzyme.

This general process can be illustrated by some of the steps in the hydrolysis of proteins by the digestive enzyme trypsin. In the following equation, the C=O of the peptide bond is that of a lysine or arginine unit, and P and P' are polypeptide chains of the protein substrate:

$$\text{P}-\overset{\overset{\displaystyle O}{\|}}{\text{C}}-\overset{\overset{\displaystyle H}{|}}{\text{N}}-\text{P}' + H_2O \longrightarrow \text{P}-\overset{\overset{\displaystyle O}{\|}}{\text{C}}\text{OH} + H_2N-\text{P}'$$

Trypsin contains more than twenty serine residues; however, one of these—serine-195—is highly reactive. (The numbering system for the amino acid residues relates to a group of proteases that operate by a common mechanism.) The active site of trypsin also contains, among other amino acids, a histidine (residue 57) and aspartic acid (residue 102). As a result of their interaction in the three-dimensional structure of the enzyme, serine-195 is rendered strongly nucleophilic (Fig. 2-13A). Consequently, after formation

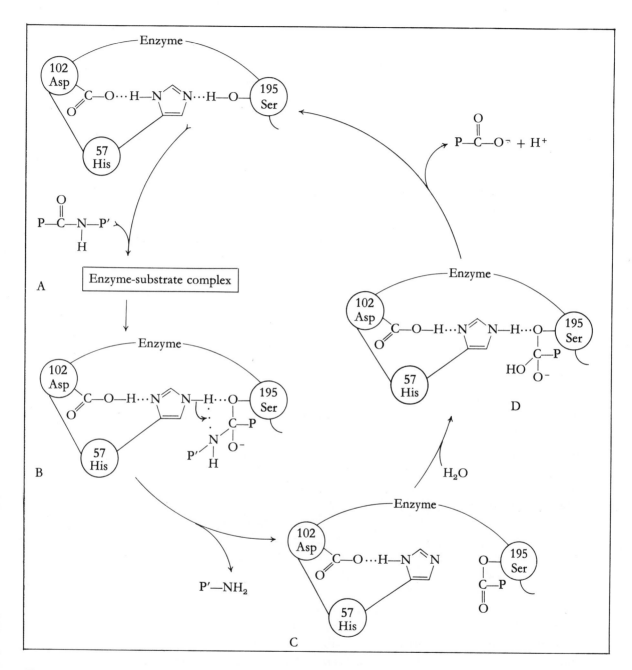

Fig. 2-13. *Mechanism for the cleavage of peptide bonds by trypsin. A. Interaction of three amino acids in the active side of the enzyme. B. Tetrahedral intermediate between serine-195 and substrate,* P—C(=O)—N(H)—P'. *C. Acyl-enzyme. D. Acyl—enzyme—H_2O intermediate.*

of the enzyme-substrate complex, serine-195 interacts with the carbonyl unit of the peptide bond in the substrate to generate a "tetrahedral intermediate" (Fig. 2-13B). Attack by a proton from histidine-57 results in the release of a polypeptide fragment (containing the amino group of the cleaved peptide bond) and formation of an "acyl enzyme" (Fig. 2-13C). Interaction of the "acyl enzyme" with a molecule of water results in another tetrahedral intermediate (Fig. 2-13D) that yields the carboxyl fragment of the substrate and the regenerated enzyme. A critical feature of this sequence is the shifting of protons between histidine-57 and aspartate-102. Hydrogen bonding between these residues and that between histidine-57 and serine-195, as well as the reversible movement of protons through these three amino acids, provides a "charge relay" system that produces the catalytic effect of the enzyme.

Although this is a highly oversimplified summary, it illustrates certain important aspects of enzyme catalysis, including (1) the interactions between amino-acid residues at the active site, (2) the formation of covalent intermediates between certain enzymes and substrates, (3) the contributions of acid-base reactions, and (4) the re-formation of the original enzyme. It can also be seen that DFP will inhibit the reaction by blocking the serine hydroxyl group, and that the process will be sensitive to the pH of the medium.

2.13. Enzyme Inhibition

The catalytic action of an enzyme can be modified, diminished, or completely nullified in a number of ways. Obviously, if the protein is denatured and the three-dimensional structure of the enzyme is damaged, it will lose its effectiveness. Similarly, any reaction that blocks functional groups at the active site will modify or nullify the activity of the enzyme. Generally, such changes are irreversible. Thus, when a substance that causes this type of inhibition is added to an enzyme, the enzyme is so modified that its complete original activity cannot be regained. For example, if the integrity of an enzyme requires the presence of free sulfhydryl groups (–SH), it will be deactivated by iodoacetate. Similarly, enzymes that depend on active serine residues are inactivated by DFP.

In contrast to the irreversible deactivation of enzymes caused by the above-mentioned reagents and a number of similar ones, enzymes may be inhibited by various substances in a reversible manner. Two major types of inhibition can be distinguished on the basis of kinetic analyses. These are termed *competitive* and *non-competitive* inhibition. A third type, which occurs comparatively rarely when only a single substrate is involved, is termed *uncompetitive* inhibition. With multiple substrate reactions, various types of mixed inhibition may occur in addition to competitive, non-competitive, or uncompetitive inhibition.

Competitive inhibition involves a reversible binding of the inhibitor with the binding site of the enzyme. Substances that produce this type of inhibition are related structurally to the substrate. They fit into the area where the substrate binds, but they do not undergo any transformations. Thus, in addition to the normal reversible binding that occurs between the substrate and enzyme, there is a concurrent reversible binding between inhibitor and enzyme. The two competing reactions may be depicted as follows:

$$E + S \rightleftharpoons ES$$
$$E + I \rightleftharpoons EI$$

where I is the competitive inhibitor.

The degree of inhibition will depend on the relative affinity of the enzyme for the substrate versus that for the inhibitor as well as on the relative concentrations of substrate and inhibitor. For a fixed concentration of inhibitor, the rate of reaction will increase as the concentration of substrate is increased. Ultimately, the same maximal velocity as in the uninhibited reaction will be attained. However, the concentration of substrate required to produce maximal velocity will be higher than that necessary when there is no inhibitor (Fig. 2-14). In other words, since two components are competing for the same site and both reactions are reversible, the inhibition can be overcome by relatively high concentrations of substrate. The equation for the reaction velocity with competitive inhibition is:

$$v_0 = \frac{V_{max}[S]}{[S] + K_M \left(1 + \frac{[I]}{K_I}\right)} \qquad (2\text{-}18)$$

where K_I is the dissociation constant of the enzyme-inhibitor complex; $K_I = [E][I]/[EI]$. Competitive inhibition is thus characterized by a change in the apparent K_M but no change in V_{max}. This is shown more clearly by comparing the reciprocal plots of the inhibited and uninhibited reactions (Fig. 2-14).

A classic example of competitive inhibition is the action of malonic acid on the activity of succinate dehydrogenase. This enzyme catalyzes the oxidation of succinic acid (section 3.4) and the reaction is inhibited competitively by malonic acid. The possibility for the binding of malonic acid at the substrate binding site of the enzyme arises from the similarity of the two structures:

COOH
|
CH$_2$
|
CH$_2$
|
COOH
Succinic acid

COOH
|
CH$_2$
|
COOH
Malonic acid

Although the enzyme binds with malonate, it does not produce any reaction. Thus, malonate competes with succinate for the enzyme and inhibits the oxidation of the natural substrate. When the concentration of succinate relative to that of malonate is increased, the inhibitor is displaced from the enzyme and oxidation of succinate can reach the maximum rate.

A *non-competitive inhibitor* binds reversibly to the enzyme at a site other than the substrate binding site. The resultant enzyme-inhibitor complex has no enzymatic activity, even though it can still bind to the substrate. Since the substrate and enzyme are not competing for the same site, the inhibition is

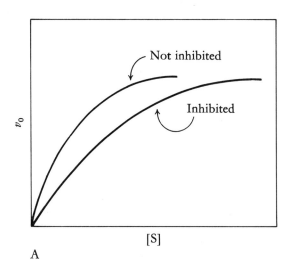
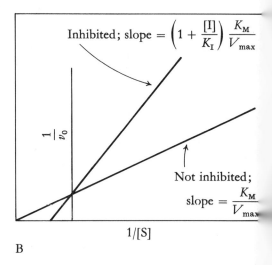

*Fig. 2-14. A. Effect of substrate concentration [S] on initial reaction velocity v_0 in the presence of a competitive inhibitor. This is compared with the uninhibited reaction.
B. Plot of reciprocal velocity versus reciprocal substrate concentration for the reaction shown in A. The equation for this plot is:*

$$\frac{1}{v_0} = \frac{K_M}{V_{max}} \left(1 + \frac{[I]}{K_I}\right) \left(\frac{1}{[S]}\right) + \frac{1}{V_{max}}$$

Note that the intercept on the ordinate is the same for the inhibited and non-inhibited reactions, namely, $1/V_{max}$. However, the slope is increased by the factor $(1 + [I]/K_I)$.

not reversed by increasing the relative amount of substrate (Fig. 2-15). The equation for the reaction velocity in the presence of a non-competitive inhibitor is:

$$v_0 = \frac{V_{max}[S]}{(K_M + [S])\left(1 + \frac{[I]}{K_I}\right)} \qquad (2\text{-}19)$$

In this situation, the inhibitor causes a decrease in the maximal reaction velocity, but it does not affect the K_M (Fig. 2-15).

Non-competitive inhibitors include certain metals; e.g., heavy metals inactivate some enzymes. Other inhibitors that may be non-competitive are agents that form complexes with essential metals that are required by enzymes as cofactors.

A third type of inhibition is termed *uncompetitive*. In this type, the inhibitor effects a change in both the apparent K_M and V_{max}. The velocity-substrate relationship is expressed by the equation:

$$v_0 = \frac{V_{max}[S]}{K_M + [S]\left(1 + \frac{I}{K_I}\right)} \qquad (2\text{-}20)$$

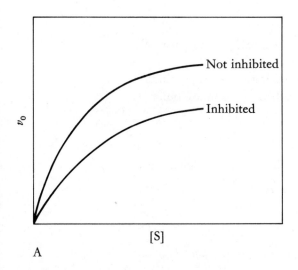
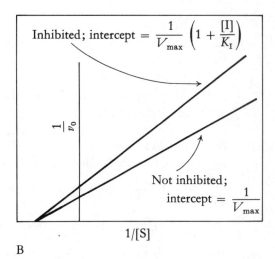

Fig. 2-15. A. Effect of substrate concentration on initial velocity in the presence of a non-competitive inhibitor. B. Double reciprocal plot for the reaction shown in A. The equation for this graph is:

$$\frac{1}{v_0} = \frac{K_M}{V_{max}}\left(1 + \frac{[I]}{K_I}\right)\left(\frac{1}{[S]}\right) + \frac{1}{V_{max}}\left(1 + \frac{[I]}{K_I}\right)$$

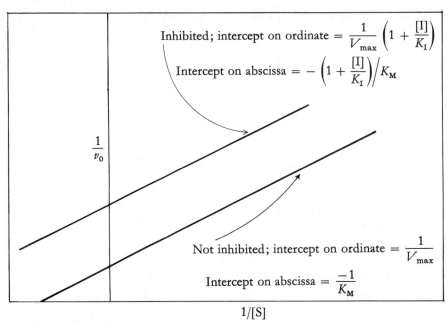

Fig. 2-16. *Double reciprocal plot for an uncompetitive inhibitor. The equation for this graph is:*

$$\frac{1}{v_0} = \frac{K_M}{V_{max}[S]} + \frac{1}{V_{max}}\left(1 + \frac{[I]}{K_I}\right)$$

A plot of $1/v_0$ versus $1/[S]$ for the reaction with uncompetitive inhibitor will be parallel to that for the uninhibited reaction (Fig. 2-16).

Several situations may give rise to this type of inhibition. One possibility is that the inhibitor interacts with the enzyme-substrate complex but not with the enzyme itself.

Suggested Reading

Bernhard, S. A. *The Structure and Function of Enzymes.* New York: Benjamin, 1968.

Chipman, D. M., and Sharon, N. Mechanism of Lysozyme Action. *Science* 165: 454–465, 1969.

Gutfreund, H. *Enzymes: Physical Principles.* New York: Wiley-Interscience, 1972.

Koshland, D. E., Jr. Protein Shape and Biological Control. *Scientific American* 229:52–64, 1973.

Phillips, D. C. The Three Dimensional Structure of an Enzyme Molecule. *Scientific American* 215:78–90, 1966.

Stroud, R. M. A Family of Protein Cutting Enzymes. *Scientific American* 231: No. 1, 24–88, 1974.

3. Catabolism of Triglycerides and the Citric Acid Cycle

The principal functions of triglycerides in cellular processes are to store energy and to provide a source of fuel. The free fatty acids that are derived from triglycerides are degraded by a series of oxidation reactions, and the energy evolved in this process can be channeled to fulfill the requirements of the organism. The triglycerides and fatty acids utilized in these transformations are derived from the diet; however, the immediate source of these substances is the material stored in the cells of various tissues.

Triglycerides are classified as a subgroup of the *lipids*. Lipids comprise a variety of substances with different chemical structures whose common characteristic is their solubility in organic solvents (e.g., chloroform, ether, or methanol) and their limited solubility in water. Since the chemistry and intermediary metabolism of the individual groups of lipids have comparatively few common features, they will be discussed in separate sections. This chapter deals exclusively with the metabolic degradation of triglycerides and the processes by which they yield energy.

3.1. Structural Chemistry

The *neutral triglycerides* are esters of glycerol and three molecules of aliphatic carboxylic acids. The general structures are depicted as follows:

$$\begin{array}{l} CH_2OH \\ | \\ CHOH \\ | \\ CH_2OH \end{array} \qquad RCH_2COOH \qquad \begin{array}{l} CH_2OCOCH_2R \\ | \\ CHOCOCH_2R \\ | \\ CH_2OCOCH_2R \end{array}$$

Glycerol   Fatty acid   Triglyceride

The two terminal carbons in glycerol or glycerol esters are called the α carbon and the α' carbon, and the center carbon is called the β carbon.

Table 3-1. Common Fatty Acids

Structure	Name	Designation*
Saturated		
$CH_3(CH_2)_{12}COOH$	Myristic	14:0
$CH_3(CH_2)_{14}COOH$	Palmitic	16:0
$CH_3(CH_2)_{16}COOH$	Stearic	18:0
$CH_3(CH_2)_{18}COOH$	Arachidic	20:0
Unsaturated		
$CH_3(CH_2)_5CH=CH(CH_2)_7COOH$	Palmitoleic	16:1 or Δ^9
$CH_3(CH_2)_7CH=CH(CH_2)_7COOH$	Oleic	18:1 or Δ^9
$CH_3(CH_2)_4CH=CHCH_2CH=CH(CH_2)_7COOH$	Linoleic	18:2 or $\Delta^{9,12}$
$CH_3CH_2CH=CHCH_2CH=CHCH_2CH=CH(CH_2)_7COOH$	Linolenic	18:3 or $\Delta^{9,12,15}$
$CH_3(CH_2)_4CH=CHCH_2CH=CHCH_2CH=CHCH_2CH=CH(CH_2)_3COOH$	Arachidonic	20:4 or $\Delta^{5,8,11,14}$

* The numerical designation defines the number of carbon atoms and the number of double bonds. Thus, 18:2 indicates that the acid is composed of 18 carbon atoms and contains two double bonds. The other notation, which uses the Greek letter delta (Δ) and a superscript, indicates the position of the double bond. For example, $\Delta^{9,12}$ means that there are double bonds between carbon-9 and carbon-10 and between carbon-12 and carbon-13.

Another nomenclature system employs numerals (i.e., 1, 2, 3) to specify the carbons on glycerol esters.

The fatty acids that are generally found in animal triglycerides are monocarboxylic acids with an even number of carbons. They contain long hydrocarbon chains, which may be either saturated or unsaturated. The structures of the fatty acids that occur most frequently are listed in Table 3-1. The most abundant of the saturated acids are palmitic and stearic acids. In addition to the acids itemized in Table 3-1, significant amounts of other fatty acids are found in animal tissues. These are listed in Table 3-2.

Over 50% of the fatty acids in triglycerides are unsaturated. The most abundant of the unsaturated fatty acids is oleic acid. Several characteristics of the structure of these acids should be noted. Except for arachidonic acid, the naturally occurring unsaturated acids generally have one unsaturation site between carbon-9 and carbon-10. (Carbons are numbered from the carboxyl end, which is designated as carbon-1.) In the polyunsaturated acids, the double bonds are spaced three carbons away from each other. The geometric configuration around the double bond is usually a *cis* relationship; the structure of oleic acid would thus be:

$$HC(CH_2)_7CH_3$$
$$\parallel$$
$$HC(CH_2)_7COOH$$

The unsaturated fatty acids are liquids at room temperature; saturated acids with ten or more carbons are solids.

Table 3-2. Other Fatty Acids of Animal Tissue

Structure	Name	Designation
Saturated		
$CH_3(CH_2)_2COOH$	Butyric	4:0
$CH_3(CH_2)_4COOH$	Caproic	6:0
$CH_3(CH_2)_8COOH$	Capric	10:0
$CH_3(CH_2)_{10}COOH$	Lauric	12:0
$CH_3(CH_2)_{22}COOH$	Lignoceric	24:0
Unsaturated		
$CH_3(CH_2)_7CH=CH(CH_2)_{11}COOH$	Erucic	22:1 or Δ^{13}
$CH_3(CH_2)_7CH=CH(CH_2)_{13}COOH$	Nervonic	24:1 or Δ^{15}
Hydroxy acids		
$CH_3(CH_2)_{21}CHOHCOOH$	Cerebronic	24:0, hydroxy
$CH_3(CH_2)_7CH=CH(CH_2)_{12}CHOHCOOH$	α-Hydroxynervonic	24:1, hydroxy

Natural triglycerides are esters of glycerol with different fatty acids. When the substituents on the terminal carbons of the glycerol differ, optical isomers result. The glycerol carbons are numbered from top to bottom as written, and the animal triglycerides have an L configuration. The following compound would thus be L-1-stearoyl-2-palmitoyl-3-oleyl-glycerol:

$$\begin{array}{c} CH_2OCO(CH_2)_{16}CH_3 \\ | \\ CH_3(CH_2)_{14}COOCH \\ | \\ CH_2OCO(CH_2)_7CH=CH(CH_2)_7CH_3 \end{array}$$

Generally, the numerals preceding names of the acid radicals can be deleted, since the order is assumed in the nomenclature. This compound can also be named α-stearyl-β-palmityl-α′-oleylglyceride.

Hydrolysis of triglycerides with acids yields glycerol and the component free fatty acids:

$$\begin{array}{c} CH_2OCOR \\ | \\ R'COOCH \\ | \\ CH_2OCOR'' \end{array} + H_2O \longrightarrow \begin{array}{c} CH_2OH \\ | \\ CHOH \\ | \\ CH_2OH \end{array} + RCOOH + R'COOH + R''COOH$$

Hydrolysis in the presence of alkali (saponification) results in the formation of the salts of the fatty acids in addition to glycerol. These salts of long-chain fatty acids are called *soaps* and have characteristic detergent activity. Since the free fatty acids are weak acids with a pK of about 5, they occur as anions in the physiologic environment, where the pH is somewhat higher than 7.

3.2. Digestion and Absorption

Digestion of dietary triglycerides occurs in the small intestine. The agents that promote this process are *pancreatic lipase* and the *bile salts*. Both of these are introduced into the duodenum via the common bile duct. The lipase is formed in the pancreas and is one of the components of pancreatic juice. This enzyme catalyzes the hydrolysis of fatty acyl moieties from the *alpha* positions in triglycerides, producing β-monoglycerides (2-monoglycerides) and fatty acids:

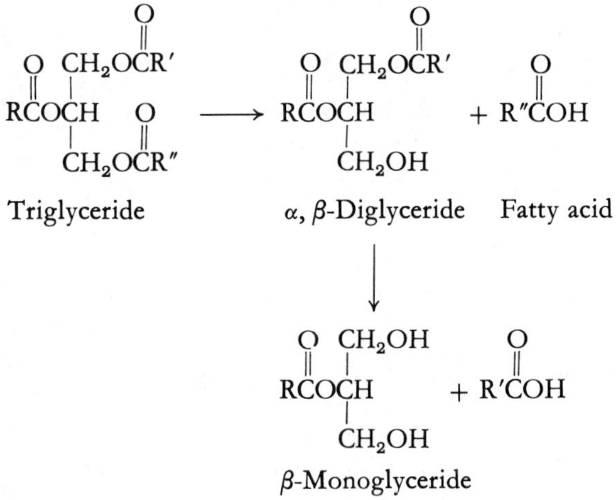

$$\underset{\text{Triglyceride}}{\begin{array}{c} \text{O} \\ \| \\ \text{RCOCH}_2\text{OCR}' \\ | \\ \text{CH}_2\text{OCR}'' \end{array}} \longrightarrow \underset{\alpha,\beta\text{-Diglyceride}}{\begin{array}{c} \text{O} \\ \| \\ \text{RCOCH}_2\text{OCR}' \\ | \\ \text{CH}_2\text{OH} \end{array}} + \underset{\text{Fatty acid}}{\text{R}''\text{COH}}$$

$$\downarrow$$

$$\underset{\beta\text{-Monoglyceride}}{\begin{array}{c} \text{O} \\ \| \\ \text{RCOCH}_2\text{OH} \\ | \\ \text{CH}_2\text{OH} \end{array}} + \text{R}'\text{COH}$$

Dietary lipids are generally in the form of oil droplets that may be visible or too fine to be seen. Since triglycerides are insoluble in the water solution of the intestinal lumen, any action of the enzyme is limited to the area of the lipid–water interface. One function of the bile salts is to emulsify the mixture and disperse the oily phase into numerous droplets. The area of the lipid–water interface is thus increased considerably, and the surface at which lipase action can occur is multiplied many times.

Bile, which contains a number of components, is secreted by the liver and is concentrated in the gallbladder. As a result of a specific stimulatory mechanism, the bile is transferred through the cystic duct and the common bile duct into the duodenum. The components involved in metabolism of lipids are the bile acids, such as glycocholic acid and taurocholic acid:

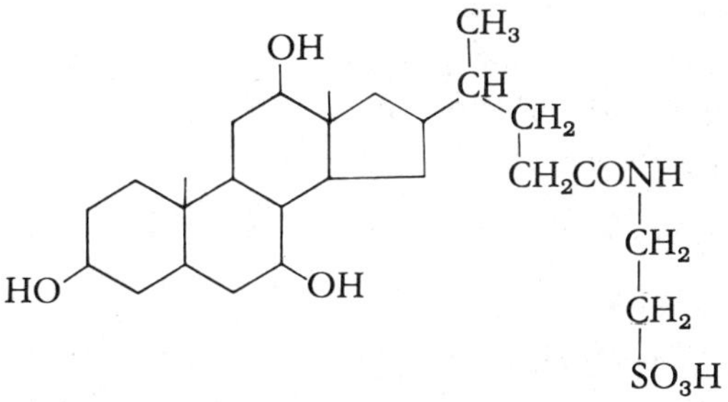

Taurocholic acid

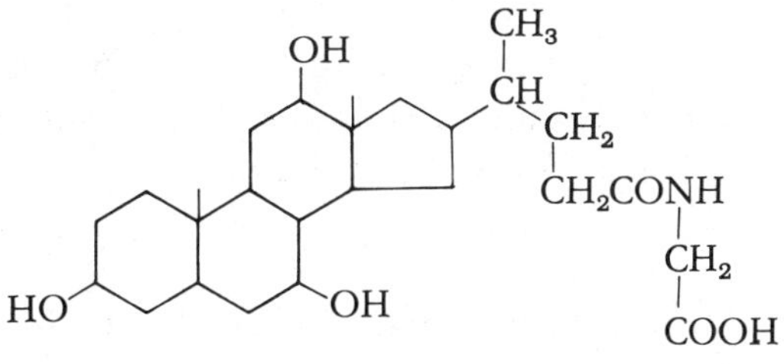

Glycocholic acid

The carboxyl and sulfate groups in these compounds are ionized at physiologic pH and in the medium of the intestine. Thus these substances are more correctly designated *bile salts*. The bile salts contain polar sections, i.e., the carboxylate or sulfate groups, that are soluble in water (hydrophilic) and non-polar hydrocarbon portions that are water-insoluble (hydrophobic). In a two-layer system involving water and lipids, such compounds accumulate at the interface, where the hydrophobic and hydrophilic portions of the molecules interact with the phase with which they are miscible. If the two phases are mixed intimately or homogenized, the result will be a comparatively stable suspension of one phase in the other. This mixture is stabilized by compounds containing hydrophobic and hydrophilic groups that interact with both phases; such compounds are called *detergents*. In addition to the bile salts, other compounds that have detergent activity are the long-chain fatty acids. Thus, the fatty acids that are released by the action of pancreatic lipase and are ionized at pH 7 serve as detergents, to aid in the further enzymatic digestion of glycerides.

The bile salts also form micelles with the fatty acids and monoglycerides released during lipolysis of triglycerides. This process increases the effective solubility of the digestion products and enhances their absorption by the mucosal cells of the intestinal villi. Within these cells, the monoglycerides are reesterified to triglycerides and combined with protein. The lipid-protein complexes, termed *chylomicrons*, are released into the lymph and transferred to the blood via the thoracic duct.

The triglycerides in circulating chylomicrons are hydrolyzed to free fatty acids and glycerol by *lipoprotein lipase*. The released fatty acids can be absorbed by adipocytes, where they are esterified and stored as triglycerides. The characteristics, means of transport, and sites of deposition of plasma lipids will be discussed in Chapter 6.

Triglycerides are stored primarily in adipose tissue. Smaller amounts are also present in liver and other tissues. In order to be metabolized, the triglycerides are hydrolyzed to glycerol and fatty acids, and the latter are transported to cells of various tissues. Oxidative degradation of fatty acids occurs in skeletal and heart muscle, as well as in liver, kidney, and other tissues. Nerve tissue and brain do not utilize fatty acids for energy.

3.3. Degradation of Fatty Acids to Acetyl-Coenzyme A

The first step in the intracellular catabolism of fatty acids is their interaction with *coenzyme A*:

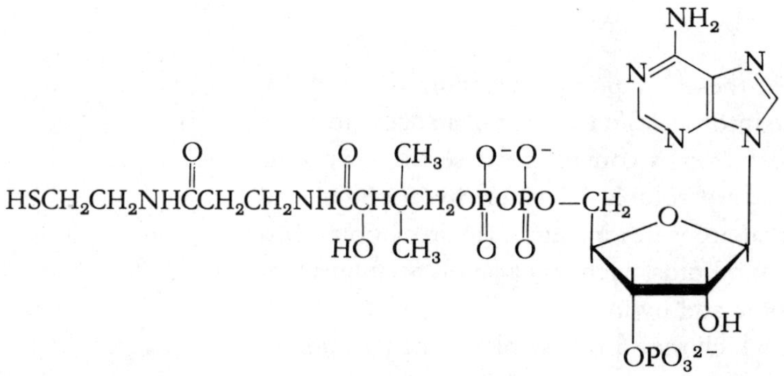

$$HSCH_2CH_2NHCCH_2CH_2NHCCHCCH_2OPOPO-CH_2\cdots$$

Coenzyme A

This compound is of special interest because it serves as a carrier of acyl units in a number of diverse reactions. It should also be noted that a section of its structure includes the vitamin pantothenic acid,

$$HOOCCH_2CH_2NHCOCHOHC(CH_3)_2CH_2OH$$

Coenzyme A combines with carboxylic acids through the sulfhydryl group to produce a thioester. In order to emphasize the importance of the sulfhydryl group in various reactions, coenzyme A is often written as *CoA–SH*, even though the SH group is actually part of the molecule. Similarly, the compound formed from coenzyme A and a carboxylic acid may be written as *RCO–SCoA* to indicate the sulfur linkage.

Long-chain fatty acids interact with coenzyme A and ATP to yield the acyl-coenzyme A plus AMP and pyrophosphate. The enzyme *thiokinase* (acyl-CoA synthetase), which is bound to the outer membrane of mitochondria, catalyzes this reaction:

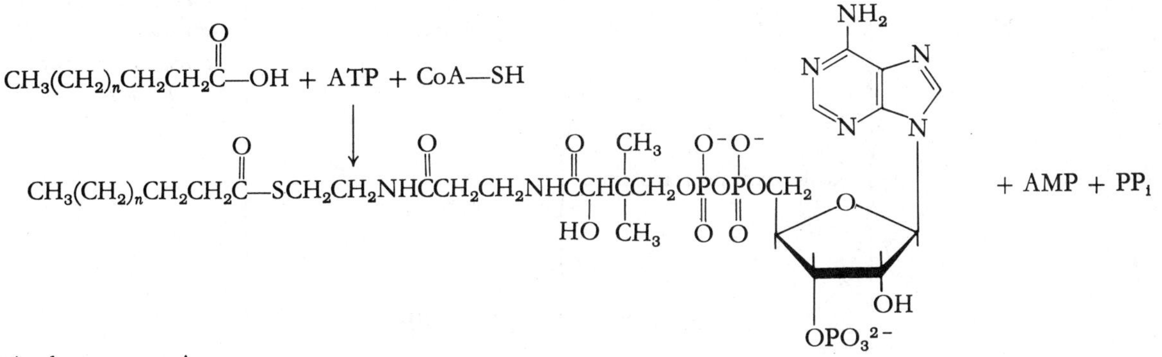

$$CH_3(CH_2)_nCH_2CH_2C(=O)-OH + ATP + CoA-SH \longrightarrow$$

$$CH_3(CH_2)_nCH_2CH_2C(=O)-SCH_2CH_2NHC(=O)CH_2CH_2NHC(=O)CH(OH)C(CH_3)_2CH_2OP(=O)(O^-)OP(=O)(O^-)OCH_2-\text{(adenosine-3'-phosphate)} + AMP + PP_i$$

Acyl-coenzyme A

The process involves the hydrolysis of a pyrophosphate group from ATP. A major fraction of the energy released from this cleavage is conserved by formation of a new energy-rich compound, namely, acyl-coenzyme A ($\Delta G^{0\prime}$ for the hydrolysis of acyl-coenzyme A is about -7 kilocalories per mole). In an isolated system, therefore, the reaction shown is reversible, and it has an equilibrium constant of about 1. In the cell, however, the process is driven toward the formation of acyl-coenzyme A by removal of one of the products; specifically, the inorganic pyrophosphate is hydrolyzed irreversibly to inorganic phosphate by the enzyme *inorganic pyrophosphatase*:

$$PP_i \longrightarrow 2P_i$$

The subsequent steps in the catabolism of acyl-coenzyme A occur within the mitochondria. Since the mitochondrial membrane is not permeable to the acyl-coenzyme A, special mechanisms are required for the transport of this substrate. A system that is instrumental in this process involves the compound *carnitine*:

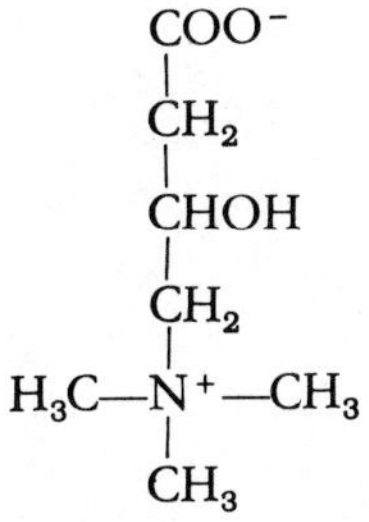

$$\begin{array}{c} COO^- \\ | \\ CH_2 \\ | \\ CHOH \\ | \\ CH_2 \\ | \\ H_3C-N^+-CH_3 \\ | \\ CH_3 \end{array}$$

Carnitine

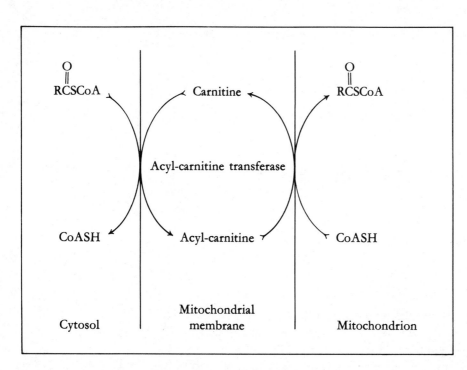

Fig. 3-1. Transport of acyl-coenzyme A through the mitochondrial membrane via carnitine carrier. The acyl-carnitine transferase is bound to the mitochondrial membranes.

The mechanism for the transfer of acyl-coenzyme A involves its interaction with carnitine to form acyl-carnitine and coenzyme A. This reaction is catalyzed by *acyl-carnitine transferase*, which is bound to the mitochondrial membrane:

$$CH_3(CH_2)_nCH_2CH_2COOCH\begin{matrix}COO^-\\|\\CH_2\\|\\\\|\\CH_2\\|\\N^+(CH_3)_3\end{matrix}$$

Acyl-carnitine

As acyl-carnitine passes through the mitochondrial membrane, it reacts with intramitochondrial coenzyme A to regenerate acyl-coenzyme A within the mitochondria. Transport processes of this type are encountered in a number of instances in intermediary metabolism. Essentially, the mechanism involves the formation of a membrane-permeating intermediate by an enzyme that is bound to the membrane (Fig. 3-1).

In addition to the thiokinase reaction, there is another process by which acyl-coenzyme A can be formed directly within the mitochondrial compartment. This involves the mitochondrial enzyme *acyl-coenzyme-A synthetase*, which catalyzes a reaction between the fatty acid, coenzyme A, and guanosine triphosphate (GTP) to yield acyl-coenzyme A, guanosine diphosphate (GDP), and inorganic phosphate:

$$CH_3(CH_2)_nCH_2CH_2COOH + CoA-SH + GTP \longrightarrow CH_3(CH_2)_nCH_2CH_2CO-SCoA + GDP + P_i$$

In both systems, the fatty acid is activated to acyl-coenzyme A. All the subsequent degradative reactions occur within the mitochondria.

Acyl-coenzyme A is dehydrogenated by a flavoprotein enzyme, *acyl-coenzyme-A dehydrogenase*. This enzyme catalyzes an oxidation that involves the transfer of two electrons and two hydrogen ions from the substrate to the enzyme's tightly bound prosthetic group, flavin adenine dinucleotide (FAD). The structures of the latter coenzyme and its reduced form (FADH$_2$) are depicted as follows:

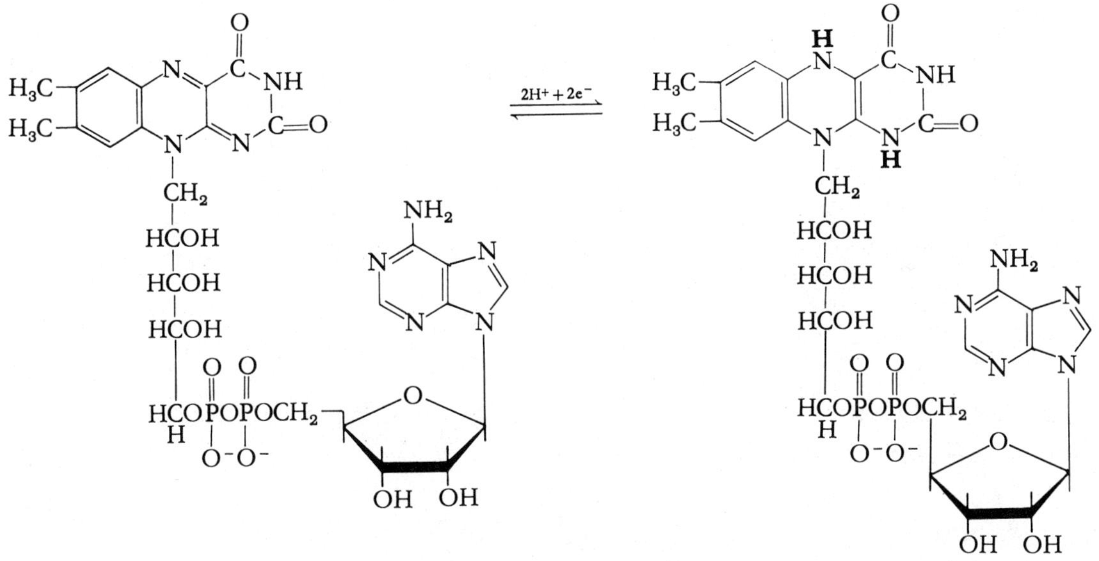

FAD   FADH$_2$

It is noteworthy that the vitamin riboflavin constitutes a portion of this coenzyme. The interaction between acyl-coenzyme A and the flavoprotein enzyme yields the α,β-unsaturated form of acyl-coenzyme A and reduced flavoprotein:

$$CH_3(CH_2)_nCH_2CH_2CO-SCoA + FAD\text{-protein} \longrightarrow CH_3(CH_2)_nCH=CHCO-SCoA + FADH_2\text{-protein}$$

The unsaturated product has a *trans* configuration, rather than the usual *cis* configuration of the naturally occurring, unsaturated fatty acids.

The enzyme *enoyl hydratase* brings about the addition of water across the double bond to yield L-3-hydroxyacyl-coenzyme A:

$$CH_3(CH_2)_nCH=CHCO-SCoA + H_2O \rightleftharpoons CH_3(CH_2)_n\overset{OH}{\underset{|}{C}}HCH_2CO-SCoA$$

α, β-Unsaturated acyl-CoA   L-3-Hydroxyacyl-CoA
                               (β-hydroxyacyl-CoA)

The hydration of the double bond is followed by oxidation of the hydroxyl group, which is catalyzed by an NAD-specific dehydrogenase, L-*3-hydroxyacyl-coenzyme-A dehydrogenase*. The products of the reaction are β-ketoacyl-coenzyme A and NADH:

$$CH_3(CH_2)_n\overset{OH}{\underset{|}{C}}HCH_2CO-SCoA + NAD^+ \rightleftharpoons CH_3(CH_2)_n\overset{O}{\overset{\|}{C}}CH_2CO-SCoA + NADH + H^+$$

β-Ketoacyl-coenzyme-A

In the next step, the enzyme *thiolase* catalyzes a reaction between free coenzyme A and the β-ketoacyl-coenzyme A in which a two-carbon unit—acetyl-coenzyme A—is cleaved from the chain. The remainder of the fatty acid becomes linked to coenzyme A to yield a fatty acyl-coenzyme A derivative with two carbons less than the original acid:

$$CH_3(CH_2)_n\overset{O}{\overset{\|}{C}}CH_2CO-SCoA + CoA-SH \longrightarrow CH_3(CH_2)_n\overset{O}{\overset{\|}{C}}-SCoA + CH_3\overset{O}{\overset{\|}{C}}-SCoA$$

Acetyl-coenzyme A

This reaction is highly exergonic; its equilibrium constant is about 6×10^4.

The new fatty acyl-coenzyme A derivative can now undergo the same sequence of reactions (flavoprotein-linked dehydrogenation, hydration, NAD-linked oxidation, and thiolase cleavage), which will result in the removal of another two-carbon unit. Ultimately, when this reaction sequence is repeated a sufficient number of times, all the original fatty acid will be converted to acetyl-coenzyme A. Thus, the total degradation of stearic acid, for example, yields 9 moles of acetyl-coenzyme A (Fig. 3-2).

Unsaturated fatty acids undergo degradative reactions similar to those of saturated acids. For example, oleic acid can be converted to oleyl-coenzyme A and oxidized by the general sequence outlined in Figure 3-2 until the twelve-carbon acid with an unsaturation site between carbon-3 and carbon-4

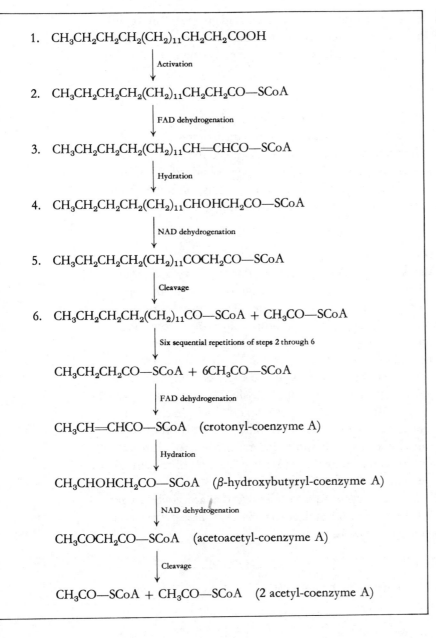

Fig. 3-2. Metabolic degradation of stearic acid to acetyl-coenzyme A.

is obtained:

cis-CH$_3$(CH$_2$)$_7$CH=CHCH$_2$CH$_2$CH$_2$CH$_2$CH$_2$CH$_2$CH$_2$CO—SCoA

cis-CH$_3$(CH$_2$)$_7$CH=CHCH$_2$CO—SCoA + 3CH$_3$CO—SCoA

Two reactions that are not required with saturated acids must occur at this point. The double bond migrates to the α,β position (between carbon-2 and carbon-3), and its original *cis* configuration is transformed to the *trans* isomer:

trans-$CH_3(CH_2)_7CH_2CH\!=\!CHCO\!-\!SCoA$

This compound may then be hydrated, dehydrogenated, and cleaved in the usual fashion. Analogous processes occur with polyunsaturated fatty acids.

Fatty acids with an odd number of carbons are degraded in the usual manner until propionyl-coenzyme A ($CH_3CH_2CO\!-\!SCoA$) is obtained. The latter is also formed in the degradation of certain amino acids. The metabolic disposition of propionyl-coenzyme A is described in Chapter 8.

3.4. Citric Acid Cycle

The *citric acid cycle* (also known as the *Krebs cycle* or the *tricarboxylic acid cycle*) consists of a series of reactions in which the acetate component in acetyl-coenzyme A is degraded to carbon dioxide and water. The process is oxygen-dependent and consequently does not occur under anaerobic conditions. The energy that is released during the specific oxidative steps of the cycle is conserved and utilized for the generation of ATP. The degradative reactions and the energy-yielding processes utilize various components of the mitochondrial apparatus. For purposes of discussion, the overall process will be divided into three topics: (1) the citric acid cycle itself, which involves the conversion of a mole of acetate to 2 moles of carbon dioxide, (2) the oxidative chain that deals with the redox reactions that ultimately lead to the reduction of oxygen to water, and (3) the utilization of the energy that is released for the generation of ATP.

The first step in the citric acid cycle is the condensation of acetyl-coenzyme A with oxaloacetate to yield citrate and coenzyme A. The enzyme that catalyzes this reaction is *citrate synthase*:

$$CH_3CO\!-\!SCoA + \underset{\underset{\displaystyle CH_2COOH}{|}}{O\!=\!CCOOH} + H_2O \longrightarrow \underset{\underset{\displaystyle CH_2COOH}{|}}{\overset{\overset{\displaystyle CH_2COOH}{|}}{HOCCOOH}} + CoA\!-\!SH$$

Acetyl-CoA   Oxaloacetic acid   Citric acid   Coenzyme A

The formulas are given as the undissociated acids; however, it should be remembered that these substances are ionized at the physiologic pH. Note that the methyl carbon of acetate is the one that links with the keto carbon of oxaloacetate. The formation of citrate as catalyzed by citrate synthase is essentially irreversible ($K = 3.2 \times 10^5$; $\Delta G^{0\prime} = -7.7$ kilocalories per mole).

Elevated levels of citrate or NADH have an inhibitory effect on the enzyme. The possible implications of these factors for metabolic regulation are discussed in a later section.

The enzyme *aconitase* then brings about the conversion of citrate to aconitate and isocitrate. The same enzyme catalyzes both reactions, i.e., both the removal and the addition of water. This results in an equilibrium mixture of the three tricarboxylic acids, citrate, aconitate, and isocitrate:

$$\begin{array}{c} CH_2COOH \\ | \\ HOCCOOH \\ | \\ CH_2COOH \end{array} \rightleftharpoons \begin{array}{c} CH_2COOH \\ | \\ CCOOH \\ \| \\ HCCOOH \end{array} + H_2O \rightleftharpoons \begin{array}{c} CH_2COOH \\ | \\ CHCOOH \\ | \\ HOCHCOOH \end{array}$$

Citric acid Aconitic acid Isocitric acid

Aconitase has a requirement for Fe^{2+}.

The next step involves the oxidation of isocitrate to α-ketoglutarate (oxoglutarate). *Isocitrate dehydrogenase*, the enzyme for this reaction, operates in conjunction with NAD^+. In addition to the oxidation that occurs in this reaction, there is a decarboxylation in which the center carboxyl group is removed as carbon dioxide:

$$\begin{array}{c} CH_2COOH \\ | \\ CHCOOH \\ | \\ HOCHOOH \end{array} + NAD^+ \longrightarrow \begin{array}{c} CH_2COOH \\ | \\ CH_2 \\ | \\ O{=}CCOOH \end{array} + CO_2 + NADH + H^+$$

Isocitric acid α-Ketoglutaric acid

This step of the citric acid cycle is of special importance since it is subject to various regulatory mechanisms (page 150).

The reaction that follows involves a complex, multienzyme system, termed the *α-ketoglutarate-dehydrogenase complex*. Operation of this system requires the coenzymes thiamine pyrophosphate (TPP), lipoate (lipoamide), FAD, NAD^+, and coenzyme A. The overall reaction may be written as follows:

$$\begin{array}{c} CH_2COOH \\ | \\ CH_2 \\ | \\ O{=}CCOOH \end{array} + CoA{-}SH + NAD^+ \xrightarrow[FAD]{TPP,\ lipoate} \begin{array}{c} CH_2COOH \\ | \\ CH_2 \\ | \\ O{=}C{-}SCoA \end{array} + CO_2 + NADH + H^+$$

α-Ketoglutaric acid Succinyl-CoA

This reaction proceeds by a series of steps catalyzed by three enzymes that are tightly associated with each other. These enzymes are *α-ketoglutarate decarboxylase, dihydrolipoate transsuccinylase*, and *dihydrolipoate dehydrogenase*. The first step, which is catalyzed by α-ketoglutarate decarboxylase, is the decarboxylation of α-ketoglutarate and the binding of the remaining group (HOOCCH$_2$CH$_2$CHOH–) to thiamine pyrophosphate:

$$\text{CH}_2\text{COOH} \atop \text{CH}_2 \atop \text{O=CCOOH}$$

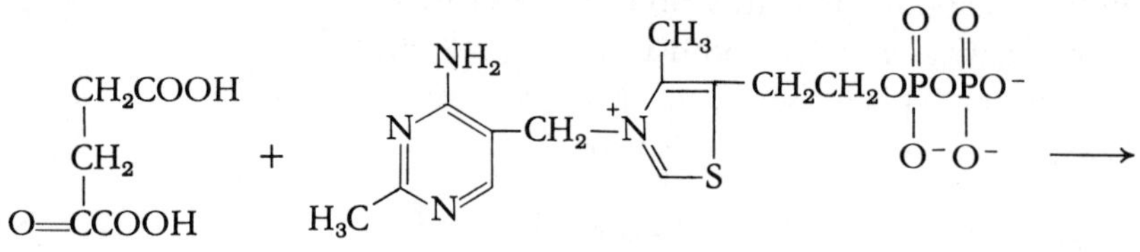

α-Ketoglutaric acid   +   Thiamine pyrophosphate   ⟶

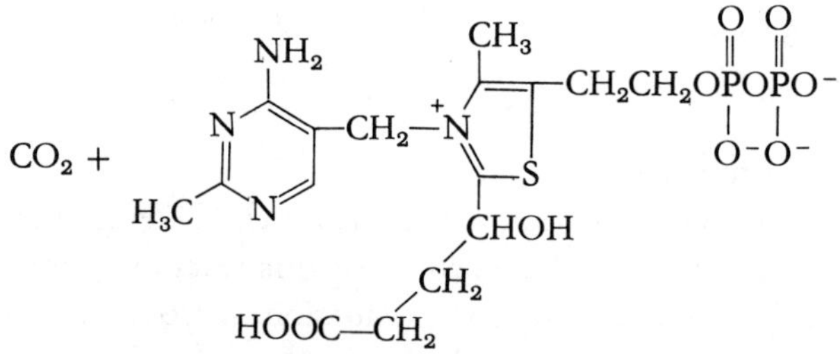

$$\text{CO}_2 \; +$$

α-Hydroxy-γ-carboxypropyl thiamine pyrophosphate

Essentially, the α-ketoglutarate is oxidized, and the thiamine pyrophosphate forms a derivative that is in a reduced state, i.e., α-hydroxy-γ-carboxypropyl thiamine pyrophosphate. The TPP derivative remains bound to the enzyme. In the second step, the α-hydroxy-γ-carboxypropyl moiety is transferred to lipoate. This reaction is catalyzed by dihydrolipoate transsuccinylase:

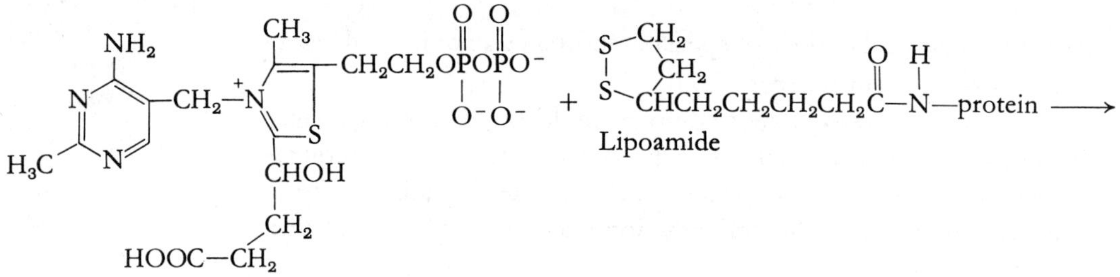

Lipoamide

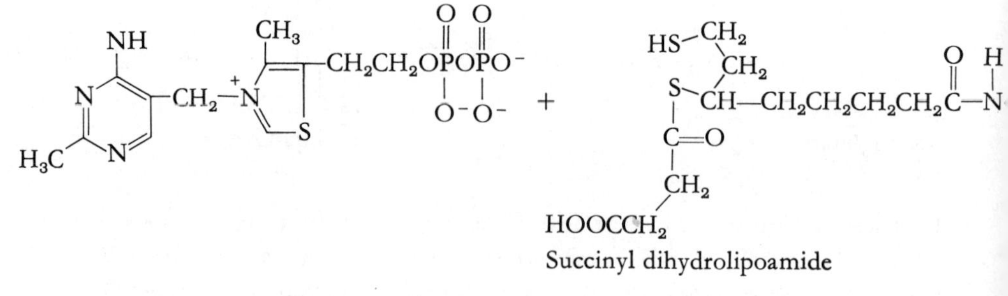

Succinyl dihydrolipoamide

As a result, thiamine pyrophosphate is regenerated and lipoate is reduced. The lipoate is not free but is linked through its carboxyl end to an ε-amino group of a lysine residue in the enzyme. (Lipoate in this form is also termed *lipoamide*.) The same enzyme, i.e., dihydrolipoate transsuccinylase, subsequently brings about the transfer of the succinyl group from dihydrolipoate to coenzyme A:

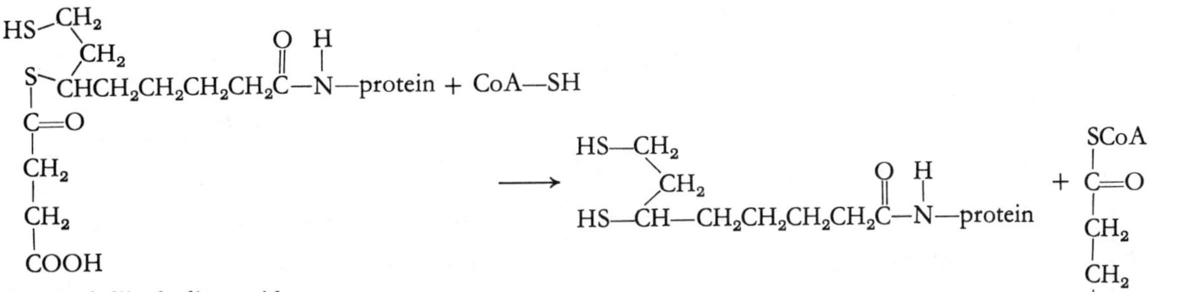

$$\text{Succinyl dihydrolipoamide} + \text{CoA—SH} \longrightarrow \text{Dihydrolipoamide} + \text{Succinyl-CoA}$$

The third enzyme, dihydrolipoate dehydrogenase, is an FAD-containing flavoprotein that catalyzes the regeneration of lipoate:

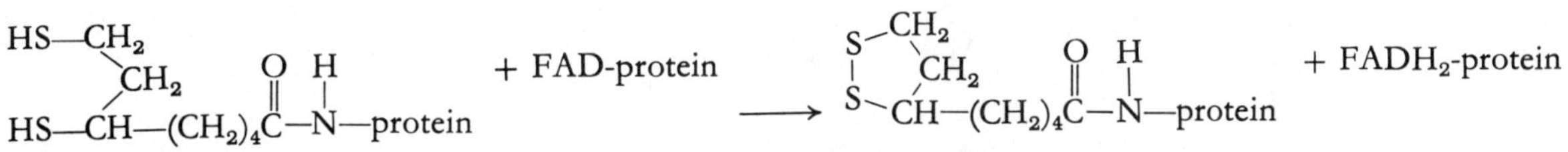

$$\text{Dihydrolipoamide} + \text{FAD-protein} \longrightarrow \text{Lipoamide} + \text{FADH}_2\text{-protein}$$

An NAD-linked dehydrogenase then effects the oxidation of the reduced flavoprotein by NAD^+:

$$\text{FADH}_2\text{-protein} + NAD^+ \longrightarrow \text{FAD-protein} + NADH + H^+$$

Hence, the net result of the reactions catalyzed by the α-ketoglutarate dehydrogenase complex is the conversion of α-ketoglutarate to succinyl-coenzyme A and the reduction of NAD. All the coenzymes in the system except NAD and coenzyme A are regenerated in the process (Fig. 3-3).

The next step in the citric acid cycle is the reaction between succinyl-coenzyme A, guanosine diphosphate (GDP), and inorganic phosphate (P_i)

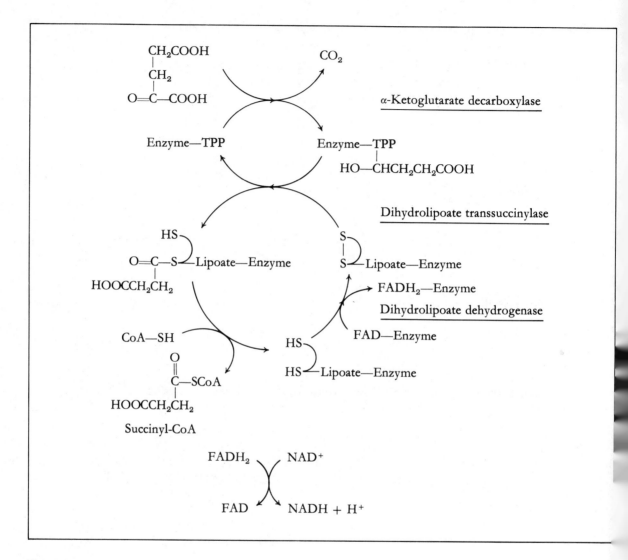

Fig. 3-3. Action of the α-ketoglutarate-dehydrogenase complex and the sequence of the intermediate steps.

to yield succinate, coenzyme A, and GTP. This reaction is catalyzed by the enzyme *succinate thiokinase*:

$$\begin{array}{c} CH_2COOH \\ | \\ CH_2 \\ | \\ O{=}C{-}SCoA \\ \text{Succinyl-CoA} \end{array} + GDP + P_i \xrightarrow{Mg^{2+}} \begin{array}{c} CH_2COOH \\ | \\ CH_2 \\ | \\ O{=}C{-}OH \\ \text{Succinic acid} \end{array} + CoA{-}SH + GTP$$

The guanosine triphosphate can yield ATP in a reaction catalyzed by nucleoside diphosphate kinase. This reaction involves the reversible transfer of phosphate from one nucleoside triphosphate to another nucleoside diphosphate:

$$GTP + ADP \rightleftharpoons GDP + ATP$$

Succinate is then oxidized by a flavoprotein-linked enzyme, *succinate dehydrogenase*. The products are fumarate and reduced flavoprotein:

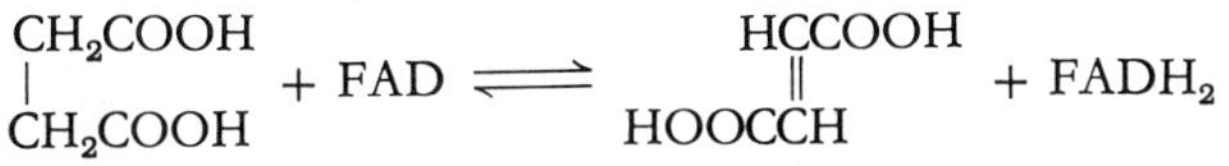

$$\begin{array}{c}CH_2COOH\\|\\CH_2COOH\end{array} + FAD \rightleftharpoons \begin{array}{c}HCCOOH\\||\\HOOCCH\end{array} + FADH_2$$

Succinic acid   Fumaric acid

It should be noted that the oxidation of succinate is the only oxidation of a citric acid cycle intermediate that is performed by a flavoprotein (the flavoprotein in the α-ketoglutarate reaction serves only to regenerate lipoate).

The enzyme *fumarase* then effects the addition of a molecule of water across the double bond of fumarate. This reaction is stereospecific and yields L-malate:

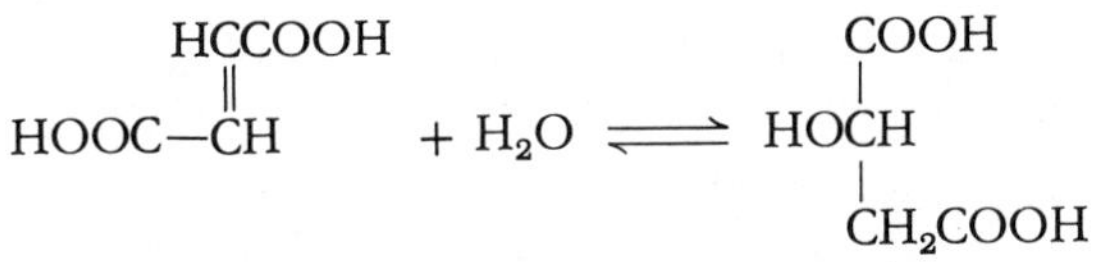

$$\begin{array}{c}HCCOOH\\||\\HOOC-CH\end{array} + H_2O \rightleftharpoons \begin{array}{c}COOH\\|\\HOCH\\|\\CH_2COOH\end{array}$$

Fumaric acid   L-Malic acid

Malate is then oxidized by an NAD-linked enzyme, *malate dehydrogenase*. This yields oxaloacetate and NADH:

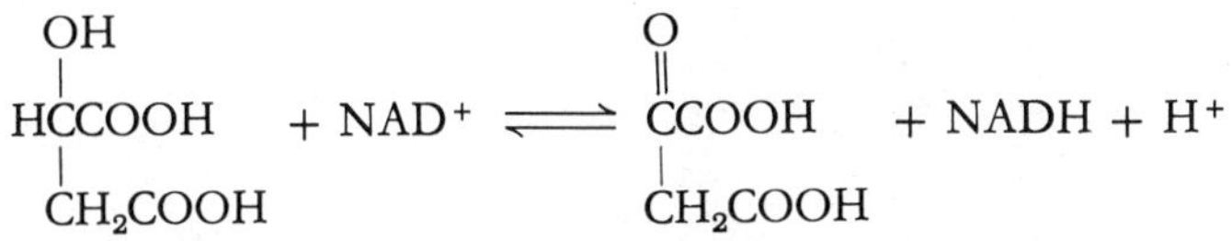

$$\begin{array}{c}OH\\|\\HCCOOH\\|\\CH_2COOH\end{array} + NAD^+ \rightleftharpoons \begin{array}{c}O\\||\\CCOOH\\|\\CH_2COOH\end{array} + NADH + H^+$$

L-Malic acid   Oxaloacetic acid

The standard free energy change for this step is about +7 kilocalories per mole. However, the reaction is driven toward the oxidation of malate by the removal and utilization of oxaloacetate. Thus, the concentration of oxaloacetate can have a regulatory effect on the cycle.

The complete sequence from the synthesis of citrate through the formation of oxaloacetate is called a *cycle* because the compound utilized in the initial step is regenerated in the final reaction; the oxaloacetate produced in the final step can react with another mole of acetyl-coenzyme A, and another turn of the cycle can begin. The complete cycle is shown in Figure 3-4.

A key function of the citric acid cycle is the degradation of acetate to two molecules of carbon dioxide. The acetate may be derived from fatty acids, pyruvate, or amino acids. The general process involves the condensation of acetate with oxaloacetate (four carbons) to yield the six-carbon compound, citrate. This is converted to the other tricarboxylic acids, aconitate and isocitrate. Isocitrate loses 1 mole of carbon dioxide to yield a five-carbon dicarboxylic acid, α-ketoglutarate. Another mole of carbon dioxide is then lost from α-ketoglutarate. At this stage, two carbons have been released as carbon dioxide, and the remaining intermediate, succinate, is a four-carbon dicarboxylic acid. (It will be shown later that the carbons are actually not those from acetate; however, this does not affect the present discussion.) The subsequent sequence—i.e., fumarate to malate to oxaloacetate—serves to regenerate the original reactant of the cycle. As oxaloacetate is formed after each turn of the cycle, it can condense with another molecule of acetyl-coenzyme A, and the same sequence of reactions may be repeated.

Presumably, if all the enzymes and cofactors of the citric acid cycle are available, and if the cycle intermediates are not utilized for other purposes, then minimal or catalytic amounts of oxaloacetate can bring about the oxidation of unlimited amounts of acetate. This implies that there is no need for *equivalent* quantities of oxaloacetate or any of the cycle intermediates in order to degrade acetate to carbon dioxide. Although this is true to a certain extent, it should be noted that specific stoichiometric quantities of certain *cofactors* are utilized in the reactions of the cycle and thus in the degradation of acetate. For example, NAD^+ is required for three of the reactions (isocitrate to α-ketoglutarate, α-ketoglutarate to succinyl-coenzyme A, and malate to oxaloacetate). As a result of these conversions, NAD^+ is removed and reduced to NADH. Similarly, the conversion of succinate to fumarate utilizes FAD. The utilization and loss of these cofactors would set limits on the activity of the cycle were it not for other reactions that bring about their regeneration. Furthermore, it is in the reoxidation of these reduced cofactors that energy is released and channeled toward the formation of ATP.

3.5. Oxidation, Reduction, and Free Energy

Before discussing biologic oxidative processes, some general aspects of oxidation-reduction will be reviewed. *Oxidation* is defined as the loss of

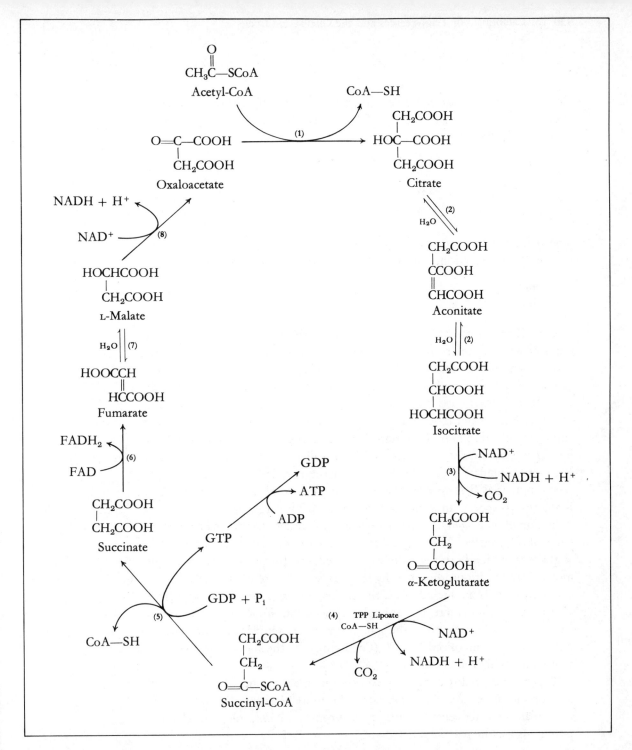

Fig. 3-4. Citric acid cycle. Enzymes: (1) citrate synthase, (2) aconitase, (3) isocitrate dehydrogenase, (4) α-ketoglutarate dehydrogenase complex, (5) succinate thiokinase, (6) succinate dehydrogenase, (7) fumarase, and (8) malate dehydrogenase.

electrons by an atom or molecule; *reduction* is the gain of electrons. Oxidation reactions may involve a transfer of hydrogen, but this is not a necessary requirement. Examples of reduction are:

$$Fe^{2+} + 2e^- \rightleftharpoons Fe^0$$

$$H^+ + e^- \rightleftharpoons \tfrac{1}{2}H_2$$

$$\tfrac{1}{2}O_2 + 2H^+ + 2e^- \rightleftharpoons H_2O$$

The reduction proceeds to a degree governed by the equilibrium constant for the specific reaction. For example, in the reduction of ferrous ion:

$$K = \frac{[Fe^0]}{[Fe^{2+}]}$$

The equilibrium constant K is a function of the relative stabilities of the oxidized and reduced forms for the specific system. It can also be considered as a measure for the ease of reduction. Actually, K is not measured directly; rather, a function of its logarithm, which is called the *electrode potential*, is the value that is determined experimentally. The value of the electrode potential of a reaction is determined relative to the oxidation–reduction of hydrogen and is expressed in *volts*. The relative value derived by comparison with the hydrogen reaction *under standard conditions* is called the *standard redox potential*, E^0, or the standard electrode potential of the reaction. In this system, the redox potential for hydrogen under standard conditions is set at zero. Reduction reactions in which the value of E^0 is larger than that for the reduction of hydrogen have positive redox potentials, and those in which the value of E^0 is smaller have negative values. *Standard conditions* here are defined as those in which both the oxidized and reduced forms are present in one-molar concentrations. For the hydrogen oxidation-reduction system, the standard conditions include that the hydrogen gas is under a pressure of 1 atmosphere and there is one-molar concentration of hydrogen ions at 25°C. For biologic reactions, the standard redox potentials are calculated for pH 7 and are symbolized by $E^{0\prime}$. (For hydrogen, the biochemical standard redox potential would then be -0.42 volt.)

The standard redox potentials express the relative tendency for an atom, ion, or molecule to attract electrons. The more positive the value of $E^{0\prime}$, the greater the tendency to attract electrons. For example, the electrode potentials for the reactions listed are:

$$Fe^{2+} + 2e^- \rightleftharpoons Fe^0; \quad E^{0\prime} = -0.44 \text{ volt}$$

$$H^+ + e^- \rightleftharpoons \tfrac{1}{2}H_2; \quad E^{0\prime} = -0.42 \text{ volt}$$

$$\tfrac{1}{2}O_2 + 2H^+ + 2e^- \rightleftharpoons H_2O; \quad E^{0\prime} = +0.82 \text{ volt}$$

Since the hydrogen system has a more positive $E^{0\prime}$ than the iron system, hydrogen ion has a greater tendency to attract electrons than does ferrous iron. By the same reasoning, molecular oxygen is a more avid electron attractor than are ferrous or hydrogen ions. Thus, if one-molar concentrations of molecular oxygen and elemental iron were combined, the oxygen would be reduced and the metallic iron would be oxidized.

Since electrons are not commonly found free in chemical reactions, the oxidation of one substance implies that another substance is reduced. In such a coupled system, the substance that is reduced is termed the *oxidizing agent* and the material that is oxidized is the *reducing agent*. Considering the value for $E^{0\prime}$, it can be seen that under standard conditions, the component with the more positive redox potential will be the better oxidizing agent; this substance will bring about the oxidation of a substance with a more negative redox potential.

Oxidation reactions that are spontaneous result in a release of energy. The relation between the difference in the standard redox potentials and the free-energy change is given by the equation:

$$\Delta G_{\text{joules}} = -nF\Delta E \tag{3-1}$$

where n is the number of electrons transferred, F is Faraday's constant (96,500 joules per volt equivalent), and ΔE is the difference in redox potentials (i.e., the value of E, in volts, of the substance that gains electrons minus the value of E of the substance that loses electrons). In order to obtain ΔG in calories, F must be expressed in calories, i.e., 23,063 calories per volt equivalent. Hence, equation 3-1 for $\Delta G^{0\prime}$ in calories is:

$$\Delta G^{0\prime}_{\text{calories}} = -23,063\, n\Delta E^{0\prime} \tag{3-2}$$

As an example, we may consider the oxidation of NADH by oxygen at pH 7 when all components are present at standard concentrations:

$$NAD^+ + 2H^+ + 2e^- \rightleftharpoons NADH + H^+; \quad E^{0\prime} = -0.32 \text{ volt}$$

$$\tfrac{1}{2}O_2 + 2H^+ + 2e^- \rightleftharpoons H_2O; \quad E^{0\prime} = +0.82 \text{ volt}$$

$$\Delta E^{0\prime} = 0.82 - (0.32) = +1.14 \text{ volts}$$

Thus, $\Delta G^{0\prime} = -23,063\,(2)\,(1.14) = -52.6$ kilocalories. Since the value of $\Delta G^{0\prime}$ is negative, this reaction is spontaneous.

The calculations given are based on standard redox potentials. At conditions other than standard, however, the redox potential is a function of the concentrations of oxidized and reduced forms of each component, according to the equation:

$$E = E^{0'} + \frac{RT}{nF} \ln \frac{[\text{Oxidized form}]}{[\text{Reduced form}]} \tag{3-3}$$

where E is the observed electrode potential, $E^{0'}$ is the standard redox potential, the value for R is 8.314 joules per degree per mole, T is the absolute temperature, and ln is the natural logarithm (2.303 log). Thus, at 25°C, equation 3-3 becomes:

$$E = E^{0'} + \frac{(8.314)(298)(2.303)}{n(96,500)} \log \frac{[\text{Oxidized form}]}{[\text{Reduced form}]}$$

or

$$E = E^{0'} + \frac{0.06}{n} \log \frac{[\text{Oxidized form}]}{[\text{Reduced form}]} \tag{3-4}$$

It is seen that increasing the concentration of the oxidized form in a mixture containing both oxidized and reduced forms results in a more positive electrode potential, E. Within the cell, the various components are not present in equimolar concentrations, and their concentrations are far below 1 M. However, some relative notions of energy release may be obtained from calculations with standard redox potentials.

3.6. Biologic Oxidations

The ultimate oxidizing agent that is required for the perpetuation of the oxidation reactions within the mitochondria is molecular oxygen. This is not apparent from the reactions described heretofore because the substrates involved do not react *directly* with oxygen. Actually, the immediate oxidant in these reactions is either NAD^+ or a flavoprotein. The resulting reduced forms of these materials are reoxidized through a stepwise series of oxidations that terminates in a reaction with oxygen. In this final oxidation, the oxygen is converted to water. This sequence of oxidations and reductions takes place on the mitochondrial membrane and is called the *respiratory chain* or the *oxidative chain*. It was indicated previously that spontaneous redox reactions proceed with a release of energy. Within the mitochondria, the energy released as a result of biologic oxidations is channeled and utilized for the synthesis of ATP.

The steps in the degradation of fatty acids that involve oxidation are listed in Table 3-3. In this list, the first two types of oxidative transformations, which are catalyzed by acyl-coenzyme-A dehydrogenase and L-3-hydroxy-acyl-coenzyme-A dehydrogenase, are repeated a number of times during the sequential release of acetyl-coenzyme A. The former enzyme is a flavoprotein

Table 3-3. Oxidation Steps in Fatty-Acid Catabolism

Reaction	Oxidant
Acyl-coenzyme A → α,β-Unsaturated acyl-coenzyme A	Flavoprotein
L-3-Hydroxyacyl-CoA → β-Ketoacyl-CoA	NAD^+
Isocitrate → α-Ketoglutarate	NAD^+
α-Ketoglutarate → Succinyl-CoA (Dihydrolipoate → Lipoate)*	NAD^+
Succinate → Fumarate	Flavoprotein
Malate → Oxaloacetate	NAD^+

* The dihydrolipoate dehydrogenase in the α-ketoglutarate dehydrogenase multienzyme complex is a flavoprotein; however, the flavoprotein is reoxidized by NAD^+.

in that it contains FAD, whereas the latter involves NAD^+ as a coenzyme. The last four reactions listed in the table are oxidations that occur in the citric acid cycle. Except for the conversion of succinate to fumarate, the oxidant in each of the citric acid cycle reactions is NAD^+. As a result of all these oxidations, NAD^+ and flavoprotein are reduced. Since the cellular concentrations of the coenzymes are fixed within narrow limits, the immediate regeneration of the oxidized forms of these coenzymes or cofactors is a vital requirement.

Reduced NAD^+ (NADH) can be reoxidized within the mitochondria by a flavoprotein enzyme termed *NADH dehydrogenase*. The effective oxidizing unit in this reaction is the prosthetic group of the dehydrogenase, i.e., flavin mononucleotide (FMN):

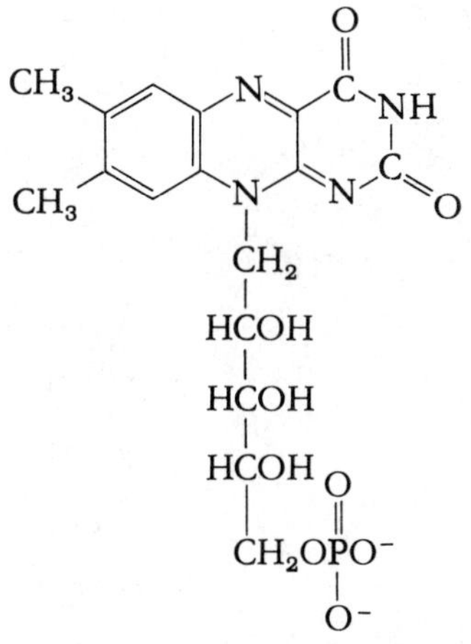

Flavin mononucleotide (FMN)

Unlike the general situation with NAD^+ in NAD-linked dehydrogenases, the FMN in flavoproteins is tightly bound to the protein. Thus, in the former instance, the NAD^+ can actually be considered one of the substrates in the dehydrogenation reaction. In the case of the flavoprotein, however, the FMN moiety is strongly bound to the protein, so the enzyme contains within itself the specific oxidizing unit.

The reaction between NADH and FMN is shown as follows:

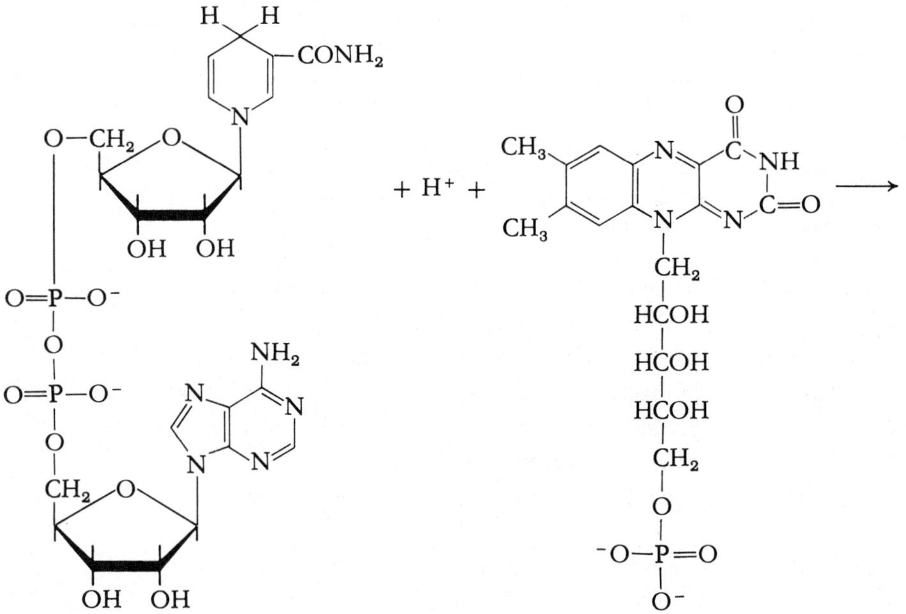

Reduced NAD + H⁺ + Oxidized FMN ⟶

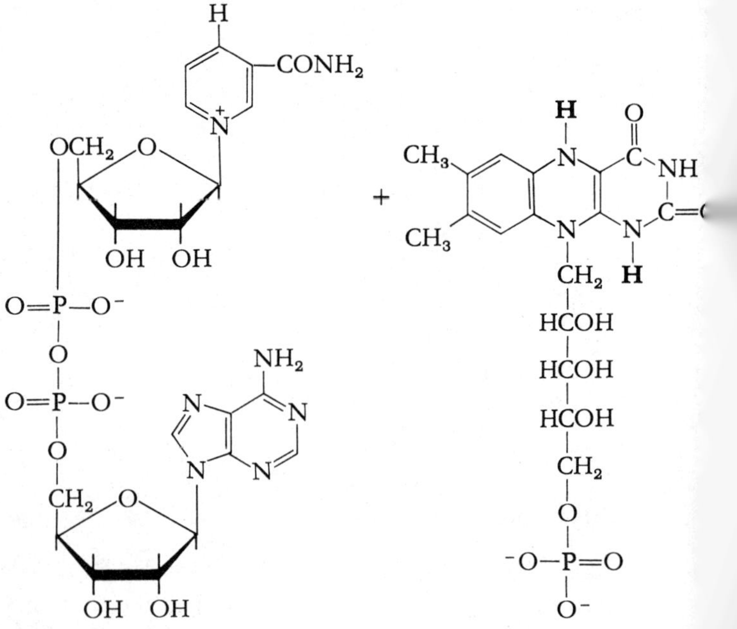

Oxidized NAD⁺ + Reduced FMN

Two electrons and a hydrogen from NADH and a proton from solution are transferred to the isoalloxazine ring of FMN to form $FMNH_2$. As a result, the pyridine nucleotide is oxidized and the flavoprotein is reduced. Starting with a specific metabolite, e.g., malate, the sequence is as follows:

$$\text{Malate} + NAD^+ \longrightarrow \text{Oxaloacetate} + NADH + H^+$$

$$NADH + H^+ + \text{FMN-protein} \longrightarrow NAD^+ + FMNH_2\text{-protein}$$

The cyclic aspect of the process may be seen more clearly when the two steps are summarized as follows:

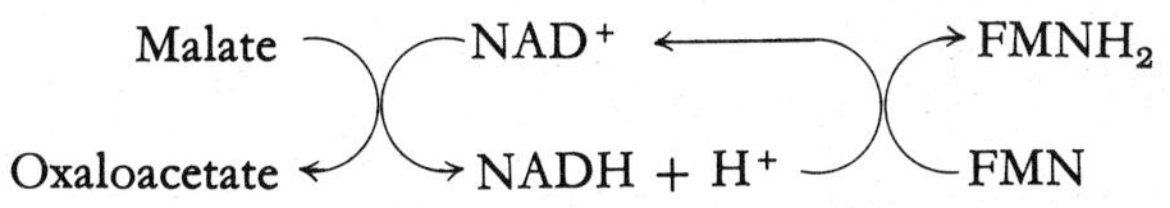

$$\begin{array}{c} \text{Malate} \\ \text{Oxaloacetate} \end{array} \rightleftarrows \begin{array}{c} NAD^+ \\ NADH + H^+ \end{array} \rightleftarrows \begin{array}{c} FMNH_2 \\ FMN \end{array}$$

In the oxidation of succinate or acyl-coenzyme A (pages 127 and 119) there is no NAD^+ involved and the substrate is oxidized directly by a flavoprotein. The prosthetic group in these dehydrogenases is flavin adenine dinucleotide (FAD); however, the actual redox reaction is similar to that of FMN in that the isoalloxazine ring is reduced. For example, the reaction with succinate may be depicted as follows:

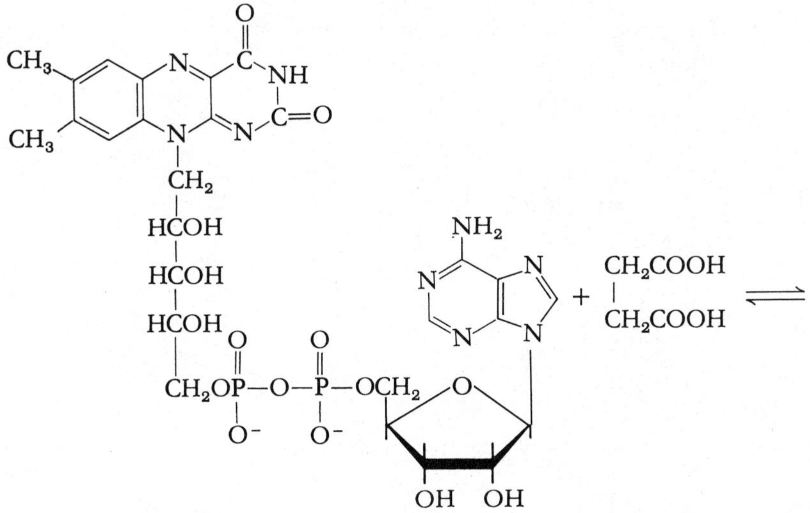

FAD + Succinic acid ⇌

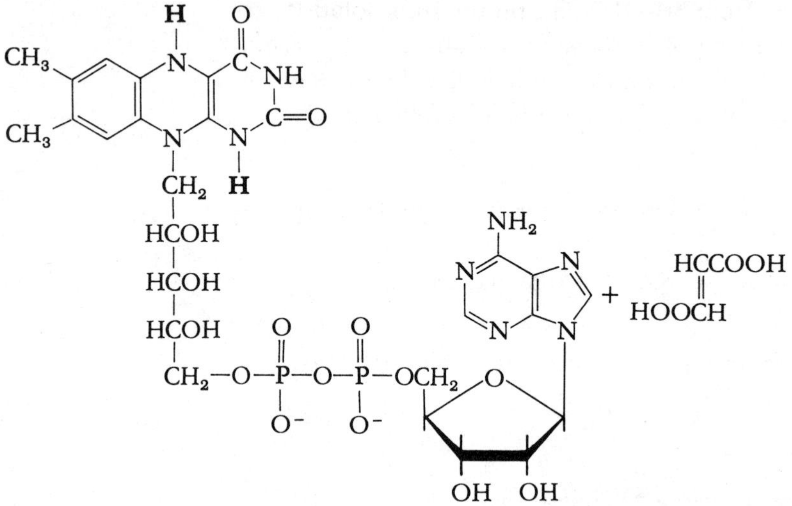

FADH$_2$ + Fumaric acid

Flavoproteins of the oxidative chain are complexed with iron, which undergoes reduction and oxidation ($Fe^{3+} + e^- \rightleftharpoons Fe^{2+}$). The total metalloprotein units are termed *iron-sulfur proteins* (FeS-proteins), because H$_2$S is released when they are decomposed with acid. The sequence of redox reactions within this mitochondrial complex appears to include a transfer of electrons from reduced flavin to the ferric ion and the release of hydrogens as protons:

$$FMNH_2 + 2Fe^{3+}\text{S-protein} \rightleftharpoons FMN + 2H^+ + 2Fe^{2+}\text{S-protein}$$

The reduced iron-sulfur proteins are reoxidized by another component of the respiratory chain, *coenzyme Q*. The structure and the redox reaction for this electron carrier are shown as follows:

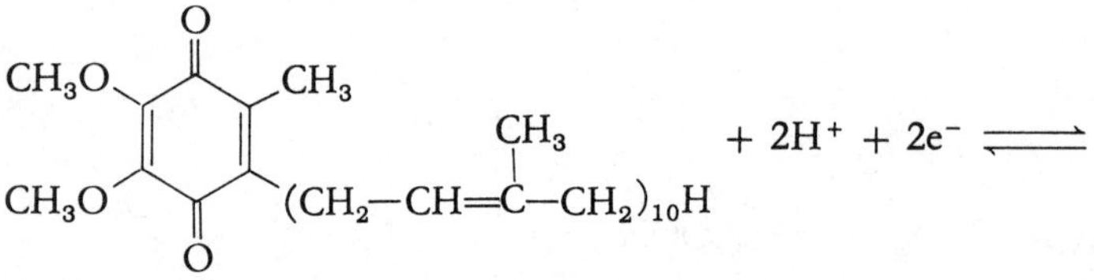

Coenzyme Q, oxidized + 2H$^+$ + 2e$^-$ ⇌

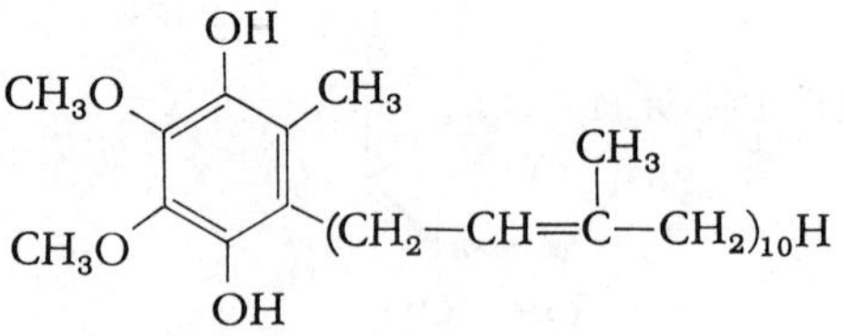

Coenzyme Q, reduced

Coenzyme Q is a derivative of benzoquinone in which one of the substituents is a polyisoprene chain; it is also known as *ubiquinone*. Some variation occurs in the number of isoprene units in the chain of coenzyme Q, which depends on the species of organism from which the coenzyme is obtained. The most common type in the mitochondria of animal tissues contains ten isoprene units and is thus called coenzyme Q_{10} or CoQ_{10}. When it functions as an oxidizing agent, the quinone is converted to a hydroquinone unit.

When the reduced iron-sulfur proteins from the oxidative chain are reoxidized by coenzyme Q, the reaction may be depicted as follows:

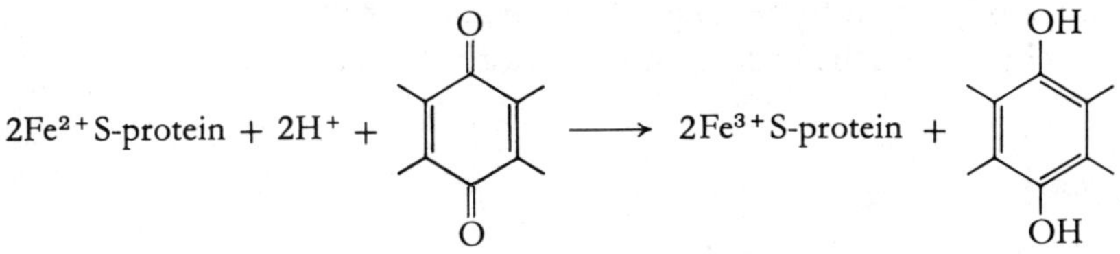

$$2Fe^{2+}\text{S-protein} + 2H^+ + \text{[quinone]} \longrightarrow 2Fe^{3+}\text{S-protein} + \text{[hydroquinone]}$$

A more abbreviated notation for outlining the electron flow is:

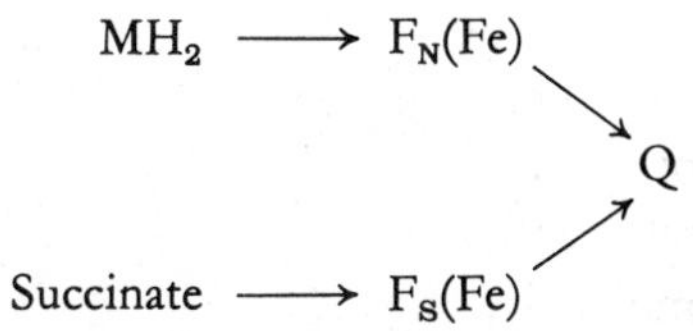

$$MH_2 \longrightarrow F_N(Fe) \searrow$$
$$Q$$
$$\text{Succinate} \longrightarrow F_S(Fe) \nearrow$$

where MH_2, F_N, F_S, and Q are the metabolite, NADH dehydrogenase, succinate dehydrogenase, and coenzyme Q, respectively, and Fe indicates that the carriers are iron-containing complexes.

Reduced coenzyme Q is reoxidized by an iron-porphyrin enzyme called *cytochrome b*. The reacting center in this redox intermediate is the iron atom in the porphyrin complex:

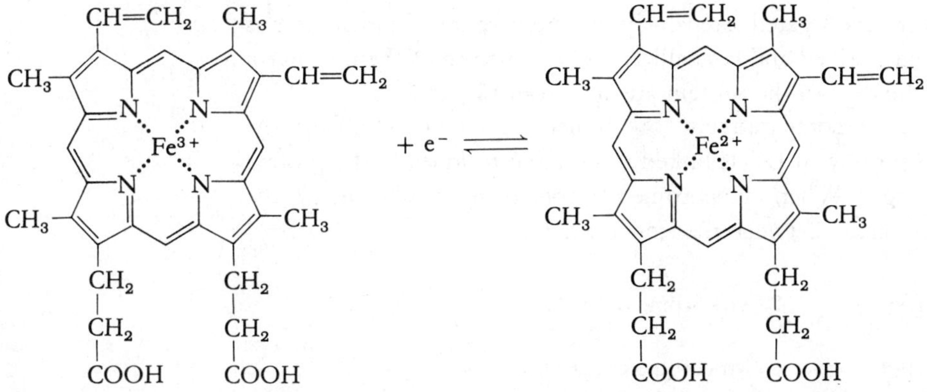

$$[\text{Fe}^{3+}\text{ porphyrin}] + e^- \rightleftharpoons [\text{Fe}^{2+}\text{ porphyrin}]$$

Porphyrins are cyclic tetrapyrroles, and numerous variations in their structure are possible, depending on the substituents on the pyrrole moieties and their specific disposition. In the case of cytochrome b, the porphyrin is classified as *protoheme IX*. The iron-porphyrin complex is bound tightly to the protein, and the complete structure is necessary for total oxidative function. (See section 10.1.)

When cytochrome b serves as an oxidizing agent, the iron is converted from Fe^{3+} to Fe^{2+}. The cytochrome thus accepts one electron. Since the oxidation of reduced coenzyme Q involves the removal of two electrons, complete oxidation of one molecule of this intermediate requires two molecules of cytochrome b. Another item to be noted at this point is that there is no transfer of hydrogen; instead, the hydrogens are released into the medium as protons:

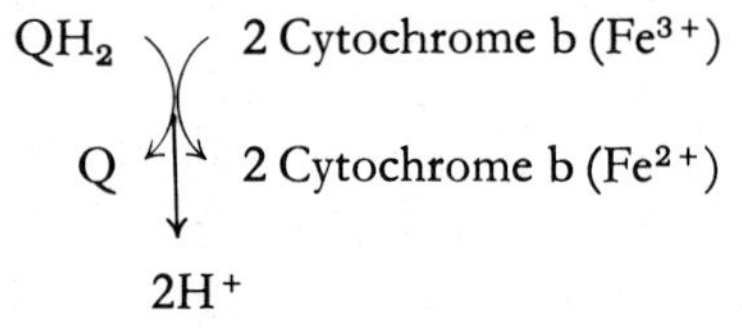

$$QH_2 \quad\quad 2\text{ Cytochrome b }(Fe^{3+})$$
$$Q \quad\quad 2\text{ Cytochrome b }(Fe^{2+})$$
$$2H^+$$

Reduced cytochrome b is subsequently oxidized by *cytochrome c_1*. This carrier, like cytochrome b and all the cytochromes, is an iron-porphyrin-protein complex. The cytochromes differ from each other structurally with respect to both the protein and the porphyrin portions of the molecule. However, the redox reactions of all the cytochromes involve the oxidation and reduction of iron:

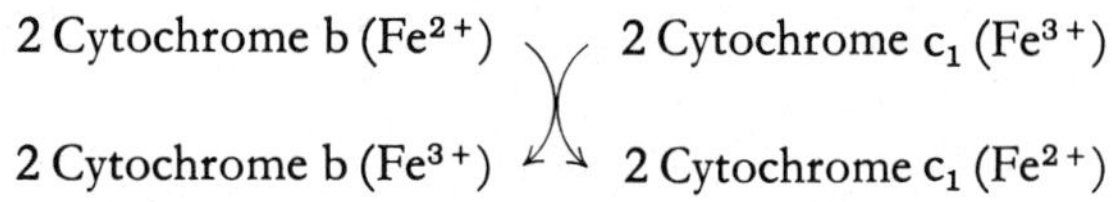

$$2\text{ Cytochrome b }(Fe^{2+}) \quad\quad 2\text{ Cytochrome } c_1\ (Fe^{3+})$$
$$2\text{ Cytochrome b }(Fe^{3+}) \quad\quad 2\text{ Cytochrome } c_1\ (Fe^{2+})$$

Cytochrome c_1, which is now in the reduced state, is oxidized by *cytochrome c*. The latter has been investigated extensively, and preparations from numerous sources have been purified and crystallized. The structure of the porphyrin and its mode of linkage to the protein are shown in Figure 3-5.

Cytochrome c has a comparatively low molecular weight (13,000); the porphyrin (protoporphyrin IX) is linked to cysteine residues of the protein by thioether linkages. When cytochrome c reacts with cytochrome c_1, the electron transfer occurs between the iron atoms:

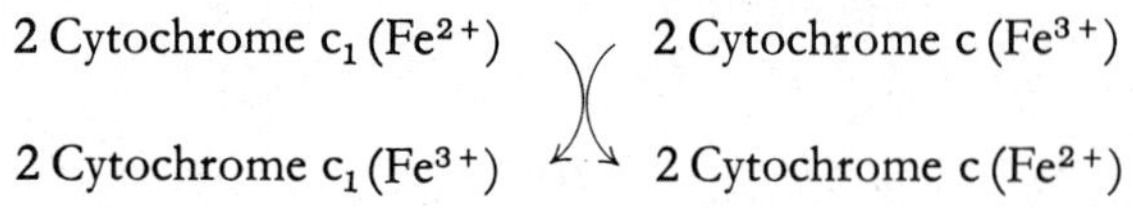

$$2\text{ Cytochrome } c_1\ (Fe^{2+}) \quad\quad 2\text{ Cytochrome c }(Fe^{3+})$$
$$2\text{ Cytochrome } c_1\ (Fe^{3+}) \quad\quad 2\text{ Cytochrome c }(Fe^{2+})$$

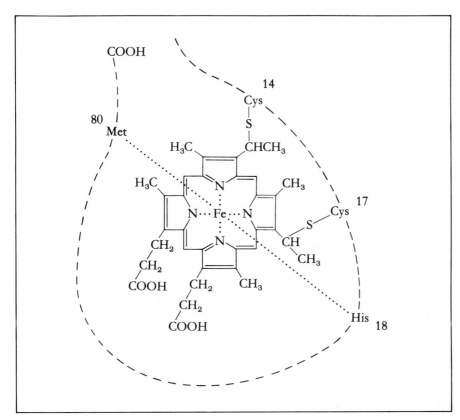

Fig. 3-5. Covalent linkages and iron binding in cytochrome c. The protein is represented by dashed lines. The porphyrin system is a planar structure in which the iron is complexed with the four nitrogen atoms. Two additional positions in the iron are coordinated with histidine-18 and methionine-80 perpendicular to the porphyrin plane as shown by the dotted line.

Reduced cytochrome c is oxidized by a complex iron-porphyrin-protein called *cytochrome oxidase*. This electron carrier is a tightly bound composite of two types of structures, termed *cytochrome a* and *cytochrome a_3*. Electron flow occurs from cytochrome a to cytochrome a_3. The complex also contains copper ions, which undergo reduction and oxidation (Cu^{2+}, Cu^+). When cytochrome oxidase reacts with cytochrome c, its iron is reduced from the ferric to the ferrous state:

Cytochrome c (Fe^{2+}) + Cytochrome oxidase (Fe^{3+}) $\rightleftharpoons$ Cytochrome c (Fe^{3+}) + Cytochrome oxidase (Fe^{2+})

Cytochrome oxidase (Fe^{2+}) is the terminal carrier in the oxidative chain. The reduced form of this enzyme reacts with oxygen in a redox reaction in which the oxygen is reduced to water and the cytochrome oxidase is reoxidized:

2 Cytochrome oxidase (Fe^{2+}) + $\frac{1}{2}O_2$ + $2H^+$ $\longrightarrow$ 2 Cytochrome oxidase (Fe^{3+}) + H_2O

The precise mechanism in this step is not clearly defined. It is known that the bound copper is critical for the final transfer of electrons to oxygen. A more complete equation of the interaction with oxygen is probably:

$$2 \text{ Cytochrome a } (Fe^{2+}) + 2 \text{ Cytochrome } a_3 (Fe^{3+}) \longrightarrow 2 \text{ Cytochrome } (Fe^{3+}) + 2 \text{ Cytochrome } a_3 (Fe^{2+})$$

$$2 \text{ Cytochrome } a_3 (Fe^{2+})(Cu^+ \rightleftharpoons Cu^{2+}) + \tfrac{1}{2}O_2 + 2H^+ \longrightarrow 2 \text{ Cytochrome } a_3 (Fe^{3+})(Cu^+ \rightleftharpoons Cu^{2+}) + H_2O$$

The reduction of oxygen to water involves the transfer of *two* electrons to each atom of oxygen. Thus, two molecules of reduced cytochrome oxidase are required for the reduction of each oxygen atom. The oxidation of the original metabolite involved the transfer of two of its electrons to NAD^+ and then to flavoprotein, and coenzyme Q. Subsequent oxidation of reduced coenzyme Q required two moles of cytochrome b. Thus, two moles of reduced cytochrome oxidase are generated in the terminal reaction; these are then available for the reduction of each atom of oxygen.

A series of equations showing the electron transfers from metabolite (MH_2) to oxygen are summarized below:

1. $MH_2 + NAD^+ \rightleftharpoons M + NADH + H^+$
2. $NADH + H^+ + FMN \rightleftharpoons NAD^+ + FMNH_2$
3. $FMNH_2 + 2Fe^{3+}\text{S-protein} \rightleftharpoons FMN + 2Fe^{2+}\text{S-protein} + 2H^+$
4. $2H^+ + 2Fe^{2+}\text{S-protein} + \text{Coenzyme Q} \rightleftharpoons 2Fe^{3+}\text{S-protein} + \text{Coenzyme } QH_2$
5. $\text{Coenzyme } QH_2 + 2 \text{ Cytochrome b } (Fe^{3+}) \rightleftharpoons \text{Coenzyme Q} + 2 \text{ Cytochrome b } (Fe^{2+}) + 2H^+$
6. $2 \text{ Cytochrome b } (Fe^{2+}) + 2 \text{ Cytochrome } c_1 (Fe^{3+}) \rightleftharpoons$
   $\qquad\qquad 2 \text{ Cytochrome b } (Fe^{3+}) + 2 \text{ Cytochrome } c_1 (Fe^{2+})$
7. $2 \text{ Cytochrome } c_1 (Fe^{2+}) + 2 \text{ Cytochrome c } (Fe^{3+}) \rightleftharpoons$
   $\qquad\qquad 2 \text{ Cytochrome } c_1 (Fe^{3+}) + 2 \text{ Cytochrome c } (Fe^{2+})$
8. $2 \text{ Cytochrome c } (Fe^{2+}) + 2 \text{ Cytochrome a } + a_3 (Fe^{3+})(Cu^+, Cu^{2+}) \rightleftharpoons$
   $\qquad\qquad 2 \text{ Cytochrome c } (Fe^{3+}) + 2 \text{ Cytochrome a } + a_3 (Fe^{2+})(Cu^+, Cu^{2+})$
9. $2 \text{ Cytochrome a} + a_3 (Fe^{2+})(Cu^+, Cu^{2+}) + \tfrac{1}{2}O_2 + 2H^+ \longrightarrow$
   $\qquad\qquad 2 \text{ Cytochrome a } + a_3 (Fe^{3+})(Cu^+, Cu^{2+}) + H_2O$

Net: $MH_2 + \tfrac{1}{2}O_2 \longrightarrow M + H_2O$

The equation for the net reaction stresses the fact that all the intermediates of the respiratory chain are regenerated and that the ultimate outcome of

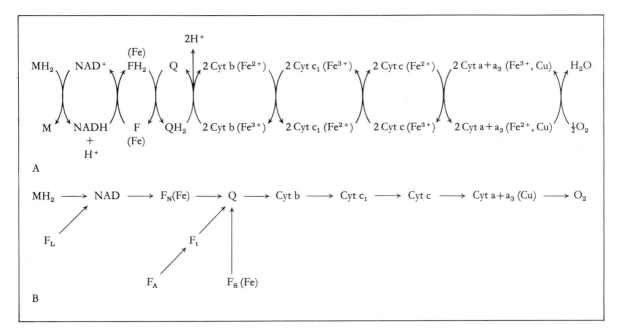

Fig. 3-6. A. Oxidative chain from intermediary metabolite (MH_2) to oxygen. This representation shows the regeneration of each electron carrier. B. Direction of electron flow from metabolite to oxygen. In addition to the direct chain, the diagram shows where other items are fed into the system. F_N, F_L, F_A, and F_S are, respectively, the flavoproteins for NADH dehydrogenase, lipoate dehydrogenase, acyl-CoA dehydrogenase, and succinate dehydrogenase. F_A does not feed directly into the chain; it reacts first with another flavoprotein called the electron-transfer flavoprotein *(F_t). In this notation, (Fe) represents non-heme iron, as distinct from the iron in the porphyrin complexes of the cytochromes.*

mitochondrial aerobic oxidations is the transfer of electrons from the metabolite to oxygen. A critical consequence of this process is that oxygen is required for the oxidation steps in the degradation of fatty acids, i.e., in the conversion of acyl-coenzyme A to the α,β-unsaturated acyl-coenzyme A, the conversion of β-hydroxyacyl-coenzyme A to β-ketoacyl-coenzyme A, and the oxidations of the citric acid cycle.

Two other ways of representing the sequence of reactions in the oxidative chain are shown in Figure 3-6.

3.7. Energy from Biologic Oxidations

The discussion to this point has been concerned with the sequence of molecular transformations that lead to the effective oxidation of metabolic intermediates by oxygen. It is important to bear in mind, however, that these reactions proceed with a release of energy and that this energy is vital for the functioning of the organism. Actually, the principal source of energy for the cell is that provided by the oxidation of various metabolites by oxygen. This section will deal with some of the quantitative aspects of energetics, i.e., with the amount of energy that can be derived from specific reactions.

The essential approach for calculating the free energy evolved in an oxidation is based on analyzing the redox potentials of the half-reactions in the oxidative process. The standard free-energy change $\Delta G^{0\prime}$ (in calories) in such reactions is related to the difference in the redox potential $\Delta E^{0\prime}$ of two reactants by equation 3-2:

$$\Delta G^{0\prime} = -nF\Delta E^{0\prime}$$
$$= -23{,}063\, n\Delta E^{0\prime} \text{ (calories)}$$

Standard redox potentials for a number of the substances involved in biologic oxidations are listed in Table 3-4. The values given are for reduction, i.e., for the electron-attracting capability of the component on the left of the arrows. Some reservations should be noted concerning the energy yields obtained from the values in the table and the use of equation 3-2. First, it is assumed that the redox potentials of the respiratory electron carriers as they occur in intact mitochondria are the same as those found for the experimentally isolated components. Second, the calculated free-energy values are

Table 3-4. Standard Redox Potentials at pH 7 and 25°C for Substances Involved in Biologic Oxidations

Reaction	Redox Potential* $E^{0\prime}$ (volts)
$\tfrac{1}{2}O_2 + 2H^+ + 2e^- \rightleftharpoons H_2O$	+0.82
Cytochrome a (Fe^{3+}) + e^- $\rightleftharpoons$ Cytochrome a (Fe^{2+})	+0.29
Cytochrome c (Fe^{3+}) + e^- $\rightleftharpoons$ Cytochrome c (Fe^{2+})	+0.25
Cytochrome c_1 (Fe^{3+}) + e^- $\rightleftharpoons$ Cytochrome c_1 (Fe^{2+})	+0.22
Crotonyl-coenzyme A + $2H^+$ + $2e^-$ $\rightleftharpoons$ Butyryl-coenzyme A	+0.19
Coenzyme Q + $2H^+$ + $2e^-$ $\rightleftharpoons$ Coenzyme QH_2	+0.10
Cytochrome b (Fe^{3+}) + e^- $\rightleftharpoons$ Cytochrome b (Fe^{2+})	+0.05
Fumarate + $2H^+$ + $2e^-$ $\rightleftharpoons$ Succinate	+0.03
Oxaloacetate + $2H^+$ + $2e^-$ $\rightleftharpoons$ Malate	−0.17
Lipoate + $2H^+$ + $2e^-$ $\rightleftharpoons$ Dihydrolipoate	−0.29
NAD^+ + $2H^+$ + $2e^-$ $\rightleftharpoons$ NADH + H^+	−0.32
α-Ketoglutarate + CO_2 + $2H^+$ + $2e^-$ $\rightleftharpoons$ Isocitrate	−0.38
Succinate + CO_2 + $2e^-$ $\rightleftharpoons$ α-Ketoglutarate	−0.67

* Values are for 1M concentrations of oxidized and reduced forms.

based on one-molar concentrations of both the oxidized and reduced states of the materials at pH 7; these concentrations are unlikely in the intact cell. Third, the fundamental equations utilized in these procedures are derived for isolated, reversible systems approaching equilibrium, whereas the actual biologic situation is that of a dynamic steady state. Nonetheless, the calculations from the differences in redox potentials give a fair estimate of the amounts of energy released, and many of the conclusions derived by this method are borne out by other approaches.

The energy released as a result of the oxidation of reduced NAD^+ via the oxidative chain is sufficient to bring about the synthesis of several moles of ATP. Since hydrolysis of 1 mole of ATP to ADP and inorganic phosphate yields 7.3 kilocalories of free energy, the reversal of this reaction,

$$ADP + P_i \longrightarrow ATP$$

should be feasible when at least this amount of energy is provided to drive it. Indeed, it has been shown that the mitochondrial oxidation of certain metabolites in the presence of ADP and inorganic phosphate yields 3 moles of ATP. For example, the oxidation of isocitrate via the respiratory chain yields three molecules of ATP. This process is called *oxidative-chain phosphorylation*, i.e., the phosphorylation of ADP to ATP as a result of the operation of the oxidative chain. However, these 3 moles of ATP are not produced in one step, since all the energy is not released in a single reaction. Instead, the oxidations and concomitant energy release are regulated to generate ATP at specific sequences in the oxidative chain. One of these is the transfer of electrons from NADH to coenzyme Q via FMN-FeS proteins (site 1). Another ATP-generating sequence is in the flow of electrons from reduced cytochrome b to cytochrome c (site 2). A third mole of ATP is produced in the oxidation of reduced cytochrome a by oxygen (site 3). Utilizing equation 3-2 and the data from Table 3-4, the free-energy change for some of the oxidations in the respiratory chain can be calculated. The results for the ATP-generating steps are shown in Table 3-5.

Table 3-5. Standard Free-Energy Changes in Reactions of the Oxidative Chain

Electron Transfer	$\Delta E^{0\prime}$ (volts)	$\Delta G^{0\prime}$ (kcal)
NADH → Coenzyme Q	0.42	−19.4
Cytochrome b → Cytochrome c	0.20	−9.2
Cytochrome a → Oxygen	0.53	−24.4

Additional details on the mechanisms of the respiratory chain and sites of generation of ATP are described in section 3.11 of this chapter.

The oxidation of succinate bypasses one of the sites for ATP synthesis (NADH-dehydrogenase), since the electrons are transferred directly from substrate to flavoprotein. As a consequence, only 2 moles of ATP are generated by the respiratory chain in the oxidation of succinate. Calculations from the difference in redox potentials between the succinate-fumarate and the oxygen-water systems (Table 3-4) show that the oxidation of succinate should produce considerably less energy than the oxidation of NADH. For the oxidation of succinate:

$$\Delta G^{0\prime} = -23{,}063\,(2)\,(0.79) = -36.4 \text{ kilocalories}$$

For the oxidation of NADH by oxygen via the respiratory chain:

$$\Delta G^{0\prime} = -23{,}063\,(2)\,(1.14) = -52.6 \text{ kilocalories}$$

The findings derived from consideration of the redox potentials thus conform with the experimental results.

The ratio of the number of moles of inorganic phosphate utilized for the generation of ATP to the number of atoms of oxygen reduced to water is called the *P-to-O ratio* (P/O). For example, the P/O ratio for the mitochondrial oxidation of malate to oxaloacetate is 3, whereas that for the oxidation of succinate is 2.

In addition to being synthesized under aerobic conditions (i.e., when oxygen is available) by oxidative phosphorylation, ATP can be formed directly. One example of this type of ATP synthesis is the result of combining the succinyl thiokinase and the nucleoside-diphosphokinase reactions:

$$\text{Succinyl-CoA} + \text{GDP} + \text{P}_i \rightleftharpoons \text{Succinate} + \text{GTP}$$

$$\text{ADP} + \text{GTP} \rightleftharpoons \text{ATP} + \text{GDP}$$

Other instances in which ATP is formed directly will be described in conjunction with glycolysis. The formation of ATP by such mechanisms is called *substrate-level phosphorylation* in distinction from its formation by oxidative phosphorylation.

3.8. Energy from Fatty-Acid Catabolism

As a result of the metabolic oxidation of fatty acids to carbon dioxide and water, a considerable amount of energy is released and conserved in the form of ATP. This production of energy occurs during the degradation of the long-chain fatty acid to acetyl-coenzyme A and as a result of several steps of the citric acid cycle.

The first step in the cellular metabolism of fatty acids—i.e., the formation of the acyl-coenzyme A derivative from the free fatty acid—consumes 1 mole of ATP (section 3-3). Subsequently, the dehydrogenation of the acyl-coenzyme A by a flavoprotein results in the formation of reduced flavoprotein. Oxidation of the latter via the respiratory chain yields 2 moles of ATP by oxidative phosphorylation. The next step, in which the α,β-unsaturated acyl-coenzyme A is converted to L-3-hydroxyacyl-coenzyme A by enoyl hydrase, does not involve any ATP. The oxidation of L-3-hydroxyacyl-coenzyme A to β-ketoacyl-coenzyme A is catalyzed by an NAD-linked enzyme, L-3-hydroxyacyl-coenzyme-A dehydrogenase. The NAD^+ that is reduced in this reaction is reoxidized through the oxidative chain and thus yields 3 moles of ATP. The subsequent step—the thiolase cleavage reaction—neither generates nor consumes ATP. The sites in which ATP is involved are summarized in Table 3-6.

As each two-carbon unit (acetyl-coenzyme A) is cleaved from a long-chain acyl-coenzyme A, 5 moles of ATP are generated. For example, the metabolism of stearic acid, which is composed of eighteen carbons, yields 9 moles of acetyl-coenzyme A. This is accomplished by repeating the degradative sequence eight times. The complete process yields 40 ATPs. The energy expended in the activation step is equivalent to that of 2 ATPs since it involves a pyrophosphate cleavage. Hence, conversion of a mole of stearate to acetyl-coenzyme A yields a net amount of 38 moles of ATP.

The complete oxidation of acetyl-coenzyme A to carbon dioxide and water via the citric acid cycle yields another 12 moles of ATP. The specific steps in the cycle at which ATP is generated are listed in Table 3-7.

Table 3-6. ATP Yield in the Degradation of Fatty Acids to Acetyl-CoA

Reaction	ATP*
Fatty acid → Acyl-CoA	$-1, \approx -2$
Acyl-CoA → α,β-Unsaturated acyl-CoA	+2
α,β-Unsaturated acyl-CoA → L-3-Hydroxyacyl-CoA	0
L-3-Hydroxyacyl-CoA → β-Ketoacyl-CoA	+3
β-Ketoacyl-CoA → Acetyl-CoA + Acyl-CoA	0

* Moles of ATP consumed (−) or produced (+) as a result of the listed reactions.

Table 3-7. ATP Yield from the Citric Acid Cycle

Reaction	ATP*
Acetyl-CoA → Citrate	0
Citrate → Aconitate	0
Aconitate → Isocitrate	0
Isocitrate → α-Ketoglutarate	3
α-Ketoglutarate → Succinyl-CoA (dihydrolipoate oxidation)	3
Succinyl-CoA → Succinate (GTP + ADP → GDP + ATP)	1
Succinate → Fumarate	2
Fumarate → Malate	0
Malate → Oxaloacetate	3
Total:	12

* Moles of ATP produced as a result of the reactions listed.

In the citric acid cycle, reduced NAD is formed in the oxidation of isocitrate (isocitrate dehydrogenase), α-ketoglutarate (α-ketoglutarate dehydrogenase), and malate (malate dehydrogenase). The NADH generated in each of these reactions is reoxidized by the reactions of the respiratory chain in which 3 moles of ATP are generated for each mole of NADH. The oxidation of succinate, which is catalyzed by a flavoprotein (succinate dehydrogenase), yields 2 moles of ATP. Another mole of ATP is produced from the GTP generated during the conversion of succinyl-coenzyme A to succinate (succinate thiokinase).

A long-chain fatty acid such as stearate, which yields 9 moles of acetyl-coenzyme A, can therefore generate 108 moles of ATP as a result of the citric acid cycle oxidation of the acetyl units. Taking into account the 38 moles of ATP provided in the conversion of stearate to acetyl-coenzyme A, it is seen that total degradation of stearate to carbon dioxide and water yields a net total of 146 moles of ATP. (The amount of ATP produced in the metabolism of other fatty acids can be determined in an analogous manner.)

The minimal value for the free energy of hydrolysis of ATP is about 7.3 kilocalories per mole. Multiplication of this value by the number of moles of ATP that are generated gives the energy yield in kilocalories. Thus, the metabolic degradation of 1 mole of stearic acid provides at least 1066 kilocalories of energy. The amount of energy derived from the catabolism of fatty acids is considerably greater than that obtained from any other nutrient. In terms of energy yield per weight and per mole, fatty acids are superior to other substances as energy-yielding materials.

3.9. Special Features of the Citric Acid Cycle

In the previous sections of this chapter, it was shown how the citric acid cycle serves to convert the acetyl-coenzyme A derived from the metabolism of fatty acids to carbon dioxide and water. It should be remembered, however, that a number of other substances—e.g., carbohydrates and certain amino acids—are also metabolized to acetyl-coenzyme A. The cycle is thus a central mechanism for the final degradation of all metabolites that yield acetyl-coenzyme A. Since the citric acid cycle occupies such an important position in intermediary metabolism, it is of interest to define some of the details of the processes and the mechanisms of some of the individual reactions.

A. ASYMMETRY OF CITRATE REACTIONS

Citric acid does not contain an asymmetric carbon atom and is thus not a chiral molecule:

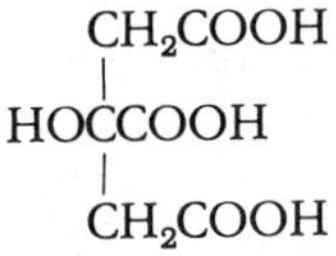

$$\begin{array}{l} CH_2COOH \\ | \\ HOCCOOH \\ | \\ CH_2COOH \end{array}$$

From considerations of organic chemistry, it would be expected that in the conversion of citrate to aconitate and isocitrate, there should be no distinction between the two $-CH_2COOH$ groups. Thus, when citrate that is formed from oxaloacetate and radioactively labeled ^{14}C-acetyl-coenzyme A undergoes the aconitase reaction, the radioactive carbon (indicated by an asterisk) should be found in either of two positions on the isocitrate:

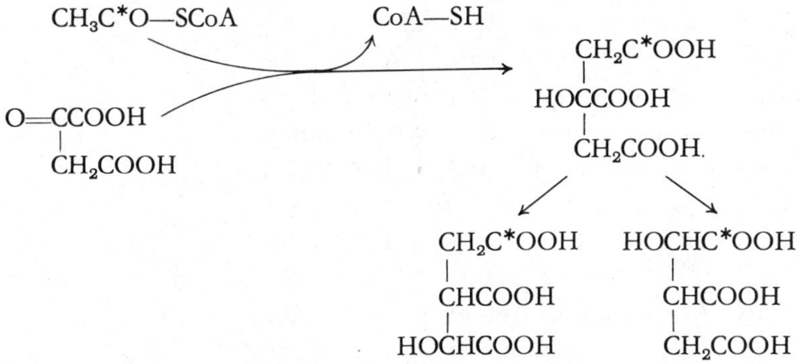

$$\begin{array}{c} CH_3C^*O-SCoA \\ \\ O=CCOOH \\ | \\ CH_2COOH \end{array} \xrightarrow{CoA-SH} \begin{array}{c} CH_2C^*OOH \\ | \\ HOCCOOH \\ | \\ CH_2COOH. \end{array}$$

$$\begin{array}{cc} CH_2C^*OOH & HOCHC^*OOH \\ | & | \\ CHCOOH & CHCOOH \\ | & | \\ HOCHCOOH & CH_2COOH \\ A & B \end{array}$$

Subsequent conversion of the isocitrate to α-ketoglutarate and succinate would then be expected to yield the following:

$$
\begin{array}{ccc}
A & & B \\
\text{CH}_2\text{C*OOH} & & \text{HOCHC*OOH} \\
| & & | \\
\text{CHCOOH} & \text{Isocitrate} & \text{CHCOOH} \\
| & & | \\
\text{HOCCOOH} & & \text{CH}_2\text{COOH} \\
\big\downarrow \!\!\searrow\! CO_2 & & \big\downarrow \!\!\searrow\! CO_2 \\
\text{CH}_2\text{C*OOH} & & \text{O=CC*OOH} \\
| & & | \\
\text{CH}_2 & \alpha\text{-Ketoglutarate} & \text{CH}_2 \\
| & & | \\
\text{O=CCOOH} & & \text{CH}_2\text{COOH} \\
\big\downarrow \!\!\searrow\! CO_2 & & \big\downarrow \!\!\searrow\! C^*O_2 \\
\text{CH}_2\text{C*OOH} & & \text{O=C-OH} \\
| & & | \\
\text{CH}_2 & \text{Succinate} & \text{CH}_2 \\
| & & | \\
\text{O=C-OH} & & \text{CH}_2\text{COOH}
\end{array}
$$

It would therefore be expected that reaction pathways A and B would occur to an equal extent, i.e., both the succinate and the CO_2 would be radioactive. However, actual experiments of this type showed that only one of these possibilities occurred: only the succinate had the radioactive carbon, as shown in pathway A.

The apparent conclusion from these findings was that citrate, since it is a symmetric molecule, could not be an intermediate in the series of reactions leading to α-ketoglutarate and succinate. Subsequent, more definitive studies, however, demonstrated that these compounds were indeed formed from citrate. The dilemma was resolved by the postulation of a novel concept, which contributed significantly to our general understanding of the stereochemistry of certain enzymatically catalyzed reactions. A molecule such as citrate, which contains *three* different groups on one of its carbons (although it is symmetric), will behave in an asymmetric fashion if it binds at *three* sites in combination with an enzyme. Since the binding sites on the enzyme are fixed geometrically, such a substrate can form an enzyme-substrate complex with only one of the two similar substituents on the central carbon (Fig. 3-7).

Although the asymmetric reactivity of citrate can be explained on the basis of a three-point attachment of the substrate to the enzyme, it is not absolutely necessary to assume this type of multiple binding. The two $-CH_2COOH$ groups of citrate are not equivalent sterically, as is seen from Figure 3-7. Thus, from the "vantage point" of each $-CH_2COOH$ group, the three other substituents on the central carbon atom have a different spatial

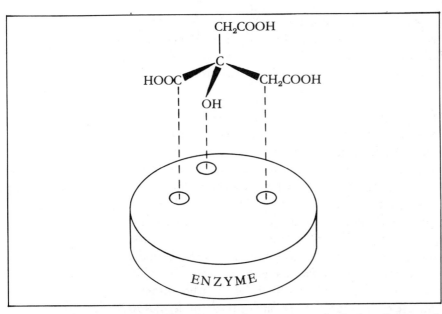

Fig. 3-7. Interaction of enzyme at three sites of the substrate. The binding sites on the enzyme for –COOH, –OH, and –CH$_2$COOH are fixed spatially; this does not allow the two –CH$_2$COOH groups to be interchanged.

order. Enzymes are essentially asymmetric reagents and are therefore sensitive to such steric differences.

It should be noted that the formation of citrate from acetyl-coenzyme A and oxaloacetate is also a stereospecific reaction. The position in citrate of the group derived from acetate is such that it is not the one that is modified in the aconitase reaction.

B. RANDOMIZATION AFTER SUCCINATE

The symmetric structure of succinate does not allow for any distinction between one methylene carbon and the other. As a result, the subsequent steps yield malate, with the isotopic carbon in two different positions:

$$\begin{array}{c}
\text{CH}_2\text{C*OOH} \\
| \\
\text{CH}_2\text{CO—COOH}
\end{array} \qquad \alpha\text{-Ketoglutarate}$$

$$\downarrow -\text{CO}_2$$

$$\begin{array}{c}
\text{CH}_2\text{C*OOH} \\
| \\
\text{CH}_2\text{COOH}
\end{array} \left(= \begin{array}{c} \text{CH}_2\text{COOH} \\ | \\ \text{CH}_2\text{C*OOH} \end{array} \right) \qquad \text{Succinate}$$

$$\downarrow$$

$$\begin{array}{c}
\text{HCC*OOH} \\
\parallel \\
\text{HOOCCH}
\end{array} \left(= \begin{array}{c} \text{HCCOOH} \\ \parallel \\ \text{HOOC*CH} \end{array} \right) \qquad \text{Fumarate}$$

$$\swarrow \qquad \searrow$$

$$\begin{array}{cc}
\text{HOCHC*OOH} & \text{CH}_2\text{C*OOH} \\
| & | \\
\text{CH}_2\text{COOH} & \text{HOCHCOOH}
\end{array} \qquad \text{Malate}$$

Half of the radioactivity will be found in the carboxyl group attached to the hydroxyl carbon of malate and the other half will be in the carboxyl group attached to the methylene carbon. Expressed in another way, the radioactive carbon from α-ketoglutarate is randomized between the two carboxyls of the malate.

C. FATE OF THE CARBONS FROM ACETATE

On the basis of the considerations discussed, it is possible to analyze the distribution of the carbon atoms from acetate among the intermediates of the citric acid cycle (Fig. 3-8). It can be seen that during one turn of the cycle, all the carbons from the acetate are in the newly formed oxaloacetate; none of the original acetate is converted to carbon dioxide. The new oxaloacetate can then condense with another mole of acetate, and the cycle is repeated. During this second turn of the cycle, one of the carbons from the first molecule of acetate is released as carbon dioxide. (Since in Figure 3-8 all the possible positions of the carbons from the first molecule of acetate are indicated, two of these carbons are shown as being liberated as carbon dioxide in the second turn of the cycle. However, either one of these molecules of carbon dioxide, but not both, contains carbons from the original acetate.) The remaining carbon from the original molecule of acetate is liberated in subsequent turns of the cycle. Thus, although the acetate carbons are not released as carbon dioxide during the first complete cycle, they are eventually converted to carbon dioxide in later cycles. In metabolic considerations, it is therefore valid to consider the cycle as a pathway by which acetate is oxidized to carbon dioxide and water.

3.10. Isocitrate Dehydrogenase and Allosteric Enzymes

The oxidation of isocitrate by *NAD-linked* isocitrate dehydrogenase (page 123) is an important control site in the operation of the citric acid cycle. The enzyme is activated by ADP and is inhibited by ATP or NADH. Isocitrate dehydrogenase is an example of an *allosteric enzyme*. Inhibitors of such enzymes are not related structurally to the substrate, as is the case with competitive inhibition. For example, ATP bears no structural similarity to isocitrate, yet it inhibits isocitrate dehydrogenase. Another characteristic of allosteric enzymes is that they have a comparatively high molecular weight and are composed of several subunits. The inhibitors bring about a change in the intramolecular binding of the units and the conformation of the enzymes, thus modifying their catalytic activity.

Allosteric enzymes may be modified by specific compounds that inhibit or *enhance* their catalytic activity, depending on the particular case. The modifier is therefore termed an *allosteric effector* rather than merely an inhibitor. In contrast to the action of a competitive inhibitor, the allosteric effector

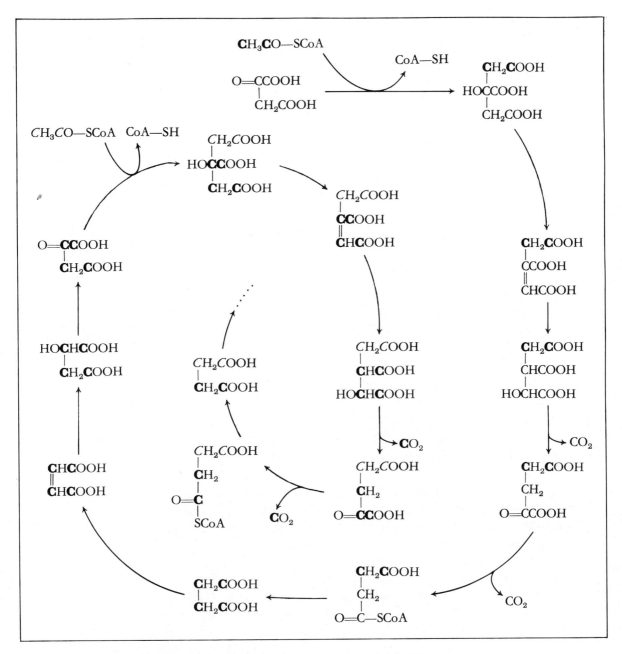

Fig. 3-8. Fate of the carbon atoms from acetate after one turn of the citric acid cycle. The positions of the carbon atoms from the initial acetate molecule are shown in boldface type. (In succinate and following intermediates, all four positions are indicated as a result of the symmetry of succinate; in fact, only one pair of positions will be occupied by the original acetate carbons.) The positions of the atoms from a second acetate molecule are shown in italic type.

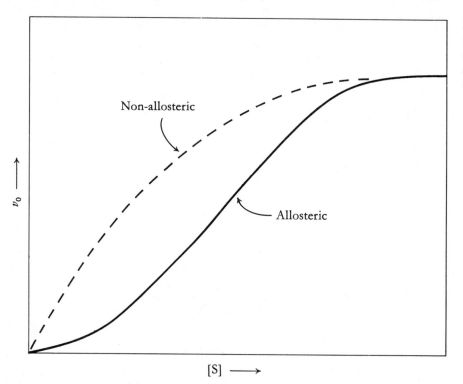

Fig. 3-9. *Initial reaction velocity* (v_0) *versus substrate concentration* ($[S]$) *for allosteric enzymes compared with non-allosteric enzymes.*

does not bind to the same site as the substrate. Rather, the enzyme contains a specific binding site for the effector, which is called the *allosteric site* or *regulatory site*. The site where the substrate is bound and transformed is termed the *catalytic site*, to distinguish it from the allosteric site. In a number of instances in which detailed studies were made of allosteric enzymes, it was found that the allosteric sites and the catalytic sites were on different subunits of the enzyme. Thus, the individual subunits may be referred to as either *catalytic subunits* or *regulatory subunits*.

Allosteric enzymes exhibit characteristic kinetic relationships between substrate concentration and initial reaction velocity (Fig. 3-9). At low concentrations of substrate, the increase in reaction velocity is comparatively small, but at a certain point there is a dramatic change in the effect of substrate concentration on the velocity. The type of curve thus obtained is called a *sigmoid curve* because of its shape. The simplest physical explanation for this type of relationship is that the substrate, in addition to undergoing a chemical transformation, functions to increase the activity of the enzyme. More specifically, the binding of the substrate to an allosteric enzyme increases the affinity of the enzyme for more substrate. This phenomenon is generally referred to as *cooperative binding*.

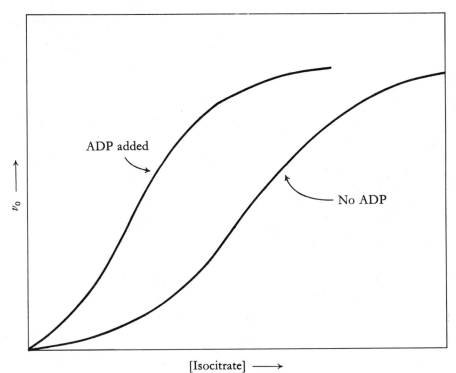

Fig. 3-10. Effect of substrate (isocitrate) concentration on initial reaction velocity for the enzyme, isocitrate dehydrogenase. A sigmoid curve is obtained in absence of ADP, an allosteric modifier that enhances this reaction. When ADP is added (1 mM), the slope of the curve increases about the point of inflection, and the apparent K_M decreases, though V_{max} remains the same.

The kinetic properties of isocitrate dehydrogenase are described in Figure 3-10. In the absence of ADP, the plot of reaction velocity versus isocitrate concentration has a sigmoidal shape that is characteristic of allosteric enzymes. The addition of ADP causes an increase in the initial reaction velocity, even when substrate concentration is low, i.e., the apparent K_M is decreased. The value of V_{max}, in this case, is unchanged. Thus, ADP binds to the enzyme and changes its conformation so that the affinity for isocitrate is increased. ATP has the opposite effect, i.e., it functions as an inhibitor of the action of isocitrate dehydrogenase.

The action of the allosteric effectors on isocitrate dehydrogenase regulate the citric acid cycle to conform with the intracellular concentrations of ATP and ADP. This is important in view of the fact that the citric acid cycle functions as a mechanism for providing ATP. When the cellular concentration of ADP is increased, the activity of isocitrate dehydrogenase is enhanced. Conversely, increased levels of ATP decrease the activity of this enzyme and, consequently, the operation of the citric acid cycle. Since there is an inverse relationship between the intracellular levels of ATP and ADP, the control of the enzyme by these two components serves to regulate the energy reserve of the cells.

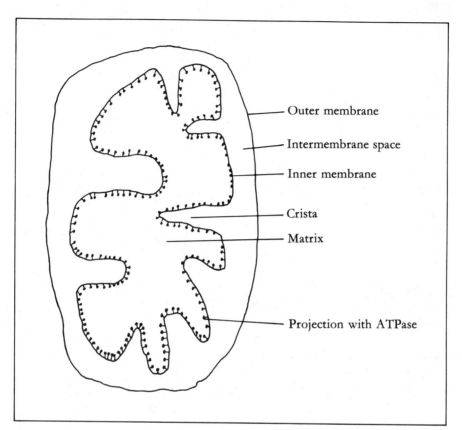

Fig. 3-11. Diagrammatic representation of a mitochondrion.

3.11. Processes of Oxidative Phosphorylation

A. MITOCHONDRIAL STRUCTURE AND ENZYMES

The aerobic metabolic processes that give rise to the formation of ATP are localized in the mitochondria. These processes include fatty acid degradation, the citric acid cycle, and oxidative phosphorylation. The mitochondria function to compartmentalize such interrelated processes and to provide structural requirements that are critical for channeling the released energy to ATP synthesis.

The *mitochondrion* is composed of outer and inner membranes that envelop a fluid matrix (Fig. 3-11). The outer membrane is relatively permeable to ions and small molecules, whereas the inner membrane is impermeable even to protons. Although various carrier proteins permit the facilitated diffusion of many materials, NAD^+ and NADH cannot cross the inner membrane. The inner membrane contains folds and invaginations that appear as protrusions into the matrix, called *cristae*. On the inner surface of the inner membrane, there are knob-like particles that project in the direction of the matrix (Fig. 3-11).

The enzymes for the oxidation of fatty acyl-coenzyme A to acetyl-coenzyme A and those of the citric acid cycle (except succinate dehydrogenase) occur in the mitochondrial matrix. The components of the respiratory chain and succinate dehydrogenase are bound to the inner membrane. *Adenosine triphosphatase* (ATPase) is found in the knob-like projections facing the matrix. This enzyme, termed the *coupling factor* 1, or F_1, is involved in linking the energy that is released by the respiratory chain to the synthesis of ATP. (Adenosine triphosphatase catalyzes the hydrolysis of ATP to ADP and inorganic phosphate. Since this reaction is reversed during oxidative phosphorylation, ATPase is considered to be the critical enzyme for the synthesis of ATP.)

B. SITES OF ATP GENERATION

On the basis of thermodynamic considerations, it was previously shown that sufficient energy for the synthesis of ATP is released at three sites in the oxidative or respiratory chain (see section 3.7); i.e., that 3 moles of ATP can be produced as a result of electron transport. The reactions at these sites are shown in Figure 3-12. In addition to theoretical calculations (which involve several assumptions), there are several other lines of evidence to implicate these sites in ATP production.

The fact that the first site—the electron transfer from NADH to coenzyme Q—contributes 1 mole of ATP is demonstrated by the finding that only 2 moles of ATP are generated during the mitochondrial oxidation of succinate, in which the step involving the oxidation of NADH by flavoprotein is bypassed. In the oxidation of succinate, the $FADH_2$ of succinate dehydrogenase

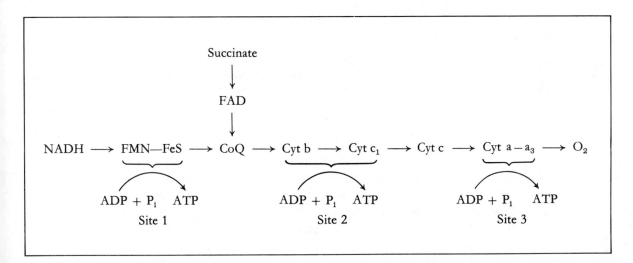

Fig. 3-12. Sequence of electron carriers and sites of energy release for synthesis of ATP.

is oxidized by coenzyme Q, and the subsequent steps are similar to the electron transport sequence of the "main chain" (Fig. 3-12). The difference of one ATP between the two pathways indicates that one energy conservation site occurs in the "main chain" before the reduction of coenzyme Q.

Similarly, the location of the second site can be inferred from the finding that 1 mole of ATP is produced upon mitochondrial oxidation of ascorbate. The reduced form of the latter transfers its electrons directly to cytochrome c, and the fact that only 1 mole of ATP is produced demonstrates that the first two energy-coupling sites occur earlier in the chain than cytochrome c oxidation. Finally, it is an obvious conclusion from these experiments that the third site of ATP production is between cytochrome c and oxygen.

Another, more direct approach for identifying the energy-conserving sites is to determine the effects of inhibitors of particular reactions of the oxidative chain. Thus, amytal (amobarbital) and rotenone, which inhibit the transfer of electrons from NADH dehydrogenase (FMN) to coenzyme Q, interfere with the production of ATP from the reactions at site 1. However, these inhibitors do not affect electron transport from succinate and the ensuing production of 2 moles of ATP in the remaining reaction sequence. Similarly, antimycin A, which inhibits the oxidation of cytochrome b by cytochrome c_1, blocks the generation of ATP from site 2. The electron flow from ascorbate to cytochrome c, a, a_3, and oxygen, and the production of 1 mole of ATP, are not blocked by antimycin A. Cyanide and carbon monoxide, which bind to cytochrome $a + a_3$, inhibit the oxidative sequence between these cytochromes and oxygen and do not allow the production of ATP at site 3. These and analogous studies with inhibitors of the oxidative chain have demonstrated that the prevention of electron flow at certain points in the sequence inhibits formation of the ATP that is normally produced from reactions at these sites.

More recent investigations on oxidative phosphorylation have led to the isolation of the constituents involved in site 1 and site 3. In all cases, mitochondrial membrane particles were necessary components for producing the oxidative sequence and the formation of ATP.

C. UNCOUPLERS OF OXIDATIVE PHOSPHORYLATION

The discussions to this point have emphasized the dependence of ATP production on electron transport in the respiratory chain. Thus, limitations in the supply of oxygen or the inhibition of any of the oxidation sites can reduce or abolish the synthesis of ATP. The converse of this relationship is also an experimentally verified fact; that is, in intact mitochondria, the oxidative chain does not operate when ATP cannot be generated. For example, if ADP is not available for production of ATP, the oxidative processes do not occur and oxygen is not consumed. Mitochondrial reactions, which exhibit this interdependence of oxidation and phosphorylation, are referred to as being tightly *coupled*. Upon aging of isolated mitochondria, the degree of coupling is decreased or completely eliminated. The addition of a metabolite to such mitochondria may cause oxidation, electron transport, and oxygen consumption to occur, but the generation of ATP is minimal. Moreover, in this system, electron transfer occurs even in the absence of ADP. In such mitochondria, oxidation and phosphorylation are uncoupled.

Certain reagents cause the uncoupling of oxidative phosphorylation in intact mitochondria. Most prominent among these is 2,4-dinitrophenol:

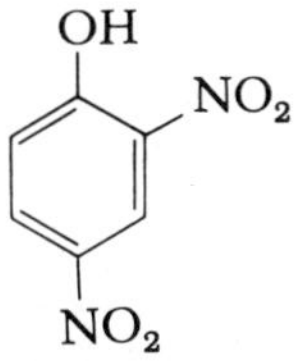

2,4-Dinitrophenol

Other substances that have this effect are dicumarol and thyroxine. In addition to effecting the uncoupling of the energy-yielding oxidation from the energy-conserving phosphorylation, dinitrophenol stimulates the respiratory chain and the hydrolysis of ATP by mitochondria. The latter effect, termed *ATPase activity*, is minimal under normal conditions. ATPase is the enzyme that catalyzes the reversible reaction:

$$ATP \rightleftharpoons ADP + P_i$$

In coupled mitochondria, the enzyme is instrumental in producing ATP from ADP and inorganic phosphate. Under conditions when the energy from the respiratory chain cannot be channeled to ATP synthesis, mitochondrial ATPase effectively causes the hydrolysis of ATP.

D. MECHANISMS OF OXIDATIVE PHOSPHORYLATION

Although the interdependence between the respiratory chain and the production of ATP is an accepted fact, all the details as to how the two processes are linked are not known. It is expected that this coupling involves a mechanism whereby the energy released in certain steps of the electron-transport sequence is conserved rather than dissipated. The energy may be trapped in certain intermediates (chemical coupling) or as a physical high-energy state (chemiosmotic coupling). In either case, the sites at which large amounts of energy are released must also be energy conservation sites, and the specific redox reactions as outlined previously may include additional steps.

According to the *chemical coupling hypothesis*, the energy from oxidation at the phosphorylation sites (sites 1, 2, and 3) is conserved by the formation of a high-energy chemical intermediate. The energy released by the oxidation of NADH, for example, would be utilized for the production of a high-energy compound composed of NAD^+ and a specific "coupling factor." If the latter is designated C, then the proposed reaction may be written as:

$$NADH + H^+ + FMN + C \rightleftharpoons NAD\sim C + FMNH_2$$

Reaction of $NAD\sim C$ with inorganic phosphate would produce an activated phosphate derivative:

$$NAD\sim C + P_i \rightleftharpoons C\sim P + NAD$$

The final step would be the utilization of the energy in the $C\sim P$ complex for the synthesis of ATP:

$$C\sim P + ADP \rightleftharpoons C + ATP$$

As a consequence of this sequence, the energy released by the oxidation reaction would be incorporated into ATP. In this view, similar series of reactions are visualized for the generation of ATP at the other phosphorylation sites.

The chemical coupling hypothesis explains a number of experimental findings; however, all attempts to isolate and identify high-energy intermediates have failed. This difficulty, as well as the demonstration that oxidative phosphorylation requires the presence of intact inner mitochondrial membrane and its particulate vesicles, led to the proposal of other hypotheses. Currently, the most attractive concept is the *chemiosmotic coupling* theory.

The chemiosmotic coupling theory is based on several overall principles, although there are differences of opinion regarding specific aspects (Fig. 3-13). The electron carriers of the respiratory chain, according to this theory, are bound to the inner mitochondrial membrane and arranged in a specific manner with respect to each other, and with respect to the side of the membrane they face. Electron flow through the carriers involves the concomitant transfer, or pumping, of protons from the inner (matrix) side of the membrane to the outer side (intermembrane space). The net effect is a

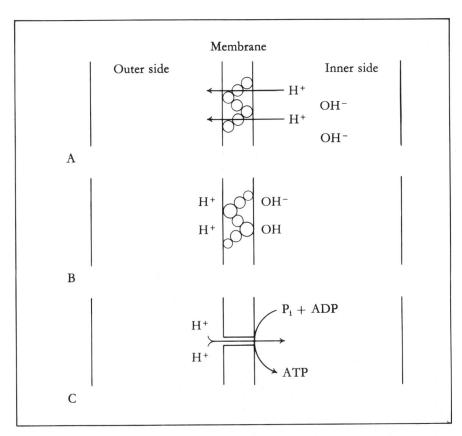

Fig. 3-13. Principal steps in chemiosmotic coupling. A. Movement of protons from inner side (matrix) to the outer side (intermembrane space). B. Resulting proton gradient. C. Return of protons to inner side and production of ATP.

difference in pH and electrical charge across the inner membrane. Thus, the free energy from the reactions of the oxidative chain is conserved as a high-energy electrochemical gradient. The free energy of the gradient is then utilized to drive the reaction between ADP and inorganic phosphate to yield ATP. This is accomplished by the return of the accumulated protons to the inner side of the membrane via a "channel" that is associated with ATPase. Hence, the energy evolved from the release of the gradient serves to drive the ATPase-catalyzed reaction in the synthetic direction.

The link between electron transport and production of ATP is thus in the movement of the protons across the inner mitochondrial membrane. The reactions of the oxidative chain are coupled with the transfer of protons from the matrix compartment to the intermembrane space; the return of the protons to the original site is linked with the synthesis of ATP. It is obvious from these considerations that oxidative chain phosphorylation depends on intact mitochondrial membranes that restrict the diffusion of protons. The action of dinitrophenol as an uncoupler of oxidative phosphorylation appears

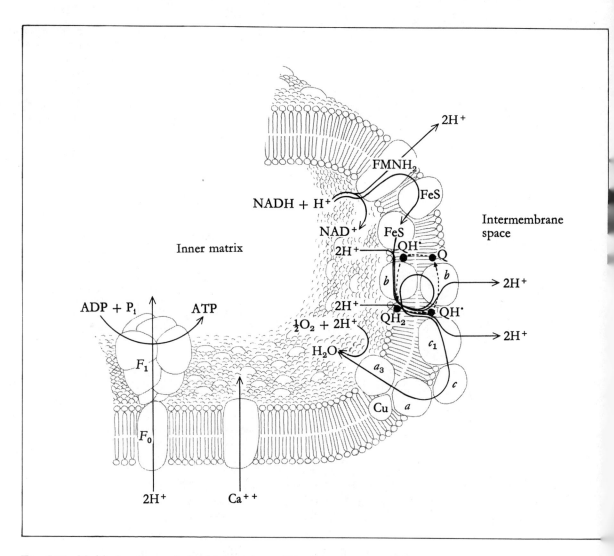

Fig. 3-14. Model of an inner mitochondrial membrane. The electron transport chain (NADH to oxygen) with possible sites for transfer of protons is shown at the right. The return of protons to the matrix side, via F_0–F_1, and the generation of ATP is represented at the far left. The energy from the oxidative chain is also utilized for translocation of Ca^{2+} as shown at the bottom. (Reprinted with permission from "How Cells Make ATP" by P. C. Hinkle and R. E. McCarty. Copyright © March 1978 by Scientific American, Inc. All rights reserved.)

to be due to its effect in altering structure of the mitochondrial membrane and, consequently, the generation of a proton gradient.

The exact disposition of the electron carriers on the inner mitochondrial membrane and the quantitative aspects of the hydrogen transfers remain to be resolved. Moreover, the precise mechanism of phosphorylation is still uncertain. The ATPase system comprises F_1, which are knob-like projections from the inner membrane into the matrix, and stalk-like structures attached to F_1, termed F_0. F_0 is imbedded in the inner membrane and appears to serve as the channel for the return of protons from the intermembrane space. The isolated F_1 component has ATPase activity; i.e., it catalyzes the hydrolysis of ATP to ADP and P_i. Promotion of the synthesis of ATP requires that F_1 be linked to F_0 on the membrane. Some of the suggested sequences of mitochondrial processes are shown in Figure 3-14. It should be noted that in addition to the synthesis of ATP, the energy from the proton gradient is used to drive the movement of various ions, e.g., calcium or sodium, across the mitochondrial membrane

Suggested Reading

Boyer, P. D., Chance, B., Ernster, L., Mitchell, P., Racker, E., and Slater, E. C. Oxidative Phosphorylation and Photophosphorylation. *Annual Review of Biochemistry* 46:955–1026, 1977.

Greville, G. D., and Tubbs, P. K. The Catabolism of Long Chain Fatty Acids in Mammalian Tissue. *Essays in Biochemistry* 4:155–212, 1968.

Hinkle, P. C., and McCarty, R. E. How Cells Make ATP. *Scientific American* 238:104–123, 1978.

Lowenstein, J. M. *Citric Acid Cycle: Control and Compartmentalization.* New York: Dekker, 1969.

Racker, E. The Two Faces of the Inner Mitochondrial Membrane. *Essays in Biochemistry* 6:1–22, 1970.

Racker, E. Inner Mitochondrial Membranes: Basic and Applied Aspects. *Hospital Practice* 9:87–93, 1974.

4. Metabolism of Carbohydrates

The principal carbohydrate that is utilized directly by the cell is glucose. In order for other sugars to be metabolized, they must be converted either to derivatives of glucose or to one of its metabolic intermediates. Consequently, when considering the metabolism of carbohydrates, the central subject is the physiologic disposition of glucose.

Living organisms utilize glucose for essentially four purposes. First, it serves as a fuel for deriving energy; the controlled cellular degradation of glucose proceeds with a release of energy that can be channeled toward the generation of ATP. Second, a number of the intermediates produced during the metabolic interconversions of glucose serve as precursors for the biosynthesis of non-carbohydrate components of cells and tissues. For example, the fragments arising from the breakdown of sugars can be used for the synthesis of amino acids, lipids, and various other substances. Third, glucose is utilized to form polysaccharides, complex glycoproteins, and nucleic acids. More specifically, these molecules include glycogen, which acts as the storage form of glucose; glycoproteins and glycolipids, which are essential components of membranes; and mucopolysaccharide complexes, which are important in maintaining the integrity of connective tissue. Various complex polysaccharides and glycoproteins are also constituents of certain secretions, blood-group substances, and a number of enzymes. Finally, certain reactions in glucose metabolism result in the reduction of nicotinamide adenine dinucleotide phosphate ($NADP^+$). This reduced coenzyme is essential as a reducing agent in the biosynthesis of a number of cellular components, e.g., fatty acids, steroids, and DNA. The metabolism of glucose thus provides the organism with energy, complex polysaccharides, intermediates for the synthesis of various non-carbohydrate molecules, and the reducing power required for specific cellular processes.

4.1. Digestion and Absorption

The principal nutrient carbohydrates of the average diet are the polysaccharide starch, and the disaccharides sucrose and lactose. Glycogen, maltose, glucose, and fructose, though present in certain foods, constitute a relatively minor fraction of ingested carbohydrates. Other polysaccharides found in foods—such as celluloses, dextrans, levans, and mannans—cannot be utilized because they are not degraded by the enzymes of the gastrointestinal tract. In order for starch and disaccharides to be used by the organism, they must first be hydrolyzed to their component monosaccharides by the enzymes of the digestive system.

Preliminary digestion of starch may take place in the mouth by an enzyme present in the salivary secretions known as *ptyalin* or *amylase*. This enzyme catalyzes the hydrolysis of starch to maltose. However, salivary amylase does not contribute significantly to the digestion of polysaccharides because the food does not remain in the mouth for very long. The enzyme has no activity when the carbohydrates reach the stomach, since its pH optimum is 7.1 and the medium in the stomach is highly acidic.

The major degradation of polysaccharides occurs in the small intestine. Pancreatic juice contains an amylase that is very similar in its enzymatic properties to salivary amylase. It is active at pH 6.9 to pH 7.2, and it catalyzes the random cleavage of the α-1,4 linkage of starch or glycogen to yield maltose. Since starch contains components having some branches through carbon-6 of the glucose units, a certain amount of isomaltose is also formed (Fig. 4-1).

The maltose and isomaltose resulting from the digestion of starch or glycogen are hydrolyzed to glucose by the intestinal disaccharidases *maltase* and *isomaltase*. Other disaccharide-splitting enzymes of the intestine are *sucrase*, which hydrolyzes sucrose to glucose and fructose, and *lactase*, which yields glucose and galactose from lactose (Fig. 4-2). The disaccharidases are enzymes of the intestinal mucosa, and the reactions that they catalyze occur at the surface of these cells. In addition, the mucosal cells have a high turnover rate, and as they break down, the enzymes are released into the intestinal lumen.

The monosaccharides are absorbed from the intestine into the portal circulation. This absorption involves an active-transport process, since its rate is not related to the concentration gradient or the molecular weight of the monosaccharide. Thus, the rates of transport of the common monosaccharides, in decreasing order, are as follows: galactose, glucose, fructose, mannose, xylose, and arabinose. The exact mechanism for the transfer of sugar from the intestinal lumen to the portal vein is not defined completely. However, the process is quite efficient; 1 gram of glucose per kilogram of body weight can be transported in 1 hour.

From the portal circulation, glucose is transferred into the cells of the liver, where it may be metabolized or stored as glycogen. The liver also channels glucose to the systemic circulation, thereby making it available to the cells of all tissues.

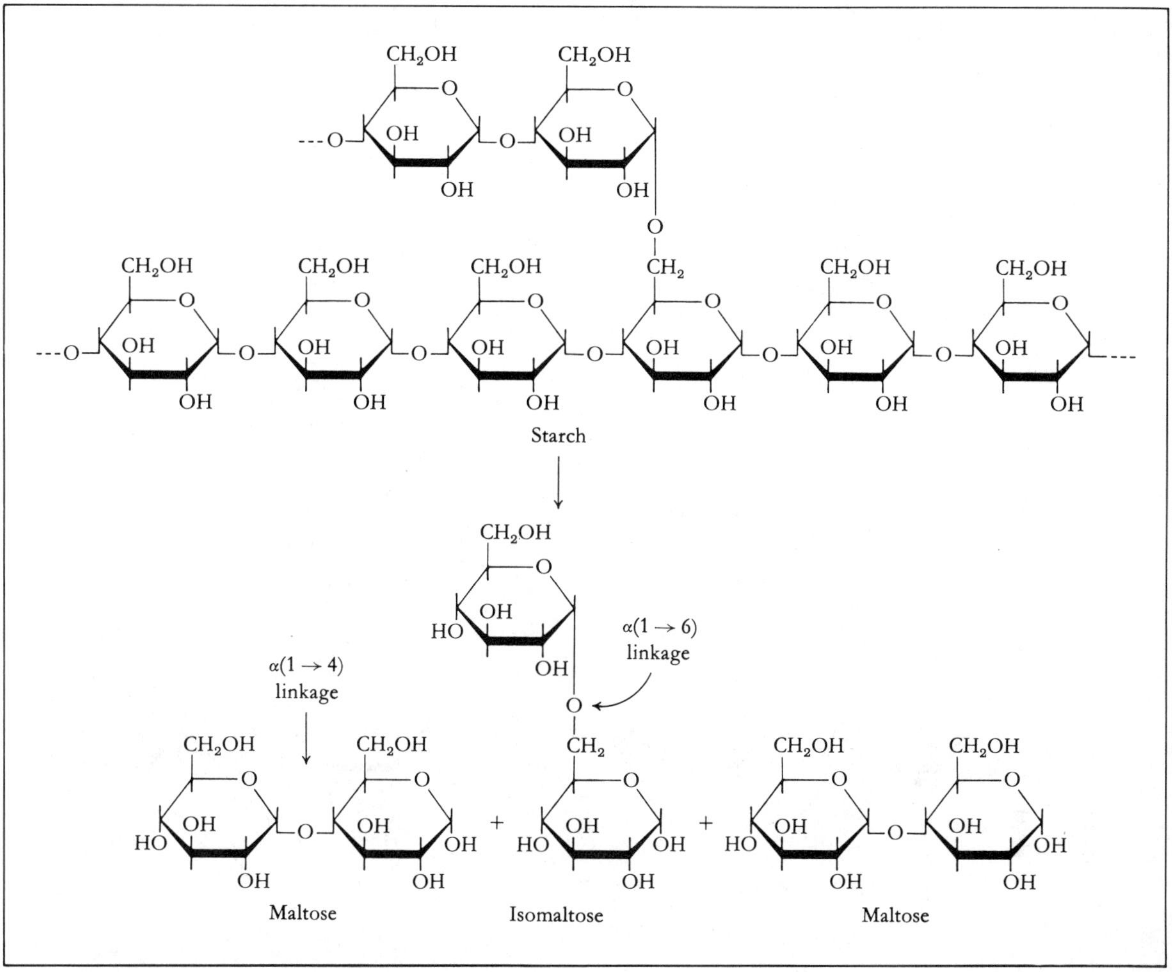

Fig. 4-1. Digestion of starch by pancreatic amylase.

The level of blood glucose is maintained within fairly narrow limits (70 to 90 milligrams per 100 milliliters) by various regulatory mechanisms. When there is an elevation of blood glucose, as occurs after the ingestion of carbohydrates, the beta cells of the pancreas are stimulated to secrete insulin. This hormone has the overall effect of decreasing the amount of glucose in the circulation. One mechanism that accomplishes this involves increasing the permeability of certain cells to glucose. Among the tissues that are dependent on insulin for making glucose available to their cells are adipose tissue, skeletal muscle, diaphragm, aorta, anterior pituitary gland, lactating mammary glands, and the lens of the eye. Other cells—such as those of nerves, brain tissue, erythrocytes, or liver—do not require insulin for glucose

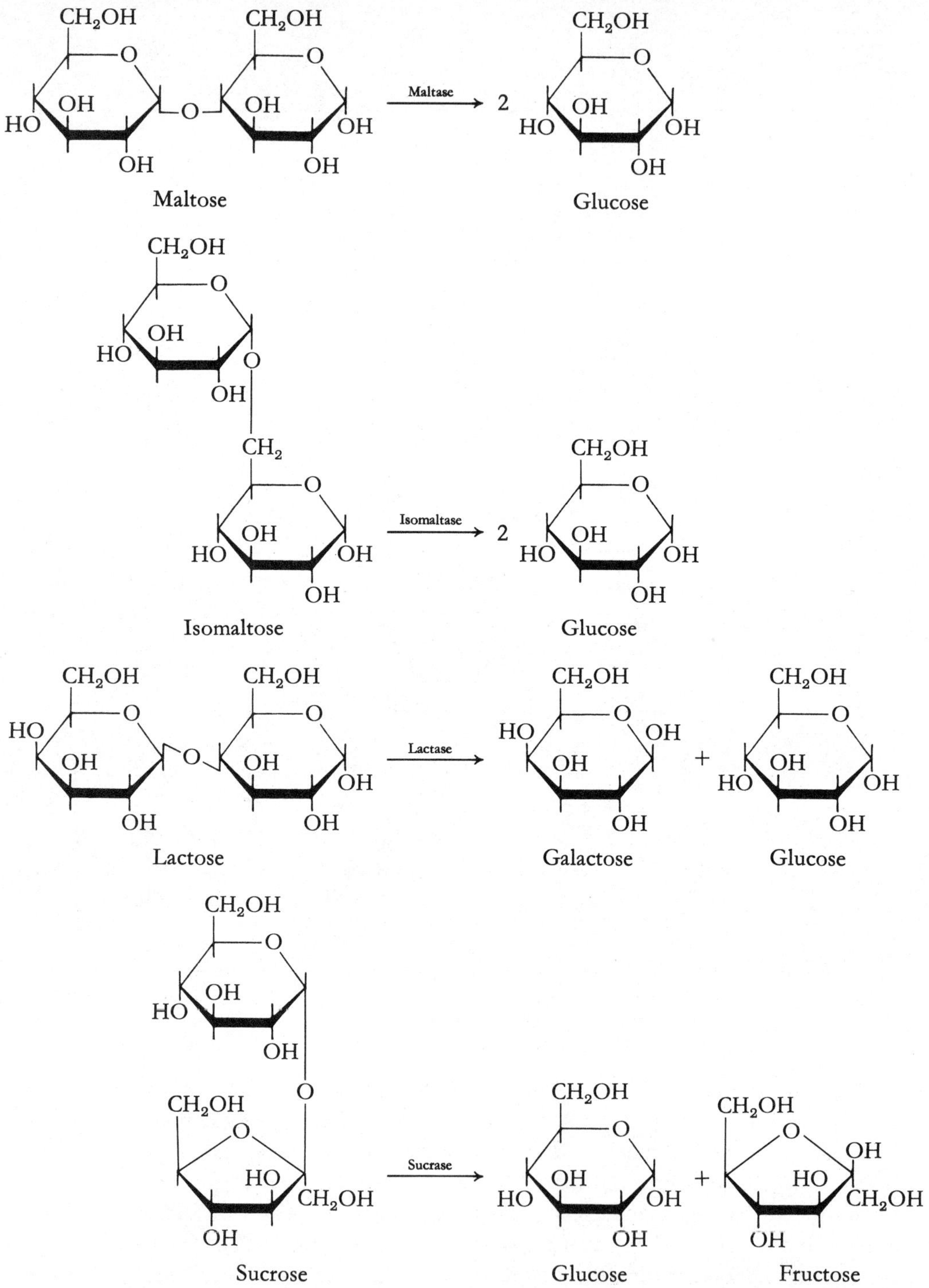

Fig. 4-2. Action of intestinal disaccharidases.

permeability. In addition to this effect of insulin, there are other actions of this hormone, and various controls by other hormones and enzymes, which are important in regulating the concentration of blood glucose. These will be discussed in Chapter 6 (section 6.3).

4.2. Glucose 6-Phosphate: Synthesis and Disposition

The first step in the utilization of glucose is its interaction with ATP in the presence of magnesium ions to yield glucose 6-phosphate and ADP. The reaction is catalyzed by the enzyme *hexokinase*:

D-Glucose + ATP (Mg^{2+}) $\xrightarrow{\text{Hexokinase}}$ D-Glucose 6-phosphate + ADP

The cell membrane is impermeable to glucose 6-phosphate, so this transformation is effective in locking glucose into the cell. It should be noted that the hexokinase reaction involves the expenditure of energy in the form of 1 mole of ATP. Since the product, glucose 6-phosphate, is not a high-energy phosphate compound ($\Delta G^{0\prime}$ of hydrolysis is -3800 calories per mole), most of the energy from the ATP is dissipated. As a consequence, the reverse reaction—i.e., the formation of ATP from glucose 6-phosphate and ADP—would be endergonic and would have a positive ΔG. The hexokinase reaction is therefore irreversible under usual physiologic conditions.

In addition to hexokinase, which is found in all cells that metabolize glucose, there is another enzyme in the liver, termed *glucokinase*, that catalyzes a similar reaction. Hexokinase and glucokinase differ from each other in their K_M values, their substrate specificities, and the effect of inhibitors. Liver glucokinase apparently plays an important role in the regulation of blood glucose. This will be described fully in the discussion of regulatory factors (Chap. 6, section 6.3).

Glucose 6-phosphate may be channeled into a number of metabolic pathways (Fig. 4-3). First, it may be converted to fructose 6-phosphate and degraded via the glycolytic pathway. Second, glucose 6-phosphate may be oxidized to gluconolactone 6-phosphate and subsequently converted to pentose phosphates. Third, it may be transformed to glucose 1-phosphate. The last reaction may be followed by the formation of uridine diphosphate (UDP) derivatives of glucose, which can be utilized for the synthesis of glycogen, uronic acids, and a host of other carbohydrates. Still another

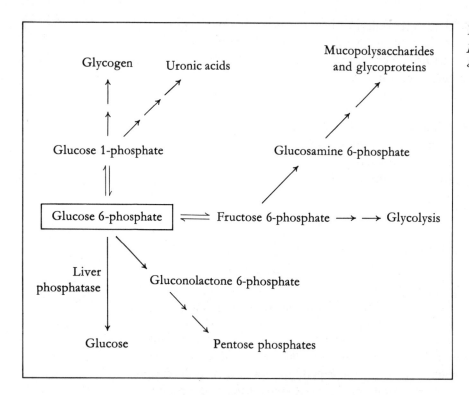

Fig. 4-3. Metabolic pathways leading from glucose 6-phosphate.

pathway of glucose 6-phosphate metabolism leads (after several steps) to the formation of glucosamine 6-phosphate. This product is an intermediate in the synthesis of mucopolysaccharides and glycoproteins. Finally, in the liver and kidney, glucose 6-phosphate can be hydrolyzed to glucose and inorganic phosphate. This reaction is catalyzed by the enzyme *glucose 6-phosphatase* (note that this reaction does not involve regeneration of ATP). The glucose formed in this hydrolysis can be delivered into the circulatory system. Thus, glucose 6-phosphate is an important keystone of metabolism; the multiplicity of pathways leading from this point provide for the derivation of energy from glucose, the formation of various metabolic intermediates, and the synthesis of a number of structural molecules for maintaining the living organism.

The cells of a number of tissues contain the enzymes for catalyzing the reactions of several metabolic pathways involving glucose 6-phosphate. The specific route that predominates depends on the metabolic state of the organism and may vary with different conditions. Other tissues with more specialized functions are geared to utilize carbohydrates for providing their specific requirements. These differences in the mode of metabolism of glucose by individual tissues will be described in detail after each of the major pathways is defined (Chap. 6, section 6.7).

4.3. Glycolysis

The term *glycolysis* refers to a series of reactions in which glucose is degraded to pyruvic acid and lactic acid. This process results in a release of energy, and it occurs even in the absence of oxygen. The enzymes that catalyze the individual reactions of this sequence are found in the soluble fraction of the cell (outside the mitochondria). Although this pathway is characteristically found in anaerobic microorganisms, the reactions of glycolysis are also fundamental steps for the degradation of glucose in higher animals, which do utilize oxygen.

Considering glucose as the initial substrate, the first step involves the formation of glucose 6-phosphate by the *hexokinase* reaction described in section 4.2.

The enzyme *glucose-phosphate isomerase* then catalyzes the conversion of glucose 6-phosphate to fructose 6-phosphate. This reaction is reversible:

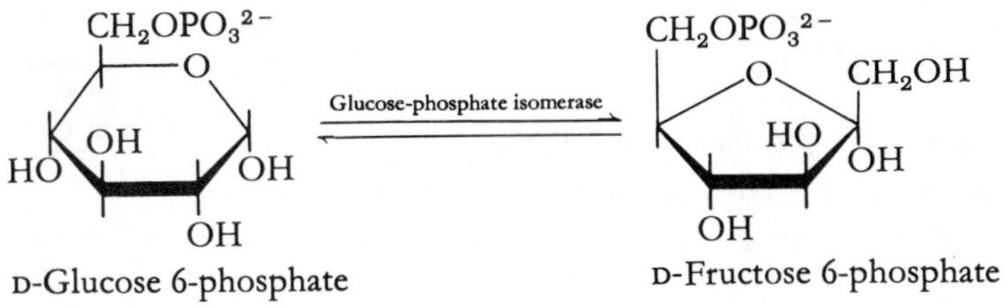

$$\text{D-Glucose 6-phosphate} \xrightleftharpoons{\text{Glucose-phosphate isomerase}} \text{D-Fructose 6-phosphate}$$

The next step in the glycolytic sequence involves the interaction of fructose 6-phosphate with ATP and Mg^{2+} to yield fructose 1,6-diphosphate. The enzyme that catalyzes this step is *phosphofructokinase*:

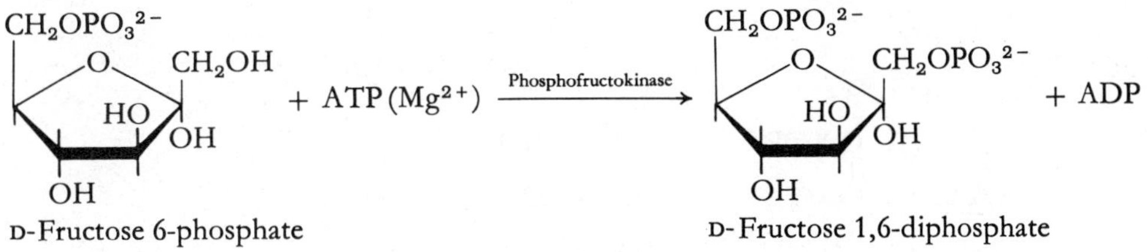

$$\text{D-Fructose 6-phosphate} + \text{ATP}(Mg^{2+}) \xrightarrow{\text{Phosphofructokinase}} \text{D-Fructose 1,6-diphosphate} + \text{ADP}$$

The reaction is irreversible since the new phosphate bond that is formed does not have a high energy of hydrolysis; the explanation for the irreversibility of this reaction is the same as that given previously for the irreversibility of the hexokinase reaction.

Phosphofructokinase is inhibited by excess ATP (substrate inhibition) and by citrate. Its activity is enhanced by AMP and ADP. Since the formation of fructose diphosphate (FDP) is the slowest of the glycolytic reactions, it is the rate-limiting step of the whole sequence. Any inhibition or activation of phosphofructokinase therefore controls the rate of glycolysis. Various characteristics of this enzyme allow for precise regulation of its activity, and it is therefore postulated that the phosphofructokinase reaction is an important site for the control of glycolysis. This subject will be discussed more fully together with other regulatory factors of carbohydrate metabolism (Chap. 6, section 6.5).

Although the phosphofructokinase reaction is irreversible, fructose 1,6-diphosphate can be converted back to fructose 6-phosphate by another enzyme, 1,6-*fructose diphosphatase*. This enzyme catalyzes the hydrolysis of fructose diphosphate to fructose 6-phosphate and inorganic phosphate; the reaction does not involve regeneration of ATP. The interconversions of fructose 6-phosphate and fructose 1,6-diphosphate may be diagrammed as follows:

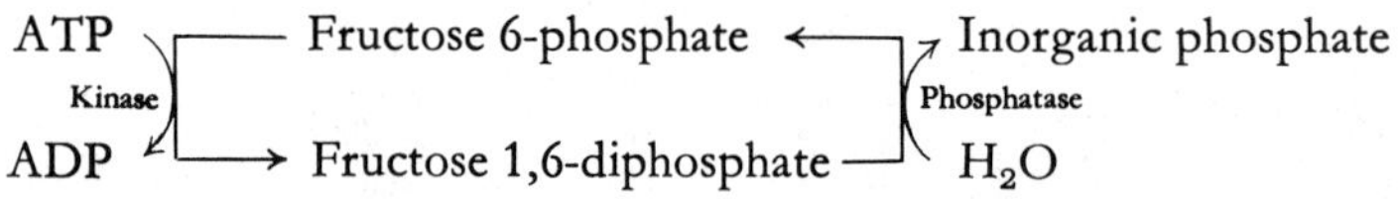

$$\text{ATP} \xrightarrow{\text{Kinase}} \text{Fructose 6-phosphate} \xleftarrow{} \text{Inorganic phosphate}$$
$$\text{ADP} \xleftarrow{} \text{Fructose 1,6-diphosphate} \xrightarrow{\text{Phosphatase}} \text{H}_2\text{O}$$

Fructose 1,6-diphosphatase is an important regulatory site for gluconeogenesis—i.e., the formation of glucose or glycogen from non-hexose structures—and will be described in detail under that topic (section 4.8, and Chap. 6, section 6.5).

The next reaction of the glycolytic pathway involves the cleavage of fructose 1,6-diphosphate to yield dihydroxyacetone phosphate and glyceraldehyde 3-phosphate. This step is catalyzed by the enzyme *aldolase*. The reaction can be visualized more clearly when fructose 1,6-diphosphate is written in the open-chain form and the individual carbons are numbered:

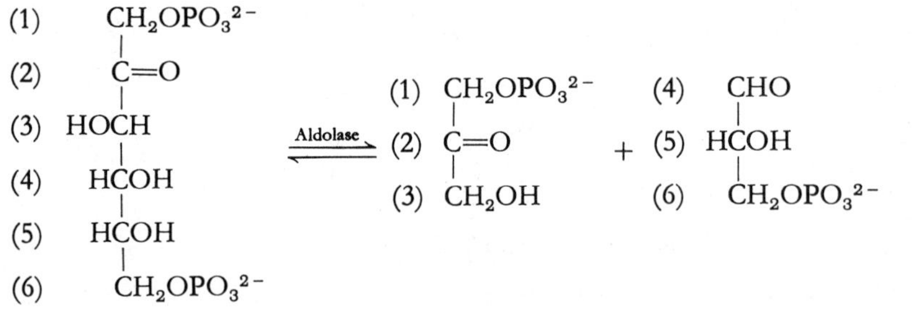

	Fructose 1,6-diphosphate		Dihydroxyacetone phosphate		D-Glyceraldehyde 3-phosphate
(1)	$CH_2OPO_3^{2-}$				
(2)	C=O	(1)	$CH_2OPO_3^{2-}$	(4)	CHO
(3)	HOCH	(2)	C=O	(5)	HCOH
(4)	HCOH	(3)	CH_2OH	(6)	$CH_2OPO_3^{2-}$
(5)	HCOH				
(6)	$CH_2OPO_3^{2-}$				

with Aldolase arrow between, and + between the two products.

The enzyme *triose-phosphate isomerase* then catalyzes the interconversion of dihydroxyacetone phosphate and glyceraldehyde 3-phosphate (the numbers refer to the specific carbon atoms in the original glucose):

$$
\begin{array}{l}
(1)\ CH_2OPO_3^{2-} \\
(2)\ C=O \\
(3)\ CH_2OH
\end{array}
\quad \underset{}{\overset{\text{Isomerase}}{\rightleftharpoons}} \quad
\begin{array}{l}
(1)\ CH_2OPO_3^{2-} \\
(2)\ HOCH \\
(3)\ CHO
\end{array}
\quad = \quad
\begin{array}{l}
(3)\ CHO \\
(2)\ HCOH \\
(1)\ CH_2OPO_3^{2-}
\end{array}
$$

Dihydroxyacetone phosphate Glyceraldehyde 3-phosphate

Since this reaction is drawn toward the formation of glyceraldehyde 3-phosphate as a result of the subsequent reaction, the original glucose 6-phosphate may be visualized as yielding, in the overall reaction to this point, 2 moles of glyceraldehyde 3-phosphate. (It should be noted that the aldehydic carbon of glyceraldehyde 3-phosphate arises from carbon-3 or carbon-4 of glucose, the center carbon from carbon-2 or carbon-5 of glucose, and the phosphorylated carbon from carbon-1 or carbon-6 of glucose.)

The next step involves the oxidation of the aldehyde group of glyceraldehyde 3-phosphate with the concomitant reduction of a mole of NAD^+. This step is catalyzed by the enzyme *glyceraldehyde 3-phosphate dehydrogenase*. The reaction requires the presence of inorganic phosphate (HPO_4^{2-}), and the product is 1,3-diphosphoglyceric acid:

$$
\begin{array}{l}
CHO \\
HCOH \\
CH_2OPO_3^{2-}
\end{array}
+ NAD^+ + HPO_4^{2-}
\quad \underset{}{\overset{\text{Glyceraldehyde 3-phosphate dehydrogenase}}{\rightleftharpoons}} \quad
\begin{array}{l}
\overset{O}{\underset{}{\|}} \\
C-OPO_3^{2-} \\
HCOH \\
CH_2OPO_3^{2-}
\end{array}
+ NADH + H^+
$$

Glyceraldehyde 3-phosphate   1,3-Diphosphoglyceric acid

This oxidation reaction is exergonic; however, the energy is conserved in the product by the formation of a high-energy phosphate compound. It should be noted that the new phosphate bond that is formed is an acyl phosphate (or an acid anhydride) rather than a phosphate ester. This bond has a considerable free energy of hydrolysis: $\Delta G^{0\prime} = -11.8$ kilocalories per mole. Since the energy of the oxidation is conserved in the compound and comparatively little is dissipated, the reaction can proceed in either direction depending on the concentration of the reacting components.

The glyceraldehyde 3-phosphate dehydrogenase reaction does not proceed in the absence of inorganic phosphate. However, the oxidation can occur when phosphate is replaced by arsenate. In this case, the product formed is not subject to the next transformation in the glycolytic sequence. Glycolysis would therefore be inhibited at this point by the arsenate.

The following step in the glycolytic pathway is the reaction between 1,3-diphosphoglyceric acid, ADP, and magnesium ions to yield 3-phosphoglyceric acid and ATP. The reaction is catalyzed by the enzyme *phosphoglycerate kinase*:

$$\begin{array}{c}\text{O}\\\|\\\text{C}-\text{OPO}_3{}^{2-}\\|\\\text{HCOH}\\|\\\text{CH}_2\text{OPO}_3{}^{2-}\end{array} + \text{ADP(Mg}^{2+}) \underset{}{\overset{\text{Kinase}}{\rightleftarrows}} \begin{array}{c}\text{COOH}\\|\\\text{HCOH}\\|\\\text{CH}_2\text{OPO}_3{}^{2-}\end{array} + \text{ATP}$$

1,3-Diphosphoglyceric acid   3-Phosphoglyceric acid

The reaction is reversible since the free energies of hydrolysis of ATP and of 1,3-diphosphoglyceric acid are within the same general range.

This is the first site in the glycolytic pathway that yields ATP. Furthermore, since 2 moles of 1,3-diphosphoglyceric acid are obtained from 1 mole of glucose, this reaction actually provides 2 moles ATP per mole of glucose.

The ATP that is generated in this step is formed by the direct transfer of phosphate from the substrate (1,3-diphosphoglyceric acid) to ADP. This mode of formation of ATP is termed *substrate-level phosphorylation*. It should be noted that this contrasts with oxidative-chain phosphorylation, where ATP is produced by the operation of the mitochondrial respiratory chain (Chap. 3, section 3.5).

The enzyme *phosphoglyceromutase* catalyzes the conversion of 3-phosphoglyceric acid to 2-phosphoglyceric acid. The reaction is reversible:

$$\begin{array}{c}\text{COOH}\\|\\\text{HCOH}\\|\\\text{CH}_2\text{OPO}_3{}^{2-}\end{array} \underset{}{\overset{\text{Phosphoglyceromutase}}{\rightleftarrows}} \begin{array}{c}\text{COOH}\\|\\\text{HCOPO}_3{}^{2-}\\|\\\text{CH}_2\text{OH}\end{array}$$

3-Phosphoglyceric acid   2-Phosphoglyceric acid

This reaction is followed by the removal of a molecule of water from 2-phosphoglyceric acid to yield phosphoenolpyruvic acid. The enzyme for this reaction is *enolase*:

$$\begin{array}{c}\text{COOH}\\|\\\text{HCOPO}_3{}^{2-}\\|\\\text{CH}_2\text{OH}\end{array} \underset{}{\overset{\text{Enolase}}{\rightleftarrows}} \begin{array}{c}\text{COOH}\\|\\\text{C}-\text{OPO}_3{}^{2-}\\\|\\\text{CH}_2\end{array} + \text{H}_2\text{O}$$

2-Phosphoglyceric acid   Phosphoenolpyruvic acid

This reaction requires the presence of magnesium. It is inhibited by fluoride ion, which forms a magnesium-fluoride-phosphate complex that prevents Mg^{2+} from binding to the enzyme.

Phosphoenolpyruvic acid is a high-energy compound, having a free energy of hydrolysis of -14.8 kilocalories per mole.

In the presence of the enzyme *pyruvate kinase* and magnesium ion, the phosphate of phosphoenolpyruvic acid is transferred to ADP. The products formed are ATP and pyruvic acid:

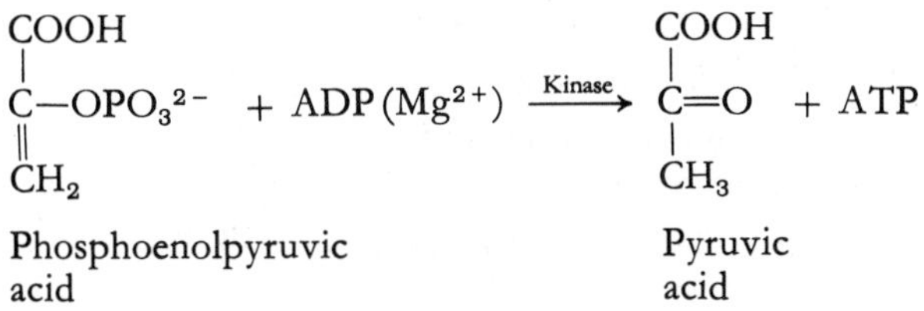

$$\begin{array}{c} COOH \\ | \\ C-OPO_3^{2-} \\ || \\ CH_2 \end{array} + ADP\,(Mg^{2+}) \xrightarrow{Kinase} \begin{array}{c} COOH \\ | \\ C=O \\ | \\ CH_3 \end{array} + ATP$$

Phosphoenolpyruvic acid　　　　　　　　　　Pyruvic acid

Since the free energy of hydrolysis of phosphoenolpyruvate is much greater than that of ATP, a considerable amount of energy is dissipated in the reaction. As expected from thermodynamic principles, this reaction is irreversible under physiologic conditions.

This is the second reaction of glycolysis that yields ATP. Again, considering that each mole of glucose yields 2 moles of phosphoenolpyruvic acid, 2 moles of ATP are generated at this stage of glycolysis for each mole glucose.

Pyruvic acid represents a central branch point in intermediary metabolism; that is, it may be converted to a number of different compounds. However, in the pathway of glycolysis, the next step is the reduction of pyruvic acid to lactic acid. This is catalyzed by the enzyme *lactate dehydrogenase* (LDH). The reducing agent in this reaction is NADH:

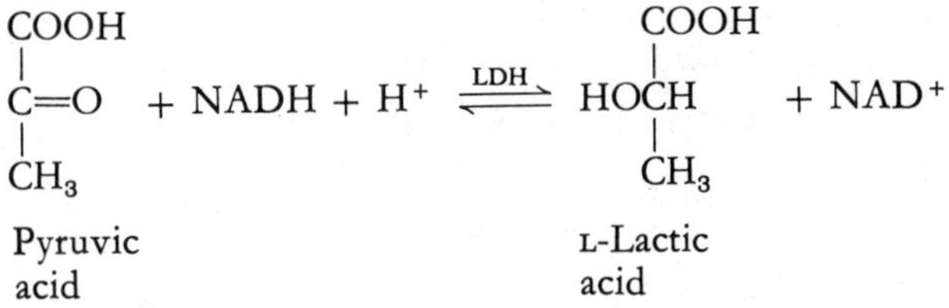

$$\begin{array}{c} COOH \\ | \\ C=O \\ | \\ CH_3 \end{array} + NADH + H^+ \underset{}{\overset{LDH}{\rightleftharpoons}} \begin{array}{c} COOH \\ | \\ HOCH \\ | \\ CH_3 \end{array} + NAD^+$$

Pyruvic acid　　　　　　　　　　L-Lactic acid

It may be noted that reduced NAD was formed in an earlier step of glycolysis, i.e., in the oxidation of glyceraldehyde 3-phosphate. The reduced NAD is thus available at this stage for the reduction of pyruvate. With respect to NAD^+, then, glycolysis may be visualized as a self-sufficient system in which this coenzyme is recycled:

Glyceraldehyde 3-phosphate + Phosphate ⇌ NAD⁺ ⇌ Lactic acid
1,3-Diphosphoglyceric acid ⇌ NADH + H⁺ ⇌ Pyruvic acid

[Reaction scheme: Glyceraldehyde 3-phosphate + Phosphate ⇌ 1,3-Diphosphoglyceric acid, coupled with NAD⁺ ⇌ NADH + H⁺, coupled with Pyruvic acid ⇌ Lactic acid]

In muscle, where lactate is formed, it cannot be metabolized further. Since it is diffusible from muscle cells, lactate moves to the bloodstream and is carried to the liver. In the liver, it may be converted to glucose (gluconeogenesis) or oxidized to pyruvate and ultimately to carbon dioxide and water.

The transformation of glucose to lactic acid via the glycolytic pathway results in the formation of a total of 4 moles of ATP. However, 2 moles of ATP are expended during the process, i.e., in the hexokinase and phosphofructokinase reactions. Hence, the net ATP yield generated in glycolysis is 2 moles of ATP per mole of glucose. The overall equation for the conversion of glucose to lactic acid is summarized as follows:

$$C_6H_{12}O_6 + 2ATP + 2NAD^+ + 2P_i + 4ADP + 2NADH + 2H^+ \longrightarrow$$
$$2CH_3CHOHCOOH + 2ADP + 2NADH + 2H^+ + 4ATP + 2NAD$$

If the substances that appear on both sides of the equation are canceled, the expression becomes:

$$C_6H_{12}O_6 + 2P_i + 2ADP \longrightarrow 2CH_3CHOHCOOH + 2ATP$$

The calculated standard free-energy change for the oxidation of 1 mole of glucose to 2 moles of lactic acid is -52.0 kilocalories. If the $\Delta G^{0\prime}$ for the hydrolysis of ATP is taken at the minimal value of 7.3 kilocalories per mole (Chap. 2, section 2.9), the glycolytic pathway, in generating 2 moles of ATP, yields a net of 14.6 kilocalories of utilizable energy. The efficiency of the process with respect to energy conservation is thus 14.6/52.0, or 28%. This value should be considered a minimum since the free energy of hydrolysis of ATP within the cell is probably more than 7.3 kilocalories per mole, as discussed previously (see Chap. 2, section 2.3).

The standard free energies of hydrolysis for some of the phosphorylated intermediates in the glycolytic pathway are listed in Table 4-1. It is seen that the compounds that transfer phosphate to ADP—i.e., 1,3-diphosphoglyceric acid and phosphoenolpyruvic acid—have a greater $-\Delta G^{0\prime}$ than does ATP. When one of these substances reacts with ADP to yield ATP, there will be a release of energy and the process will be spontaneous. In the case of phosphoenolpyruvate, the reaction is irreversible under physiologic conditions. The reaction of ADP with 1,3-diphosphoglycerate could be reversed

Table 4-1. Standard Free Energies for Phosphate Hydrolysis of Intermediates of Glycolytic Pathway

Glycolytic Intermediate	Standard Free-Energy Change, $\Delta G^{0\prime}$ (kcal/mole)
Phosphoenolpyruvate	−14.8
1,3-Diphosphoglycerate	−11.8
ATP	− 7.3
Fructose 1,6-diphosphate	− 4.0
Fructose 6-phosphate	− 3.8
Glucose 6-phosphate	− 3.3

since the energy difference between the latter and ATP can be overcome by changes in the relative concentrations of the components within the cell (Chap. 2).

It should also be noted that ATP occupies an intermediate position in the series shown in Table 4-1. Substances with higher (more negative) free energies of hydrolysis than ATP tend to channel their energy toward the generation of ATP; the latter can transfer its phosphate to effect the synthesis of phosphorylated compounds that have an even lower energy—e.g., glucose 6-phosphate and fructose diphosphate. When ATP functions in this manner, it activates the specific substrates (glucose and fructose 6-phosphate) for subsequent glycolytic reactions. ATP can therefore be conceived as an energy reserve and as an intermediary for the transfer of energy. It is generated and replenished either by oxidative phosphorylation (Chap. 3) or by substrate-level phosphorylation from 1,3-diphosphoglyceric acid and phosphoenolpyruvic acid. ATP can then be tapped and utilized for numerous functions, such as the activation of fatty acids as well as the synthesis of glucose 1,6-phosphate or fructosediphosphate.

4.4. Conversion of Pyruvate to Acetyl-Coenzyme A

Although the glycolytic sequence of reactions from glucose to pyruvate occurs in most cell types, the mode of disposition of pyruvate by these cells varies considerably. Even in the same tissue or organism, pyruvate may be metabolized by different pathways, depending on various factors and instantaneous requirements. Thus, in addition to being convertible to lactate, pyruvate may react to form alanine, oxaloacetate, malate, or acetyl-coenzyme A. In some organs—e.g., the liver—all these options are available. In cells or tissues that are more specialized, certain pathways that can serve the necessary functions of the cell or tissue will predominate.

In order for pyruvate to be utilized for the synthesis of lipids or for the derivation of energy, it must first be transferred to the mitochondrial compartment, where it is converted to acetyl-coenzyme A. This irreversible reaction is catalyzed by a multienzyme complex termed the *pyruvate dehydrogenase system*. The reactions catalyzed by this enzyme system are dependent on the operation of the respiratory chain (Chap. 3, section 3.6) and will proceed only when oxygen is available.

The pyruvate dehydrogenase complex is composed of several tightly bound enzymes and coenzymes that catalyze a series of sequential reactions leading to acetyl-coenzyme A. The coenzymes participating in this process are thiamine pyrophosphate (TPP), lipoic acid, coenzyme A, NAD^+, and FAD. The overall reaction may be depicted as follows:

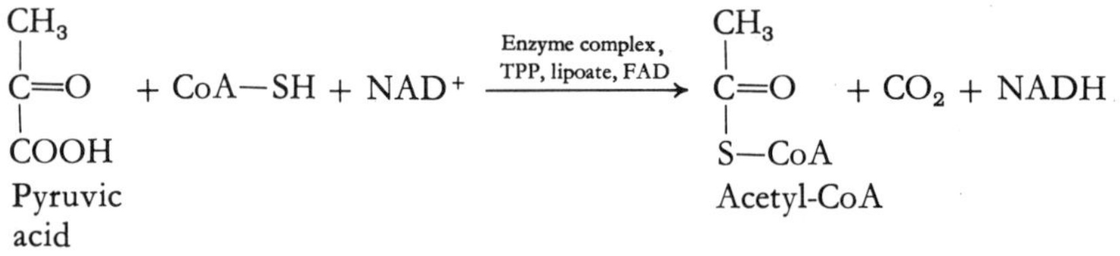

$$\underset{\substack{\text{Pyruvic}\\\text{acid}}}{\begin{array}{c}CH_3\\|\\C=O\\|\\COOH\end{array}} + CoA-SH + NAD^+ \xrightarrow[\text{TPP, lipoate, FAD}]{\text{Enzyme complex,}} \underset{\text{Acetyl-CoA}}{\begin{array}{c}CH_3\\|\\C=O\\|\\S-CoA\end{array}} + CO_2 + NADH$$

(The structure of acetyl-coenzyme A was discussed in Chapter 3, section 3.3.)

As indicated, this transformation involves several intermediate steps catalyzed by individual enzymes acting in a coordinated sequence. The first reaction is the oxidative decarboxylation of pyruvate. Carbon dioxide is released, and the acetyl fragment is linked to TPP as a hydroxyethyl group. The coenzyme TPP is also bound to *pyruvate dehydrogenase*, which catalyzes the oxidation step:

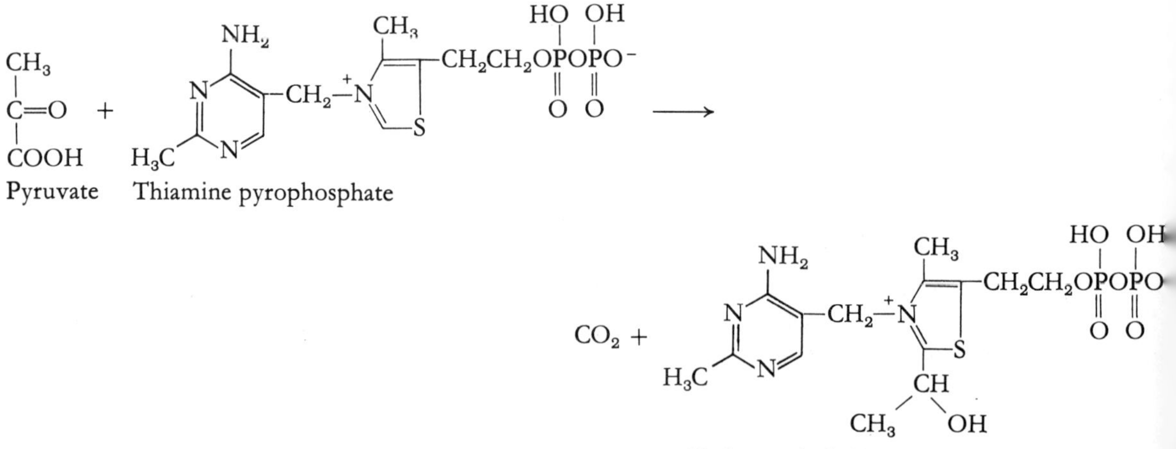

$$\underset{\text{Pyruvate}}{\begin{array}{c}CH_3\\|\\C=O\\|\\COOH\end{array}} + \underset{\text{Thiamine pyrophosphate}}{[\text{TPP structure}]} \longrightarrow$$

$$CO_2 + \underset{\alpha\text{-Hydroxyethyl thiamine pyrophosphate}}{[\text{HETPP structure}]}$$

In the next step, the hydroxyethyl group is transferred from TPP to lipoic acid. This eight-carbon acid is bound by an amide linkage to a second enzyme component of the complex, termed *dihydrolipoyl transacetylase.* As a result of this reaction, the lipoate is reduced:

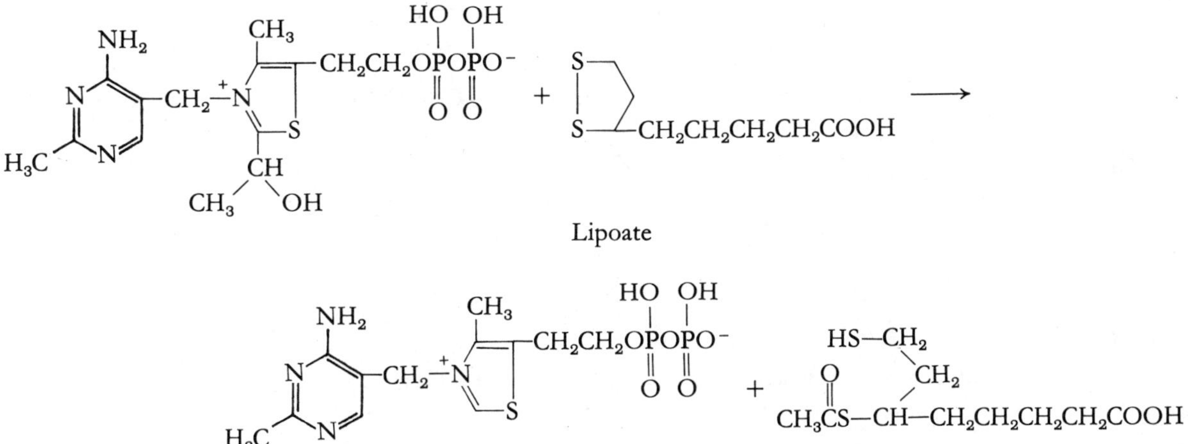

Lipoate

Acetyl lipoate

This is followed by the transfer of the acetyl group from the lipoate moiety to coenzyme A; the reaction is catalyzed by the same transacetylase:

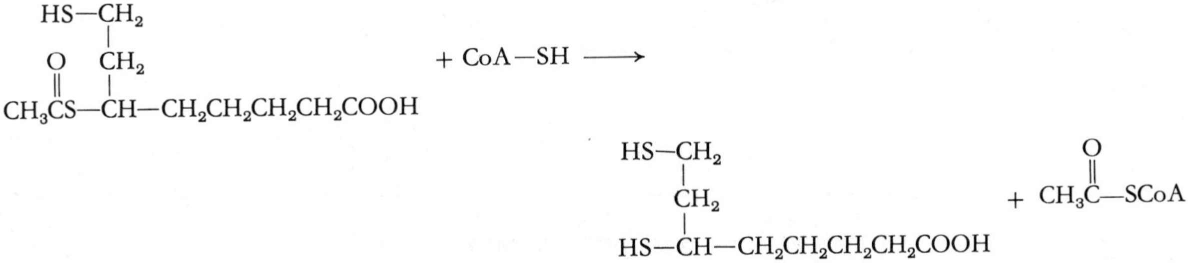

Acetyl-CoA

The enzyme-bound dihydrolipoate is then oxidized by a flavoprotein component of the enzyme *dihydrolipoyl dehydrogenase.* This brings about the regeneration of the lipoate:

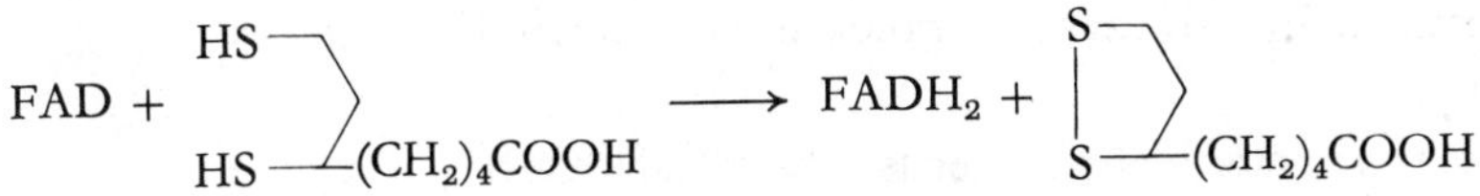

Finally, the reduced flavoprotein is reoxidized to FAD by an NAD-linked

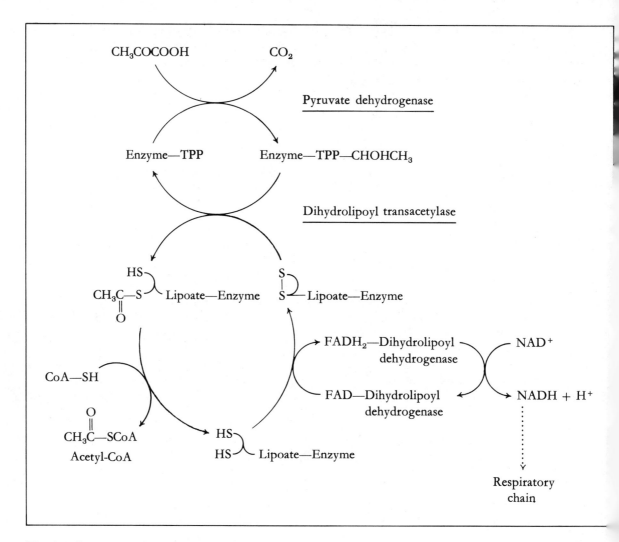

Fig. 4-4. Reactions of the pyruvate dehydrogenase system. All the intermediates are bound to an enzyme protein.

dehydrogenase, yielding reduced NAD. This NADH can then be oxidized by mitochondrial NADH dehydrogenase (Chap. 3, section 3.6) and the electrons are transferred through the respiratory chain. In terms of energy, this electron transfer yields 3 moles of ATP by oxidative phosphorylation.

The operation of the pyruvate dehydrogenase complex is summarized diagrammatically in Figure 4-4.

Multienzyme systems of the pyruvate dehydrogenase type can exert critical effects on the control of intermediary metabolism. In addition to the common enzymatic regulatory factors (e.g., inhibition or activation by various intermediates), these complexes are sensitive to conditions that affect the arrangement of the individual protein components and their intermolecular relationships. The importance of spatial organization within multienzyme systems necessitates precise coordination, and factors that effect this cooperativity will be vital control agents in the functioning of the system.

The specific regulation of the pyruvate dehydrogenase system and its role in intermediary metabolism are discussed in Chapter 6 (section 6.5).

4.5. Aerobic Oxidation of Glucose

Under conditions in which there is an ample supply of oxygen, most of the pyruvate derived from glucose is converted to acetyl-coenzyme A, which in turn may then condense with oxaloacetate and be oxidized via the citric acid cycle to carbon dioxide and water (Chap. 3, section 3.4). These processes yield considerable amounts of ATP by oxidative phosphorylation (see the following discussion). Furthermore, when oxygen is not a limiting factor, the NADH formed in glycolysis as a result of the oxidation of glyceraldehyde 3-phosphate is potentially available for oxidation by the respiratory chain. The latter reaction cannot occur directly since the NADH generated by the glyceraldehyde 3-phosphate dehydrogenase reaction is formed outside the mitochondria, whereas the reactions of the oxidative chain occur within the mitochondria. Although the mitochondrial membrane is not permeable to NADH, the electrons from the reduced NAD can be made available to the mitochondria by indirect routes that are generally termed *shuttle systems*. These will be discussed in the next section.

A. SHUTTLE SYSTEMS

One of the mechanisms by which electrons from extramitochondrial NADH can be transferred into the mitochondria involves the enzyme *glycerolphosphate dehydrogenase*. This enzyme catalyzes the reversible reaction between dihydroxyacetone phosphate (DHAP) and NADH to yield glycerol phosphate and NAD^+:

$$\begin{array}{c} CH_2OH \\ | \\ C=O \\ | \\ CH_2OPO_3H_2 \end{array} + NADH + H^+ \rightleftharpoons \begin{array}{c} CH_2OH \\ | \\ CHOH \\ | \\ CH_2OPO_3H_2 \end{array} + NAD^+$$

Dihydroxyacetone phosphate

Glycerol phosphate

When glycerol phosphate crosses the mitochondrial membrane, it is oxidized by a flavoprotein enzyme to dihydroxyacetone phosphate. The flavoprotein may then be reoxidized by the sequence of the oxidative chain. Since the mitochondrial oxidation of glycerol phosphate bypasses the NAD-FMN steps of the oxidative chain, it generates only 2 moles of ATP. The dihydroxyacetone phosphate can then be transferred back to the cytosol to react again with NADH. This shuttle system is depicted as follows:

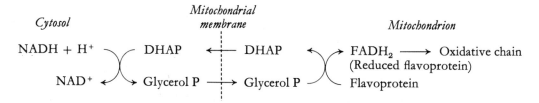

The *glycerol-phosphate shuttle* is one of several systems for transfer of electrons from NADH. It operates to a large extent within skeletal muscle.

Another transport route available to the liver and a number of other tissues is the *malate shuttle*. In this system, NADH from the soluble fraction of the cell may react with oxaloacetate in the presence of extramitochondrial malate dehydrogenase to produce malate. The malate can then be transferred into the mitochondria, where it is reoxidized to oxaloacetate by mitochondrial malate dehydrogenase. The NADH generated in this process is reoxidized via the respiratory chain, where it yields 3 moles of ATP:

Cytosol   *Membrane*   *Mitochondria*

NADH + H⁺ ⟶ Oxaloacetate ⟵ ⟵ Oxaloacetate ⟶ NADH + H⁺ ⟶ Oxidative chain
NAD⁺ ⟵ Malate ⟶ Malate ⟶ NAD⁺

The primary prerequisites for the operation of a shuttle system is that an enzyme catalyzing a particular transformation is present in both compartments and that certain products of the reaction can be transported through the mitochondrial membranes. For the glycerol-phosphate shuttle, the two glycerol-phosphate dehydrogenases fulfill the enzyme requirement. The products, dihydroxyacetone phosphate and glycerol phosphate, can be moved in and out of the mitochondria. Hence, electrons from the NADH of the cytosol are transferred through the mitochondrial membrane to the oxidative chain within the mitochondrion.

With the malate shuttle, the enzyme malate dehydrogenase is present in both compartments. Since the mitochondrial membrane transfers malate but not oxaloacetate, this system by itself could not operate as a cycle; as oxaloacetate is depleted from the cytosol, the shuttle would come to a stop.

However, this difficulty is overcome by the availability of other processes, which are analogous to shuttle systems, that in effect transfer oxaloacetate from the mitochondrion to the cytosol. These supplemental routes involve transaminases, citrate-cleaving enzyme, or malic enzyme; they are described fully in section 4.8.

The role of shuttle enzymes in metabolic regulation is discussed in Chapter 6.

B. ENERGY FROM GLUCOSE METABOLISM

The degradation of glucose to pyruvate followed by the reduction of the latter to lactate (*anaerobic glycolysis*) results in the formation of a net amount of 2 moles of ATP per mole of glucose. Under aerobic conditions (i.e., when oxygen is not a limiting factor), most of the pyruvate is oxidized to acetyl-coenzyme A with the concomitant generation of 3 moles of ATP by respiratory-chain phosphorylation, or 6 moles of ATP per mole of glucose. The conversion of each mole of acetyl-coenzyme A to carbon dioxide and water by the citric acid cycle yields 12 moles of ATP, or 24 moles of ATP per mole of glucose. In addition to the 32 moles of ATP obtained by these reactions, either 4 or 6 more moles of ATP may result from the reoxidation of NADH. The 2 moles of NADH derived from the oxidation of 2 moles of glyceraldehyde 3-phosphate in the glycolytic pathway may be reoxidized via a shuttle. When this is the glycerol-phosphate shuttle, 4 moles of ATP will be generated, whereas if the electrons from NADH are transferred by the malate-dehydrogenase shuttle, oxidation by the respiratory chain will yield 6 moles of ATP. Thus, the aerobic metabolism of glucose will yield a minimum of 36 moles of ATP.

The theoretical standard free energy released when a mole of glucose is degraded to carbon dioxide and water is 686 kilocalories. The 36 moles of ATP generated by the aerobic metabolism of glucose is equivalent to a free-energy release of 262.8 kilocalories. Hence, the total metabolic degradation of glucose to carbon dioxide and water proceeds with at least a 38% efficiency in terms of the conservation of utilizable energy. In addition to yielding eighteen times the amount of energy, the oxidation of glucose under aerobic conditions is significantly more efficient than anaerobic glycolysis.

Large amounts of lactic acid are formed from glucose when the demand for oxygen exceeds its supply. This occurs in muscle upon excessive exertion. It should be noted, however, that small amounts of lactate are formed in muscle even when there is an ample supply of oxygen.

C. PASTEUR EFFECT

When cells that are capable of metabolizing glucose both aerobically and anaerobically are maintained under aerobic conditions, they consume smaller amounts of glucose and form less lactic acid than when they are placed in an anaerobic system. This inhibition of glycolysis by oxygen is known as the *Pasteur effect*. Another related finding, termed the *Crabtree effect*, is the inhibitory action of glucose on the operation of the citric acid cycle.

Both these effects can be explained in terms of cellular energy requirements. Thus, when there is no limitation on the oxygen supply, glucose can be metabolized to pyruvate and the latter can be degraded to carbon dioxide and water. Since this pathway yields considerably more energy than does anaerobic glycolysis, the energy requirements of the cell are fulfilled by smaller amounts of glucose. On the other hand, when there are large amounts of glucose available, the energy needs can be satisfied by the glycolytic pathway, so there is a lesser demand for oxidation of pyruvate. The fact that these phenomena exist indicates that the cells have very fine control systems to regulate the mode and degree of glucose metabolism. Numerous investigations have elucidated the mechanism of the Pasteur effect, and there is good evidence that several factors are involved.

An important principle that emerges from studies of the Pasteur effect and of other regulatory mechanisms is that the cellular concentrations of certain components are limited, and the relative availability of these substances can influence the functioning of specific pathways. For example, cells of individual tissues contain fairly constant amounts of ATP, ADP, and inorganic phosphate. Under conditions in which there is considerable formation of ATP from ADP and inorganic phosphate, lesser amounts of the reactants will be available for other reactions that require these substances. As a consequence, there will be a decrease in the activity of pathways that involve steps in which ADP or inorganic phosphate are required. Hence, the ratio of ATP to ADP + P_i can be a critical factor in regulating the direction of intermediary metabolism. Another example of control by cellular cofactors is the influence of the relative amounts of NAD^+ and NADH. Reactions that result in the reduction of NAD^+ will decrease the rate of other transformations that require NAD^+. The ratio of NAD^+ to NADH can therefore have a pronounced effect on certain steps and ultimately on an entire pathway.

The Pasteur effect may be a result of such regulation; i.e., it may depend upon the availability of cofactors. Efficient operation of the glycolytic route and the respiratory chain are dependent on inorganic phosphate and ADP. Glycolysis requires inorganic phosphate for the oxidation of glyceraldehyde 3-phosphate to 1,3-diphosphoglyceric acid. It also requires ADP for the conversion of 1,3-diphosphoglyceric acid to 3-phosphoglyceric acid and ATP, and for the conversion of phosphoenolpyruvic acid to pyruvic acid and ATP. The mitochondrial chain also depends on the availability of ADP and P_i for oxidative phosphorylation. The two metabolic systems thus compete for supplies of ADP and P_i. Under anaerobic conditions when the respiratory chain is inhibited, there is a relative abundance of ADP and P_i.

The glycolytic reactions will therefore operate at a high rate, and there will be a relatively rapid consumption of glucose. However, when oxygen is introduced into the system, ADP and P_i will be removed for mitochondrial phosphorylation and glycolysis will be inhibited. The mitochondrial enzymes have a much greater affinity for ADP and P_i than do the glycolytic enzymes. Therefore, when competition exists, the mitochondrial reactions will predominate.

The low affinity (higher K_M) of the glycolytic enzymes for inorganic phosphate and ADP and the consequent subordination of this pathway by the mitochondrial phosphorylation systems can be overcome by increasing the concentration of glucose. When the level of glucose is high, P_i and ADP will be consumed by the glycolytic reactions, and the citric acid cycle, which is linked to oxidative phosphorylation, will be inhibited; i.e., the Crabtree effect will occur. One item in evidence for the relation between the Pasteur effect and the levels of ADP or P_i is the fact that the addition of 2,4-dinitrophenol relieves the inhibition of glycolysis by oxygen. Since 2,4-dinitrophenol uncouples phosphorylation of ADP from the oxidative chain, P_i and ADP are not consumed during respiration and are thus available for glycolytic reactions.

Another mechanism by which oxygen would influence the rate of glycolysis and decrease glucose consumption is that of enzyme inhibition. Citrate and excess ATP inhibit phosphofructokinase. When oxygen is available, pyruvate is oxidized to acetyl-coenzyme A, which in turn is incorporated into citrate. Significant increases in the levels of this intermediate and of ATP generated by oxidative phosphorylation, cause a decrease in the conversion of fructose 6-phosphate to fructose 1,6-diphosphate and a concomitant inhibition of the glycolytic reactions. This mechanism is essentially that of *feedback inhibition*, i.e., inhibition in which an early reaction is blocked by later products of the pathway.

As mentioned previously, an additional factor involved in these interrelationships is the level of NADH or the ratio of NAD^+ to NADH. Under aerobic conditions, when there is an ample supply of oxygen, the NADH formed during the oxidation of glyceraldehyde 3-phosphate is oxidized (via one of the shuttle reactions) by the respiratory chain. As a consequence, very little NADH is available for the reduction of pyruvate to lactate. (The major pathway for pyruvate in an aerobic system is its oxidation to acetyl-coenzyme A.) However, when the oxygen supply is limited, NADH is not removed from the system and its accumulation will consequently bring about the reduction of pyruvate to lactate. The metabolism of glucose under anaerobic conditions therefore results in the accumulation of lactate.

The regulatory systems involved in the Pasteur effect are important examples of the interaction of different metabolic pathways and their regulation by the availability of various intermediates. In the present context, these intermediates are the adenosine nucleotides (ADP and ATP), inorganic phosphate, and the pyridine nucleotide coenzymes (NAD$^+$ and NADH). In addition to the action of these substances as substrates in certain critical reactions, some of them also have regulatory effects by functioning as inhibitors or activators of key enzymes in a pathway. An example of such an effect is the inhibition of phosphofructokinase by citrate or excess ATP.

4.6. Pentose Pathway

The oxidative *pentose-phosphate pathway*, also known as the *hexose-monophosphate shunt* or *phosphogluconate pathway*, includes a group of reactions in which glucose 6-phosphate is oxidized and converted to pentose phosphates. It is important because it provides reduced nicotinamide adenine dinucleotide phosphate (NADPH) which is required for various reductive reactions, and ribose 5-phosphate, which is utilized for the synthesis of nucleotides and nucleic acids. The transformations involved in the pentose-phosphate oxidative pathway occur in the cytosol (extramitochondrial) compartment of the cell.

The first reaction in the series involves the oxidation of glucose 6-phosphate by NADP to yield gluconolactone 6-phosphate (or 6-phosphogluconolactone) and NADPH. This is catalyzed by the enzyme *glucose 6-phosphate dehydrogenase*:

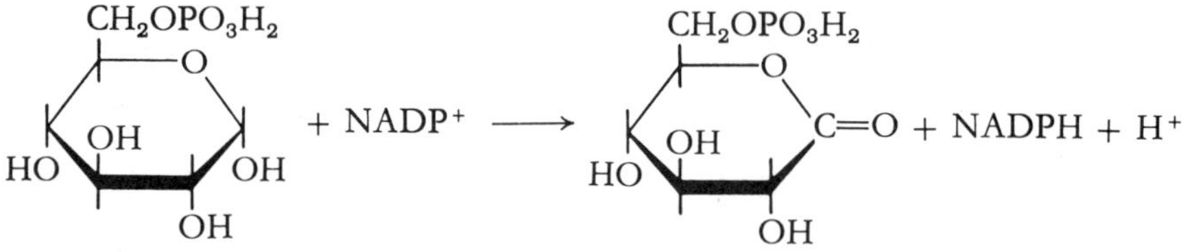

D-Glucose 6-phosphate + NADP$^+$ ⟶ D-Gluconolactone 6-phosphate + NADPH + H$^+$

Although gluconolactone 6-phosphate hydrolyzes spontaneously, it does so rather slowly. The enzyme *lactonase* catalyzes this reaction so that its velocity is increased by several orders of magnitude. The product of the hydrolysis is gluconic acid 6-phosphate (or 6-phosphogluconic acid):

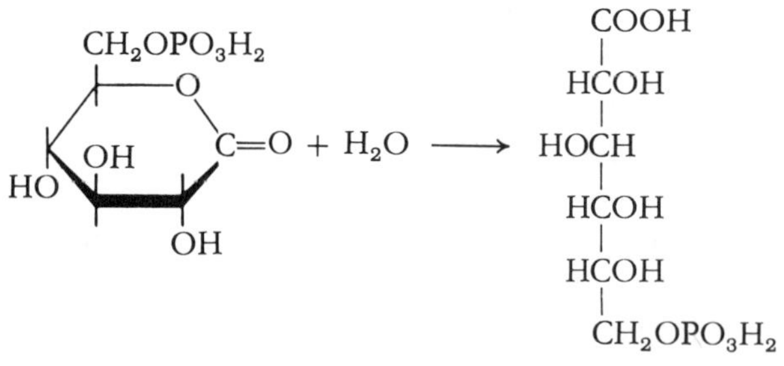

D-Gluconolactone 6-phosphate + H$_2$O ⟶ D-Gluconic acid 6-phosphate

Gluconic acid 6-phosphate is oxidized in the presence of NADP$^+$ and the enzyme *gluconic acid 6-phosphate dehydrogenase*, to D-ribulose 5-phosphate. Carbon dioxide and NADPH are also produced; the CO_2 arises from carbon-1 of the hexose:

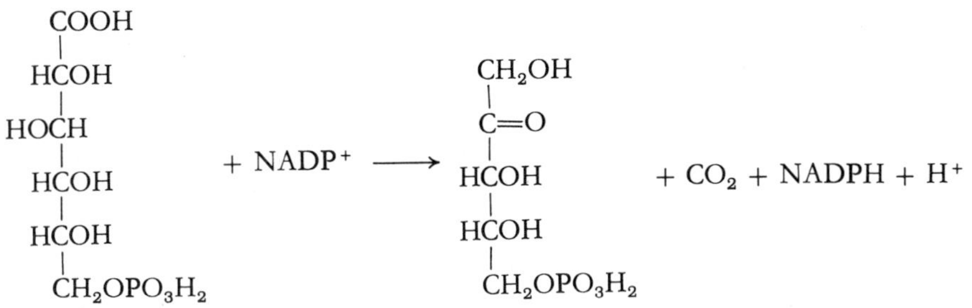

$$
\begin{array}{c}
\text{COOH} \\
|\\
\text{HCOH} \\
|\\
\text{HOCH} \\
|\\
\text{HCOH} \\
|\\
\text{HCOH} \\
|\\
\text{CH}_2\text{OPO}_3\text{H}_2 \\
\text{D-Gluconic acid} \\
\text{6-phosphate}
\end{array}
\;+\; \text{NADP}^+ \;\longrightarrow\;
\begin{array}{c}
\text{CH}_2\text{OH} \\
|\\
\text{C}=\text{O} \\
|\\
\text{HCOH} \\
|\\
\text{HCOH} \\
|\\
\text{CH}_2\text{OPO}_3\text{H}_2 \\
\text{D-Ribulose} \\
\text{5-phosphate}
\end{array}
\;+\; CO_2 \;+\; \text{NADPH} \;+\; \text{H}^+
$$

The enzyme *pentose-phosphate epimerase* catalyzes the reversible conversion of D-ribulose 5-phosphate to D-xylulose 5-phosphate:

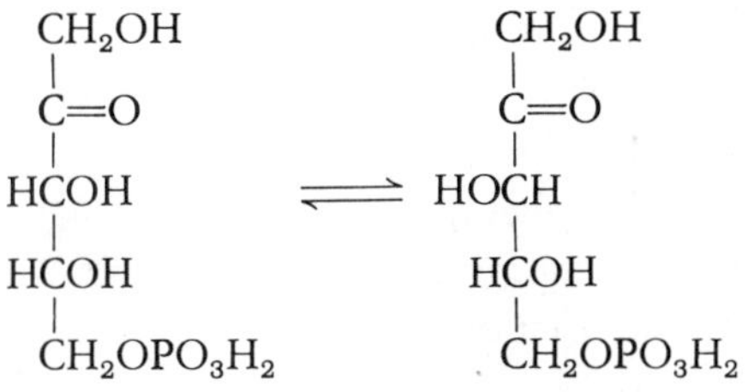

$$
\begin{array}{c}
\text{CH}_2\text{OH} \\
|\\
\text{C}=\text{O} \\
|\\
\text{HCOH} \\
|\\
\text{HCOH} \\
|\\
\text{CH}_2\text{OPO}_3\text{H}_2 \\
\text{D-Ribulose} \\
\text{5-phosphate}
\end{array}
\;\rightleftharpoons\;
\begin{array}{c}
\text{CH}_2\text{OH} \\
|\\
\text{C}=\text{O} \\
|\\
\text{HOCH} \\
|\\
\text{HCOH} \\
|\\
\text{CH}_2\text{OPO}_3\text{H}_2 \\
\text{D-Xylulose} \\
\text{5-phosphate}
\end{array}
$$

D-Ribulose 5-phosphate may also be converted to D-ribose 5-phosphate. This reaction, which is also reversible, is catalyzed by the enzyme *pentose-phosphate isomerase*:

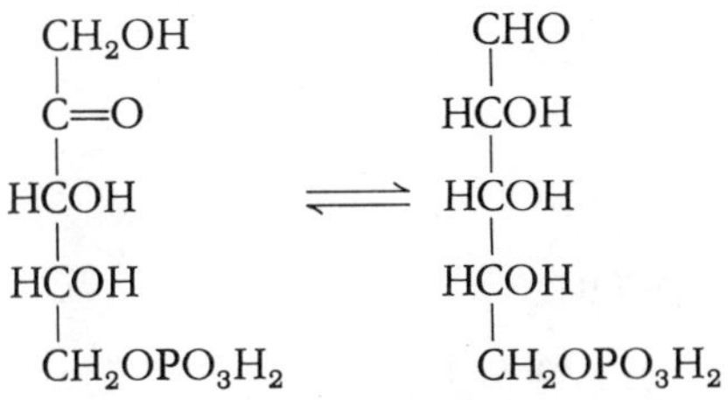

$$
\begin{array}{c}
\text{CH}_2\text{OH} \\
|\\
\text{C}=\text{O} \\
|\\
\text{HCOH} \\
|\\
\text{HCOH} \\
|\\
\text{CH}_2\text{OPO}_3\text{H}_2 \\
\text{D-Ribulose} \\
\text{5-phosphate}
\end{array}
\;\rightleftharpoons\;
\begin{array}{c}
\text{CHO} \\
|\\
\text{HCOH} \\
|\\
\text{HCOH} \\
|\\
\text{HCOH} \\
|\\
\text{CH}_2\text{OPO}_3\text{H}_2 \\
\text{D-Ribose} \\
\text{5-phosphate}
\end{array}
$$

These various interconversions give rise to a mixture of the three pentose phosphates. The latter can be utilized in a number of concurrent and sequential reversible reactions.

Xylulose 5-phosphate can react with ribose 5-phosphate to yield sedoheptulose 7-phosphate and glyceraldehyde 3-phosphate:

$$
\begin{array}{c}
\text{CH}_2\text{OH} \\
| \\
\text{C}=\text{O} \\
| \\
\text{HOCH} \\
| \\
\text{HCOH} \\
| \\
\text{CH}_2\text{OPO}_3\text{H}_2 \\
\text{D-Xylulose} \\
\text{5-phosphate}
\end{array}
+
\begin{array}{c}
\text{CHO} \\
| \\
\text{HCOH} \\
| \\
\text{HCOH} \\
| \\
\text{HCOH} \\
| \\
\text{CH}_2\text{OPO}_3\text{H}_2 \\
\text{D-Ribose} \\
\text{5-phosphate}
\end{array}
\rightleftharpoons
\begin{array}{c}
\text{CH}_2\text{OH} \\
| \\
\text{C}=\text{O} \\
| \\
\text{HOCH} \\
| \\
\text{HCOH} \\
| \\
\text{HCOH} \\
| \\
\text{HCOH} \\
| \\
\text{CH}_2\text{OPO}_3\text{H}_2 \\
\text{D-Sedoheptulose} \\
\text{7-phosphate}
\end{array}
+
\begin{array}{c}
\text{CHO} \\
| \\
\text{HCOH} \\
| \\
\text{CH}_2\text{OPO}_3\text{H}_2 \\
\text{D-Glyceraldehyde} \\
\text{3-phosphate}
\end{array}
$$

This reaction is catalyzed by the enzyme *transketolase* and requires the presence of thiamine pyrophosphate (TPP). It involves the transfer of the ketol, or glycolaldehyde group –C(=O)CH$_2$OH, from xylulose 5-phosphate to carbon-1 of ribose 5-phosphate. The ketol unit is bound to the enzyme-TPP complex and is transferred directly to the acceptor. It may be noted that the donor of the ketol group (xylulose 5-phosphate) and the product (sedoheptulose 7-phosphate) are ketoses in which carbon-3 has an "L-type" configuration. This is a steric requirement for ketol donors in transketolase reactions. Another reaction with analogous substrates will be described later.

In the next step, sedoheptulose 7-phosphate can transfer a dihydroxyacetone group (carbon-1, -2, and -3) to glyceraldehyde 3-phosphate. This reaction, which is catalyzed by *transaldolase*, yields erythrose 4-phosphate and fructose 6-phosphate:

$$
\begin{array}{c}
\text{CH}_2\text{OH} \\
| \\
\text{C}=\text{O} \\
| \\
\text{HOCH} \\
| \\
\text{HCOH} \\
| \\
\text{HCOH} \\
| \\
\text{HCOH} \\
| \\
\text{CH}_2\text{OPO}_3\text{H}_2 \\
\text{D-Sedoheptulose} \\
\text{7-phosphate}
\end{array}
+
\begin{array}{c}
\text{CHO} \\
| \\
\text{HCOH} \\
| \\
\text{CH}_2\text{OPO}_3\text{H}_2 \\
\text{D-Glyceraldehyde} \\
\text{3-phosphate}
\end{array}
\rightleftharpoons
\begin{array}{c}
\text{CH}_2\text{OH} \\
| \\
\text{C}=\text{O} \\
| \\
\text{HOCH} \\
| \\
\text{HCOH} \\
| \\
\text{HCOH} \\
| \\
\text{CH}_2\text{OPO}_3\text{H}_2 \\
\text{D-Fructose} \\
\text{6-phosphate}
\end{array}
+
\begin{array}{c}
\text{CHO} \\
| \\
\text{HCOH} \\
| \\
\text{HCOH} \\
| \\
\text{CH}_2\text{OPO}_3\text{H}_2 \\
\text{D-Erythrose} \\
\text{4-phosphate}
\end{array}
$$

This reaction is also freely reversible. Transaldolase differs from the fructose-diphosphate aldolase of the glycolytic pathway in that it does not bring about the release of free dihydroxyacetone; the latter is bound to the transaldolase enzyme and transferred directly to the acceptor.

Erythrose 4-phosphate can act as an acceptor in a *transketolase* reaction. It can thus react with xylulose 5-phosphate to yield fructose 6-phosphate and glyceraldehyde 3-phosphate:

$$
\begin{array}{c}
\text{CH}_2\text{OH} \\
| \\
\text{C}=\text{O} \\
| \\
\text{HOCH} \\
| \\
\text{HCOH} \\
| \\
\text{CH}_2\text{OPO}_3\text{H}_2 \\
\text{D-Xylulose} \\
\text{5-phosphate}
\end{array}
\; + \;
\begin{array}{c}
\text{CHO} \\
| \\
\text{HCOH} \\
| \\
\text{HCOH} \\
| \\
\text{CH}_2\text{OPO}_3\text{H}_2 \\
\text{D-Erythrose} \\
\text{4-phosphate}
\end{array}
\;\rightleftharpoons\;
\begin{array}{c}
\text{CH}_2\text{OH} \\
| \\
\text{C}=\text{O} \\
| \\
\text{HOCH} \\
| \\
\text{HCOH} \\
| \\
\text{HCOH} \\
| \\
\text{CH}_2\text{OPO}_3\text{H}_2 \\
\text{D-Fructose} \\
\text{6-phosphate}
\end{array}
\; + \;
\begin{array}{c}
\text{CHO} \\
| \\
\text{HCOH} \\
| \\
\text{CH}_2\text{OPO}_3\text{H}_2 \\
\text{D-Glyceraldehyde} \\
\text{3-phosphate}
\end{array}
$$

It can be seen that the net result of the reactions of the oxidative pentose-phosphate pathway is the production of fructose 6-phosphate and glyceraldehyde 3-phosphate. Conceivably, with the cooperation of four additional glycolytic enzymes, this pathway can result in the conversion of all the carbons of glucose to carbon dioxide. The required enzymes would be (1) *triose-phosphate isomerase* to convert glyceraldehyde 3-phosphate to dihydroxyacetone phosphate, (2) *aldolase* to yield fructose 1,6-diphosphate from glyceraldehyde 3-phosphate and dihydroxyacetone phosphate, (3) *fructose diphosphatase* to hydrolyze fructose 1,6-diphosphate to fructose 6-phosphate, and (4) *hexose-phosphate isomerase* to convert fructose 6-phosphate to glucose 6-phosphate. The glucose 6-phosphate can then be oxidized to 6-phosphogluconolactone, and the process repeated. This can be shown as a series of balanced equations starting with six molecules of glucose 6-phosphate:

1. 6 Glucose 6-phosphate + 6NADP$^+$ $\longrightarrow$ 6 6-Phosphogluconolactone + 6NADPH + 6H$^+$
2. 6 6-Phosphogluconolactone + 6H$_2$O $\longrightarrow$ 6 6-Phosphogluconic acid
3. 6 6-Phosphogluconic acid + 6NADP$^+$ $\longrightarrow$ 6 Ribulose 5-phosphate + 6CO$_2$ + 6NADPH + 6H$^+$
4. 4 Ribulose 5-phosphate $\longrightarrow$ 4 Xylulose 5-phosphate
5. 2 Ribulose 5-phosphate $\longrightarrow$ 2 Ribose 5-phosphate
6. 2 Xylulose 5-phosphate + 2 Ribose 5-phosphate $\longrightarrow$
                             2 Sedoheptulose 7-phosphate + 2 Glyceraldehyde 3-phosphate

7. 2 Sedoheptulose 7-phosphate + 2 Glyceraldehyde 3-phosphate $\longrightarrow$
  2 Erythrose 4-phosphate + 2 Fructose 6-phosphate
8. 2 Xylulose 5-phosphate + 2 Erythrose 4-phosphate $\longrightarrow$
  2 Fructose 6-phosphate + 2 Glyceraldehyde 3-phosphate
9. Glyceraldehyde 3-phosphate $\longrightarrow$ Dihydroxyacetone 3-phosphate
10. Glyceraldehyde 3-phosphate + Dihydroxyacetone 3-phosphate $\longrightarrow$ Fructose 1,6-diphosphate
11. Fructose 1,6-diphosphate $\longrightarrow$ Fructose 6-phosphate + P_i
12. 5 Fructose 6-phosphate $\longrightarrow$ 5 Glucose 6-phosphate

The sum of these reactions is:

6 Glucose 6-phosphate + 12NADP$^+$ $\longrightarrow$ 6CO$_2$ + 5 Glucose 6-phosphate + 12NADPH + 12H$^+$ + P_i

According to these considerations, 6 moles of glucose 6-phosphate are converted to 6 moles of CO$_2$ and 5 moles of glucose 6-phosphate. The latter, with the addition of another mole of glucose 6-phosphate, can then be recycled through the same series of steps.

Alternatively, the pentose-phosphate pathway may be conceived as a "shunt" for the production of fructose 6-phosphate from glucose 6-phosphate. Both the glucose 6-phosphate and the glyceraldehyde 3-phosphate produced by the pentose-phosphate pathway can be metabolized to pyruvate and finally oxidized by the mitochondrial enzyme system. (Actually, the pathway was originally called the "hexose-monophosphate shunt" because it was considered to be an alternative method in which glucose 6-phosphate is converted to fructose 6-phosphate.)

Within cells that contain the appropriate enzymes, all the routes discussed may be taken in the metabolic degradation of glucose. However, there are considerable differences in cells of various tissues as to the preferred pathway. Thus, in the liver, about 30% of the glucose is oxidized by the pentose-phosphate reactions. In adipose tissue, this pathway accounts for over 60% of glucose oxidation. Other examples of tissues that demonstrate a considerable proportion of this mode of glucose metabolism are the mammary gland, the adrenal cortex, erythrocytes, and leukocytes.

The preferred pathway for glucose metabolism in various tissues is related to their metabolic function. In general, cells that are active in synthesizing fatty acids or steroids will metabolize glucose to a greater extent by the oxidative pentose-phosphate pathway than by glycolysis. This finding is explained in terms of the need for NADPH by the involved cells; for example, the oxidative pentose-phosphate pathway provides the NADPH that is required for the synthesis of lipids. On this basis, it would appear that the primary function of the oxidative pentose-phosphate pathway is to produce reduced NADP. The latter provides a reservoir of reducing agent for the synthesis of cellular components, such as fatty acids, steroids, and DNA. NADPH also functions to maintain hemoglobin or glutathione in erythrocytes in the reduced state.

The oxidative pentose-phosphate pathway also provides the ribose 5-phosphate that is utilized in the synthesis of nucleotides and nucleic acids. However, this is not the only means of production of ribose 5-phosphate since pentose phosphates can be produced from fructose 6-phosphate by non-oxidative routes. For example, a transketolase reaction between fructose 6-phosphate and glyceraldehyde 3-phosphate yields erythrose 4-phosphate and xylulose 5-phosphate:

$$
\begin{array}{c}
CH_2OH \\
| \\
C=O \\
| \\
HOCH \\
| \\
HCOH \\
| \\
HCOH \\
| \\
CH_2OPO_3H_2
\end{array}
\;+\;
\begin{array}{c}
CHO \\
| \\
HCOH \\
| \\
CH_2OPO_3H_2
\end{array}
\;\rightleftharpoons\;
\begin{array}{c}
CHO \\
| \\
HCOH \\
| \\
HCOH \\
| \\
CH_2OPO_3H_2
\end{array}
\;+\;
\begin{array}{c}
CH_2OH \\
| \\
C=O \\
| \\
HOCH \\
| \\
HCOH \\
| \\
CH_2OPO_3H_2
\end{array}
$$

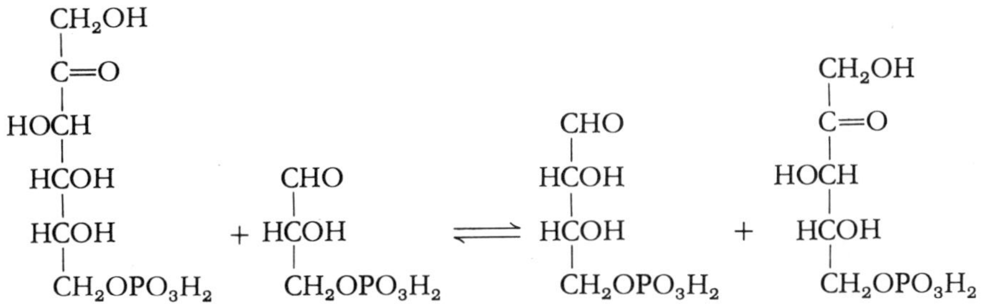

Fructose 6-phosphate   Glyceraldehyde 3-phosphate   Erythrose 4-phosphate   Xylulose 5-phosphate

The erythrose 4-phosphate may react with another molecule of fructose 6-phosphate by a transaldolase reaction:

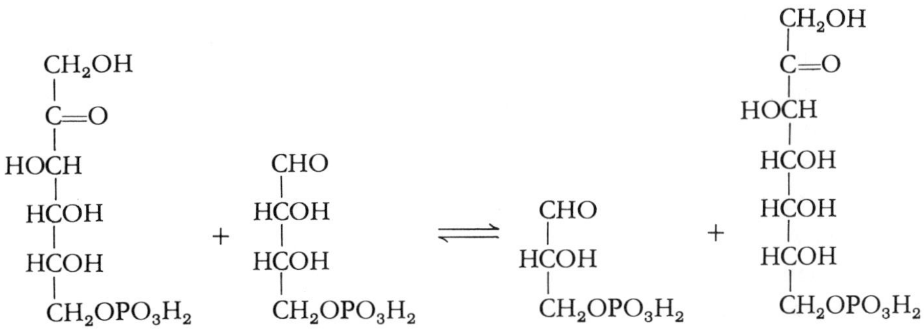

$$
\begin{array}{c}
CH_2OH \\
| \\
C=O \\
| \\
HOCH \\
| \\
HCOH \\
| \\
HCOH \\
| \\
CH_2OPO_3H_2
\end{array}
\;+\;
\begin{array}{c}
CHO \\
| \\
HCOH \\
| \\
HCOH \\
| \\
CH_2OPO_3H_2
\end{array}
\;\rightleftharpoons\;
\begin{array}{c}
CHO \\
| \\
HCOH \\
| \\
CH_2OPO_3H_2
\end{array}
\;+\;
\begin{array}{c}
CH_2OH \\
| \\
C=O \\
| \\
HOCH \\
| \\
HCOH \\
| \\
HCOH \\
| \\
HCOH \\
| \\
CH_2OPO_3H_2
\end{array}
$$

Fructose 6-phosphate   Erythrose 4-phosphate   Glyceraldehyde 3-phosphate   Sedoheptulose 7-phosphate

Reaction of the sedoheptulose 7-phosphate produced in the above reaction with another molecule of glyceraldehyde 3-phosphate yields, via another transketolase reaction, ribose 5-phosphate and xylulose 5-phosphate:

```
   CH₂OH
    |
    C=O
    |
   HOCH                              CHO                    CH₂OH
    |                                 |                      |
   HCOH                              HCOH                    C=O
    |                                 |                      |
   HCOH      +   CHO      ⇌         HCOH       +          HOCH
    |             |                   |                      |
   HCOH          HCOH                HCOH                   HCOH
    |             |                   |                      |
  CH₂OPO₃H₂    CH₂OPO₃H₂          CH₂OPO₃H₂              CH₂OPO₃H₂
 Sedoheptulose  Glyceraldehyde     Ribose                 Xylulose
  7-phosphate   3-phosphate        5-phosphate            5-phosphate
```

It should be noted that the formation of pentose phosphates by this pathway does not involve any oxidative steps. The extent to which ribose is formed by oxidative or non-oxidative routes can be evaluated by studying the disposition of glucose labeled with ^{14}C on specific carbons. Thus, glucose-2-^{14}C, when metabolized by the oxidative pentose-phosphate pathway, should yield ribose-1-^{14}C. If the ribose arises by the non-oxidative sequence, it should be labeled on carbon-2. Results of such studies have shown that the *oxidative* pentose-phosphate pathway is most prominent in cells that synthesize lipids, i.e., those that require considerable amounts of NADPH. Other tissues apparently derive their pentoses primarily by non-oxidative routes.

4.7. Glycogen Metabolism

Glycogen is a highly branched polysaccharide composed of D-glucose units joined to each other by glycosidic bonds (Fig. 4-5). The major linkages are α-1,4 glycosidic bonds. At intervals of about ten units, there are branches in the chain involving α-1,6 linkages. Each branch then continues with α-1,4 linkages. Additional α-1,6 branching then occurs on the original branch.

Although glycogen is present in many types of cells, the largest concentrations are generally found in liver and muscle. The actual amount in the liver during normal sugar intake is about 150 grams of glycogen per kilogram dry weight. During starvation, the liver glycogen is depleted rapidly and may drop to 10% of the original concentration. Muscle glycogen is not lost as quickly during fasting. Thus, in well-fed animals, muscle glycogen is about 12 grams per kilogram of tissue, whereas after a three-day fast it may drop to about 5 grams.

Glycogen serves as the metabolic storage form of carbohydrates. In order to fulfill this function efficiently, both the release of glucose from glycogen and the incorporation of glucose into glycogen must be highly responsive to the immediate requirements of the organism. Indeed, the routes for the

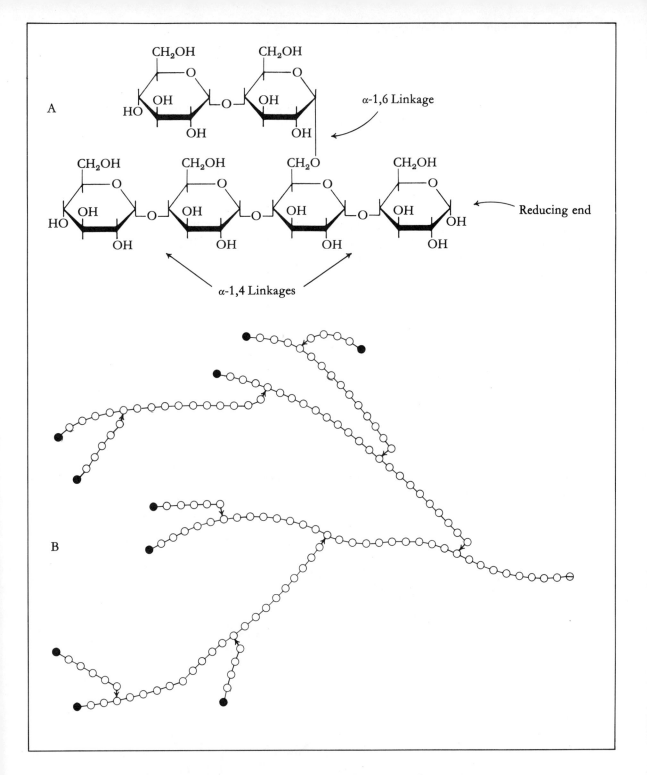

Fig. 4-5. *A. Types of linkages in glycogen. B. General structure of glycogen:* ⊖, *reducing end;* ●, *non-reducing ends;* ○—○, *α-1,4 linkage;* ○←○, *α-1,6 linkage (branch point). Each chain is extended and branched further, producing a polysaccharide of high molecular weight.*

synthesis and the degradation of glycogen are laden with intricate regulatory systems involving enzymatic and hormonal controls.

In this section, first the specific reactions and metabolic pathways in glycogen metabolism are delineated; then certain details of the enzymatic processes and control factors are described.

A. SYNTHETIC PATHWAY

In order to be incorporated into glycogen, glucose 6-phosphate must first be converted to glucose 1-phosphate. This reaction, which is reversible, is catalyzed by the enzyme *phosphoglucomutase*:

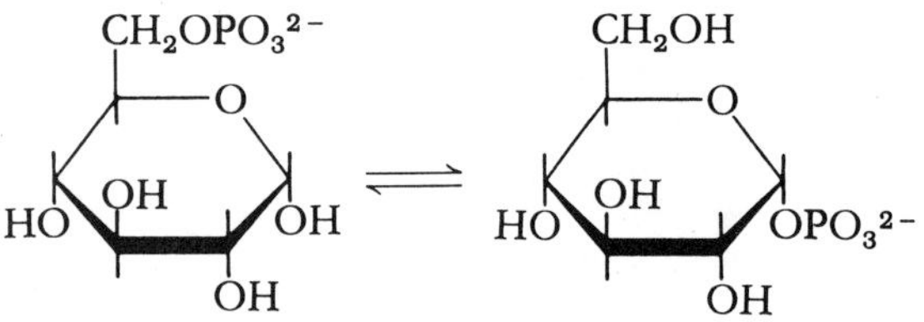

D-Glucose 6-phosphate ⇌ D-Glucose 1-phosphate

Glucose 1-phosphate then reacts with uridine triphosphate (UTP) to yield uridine diphosphoglucose (UDP-glucose) and inorganic pyrophosphate. The enzyme for this reaction is *uridine-diphosphoglucose pyrophosphorylase* or glucosyl phosphate uridylyl transferase:

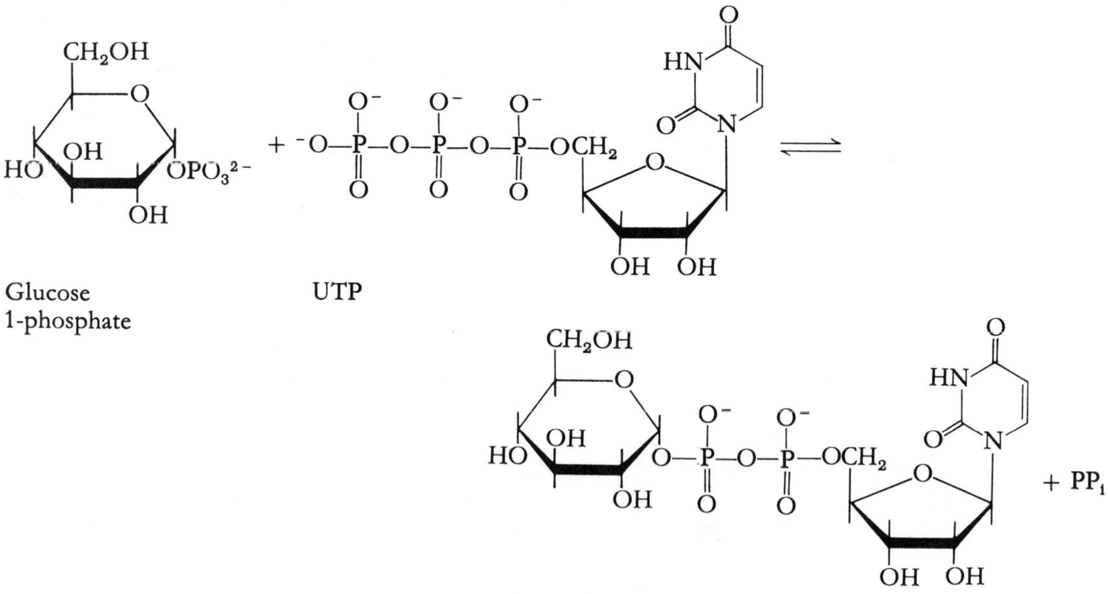

Glucose 1-phosphate + UTP ⇌ UDP-glucose + PP_i

This reaction is freely reversible in vitro. However, since the inorganic pyrophosphate is hydrolyzed to phosphate by a cellular *pyrophosphatase*, the extent of reversal of the UDP-glucose pyrophosphorylase reaction would be minimal.

The next step involves the transfer of glucose from UDP-glucose to the non-reducing end of pre-existing glycogen. This results in the attachment, via an α-1,4 linkage, of the new glucose unit to the end of the glycogen polymer. This reaction is catalyzed by the enzyme *glycogen synthetase*:

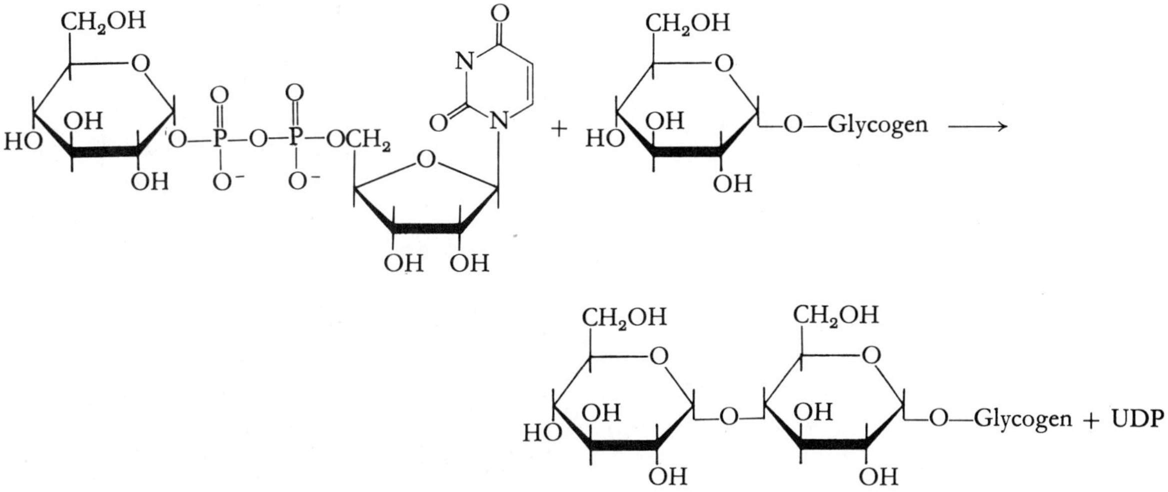

The best substrate for the reaction is glycogen itself or a large oligosaccharide. Smaller units may also serve as acceptors of glucose from UDP-glucose; however, these have much higher K_M values, and the reaction is comparatively slow.

The addition of glucose to the non-reducing end of the polysaccharide is repeated a number of times until many new residues are added. After the attachment of about ten 1,4-linked units, a portion from the non-reducing end of the chain is transferred to carbon-6 of a glucose unit in the same chain (Fig. 4-6) or in a neighboring chain (Fig. 4-7). This reaction, which produces a branch in the polysaccharide, is brought about by the enzyme *amylo-(1,4 → 1,6)-transglycosylase*; this enzyme is also known as the *branching enzyme*. It should be noted that intact chains are transferred in this process so that 1,6 linkages occur only at branch points; the rest of the polysaccharide contains 1,4 linkages. The minimal chain size for such a transfer appears to be a hexasaccharide.

After the branching has occurred, additional glucose units may be added

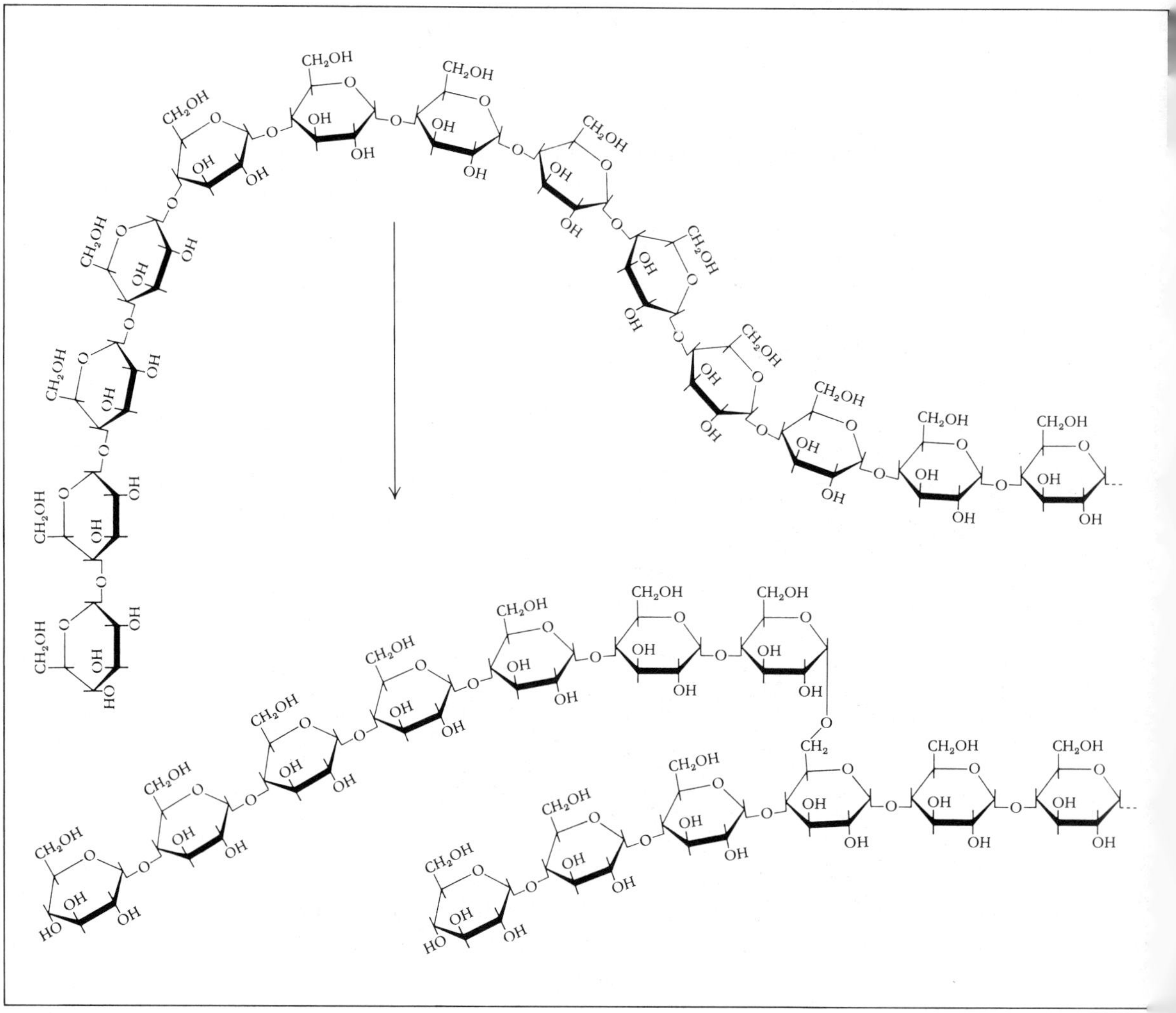

4. Metabolism of Carbohydrates

Fig. 4-6. Action of amylo-(1,4→1,6)-transglycosylase in transferring a heptasaccharide unit from the non-reducing end of a glycogen segment to produce a branch on the same chain.

from UDP-glucose to the non-reducing ends of either the original chains or the branches by means of the glycogen synthetase reaction. When a sufficient number of units are added in this manner, the branching may be repeated.

As a result of this combination of reactions, glycogen generally contains about 80% to 90% 1,4 linkages and about 10% to 20% 1,6 linkages. The branching frequency is approximately 8 to 12 glucose units.

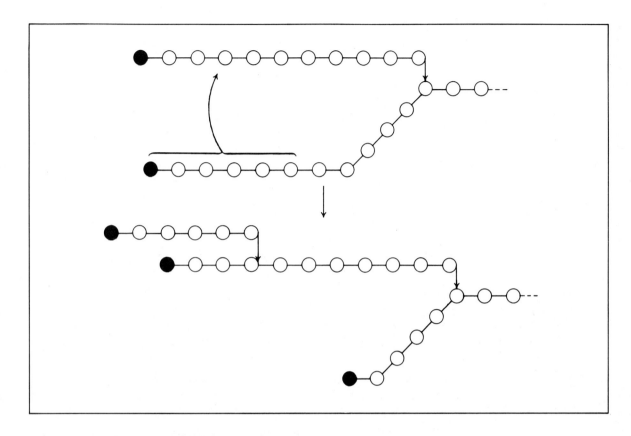

Fig. 4-7. Transfer of an oligosaccharide unit from position-4 of a glucose of one chain to position-6 of a glucose residue in another chain. ○—○, *1,4 linkage;* ○↓ , *1,6 linkage;* ●, *sugar residue at non-reducing end.*

B. GLYCOGENOLYSIS

The degradation of glycogen, or *glycogenolysis*, involves essentially two types of processes: phosphorolysis and debranching. The enzyme that brings about the removal of glucose units from glycogen is *glycogen phosphorylase*. However, this enzyme acts only on α-1,4 linkages. In order to achieve extended degradation, glycogen has to be debranched by a *debranching enzyme* so that 1,6 linkages are eliminated.

The enzyme glycogen phosphorylase catalyzes the interaction of glycogen with inorganic phosphate to yield glucose 1-phosphate. This enzyme contains pyridoxal phosphate and is specific for the α-1,4-linked glucose units on the non-reducing terminals of the polysaccharide:

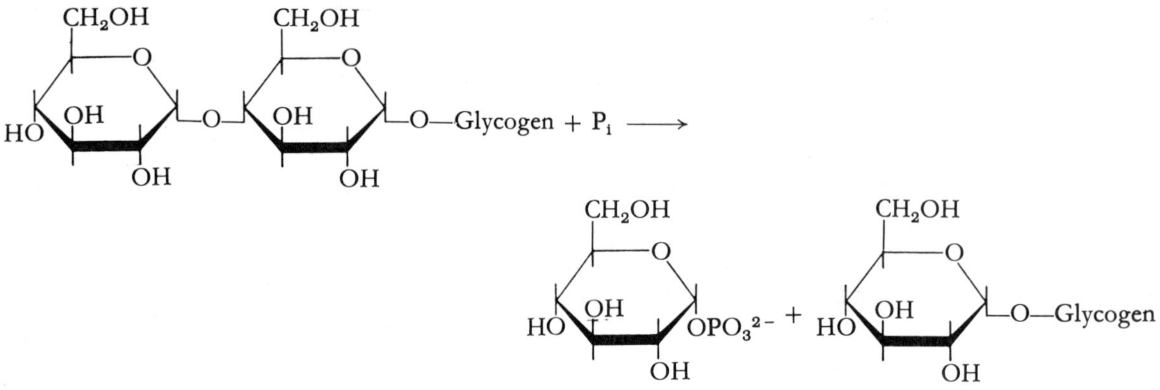

As soon as a terminal glucose is taken off the glycogen chain, the newly exposed unit itself becomes susceptible to phosphorolysis. The removal of glucose units continues until the vicinity of a branch point is reached. Glycogen at this stage of degradation is called a *limit dextrin*.

Debranching of the limit dextrin involves two separate enzymatically catalyzed reactions (Fig. 4-8). The first enzyme is termed *oligo-(1,4 → 1,4)-glucantransferase*. This brings about the transfer of a 1,4-linked oligosaccharide unit from the 1,6-linked glucose of a branch to the non-reducing end of a chain. As a result, a single 1,6-linked glucose is left at the branch point. The second enzyme in the debranching process, *amylo-1,6-glucosidase*, catalyzes the hydrolytic cleavage of the 1,6 linkage, thus removing the single glucose branch. At this stage, the unbranched glycogen is susceptible to further degradation by phosphorylase. An alternation of phosphorolysis and debranching may thus result in substantial degradation of the glycogen polysaccharide.

In addition to the above-mentioned enzymes that catalyze glycogen breakdown, the liver also contains a lysosomal *α-glucosidase* that hydrolyzes glycosidic bonds to yield glucose units. The exact significance of this enzyme is not clear; however, its absence results in a severe pathologic condition. This and other glycogen-storage diseases will be discussed in Chapter 6 (section 6.8).

C. HORMONAL EFFECTS ON GLYCOGEN METABOLISM

Glycogen phosphorylase and glycogen synthetase both have inactive (or less active) and active forms, and the activation of each of these enzymes is under hormonal control. The activity of the enzymes is also highly dependent on the availability of various intermediates and cofactors. As a consequence, both glycogenesis and glycogenolysis are highly regulated, and the amounts of glucose released for metabolism normally conform with the requirements of the organism.

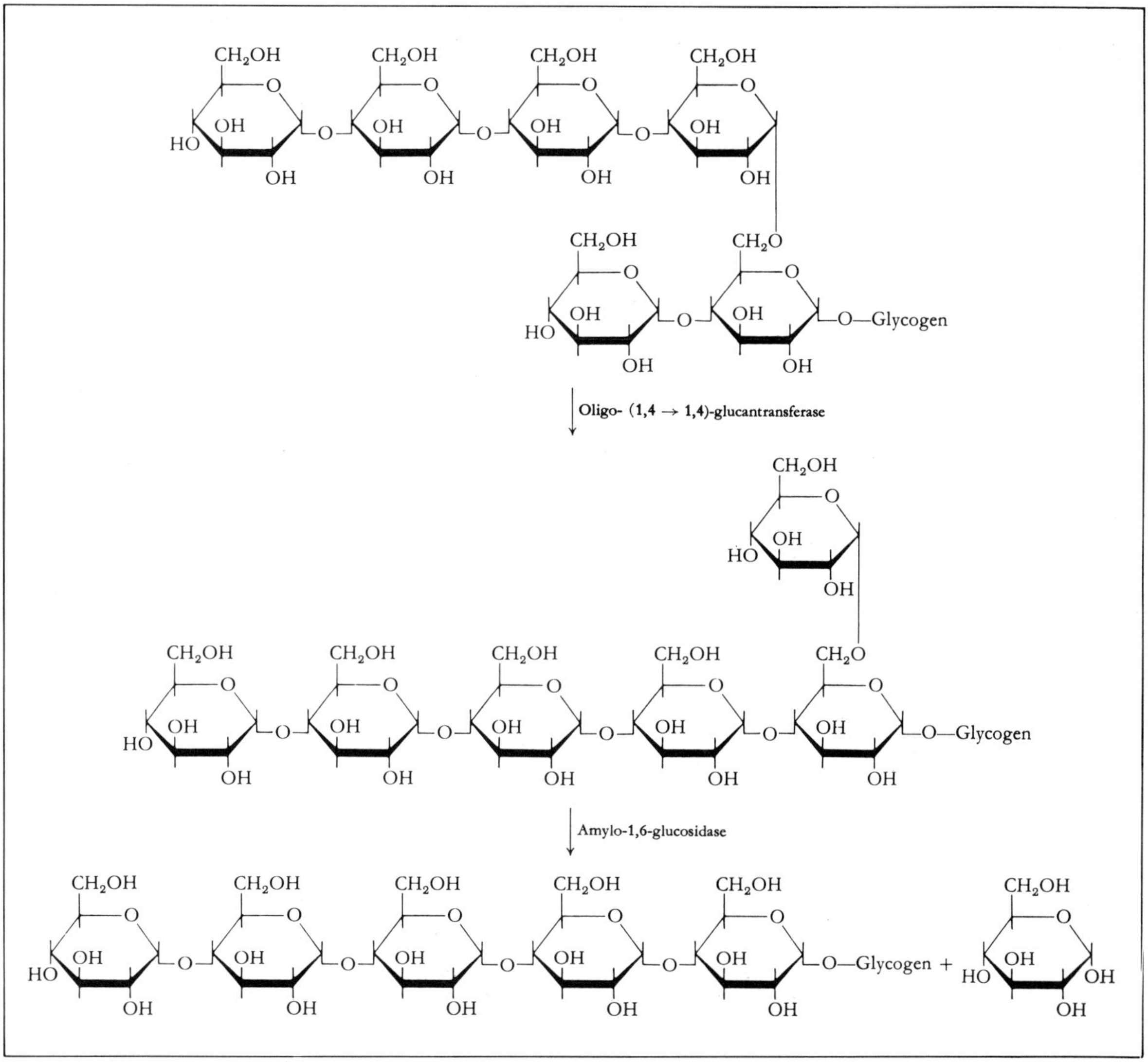

Fig. 4-8. Debranching of glycogen.

The relatively inactive form of glycogen phosphorylase in muscle is termed *phosphorylase b*. The latter functions only in the presence of comparatively high concentrations of AMP. Phosphorylase *b* is converted to its active form, *phosphorylase a*, by the active form of the enzyme *phosphorylase b kinase* and ATP. In addition to conformational changes, this transformation involves the addition of phosphate to the phosphorylase. More precisely, phosphorylase *b* is a dimer (i.e., is composed of two polypeptide units) and does not contain phosphate, whereas phosphorylase *a* is a phosphorylated tetramer. The reaction may thus be written as follows:

2 Phosphorylase *b* + nATP ⟶ Phosphorylase *a* + nADP

Phosphorylase *a* can be converted back to phosphorylase *b* by the liver enzyme *phosphorylase a phosphatase*:

Phosphorylase *a* ⟶ 2 Phosphorylase *b* + nP$_i$

The active form of phosphorylase *b* kinase also has an inactive precursor, which is activated by *protein kinase*.

The steps in the activation of phosphorylase described thus far can be summarized as follows:

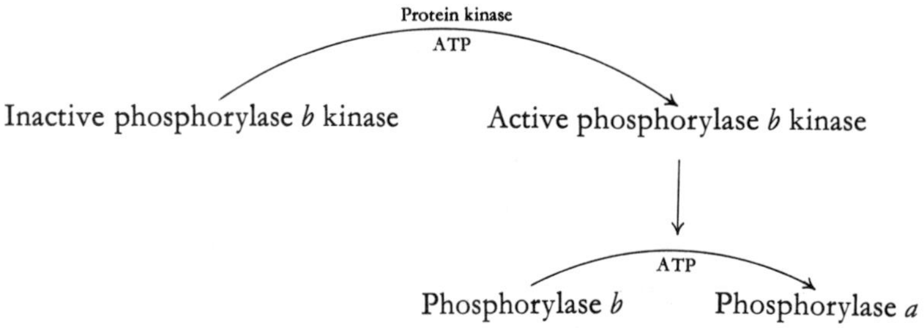

$$\text{Inactive phosphorylase } b \text{ kinase} \xrightarrow{\text{Protein kinase, ATP}} \text{Active phosphorylase } b \text{ kinase}$$

$$\text{Phosphorylase } b \xrightarrow{\text{ATP}} \text{Phosphorylase } a$$

The hormone *epinephrine* effects the sequential series of reactions leading to the activation of phosphorylase *a* by stimulating the enzyme *adenyl cyclase*, which catalyzes the conversion of ATP to cyclic AMP (3′,5′-cyclic adenylic acid):

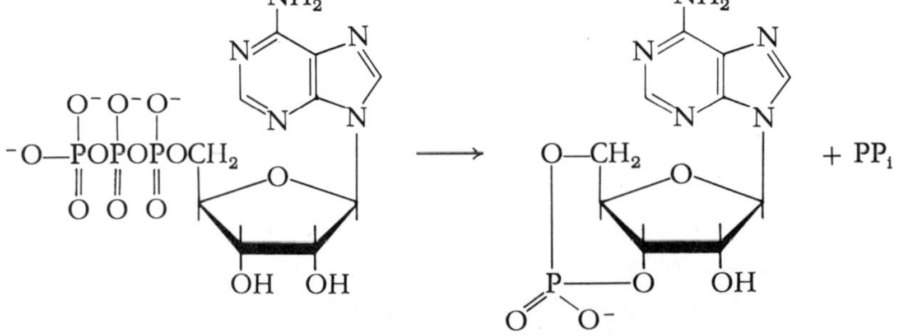

ATP ⟶ Cyclic AMP + PP$_i$

Cyclic AMP has the effect of activating protein kinase, which in turn triggers the series of steps terminating in the generation of phosphorylase *a*.

Adenyl cyclase is located in the plasma membrane of many cells. Epinephrine is secreted by the adrenal medulla and reaches particular cells via the bloodstream. The hormone binds to the plasma membrane and stimulates the cell's adenyl cyclase. The latter catalyzes the formation of cyclic AMP and sets in motion the process leading to the activation of phosphorylase. Cyclic AMP thus functions as an intracellular messenger that relays the instructions of the extracellular hormone (epinephrine) to the cellular enzyme systems. In this instance, the message of the epinephrine is to provide greater amounts of glucose for cellular metabolism.

Another hormone that stimulates glycogen phosphorolysis is *glucagon*. This is a polypeptide hormone secreted by the alpha cells of the pancreas. Glucagon, like epinephrine, stimulates adenyl cyclase and the consequent formation of cyclic AMP. Its effect is primarily on liver and heart muscle.

Glycogen synthetase also occurs in two forms. One of these is linked to phosphate and requires glucose 6-phosphate as a cofactor for enzymatic activity. It is termed *synthetase D* (for "dependent"), since it depends on glucose 6-phosphate for its activation. The other form is not phosphorylated and does not require glucose 6-phosphate; this is called the independent synthetase or *synthetase I*. Synthetase I reacts with ATP when catalyzed by *protein kinase* to yield synthetase D. Synthetase D can be converted back to synthetase I by the action of *synthetase D phosphatase*. These interconversions are depicted as follows:

$$\text{ATP} \searrow \quad \text{Synthetase I} \searrow \quad \nearrow P_i$$
$$\text{Protein kinase} \quad\quad\quad\quad\quad\quad\quad\quad \text{Phosphatase}$$
$$\text{ADP} \nearrow \quad \text{Synthetase D} \nearrow \quad \searrow H_2O$$

Epinephrine, via its effect on protein kinase through the generation of cyclic AMP, will have the overall result of inhibiting the synthesis of glycogen. Glycogen synthetase and glycogen phosphorylase are affected by phosphorylation in an opposite fashion: the activated phosphorylase is linked to phosphate, whereas the more active synthetase is in the dephospho form. Since the ultimate effect of epinephrine and cyclic-AMP generation on the two enzymes is to transform them to the phosphated form, the end result is to decrease the synthesis of glycogen (by inactivation of synthetase) and increase its rate of degradation (by activation of phosphorylase). Epinephrine, by stimulating adenyl-cyclase activity, thus makes glucose available to the cells through two mechanisms: by inhibiting its storage in the form of glycogen and by enhancing the breakdown of glycogen or withdrawal from the stores (Fig. 4-9).

Insulin is effective in decreasing the level of cyclic AMP. With respect to glycogen metabolism, the action of this hormone opposes that of epinephrine.

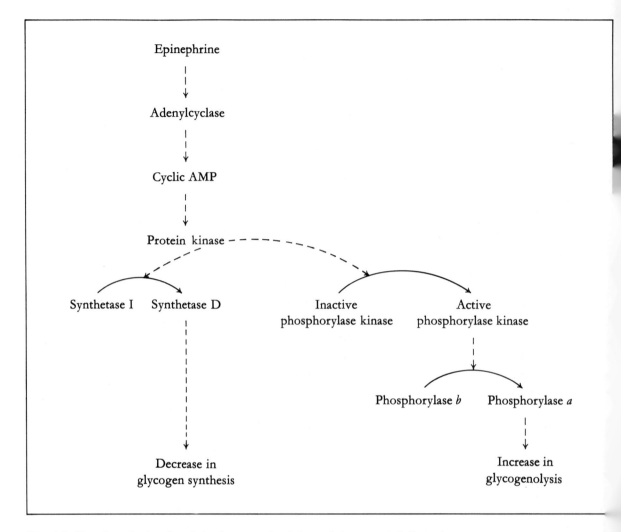

Fig. 4-9. Reactions showing the relation between epinephrine and glycogen metabolism; the effect of epinephrine is to inhibit glycogen synthesis and increase glycogenolysis. The effect of glucagon is similar to that of epinephrine. Insulin inhibits these sequences.

Insulin would therefore bring about a decrease in phosphorylase activity and an enhancement of glycogen-synthetase activity. The net result would then be the conservation of carbohydrates, rather than their utilization.

In addition to the hormonal regulation of glycogen utilization by means of the activation or deactivation of phosphorylase and synthetase, the rate of its metabolism is influenced profoundly by various intermediates and cofactors. For example, the activity of phosphorylase b is enhanced in the presence of AMP. Similarly, glycogen synthetase D is activated by glucose

6-phosphate. Another factor that can play an important role in glycogen metabolism is the inhibition of specific enzymes when certain intermediates accumulate. For example, UDP is an inhibitor of both UDP-glucose pyrophosphorylase and liver glycogen synthetase. UDP, which is formed as a result of the synthesis of glycogen from UDP-glucose, will thus serve to moderate the rate of glycogen synthesis. Glycogen itself is an inhibitor of the conversion of synthetase D to synthetase I. Consequently, an oversupply of glycogen will tend to decrease the rate of its own synthesis.

4.8. Gluconeogenesis

It is obvious from consideration of the metabolic pathways described in this chapter that in addition to being a critical source of energy, glucose is necessary for the production of a number of vital intermediates. Thus, the pentose phosphates derived from glucose metabolism are required for the synthesis of nucleotides and nucleic acids. The dihydroxyacetone phosphate formed during glycolysis yields glycerol phosphate for the formation of triglycerides. The direct oxidative pathway of glucose 6-phosphate generates reduced NADP for various synthetic processes. Glucose is also the precursor of hexosamines and uronic acids, which are building blocks for the biosynthesis of glucosaminoglycans (mucopolysaccharides), glycoproteins, and uronide conjugates (section 4.9). Hence, in order to maintain continued normal functioning of the organism, it is of utmost importance that a steady supply of glucose be available to the cells even when dietary carbohydrates are reduced or withdrawn. This is accomplished by *gluconeogenesis*, i.e., the conversion of non-carbohydrate nutrients or intermediates to glucose and glycogen. Gluconeogenesis takes place at a modest rate under all conditions; however, it is accelerated considerably during fasting and in sustained muscular activity as well as in untreated diabetes.

Gluconeogenesis occurs in the liver and kidney. The most prominent precursors for this process are certain amino acids (e.g., alanine, glutamate, aspartate, serine, and cysteine) as well as pyruvate, lactate, and glycerol. The common factor among amino acids that are glucogenic is that they can be converted to either pyruvate or one of the dicarboxylic acids of the citric acid cycle. Glycerol can be phosphorylated to glycerol phosphate and oxidized to dihydroxyacetone phosphate, and lactate is oxidizable to pyruvate.

The process of gluconeogenesis is essentially a reversal of the glycolytic pathway. The reactions leading from fructose 1,6-diphosphate to phosphoenolpyruvate are all reversible, so the enzymes that catalyze the glycolytic reactions may also serve to convert the triose phosphates back to fructose 1,6-diphosphate. The glycolytic formation of the latter from fructose 6-phosphate is not reversible; however, fructose 1,6-diphosphate can yield

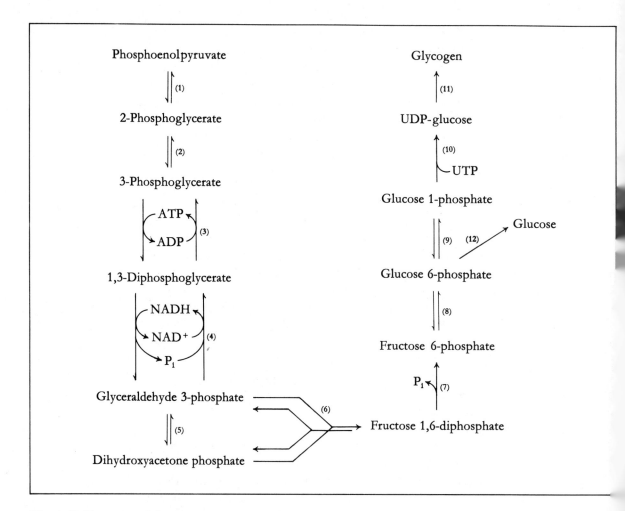

Fig. 4-10. Formation of glucose and glycogen from phosphoenolpyruvate. The enzymes indicated by the numbers in parentheses are (1) enolase, (2) phosphoglyceromutase, (3) phosphoglycerate kinase, (4) glyceraldehyde 3-phosphate dehydrogenase, (5) triose-phosphate isomerase, (6) aldolase, (7) fructose diphosphatase, (8) glucose-phosphate isomerase, (9) phosphoglucomutase, (10) UDP-glucose pyrophosphorylase, (11) glycogen synthetase, and (12) glucose 6-phosphatase.

fructose 6-phosphate through the action of *fructose diphosphatase*. Fructose 6-phosphate can then be transformed to glucose 6-phosphate by glucose-phosphate isomerase since this reaction is reversible. Glucose 6-phosphate may be either converted to glycogen or hydrolyzed to glucose by the action of *glucose 6-phosphatase* (the hexokinase reaction is also irreversible). The gluconeogenic sequence of steps from phosphoenolpyruvate is shown in Figure 4-10.

The formation of pyruvate from phosphoenolpyruvate by pyruvate kinase is not reversible under physiologic conditions (section 4.3). In order to bypass this barrier and convert pyruvate to phosphoenolpyruvate, it is necessary to utilize two additional reactions. These involve the sequence:

Pyruvate $\longrightarrow$ Oxaloacetate $\longrightarrow$ Phosphoenolpyruvate

In the first reaction, pyruvate reacts with carbon dioxide (or bicarbonate) and ATP to yield oxaloacetate and ADP plus inorganic phosphate:

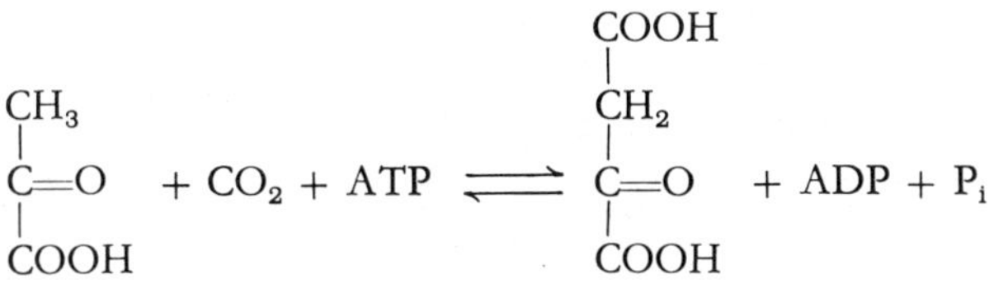

$$\begin{array}{c} CH_3 \\ | \\ C=O \\ | \\ COOH \end{array} + CO_2 + ATP \rightleftharpoons \begin{array}{c} COOH \\ | \\ CH_2 \\ | \\ C=O \\ | \\ COOH \end{array} + ADP + P_i$$

Pyruvate Oxaloacetate

This reaction is catalyzed by the mitochondrial enzyme *pyruvate carboxylase*, which contains covalently linked *biotin* (page 220). The mechanism of the reaction involves the combination of CO_2 (as a carboxyl group) with biotin, followed by the transfer of the carboxylate unit to pyruvate. Pyruvate carboxylase is an allosteric enzyme that requires acetyl-coenzyme A as a positive effector.

In the second reaction, another enzyme, *phosphoenolpyruvate carboxykinase*, which is present in the cytosol, brings about the reaction between oxaloacetate and guanosine triphosphate (GTP) or inosine triphosphate (ITP) to yield phosphoenolpyruvate, carbon dioxide, guanosine diphosphate (or inosine diphosphate), and inorganic phosphate:

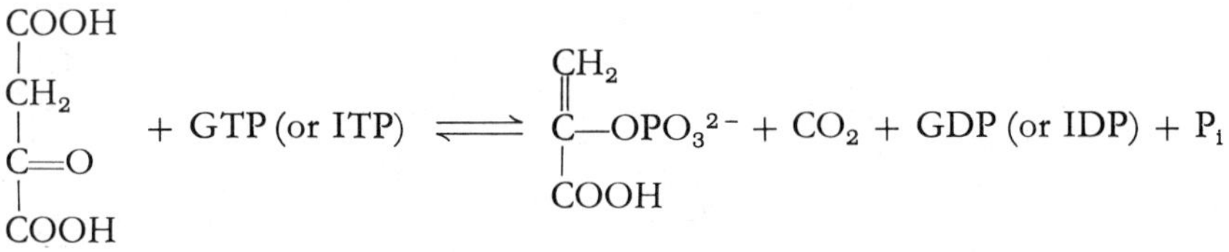

$$\begin{array}{c} COOH \\ | \\ CH_2 \\ | \\ C=O \\ | \\ COOH \end{array} + GTP\ (or\ ITP) \rightleftharpoons \begin{array}{c} CH_2 \\ \| \\ C-OPO_3^{2-} \\ | \\ COOH \end{array} + CO_2 + GDP\ (or\ IDP) + P_i$$

Oxaloacetate Phosphoenolpyruvate

The cellular location of phosphoenolpyruvate carboxykinase varies with different species. In mouse and rat liver, it is found in the cytosol.

The combination of these two reactions allows for the transformation of pyruvate (or of any intermediate that can be converted to pyruvate) to phosphoenolpyruvate. The latter may ultimately yield glucose and glycogen, as outlined in Figure 4-10. It should be noted that both the pyruvate carboxylase and the phosphoenolpyruvate carboxykinase reactions require nucleoside triphosphates, that is, they are energy-consuming reactions. Considering the other reactions of the gluconeogenic pathway, it is seen that ATP is also consumed in the formation of 1,3-diphosphoglycerate from 3-phosphoglycerate. Since the synthesis of glucose requires the formation of two triose phosphates, a net amount of 6 moles of nucleoside triphosphates are consumed for the generation of each mole of glucose from pyruvate. Gluconeogenesis is thus quite a costly process in terms of energy or ATP consumed. Under conditions when gluconeogenesis is proceeding at a high rate, more than 60% of the ATP generated in the liver may be consumed for this pathway.

It was pointed out previously (section 4.5.A) that the mitochondrial membrane is not permeable to oxaloacetate. Since oxaloacetate is formed from pyruvate in the mitochondria (by pyruvate carboxylase) and may be converted to phosphoenolpyruvate in the cytosol (by phosphoenolpyruvate carboxykinase), a special system for transporting this intermediate from one compartment to the other becomes necessary. Several reactions are available to provide the transfer of oxaloacetate. Some of these reactions are analogous to a shuttle system (section 4.5.A).

Both the mitochondrial and the cytosol compartments contain enzymes termed *aspartate-glutamate transaminases*. These enzymes will be discussed in Chapter 7 (section 7.2); however, for present purposes, it is sufficient to know that they catalyze the following reversible reaction:

$$
\begin{array}{c}
\text{COOH} \\
| \\
\text{C=O} \\
| \\
\text{CH}_2 \\
| \\
\text{COOH} \\
\text{Oxaloacetate}
\end{array}
+
\begin{array}{c}
\text{COOH} \\
| \\
\text{HCNH}_2 \\
| \\
\text{CH}_2 \\
| \\
\text{CH}_2 \\
| \\
\text{COOH} \\
\text{Glutamate}
\end{array}
\rightleftharpoons
\begin{array}{c}
\text{COOH} \\
| \\
\text{HCNH}_2 \\
| \\
\text{CH}_2 \\
| \\
\text{COOH} \\
\text{Aspartate}
\end{array}
+
\begin{array}{c}
\text{COOH} \\
| \\
\text{C=O} \\
| \\
\text{CH}_2 \\
| \\
\text{CH}_2 \\
| \\
\text{COOH} \\
\alpha\text{-Ketoglutarate}
\end{array}
$$

Since the mitochondrial inner membrane has a carrier system for transporting these two amino acids, the transaminase reactions allow for the effective transfer of oxaloacetate from one compartment to the other:

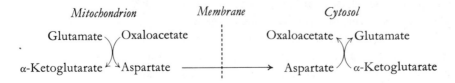

The oxaloacetate is thus made available in the cytosol and can be converted there to phosphoenolpyruvate by extramitochondrial phosphoenolpyruvate carboxykinase.

Another process that can generate oxaloacetate in the extramitochondrial compartment involves the *citrate-cleaving enzyme* (citrate lyase). The latter, which is found in the soluble fraction, catalyzes the reaction between citrate, coenzyme A, and ATP to yield oxaloacetate, acetyl-coenzyme A, ADP, and inorganic phosphate:

$$\underset{\text{Citrate}}{\begin{array}{c}\text{CH}_2\text{COOH}\\|\\\text{HOCCOOH}\\|\\\text{CH}_2\text{COOH}\end{array}} + \text{CoA—SH} + \text{ATP} \longrightarrow \underset{\text{Acetyl-CoA}}{\text{CH}_3\overset{\text{O}}{\overset{\|}{\text{C}}}\text{—S—CoA}} + \underset{\text{Oxaloacetate}}{\begin{array}{c}\overset{\text{O}}{\overset{\|}{\text{C}}}\text{COOH}\\|\\\text{CH}_2\text{COOH}\end{array}} + \text{ADP} + \text{P}_i$$

When this transfer system is employed, the oxaloacetate that is formed from pyruvate by mitochondrial pyruvate carboxylase may be converted to citrate by citrate synthetase (page 122). The citrate, which can be transported to the cytosol by a special mitochondrial-membrane carrier system, can then be cleaved to oxaloacetate, which in turn yields phosphoenolpyruvate:

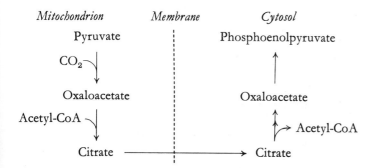

A third mechanism for the transport of oxaloacetate across the mitochondrial membrane involves a reversal of the malate shuttle described previously (section 4.5.A). Mitochondrial oxaloacetate can be reduced to malate by mitochondrial malate dehydrogenase. The malate thus formed is diffusible and can be reoxidized by the malate dehydrogenase of the cytosol:

```
     Mitochondrion          Membrane              Cytosol
NADH + H⁺ \ /Oxaloacetate              Oxaloacetate \ / NADH + H⁺
           X                                        X
NAD⁺ / \ Malate      ────────────→     Malate     / \ NAD⁺
```

The reactions involved in the synthesis of phosphoenolpyruvate from

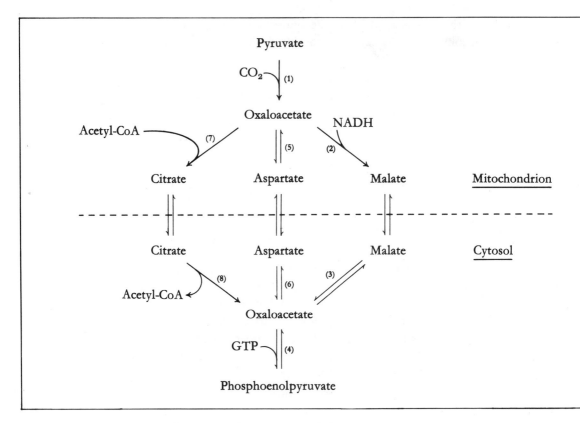

Fig. 4-11. *Synthesis of phosphoenolpyruvate from pyruvate and routes for the transport of oxaloacetate across the mitochondrial membrane. The enzymes indicated by the numbers in parentheses are (1) pyruvate carboxylase, (2) mitochondrial malate dehydrogenase, (3) cytosol malate dehydrogenase, (4) phosphoenolpyruvate carboxykinase, (5) mitochondrial transaminase, (6) cytosol transaminase, (7) citrate synthase, and (8) citrate-cleaving enzyme.*

pyruvate and the three systems for the effective mitochondrial transport of oxaloacetate are outlined in Figure 4-11.

The fact that individual metabolic intermediates may undergo different reactions that channel them into opposing pathways poses the problem as to what controls the mode of utilization of these substances. More specifically, the question may be posed: What are the regulating factors that determine whether glycolysis or glyconeogenesis should predominate in the liver cell? Obviously, if glucose were degraded to pyruvate and all of the latter were then reconverted to glucose, it would be a completely useless and wasteful exercise. Certain conditions must therefore govern whether the primary effort should be invested into the formation of glucose or into its utilization. Although a number of factors play roles in this control, the present discussion will be concerned with the characteristics of the enzymes involved and the dependence of their activity on the concentrations of specific intermediates. Some overall aspects of the hormonal control of glucose metabolism are discussed in Chapter 6.

In considering the reactions of glycolysis and gluconeogenesis, it will be noted that most of the enzymes are common to both pathways. Essentially, these are the enzymes that catalyze the reversible reactions in glycolysis. The enzymes that are specific for the degradation of glucose 6-phosphate to pyruvate are phosphofructokinase and pyruvate kinase; these catalyze irreversible reactions in glycolysis. Similarly, the enzymes catalyzing reactions involved only in the synthesis of glucose 6-phosphate are pyruvate carboxylase, phosphoenolpyruvate carboxykinase, and fructose diphosphatase; these, too, are involved only in irreversible reactions. Modulating an enzyme that catalyzes a reversible glycolytic reaction would not control the direction of the pathway, since the enzyme only decreases the time taken to reach equilibrium. The efficient regulation of an individual pathway must therefore be concerned with those enzymes that catalyze irreversible steps, i.e., those that are specific for that pathway. This is indeed found to be the case. Moreover, certain factors have an opposite effect on the enzymes of glycolysis to what they have on enzymes of the gluconeogenic route.

One of the critical factors determining the direction of glucose metabolism is the ratio of ATP to ADP. It has been pointed out previously (section 4.5.C) that excess ATP inhibits the activity of phosphofructokinase in glycolysis. Fructose diphosphatase, the enzyme that catalyzes the rate-limiting step in gluconeogenesis, is inhibited by AMP. Pyruvate kinase is also inhibited by ATP, whereas pyruvate carboxylase requires ATP. Thus, the relative levels of ATP, ADP, and AMP constitute one of the factors that regulate the direction of glucose metabolism.

Another controlling factor is the level of acetyl-coenzyme A. This intermediate activates pyruvate carboxylase, and consequently when its concentration increases, gluconeogenesis will be stimulated. This is probably one of the mechanisms for enhanced gluconeogenesis in untreated diabetics or in normal persons during starvation. In these situations, the rate of fatty-acid degradation increases and there is a concomitant rise in the cellular concentration of acetyl-coenzyme A.

The level of lactate also appears to influence the degree of gluconeogenesis. High lactate levels may explain the increased gluconeogenesis that occurs during extended muscular exertion. Lactate from muscle is transported via the bloodstream to the liver (Cori cycle; see Chap. 6, section 6.2A), where it is channeled to form glucose or glycogen. The oxidation of lactate to pyruvate generates reduced NAD, which can be utilized for the conversion of 1,3-diphosphoglyceric acid to glyceraldehyde 3-phosphate, thus facilitating gluconeogenesis.

The preceding discussion centered on the regulation of the direction of glucose metabolism by the concentration of various intermediates and cofactors. Another critical factor in such regulation comprises the levels or activities of specific enzymes. For example, the levels of pyruvate carboxylase, phosphoenolpyruvate carboxykinase, and fructose diphosphatase are elevated in diabetics, in whom the degree of gluconeogenesis is considerably elevated. This is discussed in a later chapter in conjunction with hormonal control of glycolysis and gluconeogenesis (Chap. 6, section 6.5).

It is important to note that although oxaloacetate can be converted to glycogen, acetate or acetyl-coenzyme A cannot be utilized to increase the *net weight* of liver glycogen. The primary reason for this is that the pyruvate dehydrogenase reaction (page 176) is not reversible, i.e., pyruvate cannot be formed from acetyl-CoA and carbon dioxide. Consequently, the only pathway that links acetyl-coenzyme A with carbohydrate synthesis (in mammals) is via the formation of citrate, conversion of the latter by reactions of the citric acid cycle to oxaloacetate, and transformation of oxaloacetate to phosphoenolpyruvate (Fig. 4-12). This sequence of reactions does not allow for a net increase in glucose or glycogen because two carbon atoms are eliminated when oxaloacetate is produced from citrate. Specifically, when acetate (as acetyl-CoA) is introduced into the cycle, a four-carbon compound (oxaloacetate) is converted to a six-carbon compound (citrate); the latter loses one carbon when α-ketoglutarate is formed and it loses a second carbon when succinate is formed. Thus, the compound channeled toward glucose formation—namely, oxaloacetate—is the same as that available before the introduction of acetate.

In spite of the fact that acetate cannot provide for the synthesis of *new* carbohydrate, carbon atoms from acetate can be incorporated into glucose or glycogen. The metabolic sequence that gives rise to this phenomenon is illustrated in Figure 4-12, with acetyl-coenzyme A tagged on the methyl carbon. As indicated earlier, the carbons from acetate are retained in the succinate when citrate is converted to succinate (page 148). Moreover, since succinate is a symmetric molecule, the two methylene carbons are indistinguishable. Oxaloacetate produced from the succinate can be converted to phosphoenolpyruvate, and the latter can serve as a precursor for glucose and glycogen. The tagged carbon of acetate is then localized in carbons 1, 2, 5, and 6 of glucose. However, this incorporation of carbon atoms of acetate into glucose does not increase the net production of the carbohydrate.

It is seen from the considerations just discussed that oxaloacetate functions not only as an intermediate in the citric acid cycle, but it also serves as a precursor for the biosynthesis of carbohydrates. Unlike the citric acid cycle sequence in which oxaloacetate is regenerated, the synthetic pathway removes and consumes this intermediate. Consequently, in order to maintain the

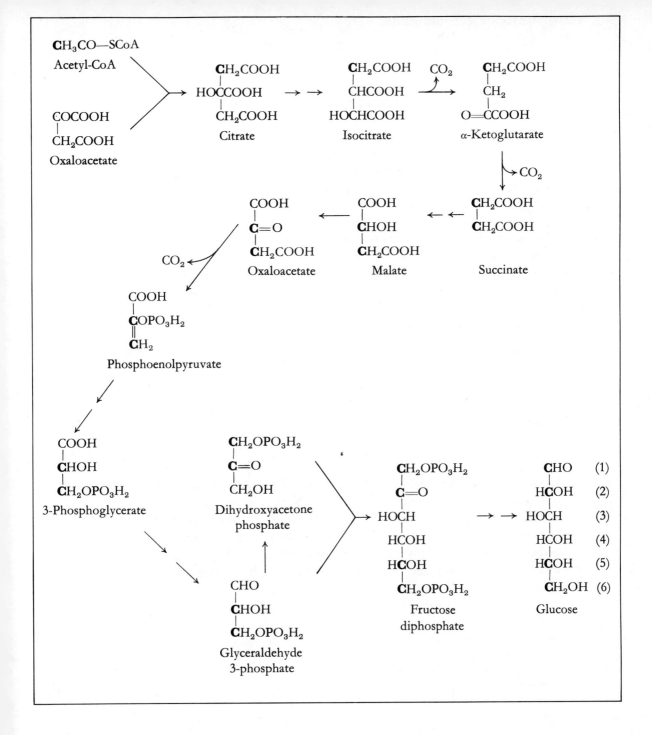

Fig. 4-12. Metabolic interrelationship between acetate (acetyl-CoA) and glucose. The tagged methyl carbon of acetyl-CoA is indicated in boldface letters to show the pathway for its incorporation into glucose. Broken arrows indicate a sequence of reactions. Note that the tagged carbon in α-ketoglutarate is distributed equally in the methylene carbons of succinate, since the latter is a symmetric compound.

normal operation of the citric acid cycle, specific processes must be available to replenish the supply of oxaloacetate or other intermediates. Reactions that fulfill this function are termed *anaplerotic* reactions. The most important of this class of reactions is the pyruvate carboxylase-catalyzed synthesis of oxaloacetate from pyruvate and carbon dioxide. Hence, this reaction serves in the metabolic replenishment of oxaloacetate, as well as in the biosynthesis of carbohydrate from pyruvate.

4.9. Nucleotide Sugar Interconversions

The synthesis of UDP-glucose and its utilization for the formation of glycogen was described previously (section 4.7.A). UDP-glucose can also undergo a number of other reactions, one of which is its conversion to UDP-galactose. This reversible reaction, which is catalyzed by *UDP-glucose 4-epimerase*, requires small amounts of NAD^+:

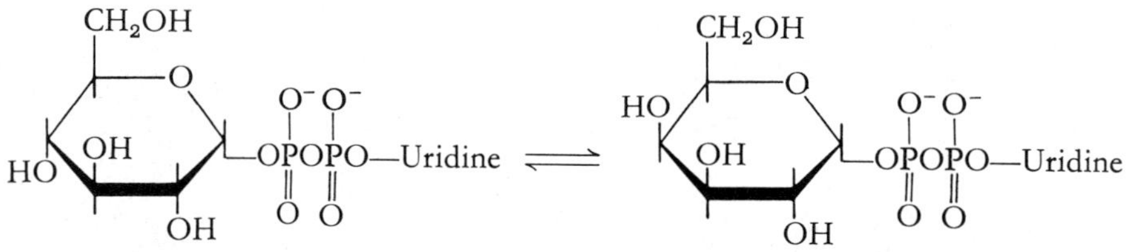

$$\text{UDP-glucose} \rightleftharpoons \text{UDP-galactose}$$

UDP-galactose is a substrate for the biosynthesis of mucopolysaccharides and glycoproteins; it will be described further in Chapter 14 (section 14.3.E). The interconversion of UDP-glucose and UDP-galactose is also important for the metabolism of galactose, which will be discussed in the next section.

Another reaction of UDP-glucose is its oxidation by NAD^+ to yield UDP-glucuronic acid. The enzyme catalyzing this reaction is *UDP-glucose dehydrogenase*:

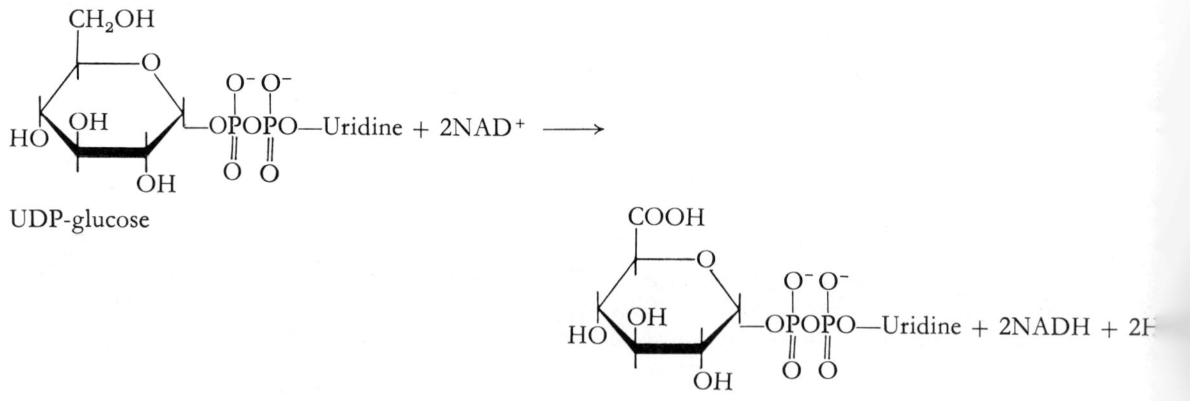

$$\text{UDP-glucose} + 2NAD^+ \longrightarrow \text{UDP-glucuronic acid} + 2NADH + 2H^+$$

UDP-glucuronic acid is another substrate for the synthesis of mucopolysaccharides (Chap. 14). In addition, it may undergo decarboxylation to UDP-xylose, which is also utilized in mucopolysaccharide structure:

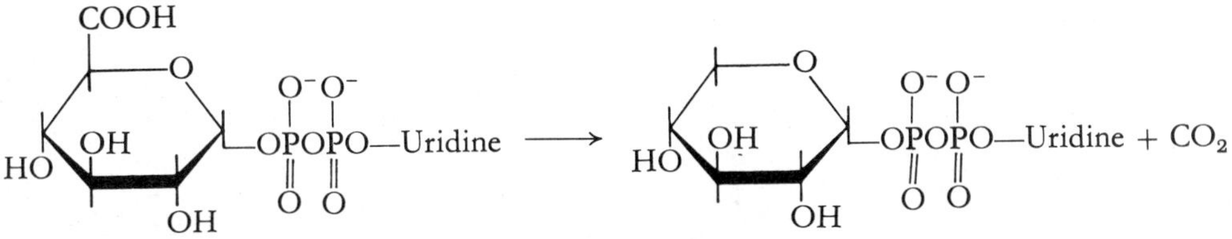

$$\text{UDP-glucuronic acid} \longrightarrow \text{UDP-xylose} + CO_2$$

UDP-glucuronic acid is a donor of glucuronic acid in the synthesis of glucuronides. The latter include glucosides composed of glucuronic acid and various compounds that contain hydroxyl groups; their general structure is:

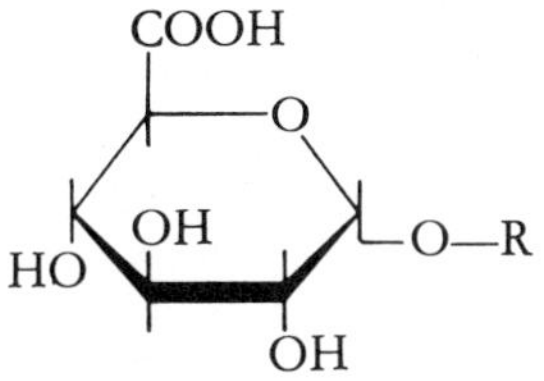

Compounds such as steroids and bilirubin are transported and excreted as glucuronides since linkage with uronic acid renders them more soluble and compatible with the physiologic environment. A number of drugs or their degradation products are converted in the liver to glucuronides and excreted in this form. The glucuronides are discussed more fully in conjunction with specific compounds (Chap. 5, section 5.5).

4.10. Metabolism of Galactose and Fructose

Milk contains the disaccharide lactose. Upon enzymatic hydrolysis in the intestinal tract, it yields glucose and galactose (section 4.1). After absorption, the galactose is transferred to the cells by the circulatory system. Within the cell, galactose is phosphorylated by ATP in the presence of the enzyme *galactokinase*:

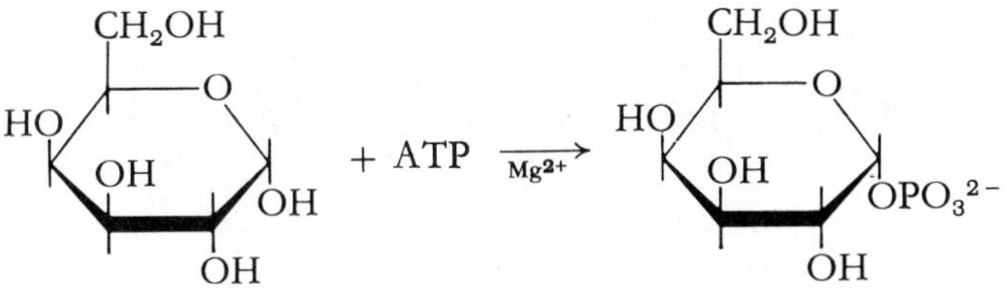

$$\text{Galactose} + ATP \xrightarrow{Mg^{2+}} \text{Galactose 1-phosphate}$$

This reaction is followed by interaction with UDP-glucose. The products are UDP-galactose and glucose 1-phosphate. The enzyme that catalyzes this reaction is *galactose 1-phosphate uridyl transferase*:

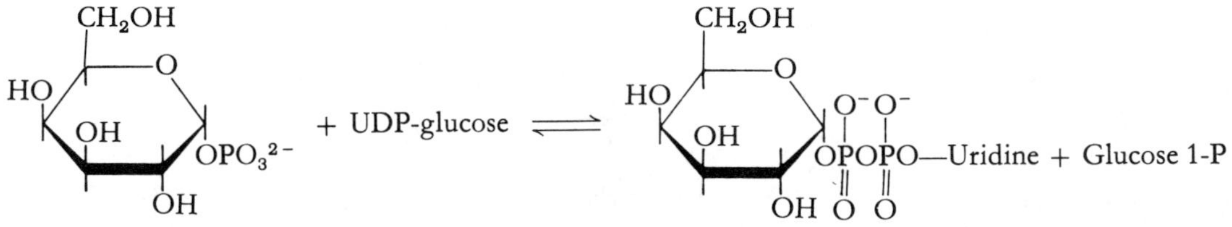

Galactose 1-phosphate + UDP-glucose ⇌ UDP-galactose + Glucose 1-P

Finally, UDP-galactose is converted to UDP-glucose by the enzyme *UDP-glucose 4-epimerase*; this reaction requires catalytic amounts of NAD^+:

$$\text{UDP-galactose} \underset{}{\overset{NAD^+}{\rightleftharpoons}} \text{UDP-glucose}$$

UDP-glucose may then undergo any of the reactions just described, such as incorporation into glycogen or conversion to UDP-glucuronic acid.

The metabolism of galactose is of special interest in that it is involved in a specific congenital disease, termed *galactosemia*. This abnormality is transmitted by an autosomal recessive gene and occurs in about 1 out of 20,000 births. The principal symptoms are cataract formation, hepatosplenomegaly, and mental retardation. The aberration in this disease involves the absence of the enzyme galactose 1-phosphate uridyl transferase. As a result, galactose cannot be metabolized and therefore accumulates as galactose 1-phosphate. The only treatment is to withdraw lactose or, more specifically, milk from the diet. This has to be done at an early stage, since the damage that results from the disorder is irreversible. Assay procedures for uridylyl transferase are available, so the disease can be diagnosed precisely.

Another inborn error of galactose metabolism involves the absence of the enzyme galactokinase. This leads to an accumulation of galactose in the tissues, and urinary excretion of the sugar occurs. The excess galactose in the tissue is converted to galactitol (dulcitol), which brings about cataract formation.

The metabolism of fructose, which is generally derived from ingested sucrose, is initiated by the enzyme *fructokinase*; this catalyzes the formation of fructose 1-phosphate from fructose and ATP:

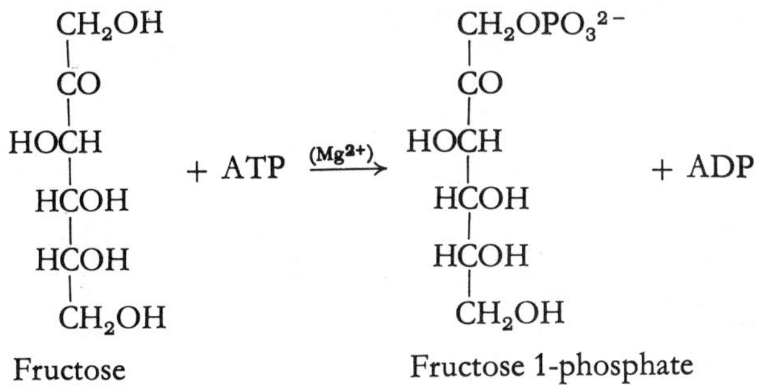

$$\underset{\text{Fructose}}{\begin{array}{c}CH_2OH\\|\\CO\\|\\HOCH\\|\\HCOH\\|\\HCOH\\|\\CH_2OH\end{array}} + ATP \xrightarrow{(Mg^{2+})} \underset{\text{Fructose 1-phosphate}}{\begin{array}{c}CH_2OPO_3^{2-}\\|\\CO\\|\\HOCH\\|\\HCOH\\|\\HCOH\\|\\CH_2OH\end{array}} + ADP$$

Fructose 1-phosphate is acted upon by a specific isozyme of aldolase that catalyzes its reaction to form dihydroxyacetone phosphate and glyceraldehyde:

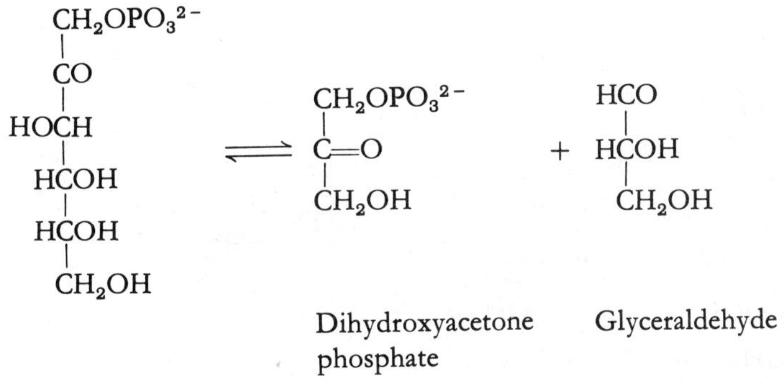

$$\begin{array}{c}CH_2OPO_3^{2-}\\|\\CO\\|\\HOCH\\|\\HCOH\\|\\HCOH\\|\\CH_2OH\end{array} \rightleftharpoons \underset{\text{Dihydroxyacetone phosphate}}{\begin{array}{c}CH_2OPO_3^{2-}\\|\\C=O\\|\\CH_2OH\end{array}} + \underset{\text{Glyceraldehyde}}{\begin{array}{c}HCO\\|\\HCOH\\|\\CH_2OH\end{array}}$$

The glyceraldehyde may be phosphorylated by ATP to yield glyceraldehyde 3-phosphate, and the latter can then follow the general pathway of glycolysis. Other pathways available to free glyceraldehyde are its reduction to glycerol or its oxidation to glyceric acid and ultimate conversion to serine.

There is a genetic defect in which fructokinase is absent. The only symptom in these individuals is the presence of fructose in the urine. Another, more serious defect entails the absence of the specific aldolase isozyme. Fructose 1-phosphate then accumulates in the liver and kidneys, resulting in damage to the kidneys and disturbance of their function. Other symptoms are abdominal pain and vomiting. Again, the only way to treat this disorder is to remove fructose and sucrose from the diet.

4.11. Hexosamines
Glucosamine and galactosamine are integral components of mucopolysaccharides and glycoproteins (Chap. 14, section 14.3). These are sugar

derivatives in which the hydroxyl groups on carbon-2 of the respective hexoses have been replaced by amino groups:

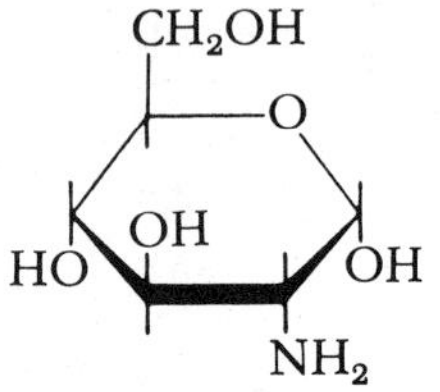

Glucosamine
(2-Amino-2-deoxy-α-D-glucose)

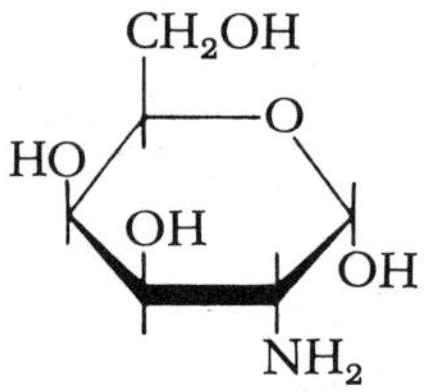

Galactosamine
(2-Amino-2-deoxy-α-D-galactose)

Glucosamine 6-phosphate is formed by the reaction of fructose 6-phosphate with glutamine. The enzyme catalyzing this reaction is an *amido transferase*:

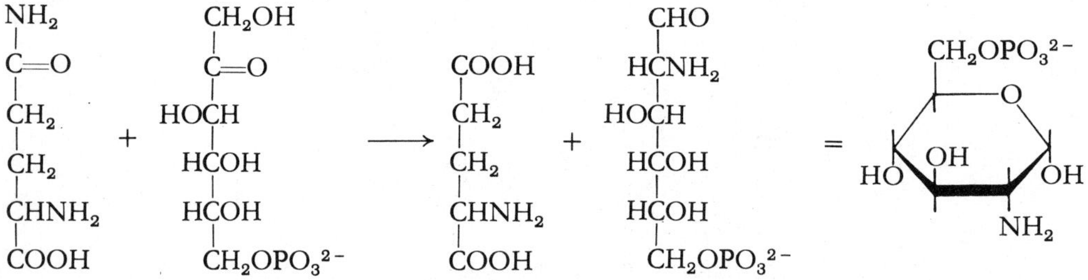

$$\begin{array}{c} NH_2 \\ | \\ C=O \\ | \\ CH_2 \\ | \\ CH_2 \\ | \\ CHNH_2 \\ | \\ COOH \end{array} + \begin{array}{c} CH_2OH \\ | \\ C=O \\ | \\ HOCH \\ | \\ HCOH \\ | \\ HCOH \\ | \\ CH_2OPO_3^{2-} \end{array} \longrightarrow \begin{array}{c} COOH \\ | \\ CH_2 \\ | \\ CH_2 \\ | \\ CHNH_2 \\ | \\ COOH \end{array} + \begin{array}{c} CHO \\ | \\ HCNH_2 \\ | \\ HOCH \\ | \\ HCOH \\ | \\ HCOH \\ | \\ CH_2OPO_3^{2-} \end{array}$$

Glutamine Fructose Glutamic Glucosamine
6-phosphate acid 6-phosphate

Note that this transfer involves the amide nitrogen of glutamine.

The amino group in glucosamine 6-phosphate may then be acetylated by acetyl-coenzyme A to yield N acetylglucosamine 6-phosphate:

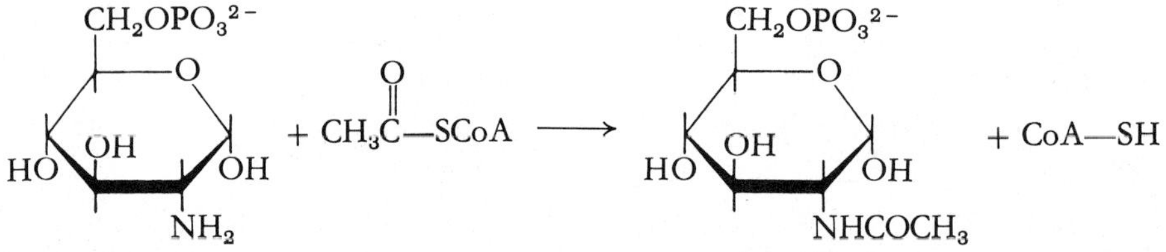

Glucosamine
6-phosphate
+ CH$_3$C—SCoA ⟶
N-acetylglucosamine
6-phosphate
+ CoA—SH

The latter is converted to N-acetylglucosamine 1-phosphate, which in turn reacts with uridine triphosphate to form UDP-N-acetylglucosamine:

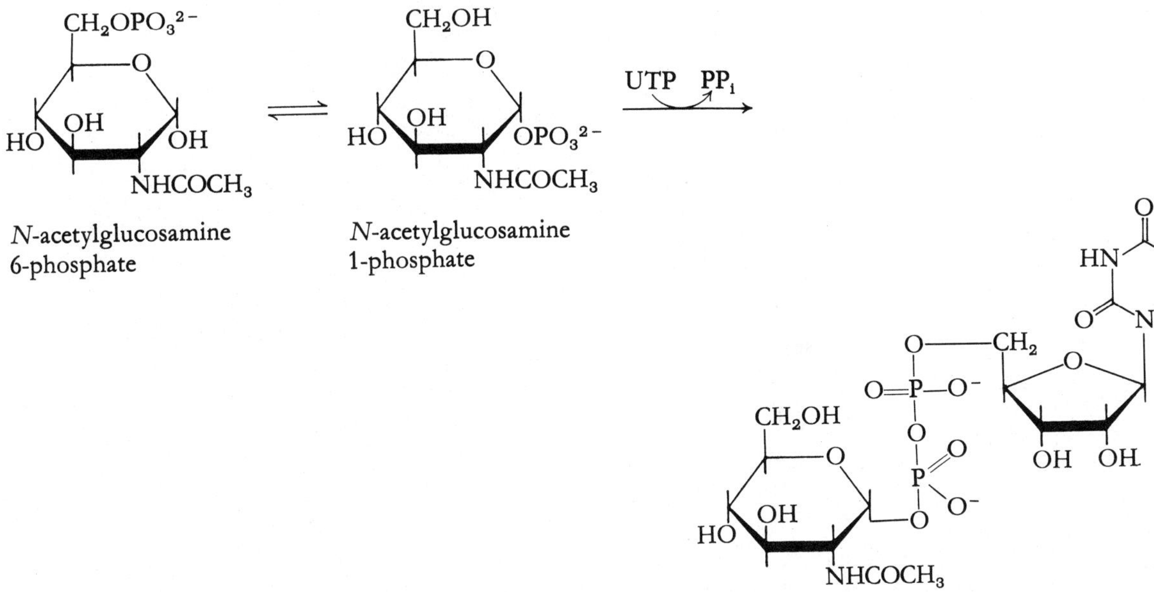

[Structures: N-acetylglucosamine 6-phosphate ⇌ N-acetylglucosamine 1-phosphate → (UTP, PP$_i$) UDP-N-acetylglucosamine]

UDP-N-acetylglucosamine may then be epimerized to UDP-N-acetylgalactosamine:

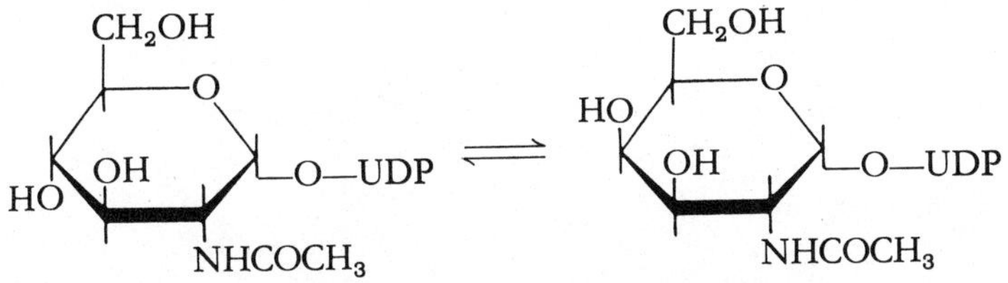

[Structures: UDP-N-acetylglucosamine ⇌ UDP-N-acetylgalactosamine]

These nucleoside-diphosphate hexosamines function to transfer the respective hexosamines for the synthesis of mucopolysaccharides such as hyaluronic acid, chondroitin sulfate, and heparin (Chap. 14, section 14.3.D).

The metabolic interconversions of carbohydrates are summarized in Figure 4-13.

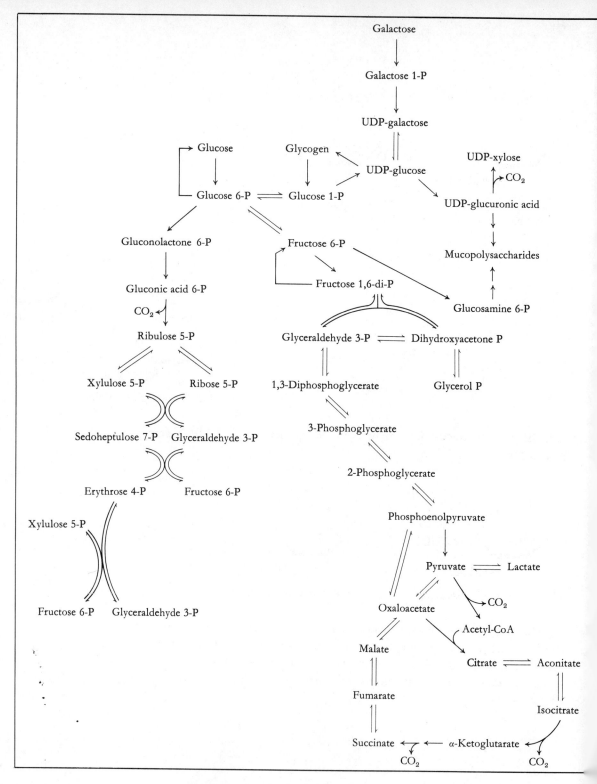

Fig. 4-13. Pathways of carbohydrate metabolism. Broken arrows indicate a sequence of several reactions; phosphate is designated as P.

Suggested Reading

Fischer, E. H., Pocker, A., and Saari, J. C. The Structure, Function and Control of Glycogen Phosphorylase. *Essays in Biochemistry* 6:23–68, 1970.

Hers, H. G. The Control of Glycogen Metabolism in Liver. *Annual Review of Biochemistry* 45:167–189, 1976.

Krebs, H. A. The Pasteur Effect and the Relation Between Respiration and Fermentation. *Essays in Biochemistry* 8:2–34, 1972.

Lai, C. Y., and Horecker, B. L. Aldolase: A Model for Enzyme Structure-Function Relationships. *Essays in Biochemistry* 8:149–178, 1972.

Pontremoli, S., and Grazi, E. Gluconeogenesis. In F. Dickens, P. J. Randle, W. J. Whelan, (Eds.), *Carbohydrate Metabolism and Its Disorders*, Vol. 1. New York: Academic Press, 1968. Pp. 259–295.

Scrutton, M. C., and Utter, M. F. The Regulation of Glycolysis and Gluconeogenesis in Animal Tissue. *Annual Review of Biochemistry* 37:249–302, 1968.

Wicks, W. D., Barnett, C. A., and McKibbin, J. B. Interaction Between Hormones and Cyclic AMP in Regulating Specific Enzyme Synthesis. *Federation Proceedings* 33:1105–1111, 1974.

5. Biosynthesis of Lipids

5.1. Intracellular Transport of Acetyl-Coenzyme A

The primary metabolic substrate for the biosynthesis of the fatty acids in triglycerides and phospholipids is *acetyl-coenzyme A*. The latter is also the building block for the formation of cholesterol and other steroids. Acetyl-coenzyme A arises primarily from the catabolism of carbohydrates and fatty acids. Considerable amounts of acetyl-coenzyme A are also formed in the degradation of amino acids. All these materials are thus potential contributors of integral units for lipid synthesis.

Acetyl-coenzyme A from catabolic reactions is generated within the mitochondria, whereas its utilization for the biosynthesis of fatty acids occurs in the cytosol. Since the mitochondrial membrane is not permeable to acetyl-coenzyme A, special mechanisms are required for making this compound available in the cytosol. One of these processes involves the action of the *citrate cleavage enzyme* (ATP-citrate lyase). This catalyzes the interaction between citrate and coenzyme A to yield oxaloacetate and acetyl-coenzyme A (Chap. 4):

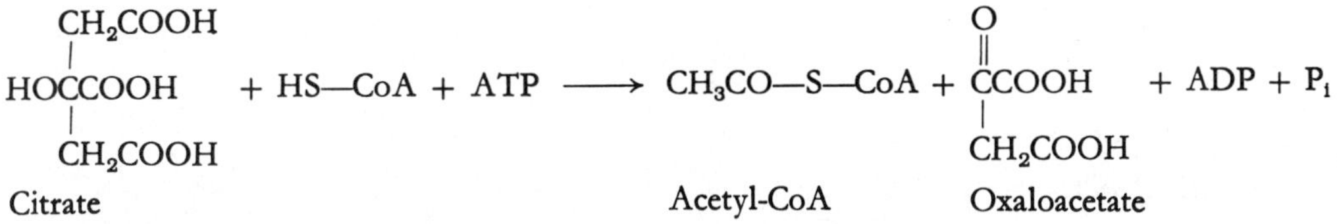

$$\underset{\text{Citrate}}{\begin{array}{c}CH_2COOH\\|\\HOCCOOH\\|\\CH_2COOH\end{array}} + HS-CoA + ATP \longrightarrow \underset{\text{Acetyl-CoA}}{CH_3CO-S-CoA} + \underset{\text{Oxaloacetate}}{\begin{array}{c}O\\\|\\CCOOH\\|\\CH_2COOH\end{array}} + ADP + P_i$$

The fact that citrate can move between the mitochondria and cytosol allows for the effective transfer of acetyl-coenzyme A by the following process:

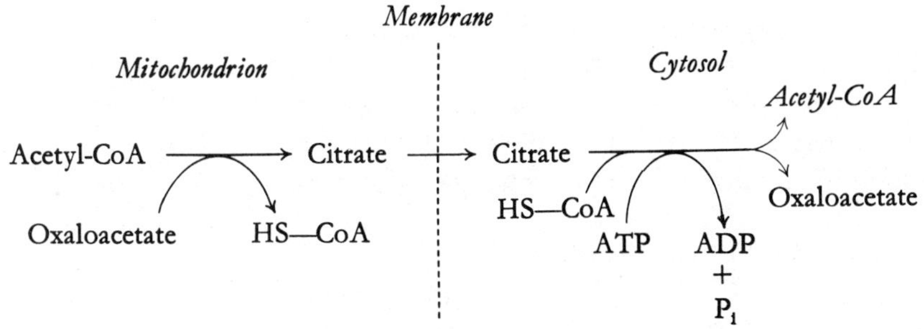

```
              Membrane
  Mitochondrion  |  Cytosol
                 |           Acetyl-CoA
                 |              ↑
Acetyl-CoA ─┐ ┌─ Citrate ─┼─→ Citrate ─┐ ┌─
            ╳              |            ╳
Oxaloacetate ┘ └─ HS—CoA  |   HS—CoA ┘ └─→ Oxaloacetate
                 |           ATP   ADP
                 |                  +
                 |                  P_i
```

(It may be noted that this process also provides for the transfer of oxaloacetate from the mitochondrion to the cytosol.)

5.2. Synthesis of Fatty Acids

A. MALONYL-COENZYME A

A pivotal intermediate in the synthesis of fatty acids is *malonyl-coenzyme A*. This is produced from acetyl-coenzyme A by a reaction between the latter and CO_2 (or bicarbonate ions) in the presence of ATP:

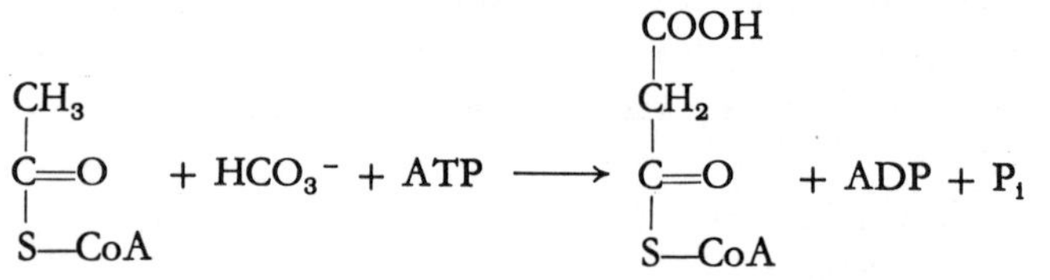

$$\begin{array}{c} CH_3 \\ | \\ C\!=\!O \\ | \\ S\!-\!CoA \end{array} + HCO_3^- + ATP \longrightarrow \begin{array}{c} COOH \\ | \\ CH_2 \\ | \\ C\!=\!O \\ | \\ S\!-\!CoA \end{array} + ADP + P_i$$

Acetyl-CoA   Malonyl-CoA

The enzyme that catalyzes this reaction, *acetyl-coenzyme A carboxylase*, contains biotin as a prosthetic group. The biotin is attached to the protein portion of the enzyme through an amide linkage between the carboxyl group of the former and the ε-amino group of a lysine residue of the protein:

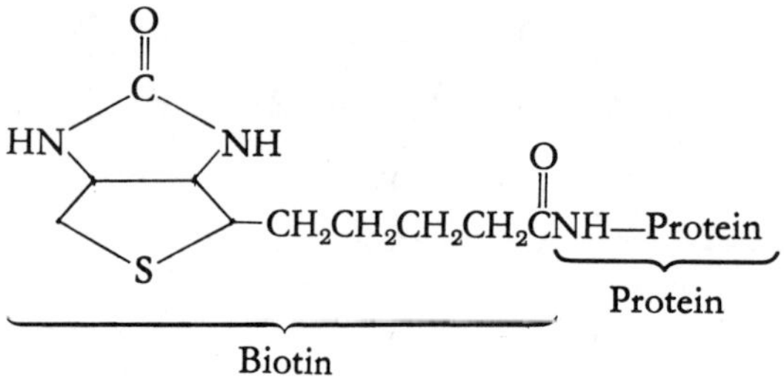

[Structure of biotin linked to protein: biotin ring system with HN–C(=O)–NH, fused to sulfur-containing ring, with side chain –CH₂CH₂CH₂CH₂C(=O)NH–Protein]

Acetyl-coenzyme A carboxylase is a complex allosteric enzyme for which citrate, isocitrate, or α-ketoglutarate function as positive modulators. The formation of malonyl-coenzyme A is stimulated about 15-fold by these intermediates. Carboxylation of acetyl-coenzyme A is the rate-limiting step in the biosynthesis of fatty acids. This reaction is therefore an important regulatory site in the overall process. The enzyme is inhibited by the coenzyme A derivatives of the long-chain fatty acids, e.g., palmitoyl-coenzyme A.

B. FATTY-ACID SYNTHETASE MULTIENZYME SYSTEM

The production of long-chain acyl derivatives from acetyl units involves a multienzyme complex called the *fatty-acid synthetase system*. The individual enzymatic proteins of this complex are bound to each other, so the steps that they catalyze operate in an efficient and regulated sequence. Thus, although each reaction will be described separately, it should be noted that they are tightly integrated with each other.

In contrast to the fatty-acid catabolic process in which the metabolites are in the form of coenzyme-A derivatives, the biosynthetic pathway requires that the intermediates be bound to *acyl-carrier protein* (ACP). The latter is a low-molecular-weight protein that is linked via the hydroxyl oxygen of a serine residue to phosphate, pantothenic acid, and mercaptoethylamine:

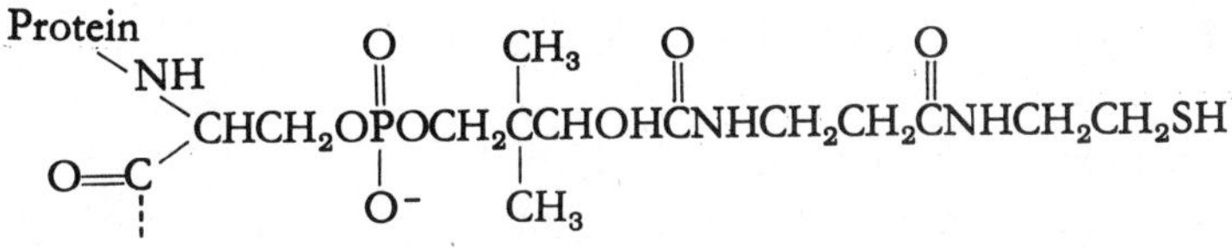

$$\text{Protein}\diagdown_{\text{NH}}\diagdown\text{CHCH}_2\text{OPOCH}_2\text{CCHOHCNHCH}_2\text{CH}_2\text{CNHCH}_2\text{CH}_2\text{SH}$$

Acyl-carrier protein (ACP)

Acetate and other fatty-acid intermediates form thioesters with acyl-carrier proteins that are similar to the thioesters formed with coenzyme A.

The malonyl-coenzyme A that is formed by the acetyl-coenzyme-A carboxylase reaction is converted to malonyl-ACP by the action of the enzyme *malonyl transacylase*:

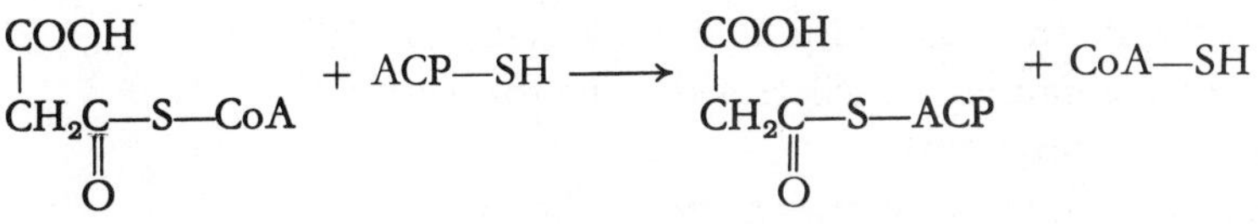

$$\underset{\text{Malonyl-CoA}}{\text{COOH}\mid\text{CH}_2\text{C}-\text{S}-\text{CoA}\parallel\text{O}} + \text{ACP}-\text{SH} \longrightarrow \underset{\text{Malonyl-ACP}}{\text{COOH}\mid\text{CH}_2\text{C}-\text{S}-\text{ACP}\parallel\text{O}} + \text{CoA}-\text{SH}$$

A similar transfer of the acetyl group from coenzyme A to acyl-carrier protein is catalyzed by *acetyl transacylase*:

$$\underset{\text{Acetyl-CoA}}{\text{CH}_3\text{C}-\text{S}-\text{CoA}} + \text{ACP}-\text{SH} \longrightarrow \underset{\text{Acetyl-ACP}}{\text{CH}_3\text{C}-\text{S}-\text{ACP}} + \text{CoA}-\text{SH}$$

The acetyl group is then transferred from ACP to a specific sulfhydryl group of a cysteine residue on the enzyme complex. This enzyme unit is termed *β-ketoacyl-ACP synthetase*:

$$\text{CH}_3\text{C}-\text{S}-\text{ACP} + \text{HS}-\text{Enz} \longrightarrow \text{CH}_3\text{C}-\text{S}-\text{Enz} + \text{ACP}-\text{SH}$$

Malonyl-ACP then condenses with the enzyme-bound acetyl group to form acetoacetyl-ACP. In the process, the free carboxyl of malonyl-ACP is released as carbon dioxide, and the reaction is rendered irreversible:

$$CH_3\overset{O}{\overset{\|}{C}}-S-Enz + \underset{COOH}{CH_2\overset{O}{\overset{\|}{C}}-S-ACP} \longrightarrow \underset{\text{Acetoacetyl-ACP}}{CH_3\overset{O}{\overset{\|}{C}}CH_2\overset{O}{\overset{\|}{C}}-S-ACP} + Enz + CO_2$$

It may be noted that the carbon dioxide incorporated during the synthesis of malonyl-coenzyme A is not incorporated in the carbon chain of the fatty acid.

The following step involves the reduction of acetoacetyl-ACP by NADPH. The enzyme for the reaction is *β-ketoacyl-ACP reductase*. This step results in the formation of D-β-hydroxybutyryl-ACP:

$$\underset{\text{Acetoacetyl-ACP}}{CH_3\overset{O}{\overset{\|}{C}}CH_2\overset{O}{\overset{\|}{C}}-S-ACP} + NADPH + H^+ \longrightarrow \underset{\text{D-β-Hydroxybutyryl-ACP}}{CH_3\overset{OH}{\overset{|}{C}}HCH_2\overset{O}{\overset{\|}{C}}-S-ACP} + NADP^+$$

(Note that the β-hydroxy acid in the biosynthetic process differs from that of the β-hydroxyacyl-CoA produced in the degradation of fatty acids in that the latter intermediate has the L configuration, whereas the one formed in the synthetic pathway has the D configuration.)

D-β-Hydroxybutyryl-ACP releases a molecule of water and is thus converted to crotonyl-ACP. This reaction is catalyzed by *enoyl-ACP hydratase*:

$$\underset{\text{D-β-Hydroxybutyryl-ACP}}{CH_3\overset{OH}{\overset{|}{C}}HCH_2\overset{O}{\overset{\|}{C}}-S-ACP} \longrightarrow \underset{\text{Crotonyl-ACP}}{CH_3CH=CH\overset{O}{\overset{\|}{C}}-S-ACP} + H_2O$$

The product has a *trans* configuration at the double bond.

The crotonyl-ACP is reduced by another mole of NADPH to yield butyryl-ACP. The enzyme that catalyzes this reduction is *enoyl-ACP reductase*:

$$\underset{\text{Crotonyl-ACP}}{CH_3CH=CH\overset{O}{\overset{\|}{C}}-S-ACP} + NADPH + H^+ \longrightarrow \underset{\text{Butyryl-ACP}}{CH_3CH_2CH_2\overset{O}{\overset{\|}{C}}-S-ACP} + NADP^+$$

This series of reactions, which is summarized in Figure 5-1, thus serves to increase the length of the carbon chain from acetate by two units. Condensation of butyryl-ACP with another molecule of malonyl-ACP followed by the repetition of the subsequent steps brings about the addition of another pair of carbon units (Fig. 5-1). The sequence starting with the addition of malonyl-ACP to acyl-ACP is repeated a number of times until a 16-carbon fatty acid (palmitoyl-ACP) is formed. The carbon chain of the latter is thus derived from one unit of acetyl-ACP and seven units of malonyl-ACP. The carbons from the starting acetyl-ACP are utilized only for carbon-15 and carbon-16 of the resulting palmitoyl-ACP.

C. ELONGATION OF FATTY ACIDS

Palmitoyl-ACP is converted to palmitoyl-coenzyme A. Further elongation of the long-chain fatty acids can occur with the latter derivative. Two systems are available for this purpose: one of these utilizes microsomal enzymes and the other involves mitochondrial reactions. In the microsomes (endoplasmic reticulum), the following sequence may occur:

1. $RCH_2\overset{O}{\underset{\|}{C}}-S-CoA + \overset{HOOC}{\underset{|}{CH_2}}\overset{O}{\underset{\|}{C}}-S-CoA \longrightarrow RCH_2\overset{O}{\underset{\|}{C}}CH_2\overset{O}{\underset{\|}{C}}-S-CoA + CO_2 + CoA$

2. $RCH_2\overset{O}{\underset{\|}{C}}CH_2\overset{O}{\underset{\|}{C}}-S-CoA + NADPH + H^+ \longrightarrow RCH_2\overset{OH}{\underset{|}{C}}HCH_2\overset{O}{\underset{\|}{C}}-S-CoA + NADP^+$

3. $RCH_2\overset{OH}{\underset{|}{C}}HCH_2\overset{O}{\underset{\|}{C}}-S-CoA \longrightarrow RCH_2CH=CH\overset{O}{\underset{\|}{C}}-S-CoA + H_2O$

4. $RCH_2CH=CH\overset{O}{\underset{\|}{C}}-S-CoA + NADPH + H^+ \longrightarrow RCH_2CH_2CH_2\overset{O}{\underset{\|}{C}}-S-CoA + NADP^+$

The product of this sequence is a fatty-acid coenzyme-A derivative with two additional carbons.

Another pathway for the elongation of coenzyme-A derivatives of long-chain fatty acids is available in the mitochondrial compartment. In this process, acetyl-coenzyme A (instead of malonyl-coenzyme A) is the two-carbon substrate.

D. NADPH REQUIREMENTS AND SOURCES

In the synthesis of fatty acids, 2 moles of NADPH are utilized each time the length of the chain is extended by two carbons. Thus, the complete synthesis of palmitic acid requires 14 moles of NADPH. Reduced $NADP^+$ can be derived from the oxidation of glucose via the pentose-phosphate pathway.

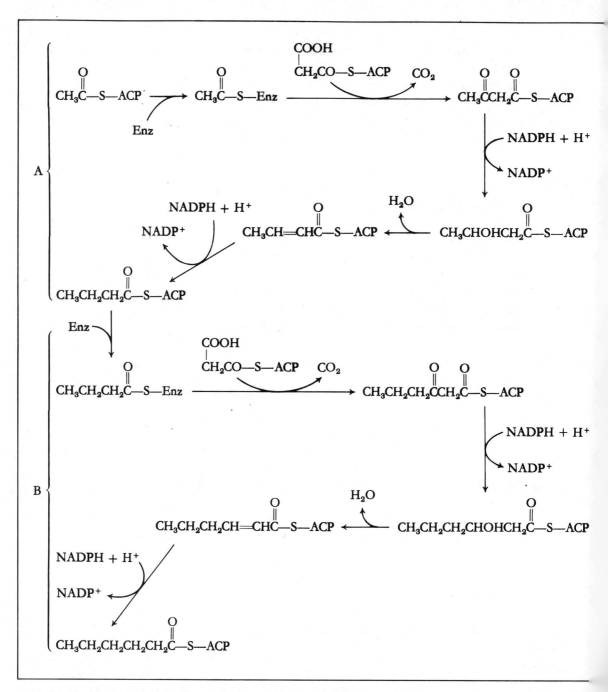

Fig. 5-1. Operations of the fatty-acid synthetase complex. A. Biosynthesis of butyryl-ACP from acetyl-ACP and malonyl-ACP. B. Conversion of butyryl-ACP to caproyl-ACP. Note that each time, the malonyl-ACP reacts with the carboxyl end of the saturated fatty acid.

Another reaction that yields this coenzyme is the conversion of malate to pyruvate by the *malic enzyme*:

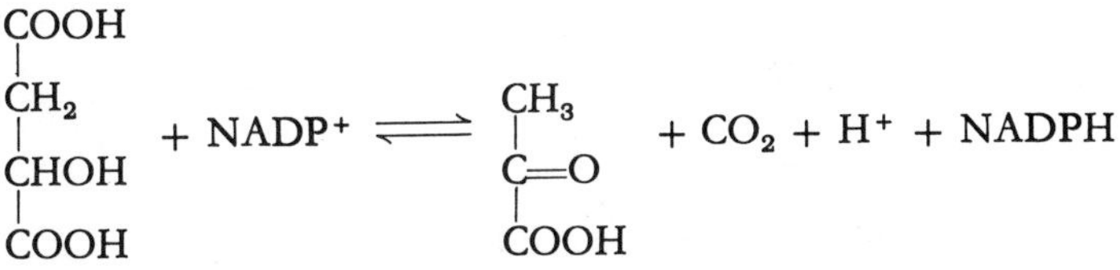

$$\begin{array}{l}\text{COOH}\\|\\\text{CH}_2\\|\\\text{CHOH}\\|\\\text{COOH}\end{array} + \text{NADP}^+ \rightleftharpoons \begin{array}{l}\text{CH}_3\\|\\\text{C}=\text{O}\\|\\\text{COOH}\end{array} + \text{CO}_2 + \text{H}^+ + \text{NADPH}$$

Malate → Pyruvate

A third reaction that may provide NADPH is the oxidation of isocitrate by an *NADP$^+$-dependent isocitrate dehydrogenase*:

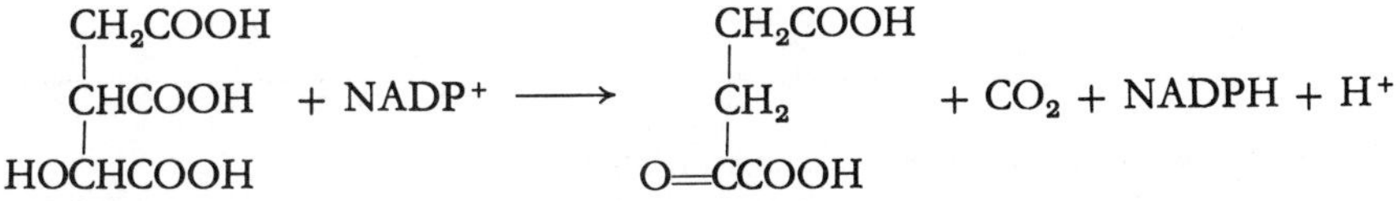

$$\begin{array}{l}\text{CH}_2\text{COOH}\\|\\\text{CHCOOH}\\|\\\text{HOCHCOOH}\end{array} + \text{NADP}^+ \longrightarrow \begin{array}{l}\text{CH}_2\text{COOH}\\|\\\text{CH}_2\\|\\\text{O}=\text{CCOOH}\end{array} + \text{CO}_2 + \text{NADPH} + \text{H}^+$$

Isocitrate → α-Ketoglutarate

The amount of NADPH contributed by the two latter reactions appears to be much less than that derived from the oxidation of glucose. It is therefore obvious why the tissues that are actively synthesizing fatty acids should metabolize glucose via the oxidative pentose pathway.

E. UNSATURATED FATTY ACIDS

The cells of liver and adipose tissue contain the necessary enzymes for converting palmitoyl-coenzyme A and stearyl-coenzyme A to the respective Δ^9-unsaturated products, palmitoleyl-coenzyme A and oleyl-coenzyme A. These enzymes, termed *mixed-function oxygenases*, catalyze the concurrent oxidations of NADPH and the fatty-acid derivatives by molecular oxygen. The overall process can be summarized by the equation:

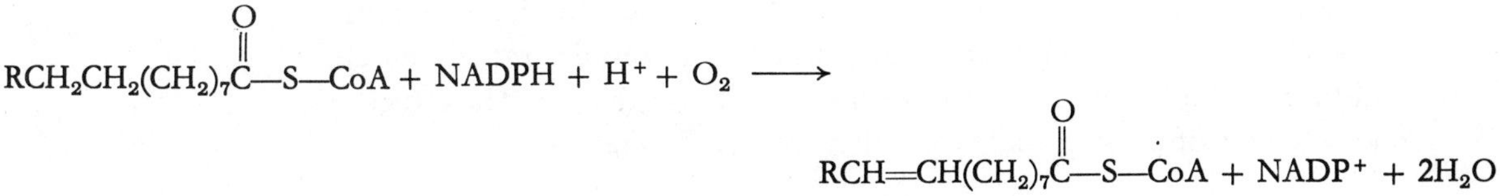

$$\text{RCH}_2\text{CH}_2(\text{CH}_2)_7\overset{\text{O}}{\underset{\|}{\text{C}}}-\text{S}-\text{CoA} + \text{NADPH} + \text{H}^+ + \text{O}_2 \longrightarrow$$

$$\text{RCH}=\text{CH}(\text{CH}_2)_7\overset{\text{O}}{\underset{\|}{\text{C}}}-\text{S}-\text{CoA} + \text{NADP}^+ + 2\text{H}_2\text{O}$$

This oxidation-reduction system, which occurs in the endoplasmic reticulum, entails the transfer of two electrons from reduced NADP$^+$ to cytochrome b$_5$. The electrons from the reduced form of the latter are accepted by the

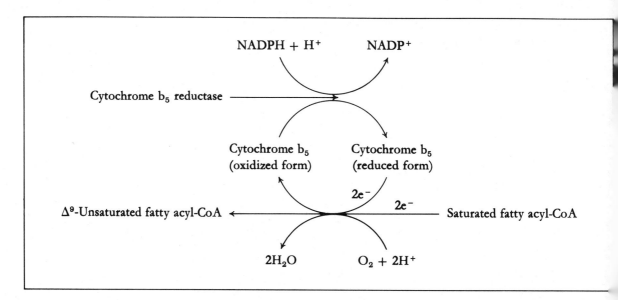

Fig. 5-2. Electron-transport sequences in the microsomal mixed oxygenase system that yields Δ^9-unsaturated fatty acyl-coenzyme A. The transfer of electrons from NADPH to cytochrome b_5 is catalyzed by a flavoprotein, cytochrome b_5 reductase.

enzyme-bound oxygen. Two additional electrons are transferred to oxygen from the fatty acyl-coenzyme A. As a result, the oxygen molecule is reduced to form two molecules of water (Fig. 5-2).

In mammals, the enzyme systems for desaturating fatty acyl-coenzyme A cannot dehydrogenate these acids between carbons 12 and 13 or carbons 15 and 16. As a consequence, mammals are unable to produce linoleic ($\Delta^{9,12}$) or linolenic ($\Delta^{9,12,15}$) acids. These two fatty acids, which must be supplied in the diet, are called *essential fatty acids*. Although the requirements for essential fatty acids are not defined, it has been shown in rats that deficiencies of these acids during early life lead to inhibited growth and abnormal skin conditions.

Linoleic acid serves as a precursor in the biosynthesis of *prostaglandins*, which are a variety of compounds with highly potent biologic activities. Prostaglandins, for example, enhance contraction of certain smooth muscles, and they are highly effective in lowering the blood pressure. The structures and activities of prostaglandins are discussed in Chapter 14.

5.3. Synthesis of Triglycerides

Saturated and unsaturated fatty acids are stored as glycerol esters. These are synthesized in the endoplasmic reticulum or microsomal components of the cells. The coenzyme-A derivatives of fatty acids react with glycerol phosphate to produce L-phosphatidic acids:

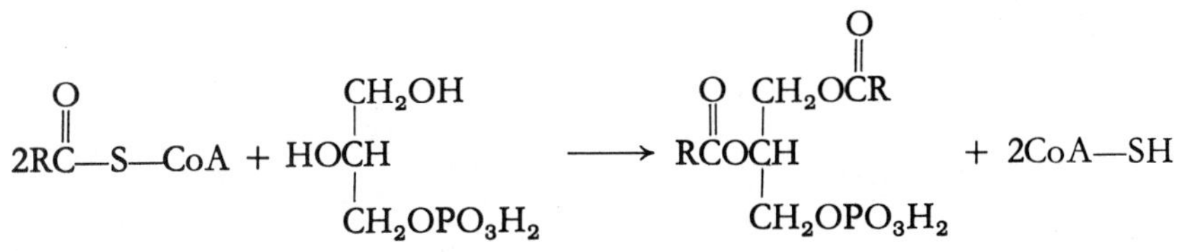

$$2RC\text{—}S\text{—}CoA + \begin{matrix}CH_2OH\\|\\HOCH\\|\\CH_2OPO_3H_2\end{matrix} \longrightarrow \begin{matrix}O\ CH_2OCR\\||\ |\\RCOCH\\|\\CH_2OPO_3H_2\end{matrix} + 2CoA\text{—}SH$$

Glycerol phosphate Phosphatidic acid

The hydrolysis of phosphatidic acids yields diglycerides (i.e., diacylglycerols) and inorganic phosphate. This reaction is catalyzed by a *phosphatase*:

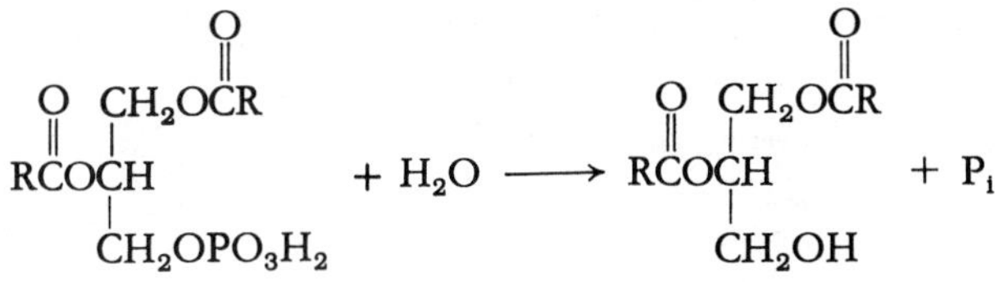

$$\begin{matrix}O\ CH_2OCR\\||\ |\\RCOCH\\|\\CH_2OPO_3H_2\end{matrix} + H_2O \longrightarrow \begin{matrix}O\ CH_2OCR\\||\ |\\RCOCH\\|\\CH_2OH\end{matrix} + P_i$$

Phosphatidic acid Diglyceride

The diglyceride may then condense with another acyl-coenzyme A unit to form a triglyceride:

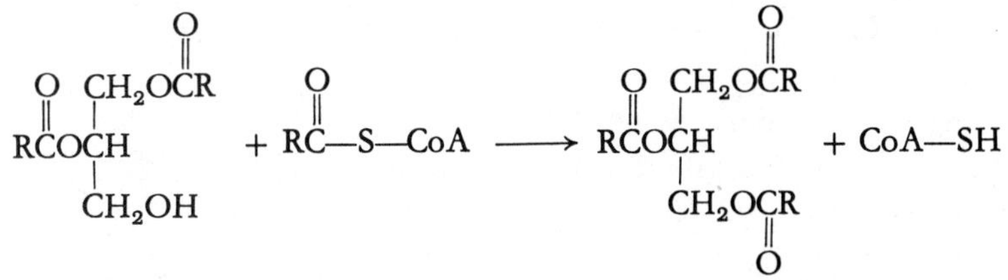

$$\begin{matrix}O\ CH_2OCR\\||\ |\\RCOCH\\|\\CH_2OH\end{matrix} + RC\text{—}S\text{—}CoA \longrightarrow \begin{matrix}O\ CH_2OCR\\||\ |\\RCOCH\\|\\CH_2OCR\\||\\O\end{matrix} + CoA\text{—}SH$$

Diglyceride Triglyceride

The glycerol phosphate required for the synthesis of triglycerides may be produced by either of two reactions. In adipose tissue, glycerol phosphate is formed by the reduction of dihydroxyacetone phosphate. This reaction is catalyzed by an NADH-linked enzyme, *glycerol-phosphate dehydrogenase* (Chap. 4):

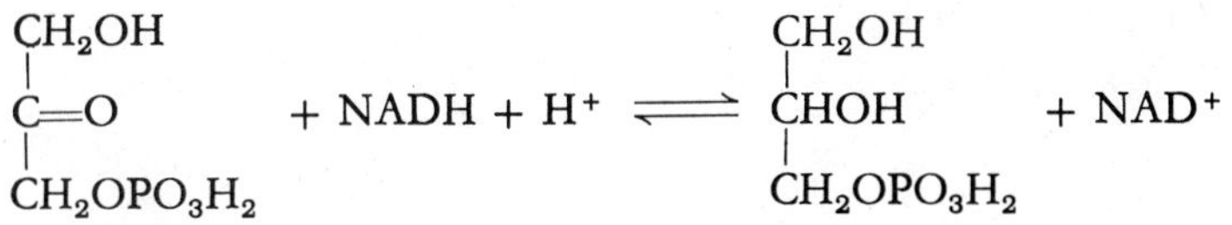

$$\begin{matrix}CH_2OH\\|\\C\!=\!O\\|\\CH_2OPO_3H_2\end{matrix} + NADH + H^+ \rightleftharpoons \begin{matrix}CH_2OH\\|\\CHOH\\|\\CH_2OPO_3H_2\end{matrix} + NAD^+$$

Dihydroxyacetone phosphate Glycerol phosphate

Liver contains an additional enzyme, *glycerol kinase,* which is not available in adipose tissue. This enzyme catalyzes the formation of glycerol phosphate from glycerol and adenosine triphosphate:

$$\begin{array}{l} CH_2OH \\ | \\ CHOH \\ | \\ CH_2OH \end{array} + ATP \longrightarrow \begin{array}{l} CH_2OH \\ | \\ CHOH \\ | \\ CH_2OPO_3H_2 \end{array} + ADP$$

Glycerol   Glycerol phosphate

This reaction allows for the conversion to glycerol phosphate of the glycerol that is released by the hydrolysis of triglycerides.

Triglycerides of adipose tissue and liver consist of a mixture of different fatty acids. Generally, over 50% of the acids are unsaturated. However, the composition depends to a significant extent on the nature of the ingested triglycerides.

5.4. Synthesis of Phospholipids

In addition to their utilization for the formation of neutral triglycerides, α,β-diglycerides (1,2-diglycerides) serve as intermediates for the biosynthesis of phospholipids. These include phosphatidylcholine, phosphatidylethanolamine, and phosphatidylserine (page 56). The first of these phospholipids is also called *lecithin* and the two latter substances are termed *cephalins.*

Lecithins and cephalins are synthesized in animal tissues from choline and ethanolamine derived from the diet or from the catabolism of endogenous phospholipids. The first step in the biosynthesis involves the conversion of choline or ethanolamine to their respective phosphates by interaction with adenosine triphosphate.

$(CH_3)_3\overset{+}{N}CH_2CH_2OH + ATP \longrightarrow (CH_3)_3\overset{+}{N}CH_2CH_2OPO_3H^- + ADP$
Choline   Choline phosphate

$H_2NCH_2CH_2OH + ATP \longrightarrow H_2NCH_2CH_2OPO_3H^- + ADP$
Ethanolamine   Ethanolamine phosphate

Choline phosphate can then react with cytidine triphosphate (CTP) to yield cytidine diphosphate choline (CDP-choline):

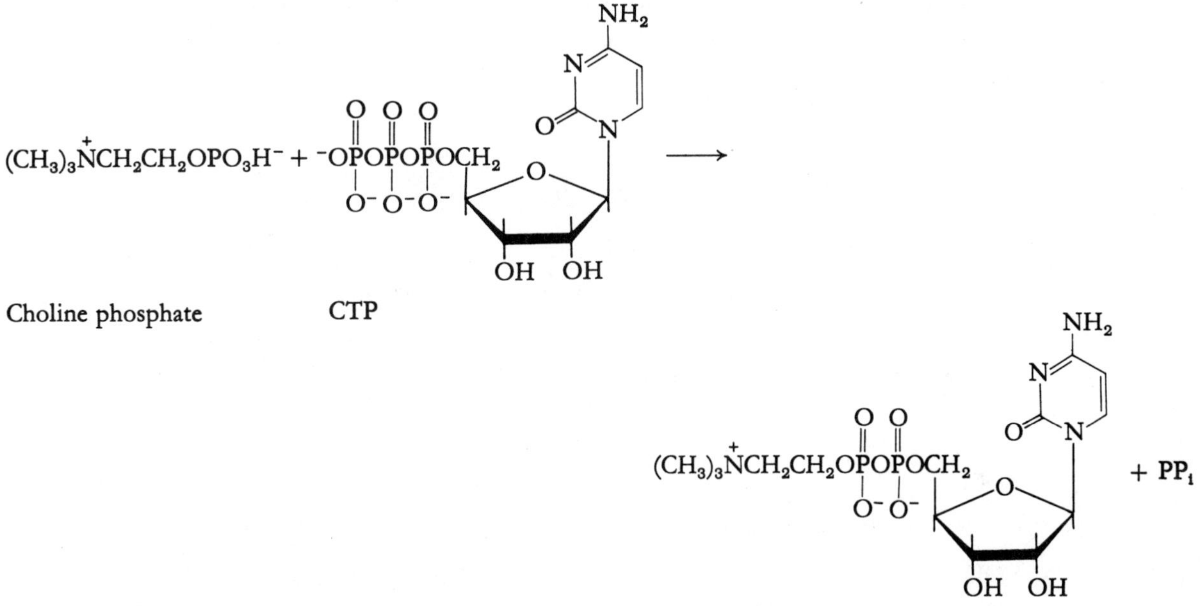

Choline phosphate + CTP → CDP-choline + PP$_i$

Similarly, cytidine diphosphate ethanolamine (CDP-ethanolamine) can be formed from ethanolamine phosphate and CTP:

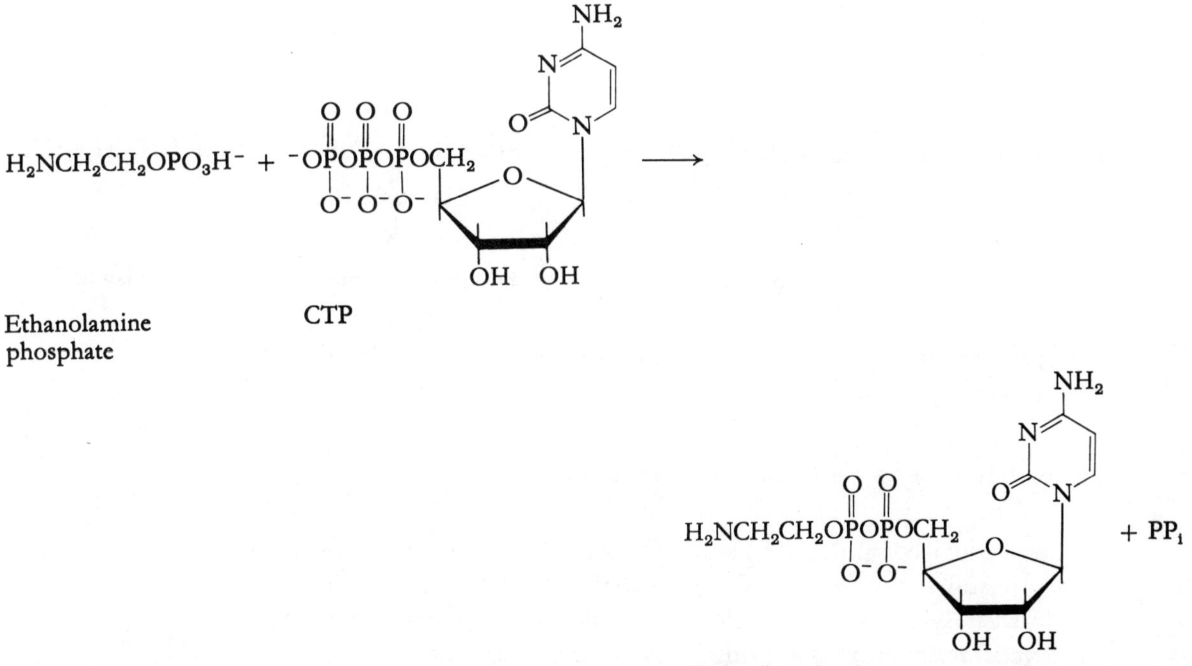

Ethanolamine phosphate + CTP → CDP-ethanolamine + PP$_i$

Subsequent reaction of CDP-choline and CDP-ethanolamine with α,β-diglycerides yields phosphatidylcholine and phosphatidylethanolamine, respectively.

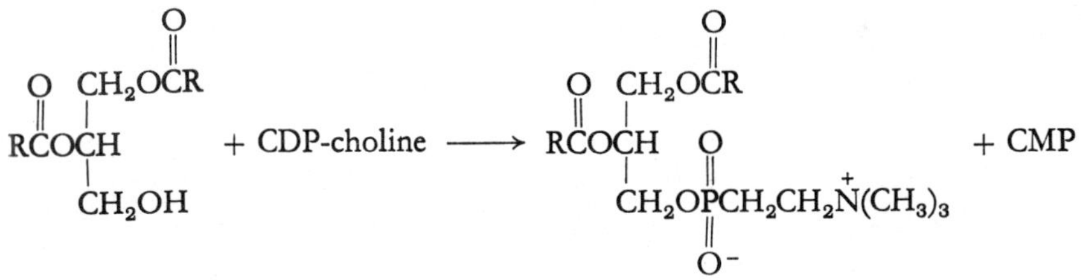

$$\begin{array}{c}\text{O}\\\parallel\\\text{O}\quad\text{CH}_2\text{OCR}\\\parallel\quad|\\\text{RCOCH}\\|\\\text{CH}_2\text{OH}\end{array} + \text{CDP-choline} \longrightarrow \begin{array}{c}\text{O}\\\parallel\\\text{O}\quad\text{CH}_2\text{OCR}\\\parallel\quad|\quad\text{O}\\\text{RCOCH}\quad\parallel\\|\quad\text{CH}_2\text{OPCH}_2\text{CH}_2\overset{+}{\text{N}}(\text{CH}_3)_3\\|\\\text{O}^-\end{array} + \text{CMP}$$

α,β-Diglyceride → Phosphatidylcholine

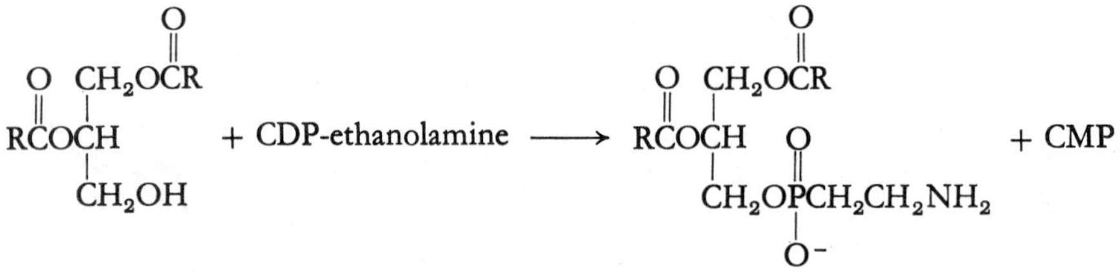

$$\begin{array}{c}\text{O}\\\parallel\\\text{O}\quad\text{CH}_2\text{OCR}\\\parallel\quad|\\\text{RCOCH}\\|\\\text{CH}_2\text{OH}\end{array} + \text{CDP-ethanolamine} \longrightarrow \begin{array}{c}\text{O}\\\parallel\\\text{O}\quad\text{CH}_2\text{OCR}\\\parallel\quad|\quad\text{O}\\\text{RCOCH}\quad\parallel\\|\quad\text{CH}_2\text{OPCH}_2\text{CH}_2\text{NH}_2\\|\\\text{O}^-\end{array} + \text{CMP}$$

α,β-Diglyceride → Phosphatidylethanolamine

Phosphatidylserine may be formed from a reaction between phosphatidylethanolamine and L-serine.

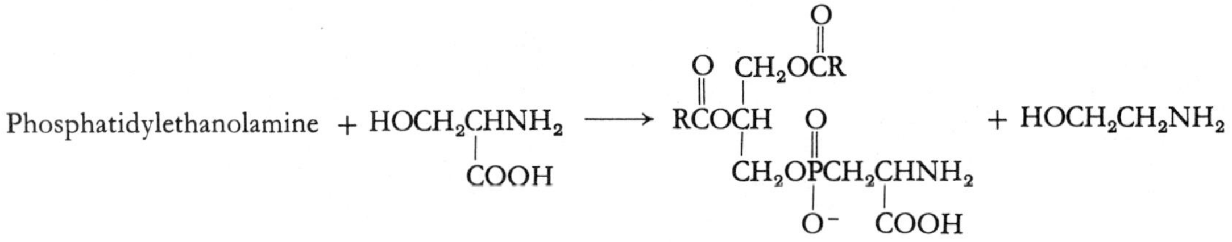

$$\text{Phosphatidylethanolamine} + \text{HOCH}_2\text{CHNH}_2 \longrightarrow \begin{array}{c}\text{O}\\\parallel\\\text{O}\quad\text{CH}_2\text{OCR}\\\parallel\quad|\quad\text{O}\\\text{RCOCH}\quad\parallel\\|\quad\text{CH}_2\text{OPCH}_2\text{CHNH}_2\\|\quad\quad|\\\text{O}^-\quad\text{COOH}\end{array} + \text{HOCH}_2\text{CH}_2\text{NH}_2$$
$$\qquad\qquad\qquad\quad|\\\qquad\qquad\qquad\text{COOH}$$

Serine → Phosphatidylserine → Ethanolamine

Another reaction that yields phosphatidylethanolamine is the decarboxylation of phosphatidylserine:

Phosphatidylserine ⟶ Phosphatidylethanolamine + CO_2

Phosphatidylcholine (lecithin) is also formed as a result of the methylation of phosphatidylethanolamine. The methylating intermediate for this reaction, S-adenosylmethionine, is discussed in detail in Chapter 7 (page 317). The overall reaction for the formation of lecithin is as follows:

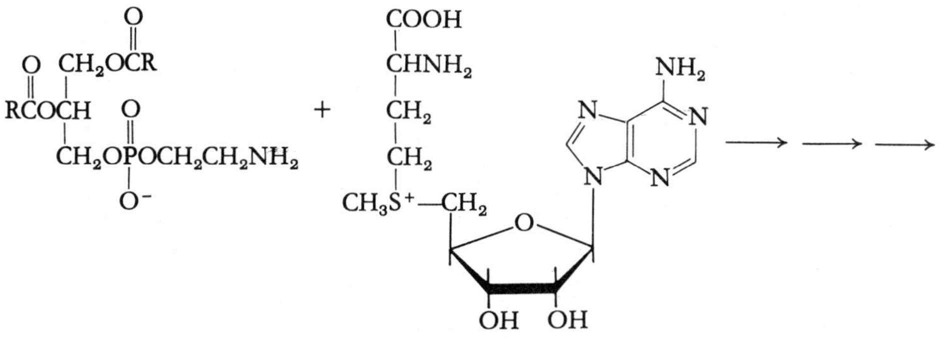

Phosphatidylethanolamine + S-Adenosylmethionine → → →

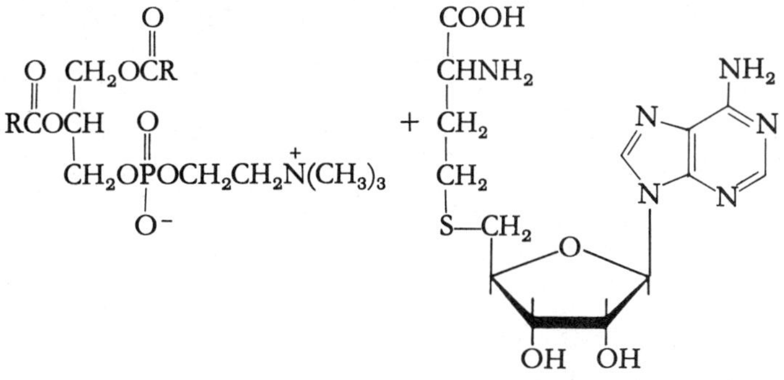

Lecithin + S-Adenosylhomocysteine

This reaction occurs in three steps. In each step, one methyl group is donated by S-adenosylmethionine.

The other glycerol phospholipids—phosphatidylinositol, phosphatidylglycerol, and diphosphatidylglycerol (cardiolipin)—are synthesized from phosphatidic acids (page 227). The common step in the synthesis of these phosphatides is a reaction between phosphatidic acid and cytidine triphosphate (CTP) to yield a cytidine diphosphate diglyceride (CDP-diacylglycerol) and inorganic pyrophosphate:

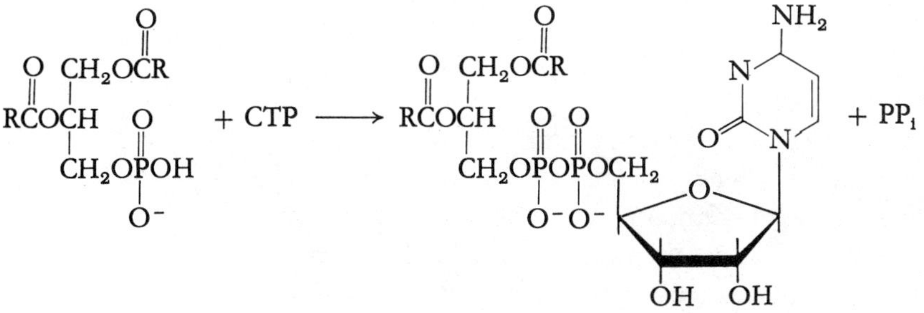

Phosphatidic acid + CTP ⟶ CDP-diacylglycerol + PP_i

Cytidine diphosphate diglyceride can react with inositol to provide phosphatidylinositol and cytidine monophosphate:

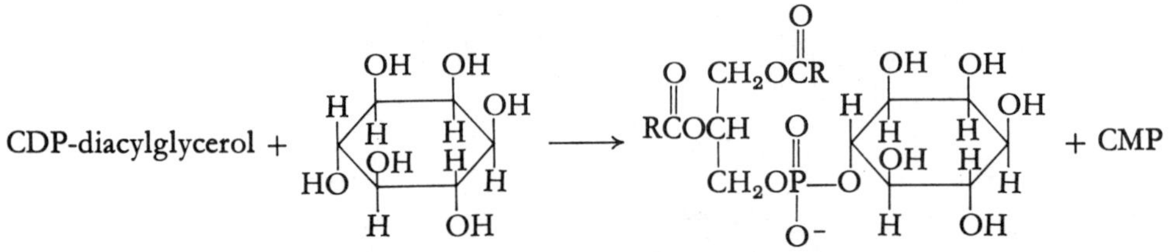

$$\text{CDP-diacylglycerol} + \text{Inositol} \longrightarrow \text{Phosphatidyl inositol} + \text{CMP}$$

By another series of reactions, CDP-diglyceride gives rise to phosphatidylglycerol and cardiolipin:

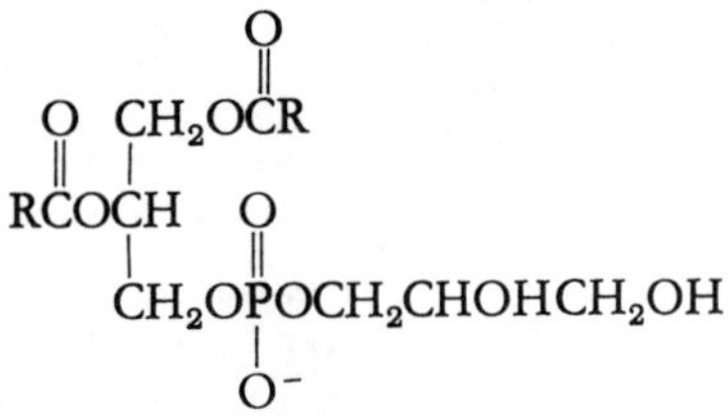

Phosphatidylglycerol:
$$\begin{array}{l} \text{O} \quad \text{CH}_2\text{OCR} \\ \text{RCOCH} \quad \text{O} \\ \quad \text{CH}_2\text{OPOCH}_2\text{CHOHCH}_2\text{OH} \\ \quad\quad\quad \text{O}^- \end{array}$$

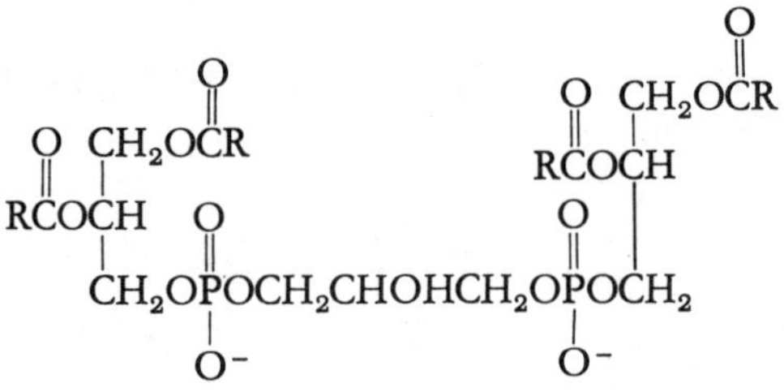

Cardiolipin:
$$\begin{array}{l} \text{O} \quad \text{CH}_2\text{OCR} \quad\quad\quad\quad\quad\quad \text{O} \quad \text{CH}_2\text{OCR} \\ \text{RCOCH} \quad \text{O} \quad\quad\quad\quad\quad\quad \text{RCOCH} \\ \quad \text{CH}_2\text{OPOCH}_2\text{CHOHCH}_2\text{OPOCH}_2 \\ \quad\quad\quad \text{O}^- \quad\quad\quad\quad\quad\quad\quad \text{O}^- \end{array}$$

Phospholipids are integral components of membrane structures. For example, glycerol phospholipids comprise about 40% of the lipid portion of the erythrocyte plasma membrane and over 95% of the lipid in mitochondrial membranes. About 20% of the phospholipid of the inner mitochondrial membrane is cardiolipin. In the circulatory system, the phospholipids occur as components of lipoproteins (Chap. 6, section 6.4).

The relationships between the pathways in the biosynthesis of neutral triglycerides and phosphatides are shown in Figure 5-3.

5.5. Cholesterol Metabolism

A. BIOSYNTHESIS OF CHOLESTEROL

Cholesterol and a number of related compounds, termed *steroids,* are synthesized from acetyl-coenzyme A. These substances serve as structural materials and have important metabolic functions. Cholesterol is an integral component of cellular membranes. It also circulates in the blood as part of a macromolecular complex with protein, phospholipids, and triglycerides (lipoproteins). Additionally, cholesterol serves as the precursor of the

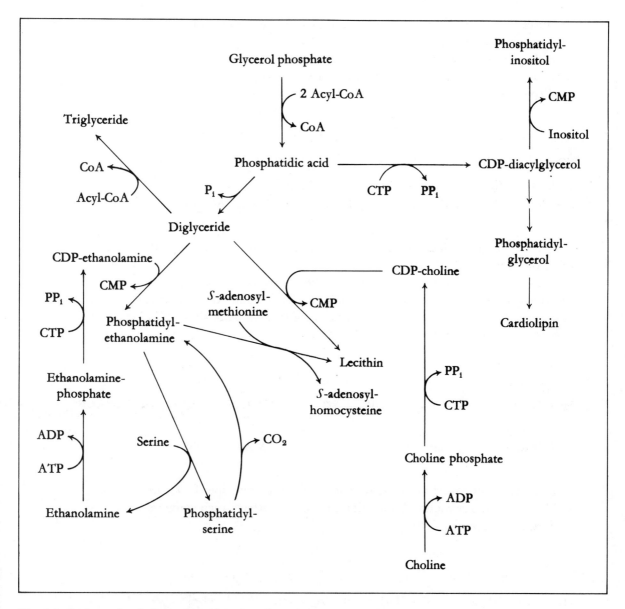

Fig. 5-3. Pathways for the biosynthesis of triglycerides and glycerol phospholipids in animal tissues.

various steroid hormones that have regulatory functions, e.g., progesterone, testosterone, and cortisol.

The major site of synthesis and catabolism of cholesterol is the liver. The following biosynthetic sequence occurs in the microsomal fraction (endoplasmic reticulum). All the carbons of cholesterol are provided by the acetate carbons of acetyl-coenzyme A. Hence, substances that are catabolized to acetyl-coenzyme A can give rise to cholesterol.

In the synthesis of cholesterol, acetyl-coenzyme A must first condense with acetoacetyl-coenzyme A (page 254) to form β-hydroxy-β-methylglutaryl-coenzyme A:

$$\underset{\text{Acetyl-CoA}}{\overset{\text{O}}{\underset{|}{\overset{\|}{\text{C}}\text{CH}_3}}\atop \text{S—CoA}} + \underset{\text{Acetoacetyl-CoA}}{\overset{\text{O}}{\overset{\|}{\text{C}}\text{CH}_2}\overset{\text{O}}{\underset{|}{\overset{\|}{\text{C}}\text{—S—CoA}}}\atop \text{CH}_3} \longrightarrow \underset{\beta\text{-Hydroxy-}\beta\text{-methylglutaryl-CoA}}{\overset{\text{O}}{\overset{\|}{\text{HOC}}\text{CH}_2}\overset{\text{OH}}{\underset{|}{\text{C}}\text{CH}_2}\overset{\text{O}}{\overset{\|}{\text{C}}\text{—S—CoA}}\atop \text{CH}_3} + \text{CoA—SH}$$

(β-Hydroxy-β-methylglutaryl-CoA is also an intermediate in the formation of ketone bodies [page 254]).

The committed step in the synthesis of cholesterol is the next reaction in which β-hydroxy-β-methylglutaryl-coenzyme A is reduced to *mevalonic acid* by an NADPH-linked reductase, termed *β-hydroxy-β-methylglutaryl-CoA reductase* (HMG-CoA reductase):

$$\underset{\beta\text{-Hydroxy-}\beta\text{-methylglutaryl-CoA}}{\overset{\text{O}}{\overset{\|}{\text{HOC}}\text{CH}_2}\overset{\text{OH}}{\underset{|}{\text{C}}\text{CH}_2}\overset{\text{O}}{\overset{\|}{\text{C}}\text{—S—CoA}}\atop \text{CH}_3} + 2\text{NADPH} + 2\text{H}^+ \longrightarrow \underset{\text{Mevalonic acid}}{\overset{\text{O}}{\overset{\|}{\text{HOC}}\text{CH}_2}\overset{\text{OH}}{\underset{|}{\text{C}}\text{CH}_2\text{CH}_2\text{OH}}\atop \text{CH}_3} + 2\text{NADP}^+ + \text{CoA—SH}$$

Three successive reactions of mevalonate with ATP and specific kinases yield 3-phospho-5-pyrophosphomevalonic acid. The latter is then converted to *3-isopentenyl pyrophosphate*:

$$\underset{\substack{\text{3-Phospho-5-pyrophospho-}\\\text{mevalonic acid}}}{\text{HOOCCH}_2\overset{\overset{\text{CH}_3}{|}}{\underset{\underset{\text{OPO}_3\text{H}_2}{|}}{\text{C}}}\text{CH}_2\text{CH}_2\text{O}\overset{\overset{\text{O}}{\|}}{\underset{\underset{\text{OH}}{|}}{\text{P}}}\text{—O—}\overset{\overset{\text{O}}{\|}}{\underset{\underset{\text{OH}}{|}}{\text{P}}}\text{OH}} \longrightarrow \underset{\text{3-Isopentenyl pyrophosphate}}{\text{CH}_2\text{=}\overset{\overset{\text{CH}_3}{|}}{\text{C}}\text{CH}_2\text{CH}_2\text{O}\overset{\overset{\text{O}}{\|}}{\underset{\underset{\text{OH}}{|}}{\text{P}}}\text{—O—}\overset{\overset{\text{O}}{\|}}{\underset{\underset{\text{OH}}{|}}{\text{P}}}\text{OH}} + \text{CO}_2 + \text{P}_i$$

3-Isopentenyl pyrophosphate isomerizes (reversibly) to form 3,3-dimethylallyl pyrophosphate, and these two isomers interact to yield *geranyl pyrophosphate*:

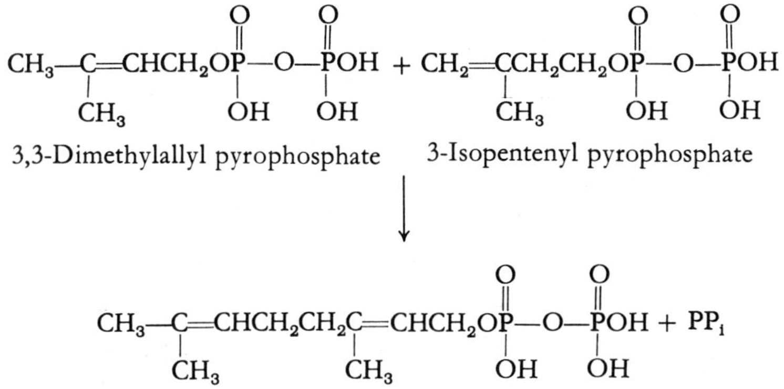

$$\text{CH}_3-\underset{\underset{\text{CH}_3}{|}}{\text{C}}=\text{CHCH}_2\text{O}\underset{\underset{\text{OH}}{|}}{\overset{\overset{\text{O}}{\|}}{\text{P}}}-\text{O}-\underset{\underset{\text{OH}}{|}}{\overset{\overset{\text{O}}{\|}}{\text{P}}}\text{OH} + \text{CH}_2=\underset{\underset{\text{CH}_3}{|}}{\text{C}}\text{CH}_2\text{CH}_2\text{O}\underset{\underset{\text{OH}}{|}}{\overset{\overset{\text{O}}{\|}}{\text{P}}}-\text{O}-\underset{\underset{\text{OH}}{|}}{\overset{\overset{\text{O}}{\|}}{\text{P}}}\text{OH}$$

3,3-Dimethylallyl pyrophosphate   3-Isopentenyl pyrophosphate

$$\downarrow$$

$$\text{CH}_3-\underset{\underset{\text{CH}_3}{|}}{\text{C}}=\text{CHCH}_2\text{CH}_2\underset{\underset{\text{CH}_3}{|}}{\text{C}}=\text{CHCH}_2\text{O}\overset{\overset{\text{O}}{\|}}{\text{P}}-\text{O}-\overset{\overset{\text{O}}{\|}}{\text{P}}\text{OH} + \text{PP}_i$$

Geranyl pyrophosphate

Reaction of geranyl pyrophosphate with another mole of 3-isopentenyl pyrophosphate yields *farnesyl pyrophosphate* and inorganic pyrophosphate:

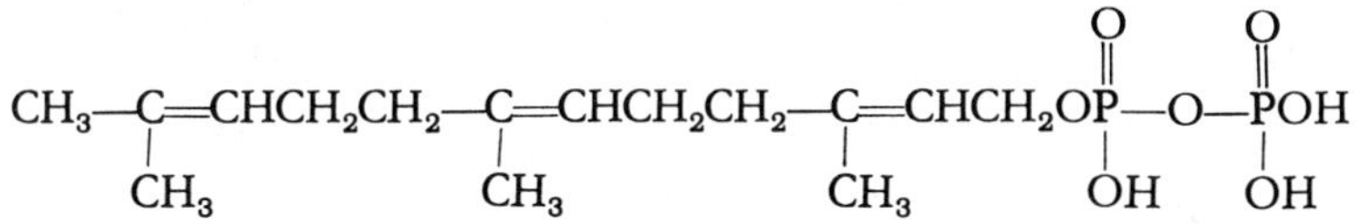

$$\text{CH}_3-\underset{\underset{\text{CH}_3}{|}}{\text{C}}=\text{CHCH}_2\text{CH}_2-\underset{\underset{\text{CH}_3}{|}}{\text{C}}=\text{CHCH}_2\text{CH}_2-\underset{\underset{\text{CH}_3}{|}}{\text{C}}=\text{CHCH}_2\text{O}\overset{\overset{\text{O}}{\|}}{\text{P}}-\text{O}-\overset{\overset{\text{O}}{\|}}{\text{P}}\text{OH}$$

Farnesyl pyrophosphate

The enzyme, squalene synthetase, catalyzes the condensation of 2 moles of farnesyl pyrophosphate, in the presence of NADPH, to produce *squalene*, NADP⁺, and 2 moles of pyrophosphate:

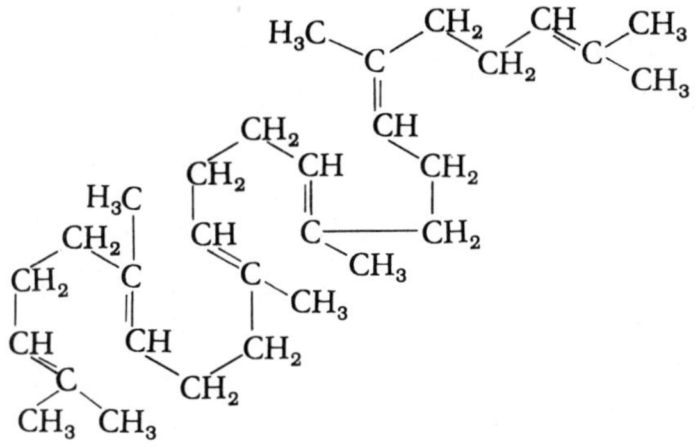

Squalene

Oxidation of squalene by oxygen is catalyzed by an NADPH-linked oxidase; this yields *squalene 2,3-epoxide*. The latter undergoes ring closure to form *lanosterol*:

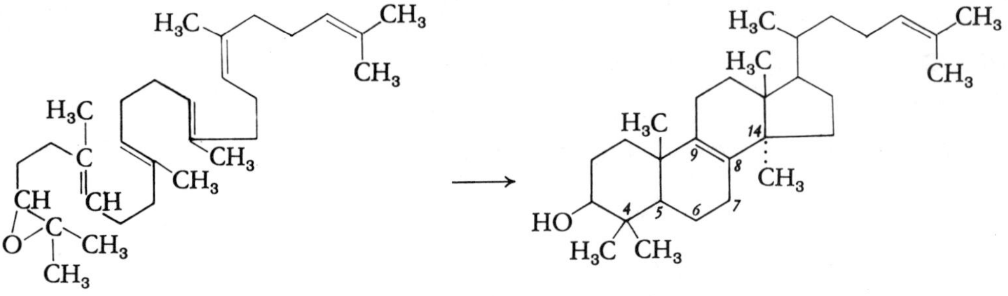

236   5. Biosynthesis of Lipids

Squalene 2,3-epoxide → **Lanosterol**

In a series of steps, the methyl groups on carbon-4 and carbon-14 are eliminated; the resulting product is *zymosterol*. The double bond between carbon-8 and carbon-9 is then shifted to carbon-5 and carbon-6, yielding *desmosterol*:

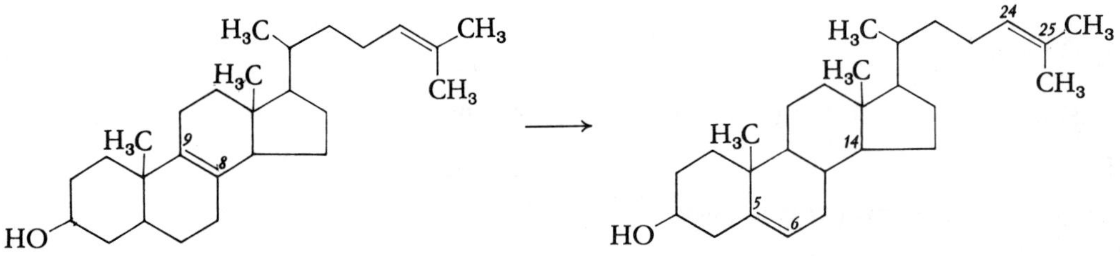

Zymosterol → **Desmosterol**

Finally, desmosterol is reduced at carbon-24 by another NADPH-linked enzyme to provide *cholesterol*:

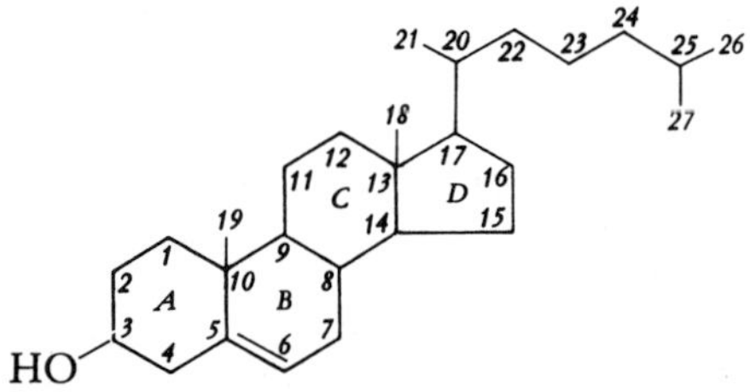

Cholesterol

The synthesis of cholesterol utilizes 18 moles of acetate. Each mole of mevalonic acid is produced from 3 moles of acetate, and 6 moles of this intermediate are required for 1 mole of cholesterol. Six moles of CO_2 are eliminated in the conversion of the mevalonic acid units to 3-isopentenyl pyrophosphate, and 3 moles of CO_2 are released in the conversion of lanosterol to cholesterol. The final product therefore contains 27 carbons. In

Table 5-1. Nomenclature of Steroids

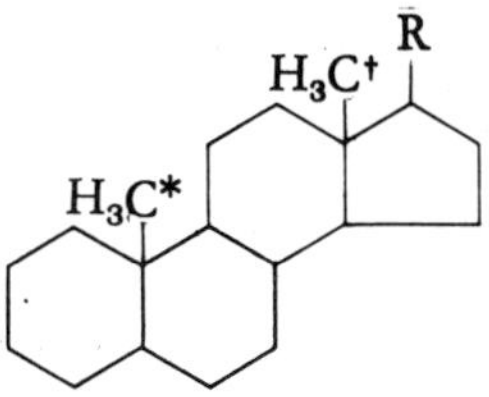

Total Carbons	Number of Carbons in R Group	Designation Based on R-Group Type	Representative Compound	Class
27–29	8–10	Cholestane	Cholesterol	Sterols
24	5	Cholane	Cholic acid	Bile acids
18	0	Estrane*	Estradiol	Estrogens
19	0	Androstane	Testosterone	Androgens
21	2	Pregnane	Progesterone	Progestogens
			Aldosterone†	Mineralocorticoids
			Cortisol	Glucocorticoids

* The estranes lack the methyl group at position 19 of cholesterol.
† Aldosterone has an aldehyde group instead of the methyl group at position 18 of cholesterol.

addition, the synthesis of 1 mole of cholesterol requires 18 moles of ATP and 13 moles of NADPH.

A critical regulatory site in the biosynthesis of cholesterol is at the committed step in the process, that is, in the conversion of β-hydroxy-β-methylglutaryl-coenzyme A to mevalonic acid. The synthesis of the enzyme (HMG-CoA reductase) that catalyzes this reaction is inhibited by elevated levels of cholesterol. This is probably an important factor in maintaining normal cholesterol concentrations and in blocking excessive synthesis when cholesterol is taken in the diet.

B. NOMENCLATURE OF STEROIDS

The numbering system for the positions of the carbons in steroids and the letter designations for the rings are indicated in the structure of cholesterol. The spatial distribution of substituents can be visualized if the ring carbons are considered to be in the plane of the paper. Atoms or groups linked to the carbons project above or below the plane of the rings. By convention, an atom or group that projects below the plane of the rings is said to be α *oriented* and its bond is shown by a dotted line. Those above the plane are designated as being β *oriented* and their bonds shown by a heavy solid line.

Rings A and B may be fused in either a *cis* or *trans* configuration. In the former case, any substituent on carbon-5 would be β oriented. Conversely, the α orientation of such a substituent would occur when the two rings are in a *trans* configuration. It may be noted that cholesterol does not have a hydrogen on carbon-5. This is also the case for many steroids. The steroid hormones that do have a hydrogen on carbon-5 have a *trans* configuration, whereas the bile salts have a *cis* configuration.

The methyl groups at positions 18 and 19 are called *angular methyl groups*. Compounds that contain hydroxyl groups but no keto (C=O) or carboxyl (COOH) groups are called *sterols*. The *steroids* include sterols as well as those compounds containing keto or carboxyl groups. In the general system of nomenclature, the compounds are further differentiated according to the number of carbons in the side chain attached to ring D (Table 5-1).

C. ESTERIFICATION OF CHOLESTEROL

About two-thirds of the cholesterol in blood is in the form of an ester. Such esters are synthesized by the transfer of a fatty-acid unit from the β position of lecithin (phosphatidylcholine) to the 3-hydroxyl group of cholesterol. The reaction is catalyzed by *lecithin-cholesterol acyltransferase* (LCAT):

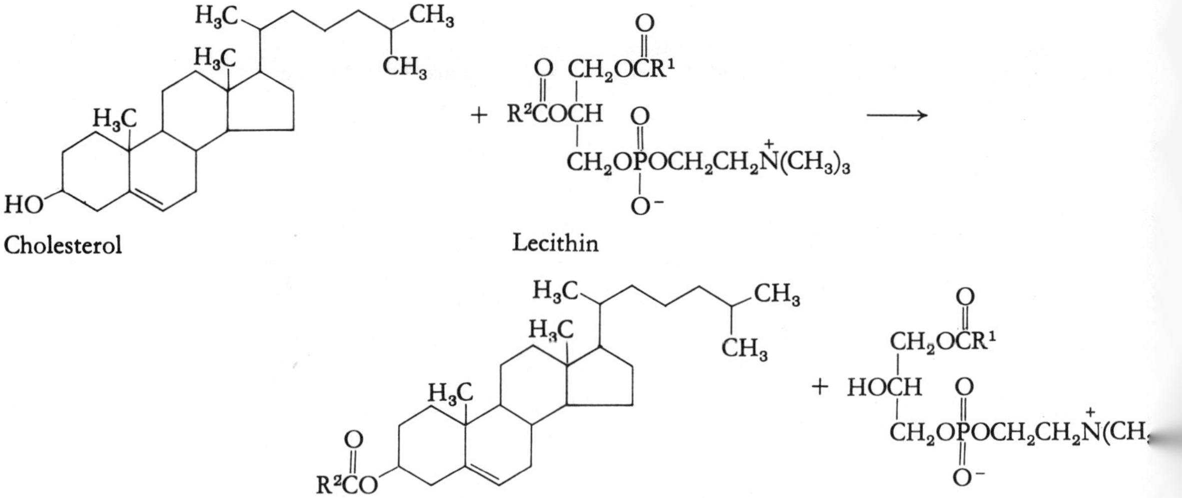

Cholesterol + Lecithin ⟶ Cholesterol ester + Lysolecithin

A specific disease in which there is a deficiency in LCAT has been described. Individuals with this abnormality have high plasma levels of lecithin and free cholesterol. Other manifestations of the disease are anemia, proteinuria, and corneal infiltration. The patients are also deficient in α_1- and pre-β-lipoproteins (page 267).

Cholesterol is also esterified in the liver by a reaction with coenzyme-A derivatives of fatty acids.

D. TURNOVER AND ELIMINATION OF CHOLESTEROL AND BILE-SALT FORMATION

The greatest amount of cholesterol in human adults is found in the adrenal glands (about 10% of the weight of the gland). The approximate concentrations in other tissues that contain relatively high levels are 0.3% in the liver and skin and 0.2% in the brain, nerves, and intestine.

The average adult normally synthesizes 1.5 to 2.0 grams of cholesterol each day. In addition, about 300 milligrams are consumed in the diet. The major catabolic pathway for cholesterol is its conversion to bile acids and ultimate elimination as fecal sterols. Relatively smaller amounts are transformed into steroid hormones.

Bile salts are formed in the liver and secreted into the gallbladder. The processes involve scission and oxidation of the side chain of cholesterol, hydroxylation at two sites, and formation of a coenzyme A derivative. The resulting intermediate, *choloyl-coenzyme A*, reacts with glycine to form glycocholic acid, which is the principal bile acid. In addition, choloyl-coenzyme A can condense with taurine to yield taurocholic acid:

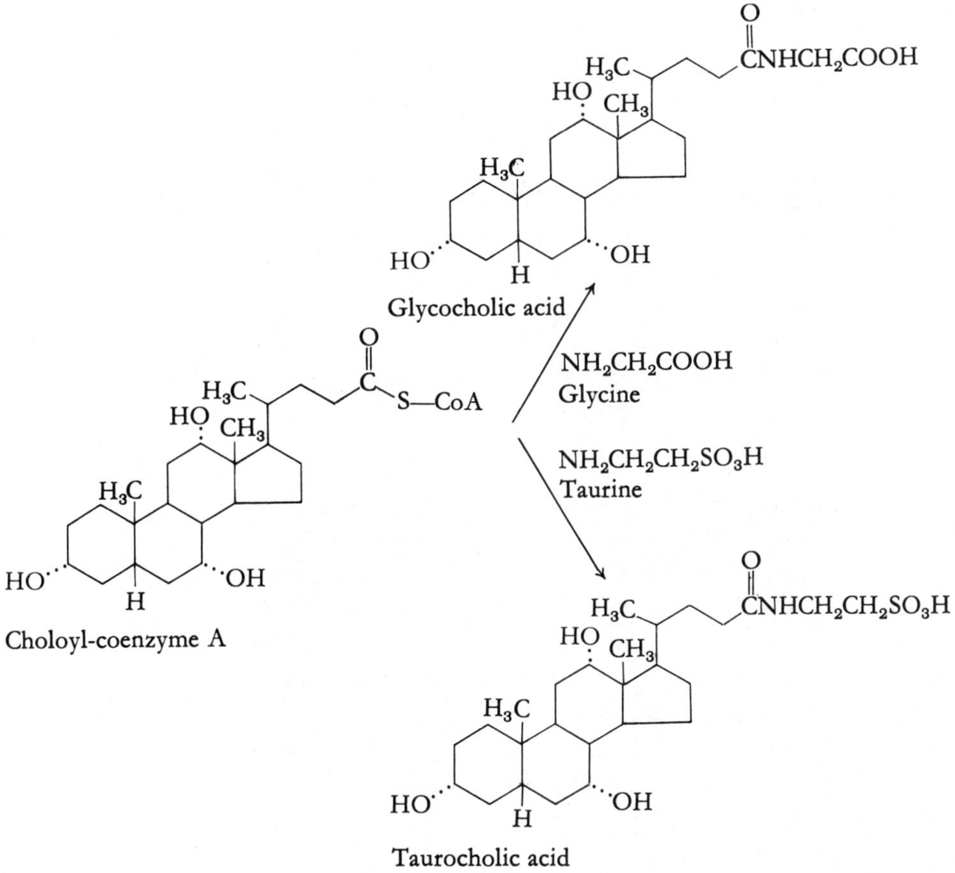

When secreted into the intestine, bile acids in their dissociated forms (i.e., their salts) serve as emulsifying agents for the disposal of water-soluble lipids (page 115). When emulsified, such lipids are susceptible to digestion by pancreatic lipase.

A large fraction of the steroid components of bile (cholesterol and bile acids) that are secreted into the intestine can be reabsorbed and returned to the liver via the portal system. The remainder is reduced to *coprostanol* and *cholestanol*:

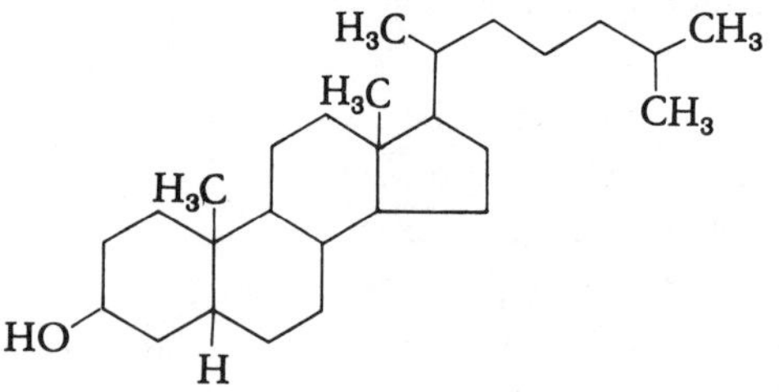

Coprostanol 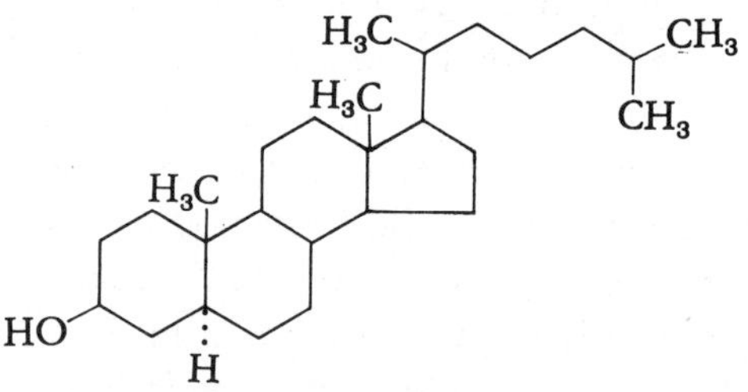   Cholestanol

Although the reduction of cholesterol to these two products is effected primarily by intestinal microorganisms, small amounts are also produced in the liver. Coprostanol and cholestanol, as well as excess cholesterol, are removed in the feces.

As a result of the movement of cholesterol from the liver to the gallbladder and the normal concentrating effect upon the bile, the level of cholesterol in this organ is relatively high. Under certain conditions—e.g., in gallbladder inflammation—the cholesterol may crystallize to form "cholesterol stones."

5.6. Steroid Hormones
A. PREGNENOLONE

Cholesterol is the common precursor of all the groups of steroid hormones that were shown in Table 5-1. These hormones include the progestogens, the glucocorticoids, the mineralocorticoids (21 carbons), the androgens (19 carbons), and the estrogens (18 carbons). A central compound in the formation of steroid hormones from cholesterol is the 21-carbon intermediate, *pregnenolone*. The process involves several oxidations and the removal of a six-carbon unit from the side chain of cholesterol.

Hydroxylation of cholesterol by mixed-function oxygenases, which utilize O_2 and NADPH, yields 20α,22-dihydroxycholesterol. The latter is also oxidized by the mixed-function oxygenase, *desmolase,* providing pregnenolone and isocaproic aldehyde:

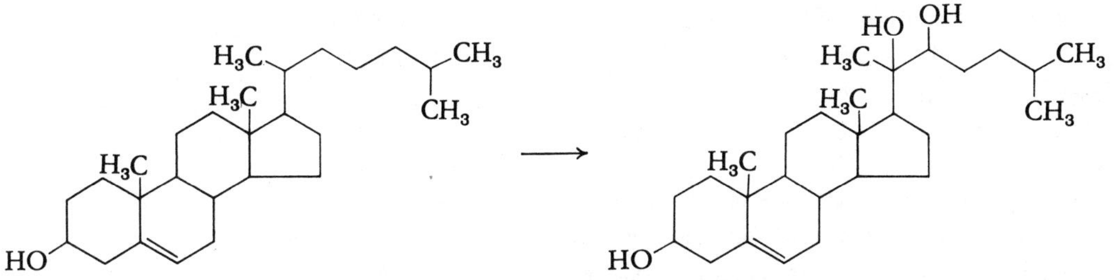

[Structures: Cholesterol → 20α,22-Dihydroxycholesterol → Pregnenolone (3β-hydroxypregn-5-ene-20-one) + Isocaproic aldehyde]

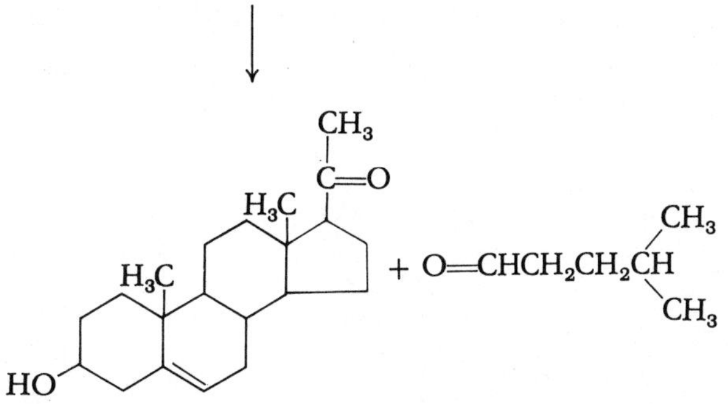

The conversion of cholesterol to pregnenolone is stimulated by the pituitary hormone, adrenocorticotropic hormone (ACTH).

B. PROGESTERONE

Progesterone is synthesized in the corpus luteum, adrenals, and placenta. Its formation from pregnenolone involves an oxidation of the hydroxyl group and a shift in the double bond from ring B to ring A:

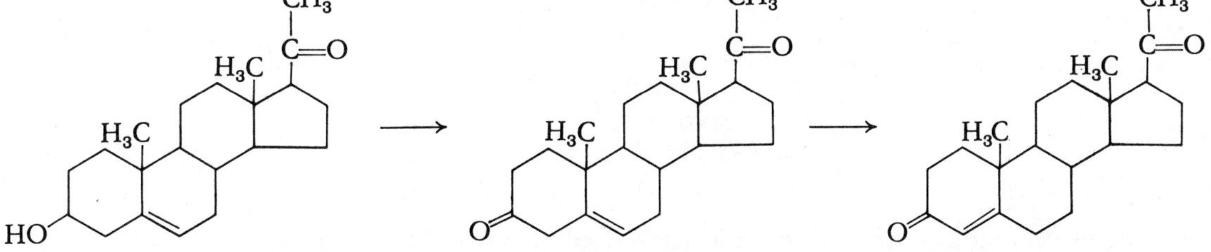

[Structures: Pregnenolone → Pregn-5-ene-3,20-dione → Progesterone (pregn-4-ene-3,20-dione)]

The secretion of progesterone by the corpus luteum in the latter half of the menstrual cycle induces the release of mucus from the endometrium. This is required for implantation of the fertilized ovum and the progress of normal pregnancy. Additional effects of this and other steroid hormones will be described in Chapter 14.

C. CORTISOL AND ALDOSTERONE

The cells of the adrenal cortex synthesize cholesterol from acetate. Cholesterol is also converted to progesterone via pregnenolone. As a result of the action of a number of mixed-function oxygenases (utilizing NADPH and O_2) on carbon-21 and carbon-11, progesterone is converted to *corticosterone*. The angular methyl group (carbon-18) is then oxidized to yield *aldosterone*. Alternatively, progesterone may be hydroxylated—first at carbon-17, then at carbon-11 and carbon-21—to provide *cortisol*.

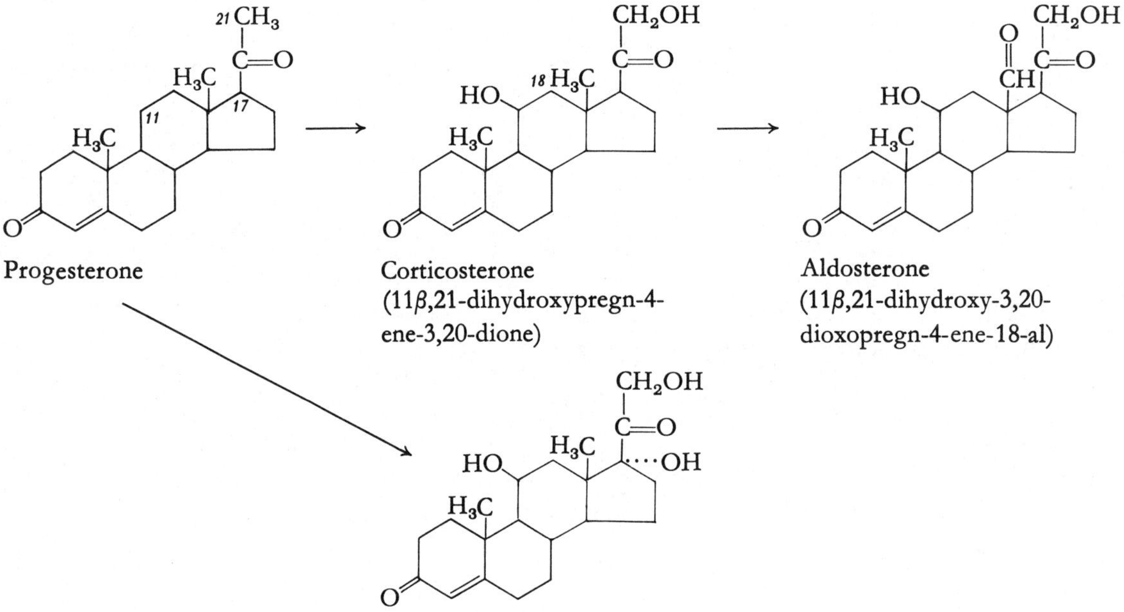

Progesterone

Corticosterone
(11β,21-dihydroxypregn-4-ene-3,20-dione)

Aldosterone
(11β,21-dihydroxy-3,20-dioxopregn-4-ene-18-al)

Cortisol
(Hydrocortisone; 11β,17α,21-trihydroxypregn-4-ene-3,20-dione)

A number of steroid hormones have a marked effect on the metabolism and utilization of carbohydrates. One overall result that some of them induce is an increase in levels of blood glucose; hormones having this effect are called *glucocorticoids*. The most potent glucocorticoid is cortisol. Among its actions are to increase gluconeogenesis, catabolism of proteins, and glycogen deposition in liver. The glucocorticoids also appear to decrease peripheral utilization of glucose. These hormones have anti-inflammatory activity as well.

Another function of certain steroids is the regulation of minerals (especially sodium and potassium ions) in blood and extracellular fluids; such steroids are termed *mineralocorticoids*. Although a number of steroids influence mineral metabolism, aldosterone is by far the most effective. The hormone is important in promoting resorption in the distal tubules of the kidney of Na^+, together with Cl^- and HCO_3^-. The maintenance of definite levels of these ions in the blood is necessary for the transfer of intracellular K^+ to the extracellular fluid (Chap. 13). By controlling the concentrations of mineral electrolytes, the mineralocorticoids have a critical effect on body-fluid volume and blood pressure. The importance of the mineralocorticoids may be realized from the fact that a deficiency in these materials results in the loss of inordinate amounts of Na^+ to the urine, a decrease in plasma volume, and an increased concentration of blood. If not treated, this condition leads to decreased cardiac output and hypotension.

D. ANDROGENS

The androgens are 19-carbon steroids. One route for their synthesis (which occurs in testis, adrenal gland, ovary, and placenta) involves the hydroxylation of pregnenolone on carbon-17 to yield 17α-hydroxypregnenolone. Removal of carbon-20 and carbon-21 provides dehydroepiandrosterone. This is followed by an oxidation on carbon-3 to yield androstenedione and a reduction of carbon-17 to produce *testosterone*:

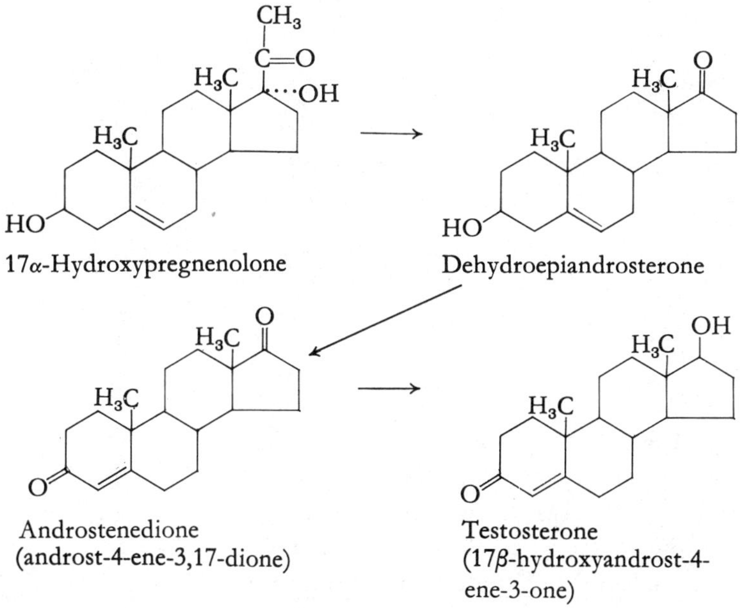

17α-Hydroxypregnenolone

Dehydroepiandrosterone

Androstenedione
(androst-4-ene-3,17-dione)

Testosterone
(17β-hydroxyandrost-4-ene-3-one)

The androgens (e.g., testosterone) are necessary for the development of secondary sex characteristics in male animals.

E. ESTROGENS

The estrogens contain 18 carbons. They differ from other steroid hormones with respect to two items: ring A is aromatic and there is no angular methyl group attached to carbon-10. The principal hormone of the ovary is *17β-estradiol*, which is synthesized from testosterone:

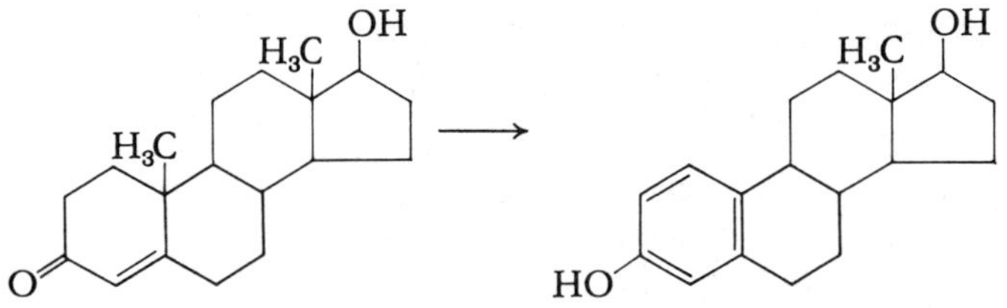

Testosterone → 17β-Estradiol

Another estrogen that is formed in the placenta, as well as in the adrenal glands and testes, is *estrone*. This can be formed from androstenedione:

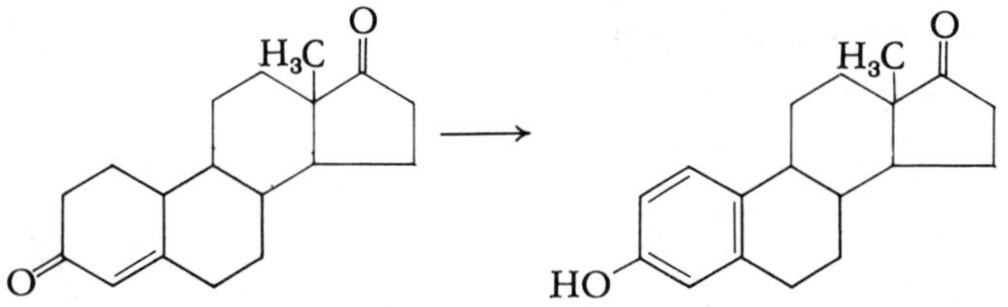

Androstenedione → Estrone

Estradiol has about 10 times the potency of estrone in terms of hormonal activity. The primary effect of these hormones is on the development of female secondary sex characteristics.

The syntheses and interconversions of steroid hormones are summarized in Figure 5-4.

5.7. Sphingolipids

Unlike the glycerol-containing lipids discussed in sections 5.3 and 5.4, the sphingolipids are derivatives of sphingosine (page 57). These materials have complex structures, and their biochemistry constitutes a highly specialized field. This discussion will be restricted to the aspects of their metabolism that are related to biologic function and clinical abnormalities.

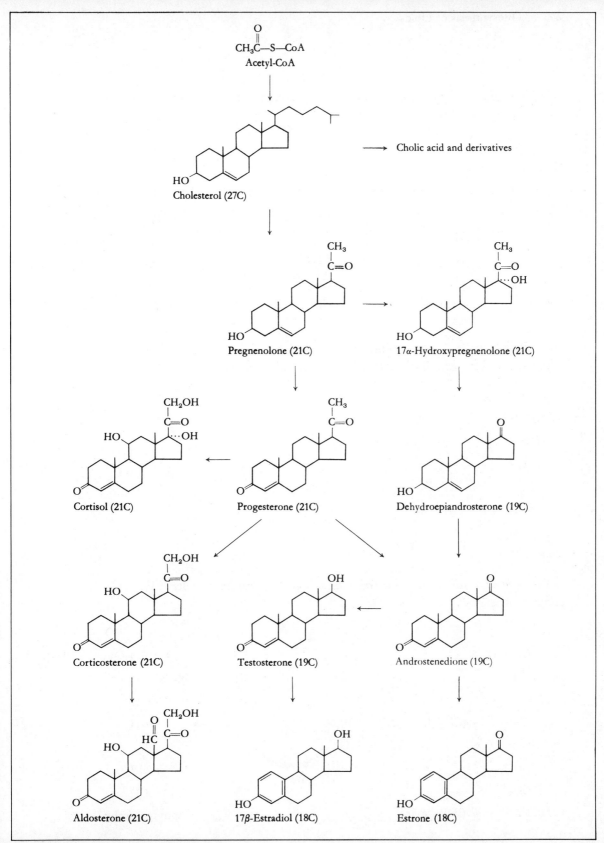

Fig. 5-4 Interconversions of steroid hormones.

Sphingosine is synthesized from serine and palmitoyl-coenzyme A according to the following sequence:

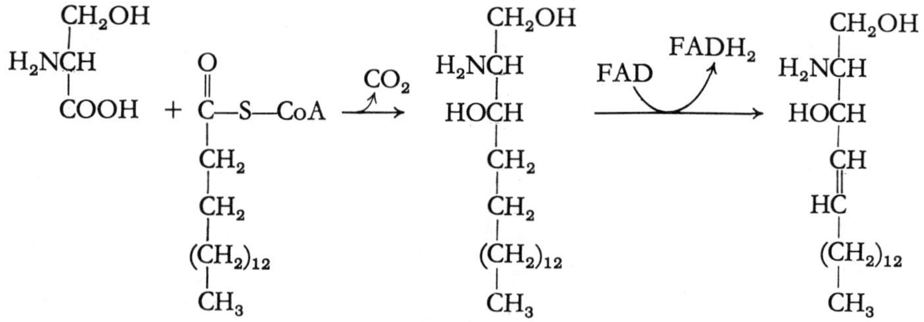

$$\underset{\text{Serine}}{\begin{array}{c}CH_2OH\\|\\H_2NCH\\|\\COOH\end{array}} + \underset{\text{Palmitoyl-CoA}}{\begin{array}{c}O\\||\\C-S-CoA\\|\\CH_2\\|\\CH_2\\|\\(CH_2)_{12}\\|\\CH_3\end{array}} \xrightarrow{CO_2} \underset{\text{Dihydro-sphingosine}}{\begin{array}{c}CH_2OH\\|\\H_2NCH\\|\\HOCH\\|\\CH_2\\|\\CH_2\\|\\(CH_2)_{12}\\|\\CH_3\end{array}} \xrightarrow[\text{FAD} \; \text{FADH}_2]{} \underset{\text{Sphingosine}}{\begin{array}{c}CH_2OH\\|\\H_2NCH\\|\\HOCH\\|\\CH\\||\\HC\\|\\(CH_2)_{12}\\|\\CH_3\end{array}}$$

In all sphingolipids, the amino group of sphingosine is acetylated with a long-chain fatty acid. Such derivatives are called *ceramides*:

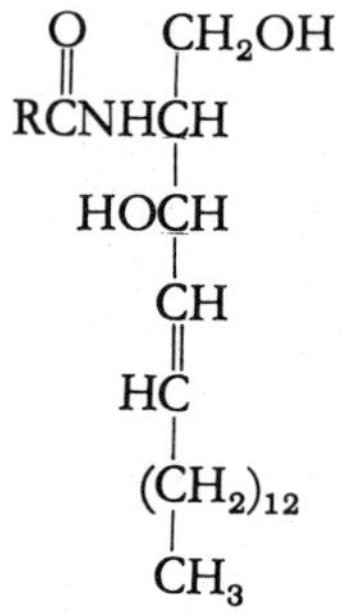

$$\begin{array}{c}O\quad CH_2OH\\||\quad|\\RCNHCH\\|\\HOCH\\|\\CH\\||\\HC\\|\\(CH_2)_{12}\\|\\CH_3\end{array}$$

Ceramide

Sphingomyelins are formed by the interaction of a ceramide with CDP-choline (page 229):

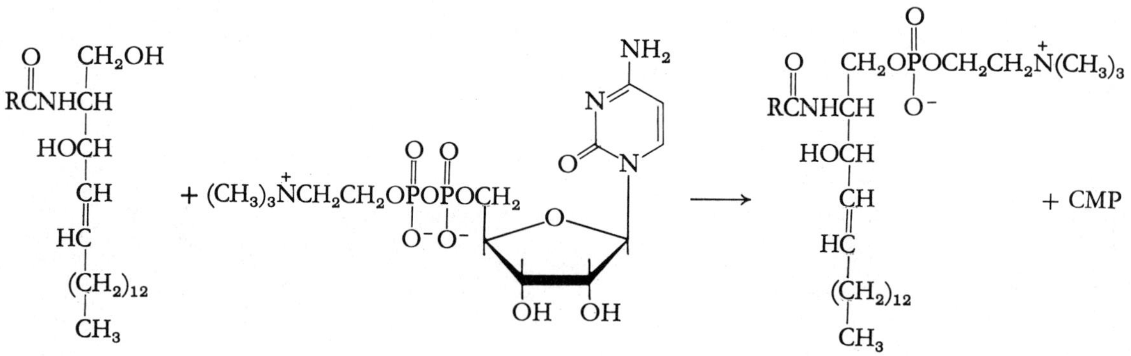

$$\underset{\text{Ceramide}}{\begin{array}{c}O\quad CH_2OH\\||\quad|\\RCNHCH\\|\\HOCH\\|\\CH\\||\\HC\\|\\(CH_2)_{12}\\|\\CH_3\end{array}} + \underset{\text{CDP-choline}}{(CH_3)_3\overset{+}{N}CH_2CH_2O\overset{O}{\underset{O^-}{P}}O\overset{O}{\underset{O^-}{P}}OCH_2\text{–cytidine}} \longrightarrow \underset{\text{Sphingomyelin}}{\begin{array}{c}O\quad CH_2O\overset{O}{\underset{O^-}{P}}OCH_2CH_2\overset{+}{N}(CH_3)_3\\||\quad|\\RCNHCH\\|\\HOCH\\|\\CH\\||\\HC\\|\\(CH_2)_{12}\\|\\CH_3\end{array}} + CMP$$

Sphingomyelins occur in membranous structures in all tissues. They are present in larger amounts in brain and nerve tissue.

Glycosphingolipids are complex molecules composed of a ceramide and one or more sugar residues. They are grouped under several types according to the number and nature of the sugar residues.

Cerebrosides are derivatives of ceramides and contain a glucose or galactose unit instead of the phosphocholine of sphingomyelins. Cerebrosides are formed from the reaction of ceramide with uridine diphosphate glucose or uridine diphosphate galactose. Specific *glycosyltransferases* are required for each type of product.

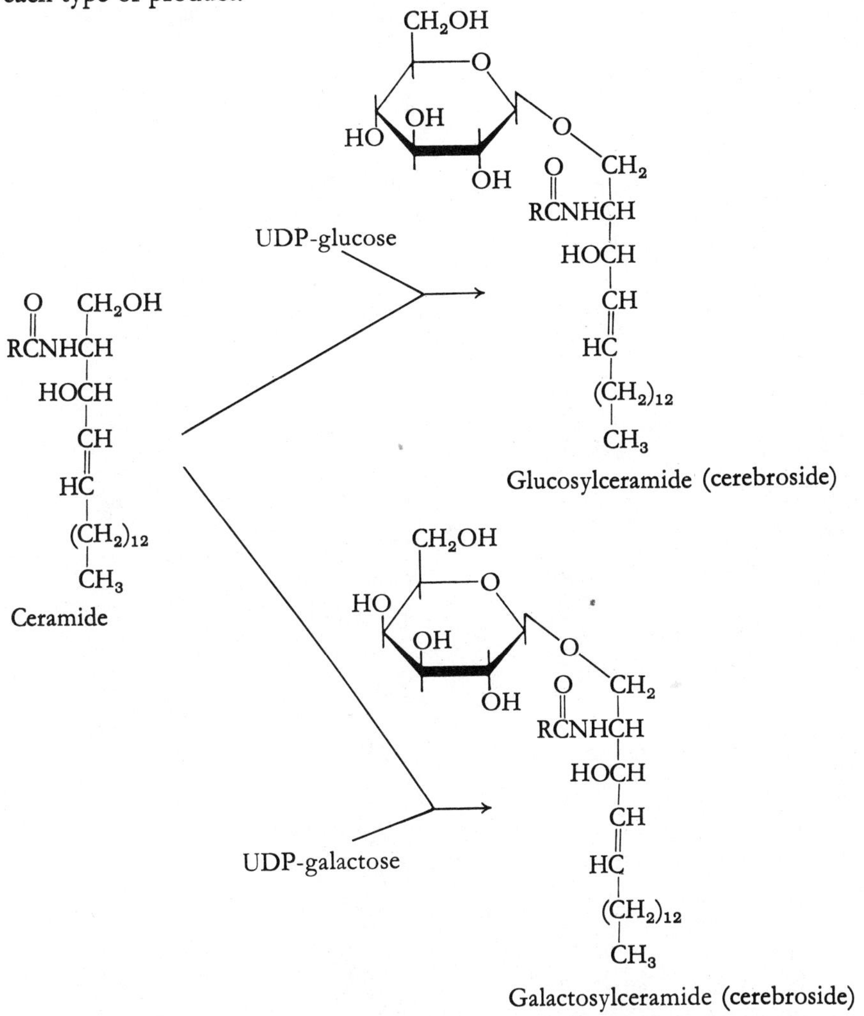

Cerebrosides of the galactosylceramide type are the predominating form in brain white matter and spinal cord. They are also important constituents of myelin. The glucosylceramides are found in extraneural tissues and fluids.

Other, more complex derivatives of ceramides (ceramide oligoglycosides) are widely distributed in animal tissues. These resemble cerebrosides but instead of having a single hexose moiety, they contain two or more sugar residues. These compounds are constituents of erythrocytes, plasma, spleen and kidney tissue, and mucous secretions of the digestive tract. Within the cellular structure, they occur as components of the plasma membranes or cell surfaces. Examples of such glycosphingolipids are diglycosylceramide, triglycosylceramide, and tetraglycosylceramide:

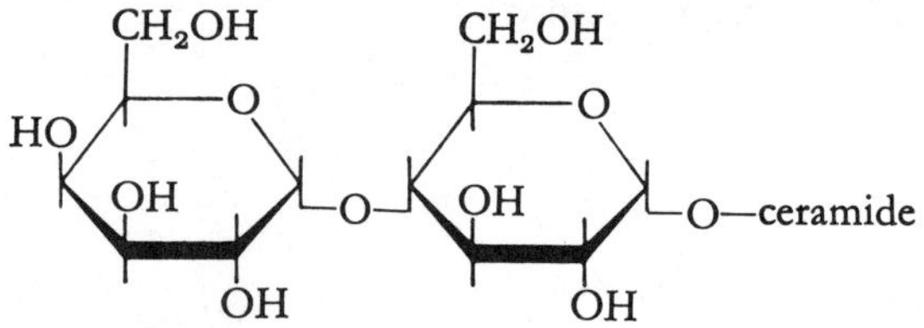

Diglycosylceramide

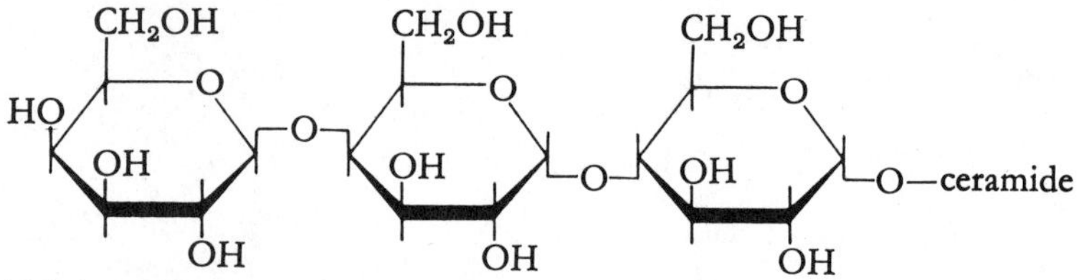

Triglycosylceramide

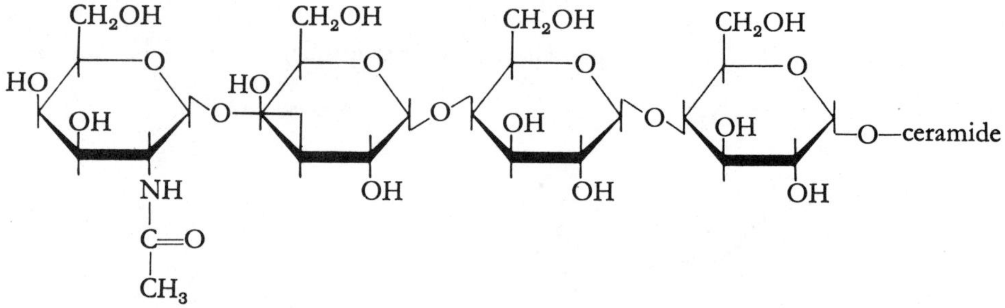

Tetraglycosylceramide

It should be noted that the tetraglycosylceramide shown contains N-acetyl-galactosamine in addition to glucose and galactose residues. Another sugar derivative that occurs in various ceramide oligosaccharides is L-*fucose* (6-deoxy-L-galactose):

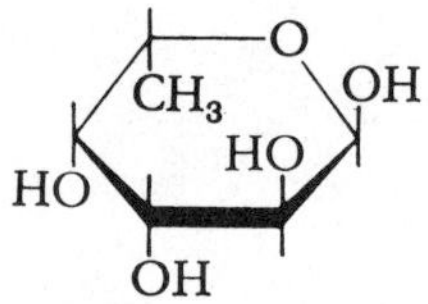

L-Fucose

Glycosphingolipids containing this sugar have been found in erythrocyte membranes and in glandular tissue.

Gangliosides are the most complex of the glycosphingolipids. In addition to containing several sugar moieties, they all contain one or more residues of sialic acid (N-acetylneuraminic acid):

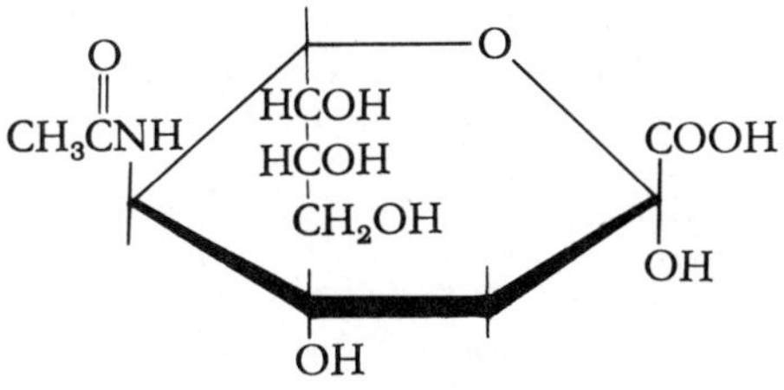

$$CH_3CNH \quad HCOH \quad HCOH \quad CH_2OH \quad COOH \quad OH \quad OH$$

N-Acetylneuraminic acid
(a sialic acid)

Over thirty different types of gangliosides have been isolated from brain tissue. The highest concentrations are found in the gray matter of brain. Examples of four types are shown as follows, where Gal = galactose, GalNAc = N-acetylgalactosamine, Glc = glucose, and NANA = N-acetylneuraminic acid:

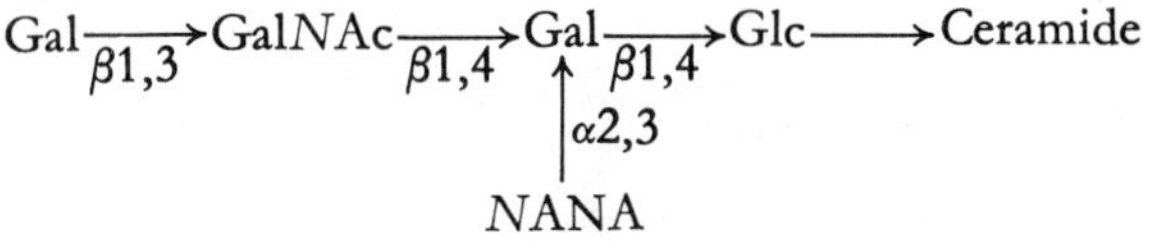

$$Gal \xrightarrow{\beta 1,3} GalNAc \xrightarrow{\beta 1,4} Gal \xrightarrow{\beta 1,4} Glc \longrightarrow Ceramide$$
$$\uparrow \alpha 2,3$$
$$NANA$$
$$G_{M1}$$

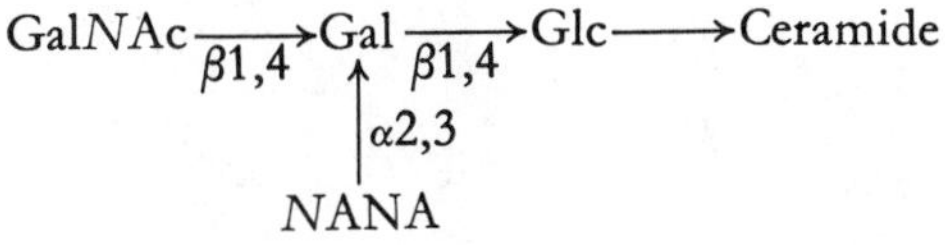

$$GalNAc \xrightarrow{\beta 1,4} Gal \xrightarrow{\beta 1,4} Glc \longrightarrow Ceramide$$
$$\uparrow \alpha 2,3$$
$$NANA$$
$$G_{M2}$$

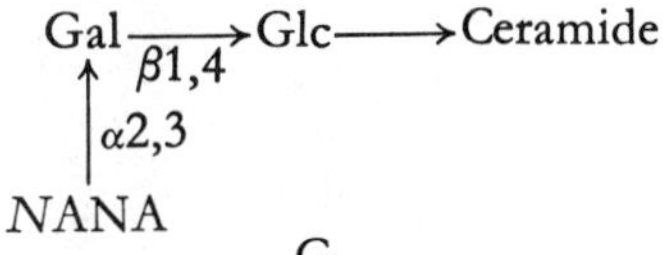

$$Gal \xrightarrow{\beta 1,4} Glc \longrightarrow Ceramide$$
$$\uparrow \alpha 2,3$$
$$NANA$$
$$G_{M3}$$

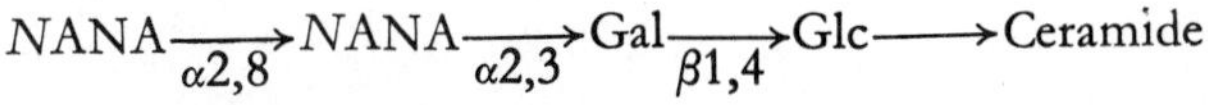

$$NANA \xrightarrow{\alpha 2,8} NANA \xrightarrow{\alpha 2,3} Gal \xrightarrow{\beta 1,4} Glc \longrightarrow Ceramide$$
$$G_{D3}$$

Gangliosides are mainly components of membranes. The sugar units and sialic acid sections of the molecule are water-soluble (i.e., hydrophilic) and negatively charged, whereas the ceramide portion is lipid-soluble (hydrophobic). The latter appears to be embedded in the membrane lipids, whereas the hydrophilic section, with its charged units, protrudes externally toward the medium. The gangliosides can therefore serve as specific membrane-binding sites (receptor sites) for circulating hormones and thereby influence various processes within the cell.

Glycosphingolipids undergo continual metabolic degradation and synthesis. Their biosynthesis involves the stepwise addition of the individual sugars via their nucleotide diphosphate derivatives; each step is catalyzed by an enzyme (glycosyltransferase) that is specific for its substrate. Similarly, catabolic processes operate via sequential hydrolysis of the sugar moieties by specific lysosomal glycosidases and neuraminidases. The absence of a specific degradative enzyme results in the accumulation of the involved glycosphingolipid or one of its intermediates. The lack of one of these enzymes is essentially the cause of a number of inborn errors of metabolism termed *sphingolipidoses*. Deficiencies of the enzymes involved in synthesis are also known; however, consideration of these requires knowledge of the details of the biosynthetic steps, which are not covered in this discussion.

The common basis of the lesions that result in sphingolipidoses may be understood by considering a representative disorder, e.g., Tay-Sachs disease. The ganglioside affected is G_{M2}, which has been shown already diagrammatically and is shown in detail as follows:

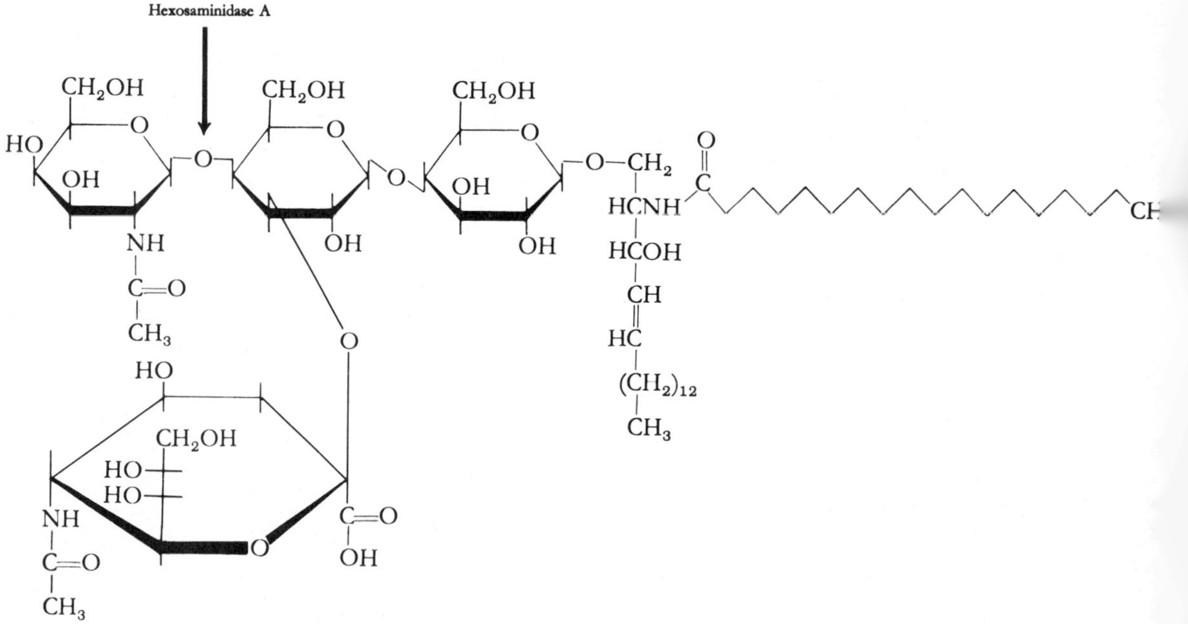

G_{M2}

Table 5-2. Lysosomal Sphingolipidoses

Disorder	Deficient Enzyme	Affected Substrate
Tay-Sachs disease	Hexosaminidase A	G_{M2} ganglioside
Sandhoff's disease	Hexosaminidase A and B	G_{M2} ganglioside
Fabry's disease	α-Galactosidase	Trihexosylceramide
Krabbe's disease	β-Galactosidase	Galactosylceramide
Lactosylceramidosis	β-Galactosidase	Lactosylceramide
G_{M1} gangliosidosis	β-Galactosidase	G_{M1} ganglioside
Gaucher's disease	β-Glucosidase	Glucosylceramide
Niemann-Pick disease	Sphingomyelinase	Sphingomyelin
Farber's disease	Ceramidase	Ceramide

Normal degradation of this ganglioside requires the action of a hydrolyzing enzyme, *hexosaminidase A,* which removes the terminal N-acetylgalactosamine residue at the site shown in the structure. Subsequently, the other components are hydrolyzed by specific enzymes. In Tay-Sachs disease, hexosaminidase A is absent, and as a consequence, G_{M2} accumulates in the affected tissues. The high concentrations of G_{M2} in the ganglia of the cerebral cortex produce abnormalities that result in mental retardation, blindness, and ultimately the death of the diseased individual before he or she reaches the age of three.

Since the breakdown of glycosphingolipids occurs in the lysosomal compartment of cells, these diseases and similar aberrations of polysaccharide hydrolysis (page 290) are classified as *disorders of lysosomal metabolism*. A number of the diseases involving glycosphingolipids are listed in Table 5-2.

Most of the sphingolipidoses are caused by autosomal recessive genetic defects. One notable exception is Fabry's disease, which is sex-linked through the X chromosome. In this case, only the mother can be a carrier in transmitting the defect to her male offspring. Unlike Tay-Sachs, Gaucher's, or Niemann-Pick disease, mental retardation does not occur in patients with Fabry's disease. Accumulation of the trihexosylceramide causes impairment of kidney function and abnormalities of the heart and eyes. The patients may live thirty or forty years. Many of the diseases can be detected prenatally by the analysis of cells obtained from amniotic fluid for the desired enzyme. Among the disorders that can be diagnosed in this way are Tay-Sachs, Gaucher's, Fabry's, and Niemann-Pick diseases. In addition, certain deficiencies can be detected in heterozygotes by assays of their levels of the specific enzyme. Theoretically, these individuals should have enzyme concentrations that are intermediate between those of normal and diseased individuals. This is essentially the basis of the screening program for Tay-Sachs disease.

Suggested Reading

Bloch, K. The Biological Synthesis of Cholesterol. *Science* 150:19–28, 1965.

Brady, R. Q. Inherited Metabolic Diseases of the Nervous System. *Science* 193:733–739, 1976.

Fishman, P. H., and Brady, R. Q. Biosynthesis and Function of Gangliosides. *Science* 194:906, 1976.

Gal, A. E., Brady, R. Q., Hibbert, S. R., and Pentchev, P. G. Sphingolipidoses. *New England Journal of Medicine* 293:632–636, 1976.

Volpe, J. J., and Vagelos, P. R. Saturated Fatty Acid Biosynthesis and Its Regulation. *Annual Review of Biochemistry* 42:21–60, 1973.

6. Tissue Disposition and Transport of Carbohydrates and Lipids

The three foregoing chapters dealt with discrete pathways in the metabolism of carbohydrates and lipids from the point of view of the interconversions of the metabolites. The primary emphasis was on basic reactions that occur within cells and the sources and disposition of various intermediates. This chapter will be concerned with some metabolic transformations that are characteristic for certain tissues or organs as well as with the integration of these reactions in the functioning of the total organism.

Since many tissues of higher animals are specialized with respect to their metabolic capabilities, they are dependent on each other for various intermediates and substrates. In addition, specific pathways for the disposition of certain products may be unique for certain tissues. Thus, a given nutrient that is stored in one tissue may be utilized and degraded in another, and some of the intermediates may be processed further in a third tissue.

In order for the animal to accomplish all the necessary metabolic transformations, various substances must be transported from one organ to another. This function is performed by the circulatory system. The components of the bloodstream are thus products of the activities of all the tissues. Various changes in the blood are therefore a reflection of transformations that occur in cells of different organs. Additionally, since a number of the parameters of blood have to be kept within fairly narrow limits for normal functioning of the organism, efficient mechanisms for regulation must be available. Several aspects of this interplay between tissues and the circulatory systems will be considered in the context of this chapter.

6.1. Ketone Bodies: Formation and Disposition

Complete metabolic degradation of fatty acids yields acetyl-coenzyme A, which in turn is oxidized to carbon dioxide and water via the citric acid cycle (Chap. 3, sections 3.2 and 3.4). For example, the ultimate transformation of the four-carbon fatty acyl-coenzyme A, butyryl-coenzyme A, involves the following sequence:

$$CH_3CH_2CH_2CO-S-CoA + FAD \longrightarrow CH_3CH=CHCO-S-CoA + FADH_2$$
Butyryl-CoA $\qquad\qquad\qquad\qquad\qquad$ Crotonyl-CoA

$$CH_3CH=CHCO-S-CoA + H_2O \longrightarrow CH_3CHOHCH_2CO-S-CoA$$
Crotonyl-CoA $\qquad\qquad\qquad\qquad\qquad$ β-Hydroxybutyryl-CoA

$$CH_3CHOHCH_2CO-S-CoA + NAD^+ \longrightarrow CH_3COCH_2CO-S-CoA + NADH + H^+$$
β-Hydroxybutyryl-CoA $\qquad\qquad\qquad\qquad\qquad$ Acetoacetyl-CoA

$$CH_3COCH_2CO-S-CoA + HS-CoA \longrightarrow 2CH_3CO-S-CoA$$
Acetoacetyl-CoA $\qquad\qquad\qquad\qquad\qquad$ Acetyl-CoA

Although the principal reaction of acetoacetyl-coenzyme A is thiolysis to 2 moles of acetyl-coenzyme A, a significant fraction is also converted to free acetoacetate. Several processes within the liver lead to the formation of acetoacetate from acetoacetyl-coenzyme A; one of these, which is of lesser importance, is a direct hydrolysis:

$$CH_3COCH_2CO-S-CoA + H_2O \longrightarrow CH_3COCH_2COOH + HS-CoA$$
Acetoacetyl-CoA $\qquad\qquad\qquad\qquad\qquad$ Acetoacetate

Another process that effectively brings about the formation of free acetoacetate involves a sequence of two reactions. Acetoacetyl-coenzyme A condenses with a molecule of acetyl-coenzyme A to yield β-hydroxy-β-methylglutaryl-coenzyme A. The latter then undergoes a cleavage that provides acetyl-coenzyme A and free acetoacetate:

$$O=\overset{\underset{\textstyle S-CoA}{|}}{C}CH_3 + \overset{\overset{\textstyle O}{\|}}{C}CH_2CO-S-CoA \longrightarrow HS-CoA + HOOCCH_2\overset{\underset{\textstyle CH_3}{|}}{\overset{\textstyle OH}{C}}CH_2CO-S-CoA$$
$\qquad\qquad\qquad\quad$ CH$_3$
Acetyl-CoA $\quad$ Acetoacetyl-CoA $\qquad\qquad\qquad\qquad\qquad$ β-Hydroxy-β-methylglutaryl-CoA

$$HOOCCH_2\overset{\underset{\textstyle CH_3}{|}}{\overset{\textstyle OH}{C}}CH_2CO-S-CoA \longrightarrow HOOCCH_2\overset{\underset{\textstyle CH_3}{|}}{\overset{\overset{\textstyle O}{\|}}{C}} + CH_3CO-S-CoA$$
β-Hydroxy-β-methylglutaryl-CoA $\qquad$ Acetoacetate $\quad$ Acetyl-CoA

(Note that β-hydroxy-β-methylglutaryl-coenzyme A is also an intermediate in the biosynthesis of cholesterol; see Chap. 5, section 5.5.A.)

Free acetoacetate formed by either of the two mechanisms outlined can be reduced to β-D-hydroxybutyrate by the NAD^+-linked enzyme *β-hydroxybutyrate dehydrogenase*:

$$CH_3COCH_2COOH + NADH + H^+ \longrightarrow CH_3CHOHCH_2COOH + NAD^+$$
Acetoacetate   β-Hydroxybutyrate

Alternatively, acetoacetate may undergo decarboxylation to yield acetone:

$$CH_3COCH_2COOH \longrightarrow CH_3COCH_3 + CO_2$$
Acetoacetate   Acetone

These three products—acetoacetate, β-hydroxybutyrate, and acetone—are formed in the mitochondria of liver and are called *ketone bodies*. Acetoacetate cannot be metabolized further in the liver; however, it can be utilized for energy by skeletal or heart muscle and, to a limited extent, by brain. As ketone bodies are formed in the liver, they are released into the bloodstream and taken up by tissues that have the necessary enzymes for their activation and oxidation.

Extrahepatic tissues—e.g., muscle or kidney—contain enzymes that catalyze the formation of acetoacetyl-coenzyme A from acetoacetate. One reaction that brings about this activation of acetoacetate involves a specific thiokinase. It utilizes adenosine triphosphate and coenzyme A and yields acetoacetyl-coenzyme A, adenosine monophosphate, and inorganic pyrophosphate:

$$CH_3COCH_2COOH + ATP + HS\text{—}CoA \longrightarrow CH_3COCH_2CO\text{—}S\text{—}CoA + AMP + PP_i$$
Acetoacetate   Acetoacetyl-CoA

Another reaction that forms activated acetoacetate is the transfer of coenzyme A from succinyl-coenzyme A to acetoacetic acid; this is catalyzed by *thiophorase*:

$$CH_3COCH_2COOH + \underset{\underset{CH_2COOH}{|}}{CH_2CO\text{—}S\text{—}CoA} \rightleftharpoons CH_3COCH_2CO\text{—}S\text{—}CoA + \underset{\underset{CH_2COOH}{|}}{CH_2COOH}$$

Acetoacetate   Succinyl-CoA   Acetoacetyl-CoA   Succinate

(The other ketone body, β-D-hydroxybutyrate, can be oxidized to acetoacetate by the NAD^+-linked dehydrogenase. The third product, acetone, is exhaled through the lungs.)

Once it is re-formed, acetoacetyl-coenzyme A can be converted to acetyl-coenzyme A (thiolase catalyst), and the acetyl-coenzyme A may then be oxidized via the citric acid cycle. Liver does not contain a system for activating acetoacetate, and thus it has no means for utilizing this intermediate.

As a result of the regular formation of ketone bodies within the liver and their subsequent transport and utilization by extrahepatic tissues, there are generally small amounts of these materials in the circulatory system. Under normal conditions, blood plasma contains about 0.3 to 2 milligrams of acetoacetate and β-hydroxybutyrate per 100 milliliters. Of the three ketone bodies, acetone is the least prominent. The amount of ketone bodies normally present in the bloodstream represents a steady state arising from the synthesis of free acetoacetic acid and its conversion to acetoacetyl-coenzyme A. In other words, the rate of production of acetoacetate in the liver is comparable to its rate of utilization by peripheral tissues. Under certain conditions, however, this relationship does not prevail; i.e., the rate of formation of acetoacetate becomes excessive. This occurs during prolonged starvation or glucose deprivation, as well as in untreated diabetics.

In situations in which the production of acetoacetic and β-hydroxybutyric acids in the liver exceeds the capacity of extrahepatic tissues for their utilization, the concentration of ketone bodies in the blood is increased, a condition termed *ketonemia*. Considerable amounts of these substances are then excreted in the urine (ketonuria). As much as 5 grams of ketone bodies have been reported to be excreted per day in the urine of uncontrolled diabetics.

The overall condition arising from an overproduction of ketone bodies is termed *ketosis*. The primary reason for the difficulties that arise from elevated levels of ketone bodies is the acidic nature of these substances. The pK_a for acetoacetic acid is 3.8, and for β-hydroxybutyric acid, it is 4.8. Since the pH of normal plasma is 7.4, these acids circulate in their ionized forms. When excreted in the urine, they are eliminated together with sodium ions. This loss of cations from the blood leads to a decrease in blood pH (metabolic acidosis) and to a loss of water from the system (dehydration). If it is not corrected, the situation results in a comatose state and the eventual death of the individual.

The common denominator in the conditions that lead to ketosis is a deprivation of glucose and a comparatively high utilization of fatty acids for energy. After extended starvation, glycogen reserves are depleted and the organism must depend on fatty acids for energy. In diabetics, either the levels of insulin are low or the activity of insulin is inhibited. As a result, glucose becomes unavailable to tissues that require the hormone for transport functions. As the intracellular carbohydrates are decreased, the utilization of fatty acids is enhanced. Insulin also functions to decrease the rate of hydrolysis of triglycerides to glycerol and fatty acids (section 6.4). Therefore, when insulin activity is minimal, greater amounts of free fatty acids become available to the liver. Thus, both during starvation and in the diabetic, degradation of fatty acids in the liver is enhanced. Concomitant with the increased catabolism of fatty acids is a more rapid formation of ketone bodies. The rate of release of ketone bodies from the liver may become much greater than the rate at which they can be utilized by peripheral tissues, and thus the ketotic condition ensues.

Additional insight into the reason for the accumulation of acetoacetate may be gained by consideration of the possible modes of disposition of acetyl-coenzyme A that are shown in Figure 6-1. One normal alternative is the utilization of acetyl-coenzyme A to produce fatty acids (pathway I of Fig. 6-1). This process, however, requires NADPH. Since the major NADPH-yielding reaction involves the oxidation of glucose by the oxidative pentose pathway, then in the absence of carbohydrates the biosynthesis of lipids is minimal. Furthermore, the storage of fatty acids involves the production of glycerol esters. Glycerol phosphate, which is necessary for this purpose, is also derived from glucose.

Pathway II of Figure 6-1, which involves the formation of cholesterol, is also dependent on the availability of NADPH (hydroxylation) and may be decreased in the absence of glucose. The third possibility, the formation of citrate (pathway III), requires the presence of oxaloacetate. Although the latter can be derived from amino acids, it is also formed from a glycolytic intermediate (pyruvate). Since the utilization of acetyl-coenzyme A in these pathways is inhibited to a significant extent when the amount of glucose is limited, a fourth route—the formation of acetoacetate—becomes most available.

Although these arguments serve to explain the formation of ketone bodies, it should be realized that the basic mechanism for the metabolic shift is extremely complex and that some of the explanations are open to criticisms. However, the fundamental concept that emerges from these considerations is that a certain amount of carbohydrate is necessary for the efficient metabolism of fatty acids.

6.2. Lactate Metabolism
A. CORI CYCLE
The cells of muscle tissue can derive energy from glucose by converting it to pyruvate (glycolysis) and then oxidizing the latter via the mitochondrial system. In the process, a small amount of the pyruvate is also reduced to lactate:

$$\text{Pyruvate} + \text{NADH} + \text{H}^+ \xrightleftharpoons{\text{Lactate dehydrogenase}} \text{Lactate} + \text{NAD}^+$$

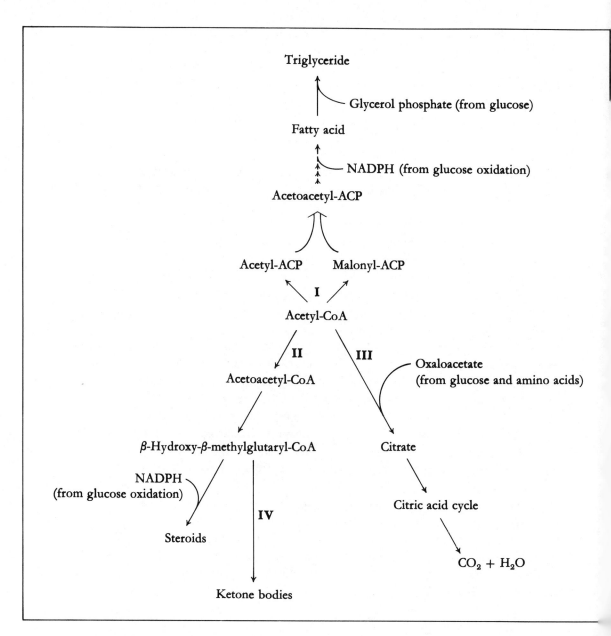

Fig. 6-1. Modes of utilization of acetyl-CoA showing the requirements for carbohydrate. (Pathways are indicated by boldface roman numerals.)

The lactate diffuses out of the cell and appears in the bloodstream. As a result, the blood of a normal individual at rest contains about 10 milligrams of lactate per 100 milliliters.

During exercise or sudden exertion, the amount of lactate produced by muscle increases considerably. This is reflected in the levels of blood lactate, which may rise to as high as 80 milligrams/100 milliliters. The reason for the increase in production of lactate is the acute demand for energy that occurs during muscle activity. This is met by an enhancement in the rate of cellular glucose consumption. However, the amount of oxygen generally available to muscle tissue is not high enough to oxidize all the pyruvate that is formed under these conditions. A considerable fraction of the excess pyruvate is therefore reduced to lactate. In addition to removing the pyruvate, the production of lactate serves to regenerate oxidized NAD^+, which is required for the maintenance of the glycolytic sequence. Although the total oxidation of glucose yields considerably higher amounts of energy than does anaerobic glycolysis, the latter process is geared to providing energy more rapidly since it is not dependent on the supply of oxygen.

Once formed, lactate cannot be metabolized unless it is reoxidized to pyruvate. Additionally, lactate that is taken up in the liver can be utilized for the synthesis of glucose and glycogen (gluconeogenesis). Glucose 6-phosphate, which is formed as a result of this process, can be hydrolyzed to glucose by the action of liver glucose 6-phosphatase. Thus, lactate formed in muscle can be converted to glucose within the liver. When reintroduced into the circulation, the glucose can be taken up by muscle and again be degraded to lactate. This cyclic series of reactions (Fig. 6-2) is known as the *Cori cycle*.

This interplay between skeletal muscle and liver in the metabolism of carbohydrates may become a critical factor when muscle activity is maximal and the supply of glucose from external sources is minimal (postabsorptive states). Under these conditions, the liver serves to maintain a constant level of glucose in the circulation via gluconeogenesis from lactate or other intermediates.

The oxygen deficit created by intense muscular exercise is normally compensated for by an immediate increase in breathing rate. This provides the necessary oxygen for lactate disposal via pyruvate and the citric acid cycle. Complete oxidation of pyruvate in the mitochondria provides for the replenishment of the ATP expended in muscle and liver.

B. ISOZYMES OF LACTATE DEHYDROGENASE

The venous blood from skeletal muscle generally has higher concentrations of lactate than does arterial blood. With heart muscle, the situation is reversed, i.e., venous blood contains less lactate than does arterial blood. This

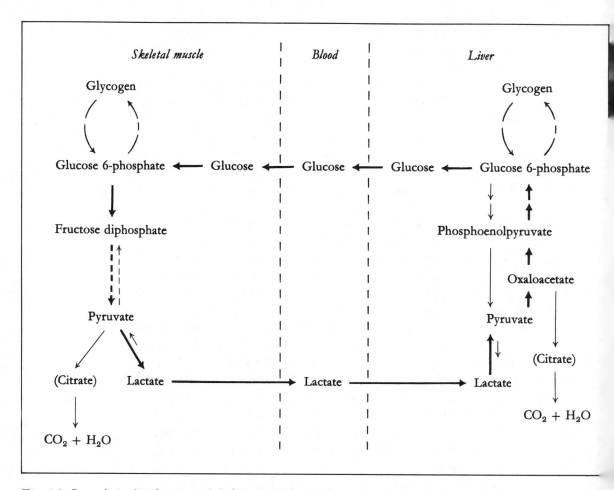

Fig. 6-2. Interrelationships between carbohydrate metabolism in liver and skeletal muscle. The Cori cycle can be followed by the darkened arrows.

phenomenon leads to the conclusion that whereas skeletal muscle generates lactate, cardiac muscle can consume it. In fact, it has been found that lactate can serve as an efficient energy source for heart muscle.

In terms of energy requirements, there is an essential difference between skeletal and heart muscle. The latter operates continuously and at a fairly steady level, whereas the activity of skeletal muscle is discontinuous and fluctuates according to momentary needs. Thus, cardiac muscle obtains energy from the aerobic oxidation of carbohydrates, fatty acids, lactate, and acetoacetate. The same substances and pathways can also serve to yield energy in skeletal muscle; however, this tissue is also equipped with mechanisms to cope with sharp, instantaneous demands for energy. One process whereby this is accomplished is the efficient and rapid derivation of energy from glucose by a pathway that is not limited by oxygen supplies, i.e., glycolysis to lactate.

An important factor controlling the degree of formation and oxidation of lactate is the nature of the enzyme that catalyzes the reaction, lactate dehydrogenase (LDH). Indeed, it was found that the lactate dehydrogenase of skeletal muscle differs in certain properties from the lactate dehydrogenase of heart muscle. The two forms of the enzyme can be distinguished by their electrophoretic mobilities, antigenic specificities, and catalytic characteristics. In addition, there are differences in certain aspects of their amino-acid composition. Enzymes that catalyze the same interconversion but exist in different forms in the same animal are called *isozymes*. A number of other enzymes also occur as isozymes; examples are phosphorylase, α-amylase, hexokinase, and alkaline phosphatase.

The molecular basis for the variation in the isozymes of lactate dehydrogenase is the difference in the subunit components of the protein. The enzymes from different tissues are composed of four subunits with independent catalytic sites. Thus, heart lactate dehydrogenase, which has a molecular weight of 140,000, can be separated into four inactive subunits, or monomers, with molecular weights of 35,000 each. The enzyme from heart muscle has a characteristic monomer that associates to form the active tetramer. If the individual subunit is designated as H, then the heart lactate-dehydrogenase tetramer can be denoted as $HHHH$. A similar situation occurs with lactate dehydrogenase from muscle. Four monomer units, termed M units, combine to form the active tetramer $MMMM$. The structures of M and H differ from each other in some aspects of their amino-acid composition. In addition to the homogeneous types, there are three hybrid forms of lactate dehydrogenase: $MHHH$, $MMHH$, and $MMMH$. The five isozymes may also be designated as M_4, M_3H, M_2H_2, MH_3, and H_4.

The M_4 isozyme of lactate dehydrogenase has a higher turnover number than does the H_4 isozyme and thus shows considerably greater catalytic activity. The heart enzyme is inhibited by excess pyruvate, and it is therefore not geared for catalyzing the reduction of pyruvate efficiently. These properties are correlated with the physiologic roles of their respective isozymes. In cardiac muscle, where the rate of metabolism is fairly constant, pyruvate is oxidized by the aerobic mitochondrial system. NADH, which is produced in glycolysis, is reoxidized by respiratory mechanisms, and the requirement for pyruvate reduction is relatively small. The primary function of lactate dehydrogenase in cardiac muscle is the oxidation of lactate to pyruvate. Metabolism in skeletal muscle, however, requires an isozyme that is more active in reducing pyruvate and is comparatively uninhibited by an excess of this substrate; these requirements are met by the characteristics of the M_4 isozyme.

Table 6-1. Average Concentration (Percentage) of LDH
Isozymes in Human Tissues

Tissue	H_4	MH_3	M_2H_2	M_3H	M_4
Skeletal muscle	4	7	17	17	55
Cardiac muscle	60	30	7	2	1
Liver	2	6	14	13	65
Erythrocytes	44	46	8	1	1

The distribution of isozymes in different tissues is of interest both from a theoretical point of view and regarding possible clinical applications. From Table 6-1, it can be seen that although the predominant isozymes in cardiac muscle are H_4 and MH_3, there are significant amounts of other forms. Similarly, skeletal muscle and liver contain small amounts of isozymes with H subunits in addition to the predominant forms. In the embryo of the human, the primary subunit is the H monomer, even in skeletal muscle. As the individual develops, the relative amount of the M type increases. It appears that the syntheses of each of the two subunits are programmed by different genes and that the hybrid patterns found in various tissues appear as a result of differing degrees of synthesis of the individual monomers.

The lactate dehydrogenase in serum is derived from the cells of various tissues. However, the principal sources are liver, skeletal muscle, and erythrocytes. Changes from the normal serum isozyme pattern occur under pathologic conditions when the permeability of certain cells is increased. Thus, following myocardial infarction, there may be an increase of the H isozyme in serum. Patients with prostatic carcinoma show a rise in the level of M_4 and M_3H isozymes. Serum isozyme patterns of numerous diseases have been investigated, and in certain cases, these patterns can be used as diagnostic tests.

6.3. Blood Glucose and Its Regulation

Since carbohydrate is a critical component for many aspects of metabolism, it is vital that a steady supply be available to all the cells. This is accomplished by the maintenance of a relatively constant concentration of glucose—an average concentration of about 90 milligrams per 100 milliliters (5×10^{-3} M)—in the circulatory system. After food intake, the concentration may rise to as much as 130 milligrams per 100 milliliters; however, this decreases to fasting levels within approximately 2 hours. The principal organ involved in the regulation of blood glucose is the liver. This is accomplished by intracellular enzymatic control processes, which in turn are highly sensitive to hormonal action.

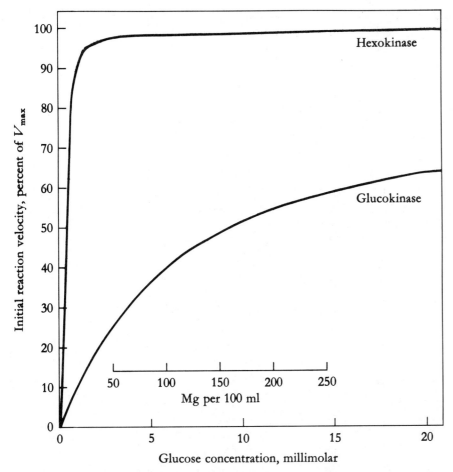

Fig. 6-3. Effects of glucose concentration on the rates of the hexokinase-catalyzed and glucokinase-catalyzed reactions. The related concentrations of glucose in milligrams per 100 milliliters are indicated on a separate scale.

One of the mechanisms through which the liver buffers sharp fluctuations in circulating glucose is based on its rate of intracellular conversion of glucose to glucose 6-phosphate. All tissues contain hexokinase; however, the activity of this enzyme is not sensitive to the usual variations in blood glucose. The K_M for hexokinase is about 4×10^{-5} M, so the enzyme will be at maximal activity even at the lowest possible blood glucose concentrations. The range of possible fluctuations in the concentration of circulating glucose is such that it would have no effect on the rate of glucose phosphorylation as catalyzed by hexokinase (Fig. 6-3). However, in addition to hexokinase, liver contains another glucose-phosphorylating enzyme, *glucokinase*. This enzyme catalyzes the same reaction as hexokinase, but certain of its characteristics endow it with important regulatory functions. Glucokinase has a comparatively low affinity for glucose (its K_M is approximately 10^{-2} M), it is not inhibited by glucose 6-phosphate, and its activity decreases to a minimum during starvation. The value of the K_M for glucokinase is such that the enzyme is not saturated by glucose in the normal blood glucose concentration range (see Fig. 6-3). Its activity would thus vary with the usual changes in circulating glucose levels. As the amount of glucose rises, glucokinase activity increases, and there is a consequent uptake of glucose and conversion to glucose 6-phosphate. When glucose levels decrease, glucokinase activity decreases, so the glucose is removed from the circulation by the liver less rapidly.

After extended periods of fasting or carbohydrate deprivation, the amount of glucokinase in liver becomes minimal. The enzyme appears to have a comparatively high turnover rate, and glucose or insulin seems to be required for its biosynthesis. When carbohydrate is decreased and insulin activity is low, the available glucokinase is insufficient to remove glucose effectively, and the enzyme is unable to exert significant regulatory action. It is of interest in this connection that in untreated diabetics, glucokinase activity is minimal. This may be one of the reasons for the failure to control blood sugar levels in this disease.

Glucokinase may also be an important enzyme in the metabolic pathway from glucose to glycogen. Increases in the levels of glucose 6-phosphate result in the inhibition of hexokinase, yet glucose 6-phosphate is required for the action of glycogen synthetase D (Chap. 4, section 4.7). Thus, maximal phosphorylation of glucose requires that glucose 6-phosphate be removed; however, as glucose 6-phosphate is removed, the subsequent step is prevented, i.e., the formation of glycogen will be diminished. In this anomalous situation, a cofactor required for a later step in a pathway acts as an inhibitor of an earlier step. This difficulty is overcome by the existence of glucokinase, which is not inhibited by glucose 6-phosphate. Thus, in addition to removing glucose by phosphorylation, glucokinase functions in the ultimate disposal of the removed glucose by facilitating its conversion to glycogen.

The liver also contains the enzyme *glucose 6-phosphatase*, which catalyzes the hydrolysis of glucose 6-phosphate to glucose and inorganic phosphate. Through the use of this enzyme, the liver functions to release glucose into the bloodstream and thus to replenish glucose supplies, even in the absence of exogenous contributions. In addition to the glucose 6-phosphate formed from glucose or glycogen, it will be remembered that this intermediate can be formed in liver as a result of gluconeogenesis from lactate, glycerol, and a number of amino acids.

6.4. Circulation and Mobilization of Lipids

In the postabsorptive state, which occurs about 12 hours after eating, the concentration of lipid in plasma is about 500 milligrams per 100 milliliters.

This lipid material is composed of approximately 125 milligrams of triglyceride per 100 milliliters of plasma, 180 milligrams of cholesterol per 100 milliliters (two-thirds of which is esterified), 180 milligrams of phospholipid per 100 milliliters (most of which is lecithin), and 15 milligrams of free fatty acids per 100 milliliters. Except for the last (which is bound to albumin), all the lipids circulate as components of lipoproteins. The levels for the plasma lipids given are only average values, and wide ranges in concentration occur even in normal individuals. This is especially true for the triglycerides and free fatty acids. For example, the concentration of triglycerides in normal individuals varies from about 30 to 160 milligrams per 100 milliliters of plasma, and the amount of free fatty acid fluctuates from 5 to 25 milligrams per 100 milliliters.

In a number of conditions, the lipid levels are higher than the usual range. These elevations may be due to direct aberrations of lipid metabolism, or they may result from organic diseases. Disorders in which plasma lipids are elevated are called *lipidemias*.

Plasma lipids arise from several sources, which include lipid absorption from the alimentary tract and lipid release from the liver and from adipose tissue. The characteristics and modes of transport of plasma lipids depend on their source.

Digested lipids are absorbed through the intestinal mucosa of the microvilli in the proximal jejunum. In the cisternae of the smooth endoplasmic reticulum, the triglycerides, cholesterol, and cholesterol esters are combined with phospholipids, small amounts of fatty acids, fat-soluble vitamins, and protein to form the *chylomicrons*. These are released into the intestinal lymph and then find their way into the thoracic duct and systemic circulation. The increase in the amount of plasma chylomicrons after a fatty meal is easily discerned by the turbid appearance of the blood. This situation, which is termed *alimentary lipemia,* is transitory due to the action of lipoprotein lipase and the processing of chylomicrons by the liver.

Lipoprotein lipase, also known as *clearing factor,* catalyzes the hydrolysis of the triglycerides in chylomicrons and lipoproteins to form fatty acids and glycerol. As a result of the action of this enzyme on the triglyceride component, the chylomicrons are decomposed, and the plasma loses its turbidity. The enzyme has almost no effect on free triglycerides. Lipoprotein lipase appears to be bound to the capillary endothelium in various tissues and thus can exert its action on circulating chylomicrons. This enzyme occurs in adipose tissue, liver, heart muscle, and a number of other tissues.

The turbidity of plasma arising after a fatty meal has been found to disappear almost instantaneously after an injection of heparin. The effectiveness of the latter in the "clearance" of alimentary lipemia has been attributed to its action on the release of bound lipoprotein lipase into the bloodstream.

Experimental results also suggest that heparin may be linked to lipoprotein lipase or may enhance the binding of the enzyme with chylomicrons.

In addition to the triglycerides that circulate with the chylomicrons, a considerable amount of triglyceride is found in the structure of the very-low-density or pre-beta lipoproteins, which will be defined later in this section. These lipoproteins are formed in the liver from the fatty acids and glycerol that are released from chylomicrons. By this action, the liver essentially functions to convert part of the chylomicron components to pre-beta lipoproteins, which are reintroduced into the bloodstream. The triglycerides of the pre-beta lipoproteins are also hydrolyzed by lipoprotein lipase.

The free fatty acids that are released by the action of lipoprotein lipase on the chylomicrons and pre-beta lipoproteins become bound to plasma albumin. They are circulated in this form and are taken up by the cells of different tissues. There, these fatty acids may be degraded and utilized for energy (in all tissues except nerve), converted to glycerides and stored (in adipose tissue), or utilized for the synthesis of phospholipids and lipoproteins (primarily in the liver). The glycerol formed by lipoprotein lipase hydrolysis can be converted in liver to glycerol phosphate and metabolized, or it may be used to form triglycerides. Liver also synthesizes families of lipoproteins termed *high-density* and *low-density lipoproteins* (see the following discussion), which serve as carriers for cholesterol and phospholipid. These lipoproteins are released from the liver and contribute significantly to the pool of circulating lipids.

The fatty acids that are stored as triglycerides in adipose tissue are made available to all the other tissues by a process called *lipid mobilization* (see also section 6.4.B). This process involves the lipase-catalyzed hydrolysis of the triglycerides to glycerol and free fatty acids, and the release of these components into the circulatory system. The lipase that functions in the intracellular hydrolysis of triglycerides is distinct from lipoprotein lipase and is usually termed *tissue lipase*. Lipid mobilization is enhanced by epinephrine and inhibited by insulin. The mechanism of the action of epinephrine in this case is analogous to its effect on glycogen phosphorylase. The sequence involves the activation of adenyl cyclase and the resultant formation of cyclic AMP. The latter brings about the conversion of lipase from an inactive to an active form. Insulin, which has an inhibitory effect on cyclic AMP, will essentially decrease the hydrolysis of triglycerides. Other hormones and intermediates—such as glucagon, ACTH, corticoids, thyroxine, norepinephrine, vasopressin, and 5-hydroxytryptamine (serotonin)—also effectively activate tissue lipase. Prostaglandins, which antagonize the action of epinephrine, cause an inhibition of the action of tissue lipase.

In the postabsorptive state, the circulating free fatty acids are derived almost exclusively from adipose-tissue triglycerides. The fatty acids are taken up by muscle and other tissues, where they are metabolized. In addition to degrading fatty acids for the derivation of energy, the liver converts these substances to triglycerides, phospholipids, and lipoproteins. Thus, whereas adipose tissue contributes lipids to the bloodstream as free fatty acids, the liver releases lipids in the form of lipoproteins.

A. LIPOPROTEINS

Since lipids are insoluble in water, they would not normally harmonize with biological fluids, which are essentially aqueous systems. This poor solubility is obviated inasmuch as the neutral lipids are associated with polar phospholipids and proteins in structures called *lipoproteins*. The individual components of lipoproteins are held together by noncovalent bonds, and the protein and phospholipid portions are concentrated on the periphery of the macromolecules. In the plasma, there are numerous types of lipoproteins, which vary according to their lipid-to-protein ratios and in the composition of the lipid and protein components. The two principal modes of classification of these substances are based on their electrophoretic mobility and on their density. First, each of these categories will be defined, and then the overall relationships between them will be described.

1. *Electrophoretic Mobility*

Electrophoresis of serum proteins at pH 8.6 results in their separation into albumin and α-, β-, and γ-globulins (Chap. 1, section 1.5). Of these, albumin has the greatest mobility toward the anode. These fractions are not homogeneous proteins but mixtures that have similar electrophoretic mobilities. The α- and β-globulin fractions contain a certain amount of lipoproteins; these are termed *α-lipoproteins* and *β-lipoproteins*. When the electrophoretic procedure is performed on a supporting medium, such as filter paper, cellulose acetate, or agarose gel, the proteins can be separated into distinct zones (Fig. 1-18, page 40). Analyses with lipid stains (e.g., Oil Red O) reveals the presence of lipoproteins in the α_1- and β-globulin sections. An additional class of lipoproteins, designated as pre-β-lipoproteins, can also be observed (Fig. 6-4). These lipoproteins differ from each other in the percentage and composition of lipids, as well as in the nature of the protein components (Tables 6-2 and 6-3). For example, the average values for the percentages of lipid in α-, β-, and pre-β-lipoproteins are 50, 75, and 90, respectively. The lipids in pre-β-lipoproteins are primarily triglycerides. In contrast, the principal lipid components in β-lipoproteins are cholesterol and its ester derivatives. α_1-Lipoproteins contain relatively high amounts of phospholipids (lecithins and sphingomyelins) and considerable amounts of cholesterol plus cholesterol esters.

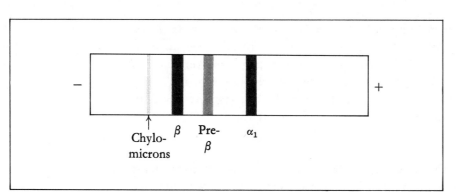

Fig. 6-4. Separation of plasma lipoproteins by agarose gel electrophoresis. The arrow shows the site of application of the sample. Chylomicrons remain at this point.

Table 6-2. Plasma Lipoprotein Classes

Properties and Composition	Chylo-microns	Lipoprotein Fractions		
		VLDL	LDL	HDL
Density, grams/milliliter	<0.95	0.95–1.006	1.006–1.063	1.063–1.210
S_f (Svedbergs)	>400	12–400	0–12	—
Electrophoretic mobility	Origin	Pre-β	β	α_1
Average composition (%)				
Protein	2	10	20	45
Total lipid	98	90	80	55
Triglyceride	88	55	8	10
Phospholipid	6	20	24	22
Cholesterol, total	4	15	48	23
Free	1	5	8	6
Esterified	3	10	40	17

Table 6-3. Apoprotein Components of Plasma Lipoproteins*

Chylomicrons	VLDL	LDL	HDL
ApoA-I	ApoB	ApoB	ApoA-I
ApoB	ApoC-I	(ApoC)	ApoA-II
ApoC-I	ApoC-II		(ApoC-I)
ApoC-II	ApoC-III		(ApoC-II)
ApoC-III	ApoE		(ApoC-III)
(ApoA-II)	(ApoA-I)		(ApoD)
(ApoE)	(ApoA-II)		(ApoE)
	(ApoD)		

* The apoproteins written in parentheses, are minor components of the indicated class of lipoproteins. (Reproduced with permission, from L. C. Smith, H. J. Pownall, and A. M. Gotto, Jr., *Annu. Rev. Biochem.*, Vol. 47. © 1978 by Annual Reviews Inc.)

Electrophoresis on paper or cellulose acetate is commonly employed for clinical analyses of lipoproteins. Various conditions that can be diagnosed by these techniques are described in section 6.9 of this chapter.

2. Density Fractionation

Since the lipids have comparatively low densities, the lipid components in lipoproteins effectively decrease the densities of the total macromolecular complexes. The densities of lipoproteins are thus inversely related to their lipid percentage. Various classes of lipoproteins can, therefore, be separated on the basis of the differences in their densities.

Plasma obtained after ingestion of lipids generally has a somewhat turbid appearance. When this plasma is centrifuged, the material that causes this turbidity separates to the top of the centrifuge tube. This top oily layer contains the *chylomicrons* (page 265). They have a density of less than 1.006 grams/milliliter, as compared with plasma, which has an average density of 1.03 grams/milliliter. By adjusting the density of the remaining plasma to different values and centrifuging the resultant solutions, various fractions with densities lower than the solutions can be separated to the top of the tube. These fractions are classified as *very-low-density lipoproteins* (VLDL), *low-density lipoproteins* (LDL), and *high-density lipoproteins* (HDL).

These lipoprotein fractions can be characterized further by ultracentrifugal analyses according to their sedimentation constants or S values (page 38). Unlike most macromolecules that travel to the bottom of the centrifuge cell during centrifugation, LDL and VLDL rise to the top of the solution. Such lipoproteins are characterized, according to the rates at which they migrate to the top of the centrifuge cell, under standardized conditions. These rates are expressed in *Svedberg flotation units,* or S_f. Accordingly, very-low-density lipoproteins have S_f values from 12 to 400, whereas low-density lipoproteins are characterized by S_f values of 0 to 12 (Table 6-2).

The electrophoretic mobilities of VLDL, LDL, and HDL depend on their net charge, their mass, and the supporting medium employed in the separation procedure. Most of the components in each density class are found to have mobilities that correspond with specific lipoprotein groups obtained by electrophoretic procedures (Table 6-2). The chylomicrons generally do not migrate from the point of origin when subjected to electrophoresis in a supporting medium, such as agarose gel.

3. Composition and Function of Lipoproteins

The general composition of each major class of plasma lipoproteins is shown in Table 6-2. HDL contain about equal amounts of protein and lipid. The major lipid components in this class of lipoproteins are cholesterol and phospholipids. Most of the cholesterol is esterified with linoleic, oleic, and palmitic acids. The phospholipid component in HDL is primarily phosphatidylcholine. The distinguishing features of LDL are that they contain about 75% to 80% lipid and that the principal lipid component is cholesterol. VLDL have an even higher proportion of lipid than LDL. The lipids in VLDL are mostly triglycerides. Chylomicrons consist primarily of triglycerides associated with small amounts of protein, cholesterol, and phospholipid.

The protein fraction in each class of lipoproteins consists of several components, termed *apolipoproteins* or *apoproteins*. The apoprotein groups are designated as *apoA*, *apoB*, *apoC*, *apoD*, and *arginine-rich apoprotein* or *apoE*. Some of the apoprotein classes can also be separated into subgroups. These are denoted by numerals, e.g., *apoA-I*, *apoA-II*, *apoC-I*, and *apoC-II* (Table 6-3). ApoA is the major apoprotein in high-density lipoproteins (HDL) and apoB is the principal apoprotein in low-density lipoproteins (LDL). ApoC, as well as considerable amounts of apoB and small amounts of apoE, occur in very-low-density lipoproteins (VLDL). In addition to the primary apoprotein constituents, each class of lipoproteins contains other proteins in lesser concentrations. For example, LDL has small amounts of apoC, and HDL contains a group of minor constituents, which include apoC, apoD, and apoE. Chylomicrons also contain apoprotein components (Table 6-3).

The chylomicrons transport triglycerides absorbed from the intestine, whereas VLDL function as the carriers of endogenous triglycerides. LDL are the major vehicle for the transport of cholesterol, and they serve in the transport of both phospholipids and cholesterol.

4. Metabolic Transformations of Lipoproteins

The very-low-density lipoproteins (VLDL) are produced in the cells of liver and intestine (Figure 6-5). ApoB that is formed in these cells is bound with the appropriate lipid components and released into the bloodstream as a nascent form of VLDL. The latter are converted to VLDL by combining with small amounts of apoC and other apoproteins that are released from HDL.

VLDL are ultimately transformed to low-density lipoproteins (LDL). The triglycerides in VLDL are hydrolyzed by the action of *lipoprotein lipase*. A certain amount of cholesterol and phospholipid is exchanged between HDL and VLDL. The cholesterol is esterified with fatty acids from carbon-2 (β-carbon) of phospholipid. This reaction is catalyzed by *lecithin-cholesterol acyltransferase* (LCAT), as described previously (Chap. 5, section 5.5). ApoA-I and apoC-I appear to promote this reaction. Within the liver, all the VLDL apoproteins are removed, except for apoB and small amounts of apoC. The product is released into the bloodstream as LDL (Fig. 6-6).

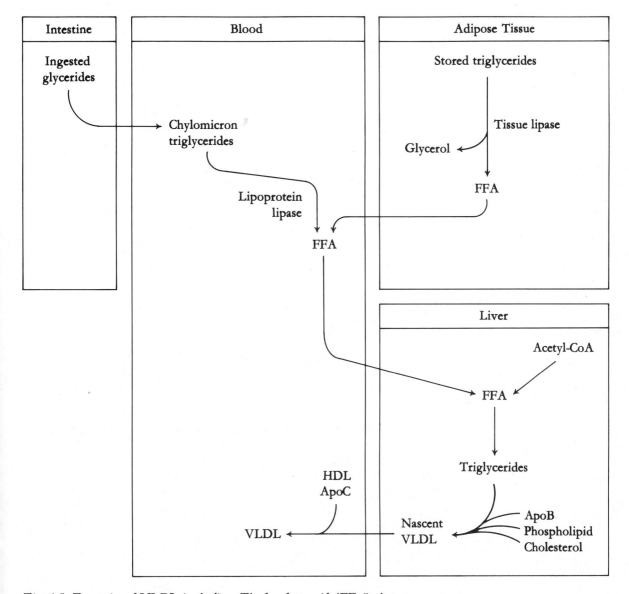

Fig. 6-5. *Formation of VLDL in the liver. The free fatty acids (FFA) that are incorporated into the triglycerides of VLDL may originate from dietary glycerides, adipose tissue stores, and biosynthesis from acetyl-CoA in the liver.*

The low-density lipoproteins (LDL) transport cholesterol to cells of extrahepatic tissues, where it is utilized for incorporation into membranes. The initial step in the utilization of LDL by these cells is their binding to specific receptors on the plasma membrane. The bound LDL pass into the cells by endocytosis and are degraded by lysosomal enzymes. The free cholesterol

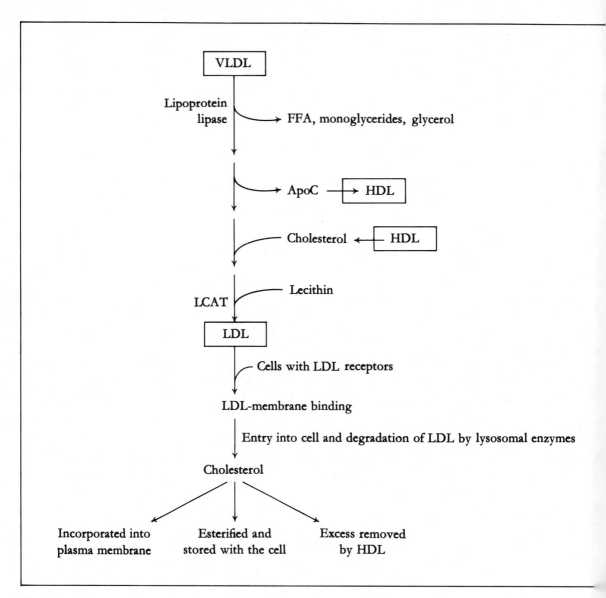

Fig. 6-6. *Conversion of VLDL to LDL and the disposition of the cholesterol from LDL. Among the functions of HDL are the removal of apoC from VLDL, and the release of cholesterol for incorporation into LDL. HDL may also function as a scavenger for removing "excess" cholesterol from extrahepatic tissues.*

that is liberated is then available for the formation of membranes. Excess cholesterol can be esterified and stored within the cells. This pathway for the utilization of LDL has been shown to occur in fibroblasts, arterial smooth muscle, endothelial cells, and lymphoid cells (Fig. 6-6).

Circulating LDL are also removed by the liver but are not reutilized for the synthesis of new VLDL.

The high-density lipoproteins (HDL) are produced in the liver and intestine. HDL from the two sources differ in their apoprotein composition; liver HDL contains apoA and apoC, whereas intestinal HDL have only apoA. The specific function of HDL is not clear, however; a defect in the production of this class of lipoproteins—Tangier disease—results in abnormal disposition of cholesterol (page 292). Conceivably, HDL serves in the transport of cholesterol from peripheral tissues to the liver.

As described previously (page 265), chylomicrons function in the transport of dietary water-insoluble lipids. Ingested cholesterol, absorbed from the small intestine, is also incorporated into chylomicrons and transferred to the liver. The major apoproteins in chylomicrons are apoA, apoB, and apoC. After partial degradation by lipoprotein lipase and transfer of apoC to HDL, the remnants are catabolized by the liver.

B. PHYSIOLOGIC SIGNIFICANCE OF PLASMA LIPID LEVELS

The nature and amounts of circulating lipids are a reflection of the metabolic state and may thus vary in accordance with various conditions and momentary requirements. During postabsorptive periods, the principal lipid components of plasma are the lipoproteins and fatty acids. Of these, the fatty acids have the highest turnover rate, since they are constantly being released from adipose tissue and transported to the tissues where they are utilized. A certain proportion is also taken up by the liver, converted to triglycerides and lipoproteins, and released into the bloodstream. Since the lipoproteins can yield free fatty acids through hydrolysis catalyzed by lipoprotein lipase, a fraction of the fatty acids that arise from adipose-tissue triglycerides may eventually find their way back to the tissue of origin.

During periods of fasting or glucose deprivation when the tissues are more dependent on fatty acids as an energy source, the rate of lipid mobilization is increased significantly. This results in an elevation in the levels of circulating free fatty acids. The plasma concentration of free fatty acids is also increased in the untreated diabetic. One consequence of this situation—i.e., ketonemia—has been discussed earlier (section 6.1).

Another result of intensive, prolonged mobilization of lipids is a condition termed *fatty liver*, in which considerable amounts of triglycerides are deposited in the liver. One explanation for the occurrence of this condition is that a major mode of disposition of fatty acids by the liver is the formation of lipoproteins (the liver does not release free fatty acids). Since the synthesis of lipoproteins is less rapid than many of the other pathways for the utilization of lipids, significant amounts of lipid accumulate in the liver. The relatively rapid mobilization of lipids from adipose tissue during glucose deprivation and in the untreated diabetic thus results in lipemia, ketosis, and fatty liver.

Two other conditions can also result in fatty liver. These are liver damage and a deficiency of choline or its precursors. Since the liver releases lipids only in the form of lipoproteins, any aberration that affects the synthesis of these macromolecules can potentially lead to the development of fatty liver. Thus, chlorinated hydrocarbons such as chloroform and carbon tetrachloride cause liver injury and produce fatty liver. This condition is also observed during chronic infections, e.g., tuberculosis.

An integral fraction of the lipoprotein macromolecule is phospholipid, which contains lecithin (phosphatidylcholine). Consequently, when choline, the building block for lecithin synthesis, is not included in the diet, then fatty liver ensues. The condition may be relieved by administering either lecithin or substances that can yield methyl groups for the synthesis of choline. Substances such as choline or methyl donors for choline are termed *lipotropic* compounds. It is of interest that materials that utilize methyl group and thus decrease their availability for choline synthesis induce fatty liver. Examples of such compounds are nicotinamide and guanidinoacetic acid; these are called *antilipotropic* substances.

6.5. Regulation of Glycolysis and Gluconeogenesis

Glucose or glycogen is constantly degraded within different tissues to provide energy and to supply certain intermediates for synthetic reactions. Concurrently, cells of the liver and kidney also synthesize sugars from various amino acids (e.g., alanine, glutamate, aspartate, serine, cysteine), from glycerol, and from lactate by gluconeogenic processes (Chap. 4, section 4.8). Although the reactions of the glycolytic and gluconeogenic pathways occur simultaneously, various regulatory mechanisms serve to enhance or inhibit specific pathways according to the condition and requirements of the organism. This control and synchronization is effected by variations in the amounts and activities of key enzymes, the concentrations of intermediates and cofactors, and the levels of certain hormones in the circulatory system. Actually, many of these modes of control are interrelated and interdependent. For example, a specific hormone may have an effect on the synthesis or activity of a certain enzyme, and the level and activity of the enzyme may determine the concentrations of various controlling intermediates. However, for purposes of clarity, each of the factors will be considered individually. The overall rate of the glycolytic pathway is regulated by the activities of phosphofructokinase and pyruvate kinase (Chap. 4, section 4.8). The former catalyzes the conversion of fructose 6-phosphate to fructose 1,6-diphosphate, and the latter produces pyruvate from phosphoenolpyruvate. The analogous controlling enzymes in the gluconeogenic sequence are fructose 1,6-diphosphatase and pyruvate carboxylase. Fructose 1,6-diphosphatase affects the hydrolysis of fructose 1,6-diphosphate to fructose 6-phosphate. Pyruvate carboxylase catalyzes the first step in the production of phosphoenolpyruvate from pyruvate, i.e., the conversion of pyruvate to oxaloacetate.

The above-mentioned key enzymes are modulated by the intracellular levels of adenosine phosphates or the ratio of the concentrations of ATP to AMP and ADP (Chap. 4, section 4.8). Adenosine monophosphate affects the activities of phosphofructokinase and fructose 1,6-diphosphatase in a reciprocal fashion; i.e., it promotes the action of the former and inhibits the latter. Hence, elevated concentrations of AMP effectively produce an enhancement in glycolysis and a decrease in the rate of the gluconeogenic sequence. Adenosine diphosphate also promotes glycolysis by enhancing the activity of phosphofructokinase. In addition, ADP is utilized as a substrate in the reaction catalyzed by pyruvate kinase. In contrast to the effects of AMP and ADP, adenosine triphosphate inhibits glycolysis and enhances gluconeogenesis. The effect on glycolysis is due to the inhibitory action of ATP on phosphofructokinase and pyruvate kinase. ATP also functions as an energy source for the pyruvate carboxylase reaction and thus promotes gluconeogenesis.

Since the production of ATP involves the phosphorylation of ADP, and the utilization of ATP results in the formation of ADP or AMP, there is an inverse relationship between the intracellular levels of ATP and those of AMP and ADP. The overall effect of the adenosine phosphates on the degrees of glucogenesis and glycolysis is, therefore, the result of the relative concentrations of these compounds or the ratio of the concentration of ATP to those of ADP and AMP. Thus, an increase in the ratios [ATP]/[AMP] and [ATP]/[ADP] results in a decrease in the rate of glycolysis. Under these conditions, glucose will be generated from other substances by an increase in the conversion of pyruvate to oxaloacetate (pyruvate carboxylase). Conversely, when these ratios are decreased, glycolysis is enhanced as a result of the activation of phosphofructokinase.

Glycolysis and gluconeogenesis are also modulated by a number of metabolic intermediates, in addition to ATP, ADP, and AMP (Table 6-4). An increase in the level of acetyl-CoA stimulates gluconeogenesis by activating pyruvate carboxylase. Citrate and fatty acids inhibit phosphofructokinase and, therefore, decrease the rate of glycolysis. Various glycolytic intermediates, e.g., glucose 6-phosphate, fructose 1,6-diphosphate, or

Table 6-4. Effects of Metabolic Intermediates on Glycolysis and Gluconeogenesis

Metabolite	Glycolysis	Gluconeogenesis
ATP	Decrease	—
ADP, AMP	Increase	Decrease
Acetyl-CoA	—	Increase
Citrate	Decrease	Increase
Fatty acids	Decrease	Increase

glyceraldehyde 3-phosphate, stimulate pyruvate kinase. Therefore, conditions that provide elevated concentrations of these substances (e.g., ample supplies of glucose) enhance the rate of glycolysis.

The rates of glycolysis and gluconeogenesis are modulated, according to the immediate requirements of the organism. Thus, when the relative concentration of ATP is decreased, the rate of the glycolytic energy-releasing pathway is enhanced. The situation is reversed when the concentration of ATP is increased. Under these conditions, the glycolytic pathway is restrained and gluconeogenesis is promoted. This also occurs when noncarbohydrate sources of energy are abundant. Hence, fatty acids decrease the rate of glucose utilization and acetyl-CoA—a product of fatty acid metabolism—promotes gluconeogenesis.

The actual amounts of most of the enzymes of carbohydrate metabolism are far higher than those normally utilized for individual reactions. Thus, the primary control of such enzymes is provided by the factors and conditions already mentioned, rather than the concentration of the enzyme itself. A significant exception is phosphoenolpyruvate carboxykinase, which brings about the formation of phosphoenolpyruvate from oxaloacetate (section 4.8). This enzyme seems to be present in limiting amounts, so variations in its concentration can have critical effects on gluconeogenesis. In fact, it has been found that the level of this enzyme increases under conditions in which the rate of gluconeogenesis is high, e.g., during glucose deprivation and in untreated diabetics. The concentration of this enzyme is under hormonal control, which apparently regulates the rate of de novo synthesis of the protein for the enzyme. Thus, regulation of gluconeogenesis by phosphoenolpyruvate carboxykinase differs from the regulatory mechanism described previously in that it involves changes in the amount of the enzyme itself.

In summary, both glycolysis and gluconeogenesis proceed under normal conditions; however, the degree to which each of these pathways contributes to carbohydrate metabolism varies with the particular conditions and requirements of the organism at any given time. Gluconeogenesis is critical for the maintenance of normal blood glucose levels. During periods when food is not ingested (e.g., between meals or in fasting) or when dietary carbohydrates are restricted, gluconeogenesis is stimulated in order to provide glucose to the circulation. When there are ample amounts of glucose available, glycolysis predominates and gluconeogenesis is restrained. Some of the major control sites regulating the flux through the two pathways are the reactions catalyzed by phosphofructokinase, pyruvate kinase, fructose diphosphatase, pyruvate carboxylase, and phosphoenolpyruvate carboxykinase. When glucose is limited, the organism mobilizes lipids and the resultant free fatty acids that are released inhibit the activity of phosphofructokinase and pyruvate kinase, which in turn retards the rate of glycolysis. Catabolism of fatty acids also provides acetyl-coenzyme A, which activates pyruvate carboxylase and inhibits the oxidation of pyruvate. The increased levels of NADH that result from fatty-acid catabolism inhibit pyruvate kinase and accelerate gluconeogenesis. NADH also yields ATP by oxidative-chain phosphorylation. The resulting increase in the [ATP]/[ADP] ratio can inhibit glycolysis and stimulate gluconeogenesis by enhancing the action of fructose diphosphatase, and pyruvate carboxylase. These interactions provide examples of the correlation of different intracellular reactions and the profound effect these may have on metabolic pathways.

Hormonal effects on gluconeogenesis are demonstrated clearly by the deviations of metabolism in diabetes and in Addison's disease. In the former, there is an increase in gluconeogenesis; in the latter, it is decreased. The explanation for the abnormalities in diabetes is essentially similar to that described for glucose deprivation; the hormonal influences will be discussed in more detail in the next section (section 6.6). In Addison's disease, there is a deficiency in the production of the adrenocortical hormones, e.g., cortisone and hydrocortisone. These hormones enhance the degradation of proteins and the utilization of amino acids for gluconeogenesis. In addition, the glucocorticoids seem to be effective in enhancing the synthesis of key gluconeogenic enzymes. The importance of such steroids is also demonstrated by the sharp decrease in gluconeogenesis that occurs after adrenalectomy.

Other hormones that have a marked effect on gluconeogenesis are epinephrine and glucagon. Since the direction of glucose metabolism is related to the activities of many diverse pathways, hormones that regulate transformations of other nutrients will also have an indirect effect on gluconeogenesis.

6.6. Hormonal Regulation of Carbohydrate and Lipid Metabolism
A. INSULIN

This hormone regulates carbohydrate and triglyceride metabolism through its action at several sites. Insulin is effective in providing cells with sources of energy by facilitating the entry of glucose into the cells and by enhancing the storage of carbohydrates and lipids. Insulin is required for the transfer of glucose from the bloodstream into the cells of certain tissues, including adipose tissue and muscle. The cell membranes of other tissues, notably those of the liver and brain, appear to be freely permeable to glucose. Since the intracellular concentration of glucose is considerably lower than that in the blood, glucose should move freely from the circulatory system into the cell. However, cells of tissues that are insulin-dependent seem to have a barrier to glucose that is apparently relieved by insulin.

By promoting the entry of glucose into the cells of muscle and adipose tissue, insulin prevents glucose accumulation in the blood (hyperglycemia). In the diabetic, there is an aberration in the functioning of this hormone, and one of the most striking effects is an increase in the level of blood glucose. It has been estimated that in a normal individual, the total blood glucose supply is turned over every 5 minutes. As glucose from the blood is consumed by the tissues, it is replenished by glucose released from the liver and by dietary carbohydrates. Injection of insulin results in the uptake of glucose from the blood by muscle and adipose tissue. When the rate of uptake exceeds the rate at which the supply of circulating glucose is replenished, there is a sharp decrease in the blood glucose level (hypoglycemia). Since glucose provides the primary energy supply for brain, its unavailability leads to a state of shock and coma. Thus, regular and balanced levels of insulin are vital for the functioning of the total organism.

Insulin also stimulates the synthesis of glucokinase. The mechanism through which this enzyme can regulate blood glucose was described previously (section 6.3). Since glucokinase is not inhibited by glucose 6-phosphate, it also performs a critical function in the transformation of glucose to glycogen, as was discussed in the earlier section.

The overall effect of insulin in glycogen metabolism is to promote the synthesis and restrain the degradation of glycogen (page 199). Insulin also seems to repress the synthesis of phosphoenolpyruvate carboxykinase, which is a critical enzyme for gluconeogenesis (section 6.5). There is also evidence that insulin inhibits the synthesis of glucose 6-phosphate phosphatase, fructose diphosphatase, and pyruvate carboxylase. Through these activities, insulin moderates the degree of gluconeogenesis. This concept is confirmed by the fact that the rate of gluconeogenesis is increased substantially in diabetics.

With respect to lipid metabolism, insulin enhances the storage of triglycerides in adipose tissue by means of two mechanisms. First, it moderates the activation of lipase in adipose tissue (page 266). Second, by promoting the entry of glucose into the cell, it provides supplies of glycerophosphate for the synthesis of triglycerides and of acetyl-coenzyme A for the synthesis of fatty acids.

There is some evidence that insulin induces the synthesis of citrate-cleaving enzyme, acetyl-coenzyme-A carboxylase, and the synthetase system for the formation of fatty acids from acetyl-coenzyme A. Its activity in this respect may be related to its general effect in enhancing the synthesis of proteins from amino acids.

The sites at which insulin affects metabolic pathways are shown in Figure 6-7. Certain of the specific actions of insulin explain many of the problems that occur in diabetes. The level of blood glucose rises, because the uptake of glucose by muscle and adipose tissue is inhibited due to a deficiency of insulin. Furthermore, the levels of glucokinase are low, so the ability to regulate the blood glucose level is impaired. The rate of glycogenolysis is increased, whereas that of glycogen synthesis is decreased. Hence, the amount of stored carbohydrate tends to become depleted. The inhibitory effect on the lipolysis of stored triglycerides becomes minimal, and free fatty acids are released into the circulation. Within the cells, these fatty acids decrease the rate of glycolysis by inhibiting phosphofructokinase and pyruvate kinase. Increased utilization of fatty acid produces a high level of acetyl-coenzyme A, which stimulates pyruvate carboxylase. These effects accelerate the rate of gluconeogenesis. The rate of synthesis of fatty acids from acetyl-coenzyme A decreases as a result of the diminished activity of acetyl-coenzyme-A carboxylase. The formation of triglycerides is reduced because of the decrease in the availability of glucose that is necessary for the production of glycerol phosphate. The increased mobilization of fatty acids can result in fatty liver, and the enhancement in utilization of fatty acids brings about a rise in the level of ketone bodies.

B. EPINEPHRINE

Epinephrine is elaborated by the adrenal medulla. Through its action in stimulating adenyl cyclase, epinephrine is effective in enhancing the activity of glycogen phosphorylase and tissue lipase. By a similar mechanism, it inhibits the formation of the more active form of glycogen synthetase or synthetase I (Chap. 4, section 4.7.C). Thus, epinephrine causes an increase in the levels of free fatty acids and a decrease in the amount of glycogen stored.

Free fatty acids have an inhibitory effect on acetyl-coenzyme-A carboxylase; this inhibition, in turn, leads to a decrease in the rate of synthesis of fatty acids from acetyl-coenzyme A. Oxidation of the fatty acids increases the levels of NADH and acetyl-coenzyme A. The former has an inhibitory action on pyruvate kinase, and the latter has a stimulatory effect on pyruvate carboxylase. Acetyl-coenzyme A also decreases the activity of the pyruvate dehydrogenase system. The sum total of these effects should be an increase in gluconeogenesis and a decrease in pyruvate utilization. The inhibitory

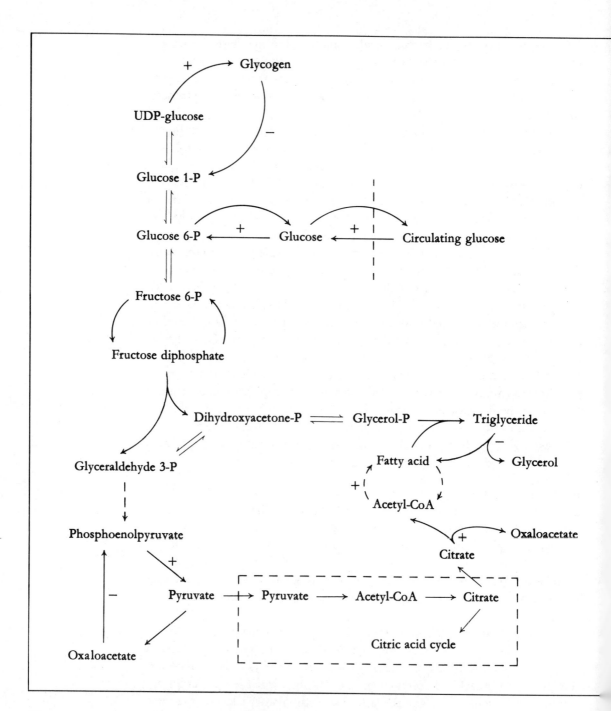

Fig. 6-7. Metabolic transformations influenced by insulin. Sites of action indicated by + (*stimulation*) and − (*inhibition*).

effect of free fatty acids on hexokinase, phosphofructokinase, and pyruvate kinase also tends to decrease the rate of glycolysis.

As a result of its action on liver phosphorylase, the injection of epinephrine causes a rise in blood sugar; i.e., epinephrine administration initiates the sequence

Glycogen → Glucose 1-phosphate → Glucose 6-phosphate → Glucose

In muscle, where there is no glucose 6-phosphatase activity, the enhancement of glycogenolysis is followed by an increase in the rate of lactic acid formation. In the blood, then, the result of epinephrine administration is an increase in the levels of glucose, lactic acid, and free fatty acids.

C. GLUCAGON
This hormone arises in the alpha cells of the pancreas. The action of glucagon is similar to that of epinephrine, except that it does not affect muscle phosphorylase. Hence, glucagon causes an increase in the concentration of circulating glucose and fatty acids. The level of lactate is not affected. Glucagon also has a stimulatory action on gluconeogenesis.

D. GLUCOCORTICOIDS
The two most important glucocorticoids are cortisol and cortisone, which are synthesized in the adrenal cortex. The glucocorticoids stimulate the release of amino acids from proteins and thus provide critical substrates for gluconeogenesis. They also induce the synthesis of pyruvate carboxylase, phosphoenolpyruvate carboxykinase, and glucose 6-phosphatase. The injection of cortisone consequently results in an increase in the rate of gluconeogenesis and a rise in the level of blood glucose.

E. THYROID HORMONE
This hormone causes an increase in blood glucose levels. It seems to increase the rate of absorption of glucose from the intestine. The hyperglycemic effect is offset to some extent by the increase in metabolic rate that is also brought about by thyroid hormone. This causes an enhancement in glucose utilization.

6.7. Metabolism of Lipids and Carbohydrates in Individual Tissues and Organs
The principal pathways of lipid and carbohydrate metabolism have been delineated previously. It will be useful to summarize here the specific functions of individual tissues and organs in intermediary metabolism.

A. LIVER

Glucose can be catabolized in the liver to pyruvate (glycolysis) and oxidized to carbon dioxide and water via the citric acid cycle. Glucose 6-phosphate may also be degraded by the oxidative pentose pathway and thus provide pentose phosphates and NADPH. It is estimated that glycolysis accounts for about 65% of the glucose that is catabolized in the liver; about 35% is catabolized via the pentose pathway. Glucose 6-phosphate is also converted to glucose 1-phosphate and then to UDP-glucose. The latter serves as the glucose donor for the synthesis of glycogen. In addition, UDP-glucose may be oxidized to UDP-glucuronic acid, which is utilized in the synthesis of mucopolysaccharides as well as in forming conjugates with various metabolites.

Hepatic cells also synthesize glucose and glycogen from pyruvate, lactate, glycerol, and various amino acids by means of gluconeogenesis. Indeed, the rate of production of glucose from lactate may exceed 60 micromoles (11 milligrams) per gram of liver per hour. Since this is the most demanding pathway in terms of energy, it consumes a considerable fraction of the nucleoside triphosphates within the liver.

The average amount of glycogen stored in the liver of a 70-kilogram man is about 100 grams. Since it can be degraded to glucose 1-phosphate and hence glucose 6-phosphate, the stored glycogen serves as a source of necessary metabolic intermediates. Through the action of glucose 6-phosphatase, the liver can maintain a constant supply of glucose in the circulatory system. The blood glucose levels are also regulated by the activity of liver glucokinase.

In addition to utilizing glucose, the liver can degrade fatty acids to provide energy. In the process, a certain amount of ketone bodies are formed and released into the blood. Liver is active in the synthesis of cholesterol and fatty acids from acetyl-coenzyme A. Glycerol can be converted to glycerol phosphate, which, alternatively, may also be formed from dihydroxyacetone phosphate. Glycerol phosphate serves in the synthesis of triglycerides and phospholipids. The desaturation of fatty acids also occurs in the liver. Cholesterol, triglycerides, and phospholipids are combined with protein to yield different types of lipoproteins, which then circulate in the bloodstream. The liver generally does not store triglycerides; instead, they are incorporated into lipoproteins and are thereby released into the circulatory system. The liver can take up triglycerides from lipoproteins without need for prior lipolysis.

The principal metabolic routes of carbohydrates and lipids in the liver are shown in Figure 6-8. The liver also functions to metabolize fructose and galactose (page 211).

When some of the enzymatic activities of the fetal liver are compared with

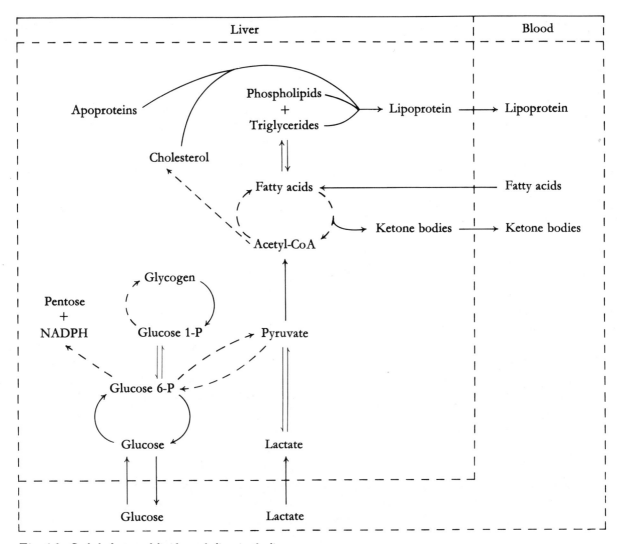

Fig. 6-8. Carbohydrate and lipid metabolism in the liver.

those of the liver during the postnatal period, certain significant changes can be detected. For instance, the gluconeogenic enzymes and glycogen synthetase are at minimal levels in the fetus. This is probably related to the fact that fetal glucose requirements are provided from the maternal circulation. The amount of glycogen stored in fetal liver is comparatively small. During the final trimester of pregnancy, the activity of glycogen synthetase increases, and there is a rise in the amount of glycogen. Immediately after birth, a large fraction of the stored glycogen is consumed; an increase in phosphorylase levels occurs, as do the activities in the gluconeogenic enzymes, phosphoenolpyruvate carboxykinase, fructose diphosphatase, and glucose 6-phosphatase. Gluconeogenesis from lactate is also enhanced by the increase in the levels of the M_4 isozyme of lactate dehydrogenase (page 261). By the time the infant is 3 weeks old, the levels of glycogen are about the same as those in the adult.

These changes are examples of the development of metabolic routes that allow the newborn to become autonomous. The activity of galactokinase also increases considerably in the newborn; this becomes necessary to deal with the lactose that the infant consumes with its milk. Thus, the levels of certain enzymes increase as the organism is presented with the specific substrates.

B. KIDNEY

The basic interconversions of carbohydrates and lipids in the kidney are similar to those that take place in the liver. Glycogen metabolism and the reactions of the pentose pathway, however, are not prominent.

C. MUSCLE

Muscle tissues take up glucose from the blood and convert it to pyruvate or to glycogen. In a 70-kilogram man, the average amount of muscle glycogen is about 245 grams (approximately 1% of the weight of skeletal muscle). This muscle glycogen can be degraded and utilized, but it cannot serve as a source of blood glucose. The pyruvate that is formed as a result of the glycolytic reactions may be oxidized to acetyl-coenzyme A and processed via the citric acid cycle to provide energy. Alternatively, the pyruvate can be reduced to lactate, and the latter will diffuse out of the cell into the bloodstream. Muscle cells also contain the necessary enzymes to degrade fatty acids and to activate and utilize acetoacetate. Muscle differs from liver with respect to the latter activity.

Resting muscle derives most of its energy from fatty acids and acetoacetate. A significant fraction of the energy required for contraction in striated muscle arises from the aerobic degradation of carbohydrates. In the heart, a considerable amount of the energy utilized for myocardial activity is derived from lactate, fatty acids, and ketone bodies. Gluconeogenesis does not occur in muscle.

The metabolic interrelationships that are prominent in muscle tissue are shown in Figure 6-9.

D. ADIPOSE TISSUE

In addition to storing triglycerides, the fat cells of adipose tissue also synthesize them and utilize them metabolically. Adipose cells are located in the abdominal cavity, beneath the skin, between muscle fibers, and in a

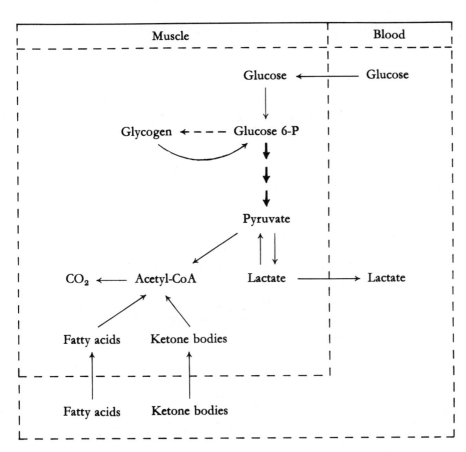

Fig. 6-9. Metabolic interconversions in muscle.

number of other sites. The stored lipids also serve to prevent rapid heat loss from the body and function to support and protect various internal organs.

The fatty acids in triglycerides are composed of both saturated and unsaturated acids of varying chain lengths. The most abundant saturated fatty acid is palmitic; it accounts for about 27% of the fatty acids in the body. Oleic acid is the most abundant unsaturated fatty acid, constituting about 53% of the total. Generally, the alpha position in the triglyceride contains a saturated fatty acid, the beta position has an unsaturated acid attached, and the α' position may be esterified with either a saturated or an unsaturated acid.

Glucose can be metabolized in adipose tissue via the glycolytic pathway and the citric acid cycle; however, the more dominant route is the oxidative pentose pathway. Fatty acids are taken up by adipose tissue and oxidized to acetyl-coenzyme A. Free fatty acids are synthesized from the latter and converted to triglycerides. Unlike liver, adipose tissue cannot produce glycerol phosphate by phosphorylation of glycerol. The only source of this intermediate in fat cells is obtained through the reduction of dihydroxyacetone phosphate yielded by glycolysis.

The hydrolysis of triglycerides in adipose tissue is catalyzed by tissue lipase. The rate of hydrolysis and the rate of synthesis of triglycerides are regulated by the activity of lipase and the availability of glucose. When the supply of glucose is adequate, the rate of hydrolysis will be matched by the rate of synthesis, thus maintaining a steady state.

The general metabolism of carbohydrates and lipids in adipose tissue is summarized in Figure 6-10.

E. BRAIN

The only metabolic fuel normally utilized by the brain is glucose. The central nervous system consumes about 115 grams of glucose per day. Conversion of glucose to glycogen is minimal; the amount of carbohydrate stored in the brain is enough to maintain it for only 10 seconds. Brain cells release lactate to the venous circulation in a similar manner as muscle. Although fatty acids do not generally serve as fuels for this organ, they can apparently be oxidized directly to aldehydes or α-hydroxy acids:

$$CH_3(CH_2)_{21}CH_2COOH + O_2 \longrightarrow CH_3(CH_2)_{21}CHOHCOOH \longrightarrow$$
$$CO_2 + CH_3(CH_2)_{21}CHO \longrightarrow CH_3(CH_2)_{21}COO$$

This process may provide α-hydroxy acids for cerebrosides.

After prolonged starvation, brain cells develop the capability to utilize ketone bodies for energy.

F. ERYTHROCYTES

Fully differentiated erythrocytes do not contain a nucleus or mitochondria. The primary source of energy for these cells is the anaerobic degradation of glucose to lactic acid. Approximately 1.5 millimoles of glucose is consumed per kilogram per hour, and, though this does not provide large amounts of ATP, it is sufficient to meet the comparatively low energy requirements of erythrocytes.

A small fraction of glucose is also oxidized by the oxidative pentose pathway. This provides NADPH, which is necessary for maintaining the integrity of the erythrocyte membrane. The mechanism for this process involves the formation of reduced glutathione from its oxidized derivative. This product, with its free sulfhydryl groups, is critical for the normal functioning of the cell.

In certain genetic diseases that involve a deficiency in glucose 6-phosphate dehydrogenase, a number of abnormalities are attributable to insufficient

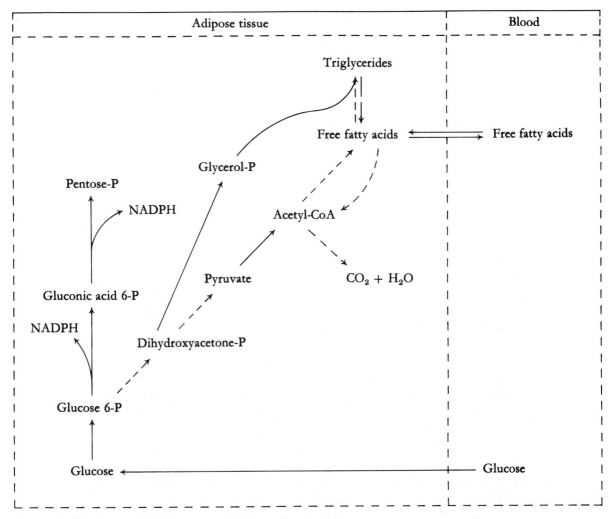

Fig. 6-10. Metabolism of carbohydrates and lipids in adipose tissue.

supplies of NADPH in the erythrocyte. These include an accumulation of oxidized glutathione, a decrease in the rate of reduction of methemoglobin, and a decrease in the lipid content of the cell. The clinical symptoms become acute when antimalarial drugs (e.g., primaquine) are administered; these drugs induce lysis of the erythrocytes or hemolytic anemia. The same effect may be brought about by eating fava beans, and glucose 6-phosphate dehydrogenase deficiency disease has also been called *favism*. The common denominator among the aggravating factors of the hemolytic anemia is that they all enhance the oxidation of glutathione. Individuals with a deficiency in glucose 6-phosphate dehydrogenase do not have ample

NADPH to regenerate the reduced glutathione. There are also a number of variants of this disease. The most common one is a genetically transmitted disorder associated with an X-linked recessive trait, which is found in African and Mediterranean people.

6.8. Glycogen Storage Diseases

A number of inborn or hereditary diseases have been shown to be due to deficiencies in the enzymes required for the interconversions of glycogen and glucose. These diseases are transmitted by autosomal recessive genes.

The types of glycogen storage diseases or glycogenoses are designated by Roman numerals; each type is described in the following discussion. The sites in glycogen metabolism that are affected by the different types of glycogenoses are indicated in Figure 6-11.

1. Type I (von Gierke's Disease)

This disease, which is transferred by an autosomal recessive inheritance, is due to a deficiency of glucose 6-phosphatase (Chap. 4, section 4.2). As a consequence, the glucose 6-phosphate that arises from glycogen phosphorolysis or from glucose phosphorylation cannot be hydrolyzed to glucose. In patients with type I glycogenosis, the glycogen content of the liver may be from 4% to 18% of the tissue. The liver becomes enlarged to as much as five times normal size (hepatomegaly). Glycogen deposits may also occur in the proximal and distal convoluted tubules of the kidney; in normal individuals, glucose 6-phosphatase activity is found in these sites. The disease is compatible with life, as is evidenced by the fact that a number of adult patients with this disorder are known.

Among the other clinical features of type I glycogenoses are a low blood glucose level (hypoglycemia) and an elevated lactic acid content. The concentration of blood glucose in fasting periods may be as low as 15 milligrams per 100 milliliters. As would be expected, the administration of epinephrine or glucagon does not bring about a marked increase in blood glucose; instead, the lactic acid level increases. Small amounts of glucose are released into the circulation as a result of the amylo-1,6-glucosidase activity of the debranching enzyme (page 196). The results of a glucose tolerance test are similar to those found in diabetics. This is attributable to the fact that in this disorder, the insulin response to glucose is about 50% that of a normal individual. Ingested galactose or fructose does not contribute to the blood glucose level.

Elevated levels of ketone bodies in the blood are also common in patients with this disease. This is explained by the high degree of lipid utilization that results because of the limitations in the supply of glucose. Hyperlipemia

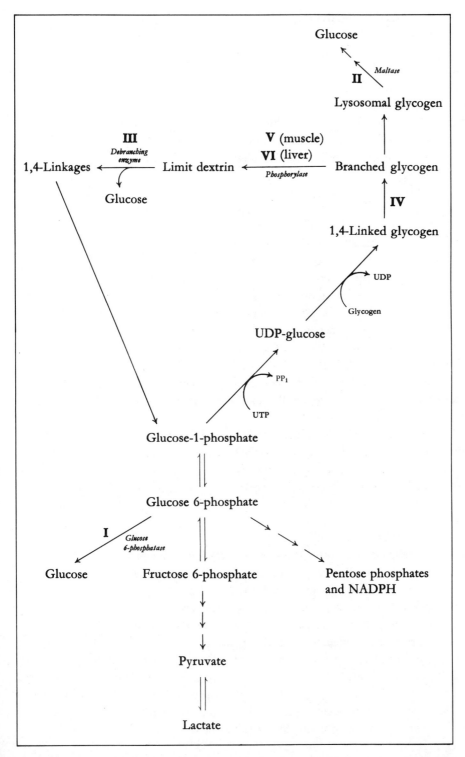

Fig. 6-11. Glycogen metabolism and the sites of deficiencies in the glycogen storage diseases. (Types of glycogenoses are shown by boldface roman numerals; the deficient enzymes are in italics.)

ensues, and lipids become stored in the liver. Blood analyses show elevated levels of triglycerides, cholesterol, and phospholipids.

About 40% of the patients also show increased urate levels in the blood and exhibit symptoms of gout (Chap. 9, section 9.11). This is probably due to the effects of lactate, β-hydroxybutyrate, and acetoacetate in decreasing the renal clearance of urate. It also seems that the hyperuricemia arises from an increased production of urate. It is conceivable that the block in the release of glucose from glucose 6-phosphate results in greater channeling of this intermediate toward the formation of ribose phosphate and nucleotides.

2. Type II (Pompe's Disease)

The deficiency in this disease is that of lysosomal α-1,4-glucosidase (acid maltase). Glycogen is found to accumulate in vacuoles of the liver and other tissues. Presumably, these vacuoles are lysosomes from which other components have been digested by normal processes, and the persistence of the glycogen is due to the absence of the specific maltase. The glycogen in this compartment is not accessible to metabolic phosphorolytic enzymes.

The organs that are most affected are the heart, liver, and muscle. Glycogen can also be shown by histologic procedures to accumulate in the glial cells of the brain. The concentrations in blood of glucose, ketone bodies, and lipids are within normal limits. The symptoms of the disease—i.e., muscle weakness, cardiomegaly, and enlarged tongue—appear within the first 2 months after birth, and the infant generally does not survive more than 6 months. Death results from cardiac failure.

During pregnancy, the presence of α-1,4-glucosidase can be detected in fibroblasts of the amniotic fluid in normal individuals. Absence of the enzyme is indicative of its deficiency in the fetus. When the disease is probable, diagnosis may be performed by amniocentesis, and abortion may be advised on the basis of the results.

3. Type III (Limit Dextrinosis, Forbe's Disease)

This disease is due to a deficiency of the glycogen debranching-enzyme system. Intracellular breakdown of glycogen proceeds only to the vicinity of the 1,6-linked branch points, yielding a product known as *limit dextrin*. The latter accumulates in the liver, muscle, heart, erythrocytes, and leukocytes. There are several variants of the disease, which are classified with respect to the severity and the tissues affected. Hepatomegaly, muscle weakness, and wasting are common features. Some patients die at an early age, while others survive to forty or older.

As expected, patients with this disease exhibit fasting hypoglycemia and do not respond to the administration of epinephrine. Injection of fructose or galactose results in a rise of blood glucose levels. The most common treatment for children is the frequent feeding of a high-protein diet. A number of patients who have reached adulthood seem to have adjusted metabolically and can function normally.

4. Type IV (Andersen's Disease)

This is a rare glycogen storage disease. The metabolic defect is a deficiency in the branching enzyme, $\alpha\text{-}(1,4 \rightarrow 1,6)$-transglycosylase. As a consequence, the glycogen that is synthesized has long outer chains with a minimal degree of branching. Glucose polymers of this nature are considerably less soluble than the more highly branched glycogen macromolecules. Deposition of such material in liver causes cirrhosis and alterations in liver function. Other tissues affected are the heart, spleen, muscle, and erythrocytes.

Children with this disease appear to be normal at birth, but symptoms such as hepatomegaly, poor weight gain, and hypotonia develop within the first few months. Glucose tolerance is normal. The effect of epinephrine is variable, and this variation may be ascribed to differences in the degree of liver damage. The individuals afflicted do not survive more than a few years.

5. Type V (McArdle's Disease)

The enzyme that is deficient in this disease of the muscles is phosphorylase. The disorder is associated with a limited ability to perform muscular exercise because of pain and weakness. The blood lactate levels do not increase significantly after exercise, as would be expected. The patient can survive and carry on a relatively normal existence, since difficulties occur only during exercise. It is of interest that the liver phosphorylase is generally normal, so the individuals are not hypoglycemic and show a normal response to epinephrine.

6. Type VI (Hers' Disease)

The patients with this disease have a high amount of liver glycogen. Although the enzyme deficiency seems to be one of liver phosphorylase, there are some questions in this regard.

6.9. Abnormalities in Lipid Disposition

Aberrations in lipoprotein metabolism and function are implicated in a number of diseases. Certain types of hyperlipidemias are known to be predisposing factors for coronary heart disease. Some disorders of lipid metabolism are clearly genetically determined. In addition to the important clinical implications, studies on lipoproteins in humans have yielded basic information on normal lipid transport. Disorders may be due to deficiencies in lipoproteins or to elevated levels of specific classes of lipoproteins. These will be discussed individually.

1. Abetalipoproteinemia

The disease is due to a defect in the formation of chylomicrons, very-low-density lipoproteins (VLDL), and low-density lipoproteins (LDL). This is an inherited disease with autosomal recessive transmission. The genetic defect appears to involve a deficiency in the synthesis or utilization of apoB (page 270). Since chylomicron formation is required for normal absorption of intestinal lipids, an absence of the required protein leads to the malabsorption of these substances, which leads to steatorrhea and abdominal distention. As might be expected, the weight gain and growth of a child with this disease is inhibited. Other clinical features in abetalipoproteinemia are acanthocytosis, ataxic neuropathy, and retinitis.

2. Familial Hypobetalipoproteinemia

The disorder is characterized by extremely low levels of plasma low-density lipoproteins and seems to arise from a block in the synthesis of the required apoprotein. The disease is transmitted as an autosomal dominant trait. The high-density lipoprotein (HDL) concentrations are normal, and only in one case have malabsorption problems been noted. The disease is not associated with any specific clinical abnormalities, which would indicate that the concentration of LDL in normal plasma is considerably greater than required for usual physiologic functions.

3. Familial High-Density Lipoprotein Deficiency (Tangier Disease)

This is a rare genetic disease, transmitted as an autosomal recessive disorder. It is characterized by minimal concentrations of high-density lipoproteins and is due to a defect in the synthesis of the required apoprotein (page 270). The levels of LDL and cholesterol are relatively low. The triglyceride concentration is generally elevated, and cholesterol esters accumulate in the reticuloendothelial system. The clinical features of this condition include orange tonsils and hepatosplenomegaly. These manifestations are presumably the results of cholesterol deposition in the affected tissues.

4. Lipoprotein Lipase Deficiency

This disease, as well as those described in the remainder of this chapter, involves higher than normal levels of various plasma lipoproteins; these abnormalities are also called *hyperlipidemias*.

The aberration that is sometimes called *type I hyperlipoproteinemia* involves the absence of lipoprotein lipase activity. The defect may be detected during the first week of life, when the infant shows symptoms of gastric distress. The abdomen may be enlarged due to hepatosplenomegaly, and xanthomas arise in the skin, generally over the trunk and extremities. High levels of chylomicrons are present in the blood, and these cannot be cleared by the administration of heparin. The blood is lipemic and thus has an intense "creamy" appearance. As expected from our knowledge of the composition of chylomicrons, the triglyceride levels are elevated and may become as high as 5% of the serum. The disease is transmitted by an autosomal recessive process. The only treatment for this condition is restriction of fat from the diet. Triglycerides of medium-chain fatty acids are also employed to replace natural triglycerides.

5. *Familial Hypercholesterolemia*
This genetic disease is inherited as an autosomal dominant disorder and is found in 0.1% to 0.5% of the population. Patients with this condition have high levels of serum cholesterol, while triglyceride concentrations are generally in the normal range. The primary defect appears to be in the processing of the low-density lipoproteins (LDL) or, more specifically, in a receptor on cell membranes that binds these materials. The normal disposition of circulating LDL, which is rich in cholesterol (page 268), involves its binding to a membrane receptor and its transfer into the cell. The intracellular degradation of the lipoprotein results in the release of its cholesterol. The latter then functions to inhibit de novo synthesis of cholesterol (page 232). This defect in the disposition of LDL thus results in uncontrolled cholesterol synthesis and hypercholesterolemia. The disease is sometimes classified as type II hyperlipidemia and is subdivided into type IIA and type IIB on the basis of the characteristic serum lipoprotein patterns (Table 6-5).

Homozygotes with this disease develop coronary heart disease as early as adolescence. About 50% of the heterozygotes have heart disease by the time they reach the age of 60 years. The levels of LDL and cholesterol may be decreased in individuals with this disorder by restricting foods that have high amounts of cholesterol and saturated fatty acids. Various agents, such as cholestyramine (a bile-sequestering resin) and nicotinic acid, appear to be effective in treating some patients.

6. *Broad Beta Disease*
Another rare, genetic disorder that results in premature vascular disease and xanthomas is known as "broad beta" disease or *type III hyperlipidemia*. The blood contains high levels of LDL and VLDL. Both cholesterol and triglyceride concentrations are elevated. The lipid levels can generally be reduced by restricting the intake of triglycerides, and drug administration (for example, clofibrate) has also been found effective.

7. *Pre-Beta Hyperlipidemia*
The abnormality known as pre-beta hyperlipidemia, also characterized as type IV hyperlipidemia, is comparatively common. Although the disorder

Table 6-5. Serum Characteristics in Hyperlipidemias

Type	Serum Abnormality*		Cholesterol Level	Triglyceride Level	Result of Glucose Tolerance Test
	Electrophoresis	Ultracentrifugation			
I	Chylo.	Chylo.	Normal	Elevated	Normal
IIA	Beta	LDL	Elevated	Normal	Normal
IIB	Beta, pre-beta	LDL, VLDL	Elevated	Moderately elevated	Varies
III	Broad beta	LDL, VLDL	Elevated	Elevated	Abnormal
IV	Pre-beta	VLDL	Moderately elevated	Elevated	Abnormal
V	Chylo., pre-beta	Chylo., VLDL	Moderately elevated	Elevated	Abnormal

* Abnormal components found utilizing either serum electrophoresis or ultracentrifugation. Chylo. = chylomicrons, beta = β-lipoproteins, pre-beta = pre-β-lipoprotein fraction, LDL = low-density lipoproteins, and VLDL = very-low-density lipoproteins.

is known to be prevalent among members of the same families, it may also arise from various secondary causes. These include diabetes, hypothyroidism, pancreatitis, and alcoholism. The direct metabolic defect, which leads to the elevation of pre-β-lipoprotein levels is not defined. The condition predisposes the afflicted individual to vascular abnormalities and coronary disease.

8. Type V Hyperlipidemia

Type V hyperlipidemia is characterized by increased plasma levels of both chylomicrons and VLDL. As expected from the composition of these blood components, the concentration of triglycerides in the blood are elevated. In addition, the level of cholesterol is elevated and glucose tolerance is abnormal. Analyses for lipoprotein lipase activity do not reveal any deficiency in this enzyme.

9. Serum Characteristics

Investigation of the serum lipid profiles in the hyperlipidemias has resulted in the system of classification shown in Table 6-5.

Suggested Reading

Cahill, G. F., Jr. Physiology of Insulin in Man. *Diabetes* 20:785–799, 1971.

Cahill, G. F., Jr. Starvation in Man. *New England Journal of Medicine* 282:668, 1970.

Exton, J. H. Progress in Endocrinology and Metabolism, Gluconeogenesis. *Metabolism* 21:945, 1972.

Fredrickson, D. S. Plasma Lipoproteins and Apolipoproteins. *Harvey Lectures* 68:185–238, 1974.

Jackson, R. L., and Gotto, A. M., Jr. Phospholipids in Biology and Medicine. *New England Journal of Medicine* 290:24–29; 87–93, 1974.

Jackson, R. L., Morrisett, J. D., and Gotto, A. M., Jr. Lipoprotein Structure and Metabolism. *Physiological Reviews* 56:259–314, 1976.

Levine, R., and Haft, D. Carbohydrate Homeostasis. *New England Journal of Medicine* 283:175 and 237, 1970.

Levy, R. I., Morganroth, J., and Rifkind, B. M. Treatment of Hyperlipidemia. *New England Journal of Medicine* 290:1295, 1974.

Neufeld, E. F., Lim, T. W., and Shapiro, L. J. Inherited Disorders of Lysosomal Metabolism. *Annual Review of Biochemistry* 45:357, 1975.

Smith, L. C., Pownall, H. J., and Gotto, A. M., Jr. The Plasma Lipoproteins: Structure and Metabolism. *Annual Review of Biochemistry* 47:751–777, 1978.

Sokoloff, L. Metabolism of Ketone Bodies by the Brain. *Annual Review of Medicine* 24:271–288, 1973.

Stanbury, J. B., Wyngaarden, J. B., and Fredrickson, D. S. *The Metabolic Basis of Inherited Diseases* (3rd ed.). New York: McGraw-Hill, 1972.

7. Metabolism of Proteins and Amino Acids

7.1. Digestion and Absorption

The fact that proteins are vital for the functioning of the total organism becomes obvious when we consider the wide variety of proteins in tissue components. The most distinctive feature of proteins in the dynamic aspects of metabolism is their capacity to function as enzymes; all enzymes have protein structure. Proteins also serve as structural materials in cells and tissues; e.g., they form cell membranes and extracellular fibrous materials. In addition, proteins participate as components of the circulatory system (i.e., albumin and globulins), and some perform intercellular regulatory functions. Like all components of living organisms, proteins undergo degradation and turnover. The ultimate source of the building blocks for their replenishment are the proteins ingested in the diet.

In order to be absorbed and utilized, dietary proteins must be digested and degraded to amino acids. This function is performed by proteolytic enzymes in the stomach and intestine. Gastric digestion of proteins is catalyzed by *pepsin*. This enzyme is an endopeptidase; that is, it hydrolyzes peptide bonds in a random fashion in the middle of the protein polymer. (Endopeptidases are to be distinguished from exopeptidases, which hydrolyze terminal amino acids. Of the latter, carboxypeptidases hydrolyze amino acids from the carboxyl end, and aminopeptidases hydrolyze those from the amino terminus of the protein.) Pepsin has an optimum pH of 2 to 3. The low pH of the gastric milieu is provided by the secretion of hydrochloric acid from the parietal cells of the stomach.

The pepsin in the stomach is formed from an inactive proenzyme or zymogen called *pepsinogen*. The latter is secreted by the chief cells of the gastric mucosa. Pepsinogen is converted to pepsin in the lumen of the stomach through the action of HCl. Pepsin itself also causes the transformation of pepsinogen to pepsin. The formation of the latter is thus an autocatalytic process. The conversion of pepsinogen to pepsin involves the hydrolysis of a polypeptide segment of the zymogen that is composed of 42 amino acid residues (the molecular weight of pepsinogen is 40,400, whereas that of pepsin is 32,700). In the process, an aspartic acid carboxyl group, which is critical in the functioning of the active site, is exposed in the functional enzyme.

Pepsin acts primarily on peptide bonds involving the carboxyl group of the aromatic amino acids phenylalanine, tyrosine, and tryptophan:

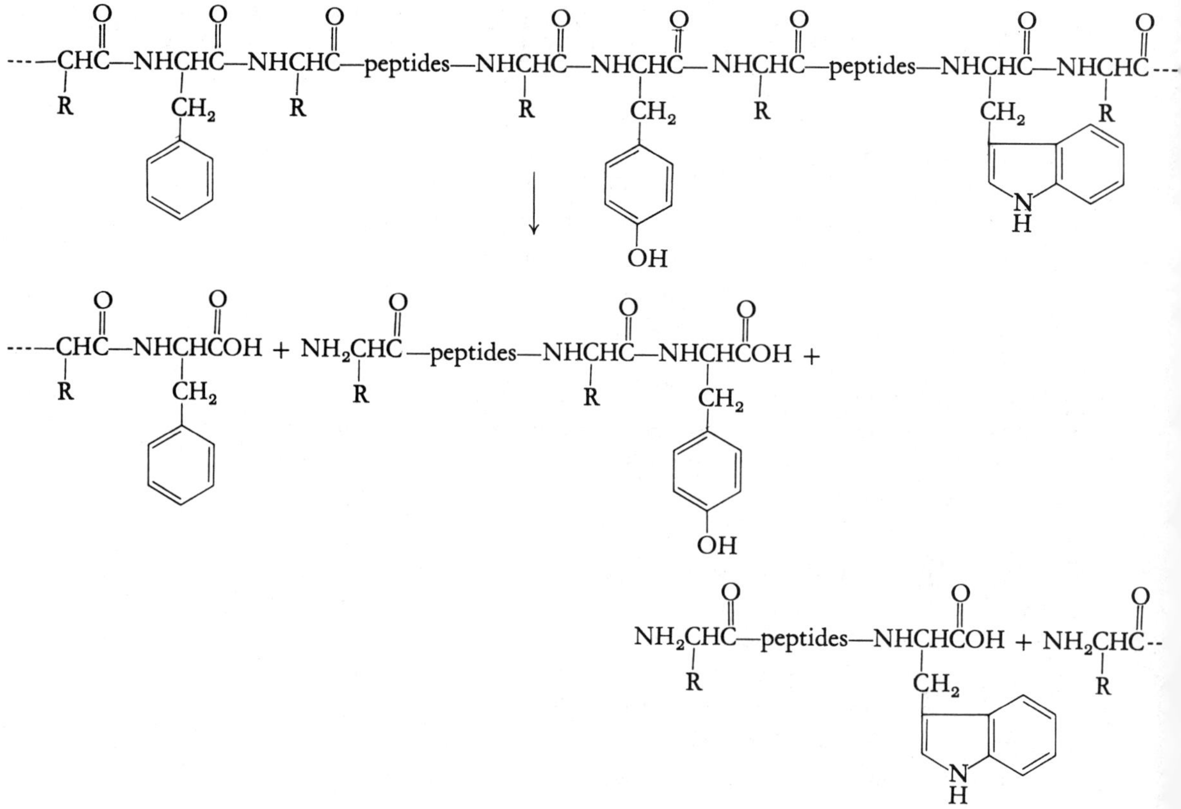

$$\cdots\text{-CHC-NHCHC-NHCHC-peptides-NHCHC-NHCHC-NHCHC-peptides-NHCHC-NHCHC}\cdots$$

$$\downarrow$$

$$\cdots\text{-CHC-NHCHCOH} + \text{NH}_2\text{CHC-peptides-NHCHC-NHCHCOH} +$$

$$\text{NH}_2\text{CHC-peptides-NHCHCOH} + \text{NH}_2\text{CHC-}\cdots$$

Other sites are affected to a comparatively lesser extent. The main products of proteolysis by pepsin are polypeptides. Relatively small amounts of free amino acids are released in the process.

In addition to the secretion of pepsinogen and hydrochloric acid by the chief cells and parietal cells respectively, a mixture of various glycoproteins are secreted by the mucous cells of the stomach. These are complex protein-polysaccharide macromolecules that contain, among other things, blood-group substances and intrinsic factor. Some aspects of these materials will be discussed in Chapters 12 and 14.

When the stomach contents, including partially digested proteins, are transferred to the small intestine, they are neutralized by the alkaline secretions of the pancreas. For the further digestion of proteins, pancreatic juice contains trypsinogen, three chymotrypsinogens, procarboxypeptidases, and proelastase.

The inactive trypsinogen is converted to the active enzyme, *trypsin,* by the action of an enzyme of the intestinal mucosa, *enterokinase.* Trypsin also catalyzes further activation of trypsinogen. In addition, trypsin converts the chymotrypsinogens, procarboxypeptidases, and proelastase to chymotrypsins, carboxypeptidases, and elastase, respectively. The activation processes involve cleavage of specific peptide bonds and the release of a peptide segment (Fig. 7-1). Trypsin catalyzes the hydrolysis of peptide bonds that involve the carboxyl groups of lysine or arginine.

Chymotrypsins cause the cleavage of proteins at peptide linkages in which the carboxyl groups of phenylalanine, tyrosine, and tryptophan are involved. *Carboxypeptidase A,* an exopeptidase, effects the release of C-terminal amino acids and is non-specific. *Carboxypeptidase B* causes the scission of C-terminal arginine or lysine:

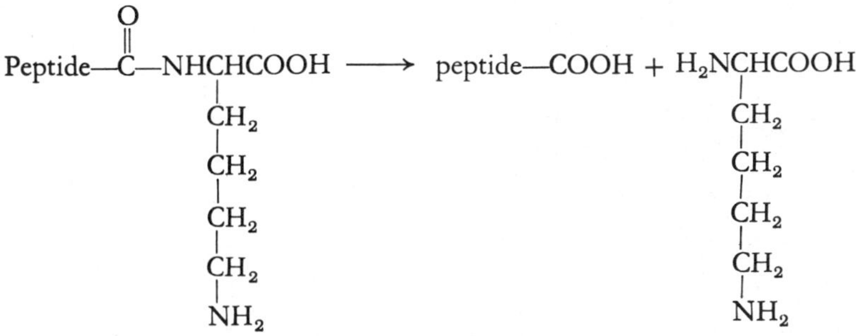

$$\text{Peptide-}\overset{\overset{O}{\|}}{C}\text{-NHCHCOOH} \longrightarrow \text{peptide-COOH} + H_2\text{NCHCOOH}$$
$$\hspace{3.2cm}|\hspace{6cm}|$$
$$\hspace{3.2cm}CH_2\hspace{5.7cm}CH_2$$
$$\hspace{3.2cm}|\hspace{6cm}|$$
$$\hspace{3.2cm}CH_2\hspace{5.7cm}CH_2$$
$$\hspace{3.2cm}|\hspace{6cm}|$$
$$\hspace{3.2cm}CH_2\hspace{5.7cm}CH_2$$
$$\hspace{3.2cm}|\hspace{6cm}|$$
$$\hspace{3.2cm}CH_2\hspace{5.7cm}CH_2$$
$$\hspace{3.2cm}|\hspace{6cm}|$$
$$\hspace{3.2cm}NH_2\hspace{5.7cm}NH_2$$

In addition to these proteolytic processes, there are several *aminopeptidases* in the intestinal mucosa. These catalyze the hydrolysis of peptide bonds from the amino terminus:

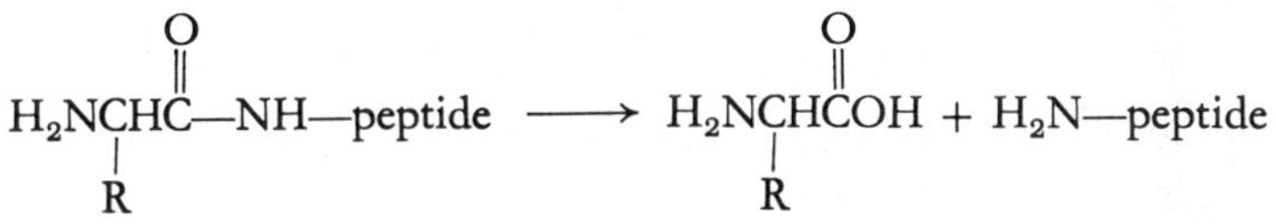

$$H_2\text{NCH}\overset{\overset{O}{\|}}{C}\text{-NH-peptide} \longrightarrow H_2\text{NCH}\overset{\overset{O}{\|}}{C}\text{OH} + H_2\text{N-peptide}$$
$$\hspace{1.5cm}|\hspace{5.5cm}|$$
$$\hspace{1.5cm}R\hspace{5.7cm}R$$

These aminopeptidases are intracellular enzymes that cleave peptides as they either enter the respective cells or are released into the intestinal lumen as the cells are turned over. The final products of all proteolytic activities are the amino acid components of the respective proteins.

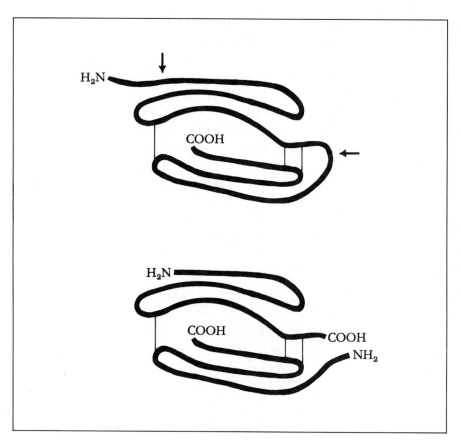

Fig. 7-1. Schematic representation of the structures of trypsinogen (top) and trypsin (bottom). The arrows show the sites where trypsinogen is cleaved. Thin lines represent disulfide linkages. Note that the transformation involves the removal of a hexapeptide unit from the N-terminal section of trypsinogen and additional hydrolysis of a specific peptide within the chain. Thus, trypsin consists of two polypeptide chains that are bridged by disulfide bonds.

Some of the characteristics of the proteolytic enzymes of the digestive system are summarized in Table 7-1.

Amino acids are absorbed from the small intestine into the portal circulation. The process involves active transport and thus requires energy. L-Amino acids are absorbed more rapidly than the D forms; neutral and hydrophilic amino acids are transferred at a higher rate than basic or hydrophobic amino acids. The rate of absorption is also influenced by other substances that may be present in the intestine. Glucose and galactose, for example, delay the absorption of amino acids; when leucine is being absorbed, there is a delay in the absorption of isoleucine and valine.

The intracellular metabolic interconversions of amino acids involve numerous reactions and thus present a highly complex and intricate system. One of the reasons for the multiplicity of reactions is that each amino acid has several specific pathways. With 20 amino acids, the total number of reactions becomes quite large, and it is difficult to visualize their metabolism as an integrated system. Several types of reactions, however, are common to

Table 7-1. Properties of Digestive Enzymes

Enzyme	pH Optimum	Molecular Weight	Isoelectric Point
Pepsin	2	32,700	1.1
Trypsin	7–8	23,000	10.5
α-Chymotrypsin	7–8	23,000	8.6
Carboxypeptidase	8	34,000	—
Leucine aminopeptidase	8–9	—	—
Pepsinogen	—	40,400	3.7
Trypsinogen	—	23,700	9.3
Chymotrypsinogen	—	23,000	9.5

most amino acids. These will be considered first, since they involve certain basic principles that can give an overall view of amino acid metabolism.

Most amino acid metabolism occurs in the liver. The kidneys also participate, but to a much lesser degree. The contribution of skeletal muscle to amino acid catabolism is minimal.

7.2. Transamination

One of the most important preliminary steps in the catabolism of amino acids is their conversion to α-ketoacids. This may be accomplished by at least three different processes: transamination, oxidative deamination, and non-oxidative deamination. Although all of these are crucial reactions and fulfill specific functions, the most general and predominant mode of nitrogen removal is *transamination*. A crucial transamination reaction is the reversible transfer of the α-amino group from an amino acid to α-ketoglutarate. The products of the reaction are the respective ketoacid and glutamic acid:

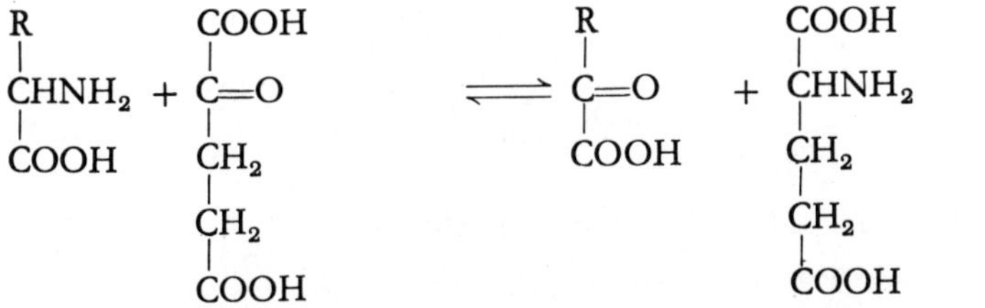

$$\begin{array}{c} R \\ | \\ CHNH_2 \\ | \\ COOH \end{array} + \begin{array}{c} COOH \\ | \\ C=O \\ | \\ CH_2 \\ | \\ CH_2 \\ | \\ COOH \end{array} \rightleftharpoons \begin{array}{c} R \\ | \\ C=O \\ | \\ COOH \end{array} + \begin{array}{c} COOH \\ | \\ CHNH_2 \\ | \\ CH_2 \\ | \\ CH_2 \\ | \\ COOH \end{array}$$

Amino acid   α-Ketoglutarate (ketoacid)   Ketoacid   Glutamate (amino acid)

All amino acids—except threonine, lysine, and proline—can undergo transamination with α-ketoglutarate.

One of the important reactions in the intermediary metabolism of amino acids is the reversible transfer of amino groups from glutamate to oxaloacetate:

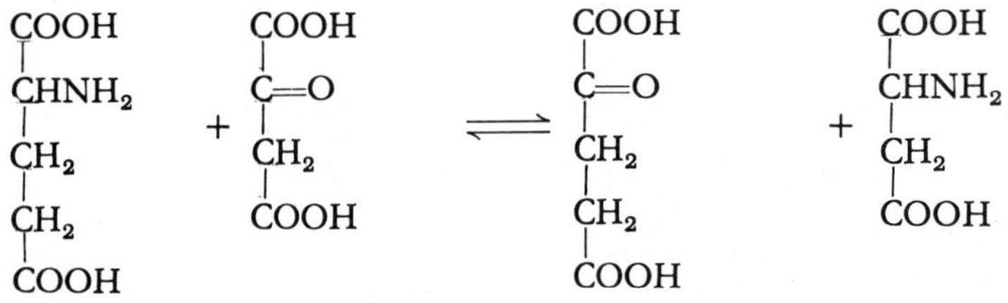

$$\begin{array}{c} \text{COOH} \\ | \\ \text{CHNH}_2 \\ | \\ \text{CH}_2 \\ | \\ \text{CH}_2 \\ | \\ \text{COOH} \end{array} + \begin{array}{c} \text{COOH} \\ | \\ \text{C=O} \\ | \\ \text{CH}_2 \\ | \\ \text{COOH} \end{array} \rightleftharpoons \begin{array}{c} \text{COOH} \\ | \\ \text{C=O} \\ | \\ \text{CH}_2 \\ | \\ \text{CH}_2 \\ | \\ \text{COOH} \end{array} + \begin{array}{c} \text{COOH} \\ | \\ \text{CHNH}_2 \\ | \\ \text{CH}_2 \\ | \\ \text{COOH} \end{array}$$

Glutamate (amino acid)   Oxaloacetate (ketoacid)   α-Ketoglutarate (ketoacid)   Aspartate (amino acid)

The enzymes that catalyze such reactions are called *transaminases*. These enzymes are usually named according to the two amino acids involved in the transamination reaction. Thus, the trivial name of the enzyme for the reaction just given is glutamate-aspartate transaminase (it is also known as glutamate-oxaloacetate transaminase, or GOT). The International Enzyme Commission name for this enzyme is L-aspartate:2-oxoglutarate aminotransferase, EC 2.6.1.1 (the first numeral refers to the fact that the enzyme is a transferase; the second indicates that nitrogens are transferred; the third specifies that they are amino groups; and the fourth is specific for the enzyme).

All the transaminases have pyridoxal phosphate as a coenzyme:

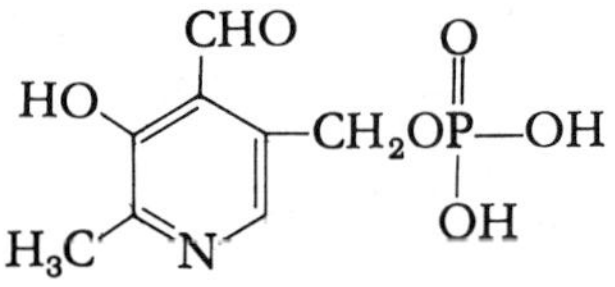

Pyridoxal phosphate

As is the situation with pyridine nucleotides and flavin coenzymes, pyridoxal phosphate is derived from a vitamin. In this case, the vitamin is pyridoxine, or vitamin B_6:

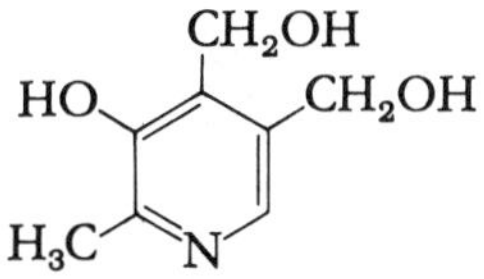

Pyridoxine (structure: pyridine ring with HO, CH₂OH, CH₂OH, H₃C, N substituents)

The mechanism for the transamination process involves the interaction of the enzyme with the amino acid substrate, RCH(NH$_2$)COOH, to form a Schiff base with the pyridoxal coenzyme:

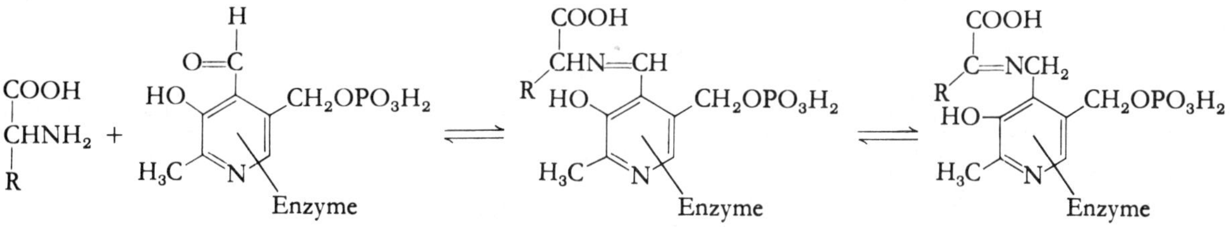

Amino acid + Pyridoxal phosphate-enzyme ⇌ Schiff base (Aldimine) ⇌ Schiff base (Ketimine)

Hydrolysis of the Schiff base yields the ketoacid and enzyme-linked pyridoxamine phosphate:

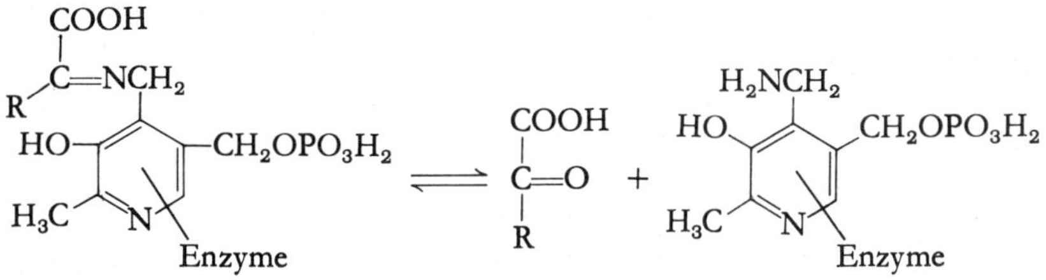

Schiff base (Ketimine) ⇌ Ketoacid + Pyridoxamine phosphate-enzyme

The ketoacid substrate, R'C(=O)COOH, of the transamination reaction then forms a Schiff base with the pyridoxamine phosphate, which finally yields the corresponding amino acid R'CH(NH$_2$)COOH and regenerated enzyme-linked pyridoxal phosphate:

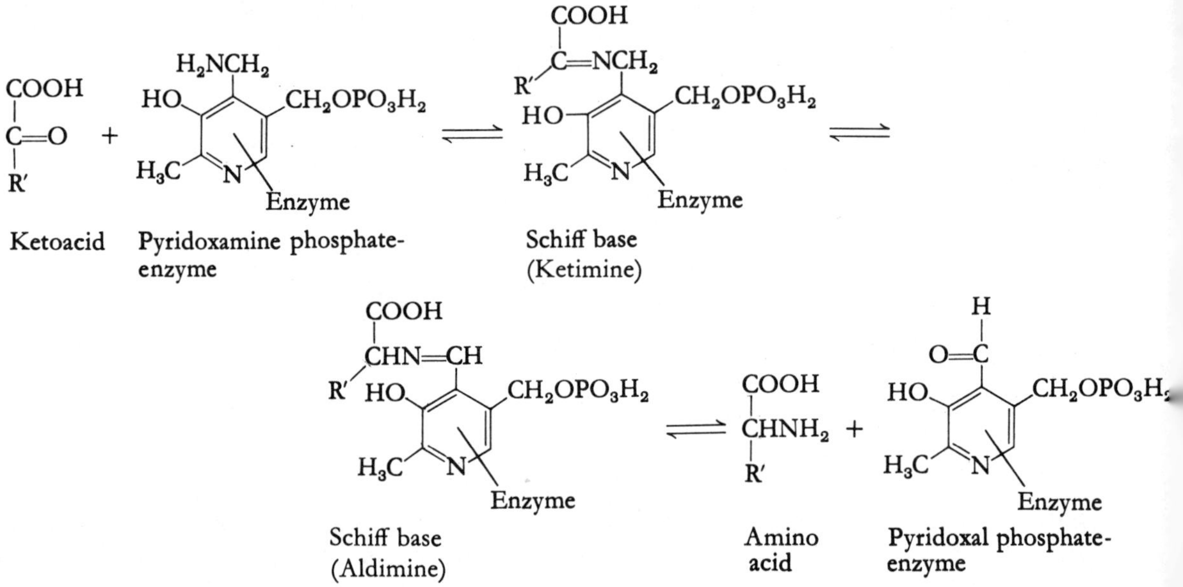

7. Metabolism of Proteins and Amino Acids

[Ketoacid + Pyridoxamine phosphate-enzyme ⇌ Schiff base (Ketimine) ⇌ Schiff base (Aldimine) ⇌ Amino acid + Pyridoxal phosphate-enzyme]

Since the transaminase-catalyzed reactions are freely reversible, they can serve to form amino acids from the corresponding ketoacids. Liver contains the required enzymes for the synthesis of all amino acids except glycine, lysine, threonine, and proline. Glutamate can serve as the amino-group donor in such syntheses. An example is the formation of alanine from pyruvate:

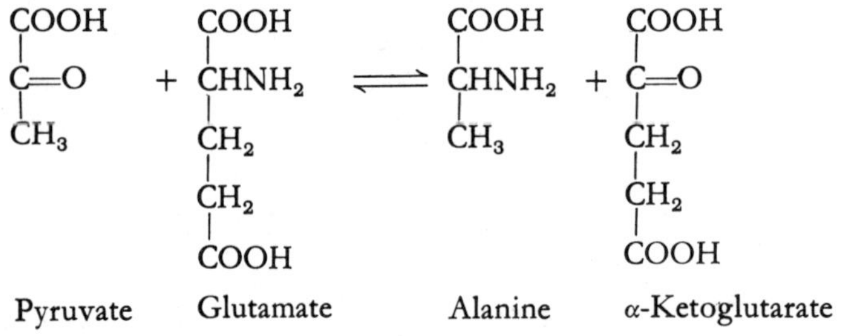

Pyruvate + Glutamate ⇌ Alanine + α-Ketoglutarate

This reaction is catalyzed by glutamate-alanine transaminase (L-alanine:2-oxoglutarate aminotransferase, EC 2.6.1.2). Alanine transaminases, in which pyruvate serves as the amino acceptor, are common in mammalian tissues.

It is seen that transaminase reactions are central to both the degradation and the synthesis of amino acids. Furthermore, since these reactions involve the interconversion of amino acids with pyruvate or dicarboxylic acids, they function as a bridge between the metabolism of amino acids and carbohydrates. These aspects will be discussed in subsequent sections; for the present, only the catabolic aspects will be considered.

Transaminases are of special significance in medical practice, since the levels of these enzymes in the blood are related to the normal functioning of liver and cardiac muscle. Glutamate-oxaloacetate transaminase (GOT) and glutamate-pyruvate transaminase (GPT) are of special clinical interest, since these enzymes are lost from damaged myocardium and liver and, as a result, their levels in the blood will increase. The rationale for these clinical determinations and their utilization will be discussed in later sections, e.g., in Chapter 12.

In addition to the reversible transaminase reactions involving glutamic acid, liver cells also contain enzymes that catalyze transamination from glutamine and asparagine. The reactions involve the transfer of the α-amino groups from these amino acids to various ketoacids. For example, transamination from glutamine can be described by the following general reaction:

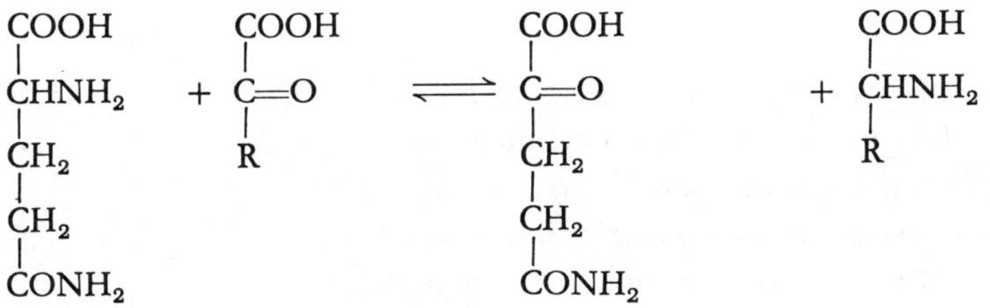

$$\begin{array}{c} COOH \\ | \\ CHNH_2 \\ | \\ CH_2 \\ | \\ CH_2 \\ | \\ CONH_2 \end{array} + \begin{array}{c} COOH \\ | \\ C=O \\ | \\ R \end{array} \rightleftharpoons \begin{array}{c} COOH \\ | \\ C=O \\ | \\ CH_2 \\ | \\ CH_2 \\ | \\ CONH_2 \end{array} + \begin{array}{c} COOH \\ | \\ CHNH_2 \\ | \\ R \end{array}$$

Glutamine Ketoacid α-Ketoglutaramate Amino acid

The α-keto-δ-amide (α-ketoglutaramate) is then hydrolyzed by a *deamidase* to α-ketoglutarate and ammonia:

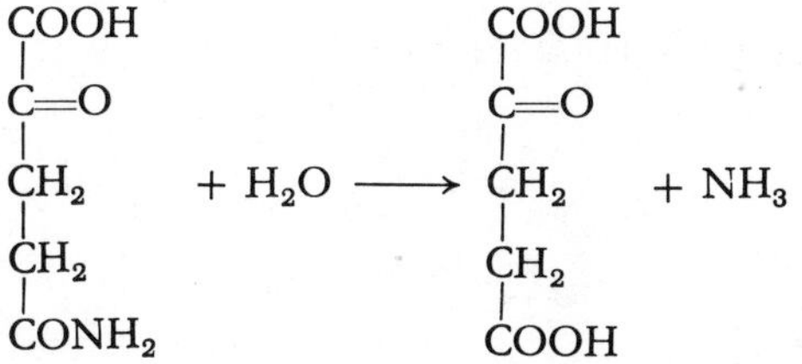

$$\begin{array}{c} COOH \\ | \\ C=O \\ | \\ CH_2 \\ | \\ CH_2 \\ | \\ CONH_2 \end{array} + H_2O \longrightarrow \begin{array}{c} COOH \\ | \\ C=O \\ | \\ CH_2 \\ | \\ CH_2 \\ | \\ COOH \end{array} + NH_3$$

α-Ketoglutarate

Transamination with asparagine follows a similar sequence to yield the amino acid, oxaloacetate, and ammonia as final products:

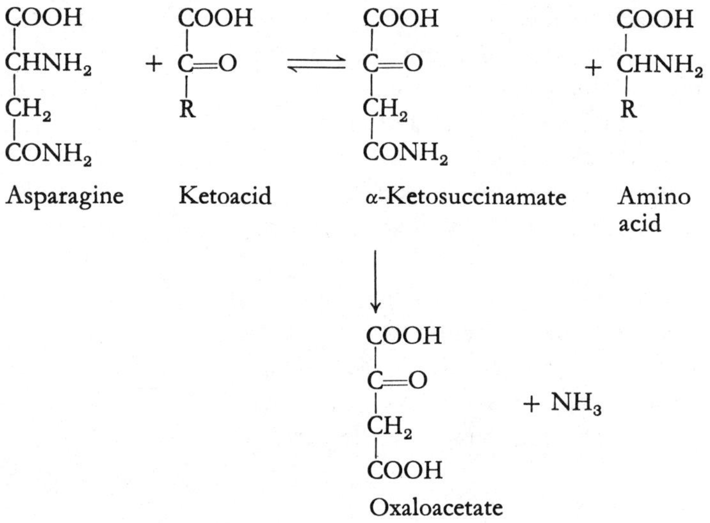

$$\begin{array}{c}\text{COOH}\\|\\\text{CHNH}_2\\|\\\text{CH}_2\\|\\\text{CONH}_2\end{array} + \begin{array}{c}\text{COOH}\\|\\\text{C=O}\\|\\\text{R}\end{array} \rightleftharpoons \begin{array}{c}\text{COOH}\\|\\\text{C=O}\\|\\\text{CH}_2\\|\\\text{CONH}_2\end{array} + \begin{array}{c}\text{COOH}\\|\\\text{CHNH}_2\\|\\\text{R}\end{array}$$

Asparagine   Ketoacid   α-Ketosuccinamate   Amino acid

$$\downarrow$$

$$\begin{array}{c}\text{COOH}\\|\\\text{C=O}\\|\\\text{CH}_2\\|\\\text{COOH}\end{array} + NH_3$$

Oxaloacetate

7.3. Deamination Reactions

In addition to following transamination pathways, some amino acids can be converted directly to their corresponding ketoacids. These deamination processes may be divided into two general categories: *oxidative deamination* and *non-oxidative deamination*. In the former, the reactions are catalyzed by NAD^+-linked enzymes or flavoproteins. Non-oxidative deamination generally involves pyridoxal phosphate-bound enzymes.

A. OXIDATIVE DEAMINATION

The most important of these reactions is the oxidation of glutamate to α-ketoglutarate, which is catalyzed by the NAD^+-linked enzyme, *glutamate dehydrogenase* (L-glutamate:NAD^+ oxidoreductase, EC 1.4.1.3):

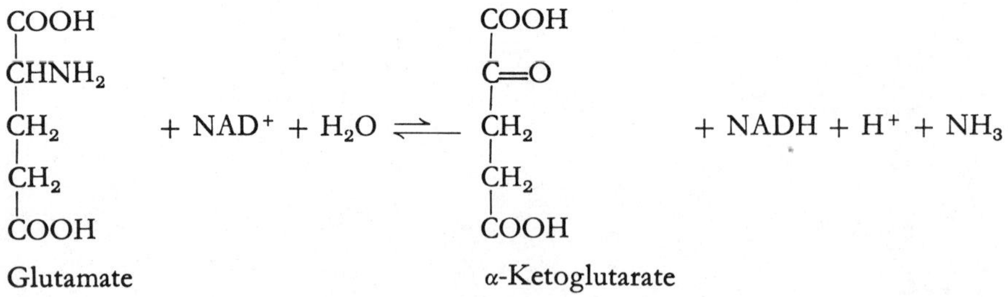

$$\begin{array}{c}\text{COOH}\\|\\\text{CHNH}_2\\|\\\text{CH}_2\\|\\\text{CH}_2\\|\\\text{COOH}\end{array} + NAD^+ + H_2O \rightleftharpoons \begin{array}{c}\text{COOH}\\|\\\text{C=O}\\|\\\text{CH}_2\\|\\\text{CH}_2\\|\\\text{COOH}\end{array} + NADH + H^+ + NH_3$$

Glutamate   α-Ketoglutarate

(A similar reaction is also catalyzed by an $NADP^+$-linked dehydrogenase.)

Since the equilibrium constant for this reaction, as written, is about 2×10^{-4}, it can be expected that the reverse reaction—i.e., the formation of glutamate—would be favored. Nonetheless, deamination of glutamate does take place, because the NADH that is generated is oxidized by the mitochondrial electron-transport system.

The glutamate dehydrogenase reaction can be visualized as a link in the general process in which amino acids are converted to ketoacids via transamination with α-ketoglutarate. The glutamate that is formed as a result of such transamination can be reoxidized by glutamate dehydrogenase to regenerate the α-ketoglutarate:

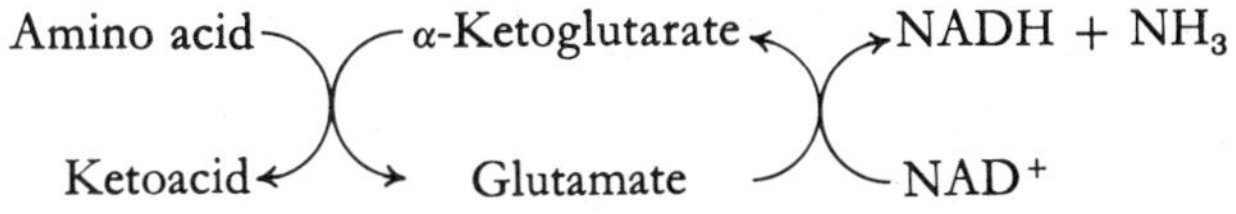

Amino acid ⟶ α-Ketoglutarate ⟶ NADH + NH₃
Ketoacid ⟵ Glutamate ⟵ NAD⁺

Conversely, the reverse reaction catalyzed by glutamate dehydrogenase can function to provide glutamate from α-ketoglutarate. The glutamate can then be utilized for the biosynthesis of amino acids from corresponding ketoacids by transamination processes.

Another type of oxidative deamination is catalyzed by L-*amino-acid oxidases*; these contain flavoproteins that are reduced in the deamination reaction and then reoxidized with oxygen:

$$\underset{\text{RCHCOOH}}{\overset{\text{NH}_2}{|}} + \text{Flavoprotein} \rightleftharpoons \underset{\text{RCCOOH}}{\overset{\text{O}}{\|}} + \text{Reduced flavoprotein} + \text{NH}_3$$

$$\text{Reduced flavoprotein} + \text{O}_2 \longrightarrow \text{Flavoprotein} + \text{H}_2\text{O}_2$$

L-Amino-acid oxidases are known to be present in the liver and kidneys. These enzymes can oxidize all amino acids except serine, threonine, and dibasic amino acids. The activity of these enzymes is comparatively low, and they probably do not play a major role in amino acid metabolism.

B. NON-OXIDATIVE DEAMINATION

Certain amino acids—e.g., serine, threonine, and cysteine—are deaminated by specific lyases that require pyridoxal phosphate. Instead of transferring the amino group to other substrates, these enzymes release ammonia into the medium. Examples of such reactions are those catalyzed by the lyases *serine dehydratase* and *histidase*. The organic products in these reactions are pyruvate and urocanic acid, respectively:

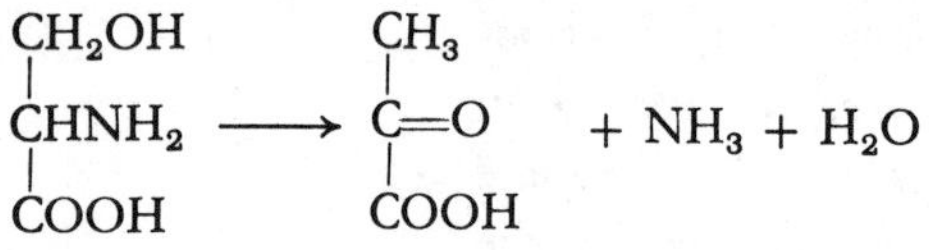

$$\begin{array}{c}\text{CH}_2\text{OH}\\|\\\text{CHNH}_2\\|\\\text{COOH}\end{array} \longrightarrow \begin{array}{c}\text{CH}_3\\|\\\text{C}=\text{O}\\|\\\text{COOH}\end{array} + \text{NH}_3 + \text{H}_2\text{O}$$

Serine → Pyruvate

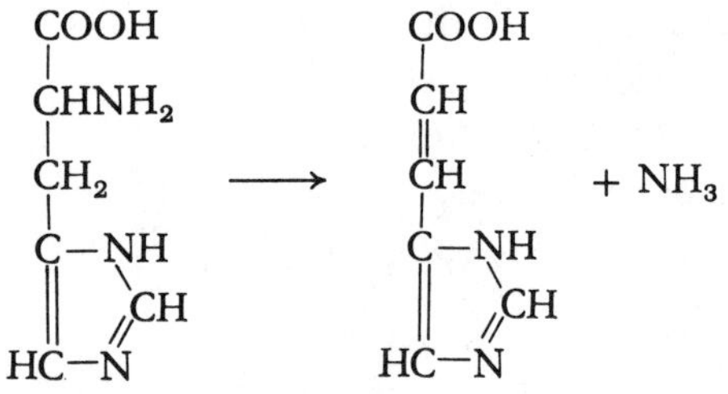

$$\begin{array}{c}\text{COOH}\\|\\\text{CHNH}_2\\|\\\text{CH}_2\\|\\\text{C}-\text{NH}\\\|\diagdown\text{CH}\\\text{HC}-\text{N}\end{array} \longrightarrow \begin{array}{c}\text{COOH}\\|\\\text{CH}\\\|\\\text{CH}\\|\\\text{C}-\text{NH}\\\|\diagdown\text{CH}\\\text{HC}-\text{N}\end{array} + \text{NH}_3$$

Histidine → Urocanate

These reactions and additional examples will be discussed in the metabolism of individual amino acids (Chap. 8).

7.4. Ammonia Metabolism

The transfer of amino groups from most amino acids to α-ketoglutarate, and the action of glutamate dehydrogenase on the resultant glutamate, leads to the release of considerable amounts of ammonia. Smaller amounts of ammonia also arise from the action of amino-acid oxidase, non-oxidative deaminases, and the oxidation of various catecholamines. Other reactions that yield ammonia include the hydrolysis of α-ketoglutaramic and ketosuccinamic acids (section 7.2) and the *glutaminase*-catalyzed hydrolysis of glutamine:

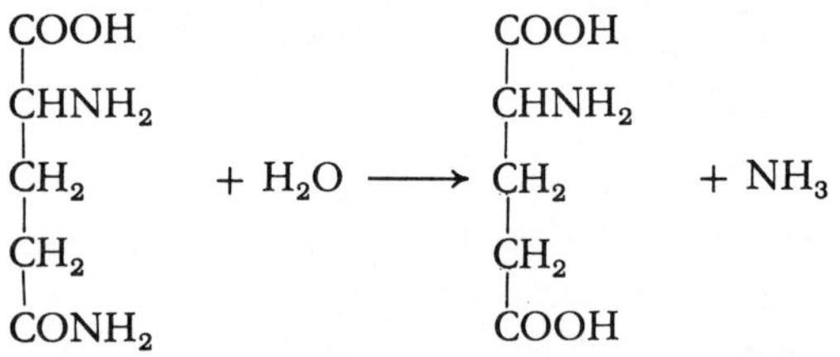

$$\begin{array}{c}\text{COOH}\\|\\\text{CHNH}_2\\|\\\text{CH}_2\\|\\\text{CH}_2\\|\\\text{CONH}_2\end{array} + \text{H}_2\text{O} \longrightarrow \begin{array}{c}\text{COOH}\\|\\\text{CHNH}_2\\|\\\text{CH}_2\\|\\\text{CH}_2\\|\\\text{COOH}\end{array} + \text{NH}_3$$

Glutamine → Glutamate

The potentially lethal action of ammonia is prevented by its incorporation into various metabolic intermediates. Essentially, there are three mechanisms for the disposal of ammonia: (1) interaction with α-ketoglutarate and NADH to yield glutamate (section 7.3.A), (2) synthesis of glutamine from glutamate, and (3) synthesis of carbamoyl phosphate. The latter two mechanisms will be discussed individually.

A. SYNTHESIS OF GLUTAMINE

Ammonia reacts with glutamic acid and ATP to provide glutamine, ADP, and inorganic phosphate. The process involves two steps: first, the formation of γ-glutamyl phosphoric acid and, second, the interaction of this intermediate with ammonia:

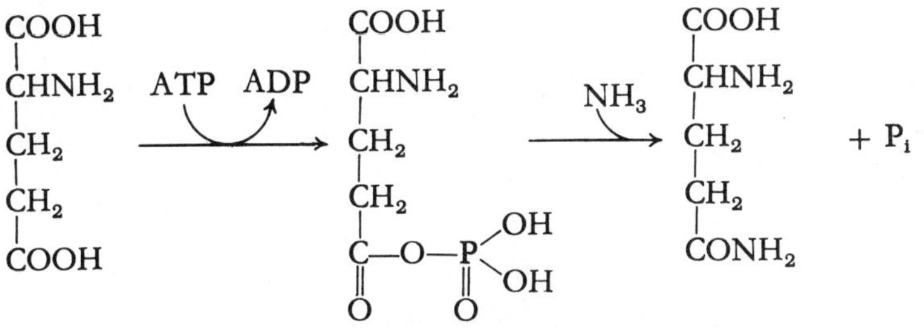

$$\begin{array}{c}\text{COOH}\\|\\ \text{CHNH}_2\\|\\ \text{CH}_2\\|\\ \text{CH}_2\\|\\ \text{COOH}\end{array} \xrightarrow{\text{ATP} \quad \text{ADP}} \begin{array}{c}\text{COOH}\\|\\ \text{CHNH}_2\\|\\ \text{CH}_2\\|\\ \text{CH}_2\\|\\ \text{C}-\text{O}-\text{P}(\text{OH})_2\\ \| \quad \|\\ \text{O} \quad \text{O}\end{array} \xrightarrow{\text{NH}_3} \begin{array}{c}\text{COOH}\\|\\ \text{CHNH}_2\\|\\ \text{CH}_2\\|\\ \text{CH}_2\\|\\ \text{CONH}_2\end{array} + P_i$$

Glutamate $\qquad$ γ-Glutamyl phosphoric acid $\qquad$ Glutamine

This reaction is catalyzed by *glutamine synthetase*, which is a highly complex enzyme and subject to various regulatory mechanisms. It has been purified from liver and brain and shown to be located in the microsomal fraction (endoplasmic reticulum) of the cells. Excess ammonia arising from the intake of ammonium salts or large amounts of amino acids appears to be removed primarily by means of the glutamine-synthetase reaction.

B. SYNTHESIS OF CARBAMOYL PHOSPHATE

Carbamoyl phosphate is formed by the reaction of carbon dioxide, ammonia, and 2 moles of ATP:

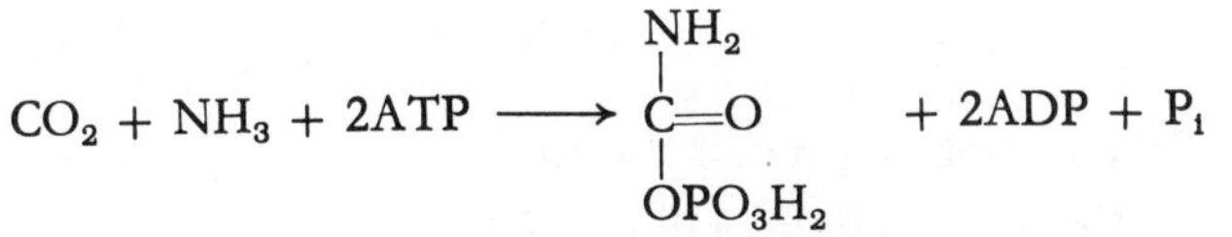

$$CO_2 + NH_3 + 2ATP \longrightarrow \begin{array}{c}\text{NH}_2\\|\\ \text{C}=\text{O}\\|\\ \text{OPO}_3\text{H}_2\end{array} + 2ADP + P_i$$

Carbamoyl phosphate

The enzyme for this reaction, *carbamoyl-phosphate synthetase*, requires N-acetylglutamate as an allosteric activator:

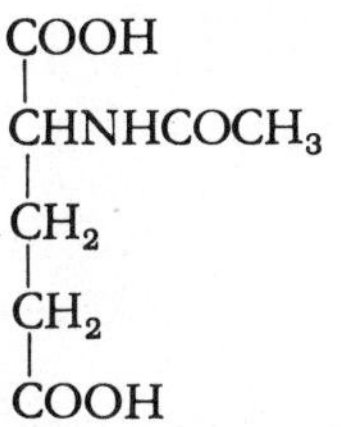

COOH
|
CHNHCOCH$_3$
|
CH$_2$
|
CH$_2$
|
COOH

N-Acetylglutamate

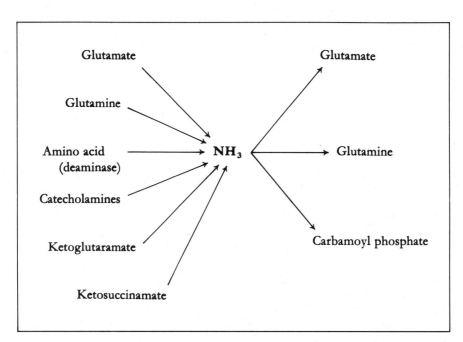

Fig. 7-2. Processes involved in the formation and utilization of ammonia.

Carbamoyl phosphate is formed in the liver and, to a lesser extent, in other cells. Its principal mode of utilization in liver is in the synthesis of urea.

The different processes for the generation and disposition of ammonia are summarized in Figure 7-2.

7.5. Formation of Urea

The principal end product of protein-nitrogen metabolism (in man) is urea, which is finally excreted in the urine. Normally, the amount of urea produced is dependent on the protein intake. The reaction leading to the formation of urea is the breakdown of arginine as catalyzed by the enzyme *arginase*. Urea is split off from the arginine molecule to leave ornithine:

$$
\begin{array}{c}
\text{COOH} \\
| \\
\text{CHNH}_2 \\
| \\
\text{CH}_2 \\
| \\
\text{CH}_2 \quad\quad \text{NH} \\
| \quad\quad\quad\quad \| \\
\text{CH}_2\text{—NHCNH}_2
\end{array}
\;+\; H_2O \longrightarrow
\begin{array}{c}
\text{COOH} \\
| \\
\text{CHNH}_2 \\
| \\
\text{CH}_2 \\
| \\
\text{CH}_2 \\
| \\
\text{CH}_2\text{NH}_2
\end{array}
\;+\;
\begin{array}{c}
\text{NH}_2 \\
| \\
\text{C}=\text{O} \\
| \\
\text{NH}_2
\end{array}
$$

Arginine   Ornithine   Urea

This reaction occurs in the liver, which seems to be the only tissue that contains arginase. Amino groups arising from the catabolism of amino acids are utilized in the synthesis of arginine, which in turn releases urea. The total sequence of reactions that ultimately yield urea is known as the *urea cycle*.

Carbamoyl phosphate, which is formed from ammonia, carbon dioxide, and ATP, condenses with ornithine to yield citrulline. The enzyme that catalyzes the reaction is *ornithine transcarbamoylase*:

$$\begin{array}{c}\text{COOH}\\|\\\text{CHNH}_2\\|\\\text{CH}_2\\|\\\text{CH}_2\\|\\\text{CH}_2\text{NH}_2\end{array} + \begin{array}{c}\text{NH}_2\\|\\\text{C}=\text{O}\\|\\\text{OPO}_3\text{H}_2\end{array} \longrightarrow \begin{array}{c}\text{COOH}\\|\\\text{CHNH}_2\\|\\\text{CH}_2\\|\\\text{CH}_2\\|\\\text{CH}_2\text{NHCNH}_2\\\|\\\text{O}\end{array} + P_i$$

Ornithine   Carbamoyl phosphate   Citrulline

This reaction is followed by the condensation of citrulline with aspartate to form argininosuccinate. The condensation reaction requires ATP and is catalyzed by *argininosuccinate synthetase*; note that the ATP undergoes pyrophosphate removal:

$$\begin{array}{c}\text{COOH}\\|\\\text{CHNH}_2\\|\\\text{CH}_2\\|\\\text{CH}_2\\|\\\text{CH}_2\text{NHCNH}_2\\\|\\\text{O}\end{array} + \begin{array}{c}\text{COOH}\\|\\\text{CH}_2\\|\\\text{H}_2\text{NCHCOOH}\end{array} + \text{ATP} \xrightarrow{\text{Mg}^{2+}} \begin{array}{c}\text{COOH}\\|\\\text{CHNH}_2\\|\\\text{CH}_2\\|\\\text{CH}_2\\|\\\text{CH}_2\text{NHCNHCHCOOH}\\\|\\\text{NH}\end{array}\begin{array}{c}\text{COOH}\\|\\\text{CH}_2\end{array} + \text{AMP} + \text{PP}_i$$

Citrulline   Aspartate   Argininosuccinate

The enzyme *argininosuccinate lyase* (*argininosuccinase*) can then effect the cleavage of argininosuccinate to provide arginine and fumarate:

$$\begin{array}{c}\text{COOH}\\|\\\text{CHNH}_2\\|\\\text{CH}_2\\|\\\text{CH}_2\\|\\\text{CH}_2\text{NHCNHCHCOOH}\\\|\\\text{NH}\end{array}\begin{array}{c}\text{COOH}\\|\\\text{CH}_2\end{array} \longrightarrow \begin{array}{c}\text{COOH}\\|\\\text{CNNH}_2\\|\\\text{CH}_2\\|\\\text{CH}_2\\|\\\text{CH}_2\text{NHCNH}_2\\\|\\\text{NH}\end{array} + \begin{array}{c}\text{HCCOOH}\\\|\\\text{HOOCCH}\end{array}$$

Argininosuccinate   Arginine   Fumarate

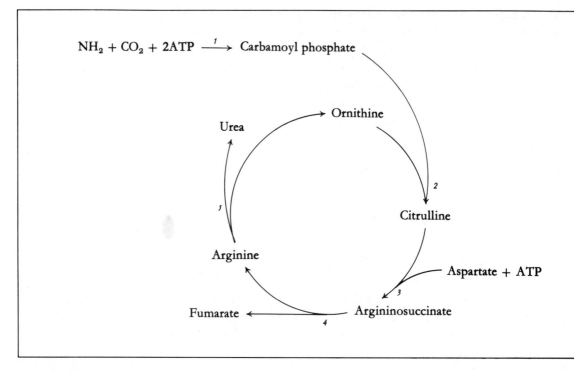

Fig. 7-3. Urea cycle. The enzymes for the reactions are 1, carbamoyl-phosphate synthetase; 2, ornithine transcarbamoylase; 3, argininosuccinate synthetase; 4, argininosuccinate lyase; 5, arginase.

Thus, the series of reactions constitutes a cyclic process (Fig. 7-3) that yields fumarate and urea from carbamoyl phosphate and aspartate. Since the intermediates of the cycle—arginine, citrulline, and ornithine—are regenerated, only small amounts of them are required to maintain the process.

It can be seen that the production of urea from NH_3 and CO_2 utilizes 3 ATPs. However, since the argininosuccinate reaction involves the release of a pyrophosphate unit from ATP, the energy expended in this step is that of 2 ATPs (pages 145 and 378). In terms of energy, therefore, the total synthesis of urea consumes the equivalent of 4 ATPs.

Most of the ammonia that is incorporated into carbamoyl phosphate is derived from the action of mitochondrial, NAD^+-linked, glutamate dehydrogenase in the liver (page 306). Glutamate is also the immediate donor of the amino group of aspartate (aspartate-glutamate transaminase reaction), and this aspartate enters the urea cycle at the citrulline condensation step. Thus, both of the nitrogen atoms of urea can arise from glutamate. Since the amino group of glutamate can be obtained from most amino acids, the urea cycle functions to transport nitrogen from amino acids to urea.

The removal of ammonia via carbamoyl phosphate and the urea cycle is of special importance, because ammonia can have poisonous effects on intermediary metabolism. Should ammonia accumulate, it is reasonable to expect that it will interact with α-ketoglutarate and NADH to form glutamate (glutamate dehydrogenase reaction). Such uptake of α-ketoglutarate may interfere with the normal functioning of the citric acid cycle and thus inhibit respiration. Furthermore, the oxidative disposal of acetate, which is also dependent on the citric acid cycle, would be depressed, and there would be a consequent increase in ketone body formation. Since α-ketoglutarate is also an intermediate in numerous other processes (e.g., gluconeogenesis and transamination), its depletion may alter the normal metabolism of the cell with respect to these processes. Several factors serve to determine that the required metabolic pathway will predominate; these include control of enzyme activity, enzyme compartmentalization, and hormonal effects.

Some interrelationships among amino acid, lipid, and carbohydrate metabolism are shown in Figure 7-4.

7.6. Catabolism of the Carbon Skeleton of Amino Acids

The hydrocarbon portions of amino acids can be carried through a wide variety of pathways and utilized in the syntheses of numerous intermediates. A number of these reactions will be described subsequently. This section is concerned with the transformations that lead to the complete oxidation of the carbon skeleton and the concomitant release of energy.

All the amino acids are ultimately convertible to either acetyl-coenzyme A, pyruvate, or intermediates of the citric acid cycle. Since the pathways for the conversion of these metabolites to carbon dioxide and water have already been described, it remains only to show how they are formed from amino acids. Some reactions in which amino acids form the requisite ketoacids directly were described in the foregoing sections. Thus, transamination from alanine, glutamate, and aspartate yields pyruvate, α-ketoglutarate, and oxaloacetate, respectively. The carbon atoms of aspartate also provide fumarate as a result of the reactions of the urea cycle. Somewhat more complex sequences lead to the formation of the relevant intermediates from the other amino acids (Chap. 8). The overall relationships are indicated in Figure 7-5.

The degree of utilization of amino acids from either dietary protein, or tissue protein that is turning over, for fulfilling energy needs depends on the nutritional state of the individual. When the dietary intake is normal,

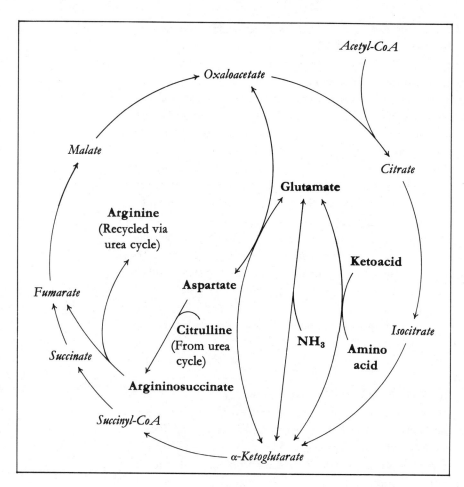

Fig. 7-4. Some interactio between the metabolism amino acids and that of lipids and carbohydrates (Citric acid cycle intermediates are indica italics; components of amino acid metabolism, boldface type.)

only about 10% of the total energy requirement is derived from protein. When the ingestion of lipids or carbohydrates is restricted, increased proportions of protein will be utilized for energy. Conversely, increased carbohydrate ingestion will have a sparing effect on the rate of amino acid degradation.

The principal pathway for the complete degradation of the carbon skeletons of amino acids is their transformation to either pyruvate, acetyl-coenzyme A, or citric acid-cycle intermediates, which in turn are oxidized to carbon dioxide and water.

7.7. Glucogenic and Ketogenic Amino Acids

In addition to being oxidized to carbon dioxide and water, the metabolic intermediates pyruvate, oxaloacetate, α-ketoglutarate, succinate, and fumarate can be utilized for the synthesis of glucose (gluconeogenesis). Amino

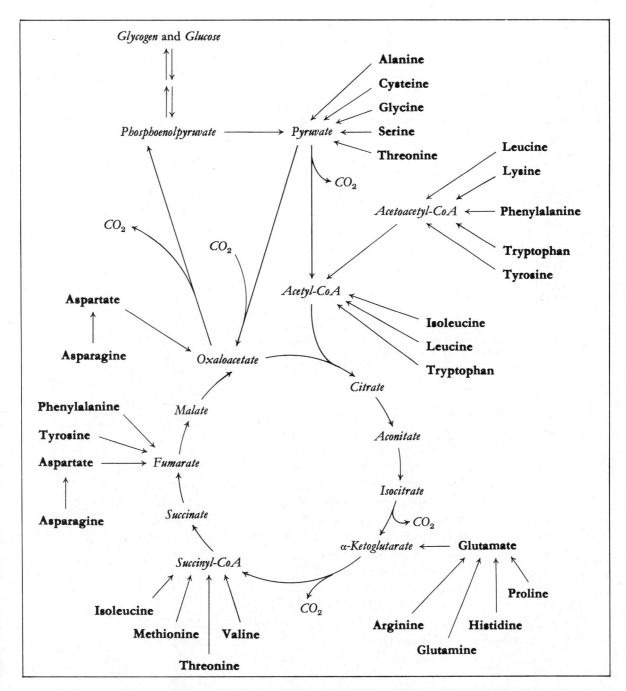

Fig. 7-5. Pyruvate, acetate, and citric acid cycle components (italics) as intermediates in the metabolism of amino acids (boldface). Phosphoenolpyruvate is shown in order to indicate the pathway to glucose and glycogen (gluconeogenesis). Note that some amino acids lead into more than one site.

Table 7-2. Glucogenic and Ketogenic Amino Acids

Glucogenic	Ketogenic	Glucogenic and Ketogenic
Alanine	Leucine	Isoleucine
Arginine	Lysine*	Phenylalanine
Asparagine		Tryptophan*
Aspartic acid		Tyrosine
Cysteine		
Glutamic acid		
Glutamine		
Glycine		
Histidine		
Proline		
Hydroxyproline		
Methionine		
Serine		
Threonine		
Valine		

* The original "feeding" studies led to the classification of tryptophan as glucogenic, and lysine as both glucogenic and ketogenic. However, experiments on intermediary metabolism of amino acids indicate that lysine is essentially ketogenic and that tryptophan is mainly ketogenic and only slightly glucogenic.

acids that can be converted to any of these intermediates are therefore potential sources of glucose; such amino acids are termed *glucogenic*. Other amino acids, which ultimately form acetoacetate or acetyl-coenzyme A, are called *ketogenic*. The experimental basis for the classification is based on observations after administering particular amino acids to diabetic animals. It was found that the administration of certain amino acids resulted in an increased urinary excretion of glucose, whereas others caused an elevation of ketone bodies. The specific amino acids in each category are shown in Table 7-2, and the rationale for this classification may be understood upon consideration of the relationships in Figure 7-5.

Some amino acids are both glucogenic and ketogenic, since they give rise to both dicarboxylic acids and acetate. For example, isoleucine can be transformed both to succinyl-coenzyme A and acetyl-coenzyme A. Similarly, tyrosine and phenylalanine yield both fumarate and acetoacetyl-coenzyme A.

7.8. Transmethylation

Carbon atoms from various amino acids can be utilized for the synthesis of numerous metabolic intermediates. Thus, in addition to providing acetate, pyruvate, and dicarboxylic acids on degradation, specific sections of amino acids can yield building blocks for the synthesis of a diversity of molecules that serve as cellular components. The reactions of certain amino acids can provide methyl groups (–CH$_3$) for such synthetic processes; these are termed transmethylation reactions.

Transmethylation, as the name implies, involves the transfer of an intact methyl group from one intermediate to another. The sources for such methyl groups are methionine and betaine. The latter is derived from choline:

$$(CH_3)_3N^+CH_2CH_2OH \longrightarrow (CH_3)_3N^+CH_2COO^-$$
Choline → Betaine

The first step in the action of methionine as a methyl donor is the formation of S-adenosylmethionine:

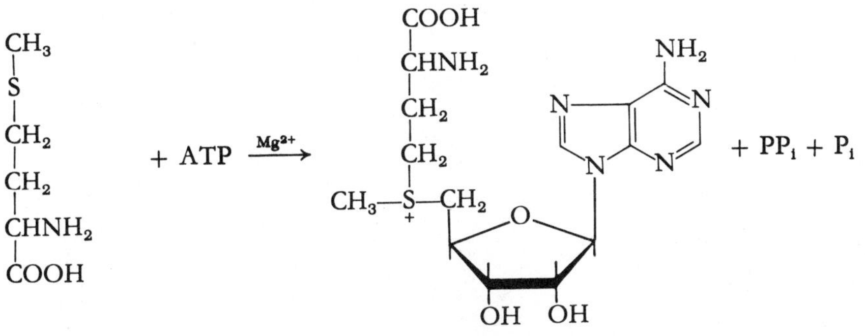

Methionine + ATP $\xrightarrow{Mg^{2+}}$ S-Adenosylmethionine + PP$_i$ + P$_i$

This reaction is catalyzed by methyl-adenosyl transferase and requires the presence of reduced glutathione. The methyl group attached to the sulfur atom of S-adenosylmethionine can then be transferred to various acceptors by specific methyltransferases. Prominent among these acceptor compounds are phosphatidylethanolamine, guanidinoacetic acid, nicotinamide, norepinephrine, and carnosine (Fig. 7-6). As a result of these reactions, S-adenosylmethionine is converted to S-adenosylhomocysteine.

The importance of transmethylation can be realized from the variety of metabolic processes that it affects. Adequate levels of phosphatidylcholine (lecithin) are required for the synthesis of lipoproteins and normal lipid metabolism. (Phosphatidylcholine can also be provided by dietary phospholipids.) Methyl groups are also necessary for the formation of epinephrine and the metabolism of nicotinic acid (niacin) and pyridine nucleotides. The role of methyl-group transfer in the formation of creatine will be described in more detail in the next section.

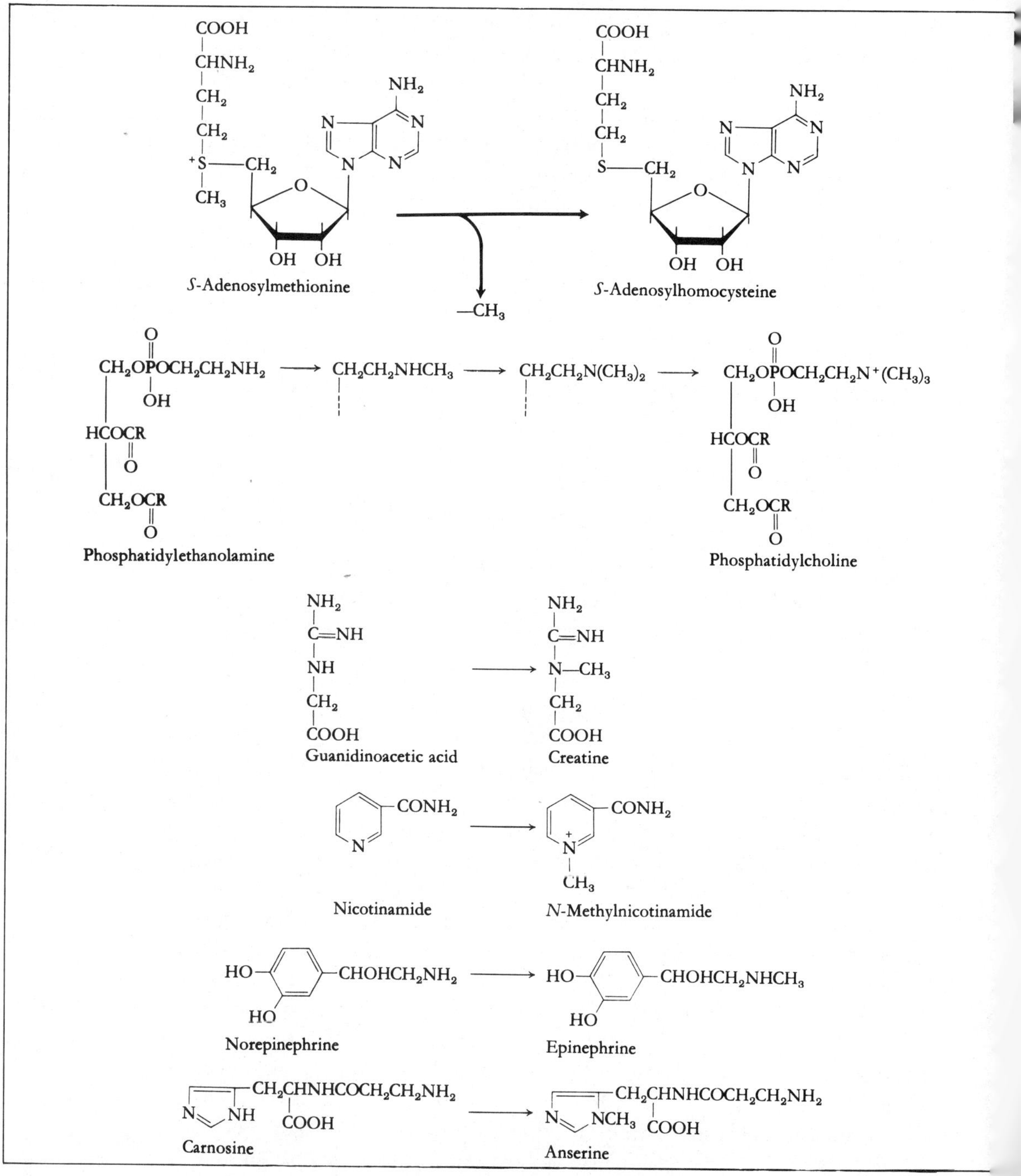

Fig. 7-6. *Transfer of methyl groups from S-adenosylmethionine. Note that the formation of phosphatidylcholine (lecithin) from phosphatidylethanolamine involves the successive transfer of three methyl groups.*

7.9. Creatine Metabolism

Creatine is vital to muscle metabolism in that it serves in the storage of energy. This compound is formed from arginine, glycine, and methionine; the latter functions as the methyl donor. The first step in creatine formation is a reaction between glycine and arginine that is catalyzed by a transamidinase. Ornithine and guanidinoacetic acid are the products of the reaction:

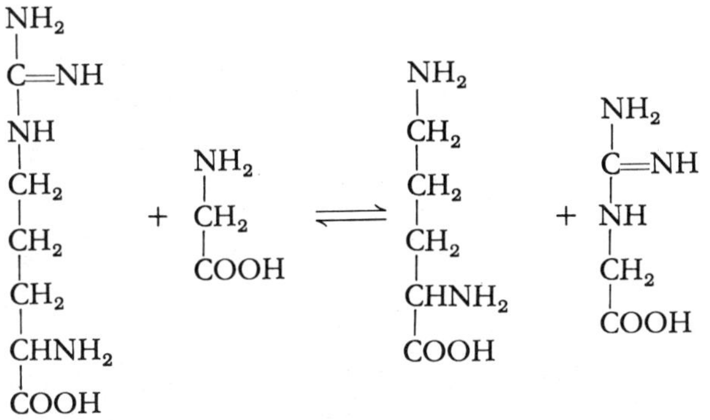

$$\text{Arginine} + \text{Glycine} \rightleftharpoons \text{Ornithine} + \text{Guanidinoacetic acid}$$

This amidine transfer reaction occurs in human kidney, liver, and pancreas. A methyl group is then transferred from methionine (via S-adenosylmethionine) to guanidinoacetic acid to provide creatine as shown (Fig. 7-6).

Creatine can be converted to phosphocreatine by the action of *creatine kinase* and ATP:

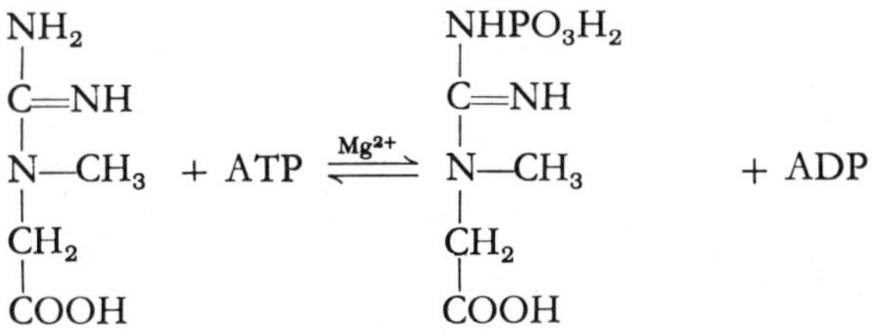

$$\text{Creatine} + \text{ATP} \xrightleftharpoons{Mg^{2+}} \text{Phosphocreatine} + \text{ADP}$$

Phosphocreatine has a large negative free energy of hydrolysis (about −9 kilocalories per mole), similar to that of ATP. Because of this, phosphocreatine functions as an energy storage form in muscle. Its availability allows for the rapid conversion of ADP to ATP and the utilization of the latter for muscle contraction.

Muscle contains about 40 micromoles of creatine per gram, or 0.5%. In resting muscle, it is present primarily as phosphocreatine. However, the latter has limited stability and is easily transformed to creatinine:

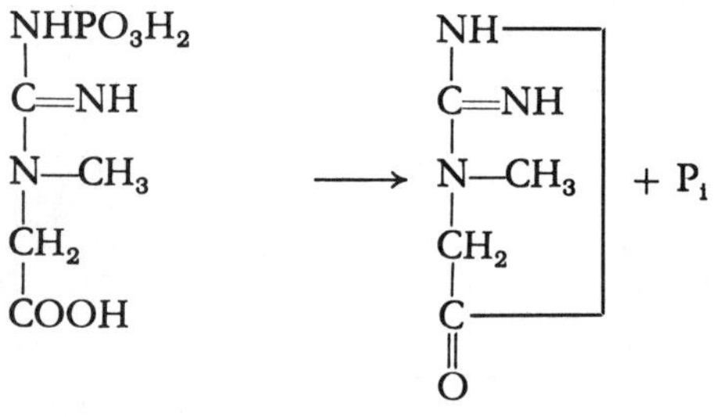

$$\begin{array}{c} NHPO_3H_2 \\ | \\ C\!=\!NH \\ | \\ N\!-\!CH_3 \\ | \\ CH_2 \\ | \\ COOH \end{array} \longrightarrow \left[\begin{array}{c} NH\!-\!\!\!\!\!\!\!\!\!\rule[0.5ex]{0.8em}{0.4pt} \\ | \\ C\!=\!NH \\ | \\ N\!-\!CH_3 \\ | \\ CH_2 \\ | \\ C\!\!\!\!\!\rule[0.5ex]{0.8em}{0.4pt} \\ \| \\ O \end{array} \right] + P_i$$

Phosphocreatine   Creatinine

Creatinine diffuses from the muscle to the blood and is ultimately excreted in the urine. As a consequence of the steady loss of creatine in the form of creatinine, it must be replenished continually. The fraction of phosphocreatine degraded daily is dependent on the muscle mass. In the average adult, about 15 millimoles or 1.7 grams of creatinine are excreted each day. The daily urinary creatinine value is fairly constant, and normally, it is affected only by the amount of ingested meat (i.e., muscle tissue). In diseases that involve muscle degeneration (e.g., dystrophy or paralysis), there is a decrease in the level of urinary creatinine. On the other hand, elevated blood levels are indicative of kidney damage.

A regulatory mechanism in creatine metabolism is the inhibition of the synthesis of guanidinoacetic acid by creatine. Such feedback inhibition functions to regulate the synthesis of creatine and its metabolites. Since the formation of creatine from guanidinoacetic acid consumes available methyl groups, fine control of the process is especially critical.

7.10. Single-Carbon Transfers Other than Methylation

Certain amino acids provide single-carbon units other than methyl groups for various metabolic syntheses. The carbon carrier in such processes is *reduced folic acid*, abbreviated as FH_4 or THFA.

Reduced folic acid, a cofactor, is derived from the vitamin folic acid. Structurally, this vitamin may be visualized as composed of glutamic acid, *p*-aminobenzoic acid, and 2-amino-4-hydroxy-6-methylpteridine. The combination of the two latter units is called *pteroic acid*. Hence, folic acid is also designated as pteroylglutamic acid:

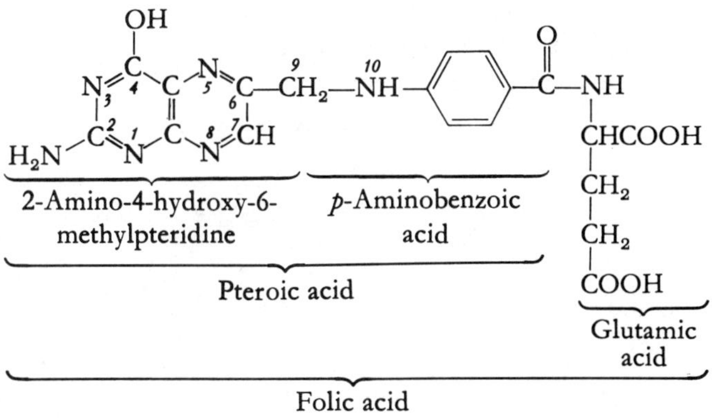

$$\underbrace{\underbrace{\overset{OH}{\underset{H_2N}{\overset{|}{\underset{|}{\overset{N^{\nwarrow^4}}{\underset{C_2}{\overset{C}{\underset{N}{\overset{|}{C}}}}}\overset{C}{\underset{|}{\overset{N^{\nwarrow^5}}{\underset{6}{C}}}}\overset{|}{\underset{7}{C}}\overset{|}{\underset{CH}{}}}}}_{\text{2-Amino-4-hydroxy-6-methylpteridine}} \underbrace{-\overset{9}{CH_2} - \overset{10}{NH} - \underset{}{\bigcirc} -}_{\text{p-Aminobenzoic acid}}}_{\text{Pteroic acid}} \underbrace{\overset{O}{\underset{}{\overset{\|}{C}}} - NH}_{\substack{| \\ CHCOOH \\ | \\ CH_2 \\ | \\ CH_2 \\ | \\ COOH}}}_{\text{Glutamic acid}}$$

Folic acid

The biologically active form of the vitamin is tetrahydrofolic acid, in which carbons 6 and 7 and nitrogens 5 and 8 are reduced:

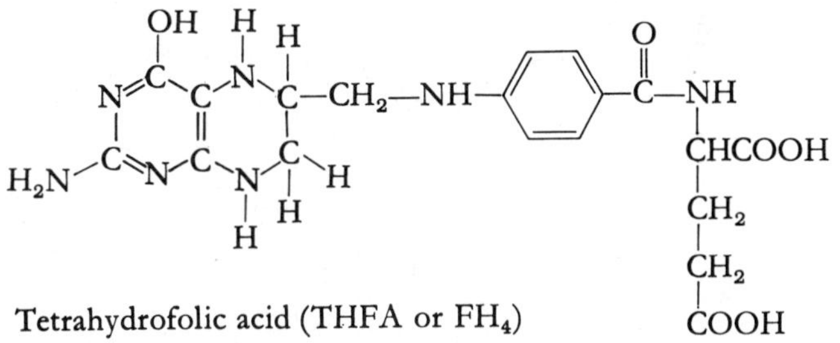

$$\overset{OH}{\underset{H_2N}{\overset{|}{\underset{|}{\overset{N}{\underset{C}{\overset{C}{\underset{N}{\overset{|}{C}}}}}\overset{|}{\underset{|}{\overset{H}{\underset{|}{\overset{N}{\underset{H}{\overset{|}{C}}}}}}}\overset{H}{\underset{|}{\overset{|}{\underset{H}{\overset{C}{\underset{H}{\overset{|}{}}}}}}}}} - CH_2 - NH - \bigcirc - \overset{O}{\underset{}{\overset{\|}{C}}} - NH \\ | \\ CHCOOH \\ | \\ CH_2 \\ | \\ CH_2 \\ | \\ COOH$$

Tetrahydrofolic acid (THFA or FH$_4$)

Table 7-3. Donors of Single-Carbon Units to Tetrahydrofolic Acid

Donor	Mechanism
Serine	Conversion to glycine with transfer of methylene
Formaldehyde	Direct combination
Methionine	Oxidation of methyl group, which is transferred as —HC=O
Choline	Conversion to betaine; methyl groups are oxidized and transferred
Sarcosine	Oxidation of methyl group, which is transferred as H$_2$C=O
Glycine	Oxidation to formate, which is transferred
δ-Aminolevulinic acid	Oxidation to formate, which is transferred
Tryptophan	Oxidation to N-formylkynurenine, which donates formate
Formate	Direct combination
Histidine	Formation of formiminoglutamic acid and transfer of formimino (—HC=NH) group; (formiminoglycine can function in a similar manner)

Tetrahydrofolic acid can accept one-carbon units from various amino acids or their metabolites. A prominent example is the removal of such a group—the hydroxymethyl group—from serine and the consequent formation of glycine. Tetrahydrofolic acid is thus transformed to N^5,N^{10}-methylenetetrahydrofolic acid:

CH2OH
CHNH2
COOH
Serine
CH2NH2
COOH
Glycine
Pyridoxal phosphate
Tetrahydrofolic acid
OH
CH2
N
C
CH–CH2
H2N
H
CH2
O
NH
CHCOOH
CH2
CH2
COOH

N^5,N^{10}-Methylenetetrahydrofolic acid

Single-carbon groups may be linked to tetrahydrofolic acid in a number of interconvertible forms that serve to transfer these groups in various synthetic reactions (Fig. 7-7). In addition to serine, other intermediates can donate single-carbon units to tetrahydrofolic acid (Table 7-3). (The metabolism of some of these substances will be described in later sections.)

The single-carbon units carried by tetrahydrofolate may be utilized in nucleotide synthesis. For example, two steps in the intracellular synthesis of purine nucleotides involve the incorporation of one-carbon units from tetrahydrofolate derivatives. Also, when a methyl group is introduced into uracil to form thymine, the reaction involves a similar one-carbon transfer.

Other examples of transfers of single-carbon groups from tetrahydrofolate are provided by the formation of serine from glycine and of methionine from homocysteine.

7.11. Essential and Non-essential Amino Acids

A. DEFINITION

A number of the metabolic reactions described in the foregoing sections

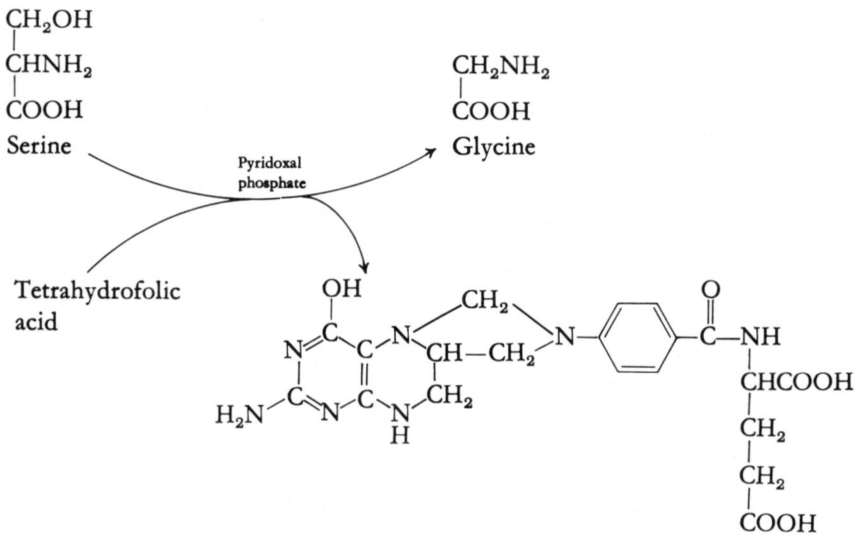

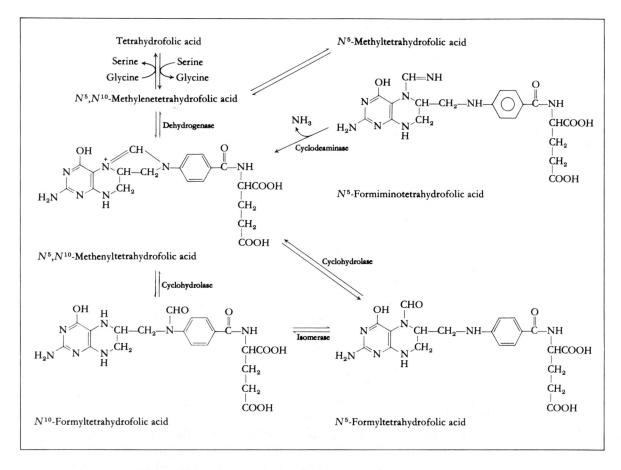

Fig. 7-7. *Single-carbon group derivatives of tetrahydrofolic acid and their interconversion.*

involved the synthesis of amino acids from components of the glycolytic pathway and the citric acid cycle. In addition, several of the amino acids can be formed from other amino acids. Such amino acids are termed *non-essential*; that is, they can be made available to the cells even though they are not included in the diet. These are distinguished from *essential* amino acids, which cannot be synthesized by the mammalian organism but must be ingested. The amino acids included in each of these categories are shown in Table 7-4.

The biosynthetic routes for the non-essential amino acids are described later in this section. In considering these processes, it should be realized

Table 7-4. Essential and Non-essential Amino Acids in Man

Essential	Non-essential
Histidine*	Alanine
Isoleucine	Arginine†
Leucine	Aspartic acid
Lysine	Asparagine
Methionine	Cysteine
Phenylalanine	Glutamic acid
Threonine	Glutamine
Tryptophan	Glycine
Valine	Hydroxyproline
	Proline
	Serine
	Tyrosine

* Histidine is essential, at least, until age 12.
† Arginine may be essential for infants.

that if any of these amino acids are excluded from the diet, there may be a higher demand for the essential amino acids, since a certain portion of the latter will be utilized for the synthesis of non-essential amino acids. For example, tyrosine is classified as non-essential because it can be formed from phenylalanine. However, in the absence of exogenous tyrosine, the amount of phenylalanine that is required is increased significantly.

A number of the non-essential amino acids are formed during the metabolic degradation of other amino acids. According to strict definition, these are categorized as non-essential, although the amounts of them contributed by metabolism may be comparatively small (these amino acids will be indicated in the subsequent discussion). The amino acids that are formed with ease and are always available include alanine, aspartate, glutamate, serine, and glycine. These amino acids are also found in greatest abundance in proteins.

B. SYNTHESIS OF NON-ESSENTIAL AMINO ACIDS

Alanine, aspartate, and *glutamate* can be formed by transamination from pyruvate, oxaloacetate, and α-ketoglutarate:

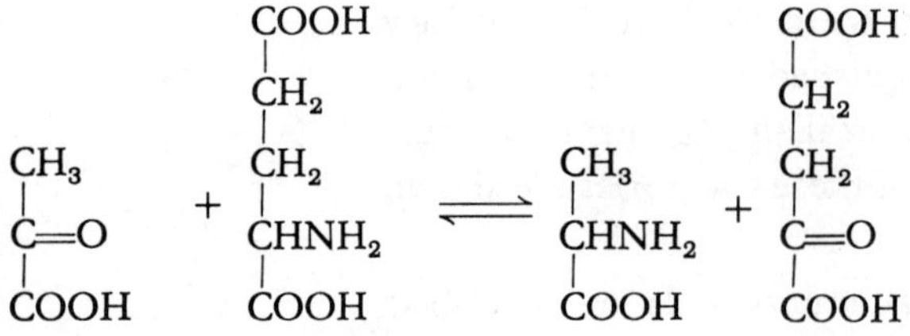

$$\begin{array}{c} CH_3 \\ | \\ C=O \\ | \\ COOH \end{array} + \begin{array}{c} COOH \\ | \\ CH_2 \\ | \\ CH_2 \\ | \\ CHNH_2 \\ | \\ COOH \end{array} \rightleftharpoons \begin{array}{c} CH_3 \\ | \\ CHNH_2 \\ | \\ COOH \end{array} + \begin{array}{c} COOH \\ | \\ CH_2 \\ | \\ CH_2 \\ | \\ C=O \\ | \\ COOH \end{array}$$

Pyruvate   Glutamate   Alanine   α-Ketoglutarate

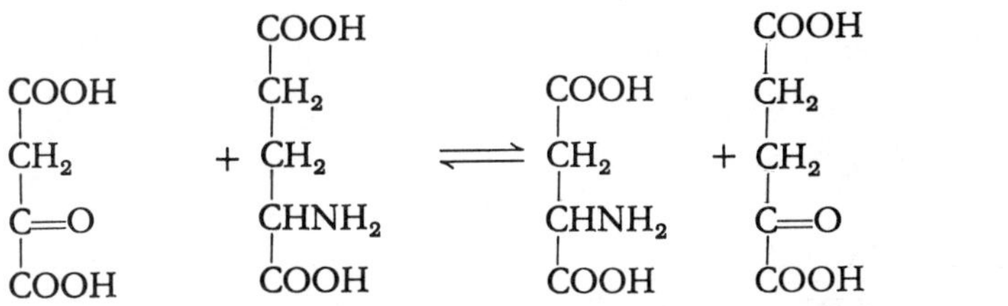

$$\underset{\text{Oxaloacetate}}{\begin{array}{c}\text{COOH}\\|\\\text{CH}_2\\|\\\text{C=O}\\|\\\text{COOH}\end{array}} + \underset{\text{Glutamate}}{\begin{array}{c}\text{COOH}\\|\\\text{CH}_2\\|\\\text{CH}_2\\|\\\text{CHNH}_2\\|\\\text{COOH}\end{array}} \rightleftharpoons \underset{\text{Aspartate}}{\begin{array}{c}\text{COOH}\\|\\\text{CH}_2\\|\\\text{CHNH}_2\\|\\\text{COOH}\end{array}} + \underset{\alpha\text{-Ketoglutarate}}{\begin{array}{c}\text{COOH}\\|\\\text{CH}_2\\|\\\text{CH}_2\\|\\\text{C=O}\\|\\\text{COOH}\end{array}}$$

Glutamate is produced when either of these reactions is reversed. Alanine is also a product of tryptophan metabolism.

In addition to its formation via transamination, glutamate is synthesized from α-ketoglutarate by the glutamate-dehydrogenase reaction:

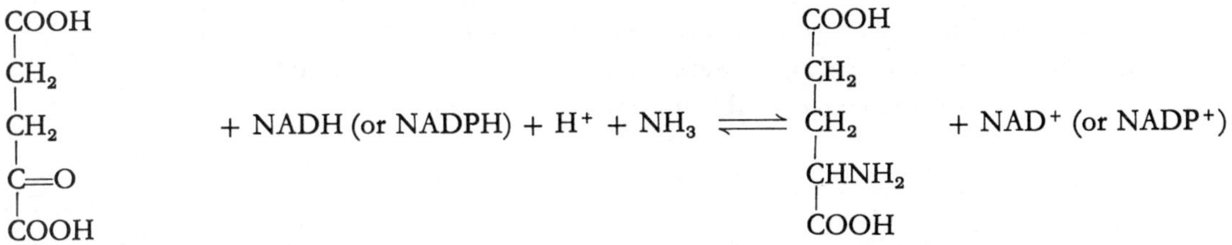

$$\underset{\alpha\text{-Ketoglutarate}}{\begin{array}{c}\text{COOH}\\|\\\text{CH}_2\\|\\\text{CH}_2\\|\\\text{C=O}\\|\\\text{COOH}\end{array}} + \text{NADH (or NADPH)} + \text{H}^+ + \text{NH}_3 \rightleftharpoons \underset{\text{Glutamate}}{\begin{array}{c}\text{COOH}\\|\\\text{CH}_2\\|\\\text{CH}_2\\|\\\text{CHNH}_2\\|\\\text{COOH}\end{array}} + \text{NAD}^+ \text{ (or NADP}^+\text{)}$$

The α-ketoglutarate utilized in these processes can be derived from either the citric acid cycle or the metabolic degradation of histidine, proline, hydroxyproline, or glutamine.

Aspartate and glutamate are also formed as a result of the hydrolyses of asparagine (asparaginase reaction) and glutamine (glutaminase reaction), respectively.

Glutamine is synthesized from glutamate, ATP, and ammonia, as described on page 309. Another enzyme, *asparagine synthetase*, catalyzes a reaction of aspartic acid with ATP and ammonia to yield *asparagine*. This process appears to be analogous to that of glutamine synthesis.

It should be noted that alanine, aspartate, glutamate, and glutamine are the most abundant free amino acids within the cells.

Proline is categorized as a non-essential amino acid, because it can be produced from glutamic acid. This involves reduction of the latter to glutamic semialdehyde, cyclization to Δ^1-pyrroline-5-carboxylic acid, and further reduction to proline:

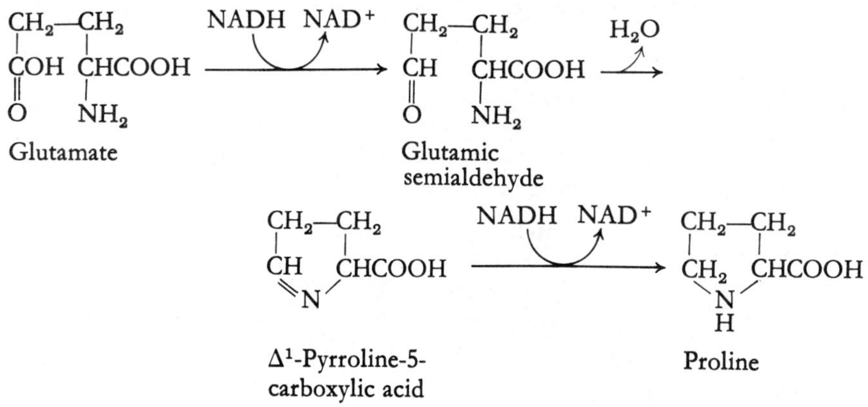

$$\underset{\text{Glutamate}}{\begin{array}{c}CH_2-CH_2\\|\quad\quad\;\;|\\COH\;\;CHCOOH\\\|\quad\quad\;\;|\\O\quad\;\;NH_2\end{array}}\xrightarrow{NADH\;\;NAD^+}\underset{\begin{array}{c}\text{Glutamic}\\\text{semialdehyde}\end{array}}{\begin{array}{c}CH_2-CH_2\\|\quad\quad\;\;|\\CH\;\;\;CHCOOH\\\|\quad\quad\;\;|\\O\quad\;\;NH_2\end{array}}\xrightarrow{H_2O}$$

$$\underset{\begin{array}{c}\Delta^1\text{-Pyrroline-5-}\\\text{carboxylic acid}\end{array}}{\begin{array}{c}CH_2-CH_2\\|\quad\quad\;\;|\\CH\;\;\;CHCOOH\\\diagdown\;\;\diagup\\N\end{array}}\xrightarrow{NADH\;\;NAD^+}\underset{\text{Proline}}{\begin{array}{c}CH_2-CH_2\\|\quad\quad\;\;|\\CH_2\;\;CHCOOH\\\diagdown\;\;\diagup\\N\\|\\H\end{array}}$$

Serine is formed from 3-phosphoglyceric acid by several routes. The principal pathway depends on the oxidation of this glycolytic intermediate to 3-phosphohydroxypyruvic acid, transamination from glutamate, and phosphate hydrolysis by serine phosphatase. An alternative pathway involves the hydrolysis of 3-phosphoglyceric acid to glyceric acid, oxidation of the latter to hydroxypyruvic acid, and transamination to yield serine:

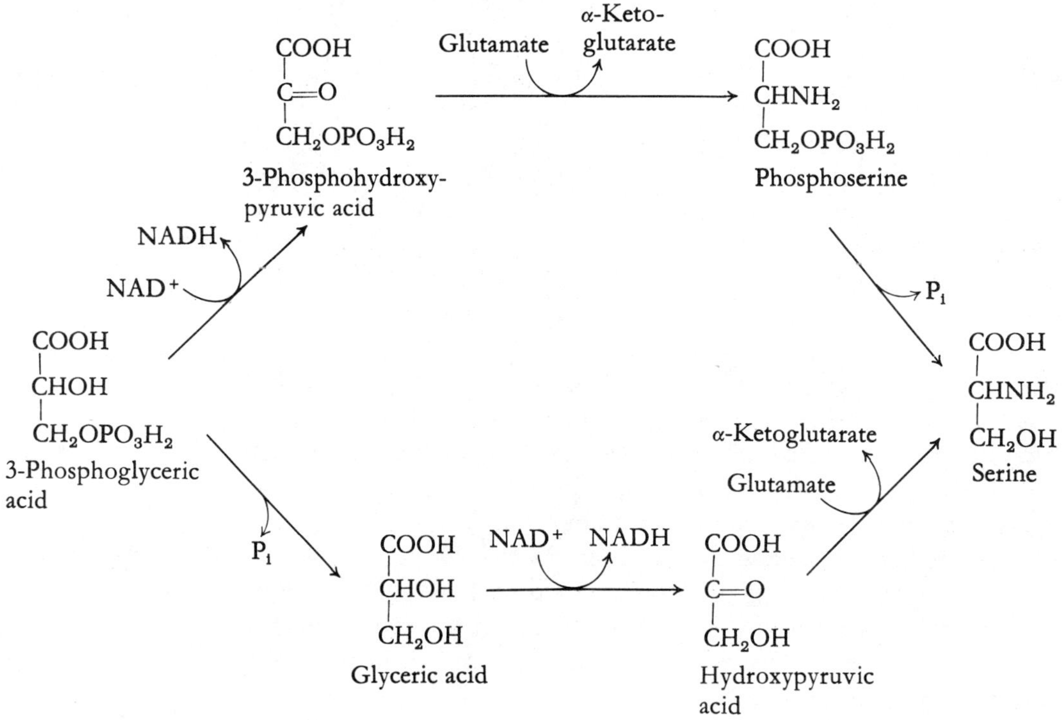

Glycine is generated intracellularly by a single-carbon unit transfer from serine to tetrahydrofolic acid (section 7.10). The reaction is catalyzed by *serine hydroxymethyltransferase* and requires the presence of tetrahydrofolic acid (THFA) and pyridoxal phosphate:

$$\begin{array}{c} \text{COOH} \\ | \\ \text{CHNH}_2 \\ | \\ \text{CH}_2\text{OH} \end{array} + \text{THFA} \underset{\text{Mn}^{2+}}{\overset{\text{Pyridoxal phosphate}}{\rightleftharpoons}} \begin{array}{c} \text{COOH} \\ | \\ \text{CH}_2\text{NH}_2 \end{array} + N^5,N^{10}\text{-methylene-THFA}$$

Serine   Glycine

Although the serine-glycine interconversion is a reversible reaction (page 322), it appears that in vivo the reaction leads mainly to the synthesis of glycine. As already indicated (section 7.10), the one-carbon unit carried by tetrahydrofolate can be utilized in a number of other synthetic processes.

The carbon skeleton of serine can be utilized for the synthesis of *cysteine*. The reaction sequence involves the interaction of serine with homocysteine to form cystathionine and the cleavage of the latter to yield cysteine, α-ketobutyrate, and ammonia. These reactions are catalyzed by *cystathionine synthase* and *cystathionase*, respectively:

$$\begin{array}{c} \text{CH}_2\text{SH} \\ | \\ \text{CH}_2 \\ | \\ \text{CHNH}_2 \\ | \\ \text{COOH} \end{array} + \begin{array}{c} \text{CH}_2\text{OH} \\ | \\ \text{CHNH}_2 \\ | \\ \text{COOH} \end{array} \longrightarrow \begin{array}{c} \text{CH}_2\text{---S---CH}_2 \\ | \quad\quad\quad\quad | \\ \text{CH}_2 \quad\quad \text{CHNH}_2 \\ | \quad\quad\quad\quad | \\ \text{CHNH}_2 \quad \text{COOH} \\ | \\ \text{COOH} \end{array} \longrightarrow \begin{array}{c} \text{CH}_3 \\ | \\ \text{CH}_2 \\ | \\ \text{C=O} \\ | \\ \text{COOH} \end{array} + \begin{array}{c} \text{CH}_2\text{SH} \\ | \\ \text{CHNH}_2 \\ | \\ \text{COOH} \end{array} + \text{NH}_3$$

Homocysteine   Serine   Cystathionine   α-Keto-   Cysteine
                                                                           butyrate

The homocysteine is derived from methionine via the formation of the intermediate, *S*-adenosylmethionine (section 7.8). The transfer of the methyl group from the latter leaves *S*-adenosylhomocysteine, which is cleaved to yield homocysteine. Although the carbons of cysteine can be derived from serine, the sulfur is derived from methionine. Dietary restriction of cysteine must therefore be compensated by an increased intake of methionine.

Cystine residues of proteins are formed by oxidation of cysteine in intact proteins or polypeptides. There is little or no free cystine present in the cells.

Tyrosine is formed from phenylalanine and oxygen by the NADPH-dependent enzyme phenylalanine hydroxylase:

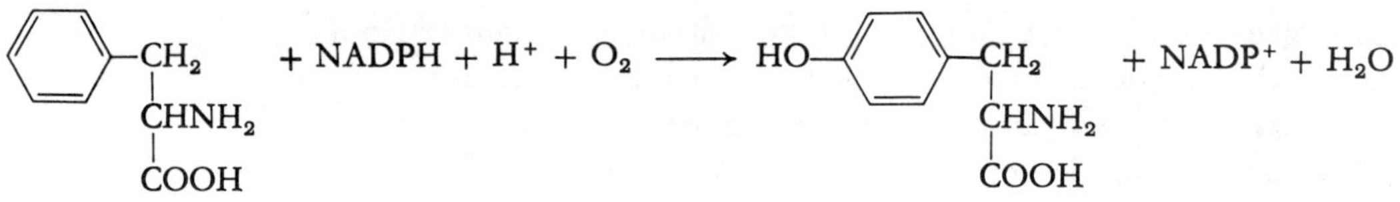

$$\text{Phenylalanine} + \text{NADPH} + \text{H}^+ + \text{O}_2 \longrightarrow \text{Tyrosine} + \text{NADP}^+ + \text{H}_2\text{O}$$

Phenylalanine → Tyrosine

The amount of ingested phenylalanine, which is an essential amino acid (Table 7-4), must be sufficient to meet the requirement for both phenylalanine and tyrosine if the latter is restricted.

The genetic disease, phenylketonuria, is due to a deficiency in the enzyme phenylalanine hydroxylase. The phenylalanine that accumulates in this disorder is converted by a transaminase to phenylpyruvate, which is excreted in the urine. The excessive phenylpyruvate causes mental retardation during development, which will ultimately become irreversible if not treated. Prevention of the symptoms requires early detection and treatment through restriction of phenylalanine intake.

Arginine is considered non-essential, since it is formed as a result of the reactions of the urea cycle. However, this amount may not be sufficient for normal nutritional requirements.

Hydroxyproline is not a precursor for the synthesis of proteins. The hydroxyproline residues in collagen are formed from proline units after the latter have been incorporated into the protein. A similar situation exists with hydroxylysine.

The precursors and principal mechanisms for the syntheses of non-essential amino acids are summarized in Table 7-5.

Table 7-5. Sources of Non-essential Amino Acids

Amino Acid	Precursor	Mechanism
Alanine	Pyruvate	Transamination
Asparagine	Aspartate	Amide formation
Aspartate	Oxaloacetate	Transamination
Cysteine (Cystine)	Methionine, serine	Transmethylation, transsulfuration
Glutamate	α-Ketoglutarate	Transamination
Glutamine	Glutamate	Amide formation
Glycine	Serine	Single-carbon transfer
Hydroxyproline	Proline (peptide-linked)	Hydroxylation
Proline	Glutamate	Reduction, cyclization
Serine	Phosphoglycerate	Oxidation, transamination
Tyrosine	Phenylalanine	Hydroxylation

Suggested Reading

Eisenberg, H. Glutamate Dehydrogenase: Anatomy of a Regulatory Enzyme. *Accounts of Chemical Research* 4:379–385, 1971.

Felig, P., and Washren, J. Protein Turnover and Amino Acid Metabolism in the Regulation of Gluconeogenesis. *Federation Proceedings* 33:1092–1097, 1974.

Kassell, B., and Kay, J. Zymogens of Proteolytic Enzymes. *Science* 180:1022–1027, 1973.

Ratner, S. Enzymes of Arginine and Urea Synthesis. *Advances in Enzymology and Related Areas of Molecular Biology* 39:1–62, 1973.

Snell, E. E., and Di Mari, S. J. Schiff Base Intermediates in Enzyme Catalysis. In P. D. Boyer (Ed.), *The Enzymes* (3rd ed.), Vol. 2, pp. 335–370. New York: Academic Press, 1970.

8. Metabolism of Specific Amino Acids

This chapter describes the metabolic disposition of each amino acid. Although the amino acids undergo a wide variety of reactions and are involved in numerous processes, the focus of discussion will be on those aspects of amino acid metabolism that are necessary for visualizing the interrelationships of metabolic pathways. Each amino acid will be considered essentially from two points of view: (1) its conversion to pyruvate, acetate, or a citric acid cycle intermediate and (2) its utilization for the synthesis of components other than glycogen or triglycerides. Some of the consequences of metabolic aberrations will also be noted.

The amino acids that are convertible to pyruvate (alanine, serine, glycine, threonine, and cysteine) will be discussed first, followed by discussions of those that yield oxaloacetate (aspartate and asparagine), α-ketoglutarate (glutamate, glutamine, arginine, histidine, and proline), succinyl-CoA (methionine, valine, and isoleucine), and so on (see Fig. 7-5).

8.1. Alanine

Alanine can be converted directly to pyruvate by transamination with α-ketoglutarate (alanine-glutamate transaminase):

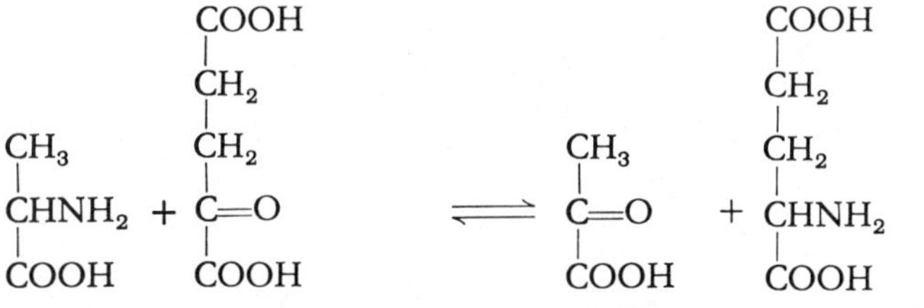

$$\underset{\text{Alanine}}{\begin{array}{c}CH_3\\|\\CHNH_2\\|\\COOH\end{array}} + \underset{\text{α-Ketoglutarate}}{\begin{array}{c}COOH\\|\\CH_2\\|\\CH_2\\|\\C=O\\|\\COOH\end{array}} \rightleftharpoons \underset{\text{Pyruvate}}{\begin{array}{c}CH_3\\|\\C=O\\|\\COOH\end{array}} + \underset{\text{Glutamate}}{\begin{array}{c}COOH\\|\\CH_2\\|\\CH_2\\|\\CHNH_2\\|\\COOH\end{array}}$$

Since pyruvate can be transformed to glucose and glycogen (pyruvate → oxaloacetate → phosphoenolpyruvate), alanine is classified as a glucogenic amino acid. Pyruvate can also be oxidized to acetyl-coenzyme A and ultimately be degraded to carbon dioxide and water via the citric acid cycle.

8. Metabolism of Specific Amino Acids

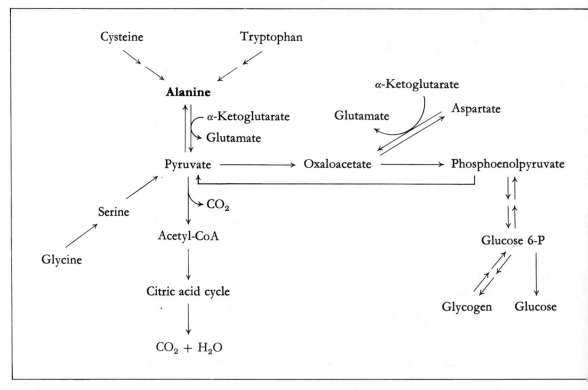

Fig. 8-1. Alanine metabolism.

Alanine is a non-essential amino acid, because it can be formed from pyruvate by a reversal of the transamination shown above. Alanine is also a product in the metabolism of tryptophan (section 8.18). These aspects of the metabolism of alanine are summarized in Figure 8-1.

8.2. Serine

Serine can be degraded by several pathways. The major route involves a non-oxidative deamination (section 7.3.B) catalyzed by *serine dehydratase*, which yields pyruvate:

$$\underset{\text{Serine}}{\begin{array}{c} CH_2OH \\ | \\ CHNH_2 \\ | \\ COOH \end{array}} \xrightarrow[\text{phosphate}]{\text{Pyridoxal}} \underset{\text{Pyruvate}}{\begin{array}{c} CH_3 \\ | \\ C{=}O \\ | \\ COOH \end{array}} + H_2O + NH_3$$

Serine is a non-essential amino acid, because it can be formed from 3-phosphoglyceric acid (section 7.11.B). To a minimal extent, it may also be produced from glycine by single-carbon unit transfer (sections 7.10 and 7.11.B).

The serine → pyruvate → alanine pathway provides for the conversion of serine to alanine. The carbon skeleton of serine can also be utilized for the synthesis of cysteine (serine → cystathionine → cysteine; see section 7.11.B). When serine functions as a single-carbon donor, it is converted to glycine. The single-carbon unit derived from serine can be incorporated into a number of cellular components, e.g., purine nucleotides and the methyl groups of methionine. In addition to these reactions, serine also serves as a building block in the formation of phosphatidylserine, cephalins and lecithins (page 230), and sphingolipids (page 246). The interconversions of serine are summarized in Figure 8-2.

8.3. Glycine

One pathway for the disposition of glycine involves its conversion to pyruvate via serine; it is thus glucogenic. The formation of serine from glycine is catalyzed by serine hydroxymethyltransferase (section 7.11.B). In this reaction, which is reversible, glycine functions as the acceptor of a single-carbon unit from N^5,N^{10}-methylene-tetrahydrofolate:

$$\begin{array}{c} CH_2NH_2 \\ | \\ COOH \end{array} + N^5,N^{10}\text{-methylene-THFA} \rightleftharpoons \begin{array}{c} CH_2OH \\ | \\ CHNH_2 \\ | \\ COOH \end{array} + THFA$$

Glycine   Serine

Glycine also undergoes oxidative deamination (section 7.3.A) to form glyoxylate and ammonia:

$$\begin{array}{c} CH_2NH_2 \\ | \\ COOH \end{array} \longrightarrow \begin{array}{c} CHO \\ | \\ COOH \end{array} + NH_3$$

Glycine   Glyoxylic acid

Glyoxylate is oxidized to carbon dioxide and formic acid, and the latter can be utilized as a transferable single-carbon unit in the formation of N^{10}-formyl-tetrahydrofolate. Hence, in addition to being an acceptor of single-carbon units, glycine can also serve as their source.

Since glycine can be formed from serine, it is classified as a non-essential amino acid.

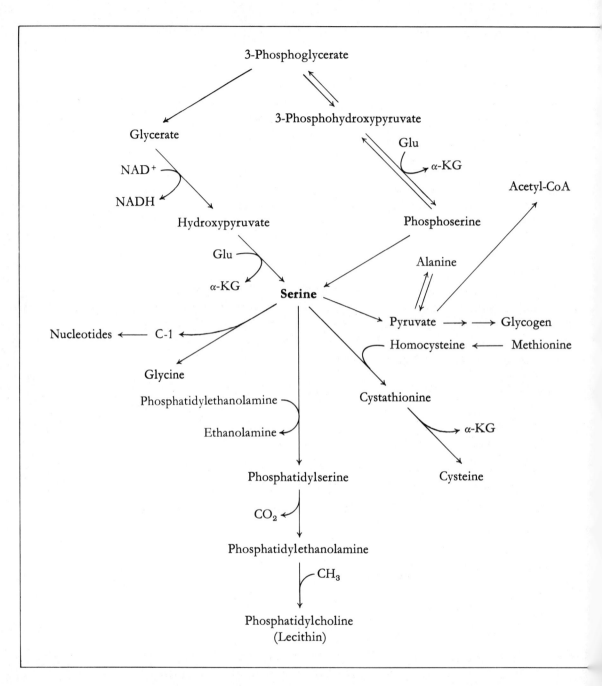

Fig. 8-2. Metabolism of serine. Transamination reactions are indicated by Glu (glutamate) and α-KG (α-ketoglutarate).

Glycine functions as a precursor for the synthesis of a wide variety of molecules. Its role in the production of creatine has already been discussed (page 319). It is also utilized in the biosynthesis of purine nucleotides (page 375) and porphyrins (page 406).

In addition, glycine functions in the synthesis of glutathione, a tripeptide. This involves the reaction of glycine with ATP and γ-glutamylcysteine. The latter is formed from glutamate and cysteine in the presence of ATP and magnesium ion:

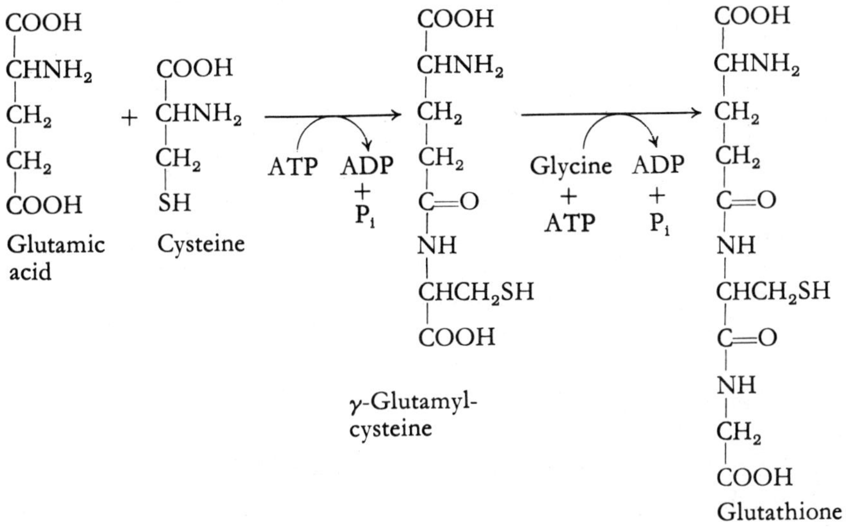

$$
\begin{array}{l}
\text{COOH} \\
| \\
\text{CHNH}_2 \\
| \\
\text{CH}_2 \\
| \\
\text{CH}_2 \\
| \\
\text{COOH} \\
\text{Glutamic acid}
\end{array}
+
\begin{array}{l}
\text{COOH} \\
| \\
\text{CHNH}_2 \\
| \\
\text{CH}_2 \\
| \\
\text{SH} \\
\text{Cysteine}
\end{array}
\xrightarrow[\text{ATP} \quad \text{ADP} + P_i]{}
\begin{array}{l}
\text{COOH} \\
| \\
\text{CHNH}_2 \\
| \\
\text{CH}_2 \\
| \\
\text{CH}_2 \\
| \\
\text{C}=\text{O} \\
| \\
\text{NH} \\
| \\
\text{CHCH}_2\text{SH} \\
| \\
\text{COOH} \\
\gamma\text{-Glutamyl-cysteine}
\end{array}
\xrightarrow[\text{Glycine} \quad \text{ADP} + P_i]{\text{ATP}}
\begin{array}{l}
\text{COOH} \\
| \\
\text{CHNH}_2 \\
| \\
\text{CH}_2 \\
| \\
\text{CH}_2 \\
| \\
\text{C}=\text{O} \\
| \\
\text{NH} \\
| \\
\text{CHCH}_2\text{SH} \\
| \\
\text{C}=\text{O} \\
| \\
\text{NH} \\
| \\
\text{CH}_2 \\
| \\
\text{COOH} \\
\text{Glutathione}
\end{array}
$$

Glutathione (G–SH) is important in maintaining the sulfhydryl groups of certain proteins in their reduced state by reducing disulfide bonds (R–SS–R′ = protein with disulfide bridge):

R—SS—R′ + G—SH $\rightleftharpoons$ G—SS—R + R′—SH

G—SS—R + G—SH $\rightleftharpoons$ G—SS—G + R—SH

Reduced glutathione is regenerated by interaction with NADPH:

$$\text{G—SS—G} \xrightarrow[\text{NADPH} \quad \text{NADP}^+]{} 2\text{G—SH}$$

The enzyme for this reaction is *glutathione reductase*. Glutathione functions in the uptake of amino acids by mammalian cells. The transport system involves the binding of amino acids by a membrane-bound *γ-glutamyl transpeptidase* and their reaction with intracellular glutathione to yield a γ-glutamyl-amino acid and cysteinylglycine:

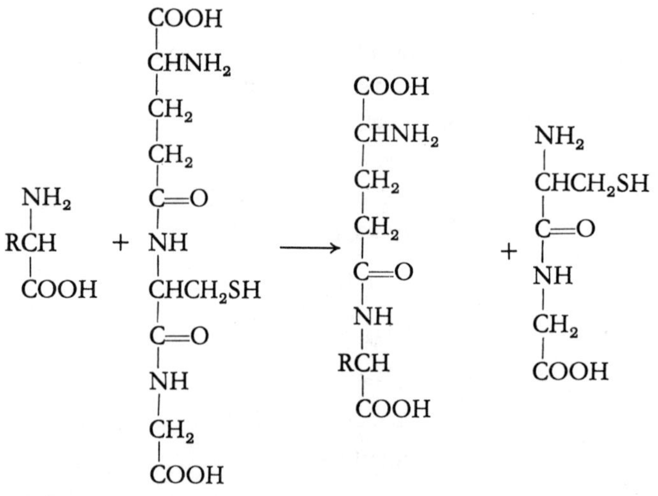

$$\begin{array}{c}\text{NH}_2\\|\\\text{RCH}\\|\\\text{COOH}\end{array} + \begin{array}{c}\text{COOH}\\|\\\text{CHNH}_2\\|\\\text{CH}_2\\|\\\text{CH}_2\\|\\\text{C=O}\\|\\\text{NH}\\|\\\text{CHCH}_2\text{SH}\\|\\\text{C=O}\\|\\\text{NH}\\|\\\text{CH}_2\\|\\\text{COOH}\end{array} \longrightarrow \begin{array}{c}\text{COOH}\\|\\\text{CHNH}_2\\|\\\text{CH}_2\\|\\\text{CH}_2\\|\\\text{C=O}\\|\\\text{NH}\\|\\\text{RCH}\\|\\\text{COOH}\end{array} + \begin{array}{c}\text{NH}_2\\|\\\text{CHCH}_2\text{SH}\\|\\\text{C=O}\\|\\\text{NH}\\|\\\text{CH}_2\\|\\\text{COOH}\end{array}$$

Amino   Glutathione   γ-Glutamyl-   Cysteinylglycine
acid   amino acid

The γ-glutamyl-amino acid within the cell is hydrolyzed to the amino acid and (by a series of steps) glutathione is regenerated. This sequence of reactions, termed the *γ-glutamyl cycle,* thus serves as a mechanism for the transfer of extracellular amino acids into the cells.

The glutathione that is present in erythrocytes is also necessary for maintaining the integrity of the plasma membrane (page 286).

Glycine is also utilized for the production of the bile salt, glycocholate:

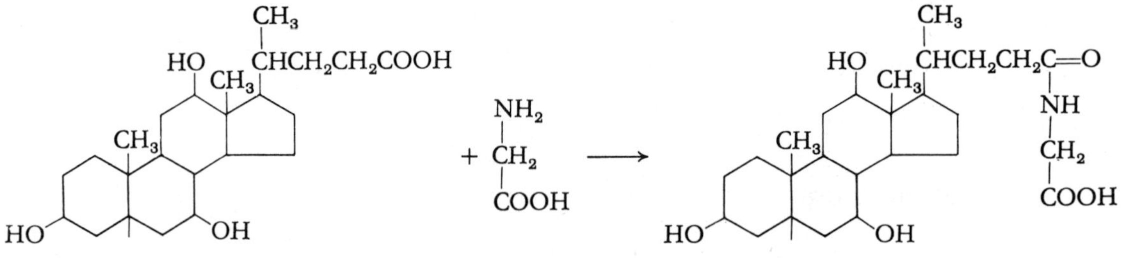

Cholic acid   + Glycine   ⟶   Glycocholic acid

Glycine also reacts with other compounds to form *N*-acylamino conjugates, i.e., it interacts with certain carboxylic compounds to yield glycine derivatives. Such conjugation reactions provide one mechanism through which the organism can dispose of water-insoluble substances. By forming the amino acid derivative, the material becomes soluble and can be excreted by the kidneys into the urine. The most prominent of these reactions is the conversion of benzoate to hippurate in the liver:

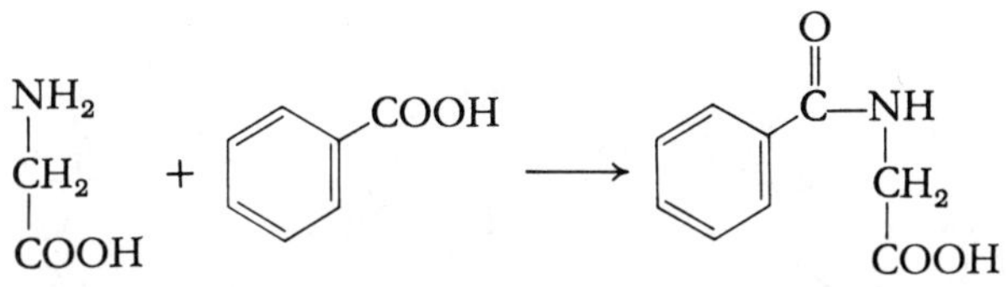

Glycine Benzoic acid Hippuric acid

Hippuric acid is found in small amounts in normal urine. Administration of benzoic acid results in a marked increase in urinary hippuric-acid levels.

The metabolism of glycine is summarized in Figure 8-3.

8.4. Threonine

Threonine is an essential amino acid. Unlike most amino acids, it is not a substrate for transaminase reactions.

Threonine is metabolized to pyruvate via glycine. Glycine and acetaldehyde are also produced from threonine by the action of the pyridoxal phosphate-dependent enzyme, threonine aldolase (serine hydroxymethyltransferase):

$$\underset{\text{Threonine}}{CH_3-CHOH-CHNH_2-COOH} \longrightarrow \underset{\text{Acetaldehyde}}{CH_3-CHO} + \underset{\text{Glycine}}{CH_2NH_2-COOH}$$

The acetaldehyde can be oxidized to acetate and converted to acetyl-coenzyme A. Threonine ultimately yields pyruvate by the following sequence:

$$\text{Threonine} \longrightarrow \text{Glycine} \longrightarrow \text{Serine} \longrightarrow \text{Pyruvate}$$

Since it yields pyruvate it is thus glucogenic.

Another reaction of threonine is its deamination by threonine deaminase (threonine dehydratase) to yield α-ketobutyrate and ammonia:

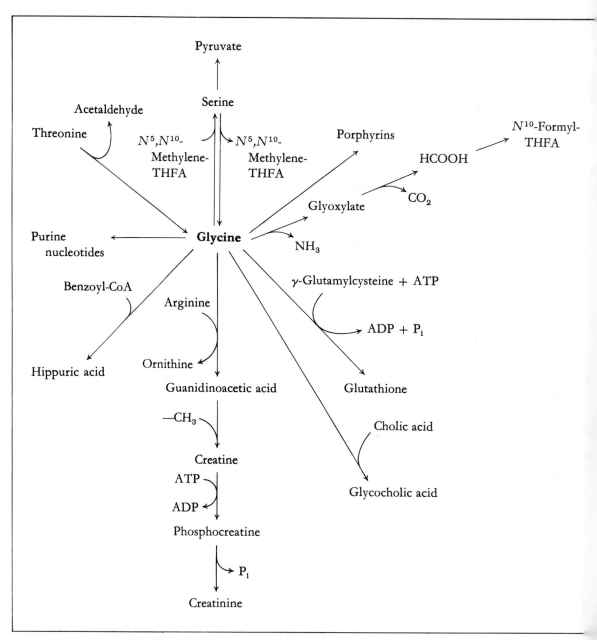

Fig. 8-3. *Metabolism of glycine.*

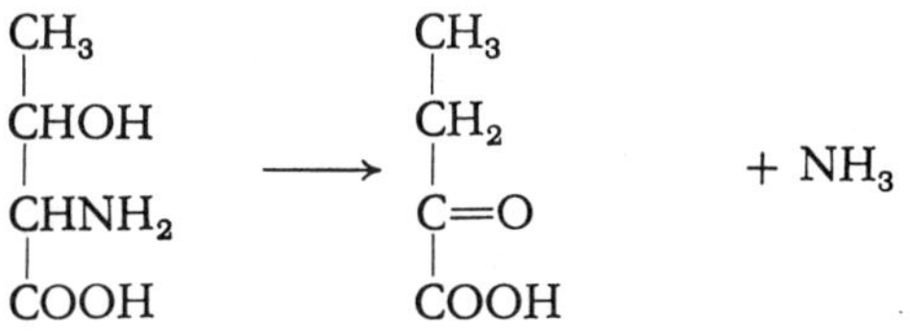

$$\begin{array}{c}\text{CH}_3\\|\\\text{CHOH}\\|\\\text{CHNH}_2\\|\\\text{COOH}\\\text{Threonine}\end{array} \longrightarrow \begin{array}{c}\text{CH}_3\\|\\\text{CH}_2\\|\\\text{C}=\text{O}\\|\\\text{COOH}\\\alpha\text{-Ketobutyrate}\end{array} + \text{NH}_3$$

α-Ketobutyrate provides succinyl-coenzyme A by the following sequence:

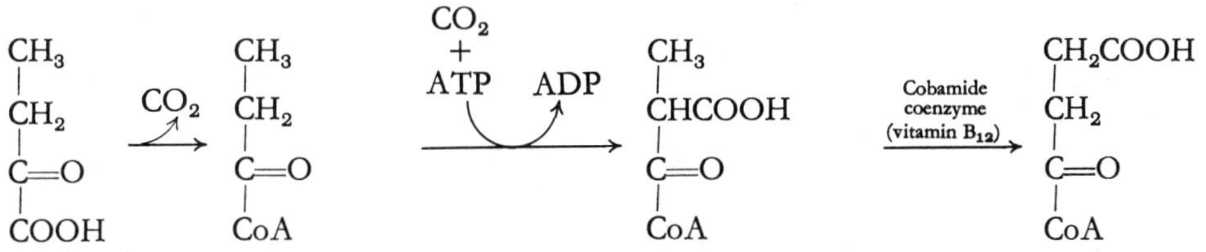

$$\begin{array}{c}\text{CH}_3\\|\\\text{CH}_2\\|\\\text{C}=\text{O}\\|\\\text{COOH}\\\alpha\text{-Keto-}\\\text{butyrate}\end{array} \xrightarrow{\text{CO}_2} \begin{array}{c}\text{CH}_3\\|\\\text{CH}_2\\|\\\text{C}=\text{O}\\|\\\text{CoA}\\\text{Propionyl-}\\\text{coenzyme A}\end{array} \xrightarrow[\text{ATP ADP}]{\text{CO}_2} \begin{array}{c}\text{CH}_3\\|\\\text{CHCOOH}\\|\\\text{C}=\text{O}\\|\\\text{CoA}\\\text{Methylmalonyl-}\\\text{coenzyme A}\end{array} \xrightarrow[\text{(vitamin B}_{12})]{\text{Cobamide coenzyme}} \begin{array}{c}\text{CH}_2\text{COOH}\\|\\\text{CH}_2\\|\\\text{C}=\text{O}\\|\\\text{CoA}\\\text{Succinyl-}\\\text{coenzyme A}\end{array}$$

Consequently, this pathway may provide an intermediate of gluconeogenesis by the route:

Succinate ⟶ Fumarate ⟶ Malate ⟶ Oxaloacetate ⟶ Phosphoenolpyruvate

Succinyl-CoA is also utilized in the biosynthesis of porphyrins (page 406); thus, some of the carbons of threonine can be incorporated into this structure.

A third pathway for the catabolism of threonine leads to the formation of aminoacetone:

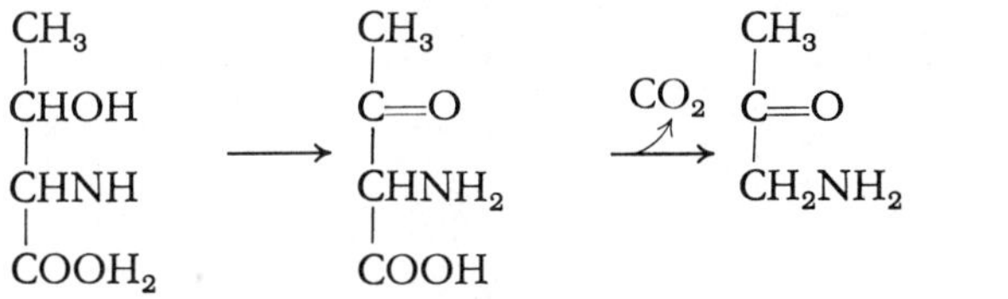

$$\begin{array}{c}\text{CH}_3\\|\\\text{CHOH}\\|\\\text{CHNH}\\|\\\text{COOH}_2\\\text{Threonine}\end{array} \longrightarrow \begin{array}{c}\text{CH}_3\\|\\\text{C}=\text{O}\\|\\\text{CHNH}_2\\|\\\text{COOH}\\\text{2-Amino-}\\\text{3-keto-}\\\text{butyric acid}\end{array} \xrightarrow{\text{CO}_2} \begin{array}{c}\text{CH}_3\\|\\\text{C}=\text{O}\\|\\\text{CH}_2\text{NH}_2\\\\\text{Aminoacetone}\end{array}$$

A portion of aminoacetone is excreted in the urine. Aminoacetone may also be oxidized to methylglyoxal, which, in turn, is converted to lactic acid.

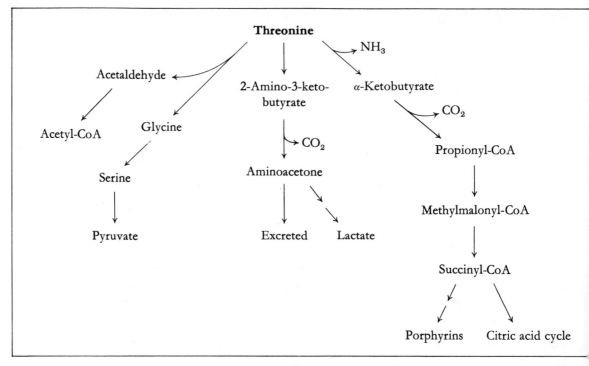

Fig. 8-4. *Metabolism of threonine.*

```
CH₃              CH₃
|                |
C=O      ⟶     CHOH
|                |
CHO              COOH
Methylglyoxal    Lactic acid
```

The metabolism of threonine is summarized in Figure 8-4.

8.5. Cysteine

A considerable amount of the cysteine in animal organisms is oxidized to cysteine sulfinic acid. Transamination of cysteine sulfinic acid yields sulfinyl pyruvic acid, which then forms pyruvic acid:

$$\underset{\text{Cysteine}}{\begin{array}{c}SH\\|\\CH_2\\|\\CHNH_2\\|\\COOH\end{array}} \xrightarrow{\underset{O_2}{\overset{NADPH}{+}} \; NADP^+} \underset{\substack{\text{Cysteine}\\\text{sulfinic acid}}}{\begin{array}{c}SO_2H\\|\\CH_2\\|\\CHNH_2\\|\\COOH\end{array}} \xrightarrow{\text{Transamination}} \underset{\substack{\text{Sulfinyl}\\\text{pyruvic acid}}}{\begin{array}{c}SO_2H\\|\\CH_2\\|\\C=O\\|\\COOH\end{array}} \longrightarrow \underset{\text{Pyruvic acid}}{\begin{array}{c}CH_3\\|\\C=O\\|\\COOH\end{array}} + SO_3^{2-}$$

The sulfite released in the process is oxidized to sulfate by hepatic *sulfite oxidase*. Absence of this enzyme is associated with mental retardation, dislocation of the lens, liver damage, and early death.

Cysteine is also converted to pyruvate by two other pathways. One of these is a transamination and removal of the sulfur:

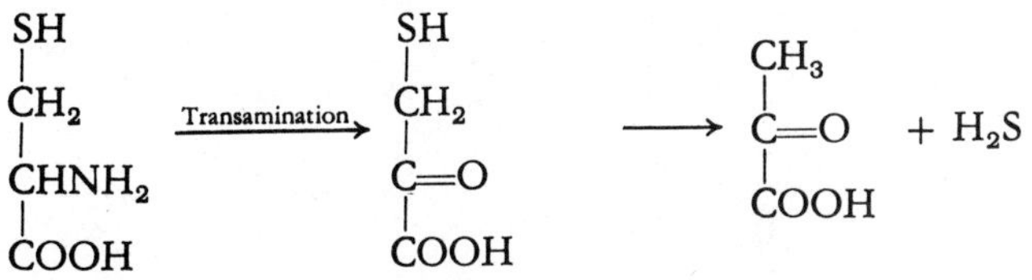

$$\begin{array}{c} SH \\ | \\ CH_2 \\ | \\ CHNH_2 \\ | \\ COOH \end{array} \xrightarrow{\text{Transamination}} \begin{array}{c} SH \\ | \\ CH_2 \\ | \\ C=O \\ | \\ COOH \end{array} \longrightarrow \begin{array}{c} CH_3 \\ | \\ C=O \\ | \\ COOH \end{array} + H_2S$$

Cysteine   β-Mercapto-   Pyruvic acid
              pyruvic acid

The other reaction, catalyzed by *cysteine desulfhydrase*, involves the release of hydrogen sulfide and ammonia:

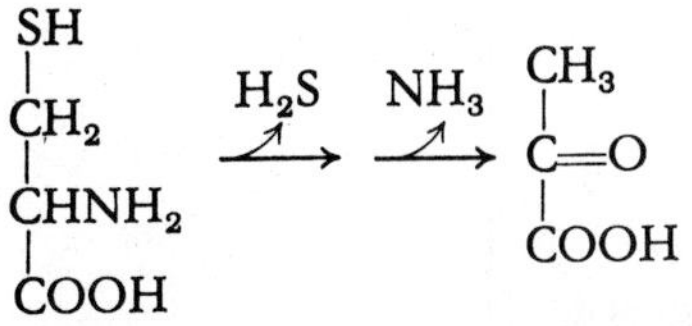

$$\begin{array}{c} SH \\ | \\ CH_2 \\ | \\ CHNH_2 \\ | \\ COOH \end{array} \xrightarrow{H_2S \quad NH_3} \begin{array}{c} CH_3 \\ | \\ C=O \\ | \\ COOH \end{array}$$

Cysteine   Pyruvic acid

In addition to providing pyruvate, cysteine can be oxidized and decarboxylated to yield taurine. The latter is necessary for the synthesis of the bile salt, taurocholate:

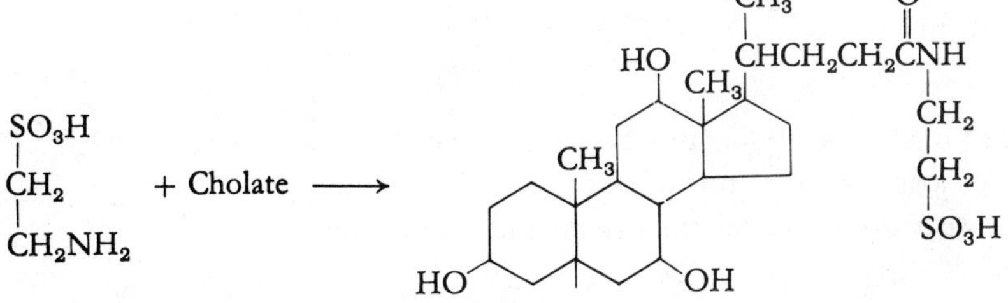

$$\begin{array}{c} SO_3H \\ | \\ CH_2 \\ | \\ CH_2NH_2 \end{array} + \text{Cholate} \longrightarrow$$

Taurine   Taurocholic acid

Since cysteine can be formed from methionine and serine (section 7.11.B), it is classified as a non-essential amino acid.

The utilization of cysteine for the synthesis of glutathione has already been described (section 8.3).

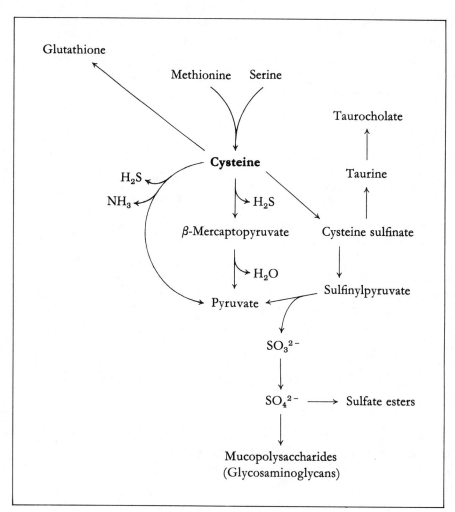

Fig. 8-5. Metabolism of cysteine.

Most of the sulfate produced in the metabolism of cysteine is excreted in the urine. A certain amount is converted to phosphoadenosine phosphosulfate and employed in the formation of sulfate esters and mucopolysaccharides (page 547).

The metabolic interconversions of cysteine are outlined in Figure 8-5.

8.6. Aspartate and Asparagine

Aspartate can be formed from oxaloacetate by transamination with glutamate; consequently, it is a non-essential amino acid:

$$\underset{\text{Oxaloacetate}}{\begin{array}{c}\text{COOH}\\|\\\text{CH}_2\\|\\\text{C}=\text{O}\\|\\\text{COOH}\end{array}} + \underset{\text{Glutamate}}{\begin{array}{c}\text{COOH}\\|\\\text{CH}_2\\|\\\text{CH}_2\\|\\\text{CHNH}_2\\|\\\text{COOH}\end{array}} \rightleftharpoons \underset{\text{Aspartate}}{\begin{array}{c}\text{COOH}\\|\\\text{CH}_2\\|\\\text{CHNH}_2\\|\\\text{COOH}\end{array}} + \underset{\alpha\text{-Ketoglutarate}}{\begin{array}{c}\text{COOH}\\|\\\text{CH}_2\\|\\\text{CH}_2\\|\\\text{C}=\text{O}\\|\\\text{COOH}\end{array}}$$

Reversal of this reaction results in the conversion of aspartate to oxaloacetate. Since the latter may be channeled toward formation of glycogen (gluconeogenesis), aspartate is a glucogenic amino acid.

Aspartate is also released by the action of the enzyme asparaginase on asparagine:

$$\underset{\text{Asparagine}}{\begin{array}{c}\text{CONH}_2\\|\\\text{CH}_2\\|\\\text{CHNH}_2\\|\\\text{COOH}\end{array}} \longrightarrow \underset{\text{Aspartate}}{\begin{array}{c}\text{COOH}\\|\\\text{CH}_2\\|\\\text{CHNH}_2\\|\\\text{COOH}\end{array}} + \text{NH}_3$$

One of the reactions of the urea cycle (section 7.5) is the condensation of aspartate with citrulline to yield argininosuccinate. In the next step of the cycle, the latter is cleaved to provide arginine and fumarate. The nitrogen of aspartate is thus incorporated into arginine, while its carbon skeleton is converted to fumarate:

$$\underset{\text{Aspartate}}{\begin{array}{c}\text{COOH}\\|\\\text{CH}_2\\|\\\text{CHNH}_2\\|\\\text{COOH}\end{array}} \xrightarrow{\text{Citrulline}} \underset{\text{Argininosuccinate}}{\begin{array}{c}\text{COOH} \quad \text{CHNH}_2\\|\qquad\qquad|\\\text{CH}_2 \; \text{NH} \; \text{CH}_2\\|\quad\;\;\|\quad|\\\text{CHNHCNHCH}_2\\|\\\text{COOH}\end{array}} \longrightarrow \underset{\text{Fumarate}}{\begin{array}{c}\text{COOH}\\|\\\text{CH}\\\|\\\text{CH}\\|\\\text{COOH}\end{array}} + \underset{\text{Arginine}}{\begin{array}{c}\text{COOH}\\|\\\text{CHNH}_2\\|\\\text{CH}_2\\|\\\text{CH}_2\\|\\\text{CH}_2\text{NHC}=\text{NH}\\|\\\text{NH}_2\end{array}}$$

Aspartate is also used in the formation of carbamoyl aspartate and the biosynthesis of pyrimidines (page 381), and its nitrogen may be incorporated into purines and pyrimidine nucleotides (page 377).

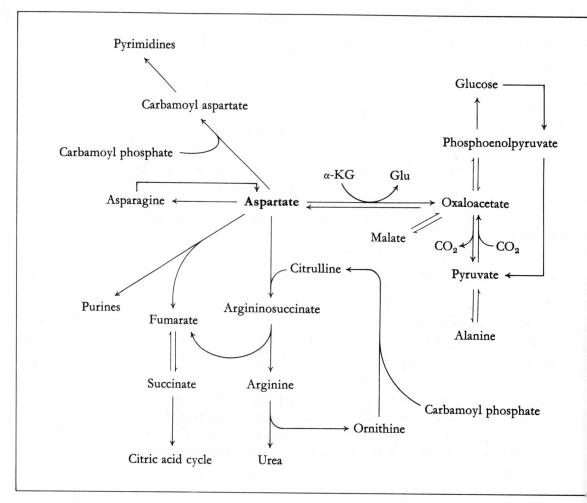

Fig. 8-6. *Metabolism of aspartate.*

Asparagine can be produced from aspartate, ammonia, and ATP in a reaction that is analogous to the synthesis of glutamine from glutamate (page 309).

The metabolism of aspartate is diagrammed in Figure 8-6.

8.7. Glutamate and Glutamine

Glutamate is a non-essential amino acid, because it can be derived from α-ketoglutarate by the action of either a transaminase or glutamate dehydrogenase (page 306). Glutamate is also a product of proline metabolism (section 8.10). Since the dehydrogenase and transaminase reactions are reversible, available glutamate can be transformed to α-ketoglutarate:

$$\begin{array}{c}\text{COOH}\\|\\\text{CH}_2\\|\\\text{CH}_2\\|\\\text{C}=\text{O}\\|\\\text{COOH}\\\alpha\text{-Ketoglutarate}\end{array} \quad\xrightleftharpoons[]{\substack{\text{NAD}^+\text{ or}\\\text{NADPH}\\+\\\text{NH}_3}\;\;\substack{\text{NAD}^+\text{ or}\\\text{NADP}^+}}\quad \begin{array}{c}\text{COOH}\\|\\\text{CH}_2\\|\\\text{CH}_2\\|\\\text{CHNH}_2\\|\\\text{COOH}\\\text{Glutamate}\end{array}\quad\xrightleftharpoons[]{\substack{\text{Keto- Amino}\\\text{acid\;\;acid}}}\quad\begin{array}{c}\text{COOH}\\|\\\text{CH}_2\\|\\\text{CH}_2\\|\\\text{C}=\text{O}\\|\\\text{COOH}\\\alpha\text{-Ketoglutarate}\end{array}$$

Formation of glutamate from α-ketoglutarate and ammonia serves as one of the mechanisms for the removal of the latter from tissues. Hydrolysis of glutamine in the kidney provides ammonium ions, which function to conserve other cations in the circulatory system.

The α-ketoglutarate that is derived from glutamate can be channeled toward glycogen synthesis. Alternatively, it can be oxidized to carbon dioxide and water via the citric acid cycle and thus serve as a source of energy. Since the brain has a comparatively minute glycogen reserve, the utilization of free amino acids becomes a significant energy source in this tissue. Glutamate especially is utilized for this purpose, because its concentration in brain is about 100 to 150 milligrams per gram of tissue, or 0.01 M.

The importance of glutamate in the formation of amino acids from keto-acids has been discussed previously (section 7.2).

Glutamate may also be reduced to glutamic semialdehyde, which may be either converted to ornithine by transamination or utilized for the synthesis of proline (section 7.11.B):

$$\begin{array}{c}\text{COOH}\\|\\\text{CH}_2\\|\\\text{CH}_2\\|\\\text{CHNH}_2\\|\\\text{COOH}\\\text{Glutamate}\end{array}\quad\xrightarrow{\text{NADH}\;\;\text{NAD}^+}\quad\begin{array}{c}\text{CHO}\\|\\\text{CH}_2\\|\\\text{CH}_2\\|\\\text{CHNH}_2\\|\\\text{COOH}\\\text{Glutamic acid}\\\text{semialdehyde}\end{array}\quad\xrightarrow{\text{Glu}\;\;\alpha\text{-KG}}\quad\begin{array}{c}\text{CH}_2\text{NH}_2\\|\\\text{CH}_2\\|\\\text{CH}_2\\|\\\text{CHNH}_2\\|\\\text{COOH}\\\text{Ornithine}\end{array}$$

$$\downarrow$$

$$\begin{array}{c}\text{CH}_2\text{—CH}_2\\|\quad\quad|\\\text{HC}\diagdown_{\text{N}}\diagup\text{CHCOOH}\\\Delta^1\text{-Pyrroline-5-}\\\text{carboxylic acid}\end{array}\quad\xrightarrow{\text{NADH}\;\;\text{NAD}^+}\quad\begin{array}{c}\text{CH}_2\text{—CH}_2\\|\quad\quad|\\\text{H}_2\text{C}\diagdown_{\text{N}}\diagup\text{CHCOOH}\\\;\;\;\;\;\;\;\;\;\;|\\\;\;\;\;\;\;\;\;\;\;\text{H}\\\text{Proline}\end{array}$$

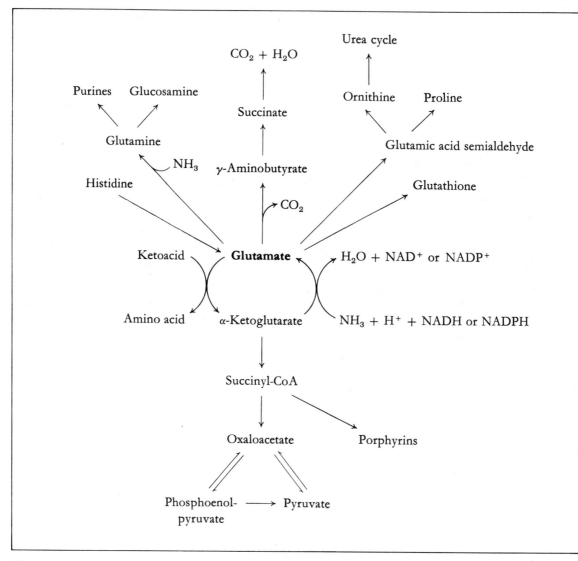

Fig. 8-7. *Metabolism of glutamate.*

In addition to these reactions, glutamate serves as a precursor for the formation of glutathione (section 8.3). The amidonitrogen atom of glutamine may also be utilized in the synthesis of purines (page 374).

The brain contains a decarboxylase that catalyzes the conversion of glutamate to γ-aminobutyrate (GABA). The latter undergoes transamination to yield succinic semialdehyde, which in turn can be oxidized to succinate:

$$\underset{\text{Glutamate}}{\begin{array}{c}\text{COOH}\\|\\\text{CH}_2\\|\\\text{CH}_2\\|\\\text{CHNH}_2\\|\\\text{COOH}\end{array}} \xrightarrow{CO_2} \underset{\substack{\gamma\text{-Amino-}\\\text{butyrate}}}{\begin{array}{c}\text{COOH}\\|\\\text{CH}_2\\|\\\text{CH}_2\\|\\\text{CH}_2\text{NH}_2\end{array}} \xrightarrow{\alpha\text{-KG} \ \ \text{Glu}} \underset{\substack{\text{Succinic}\\\text{semialdehyde}}}{\begin{array}{c}\text{COOH}\\|\\\text{CH}_2\\|\\\text{CH}_2\\|\\\text{CHO}\end{array}} \longrightarrow \underset{\text{Succinate}}{\begin{array}{c}\text{COOH}\\|\\\text{CH}_2\\|\\\text{CH}_2\\|\\\text{COOH}\end{array}}$$

Through its action as an inhibitor of synaptic transmission, γ-aminobutyric acid is apparently a regulator of brain processes.

Glutamine is converted to glutamate and α-ketoglutarate. The details of these processes were described on pages 305 and 306.

Figure 8-7 shows the metabolic interconversions of glutamate.

8.8. Arginine

Arginine can be synthesized from glutamate and aspartate through the urea-cycle reactions:

$$\underset{\text{Glutamate}}{\begin{array}{c}\text{COOH}\\|\\\text{CH}_2\\|\\\text{CH}_2\\|\\\text{CHNH}_2\\|\\\text{COOH}\end{array}} \xrightarrow{\text{NADH} \ \text{NAD}^+} \underset{\substack{\text{Glutamic}\\\text{semialdehyde}}}{\begin{array}{c}\text{CHO}\\|\\\text{CH}_2\\|\\\text{CH}_2\\|\\\text{CHNH}_2\\|\\\text{COOH}\end{array}} \xrightarrow{\text{Glu} \ \ \alpha\text{-KG}} \underset{\text{Ornithine}}{\begin{array}{c}\text{CH}_2\text{NH}_2\\|\\\text{CH}_2\\|\\\text{CH}_2\\|\\\text{CHNH}_2\\|\\\text{COOH}\end{array}} \xrightarrow{\substack{\text{Carbamoyl}\\\text{phosphate}}} \underset{\text{Citrulline}}{\begin{array}{c}\text{CH}_2\text{NHCNH}_2\\\ \ \ \ \|\\\ \ \ \ \text{O}\\|\\\text{CH}_2\\|\\\text{CH}_2\\|\\\text{CHNH}_2\\|\\\text{COOH}\end{array}} \xrightarrow{\substack{\text{Aspartate}\\+\text{ ATP}}}$$

$$\underset{\text{Argininosuccinate}}{\begin{array}{c}\ \ \ \ \ \text{NH}\\\ \ \ \ \ \|\\\text{CH}_2\text{NHCNHCHCOOH}\\|\ \ \ \ \ \ \ \ \ \ \ \ \ \ \ \ |\\\text{CH}_2\ \ \ \ \ \ \ \ \ \ \text{CH}_2\\|\ \ \ \ \ \ \ \ \ \ \ \ \ \ \ \ |\\\text{CH}_2\ \ \ \ \ \ \ \ \ \ \text{COOH}\\|\\\text{CHNH}_2\\|\\\text{COOH}\end{array}} \longrightarrow \underset{\text{Arginine}}{\begin{array}{c}\ \ \ \ \ \text{NH}\\\ \ \ \ \ \|\\\text{CH}_2\text{NHCNH}_2\\|\\\text{CH}_2\\|\\\text{CH}_2\\|\\\text{CHNH}_2\\|\\\text{COOH}\end{array}} + \underset{\text{Fumarate}}{\begin{array}{c}\text{HCCOOH}\\\|\\\text{HOOCCH}\end{array}}$$

However, this source does not provide the necessary amount of arginine for normal growth. Hence, although dietary arginine might not be required for maintaining the nitrogen balance in adults, it seems to be necessary for growing children.

When it is metabolized, arginine is converted to ornithine. In addition to its function in the urea cycle (section 7.5), ornithine may be converted to glutamate and hence to α-ketoglutarate by transamination:

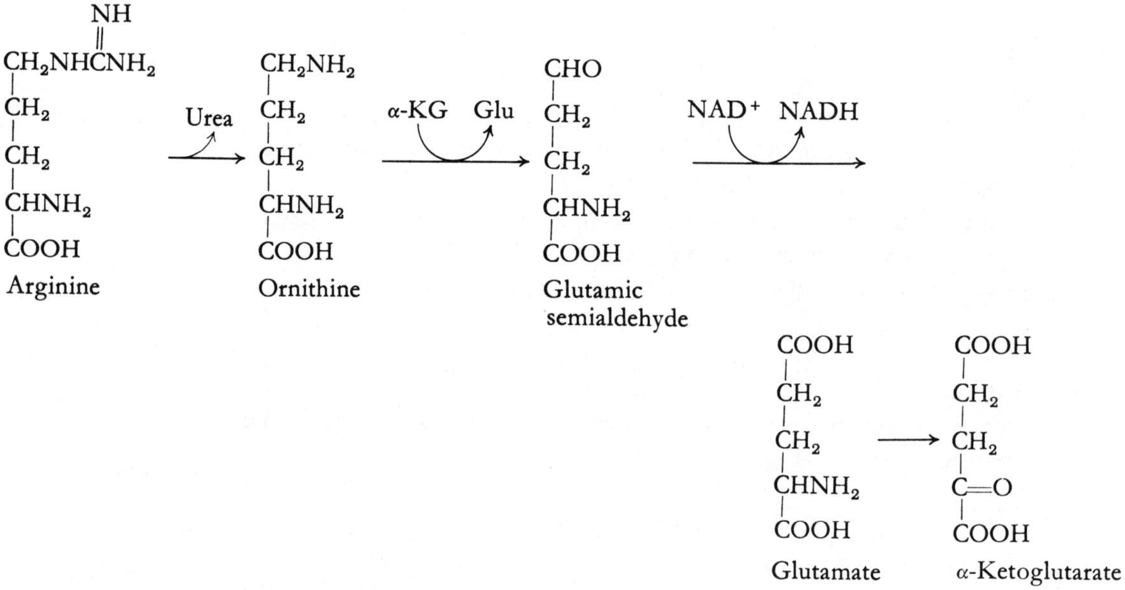

$$\underset{\text{Arginine}}{\begin{array}{c}CH_2NHCNH_2 \\ \| \\ NH \\ | \\ CH_2 \\ | \\ CH_2 \\ | \\ CHNH_2 \\ | \\ COOH\end{array}} \xrightarrow{\text{Urea}} \underset{\text{Ornithine}}{\begin{array}{c}CH_2NH_2 \\ | \\ CH_2 \\ | \\ CH_2 \\ | \\ CHNH_2 \\ | \\ COOH\end{array}} \xrightarrow{\alpha\text{-KG} \quad \text{Glu}} \underset{\substack{\text{Glutamic}\\ \text{semialdehyde}}}{\begin{array}{c}CHO \\ | \\ CH_2 \\ | \\ CH_2 \\ | \\ CHNH_2 \\ | \\ COOH\end{array}} \xrightarrow{NAD^+ \quad NADH} \underset{\text{Glutamate}}{\begin{array}{c}COOH \\ | \\ CH_2 \\ | \\ CH_2 \\ | \\ CHNH_2 \\ | \\ COOH\end{array}} \rightarrow \underset{\alpha\text{-Ketoglutarate}}{\begin{array}{c}COOH \\ | \\ CH_2 \\ | \\ CH_2 \\ | \\ C=O \\ | \\ COOH\end{array}}$$

Thus, arginine is a glucogenic amino acid.

Arginine also serves as a donor of an amidine group to glycine. This provides guanidinoacetic acid and ornithine by an alternative mechanism (section 7.9). Guanidinoacetic acid accepts a methyl group from S-adenosylmethionine to yield creatine (section 7.8).

The metabolism of arginine as well as its interrelationships with that of other amino acids and the urea and citric acid cycles are shown in Figure 8-8.

8.9. Histidine

Histidine is essential for normal growth. The metabolic degradation of histidine provides single-carbon units and glutamate. The first step in this route involves a non-oxidative deamination (section 7.3.B) catalyzed by the enzyme histidase; the product is urocanic acid:

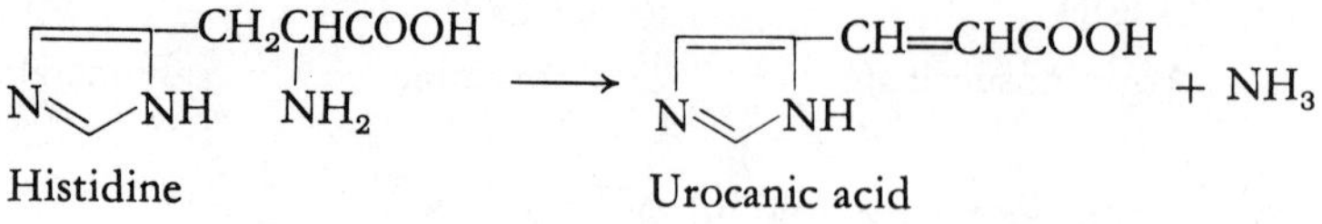

Histidine → Urocanic acid + NH_3

Urocanate is converted to imidazolone 5-propionic acid, which in turn is

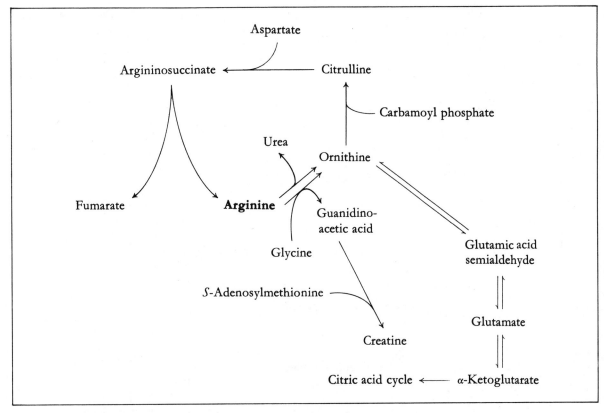

Fig. 8-8. Arginine metabolism.

transformed to *N*-formiminoglutamic acid (FIGLU):

$$\text{Urocanic acid} \xrightarrow[\text{Pyridoxal phosphate}]{\text{Urocanase}} \text{Imidazolone 5-propionic acid} \xrightarrow{\text{Hydrolase}} \text{N-Formiminoglutamic acid (FIGLU)}$$

By the action of a transferase, the formimino group is transferred to position N-5 of tetrahydrofolate (section 7.10). The remainder of the molecule is thus glutamic acid:

$$\text{FIGLU} + \text{THFA} \longrightarrow \text{Glutamic acid} + N^5\text{-Formimino-THFA}$$

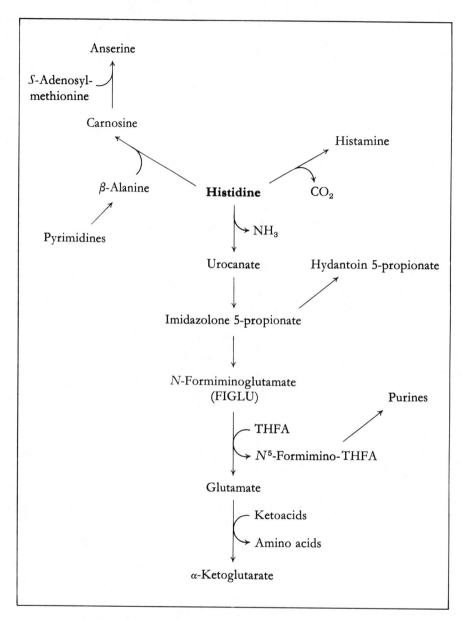

Fig. 8-9. Histidine metabolism.

Since glutamate yields α-ketoglutarate on transamination, this pathway can provide glucose from histidine; i.e., histidine is glucogenic. The single-carbon units obtained in the process are available for the synthesis of nucleotides and other intermediates requiring tetrahydrofolate-linked carbon transfers. A small fraction of the imidazolone 5-propionic acid is oxidized to hydantoin 5-propionate, which is excreted in the urine.

Another important reaction of histidine is its decarboxylation to histamine:

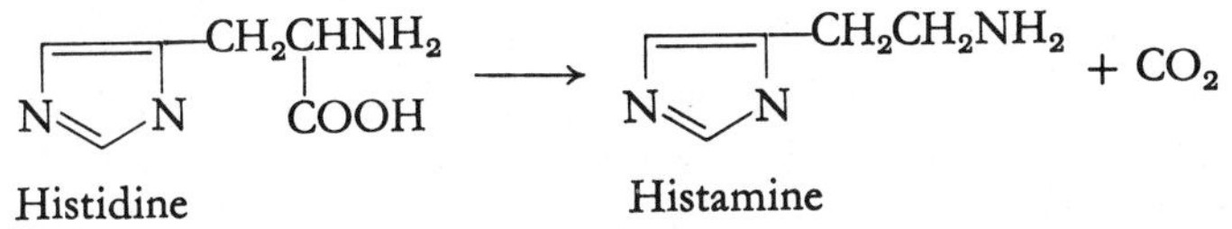

$$\text{Histidine} \longrightarrow \text{Histamine} + CO_2$$

Histidine: imidazole-CH$_2$CHNH$_2$COOH
Histamine: imidazole-CH$_2$CH$_2$NH$_2$

This reaction is especially prominent in mast cells, where histamine is stored together with serotonin and heparin. Histamine is found in significant concentrations in the liver, lung, skin, muscle, and gastric mucosa. It is released from these tissues during shock, by antigens and allergens, and by various chemical agents. Otherwise, it is tightly bound in the tissues where it is formed or stored. Histamine is a powerful vasodilator. Excessive amounts cause capillary dilation and an increase in vascular permeability. Such action results in the loss of plasma protein to the extracellular space and diminished circulatory volume. To compensate for the vasodilation, the heart rate and cardiac output increase and the systemic blood pressure drops. The acute form of these reactions is known as *histamine shock*. Histamine also stimulates the gastric secretion of pepsin and hydrochloric acid. Other effects include constriction of the bronchioles, uterine stimulation, and dilation of the vessels of the brain and meninges.

In addition to its degradation to glutamate and the formation of histamine, histidine is also utilized for the formation of carnosine and anserine, which are found in muscle.

The metabolism of histidine is summarized in Figure 8-9.

8.10. Proline

Proline is a non-essential amino acid, because it can be synthesized from glutamate. Metabolic degradation of proline also yields glutamate:

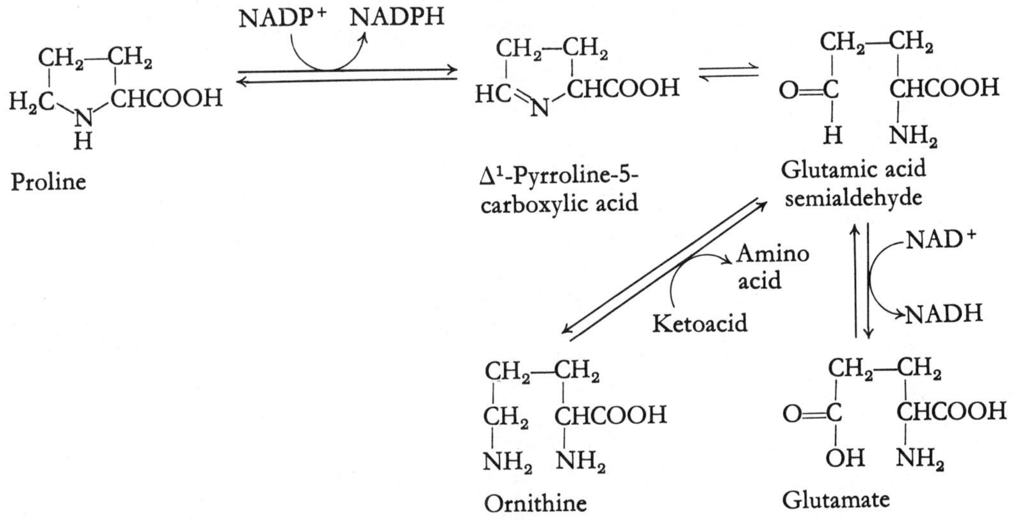

Proline $\xrightarrow{NADP^+ / NADPH}$ Δ^1-Pyrroline-5-carboxylic acid $\rightleftharpoons$ Glutamic acid semialdehyde

Glutamic acid semialdehyde $\xrightarrow{\text{Amino acid / Ketoacid}}$ Ornithine

Glutamic acid semialdehyde $\xrightarrow{NAD^+ / NADH}$ Glutamate

Conversion of the glutamate that is formed to α-ketoglutarate provides a means whereby proline can contribute to gluconeogenesis. Glutamic acid semialdehyde, which is an intermediate in the metabolism of proline, can also serve as a precursor for the synthesis of ornithine.

Proline residues in collagen and certain other fibrous proteins are oxidized to hydroxyproline. The reaction is catalyzed by proline hydroxylase:

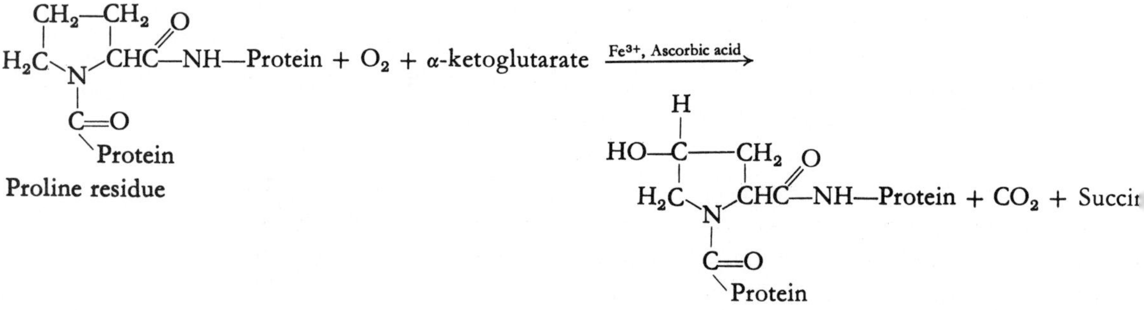

$$\text{Proline residue} + O_2 + \alpha\text{-ketoglutarate} \xrightarrow{Fe^{3+}, \text{Ascorbic acid}} \text{Hydroxyproline residue} + CO_2 + \text{Succinate}$$

Free proline does not undergo this reaction.

The metabolic pathways involving proline are summarized in Figure 8-10.

8.11. Methionine

Methionine is an essential amino acid. Metabolic degradation of methionine results in the conversion of its carbon skeleton to succinyl-coenzyme A. Since the latter can be utilized for the synthesis of glycogen (via succinate, malate, oxaloacetate, and phosphoenolpyruvate), methionine is classified as a glucogenic amino acid.

The initial steps in the pathway from methionine to succinyl-coenzyme A involve the conversion of the former to homocysteine (page 317). Interaction of homocysteine with serine yields cystathionine:

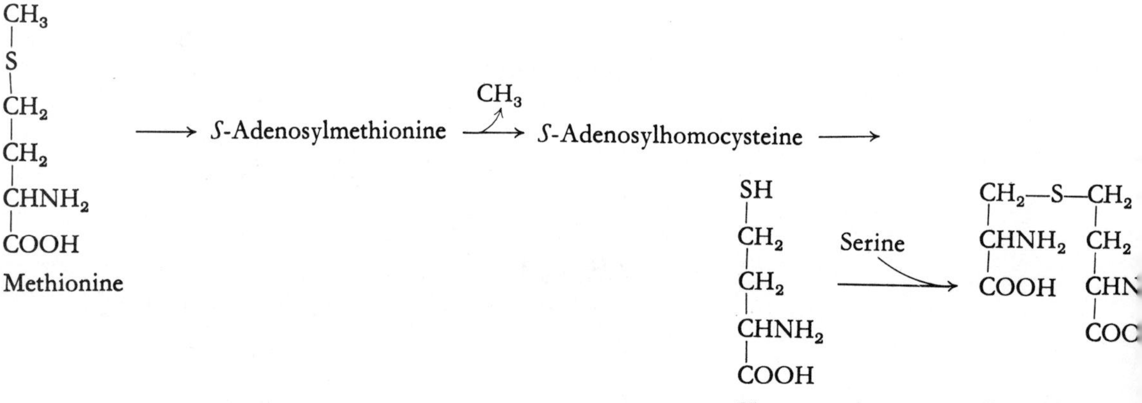

Methionine ⟶ S-Adenosylmethionine $\xrightarrow{CH_3}$ S-Adenosylhomocysteine ⟶ Homocysteine $\xrightarrow{\text{Serine}}$ Cystathionine

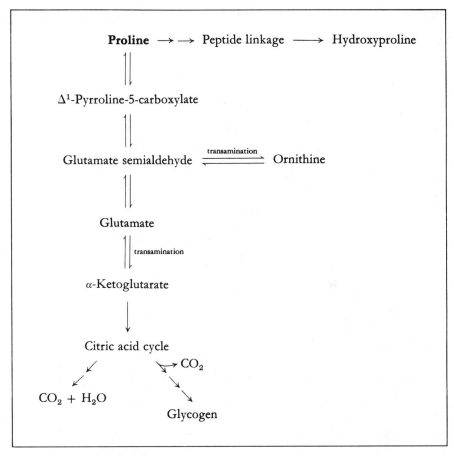

Fig. 8-10. *Metabolism of proline.*

Cystathionine is hydrolyzed to cysteine, α-ketobutyrate, and ammonia:

$$\text{Cystathionine} \longrightarrow \begin{array}{c} CH_2SH \\ | \\ CHNH_2 \\ | \\ COOH \end{array} + \begin{array}{c} CH_3 \\ | \\ CH_2 \\ | \\ C{=}O \\ | \\ COOH \end{array} + NH_3$$

Cysteine   α-Ketobutyrate

α-Ketobutyrate is converted to succinyl-coenzyme A in the reaction sequence propionyl-coenzyme A → methylmalonyl-coenzyme A → succinyl-coenzyme A, as described under the metabolism of threonine (section 8.4).

Methionine functions as the methyl donor in a variety of transmethylation reactions (section 7.8). In addition, the methyl groups derived from

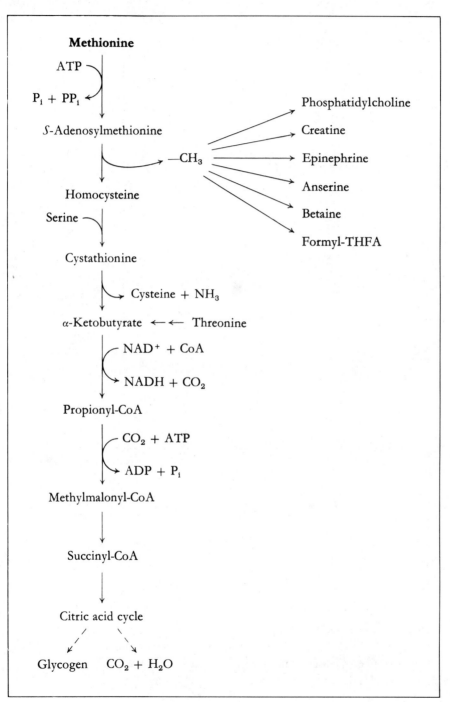

Fig. 8-11. *Metabolism of methionine.*

methionine can serve as a source of formate in reactions involving transfers via tetrahydrofolic acid (section 7.10).

The principal metabolic pathways for methionine are summarized in Figure 8-11.

8.12. Valine

Valine is an essential amino acid. The initial step in its catabolism is transamination to yield α-ketoisovaleric acid. Oxidative decarboxylation of the latter results in the formation of isobutyryl-coenzyme A:

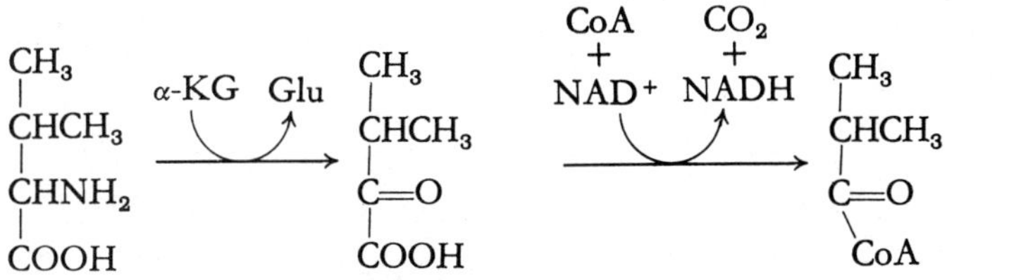

$$\begin{array}{c} CH_3 \\ | \\ CHCH_3 \\ | \\ CHNH_2 \\ | \\ COOH \end{array} \xrightarrow{\alpha\text{-KG} \quad \text{Glu}} \begin{array}{c} CH_3 \\ | \\ CHCH_3 \\ | \\ C{=}O \\ | \\ COOH \end{array} \xrightarrow[\text{NAD}^+ \quad \text{NADH}]{\text{CoA} \quad CO_2} \begin{array}{c} CH_3 \\ | \\ CHCH_3 \\ | \\ C{=}O \\ \backslash \\ CoA \end{array}$$

Valine → α-Ketoisovaleric acid → Isobutyryl-CoA

Isobutyryl-coenzyme A is oxidized to methacrylyl-coenzyme A, which is then hydrated to β-hydroxyisobutyryl-coenzyme A. Removal of coenzyme A yields β-hydroxyisobutyrate:

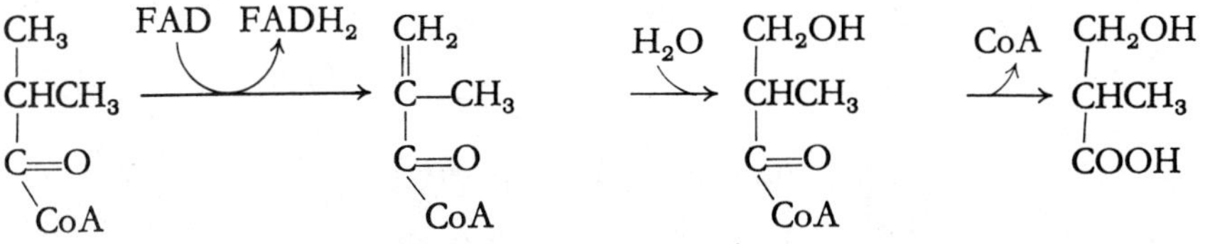

$$\begin{array}{c} CH_3 \\ | \\ CHCH_3 \\ | \\ C{=}O \\ \backslash \\ CoA \end{array} \xrightarrow{\text{FAD} \quad \text{FADH}_2} \begin{array}{c} CH_2 \\ \| \\ C{-}CH_3 \\ | \\ C{=}O \\ \backslash \\ CoA \end{array} \xrightarrow{H_2O} \begin{array}{c} CH_2OH \\ | \\ CHCH_3 \\ | \\ C{=}O \\ \backslash \\ CoA \end{array} \xrightarrow{\text{CoA}} \begin{array}{c} CH_2OH \\ | \\ CHCH_3 \\ | \\ COOH \end{array}$$

Isobutyryl-CoA → Methylacrylyl-CoA → β-Hydroxyisobutyryl-CoA → β-Hydroxyisobutyrate

Two successive oxidations yield, respectively, methylmalonate semialdehyde and methylmalonyl-coenzyme A. The latter is then converted to succinyl-coenzyme A:

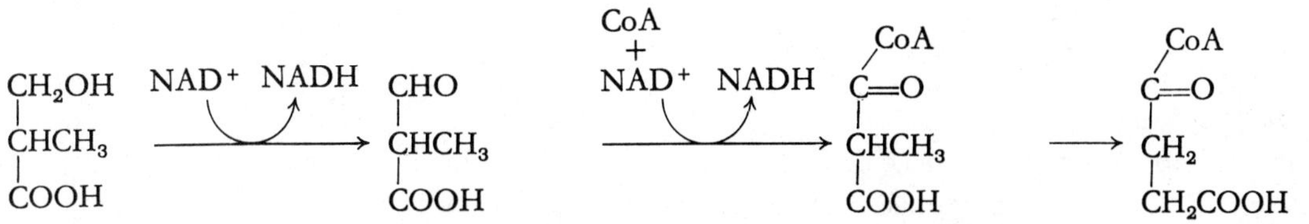

$$\begin{array}{c} CH_2OH \\ | \\ CHCH_3 \\ | \\ COOH \end{array} \xrightarrow{\text{NAD}^+ \quad \text{NADH}} \begin{array}{c} CHO \\ | \\ CHCH_3 \\ | \\ COOH \end{array} \xrightarrow[\text{NAD}^+ \quad \text{NADH}]{\text{CoA}} \begin{array}{c} CoA \\ \backslash \\ C{=}O \\ | \\ CHCH_3 \\ | \\ COOH \end{array} \longrightarrow \begin{array}{c} CoA \\ \backslash \\ C{=}O \\ | \\ CH_2 \\ | \\ CH_2COOH \end{array}$$

β-Hydroxyisobutyrate → Methylmalonate semialdehyde → Methylmalonyl-coenzyme A → Succinyl-coenzyme A

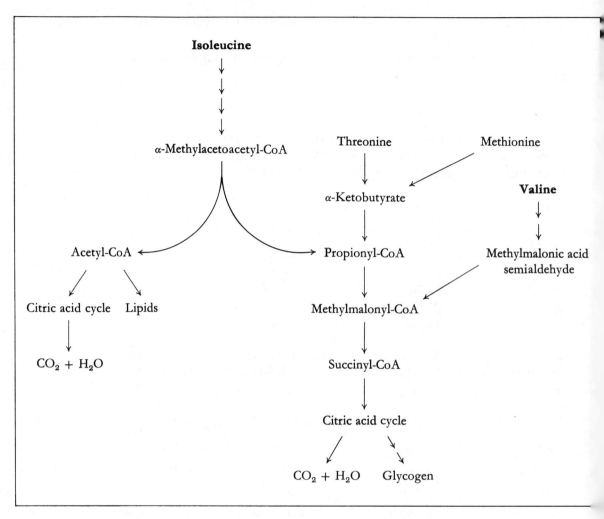

Fig. 8-12. Metabolism of valine and isoleucine and relation to metabolism of threonine and methionine.

The metabolism of valine and its relationship with that of isoleucine (which is also converted to succinate as described in the following section) is outlined in Figure 8-12.

8.13. Isoleucine

Isoleucine is an essential amino acid. It is glucogenic and ketogenic, since its metabolic degradation yields succinyl-coenzyme A (via propionyl-coenzyme A) and acetyl-coenzyme A.

The initial step in the catabolism of isoleucine is a transamination reaction in which it is converted to α-keto-β-methylvaleric acid. Oxidative decarboxylation of the latter provides α-methylbutyryl-coenzyme A:

$$\underset{\text{Isoleucine}}{\begin{array}{c}CH_3\\|\\CH_2\\|\\CHCH_3\\|\\CHNH_2\\|\\COOH\end{array}} \xrightarrow{\alpha\text{-KG} \quad \text{Glu}} \underset{\alpha\text{-Keto-}\beta\text{-methyl-}\atop\text{valeric acid}}{\begin{array}{c}CH_3\\|\\CH_2\\|\\CHCH_3\\|\\C=O\\|\\COOH\end{array}} \xrightarrow[\text{NAD}^+ \quad \text{NADH}]{\text{CoA} \quad CO_2} \underset{\alpha\text{-Methylbutyryl-}\atop\text{coenzyme A}}{\begin{array}{c}CH_3\\|\\CH_2\\|\\CHCH_3\\|\\C=O\\\diagdown\\CoA\end{array}}$$

Another oxidation by a flavoprotein-linked enzyme yields tiglyl-coenzyme A, which is hydrated to yield α-methyl-β-hydroxybutyryl-coenzyme A:

$$\underset{\alpha\text{-Methylbutyryl-}\atop\text{coenzyme A}}{\begin{array}{c}CH_3\\|\\CH_2\\|\\CHCH_3\\|\\C=O\\\diagdown\\CoA\end{array}} \xrightarrow{\text{FAD} \quad \text{FADH}_2} \underset{\text{Tiglyl-}\atop\text{coenzyme A}}{\begin{array}{c}CH_3\\|\\CH\\\|\\CCH_3\\|\\C=O\\\diagdown\\CoA\end{array}} \xrightarrow{H_2O} \underset{\alpha\text{-Methyl-}\beta\text{-hydroxy-}\atop\text{butyryl-coenzyme A}}{\begin{array}{c}CH_3\\|\\CHOH\\|\\CHCH_3\\|\\C=O\\\diagdown\\CoA\end{array}}$$

Oxidation of α-methyl-β-hydroxybutyryl-coenzyme A yields α-methylacetoacetyl-coenzyme A. Cleavage of the latter by thiolase gives rise to acetyl-coenzyme A and propionyl-coenzyme A:

$$\underset{\alpha\text{-Methyl-}\beta\text{-hydroxy-}\atop\text{butyryl-coenzyme A}}{\begin{array}{c}CH_3\\|\\CHOH\\|\\CHCH_3\\|\\C=O\\\diagdown\\CoA\end{array}} \xrightarrow{\text{NAD}^+ \quad \text{NADH}} \underset{\alpha\text{-Methylaceto-}\atop\text{acetyl-coenzyme A}}{\begin{array}{c}CH_3\\|\\C=O\\|\\CHCH_3\\|\\C=O\\\diagdown\\CoA\end{array}} \xrightarrow{\text{CoA}} \underset{\text{Acetyl-}\atop\text{coenzyme A}}{\begin{array}{c}CH_3\\|\\C=O\\\diagdown\\CoA\end{array}} + \underset{\text{Propionyl-}\atop\text{coenzyme A}}{\begin{array}{c}CH_2CH_3\\|\\C=O\\\diagdown\\CoA\end{array}}$$

Thus, the six-carbon amino acid is degraded to carbon dioxide, acetyl-coenzyme A, and propionyl-coenzyme A. The propionyl-coenzyme A is converted to succinyl-coenzyme A, as described in the discussion of threonine metabolism (section 8.4).

The disposition of isoleucine and valine is summarized in Figure 8-12, and the relationship of their metabolism to that of threonine and methionine is shown.

8.14. Leucine

As are the other branched-chain amino acids, leucine, too, is essential. Its degradation yields acetyl-coenzyme A and acetoacetate. Since neither of these contribute to a net gain of glycogen, leucine is classified as a ketogenic, non-glucogenic amino acid.

Leucine undergoes transamination with α-ketoglutarate to provide α-ketoisocaproic acid. Oxidative decarboxylation of this intermediate yields isovaleryl-coenzyme A. As a result of oxidation by a flavoprotein-linked enzyme, the latter is converted to 3-methylcrotonyl-coenzyme A:

$$\begin{array}{c}CH_3\\|\\CHCH_3\\|\\CH_2\\|\\CHNH_2\\|\\COOH\end{array}\ \xrightarrow{\alpha\text{-KG}\ \ \text{Glu}}\ \begin{array}{c}CH_3\\|\\CHCH_3\\|\\CH_2\\|\\C=O\\|\\COOH\end{array}\ \xrightarrow{\substack{CoA\ \ CO_2\\+\ \ \ +\\NAD^+\ \ NADH}}\ \begin{array}{c}CH_3\\|\\CHCH_3\\|\\CH_2\\|\\C=O\\|\\CoA\end{array}\ \xrightarrow{FAD\ \ FADH_2}\ \begin{array}{c}CH_3\\|\\CCH_3\\||\\CH\\|\\C=O\\|\\CoA\end{array}$$

Leucine   α-Ketoiso-   Isovaleryl-   3-Methyl-
              caproic acid   coenzyme A   crotonyl-
                                                                                                                                                       coenzyme A

In the presence of a carboxylase, 3-methylcrotonyl-coenzyme A reacts with carbonate and ATP to form 3-methylglutaconyl-coenzyme A and ADP. This intermediate is hydrated to yield 3-hydroxy-3-methylglutaryl-coenzyme A. The latter is cleaved to yield acetoacetate and acetyl-coenzyme A:

$$\begin{array}{c}CH_3\\|\\CCH_3\\||\\CH\\|\\C=O\\|\\CoA\end{array}\ \xrightarrow{\substack{ATP\ \ ADP\\+\ \ \ +\\HCO_3^-\ \ P_i}}\ \begin{array}{c}COOH\\|\\CH_2\\|\\CCH_3\\||\\CH\\|\\C=O\\|\\CoA\end{array}\ \xrightarrow{H_2O}\ \begin{array}{c}COOH\\|\\CH_2\\|\\HOCCH_3\\|\\CH_2\\|\\C=O\\|\\CoA\end{array}\ \longrightarrow\ \begin{array}{l}CH_2COOH\\|\\O=CCH_3\\ \text{Acetoacetate}\\\\CH_3CO-CoA\\ \text{Acetyl-coenzyme A}\end{array}$$

3-Methyl-   3-Methyl-   3-Hydroxy-
crotonyl-   glutaconyl-   3-methyl-
coenzyme A   coenzyme A   glutaryl-
                                                  coenzyme A

It should be remembered that 3-hydroxy-3-methylglutaryl-coenzyme A is an intermediate in the catabolism of fatty acids (page 254) and a precursor in the biosynthesis of cholesterol (page 234).

Two common aspects in the degradative pathways of the branched-chain amino acids are noteworthy. First, the catabolism of the keto-acid metabolites (which are produced from amino acids after removal of the amino group) follows a route analogous to that found in the degradation of fatty acids (page 121). The steps include dehydrogenations, hydrations, and hydrolyses. Second, oxidative decarboxylation of the ketoacids derived from leucine, isoleucine, and valine appears to be catalyzed by the same enzyme. This enzyme is absent in an inborn error of metabolism, *maple-syrup urine disease*. The resulting block in the catabolism of the branched-chain amino acids at the α-ketoacid stage leads to the excretion of relatively large quantities of these ketoacids in the urine. The disease owes its name to the odor of the resulting urine. The biochemical defect is accompanied by mental retardation; the reasons for the degenerative changes that occur in the central nervous system are not known.

The metabolism of leucine is summarized in conjunction with that of lysine in Figure 8-13 of the next section.

8.15. Lysine

Lysine is an essential and ketogenic amino acid. It does not undergo transamination. The first step in the predominant catabolic pathway for L-lysine in mammals is the reaction of the amino acid with α-ketoglutarate to yield saccharopine:

$$\begin{array}{c} CH_2NH_2 \\ | \\ CH_2 \\ | \\ CH_2 \\ | \\ CH_2 \\ | \\ HCNH_2 \\ | \\ COOH \end{array} + \begin{array}{c} COOH \\ | \\ C=O \\ | \\ CH_2 \\ | \\ CH_2 \\ | \\ COOH \end{array} \xrightarrow{H_2O} \left[\begin{array}{cc} & COOH \\ & | \\ CH_2N=C & \\ | & | \\ CH_2 & CH_2 \\ | & | \\ CH_2 & CH_2 \\ | & | \\ CH_2 & COOH \\ | & \\ HCNH_2 & \\ | & \\ COOH & \end{array} \right] \xrightarrow{NADH\ NAD^+} \begin{array}{cc} H & COOH \\ | & | \\ CH_2N-CH & \\ | & | \\ CH_2 & CH_2 \\ | & | \\ CH_2 & CH_2 \\ | & | \\ CH_2 & COOH \\ | & \\ HCNH_2 & \\ | & \\ COOH & \end{array}$$

L-Lysine α-Ketoglutarate   Schiff base intermediate   Saccharopine

Subsequent oxidation and hydrolysis of saccharopine provides α-aminoadipic semialdehyde and glutamate. This is followed by oxidation of α-aminoadipic semialdehyde to aminoadipic acid, and transamination to provide α-ketoadipic acid:

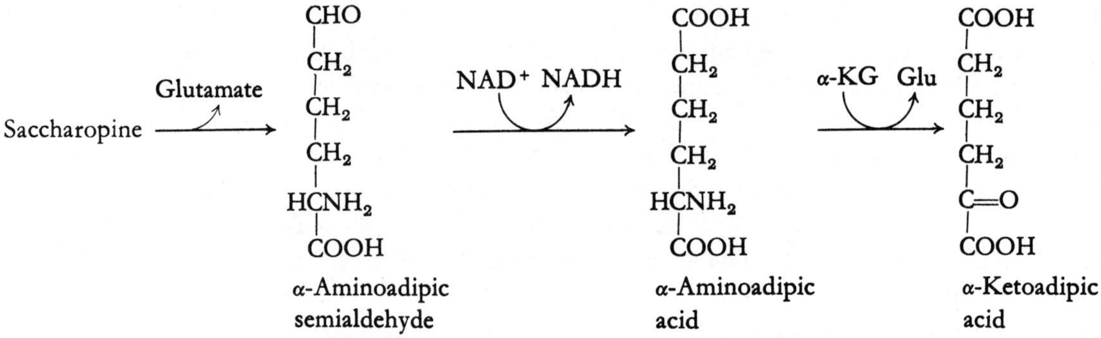

Saccharopine →[Glutamate] α-Aminoadipic semialdehyde →[NAD⁺ / NADH] α-Aminoadipic acid →[α-KG / Glu] α-Ketoadipic acid

Oxidative decarboxylation of α-ketoadipic acid in the presence of coenzyme A and NAD⁺ yields glutaryl-coenzyme A. This is oxidized to glutaconyl-coenzyme A. Decarboxylation of the latter provides crotonyl-coenzyme A, which is converted to acetoacetyl-coenzyme A:

α-Ketoadipic acid →[CoA + NAD⁺ / CO₂ + NADH] Glutaryl-coenzyme A → Glutaconyl-coenzyme A →[CO₂] Crotonyl-coenzyme A → β-Hydroxybutyryl-coenzyme A → Acetoacetyl-coenzyme A

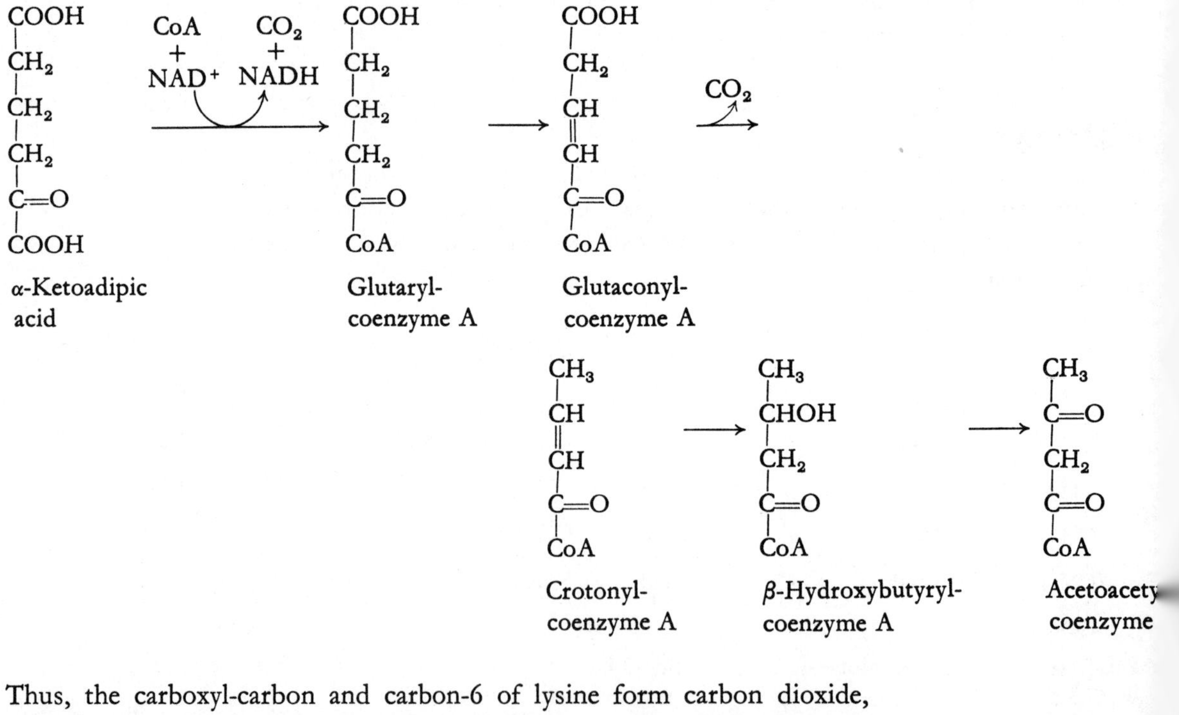

Thus, the carboxyl-carbon and carbon-6 of lysine form carbon dioxide, and the other four carbons are ultimately incorporated into acetoacetyl-coenzyme A.

The interrelationships in the metabolic pathways of leucine and lysine are shown in Figure 8-13.

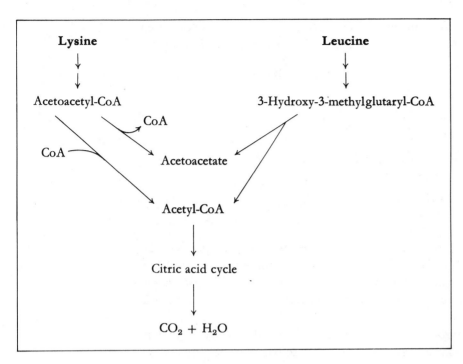

Fig. 8-13. *Catabolism of leucine and lysine.*

8.16. Tyrosine

Tyrosine is a non-essential amino acid, since it is produced from phenylalanine. Tyrosine is both glucogenic and ketogenic.

The initial step in the catabolism of tyrosine is its conversion to *p*-hydroxyphenylpyruvate by transamination. Oxidation of this ketoacid yields homogentisate:

The last reaction, which is catalyzed by *p*-hydroxyphenylpyruvate oxidase, involves hydroxylation, a shift of the side chain, and decarboxylation. Homogentisate is oxidized by molecular oxygen in the presence of homogentisic-acid oxidase to form maleylacetoacetate. This in turn is converted to fumarylacetoacetate, which is hydrolyzed to acetoacetate and fumarate:

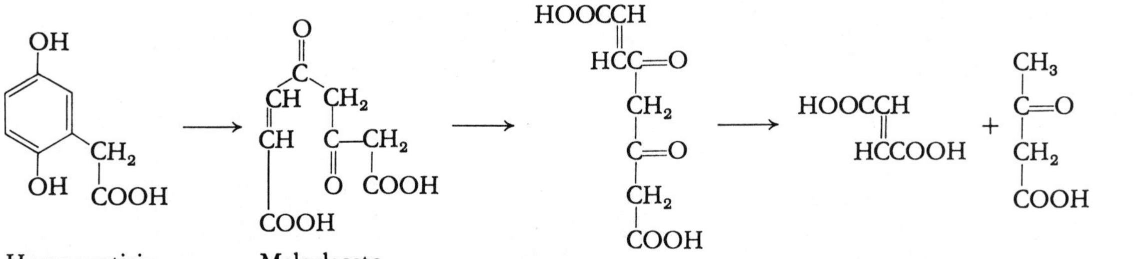

Homogentisic acid → Maleylacetoacetic acid → Fumarylacetoacetic acid → Fumaric acid + Acetoacetic acid

The fumarate thus formed can serve as a precursor of glycogen via the sequence fumarate → malate → oxaloacetate → phosphoenolpyruvate. The other product in tyrosine catabolism—acetoacetate—may give rise to the other ketone bodies or be metabolized to provide energy in extrahepatic tissues.

Several steps in tyrosine degradation are of special interest in human metabolism. The enzyme that catalyzes the conversion of tyrosine to *p*-hydroxyphenylpyruvate—i.e., tyrosine-glutamate transaminase—appears to be an *inducible enzyme*; that is, its synthesis is enhanced after ingestion of tyrosine. The levels of the enzyme are also increased after administration of adrenal steroids, insulin, or glucagon. Injection of hypophyseal somatotropin (growth hormone, GH) causes a decrease in enzymatic activity.

p-Hydroxyphenylpyruvate oxidase, which catalyzes the next reaction, requires ascorbic acid (vitamin C). When there is a deficiency in this vitamin, considerable amounts of *p*-hydroxyphenylpyruvate are excreted in the urine. Of additional interest is the finding that *p*-hydroxyphenylpyruvate-oxidase activity is low in fetal liver and becomes more prominent after birth. During infancy, especially in premature infants, a significant fraction of *p*-hydroxyphenylpyruvate is converted to *p*-hydroxyphenyllactate, and the latter and its metabolites are excreted in the urine:

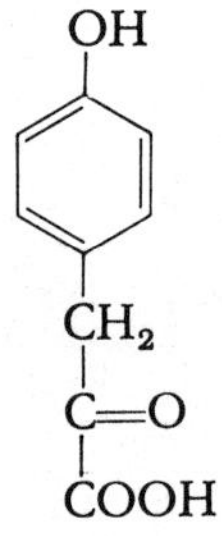

p-Hydroxyphenylpyruvate 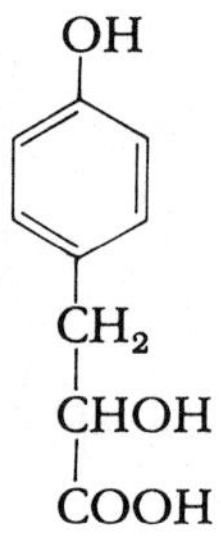   *p*-Hydroxyphenyllactate

The absence of *p*-hydroxyphenylpyruvate oxidase has been implicated as the cause of a very rare inborn error of metabolism termed *tyrosinemia I*. This disease is characterized by the excretion of *p*-hydroxyphenylpyruvate in the urine. The physical features seen in this disorder include enlargement of the spleen and liver and multiple defects of the renal tubules.

Another defect of tyrosine metabolism, called tyrosinosis or *tyrosinemia II*, has been ascribed to a defect either in the same enzyme or in tyrosine-glutamate transaminase.

Alkaptonuria is another rare hereditary disease of tyrosine metabolism. The biochemical defect is a lack of homogentisic-acid oxidase. As a consequence, homogentisic acid accumulates and is eliminated in the urine. The urine becomes dark on exposure to air, especially when alkaline, as a result of the oxidation of homogentisic acid. No physical abnormality is discernible in individuals with alkaptonuria during early life, but later, abnormal pigmentation of cartilage and other connective tissue (ochronosis) occurs due to the deposition of homogentisic acid.

Concurrent with the transformations leading to acetoacetate and fumarate, tyrosine is utilized to provide various major tissue components, such as melanin, epinephrine, and thyroxine. *Melanin* is a dark pigment present in skin and a number of other tissues (e.g., the retina, adrenal medulla, ciliary body, substantia nigra of the brain, and choroid). The production of melanin from tyrosine occurs in the melanoblasts found in the basal layer of the epidermis. The copper-containing enzyme, tyrosinase, catalyzes the oxidation of tyrosine by oxygen to produce dihydroxyphenylalanine (DOPA) and its quinone derivative:

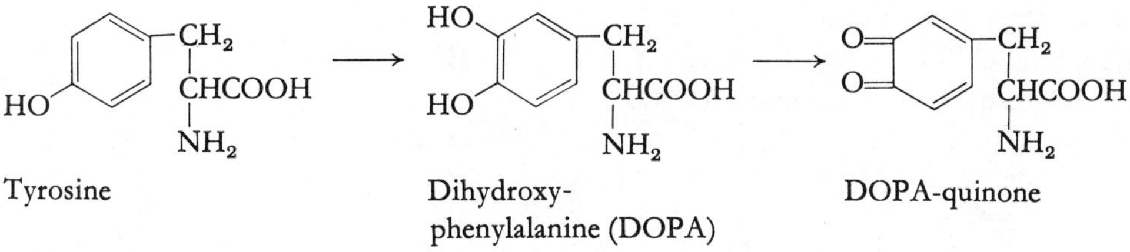

Tyrosine → Dihydroxyphenylalanine (DOPA) → DOPA-quinone

By a complex series of steps, these intermediates are transformed to indole derivatives, e.g.:

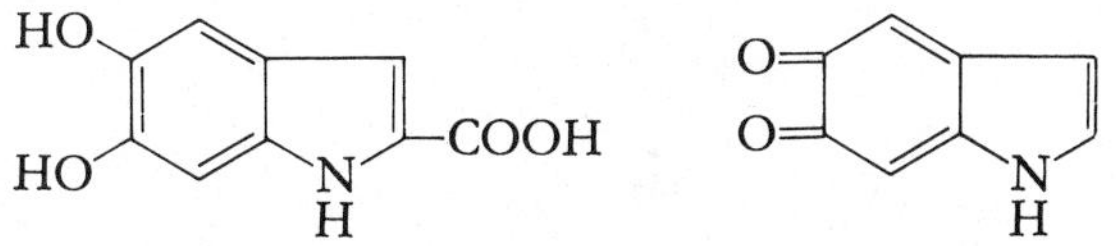

These derivatives in turn polymerize to form melanin. Melanin is a pigment, and its relative amount determines the characteristic coloration of the skin.

Albinism, the genetic defect due to the absence of tyrosinase in melanocytes, is characterized by a decrease in the melanin content of the skin, hair, and eyes. In Caucasians with this disease, the skin and hair are white, and the ocular fundus is orange-red. Neoplastic proliferation of melanoblasts gives rise to melanomas, which are generally highly malignant.

Norepinephrine and *epinephrine* are formed in the adrenal medulla via the oxidation of tyrosine to dihydroxyphenylalanine. Decarboxylation of the latter yields 3,4-dihydroxyphenylethylamine (dopamine; hydroxytyramine). Subsequent hydroxylation of the side chain results in the formation of norepinephrine, which is converted to epinephrine by a transmethylation reaction (section 7.8).

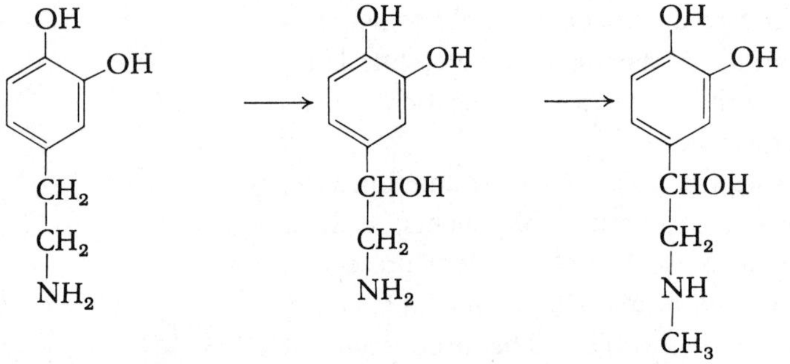

3,4-Dihydroxy-phenylethylamine (Dopamine) → Norepinephrine → Epinephrine

Tyrosine also serves as a precursor for the biosynthesis of the thyroid hormone *thyroxine*. This hormone is formed by the iodination of tyrosine residues in thyroglobulin. The iodinating agent, active iodine (I^+ or iodine free radical), is formed by a reaction between hydrogen peroxide and iodide ions:

$$2I^- + H_2O_2 + 2H^+ \xrightarrow{\text{Iodine peroxidase}} 2 \text{ Active iodine} + 2H_2O$$

The enzyme iodinase then catalyzes the iodination of tyrosine units in the protein to yield monoiodotyrosine and diiodotyrosine:

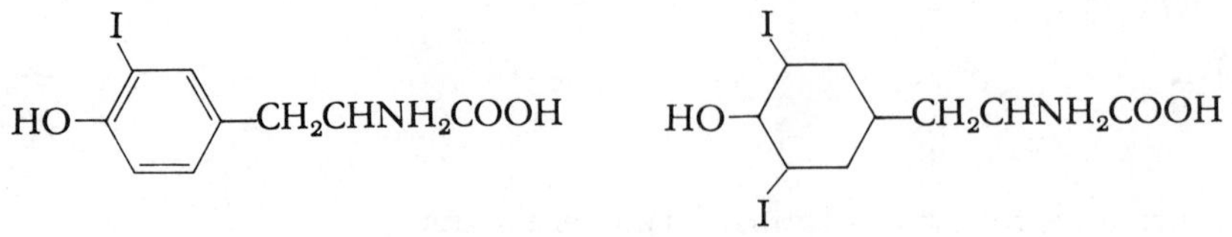

Monoiodotyrosine (protein-bound)   Diiodotyrosine (protein-bound)

Subsequent coupling of these derivatives yields thyroglobulin-bound triiodothyronine and thyroxine (tetraiodothyronine):

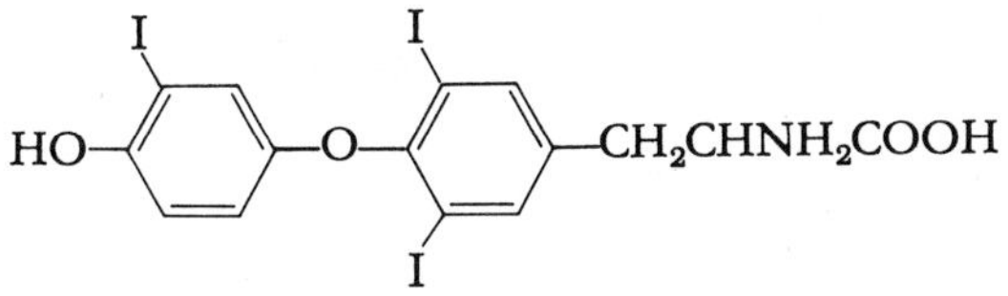

$$HO-\underset{}{\underset{I}{\bigcirc}}-O-\underset{I}{\overset{I}{\bigcirc}}-CH_2CHNH_2COOH$$

Triiodothyronine

$$HO-\underset{I}{\overset{I}{\bigcirc}}-O-\underset{I}{\overset{I}{\bigcirc}}-CH_2CHNH_2COOH$$

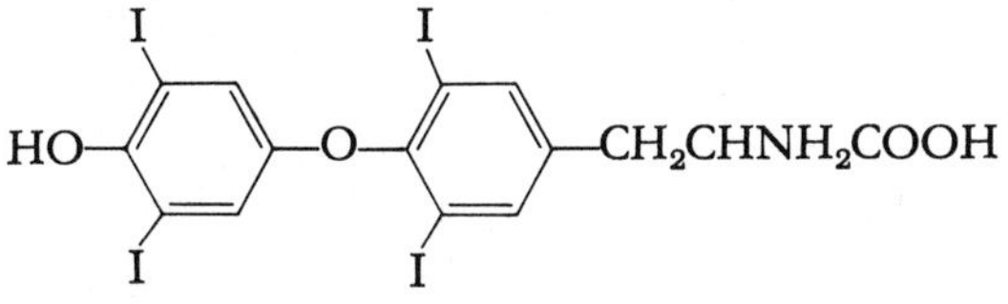

Thyroxine

Both these thyroid hormones are stored bound to thyroglobulin, a glycoprotein with molecular weight 680,000. The hormones are released from the thyroid gland into the bloodstream upon proteolysis of the thyroglobulin.

The metabolism of tyrosine as well as that of phenylalanine are outlined in Figure 8-14.

8.17. Phenylalanine

The principal metabolic fate of phenylalanine is its conversion to tyrosine in the liver. The reaction is catalyzed by *phenylalanine hydroxylase* and requires reduced biopterin (tetrahydrobiopterin) as a cofactor:

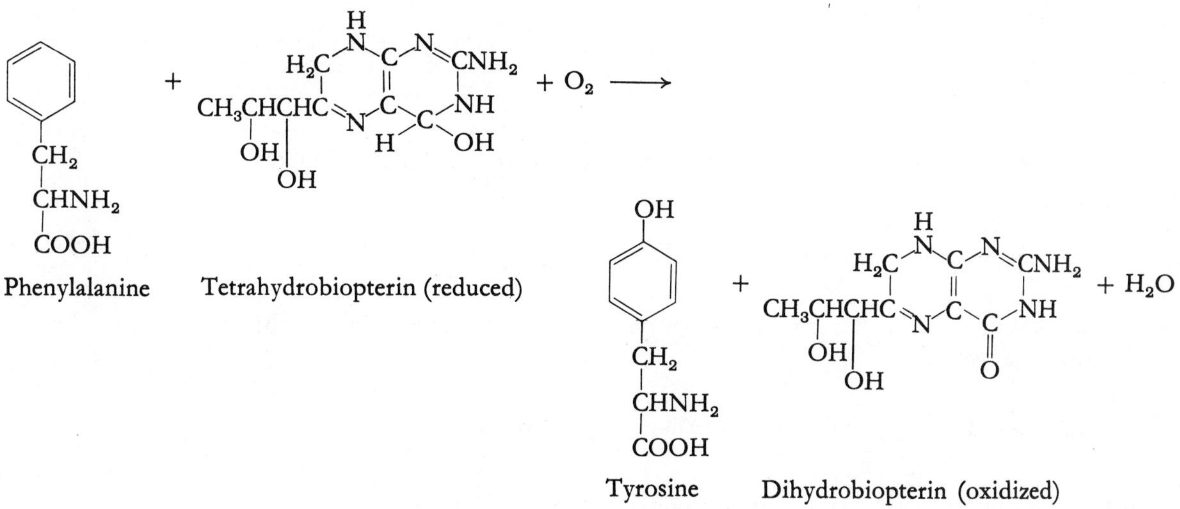

Phenylalanine + Tetrahydrobiopterin (reduced) + O_2 ⟶ Tyrosine + Dihydrobiopterin (oxidized) + H_2O

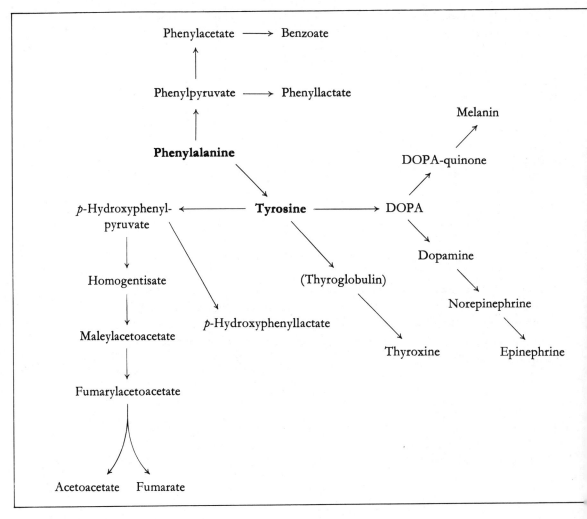

Fig. 8-14. Metabolism of phenylalanine and tyrosine.

Tetrahydrobiopterin is regenerated by the action of the NADPH-linked enzyme tetrahydropteridine dehydrogenase:

Dihydrobiopterin + NADPH + H$^+$ ⇌ Tetrahydrobiopterin + NADP$^+$

The total process for the production of tyrosine from phenylalanine can be written as follows:

NADPH + H$^+$ ⟶ Dihydrobiopterin ⟵ Tyrosine
NADP$^+$ ⟵ Tetrahydrobiopterin ⟶ Phenylalanine

The catabolic pathway of phenylalanine follows that of tyrosine (section 8.16); that is, it is ultimately degraded to acetoacetate and fumarate. In addition, a small amount of phenylalanine undergoes transamination to yield phenylpyruvate and its reduction product, phenyllactate:

$$\underset{\text{Phenylalanine}}{\underset{\text{COOH}}{\underset{|}{\overset{\text{CH}_2}{\underset{|}{\text{CHNH}_2}}}}\text{—C}_6\text{H}_5} \xrightarrow{\alpha\text{-KG} \;\; \text{Glu}} \underset{\text{Phenylpyruvic acid}}{\underset{\text{COOH}}{\underset{|}{\overset{\text{CH}_2}{\underset{|}{\text{C}=\text{O}}}}}\text{—C}_6\text{H}_5} \underset{\text{NAD}^+}{\overset{\text{NADH}}{\rightleftharpoons}} \underset{\text{Phenyllactic acid}}{\underset{\text{COOH}}{\underset{|}{\overset{\text{CH}_2}{\underset{|}{\text{CHOH}}}}}\text{—C}_6\text{H}_5}$$

Normally, the urine contains only small amounts of phenyllactate and phenylpyruvate. However, in the inborn error of metabolism, *phenylketonuria,* the enzyme phenylalanine hydroxylase is missing or defective. Since the principal pathway for phenylalanine metabolism is blocked, the serum level of phenylalanine is elevated, and high amounts of this amino acid as well as of phenylpyruvate and phenyllactate are excreted in the urine. The disease is so named because of the high concentration of phenylpyruvate in the urine. Phenylketonuria is generally linked to severe mental retardation. This impairment can be prevented if the disease is recognized in early infancy and phenylalanine is restricted from the diet.

The metabolism of phenylalanine and tyrosine was summarized in Figure 8-14.

8.18. Tryptophan

Tryptophan is an essential amino acid. The catabolism of tryptophan yields pyruvate (via alanine) and acetoacetate. On this basis, it is both glucogenic and ketogenic; however, the glucogenic aspect is minimal because of the predominance of other pathways. Tryptophan is also utilized for the synthesis of serotonin, melatonin, and nicotinic acid; the formation of each of these is considered as follows.

The initial step in the degradation of tryptophan is the opening of the five-member ring to yield N-formylkynurenine. This reaction is catalyzed by the enzyme *tryptophan pyrrolase*:

Tryptophan + O_2 → N-Formylkynurenine

Subsequent transfer of the formyl group to tetrahydrofolic acid provides kynurenine. Another oxidation results in the conversion of the latter to 3-hydroxykynurenine:

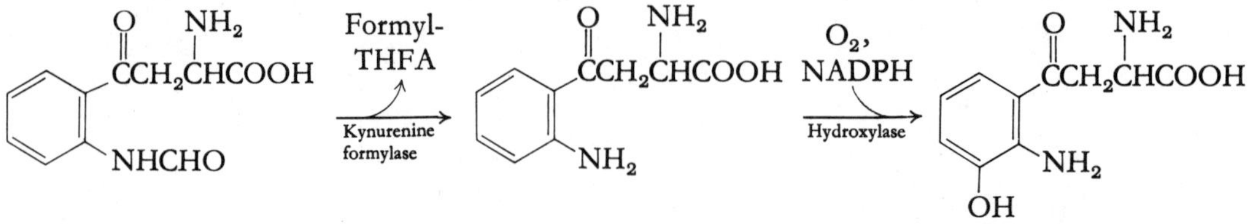

[Structure: N-Formylkynurenine] — Formyl-THFA, Kynurenine formylase → [Structure: Kynurenine] — O_2, NADPH, Hydroxylase → [Structure: 3-Hydroxykynurenine]

Action of the enzyme kynureninase on the latter yields 3-hydroxyanthranilic acid and alanine:

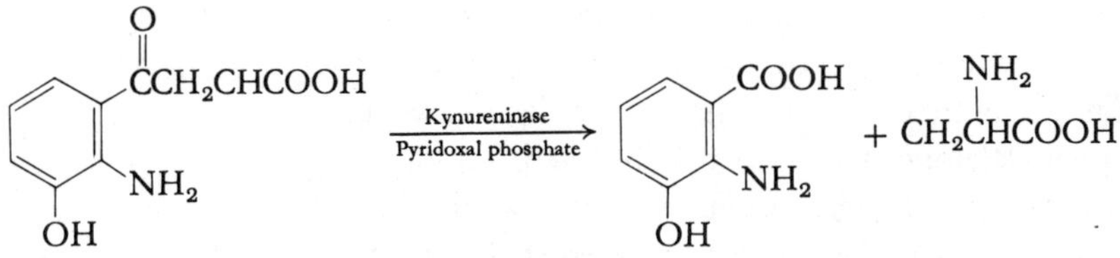

[Structure: 3-Hydroxykynurenine] — Kynureninase, Pyridoxal phosphate → [Structure: 3-Hydroxyanthranilic acid] + $CH_2CHCOOH\ |\ NH_2$ (Alanine)

3-Hydroxyanthranilic acid is degraded further by enzymes of the liver and kidney to yield α-ketoadipic acid:

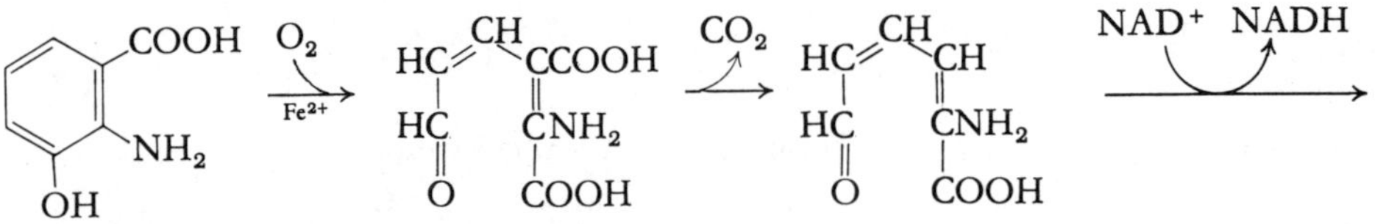

[Structure: 3-Hydroxyanthranilic acid] — O_2, Fe^{2+} → [Structure: 2-Amino-3-carboxy-muconic acid semialdehyde] — CO_2 → [Structure: 2-Aminomuconic acid semialdehyde] — NAD^+ / NADH →

$$\begin{array}{c} COOH \\ | \\ CH \\ \| \\ CH \\ | \\ CH \\ \| \\ CNH_2 \\ | \\ COOH \end{array}$$

2-Aminomuconic acid

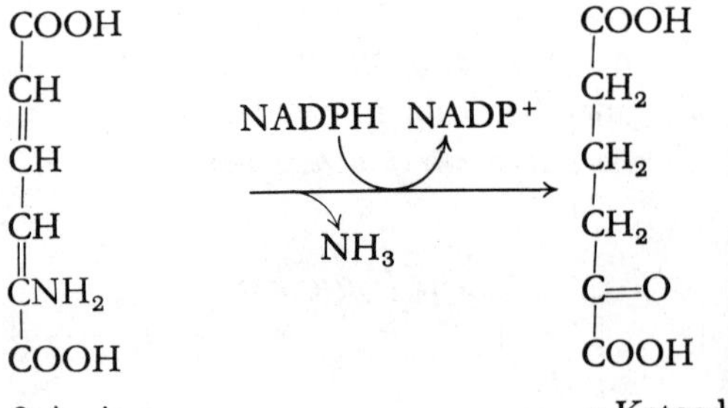

— NADPH / $NADP^+$, NH_3 →

$$\begin{array}{c} COOH \\ | \\ CH_2 \\ | \\ CH_2 \\ | \\ CH_2 \\ | \\ C=O \\ | \\ COOH \end{array}$$

α-Ketoadipic acid

α-Ketoadipic acid is converted to acetoacetyl-coenzyme A as described in the discussion of lysine (section 8.15).

One of the intermediates in the sequence just given, 2-amino-3-carboxy-muconic acid semialdehyde (or 2-acroleyl-3-aminofumaric acid), can also be utilized for the production of *nicotinic acid* and its ribosyl-phosphate derivatives:

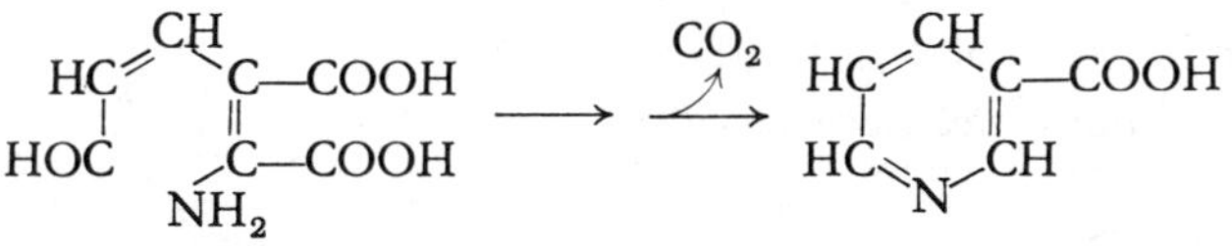

$$\text{2-Amino-3-carboxy-muconic acid semialdehyde} \xrightarrow{CO_2} \text{Nicotinic acid}$$

The amide of nicotinic acid is a B vitamin, nicotinamide, which is utilized by cells to form NAD^+ and $NADP^+$. Tryptophan metabolism does not provide the required amount of nicotinic acid for human beings. Thus, humans require additional intake of this vitamin in the diet. Only about 2% of ingested tryptophan is converted to nicotinic acid; however, a deficiency of this amino acid can exacerbate a deficiency of nicotinic acid.

The first enzyme in the metabolic pathway from tryptophan to acetoacetate and nicotinic acid is *tryptophan pyrrolase*. This is a heme-containing enzyme found in the cytosol of liver cells. Results of numerous studies indicate that there is considerable substrate and hormonal regulation of tryptophan pyrrolase. Its activity is stimulated by tryptophan and cortical steroids; NADPH appears to be an allosteric inhibitor of the enzyme. Tryptophan has a stabilizing effect on the synthesis of the enzyme, while cortical steroids induce its synthesis.

The enzyme *kynureninase* that catalyzes the conversion of 3-hydroxykynurenine to 3-hydroxyanthranilic acid (or kynurenine to anthranilic acid, as shown in Figure 8-15) requires pyridoxal phosphate as a cofactor. The enzyme is extremely sensitive to decreases in the levels of pyridoxal phosphate, and the reactions that it catalyzes will not occur when there is a deficiency in vitamin B_6. Under such conditions, large amounts of kynurenine and xanthurenic acid (an alternative metabolite of 3-hydroxykynurenine) are excreted in the urine.

Another pathway in the metabolism of tryptophan yields *serotonin* (5-hydroxytryptamine) as the principal product. This involves the action of tryptophan hydroxylase, which provides 5-hydroxytryptophan, and 5-hydroxytryptophan decarboxylase, which converts the latter to serotonin:

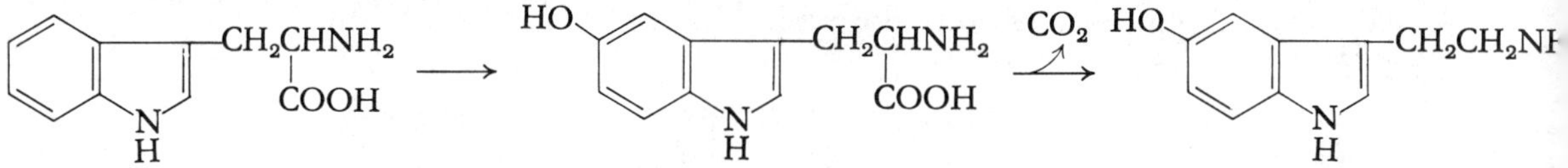

Tryptophan → 5-Hydroxytryptophan → (−CO_2) → Serotonin (5-hydroxytryptamine)

The first reaction requires tetrahydrobiopterin as a cofactor (similar to phenylalanine hydroxylase, section 8.1), and the second utilizes pyridoxal phosphate. Both enzymes are found in the brain; the decarboxylase also occurs in kidney.

Metabolic degradation of serotonin yields 5-hydroxyindoleacetic acid (5-HIAA):

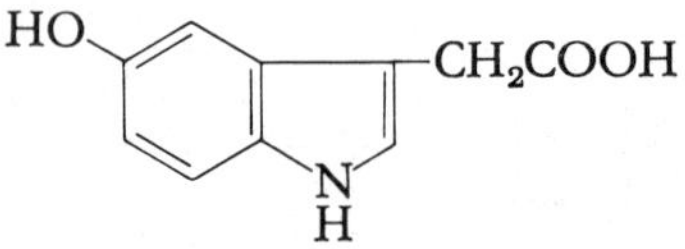

5-Hydroxyindoleacetic acid (5-HIAA)

5-HIAA can also be formed from 5-hydroxytryptophan via an alternative pathway. Low levels of 5-HIAA acid are normally excreted in the urine. When there is a deficiency in vitamin B_6 (pellagra), the amounts of 5-HIAA in the urine are increased. Similarly, in patients with Hartnup's disease (in which the pathway toward nicotinic acid is inhibited), the urinary levels of 5-HIAA, as well as those of kynurenine and xanthurenic acid, are increased markedly.

Serotonin is a powerful vasoconstrictor and stimulates the contraction of smooth muscle. It is found in the brain, intestinal tissue, mast cells, and blood platelets. In the brain, it apparently functions as a neurohormonal agent that enhances nerve activity. Some agents that have a depressing effect on the brain—e.g., reserpine—cause an increase in the urinary levels of serotonin and 5-HIAA.

Serotonin may also be converted to its *N*-acetyl-5-methoxy derivative, *melatonin*:

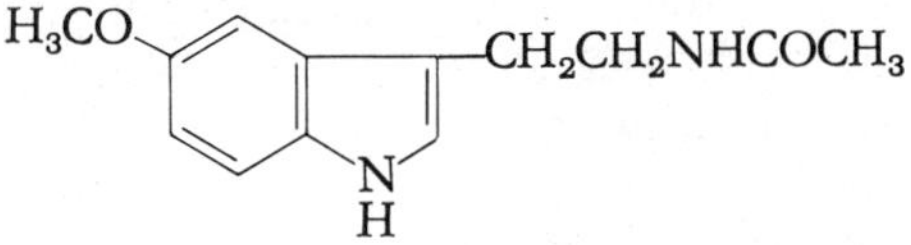

Melatonin

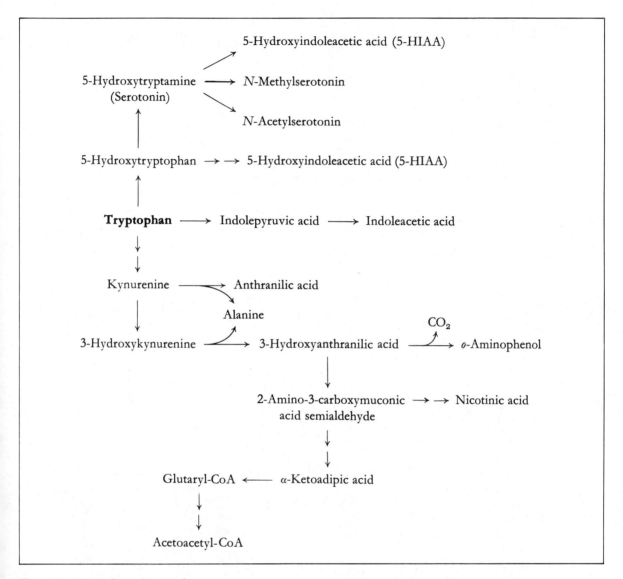

Fig. 8-15. *Metabolism of tryptophan.*

Melatonin is found in the pineal body and peripheral nerves. Although its exact function is not known, melatonin is of considerable interest because it has been implicated in the control of the circadian rhythm. This effect has been inferred from the finding that melatonin synthesis is inhibited by light. Melatonin also has been found to have an inhibitory effect on the ovary. Animals maintained in a continuously lighted room have persistent estrus, possibly as a result of the absence of the inhibitory action of melatonin.

The metabolism of tryptophan is outlined in Figure 8-15.

Suggested Reading

Axelrod, J. The Pineal Gland. A Neurochemical Transducer. *Science* 184:1341–1348, 1974.

Meister, A. *Biochemistry of the Amino Acids* (2nd ed.), Vols. 1 and 2. New York: Academic Press, 1965.

Meister, A. On the Enzymology of Amino Acid Transport. *Science* 180:33–39, 1973.

Stanbury, J. B., Wyngaarden, J. B., and Fredrickson, D. S. *The Metabolic Bases of Inherited Diseases* (3rd ed.). New York: McGraw-Hill, 1972.

9. Metabolism of Nucleotides

Nucleotides are synthesized within animal cells from relatively simple precursors. Although pre-formed purines and pyrimidines can be utilized for the biosynthesis of nucleotides and nucleic acids, most cells are not dependent on exogenous supplies of these materials. In fact, ingested or otherwise administered purines and pyrimidines are utilized to a comparatively small extent.

The general pathways for the formation of the purine nucleotides (adenosine triphosphate, ATP, and guanosine triphosphate, GTP) are different from those of the pyrimidine nucleotides (cytidine triphosphate, CTP; uridine triphosphate, UTP; and thymidine triphosphate, TTP). Each of these processes will therefore be considered separately in the following discussions of ribonucleotide synthesis.

9.1. Biosynthesis of Purine Ribonucleotides

The primary metabolite for the synthesis of nucleotides is ribose 5-phosphate. This pentose phosphate is derived from glucose via the oxidative and non-oxidative pentose pathways (section 4.6).

Ribose 5-phosphate interacts with ATP to yield 5-phospho-α-D-ribosyl-1-pyrophosphate (PRPP):

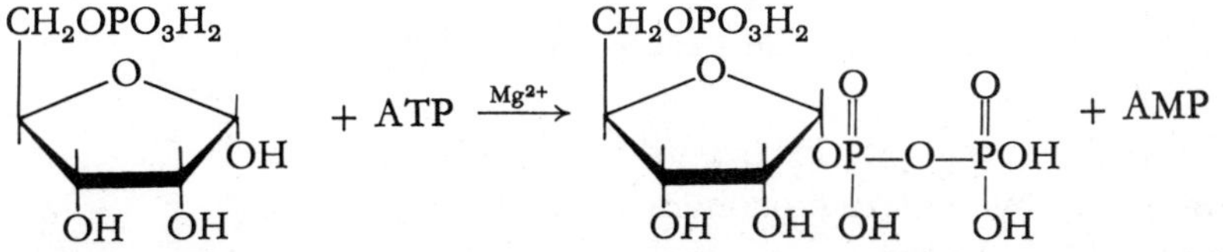

Ribose 5-phosphate + ATP $\xrightarrow{Mg^{2+}}$ 5-Phospho-α-D-ribosyl-1-pyrophosphate + AMP

The enzyme that catalyzes this reaction is *ribose-phosphate pyrophosphokinase*. It should be noted that a *pyrophosphate* group is transferred from ATP to carbon-1 of the ribose ring. 5-Phosphoribosyl-1-pyrophosphate is a common intermediate for the synthesis of both purine and pyrimidine nucleotides. It is also utilized for the synthesis of nucleotides by the so-called salvage pathways (section 9.6).

The next step toward the synthesis of purine nucleotides is catalyzed by *5-phosphoribosyl-pyrophosphate amidotransferase*. This reaction involves the transfer of the amido-nitrogen atom from glutamine to carbon-1 of ribose:

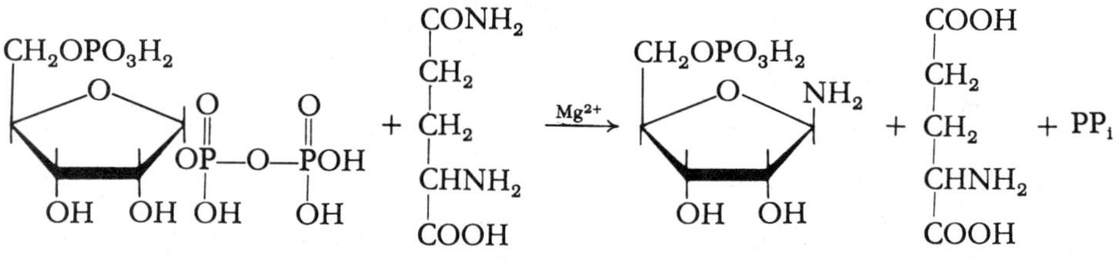

$$PRPP + Glutamine \xrightarrow{Mg^{2+}} \text{5-Phospho-}\beta\text{-D-ribosylamine} + Glutamate + PP_i$$

The 5-phosphoribosylamine produced in this reaction has the beta configuration. Thus, in addition to the displacement of pyrophosphate by an amino group, there is an inversion of configuration at carbon-1. The inorganic pyrophosphate that is released is hydrolyzed to inorganic phosphate by the action of intracellular *pyrophosphatase*. This makes the cellular synthesis of 5-phosphoribosylamine in effect an irreversible process.

The formation of 5-phosphoribosylamine is considered to be the *committed* step in the biosynthesis of purine nucleotides, because this intermediate is not utilized for any other function. The amidotransferase catalyzing this reaction is inhibited by adenosine and guanosine phosphates. Since purine nucleotides are the final products of the pathway, inhibition by nucleoside phosphates serves as a regulatory mechanism in the biosynthesis of nucleotides and nucleic acids. Control systems in nucleotide metabolism will be discussed in detail in sections 9.2 and 9.4.

The amidotransferase is also inhibited by azaserine (an antibiotic isolated from a species of *Streptomyces*) and by 6-diazo-5-oxonorleucine (DON). Inhibitors of this type are utilized in biochemical research as well as in chemotherapy (section 9.7).

In the next step toward nucleotide synthesis, 5-phosphoribosylamine interacts with glycine and ATP to yield 5-phosphoribosylglycinamide. The enzyme that catalyzes the reaction is *phosphoribosylglycinamide synthetase*:

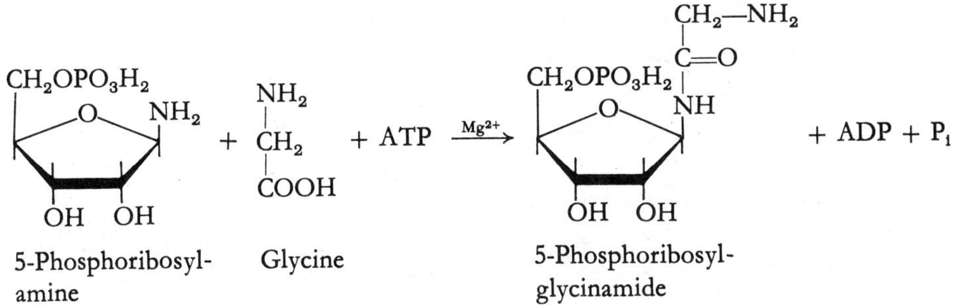

[Reaction scheme: 5-Phosphoribosylamine + Glycine + ATP →(Mg²⁺) 5-Phosphoribosylglycinamide + ADP + P$_i$]

Phosphoribosylglycinamide formyltransferase effects the transfer of a formyl group from N^5,N^{10}-methenyltetrahydrofolate to 5-phosphoribosylglycinamide. The product of the reaction is 5-phosphoribosyl-N-formylglycinamide:

[Reaction scheme: 5-Phosphoribosylglycinamide + N^5,N^{10}-Methenyl-THFA → 5-Phosphoribosyl-N-formylglycinamide + THFA]

(RP = ribose phosphate)

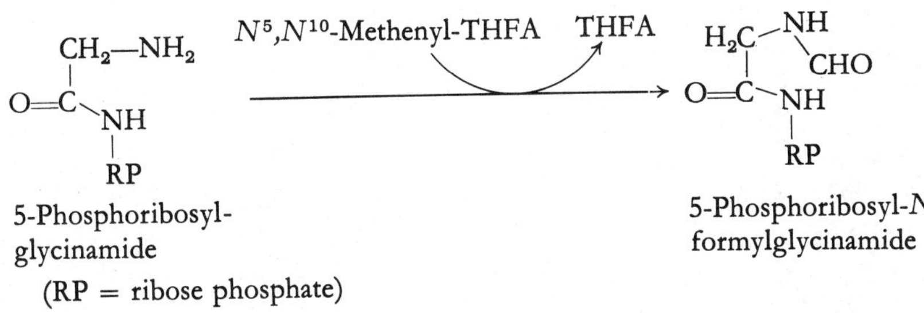

Potential donors of the single-carbon methenyl unit for this reaction include glycine, serine, and histidine (section 7.10).

The next reactions in the pathway involve the transfer of another amido-nitrogen from glutamine to 5-phosphoribosyl-N-formylglycinamide and cyclization of the product to form 5'-phosphoribosyl-5-aminoimidazole (note that once the second ring is formed, superscript prime is used to indicate the position of substitution on the ribose ring). Both steps require ATP as an energy source. The enzymes for the two steps are *phosphoribosylformylglycinamide synthetase* and *phosphoribosylaminoimidazole synthetase*:

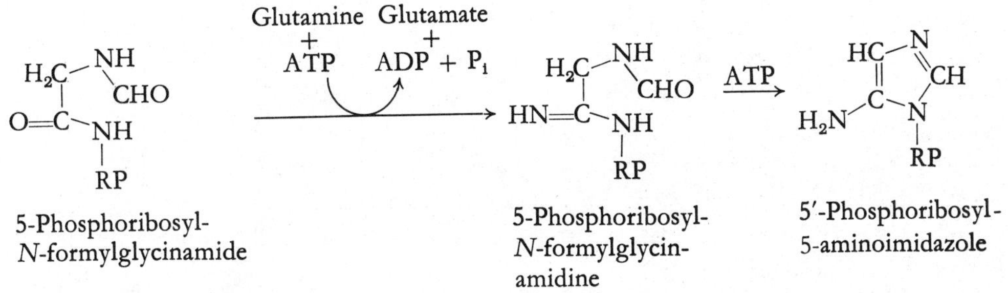

[Reaction scheme: 5-Phosphoribosyl-N-formylglycinamide + Glutamine + ATP → 5-Phosphoribosyl-N-formylglycinamidine + Glutamate + ADP + P$_i$ →(ATP) 5'-Phosphoribosyl-5-aminoimidazole]

The first synthetase involved in these reactions is similar to the previously mentioned amidotransferase in that it also transfers an amido group from glutamine and it, too, is inhibited by azaserine and 6-diazo-5-oxonorleucine.

In the following step, 5′-phosphoribosyl-5-aminoimidazole interacts with carbon dioxide (or carbonate) to yield 5′-phosphoribosyl-5-aminoimidazole-4-carboxylate. The latter combines with aspartate in the presence of ATP to form 5′-phosphoribosyl-4-(N-succinocarboxamide)-5-aminoimidazole. This is then cleaved to release fumarate and 5′-phosphoribosyl-4-carboxamide-5-aminoimidazole:

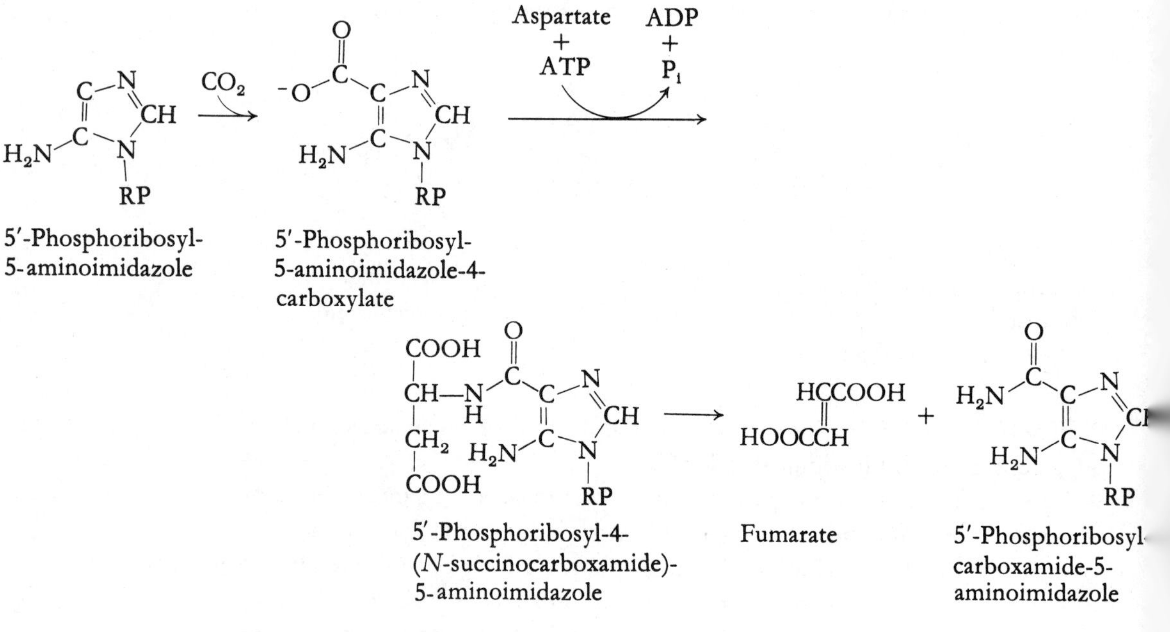

The enzymes for these reactions are *phosphoribosylaminoimidazole carboxylase*, *phosphoribosylaminoimidazole-succinocarboxamide synthetase*, and *adenylosuccinate lyase*, respectively.

Another carbon atom is introduced into the growing purine ring at this stage via N^{10}-formyltetrahydrofolate. The product of the reaction is 5′-phosphoribosyl-4-carboxamide-5-formamidoimidazole, and the enzyme catalyzing it is *phosphoribosylaminoimidazolecarboxamide formyltransferase*. Elimination of a molecule of water from the formamido compound and ring closure, which are catalyzed by *IMP cyclohydrolase*, yield inosinic acid (IMP, or hypoxanthine ribonucleotide):

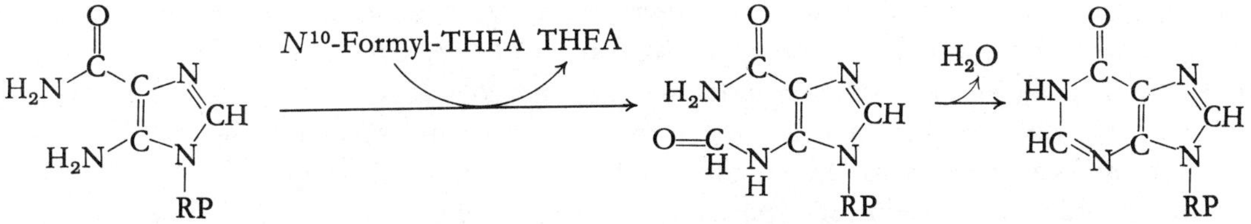

5′-Phosphoribosyl-4-carboxamide-5-aminoimidazole →(N^{10}-Formyl-THFA → THFA)→ **5′-Phosphoribosyl-4-carboxamide-5-formamidoimidazole** →(H_2O)→ **Inosinic acid (IMP)**

The biosynthesis of the purine skeleton may be summarized by considering the source of each of its components. Each of the atoms in the purine ring may be numbered:

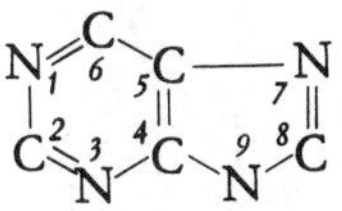

Thus, N-1 is derived from aspartate; N-3 and N-9 from the amido-nitrogen of glutamine; C-4, C-5, and N-7 from glycine; C-2 and C-8 from single-carbon donors via tetrahydrofolate; and C-6 from carbonate.

Inosinic acid, or IMP, serves as the intermediate from which both adenylic acid (AMP) and guanylic acid (GMP) can be produced. The synthesis of adenylic acid involves a reaction between inosinic acid and aspartate in the presence of GTP to form adenylosuccinate, which is catalyzed by *adenylosuccinate synthetase*. This is followed by the elimination of fumarate and the concomitant formation of adenylic acid:

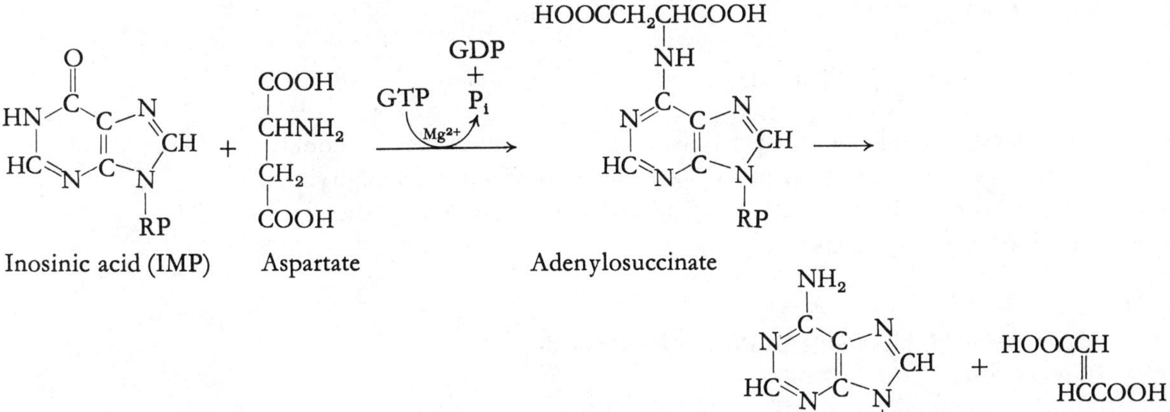

Inosinic acid (IMP) + **Aspartate** →(Mg^{2+}, GTP → GDP + P_i)→ **Adenylosuccinate** → **Adenylic acid (AMP)** + **Fumarate**

The enzyme that catalyzes the last reaction, *adenylosuccinate lyase,* appears to be identical with the one that catalyzes the elimination of fumarate from 5′-phosphoribosyl-4-(*N*-succinocarboxamide)-5-aminoimidazole. It is also of interest that the enzyme that catalyzes the formation of adenylosuccinic acid—*adenylosuccinate synthetase*—is inhibited by AMP. The implications of this inhibitory action on the control of purine biosynthesis will be discussed in the next section.

Guanylic acid (GMP) is produced from inosinic acid by two successive reactions. First, inosinic acid is oxidized by inosinic acid dehydrogenase to form xanthylic acid. Second, an amido-nitrogen is transferred to the latter from glutamine to yield guanylic acid and glutamate; this transfer is catalyzed by *GMP synthetase* and requires ATP as an energy source:

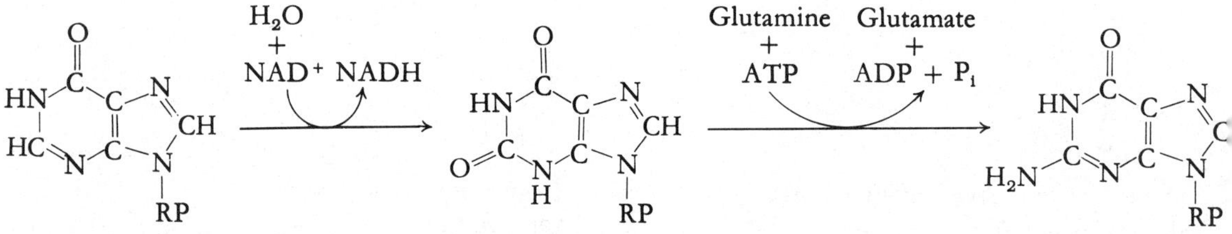

Inosinic acid (IMP) → Xanthylic acid → Guanylic acid (GMP)

The enzyme *inosinic acid dehydrogenase* is inhibited by GMP.

GMP can be converted to guanosine diphosphate (GDP) by the action of *nucleoside-monophosphate kinase* and ATP:

$$GMP + ATP \rightleftharpoons GDP + ADP$$

The enzyme *nucleoside-diphosphate kinase* catalyzes the formation of GTP from GDP and ATP:

$$GDP + ATP \rightleftharpoons GTP + ADP$$

These nucleoside phosphates are thus easily interconvertible. It should be remembered that ATP is formed from ADP by oxidative phosphorylation, and that both ATP and GTP are also produced by specific substrate-level phosphorylation reactions (Chap. 3).

9.2. Control Mechanisms in Purine Nucleotide Biosynthesis

Purine nucleotides, as well as their pyrimidine counterparts, are involved in a number of vital functions in the maintenance of the organism. In addition to serving as the building blocks of nucleic acids, they are utilized as enzymatic cofactors, energy-transferring agents, and intracellular messengers. Like all constituents of living cells, they are constantly being degraded and replaced. Unlike carbohydrates and triglycerides, however, excessive amounts of nucleotides do not appear to be stored. It is therefore critical that steady-state concentrations of the nucleotides be maintained through sensitive control of their biosynthetic pathways. Some of these regulatory mechanisms will be discussed in this section.

The committed step in the pathway leading to the synthesis of purine nucleotides is the formation of 5-phosphoribosylamine from glutamine and 5-phosphoribosyl-pyrophosphate (see the preceding section). The enzyme that catalyzes this reaction—an amidotransferase—is inhibited by any one of the purine nucleotides, i.e., by AMP, ADP, ATP, GMP, GDP, or GTP. As a consequence of this feedback inhibition, the rate of purine-nucleotide synthesis is diminished as the quantities of these nucleoside phosphates become excessive.

Another control mechanism is the allosteric inhibition exerted by AMP on the activity of the enzyme that catalyzes the conversion of inosinic acid to adenylosuccinic acid. Thus, AMP inhibits the activity of the enzyme that initiates the reaction sequence for the synthesis of AMP. A similar type of regulation in the synthesis of GMP is the inhibition of inosinic acid dehydrogenase by GMP.

For the synthesis of ribonucleic acid (RNA), the presence of four nucleoside triphosphates—ATP, GTP, cytidylic acid (CTP), and uridylic acid (UTP)—is required (page 423). It is therefore important that the concentrations of these materials be controlled relative to each other. One way in which this control is achieved is based on the fact that a particular nucleoside triphosphate is usually required for the synthesis of another. ATP, for example, is utilized in the formation of GMP from xanthylic acid. As a consequence, an increase in the concentration of the adenine nucleotide will cause an increase in the rate of formation of GMP. Similarly, GTP is required for the synthesis of adenylosuccinic acid. Thus, as the concentration of the guanine nucleotide increases, the rate of AMP synthesis increases. This interaction between the concentrations of the guanine and adenine nucleotides serves as a control factor to keep the relative amounts of these components in balance (Fig. 9-1).

9.3. Biosynthesis of Pyrimidine Ribonucleotides

The precursors for the cellular synthesis of the pyrimidine skeleton are aspartate and carbamoyl phosphate. The formation of the latter from ammonia, carbon dioxide, and 2 moles of ATP was described previously (section 7.4.B); the enzyme involved in this reaction is *carbamoyl-phosphate*

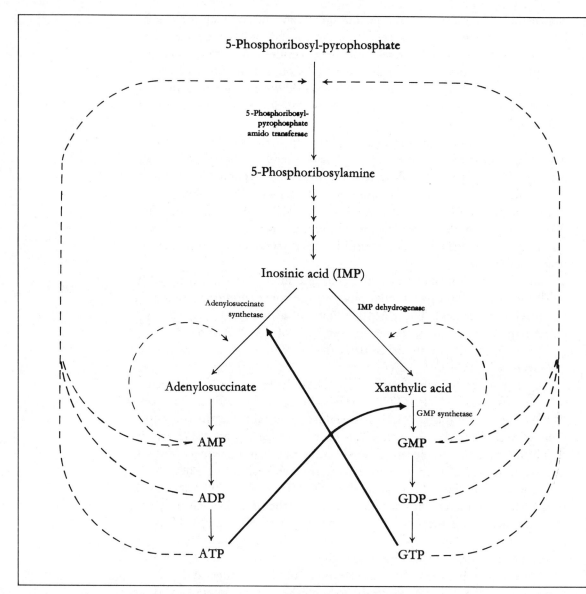

Fig. 9-1. Control of purine nucleotide biosynthesis. Dashed arrows indicate inhibition; bold arrows indicate enhancement of the reaction.

synthetase. Current studies indicate that this mode of carbamoyl-phosphate synthesis is characteristic of liver mitochondria, and it occurs primarily in conjunction with the reactions of the urea cycle (section 7.5). Another enzyme, which is found in the extramitochondrial compartment of cells, catalyzes the formation of carbamoyl phosphate from carbonate, ATP, and the amido-nitrogen of glutamine:

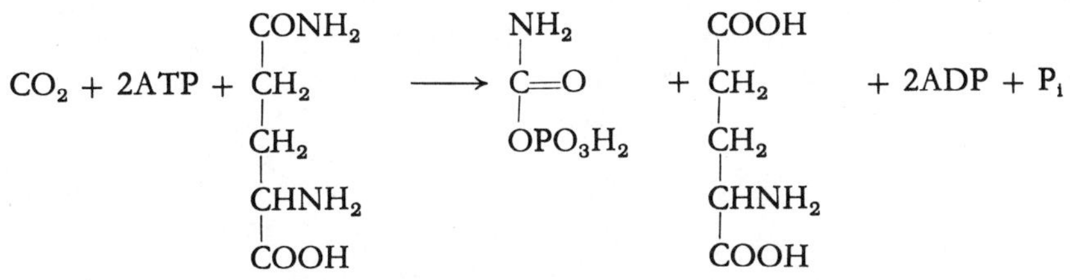

$$CO_2 + 2ATP + \begin{matrix} CONH_2 \\ | \\ CH_2 \\ | \\ CH_2 \\ | \\ CHNH_2 \\ | \\ COOH \end{matrix} \longrightarrow \begin{matrix} NH_2 \\ | \\ C=O \\ | \\ OPO_3H_2 \end{matrix} + \begin{matrix} COOH \\ | \\ CH_2 \\ | \\ CH_2 \\ | \\ CHNH_2 \\ | \\ COOH \end{matrix} + 2ADP + P_i$$

Glutamine Carbamoyl Glutamate
           phosphate

Since this specific enzyme is closely associated in the cytosol, with the one for the next step in the pathway for pyrimidine biosynthesis, it is reasonable to assume that it is the one that produces carbamoyl phosphate for the synthesis of these nucleotides.

The enzyme *aspartate transcarbamoylase (aspartate carbamoyltransferase)* catalyzes the condensation of carbamoyl phosphate with aspartate to form N-carbamoyl aspartate. This reaction is followed by ring closure catalyzed by dihydroorotase to produce dihydroorotate:

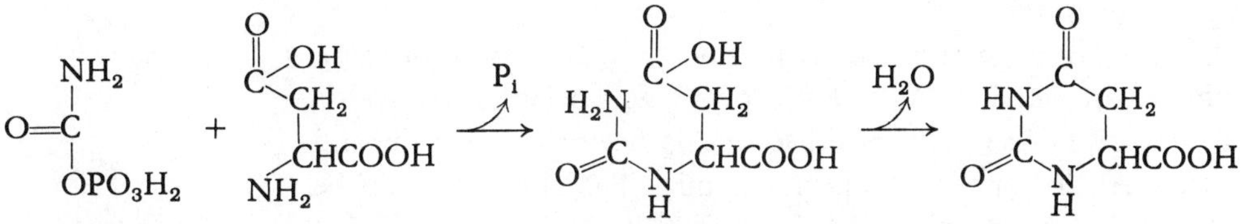

Carbamoyl phosphate + Aspartic acid $\xrightarrow{P_i}$ N-Carbamoyl aspartate $\xrightarrow{H_2O}$ L-Dihydroorotic acid

The iron-containing flavoprotein enzyme, dihydroorotate dehydrogenase, catalyzes the oxidation of dihydroorotate to orotate. The reduced enzyme is regenerated by interaction with NAD^+. The overall reaction is:

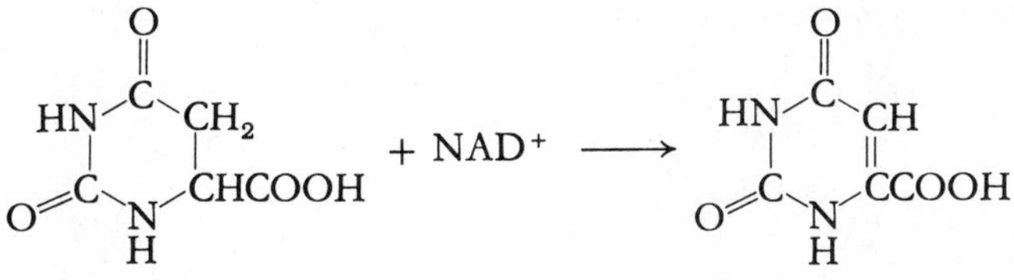

Dihydroorotic acid $+ NAD^+ \longrightarrow$ Orotic acid $+ NADH + H^+$

It may be noted that unlike the pathway for purine nucleotide synthesis, the pyrimidine ring is formed prior to the attachment of ribose phosphate.

After orotate is formed, it reacts with 5-phosphoribosyl-pyrophosphate (PRPP; section 9.1) to yield orotidine 5′-phosphate. The subsequent step involves the action of orotidine 5′-phosphate decarboxylase to yield uridylic acid (UMP):

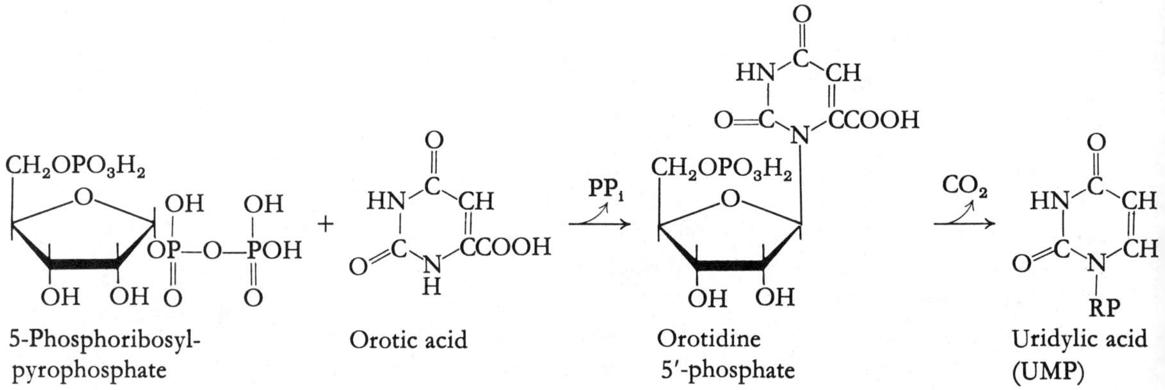

5-Phosphoribosyl-pyrophosphate + Orotic acid $\xrightarrow{PP_i}$ Orotidine 5′-phosphate $\xrightarrow{CO_2}$ Uridylic acid (UMP)

The activities of the enzymes catalyzing both of these reactions—i.e., orotate-phosphoribosyl transferase and orotidine 5′-phosphate decarboxylase—appear to reside in the same protein.

In the inborn error of metabolism known as *orotic aciduria*, both of these enzymatic activities are missing. As a result, orotic acid accumulates and is excreted in the urine. Children with this disorder do not grow normally and exhibit the characteristics of megaloblastic anemia. The condition can be corrected by the administration of uridine or cytidine. The fact that this treatment also reduces the urinary excretion of orotic acid indicates that pyrimidine nucleosides or nucleotides inhibit a step in the synthesis of orotic acid. This regulatory aspect will be discussed further in the next section.

Cytidine nucleotides are formed from UMP; however, the latter must first be converted to the triphosphate. This is accomplished by two sequential reactions with ATP, which are catalyzed by the enzymes *nucleoside-monophosphate kinase* and *nucleoside-diphosphate kinase*, respectively:

UMP + ATP $\rightleftharpoons$ UDP + ADP
UDP + ATP $\rightleftharpoons$ UTP + ADP

Uridine triphosphate reacts with glutamine, when catalyzed by *CTP synthetase*, to yield cytidine triphosphate (CTP) and glutamate:

$$\underset{\substack{\text{RPPP} \\ \text{Uridine 5'-} \\ \text{triphosphate (UTP)} \\ \text{(RPPP = ribose triphosphate)}}}{\text{UTP structure}} + \underset{\text{Glutamine}}{\begin{array}{c} CONH_2 \\ | \\ CH_2 \\ | \\ CH_2 \\ | \\ CHNH_2 \\ | \\ COOH \end{array}} + ATP \xrightarrow{Mg^{2+}} \underset{\substack{\text{RPPP} \\ \text{Cytidine 5'-} \\ \text{triphosphate (CTP)}}}{\text{CTP structure}} + \underset{\substack{\text{Glutamic} \\ \text{acid}}}{\begin{array}{c} COOH \\ | \\ CH_2 \\ | \\ CH_2 \\ | \\ CHNH_2 \\ | \\ COOH \end{array}} + ADP + P_i$$

The biosynthesis of thymidine phosphate will be discussed in the section on deoxyribonucleotides.

9.4. Control Mechanisms in Pyrimidine Nucleotide Biosynthesis

The committed step in the synthetic pathway for pyrimidine nucleotides in a number of organisms is the formation of N-carbamoyl aspartate from carbamoyl phosphate and aspartate. The enzyme that catalyzes this reaction—i.e., aspartate transcarbamoylase—is an allosteric enzyme (Chap. 3, section 3.10). Thus, the aspartate transcarbamoylase-catalyzed reaction exhibits a kinetic relationship between substrate concentration and initial reaction velocity (v_0) that is characterized by a sigmoid curve (Fig. 9-2). The activity of the enzyme is regulated by the pyrimidine nucleotide requirements of the organism. In certain bacteria, e.g., *E. coli*, the activity of aspartate transcarbamoylase is inhibited by CTP, whereas in other organisms, the enzyme is inhibited by UTP. In other words, the pyrimidine nucleoside triphosphates are *negative effectors* (page 150) of the allosteric enzyme.

The effects of CTP and UTP on aspartate transcarbamoylase are examples of *feedback inhibition*. This involves the inhibition of an enzyme that catalyzes an early step in a metabolic sequence by a product formed in a later, or final, reaction of the pathway. The usefulness of this type of inhibition in the economy of the cell is obvious. If the final product of a series of reactions is not needed, it is wasteful for the organism to perform any of the intermediate steps for its synthesis. Such control is effected most efficiently when the enzyme catalyzing the committed step in the sequence is regulated by the concentration of the ultimate product. (See also section 9.2 in this chapter.)

Aspartate transcarbamoylase from *E. coli* has a molecular weight of about 310,000. It consists of six catalytic subunits, each with a molecular weight of 33,500, and six regulatory subunits with molecular weights of 17,000. Thus, although the binding sites for the substrates are not on the same

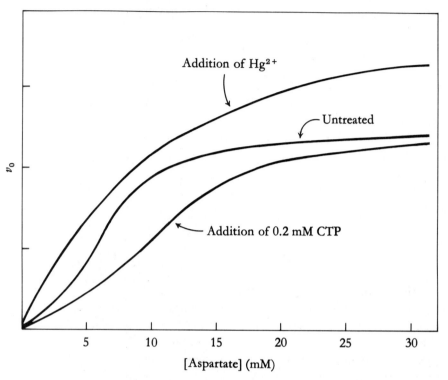

Fig. 9-2. Relationships of initial reaction velocity to substrate (aspartate) concentration for the allosteric enzyme, aspartate transcarbamoylase. Carbamoyl phosphate was present in excess. The effect of CTP, an allosteric modifier, is to depress the sigmoid curve (inhibition). Addition of mercury salts, which dissociate the enzyme, results in a hyperbolic relationship (Michaelis-Menten kinetics).

subunits as those for CTP, the binding of the latter has a profound effect on the catalytic action of the enzyme (Fig. 9-2). The sigmoid-curve function describing the relationship between substrate concentration and reaction velocity will no longer hold if the regulatory subunits are separated from the catalytic subunits. This can be accomplished by treating the enzyme with *p*-hydroxymercuribenzoate. The dissociated enzyme exhibits an increase in activity, and, as shown in Figure 9-2, its substrate-velocity relationship follows Michaelis-Menten kinetics (Chap. 2). Another finding is that the dissociated enzyme is not inhibited by CTP. When the dissociating agent (i.e., *p*-hydroxymercuribenzoate) is removed, the allosteric inhibition by CTP and the sigmoidal substrate-velocity relationship are restored.

Allosteric modulation is one of the important mechanisms for regulating metabolic pathways and for maintaining steady-state concentrations of cellular components. Although some of the experimental data discussed are characteristic for aspartate transcarbamoylase from *E. coli,* analogous systems are known to be prominent in animal tissues.

One site of regulation of pyrimidine nucleotide biosynthesis in animal tissues appears to be in the formation of carbamoyl phosphate (page 381). The *glutamine-dependent carbamoyl-phosphate synthetase* may be visualized as the enzyme that catalyzes the committed step in the synthetic pathway in animals, since carbamoyl phosphate utilized for the urea cycle is formed by the action of another enzyme. The inhibitory effect of UTP on the glutamine-dependent carbamoyl-phosphate synthetase can, therefore, serve to modulate pyrimidine nucleotide synthesis. Other sites in the pathway that are subject to control by increased levels of nucleotides are the reactions involving dihydroorotate dehydrogenase (page 381) and orotidine 5′-phosphate decarboxylase. These enzymes are inhibited when the concentrations of nucleotides are increased.

9.5. Biosynthesis of Deoxyribonucleotides

Deoxyribonucleotides are synthesized from ribonucleotides by reduction of the hydroxyl group at carbon-2′ of the ribose moiety. Although most current information concerning this process is derived from studies with microorganisms, it is apparent that the general aspects of the mechanisms are similar in higher animals.

The immediate hydrogen source for ribose reduction in *E. coli* is a protein called *thioredoxin*. The functional components of this protein are two sulfhydryl groups contained in particular cysteine residues. As a result of the action of the enzyme ribonucleoside-diphosphate reductase, electrons and hydrogen atoms are transferred from thioredoxin to the ribose unit of ADP, GDP, CDP, or UDP. Thus, the nucleoside diphosphate is reduced while the thioredoxin is oxidized:

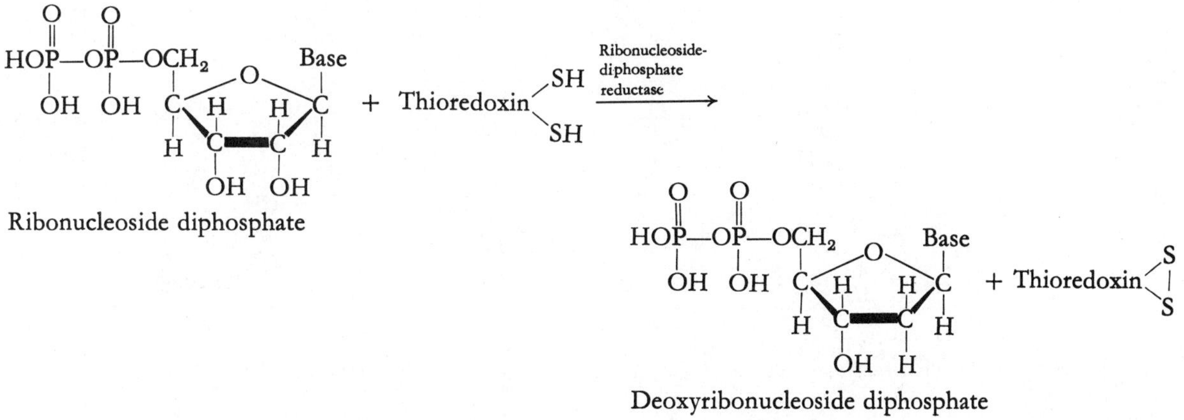

Ribonucleoside diphosphate + Thioredoxin(SH)$_2$ $\xrightarrow{\text{Ribonucleoside-diphosphate reductase}}$ Deoxyribonucleoside diphosphate + Thioredoxin(S–S)

Reduced thioredoxin is regenerated by the action of the NADPH-linked enzyme, thioredoxin reductase:

Thioredoxin(S–S) + NADPH + H$^+$ $\longrightarrow$ Thioredoxin(SH)$_2$ + NADP$^+$

The combined reactions may be written as follows:

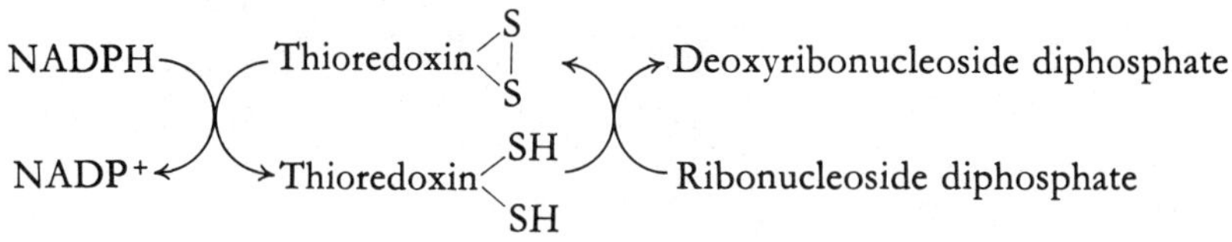

$$\text{NADPH} \searrow \quad \text{Thioredoxin} \begin{smallmatrix} S \\ | \\ S \end{smallmatrix} \quad \leftarrow \quad \text{Deoxyribonucleoside diphosphate}$$
$$\text{NADP}^+ \nearrow \quad \text{Thioredoxin} \begin{smallmatrix} SH \\ \\ SH \end{smallmatrix} \quad \rightarrow \quad \text{Ribonucleoside diphosphate}$$

Differences have been found among species with regard to the nucleotide substrate and the hydrogen donor utilized for deoxyribonucleotide synthesis. Vitamin B_{12} has been shown to be involved as a cofactor in certain microorganisms. The precise reaction pathway in humans has not been defined.

It should be noted that these reactions provide another example of a system requiring reduced NADPH. The latter is generated primarily by glucose metabolism via the oxidative-pentose pathway.

Deoxythymidine 5′-phosphate (dTMP) is formed from deoxyuridine 5′-phosphate (dUMP) and N^5,N^{10}-methylene-tetrahydrofolate. The enzyme for this reaction is thymidylate synthetase. In this process, tetrahydrofolate fulfills two functions: it serves as the carrier for the one-carbon methylene unit, which is added to carbon-5 of uracil, and it reduces the methylene unit to a methyl group. As a consequence of the reduction reaction, tetrahydrofolate (THFA) is oxidized to dihydrofolate (DHFA):

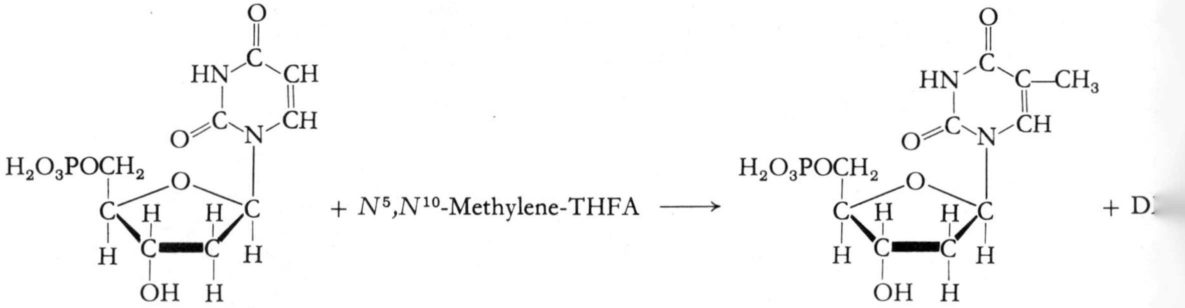

Deoxyuridylic acid (dUMP) + N^5,N^{10}-Methylene-THFA ⟶ Deoxythymidylic acid (dTMP) + D!

THFA is regenerated through a reaction of DHFA with NADPH that is catalyzed by *dihydrofolate reductase*:

DHFA + NADPH + H⁺ ⟶ THFA + NADP⁺

Deoxynucleoside monophosphates are converted to the corresponding diphosphates and triphosphates in reactions with ATP that are catalyzed by nucleoside-monophosphate kinase and nucleoside-diphosphate kinase, respectively. For example, dTTP is formed from dTMP in the following sequence:

$$dTMP + ATP \rightleftharpoons dTDP + ADP$$

$$dTDP + ATP \rightleftharpoons dTTP + ADP$$

The processes described thus far in this chapter provide the most common mechanisms for the synthesis of nucleoside and deoxynucleoside triphosphates from relatively simple precursors. In addition to their roles as cofactors and in energy transfer, the nucleoside triphosphates function as immediate precursors for the synthesis of nucleic acids. It is therefore reasonable to expect that inhibition of nucleotide biosynthesis will result in interference with the vital functions of the cell and ultimately in its death. This is the rationale for the use of certain antibiotics for the destruction of infectious organisms and for antimetabolite therapy of malignant tumors. Some aspects of the inhibitory effects of such substances on nucleotide synthesis will be discussed in section 9.7.

9.6. Salvage Pathways

From a quantitative point of view, the most important pathways for the generation of nucleotides are those involving de novo syntheses of the purine and pyrimidine rings, as described. In addition to these processes, various cells and tissues have enzyme systems that catalyze the incorporation of nucleosides and pre-formed intact bases into nucleotides and nucleic acids. The metabolic routes for such syntheses are termed *salvage pathways,* because the substrates arise primarily from the catabolism of nucleic acids or nucleotides (sections 9.9 and 9.10). Purine or pyrimidine bases and nucleosides are also absorbed to some degree from the intestine and are available for salvage processes.

One of the salvage pathways for the synthesis of nucleotides involves the utilization of the free base:

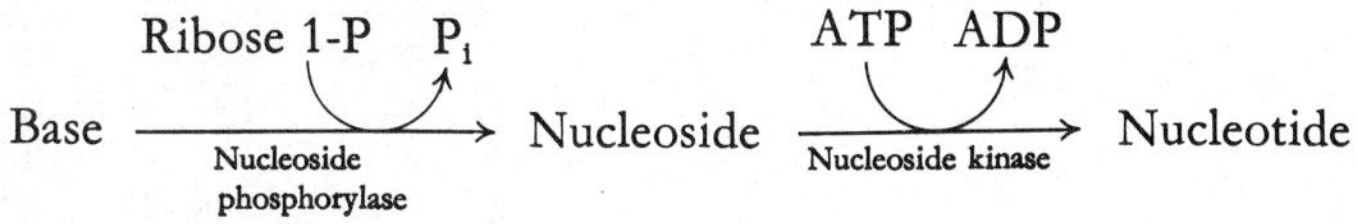

$$\text{Base} \xrightarrow[\text{Nucleoside phosphorylase}]{\text{Ribose 1-P} \quad P_i} \text{Nucleoside} \xrightarrow[\text{Nucleoside kinase}]{\text{ATP} \quad \text{ADP}} \text{Nucleotide}$$

These reactions occur to a minimal or negligible degree in most tissues.

However, in highly proliferative cells, such as those of the intestinal mucosa and certain malignant tissues, this pathway is quite significant.

Another reaction in which purine bases are converted to the respective nucleotides utilizes 5-phosphoribosyl-pyrophosphate (section 9.1). The following reaction, for example, occurs with guanine:

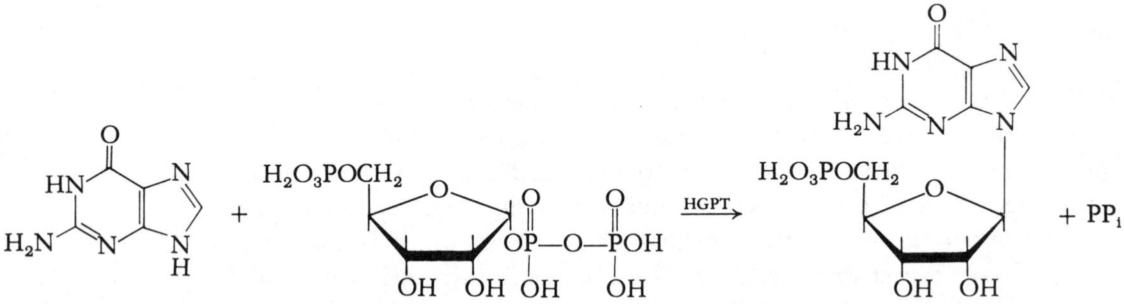

$$\text{Guanine} + \text{PRPP} \xrightarrow{\text{HGPT}} \text{GMP} + PP_i$$

The enzyme for this reaction—hypoxanthine-guanine phosphoribosyl transferase (HGPT)—also catalyzes the conversion of xanthine and hypoxanthine to xanthylic and inosinic acids, respectively:

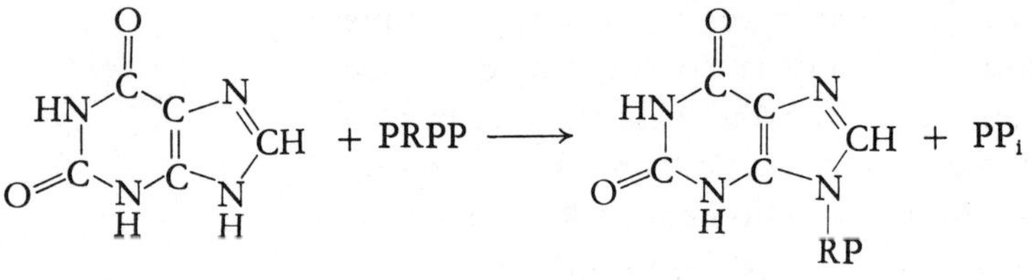

$$\text{Xanthine} + \text{PRPP} \longrightarrow \text{Xanthylic acid} + PP_i$$

$$\text{Hypoxanthine} + \text{PRPP} \longrightarrow \text{Inosinic acid (IMP)} + PP_i$$

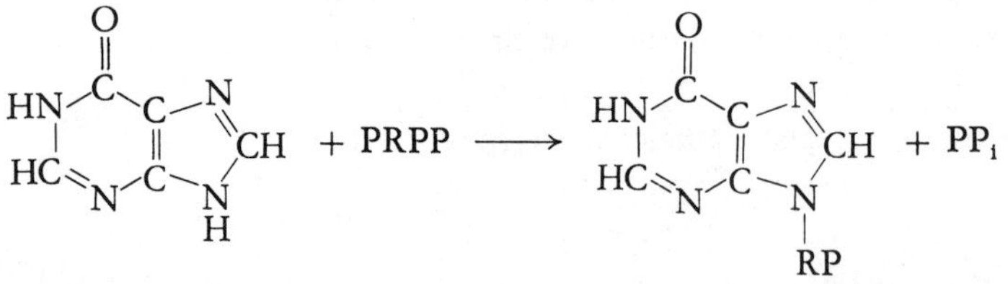

Similarly, the enzyme adenine phosphoribosyl transferase catalyzes the formation of AMP from adenine and PRPP:

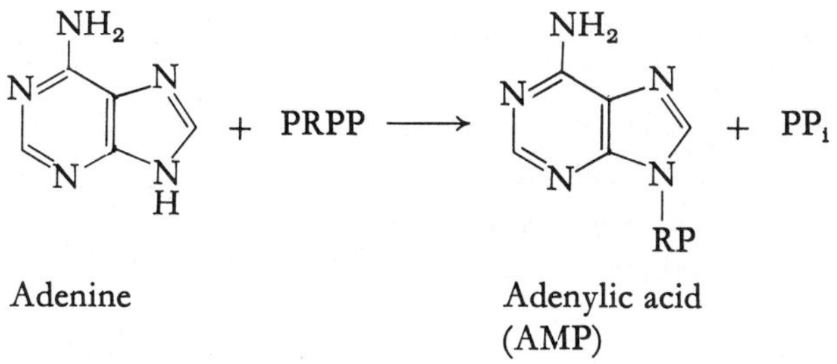

Adenine + PRPP → Adenylic acid (AMP) + PP$_i$

These transferase reactions are important in nucleotide metabolism in brain tissue, and specific hereditary diseases are attributable to the absence of the enzymes. This subject will be discussed in more detail in section 9.11.

9.7. Inhibitors of Nucleotide Synthesis

Certain substances known to interfere with specific steps of nucleotide biosynthesis have been employed for elucidating the mechanisms of individual enzymes and for defining the sequences in the biosynthetic pathways. Some of these inhibitors are also useful in chemotherapeutic treatment of diseases in which it is desirable to arrest cellular growth or to destroy invading organisms. The compounds to be described are also called *antimetabolites,* since they are analogs of specific metabolic intermediates and act by competing with or replacing the normal component in the synthetic pathway.

One type of inhibitor of nucleotide biosynthesis functions by interfering with the amidotransferase that utilizes the amido-nitrogen of glutamine. Examples of such inhibitors are O-diazoacetyl-L-serine (azaserine) and 6-diazo-5-oxo-L-norleucine (DON). These compounds probably exert their inhibitory effect because of their structural similarities to glutamine:

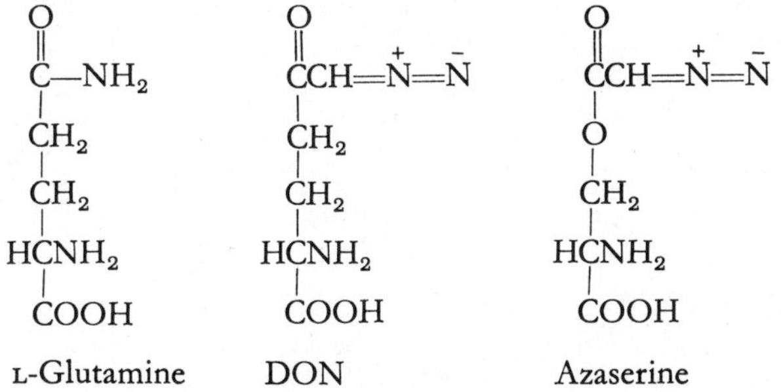

L-Glutamine   DON   Azaserine

Both azaserine and DON interfere with the synthesis of purine nucleotides by inhibiting the formation of 5-phosphoribosylamine from 5-phosphoribosyl-pyrophosphate, N-formylglycinamidine ribose-phosphate from N-formylglycinamide ribose-phosphate, and GMP from xanthylic acid (section 9.1). In addition, they affect the formation of CTP from UTP, which also involves glutamine (section 9.3).

Bacteria that normally synthesize folic acid are inhibited by sulfanilamide and other sulfonamides. These substances inhibit the incorporation of *p*-aminobenzoic acid into folic acid by the microorganisms, again because of their structural similarity to this metabolic intermediate:

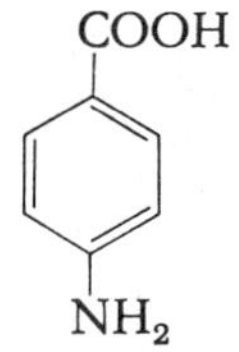

p-Aminobenzoic acid 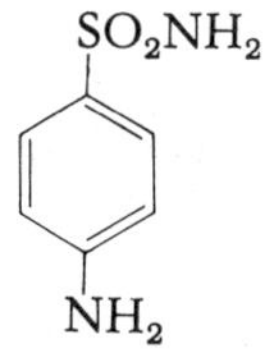   Sulfanilamide

The normal functioning of folic acid in the steps involving single-carbon transfers is inhibited by these agents so the synthesis of nucleotides cannot occur. Sulfonamide drugs are therefore potent bacteriostatic agents. They do not affect animals, which depend on exogenous folic acid for nucleotide biosynthesis.

Another group of compounds, termed *folic-acid antagonists*, also interfere with the activity of folic acid in single-carbon transfers. Examples of such drugs are aminopterin and amethopterin (methotrexate):

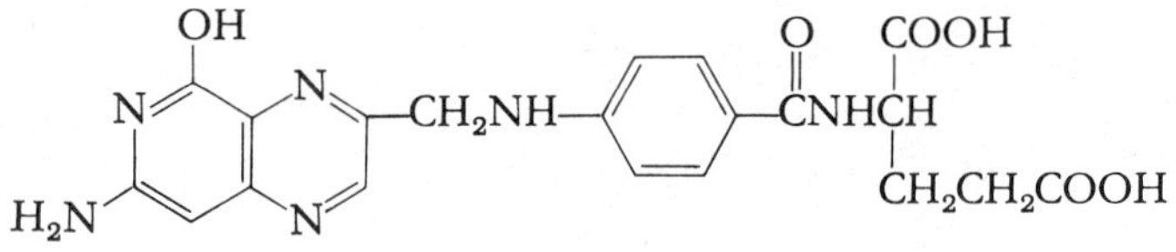

Folic acid

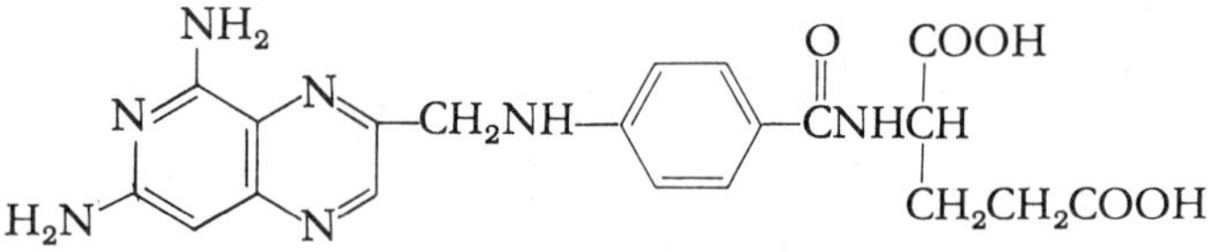

Aminopterin

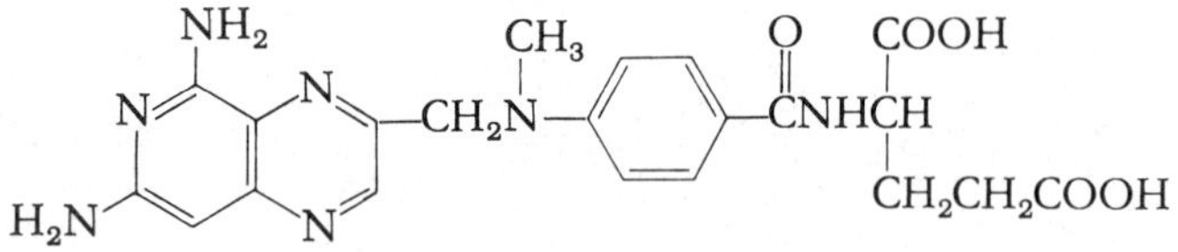

Amethopterin (methotrexate)

These substances prevent the conversion of folic acid to tetrahydrofolic acid by interfering with the action of dihydrofolate reductase (page 386). As a result, nucleotide synthesis is inhibited at the steps involving the formation of 5-phosphoribosyl-*N*-formylglycinamide from 5-phosphoribosylglycinamide, 5′-phosphoribosyl-4-carboxamide-5-formamidoimidazole from 5′-phosphoribosyl-4-carboxamide-5-aminoimidazole (section 9.1), and dTMP from dUMP (section 9.5).

Folic-acid antagonists are employed in the chemotherapy of various malignancies and are effective in inhibiting cell division. As expected, these materials also affect normal tissues, especially those that have a comparatively high rate of cell division, e.g., bone marrow or intestinal mucosa. In addition to their toxicity, another shortcoming of treatment with these drugs is that the tumor cells become resistant. Although the mechanism for this resistance has not been elucidated completely, it is known that the new cell populations that emerge show elevated levels of dihydrofolate reductase and an increase in the amounts of the enzymes involved in the synthesis of nucleotides by salvage pathways.

Other drugs employed in cancer chemotherapy are analogs of purines and pyrimidines:

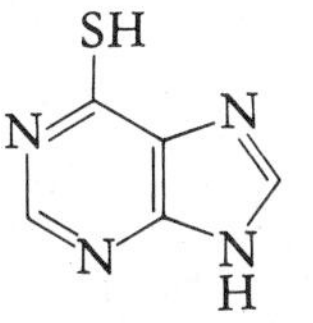

6-Mercaptopurine 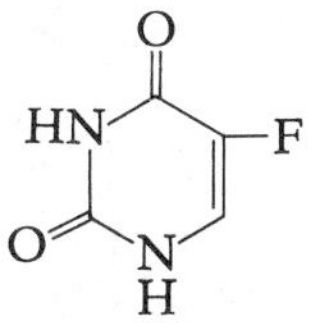   5-Fluorouracil (5-FU) 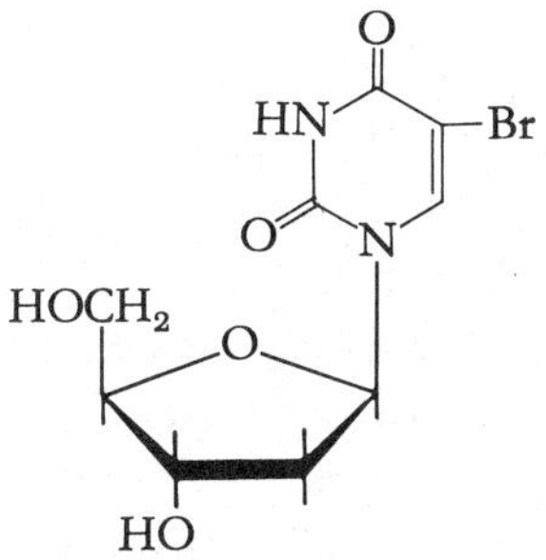   Bromodeoxyuridine 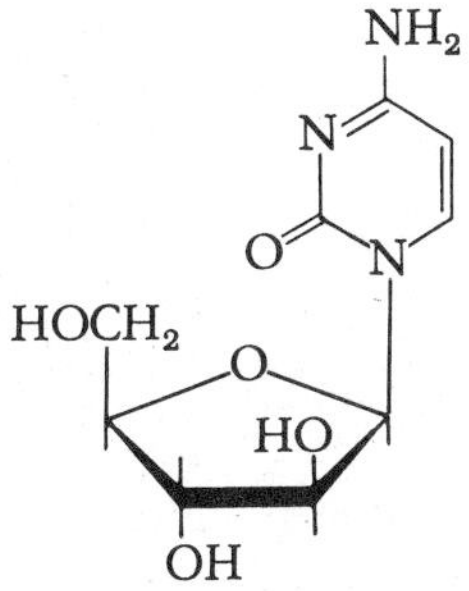   Arabinosylcytosine

These analogs are converted by enzymes of the salvage pathways to their respective nucleotides and thus inhibit normal nucleotide metabolism. For example, 6-mercaptopurine reacts with 5-phosphoribosyl pyrophosphate (PRPP) to form the mercaptopurine ribonucleotide and inorganic pyrophosphate; the reaction is catalyzed by hypoxanthine-guanine phosphoribosyl transferase (section 9.6). The mercaptopurine ribonucleotide inhibits the synthesis of AMP and GMP at several sites. Like the natural nucleotides, it inhibits the transfer of amido groups from glutamine to PRPP and thus inhibits the formation of 5-phosphoribosylamine (section 9.1). In addition, 6-mercaptopurine interferes with the conversion of inosinic acid to adenylosuccinic acid or xanthylic acid, as well as with the formation of adenylic acid from adenylosuccinic acid (section 9.1).

The overall effect of 6-mercaptopurine is to inhibit normal nucleotide biosynthesis and, consequently, cellular growth and division. Its use in the treatment of acute leukemias is based on this property. Unfortunately, as is the case with many drugs employed in cancer therapy, the cells eventually become resistant to the agent.

Injected *5-fluorouracil* (5-FU) is converted within the cell to the corresponding substituted nucleotide by the sequential action of uridine phosphorylase and uridine kinase (see also section 9.6):

$$\text{5-Fluorouracil + ribose 1-phosphate} \underset{}{\overset{\text{Uridine phosphorylase}}{\rightleftharpoons}} \text{5-Fluorouridine} + P_i$$

$$\text{5-Fluorouridine + ATP} \xrightarrow{\text{Uridine kinase}} \text{5-Fluorouridine 5'-phosphate + ADP}$$

5-Fluorouridine 5'-phosphate is transformed to the 5-fluoro-2'-deoxyuridine nucleotide, an analog of dUMP, which inhibits thymidylate synthetase (section 9.5) and thus blocks the conversion of dUMP to dTMP. 5-Fluorouracil and the corresponding deoxynucleoside (FUdR) are employed in the therapy of various malignancies, e.g., carcinomas of the colon, ovary, breast, and skin.

Bromodeoxyuridine is also phosphorylated within the cell and incorporated into DNA. As a result, normal DNA activity and cell growth are impaired.

Arabinosylcytosine (1-β-D-arabinofuranosylcytosine) inhibits the formation of dCMP and the biosynthesis of DNA. Because of this action, it is employed in the therapy of lymphocytic and acute myelocytic leukemias.

The drugs just described represent some examples of the nucleoside derivatives employed in cancer chemotherapy. Numerous compounds with modifications of the purine, pyrimidine, or pentose rings have been synthesized and tested for their therapeutic effectiveness. Most of these are highly toxic to normal cells and cannot be used in clinical practice. Even the compounds that are useful cannot be administered in high doses. Nonetheless, research on such synthetic analogs and antimetabolites has provided a number of drugs that cause remission of tumors and allow considerable prolongation of life.

A derivative of 6-mercaptopurine, *Immuran* or azathioprine, is useful as an immunosuppressive agent:

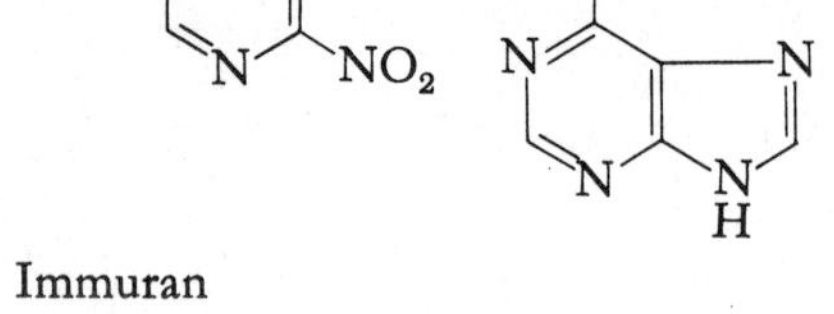

Immuran

It inhibits cellular proliferation and is employed to prevent undesirable antibody reactions.

9.8. Digestion and Absorption of Nucleotides

Digestion of nucleic acids from foods occurs primarily in the duodenal section of the small intestine. Pancreatic juice contains ribonuclease and deoxyribonuclease that hydrolyze the respective substrates to oligonucleotides. The latter are degraded to nucleosides and free bases by nucleases, phosphodiesterases, and phosphatases of the intestinal mucosa. Most of these products are degraded by intestinal microorganisms, and only a minute amount of them is absorbed. In addition, the cells of the intestinal mucosa contain a significant amount of xanthine oxidase (page 396) so most of the purines are oxidized directly. The fraction of the ingested nucleotides that is utilized by the cells for synthesis of nucleic acids is extremely small, so the purines in a normal diet will not affect the levels of uric acid in plasma.

Although ingested nucleotides are not utilized to an appreciable degree, some incorporation of bases does occur when certain of their derivatives are injected. Nucleosides, for example, are incorporated to a small extent; the injection of labeled uridine has been shown to result in the formation of labeled RNA and DNA, and labeled thymidine is incorporated specifically into DNA. This difference between the degree of utilization of nucleosides and that of the free bases has been exploited in the treatment of tumors with specific analogs.

9.9. Catabolism of Pyrimidines

Pyrimidine nucleotides are degraded to the respective nucleosides by the action of intracellular phosphatases. Nucleoside phosphorylases then catalyze the reaction between inorganic phosphate and the nucleoside—e.g., uridine—to release the free base and ribose 1-phosphate:

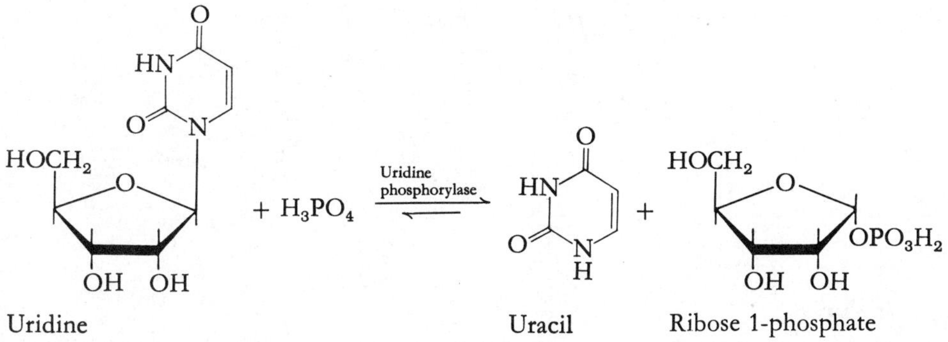

Uridine + H_3PO_4 $\xrightleftharpoons{\text{Uridine phosphorylase}}$ Uracil + Ribose 1-phosphate

Uracil also arises from the deamination of cytosine by cytosine deaminase:

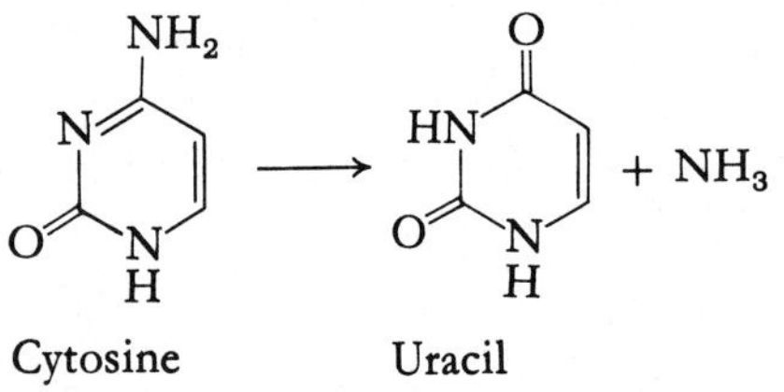

Cytosine → Uracil + NH_3

Free uracil is reduced to 5,6-dihydrouracil by dihydrouracil dehydrogenase. The reduced uracil, in turn, is converted, via β-ureidopropionic acid, to ammonia, carbon dioxide, and β-alanine:

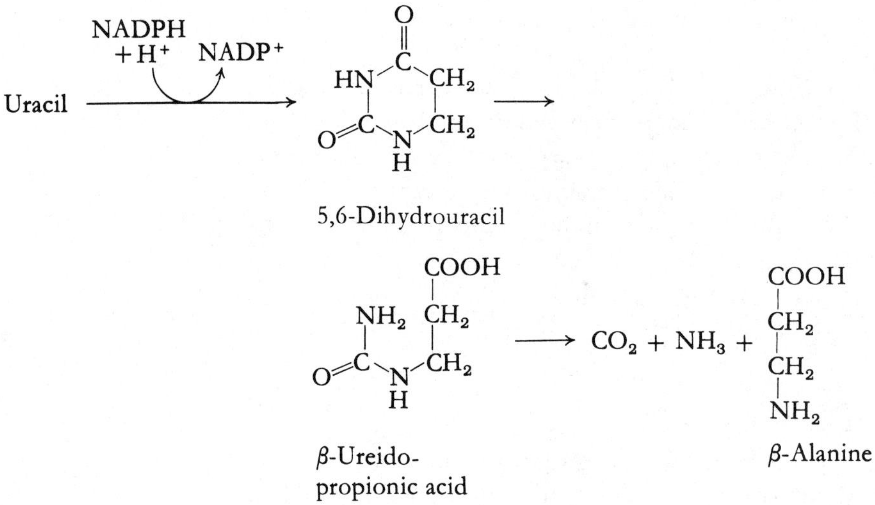

Uracil $\xrightarrow{NADPH + H^+ \to NADP^+}$ 5,6-Dihydrouracil → β-Ureidopropionic acid → CO_2 + NH_3 + β-Alanine

Some of the β-alanine participates in the formation of carnosine, anserine, and coenzyme A.

Thymine undergoes analogous degradative reactions that yield β-aminoisobutyrate:

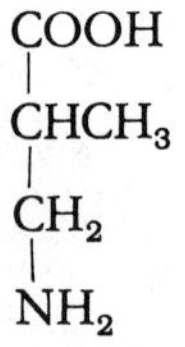

COOH
|
CHCH₃
|
CH₂
|
NH₂

β-Aminoisobutyrate (BAIB)

Urinary levels of this compound are elevated when there is excessive tissue destruction, e.g., in certain leukemias or after total x-ray irradiation. There are also reports of increased urinary excretion of BAIB in certain families of Chinese and Japanese descent. The significance of this genetic trait has not been elucidated. Since both β-alanine and β-aminoisobutyric acid undergo transamination to malonic acid semialdehyde, $HOOCCH_2CHO$, and methylmalonic acid semialdehyde, $HOOCCH(CH_3)CHO$, respectively, then an elevated excretion of BAIB may be related to a defect in the transaminases.

9.10. Catabolism of Purines

Intracellular purine nucleotides arise from the degradation of nucleic acids and, to a lesser extent, from that of nucleoside polyphosphates and coenzymes. Appropriate phosphatases catalyze the removal of phosphate from nucleotides to provide the nucleosides.

Adenylic acid (AMP) is deaminated to form inosinic acid (IMP), which in turn is hydrolyzed to inosine:

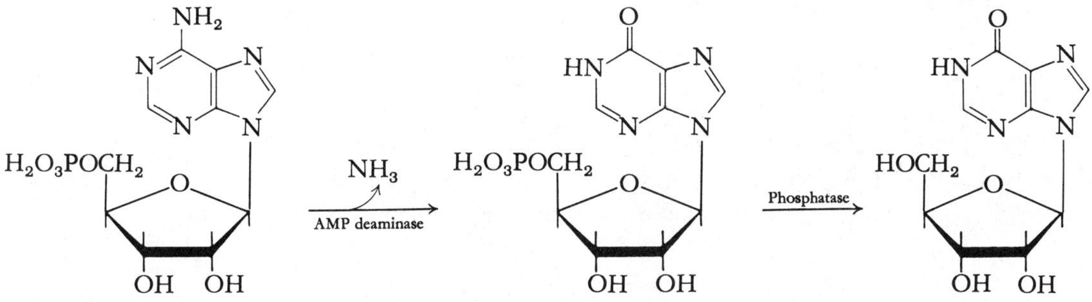

Adenylic acid (AMP) Inosinic acid (IMP) Inosine

The nucleosides adenosine, inosine, and guanosine are degraded by the respective nucleoside phosphorylases to the free bases, adenine, hypoxanthine, and guanine:

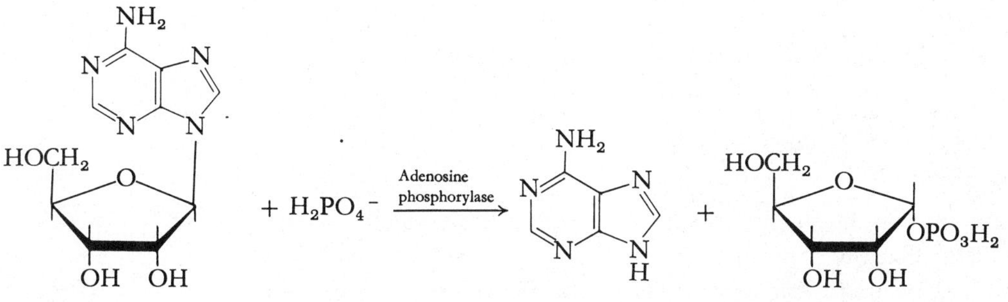

Adenosine Adenine Ribose 1-phosphate

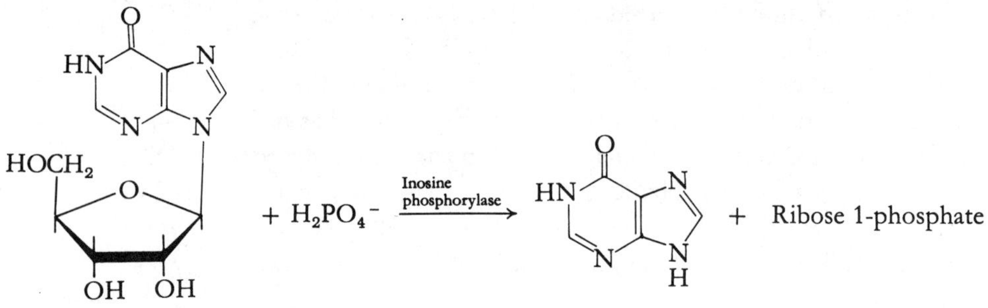

Inosine + H$_2$PO$_4^-$ $\xrightarrow{\text{Inosine phosphorylase}}$ Hypoxanthine + Ribose 1-phosphate

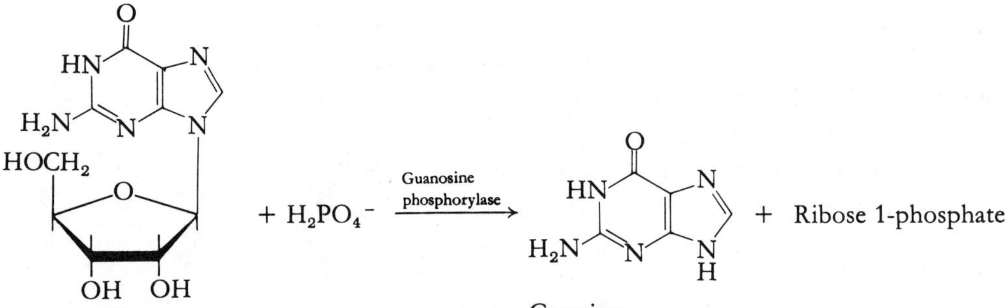

Guanosine + H$_2$PO$_4^-$ $\xrightarrow{\text{Guanosine phosphorylase}}$ Guanine + Ribose 1-phosphate

Adenine is converted to hypoxanthine by the action of the enzyme *adenase*. Oxidation of hypoxanthine by oxygen, which is catalyzed by the flavoprotein enzyme *xanthine oxidase*, yields xanthine and hydrogen peroxide:

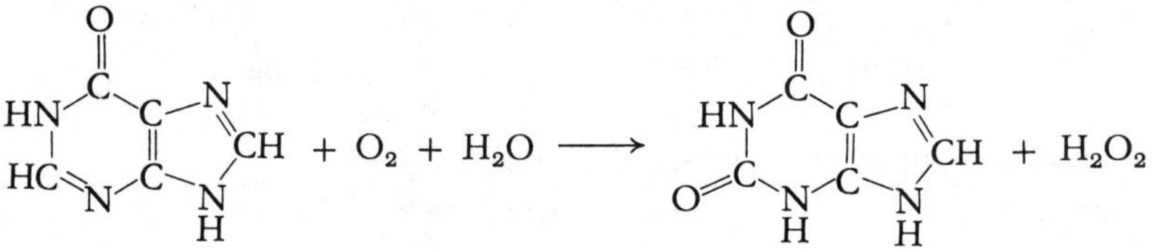

Hypoxanthine + O$_2$ + H$_2$O ⟶ Xanthine + H$_2$O$_2$

Xanthine is also formed by the action of a deaminase on guanine:

Guanine + H$_2$O ⟶ Xanthine + NH$_3$

Thus, all the purine nucleosides—adenosine, inosine, and guanosine—are ultimately metabolized to xanthine.

Xanthine is oxidized to uric acid by oxygen in the presence of xanthine oxidase. Both xanthine and uric acid exist as keto-enol tautomers:

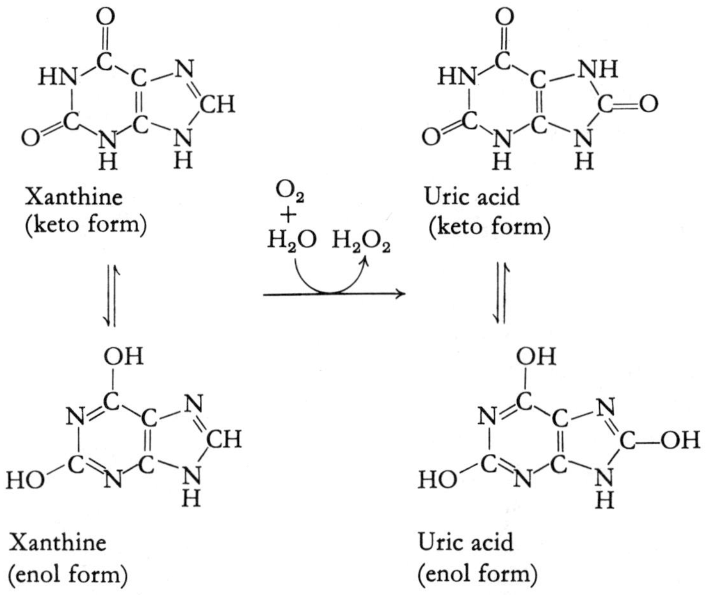

Xanthine (keto form)

O_2 + H_2O → H_2O_2

Uric acid (keto form)

Xanthine (enol form)

Uric acid (enol form)

Uric acid is the final product of purine metabolism in humans. The uric acid thus formed is transferred to the circulatory system and excreted in the urine.

Xanthine oxidase, the enzyme that catalyzes the terminal step in purine catabolism, has several interesting characteristics. It is a flavoprotein that contains two FAD units, eight non-heme iron atoms, and two molybdenum atoms. The fact that xanthine oxidase catalyzes the oxidation of both hypoxanthine and xanthine indicates that its specificity is comparatively low. This low specificity is also evidenced by the fact that xanthine oxidase also catalyzes the oxidation of acetaldehyde to acetic acid:

$$CH_3CHO + H_2O + O_2 \longrightarrow CH_3COOH + H_2O_2$$

Other interesting features of xanthine oxidase are that it catalyzes an oxidation with molecular oxygen and that one of the products of this reaction is hydrogen peroxide. The latter, which is highly toxic, is decomposed by the heme-containing enzyme, *catalase*:

$$2H_2O_2 \longrightarrow 2H_2O + O_2$$

9.11. Uric Acid and Hyperuricemic Disorders

Nucleic acids and nucleotides, like most other body constituents, undergo continuous metabolic degradation. The supplies of these substances are replenished by biosynthetic processes. The overall rate of nucleotide formation is adapted to the physiologic requirements for specific nucleotides. As a result of the regulated turnover of purine-containing substances, constant amounts of uric acid are produced and excreted. The level of urate in the circulatory system is maintained within relatively narrow limits (2 to 6 milligrams per 100 milliliters of plasma). This level represents a balance between its rate of formation and its rate of disposition via the kidneys and intestinal tract. In normal adults, the average plasma urate concentration is 5.0 milligrams per 100 milliliters for males and 4.1 milligrams per 100 milliliters for females. Although about 700 milligrams of uric acid is produced each day, only about 500 milligrams (420 ± 75 milligrams) is excreted in the urine. The rest of the uric acid reaches the intestine through the bile, and it is degraded (uricolysis) and eliminated with the feces. As already indicated (section 9.8), the contribution of dietary nucleic acids to the level of circulating uric acid is comparatively small. Most of the nucleosides arising from ingested nucleic acids are degraded in the intestine by mucosal xanthine oxidase, and only minimal amounts are absorbed and channeled through intracellular metabolic processes.

At physiologic pH, circulating urate is in the form of the sodium salt, monosodium urate. Since the latter has a limited solubility, it tends to precipitate when its levels in the plasma are elevated. This may result in urate deposition in the kidneys (renal calculi) or in the cartilage or joints (tophi).

A number of diseases are characterized by elevated concentrations of urate in the blood (hyperuricemia) and urine (hyperuricuria). These disorders may due to a primary defect in urate metabolism, or they may be induced by other disorders that interfere with the normal disposal of uric acid. Some of these abnormalities are discussed next.

Primary gout is a genetic disorder that manifests itself in sustained hyperuricemia and, ultimately, in recurrent attacks of acute arthritis. Nodules containing deposits of monosodium urate (tophi) are common in the joints; a characteristically affected site is the metatarsophalangeal joint of the great toe. These deposits give rise to inflammatory reactions (swelling, redness, and so on), extreme pain, and eventually various deformities. Although in many instances the hyperuricemia was shown to be due to overproduction of urate, this does not seem to be its only cause. Some cases of gout may be caused by a defect in the mechanism for the renal excretion of urate. The latter disorder (*primary renal gout*) should thus be differentiated from the patients in whom there is an abnormality in the control of urate production (*primary metabolic gout*). Patients with renal failure due to glomerulonephritis also develop hyperuricemia (*secondary renal gout*) as a result of deficiency in urate excretion.

Hyperuricemia and arthritic complications also occur in diseases in which there is an increased turnover of nucleic acids; these include certain leukemias and polycythemia. The clinical manifestation in such cases is termed *secondary gout*.

Increased levels of urate may arise in patients with untreated diabetes and during starvation. In these situations, the hyperuricemia is caused by the relatively high concentrations of circulating acetoacetate and β-hydroxybutyrate (ketone bodies), which compete with urate for renal secretion. Removal of urate through the kidney is also inhibited when there is an overproduction of lactate, as is the case in von Gierke's disease (glucose 6-phosphatase deficiency) and various other glycogen-storage diseases. The hyperuricemia in patients with von Gierke's disease may be due to enhanced utilization of glucose 6-phosphate and the resultant overproduction of ribose 5-phosphate. Elevated levels of the latter may cause an enhancement in the rate of synthesis of 5-phosphoribosyl pyrophosphate, which will ultimately lead to overproduction of purine nucleotides and uric acid.

Prolonged intake of alcoholic beverages may produce an increase in the blood levels of urate. Oxidation of ethanol to acetaldehyde is catalyzed by the NAD^+-linked enzyme alcohol dehydrogenase and yields acetaldehyde and NADH. The former is oxidized to acetate, which is converted to acetyl-coenzyme A. Both NADH and acetyl-coenzyme A inhibit glycolysis, and glucose 6-phosphate is thus channeled to the production of ribose 5-phosphate. Moreover, NADH stimulates the formation of lactate, which decreases the rate of removal of urate via the kidneys. Alcohol thus causes an elevation in circulating urate by enhancing its synthesis and inhibiting its disposal.

Another abnormality that can overstimulate the operation of the oxidative pentose pathway is an elevation in the activity of glutathione reductase. This enzyme catalyzes the reduction of oxidized glutathione by NADPH. Depletion of NADPH due to this reaction will cause an increase in the activity of the oxidative pentose pathway and the concurrent formation of ribose 5-phosphate.

The supply of glutamine is increased in patients with glutamate dehydrogenase deficiency. This, too, may lead to excessive production of urate by increasing the rate of the two reactions in purine synthesis that require glutamine (section 9.1). A similar effect is encountered in disorders in which there is increased activity of the amidotransferase that catalyzes the reaction of glutamine with PRPP to yield 5-phosphoribosylamine (section 9.1). The increase in urate levels in individuals with the Lesch-Nyhan syndrome will be discussed in subsequent paragraphs.

The causes of secondary gout are outlined in Figure 9-3.

The symptoms of gout can be relieved by the alkaloid, colchicine. This

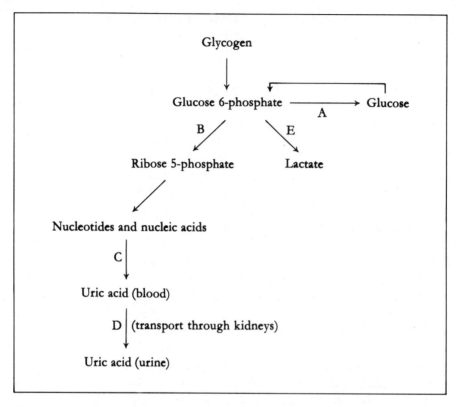

Fig. 9-3. Reaction sites associated with the causes of various types of secondary gout. If the reaction at site A does not occur (von Gierke's disease), a greater amount of glucose 6-phosphate is channeled to lactate and ribose 5-phosphate formation. Thus, pathway B is stimulated, resulting in overproduction of urate. Step D is inhibited by lactate, acetoacetate, and β-hydroxybutyrate, as well as during renal failure. Glycolysis (E) is inhibited by the NADH and acetyl-CoA formed as a result of alcohol intake, thus promoting the operation of pathway B. The activity of pathway B is also increased by the excess glutamine that arises in glutamate dehydrogenase deficiency and by the increased activity of glutamine 5-phosphoribosyl-pyrophosphate amidotransferase. Pathways B and C are activated when there is increased turnover of nucleic acid, as occurs in leukemia.

agent has an antiinflammatory effect on the arthritic joints, but it does not decrease the urate levels. Its principal clinical use is in the diagnosis of gout, rather than in its treatment. Other agents that have antiinflammatory effects (e.g., phenylbutazone and corticoids) have also been found to be useful in relieving acute arthritic attacks. Drugs that prevent renal tubular resorption (probenecid and sulfinpyrazone) are employed to decrease blood urate levels, but their use may give rise to renal calculi. The current drug of choice for decreasing hyperuricemia as well as the clinical manifestations of gout is *allopurinol* (Zyloprim):

Allopurinol

This substance is an analog of hypoxanthine and acts as an inhibitor of xanthine oxidase. Treatment with allopurinol reduces the levels of urate in the blood as well as the amount excreted in the urine. As would be expected, the concentrations of hypoxanthine and xanthine are increased. Since these are more soluble than urate, however, they do not cause the difficulties that are incurred by excessive levels of urate.

Studies of patients treated with allopurinol have revealed that the increase in the excretion of xanthine plus hypoxanthine amounts to only about two-thirds of the decrease in urate excretion. This suggests that administration of allopurinol causes a decrease in the total amount of purines produced. This finding can be explained in view of the expected conversion of some of the accumulated hypoxanthine to inosinic acid (IMP; section 9.6). Elevated levels of IMP inhibit formation of 5-phosphoribosylamine (page 379) and thus reduce the rate of purine nucleotide synthesis. Hence, in addition to decreasing the formation of urate directly, allopurinol also reduces the rate of purine nucleotide production, thereby indirectly inhibiting urate formation (Fig. 9-4).

Individuals with gout are also advised to maintain a low-protein diet. Since several amino acids are required for purine synthesis, this regimen presumably serves to decrease the rate of urate production.

In addition to the conditions described above, there is an inborn error of metabolism, *Lesch-Nyhan syndrome*, which is transmitted as an X-linked recessive trait (found only in male children). This disorder is associated with an overproduction of urate and gouty manifestations, in addition to a basic neurologic disorder involving mental retardation and destructive biting or self-mutilation. Analyses of brain, liver, red blood cells, and skin fibroblasts of these patients have revealed the absence of the enzyme hypoxanthine-guanine phosphoribosyl transferase (HGPT). Thus, individuals suffering from the Lesch-Nyhan syndrome cannot form purine nucleotides by the salvage pathways. Administration of allopurinol to these patients results in a decrease in urate and an increase in hypoxanthine levels. As would be expected, the amount of hypoxanthine plus xanthine is equal to the decrease in urate

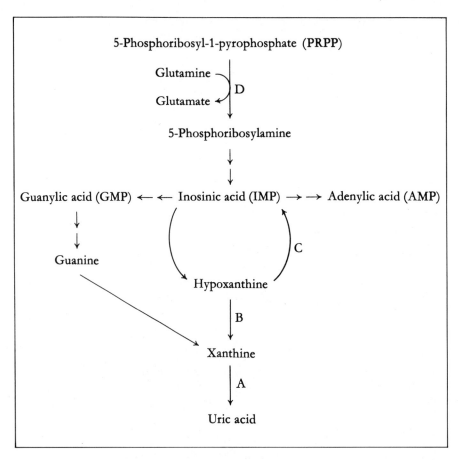

Fig. 9-4. Sites of action of allopurinol. Steps A and B are inhibited directly. This results in the stimulation of the reaction at step C, which is catalyzed by hypoxanthine-guanine phosphoribosyl transferase (HGPT). The IMP, AMP and GMP that are thus formed inhibit step D and the overall production of purine nucleotides.

production. The overall production of purines is not decreased, as in the case of primary gout, because, due to the enzyme defect, inosinic acid cannot be generated from the excess hypoxanthine (see Fig. 9-4).

Allopurinol is also employed as an auxiliary agent in the treatment of certain cancer patients with 6-mercaptopurine. The latter is partially converted to 6-thiouric acid by the action of xanthine oxidase, thus diverting 6-mercaptopurine from its purpose as a purine analog. The simultaneous administration of allopurinol maximizes the effectiveness of 6-mercaptopurine by inhibiting the activity of the enzyme, xanthine oxidase, that catalyzes the conversion of 6-mercaptopurine to the undesired by-product.

Suggested Reading
Cartier, P., and Hamet, M. The Normal Metabolism of Uric Acid. *Advances in Nephrology* 3:3, 1974.

Henderson, J. F., and Paterson, A. R. P. *Nucleotide Metabolism.* New York: Academic Press, 1973.

Seegmiller, J. E. Biochemical and Genetic Studies of an X-Linked Neurological Disease (The Lesch-Nyhan Syndrome). *Harvey Lectures* 65:175, 1971.

Smith, L. H. Pyrimidine Metabolism in Man. *New England Journal of Medicine* 15:764, 1973.

Weber, G. Enzymology of Cancer Cells. *New England Journal of Medicine* 296:486 and 541, 1977.

Wyngaarden, J. B. Defect of Primary Hyperuricemia and Gout. *American Journal of Medicine* 56:651, 1974.

Yu, T. Milestones in the Treatment of Gout. *American Journal of Medicine* 56:677, 1974.

10. Metabolism of Porphyrins

10.1. Structure and Functions

The distinguishing feature of a porphyrin is its cyclic tetrapyrrole structure:

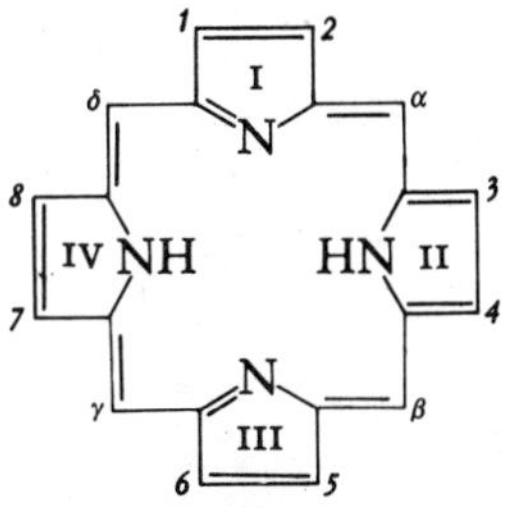

Porphyrin nucleus

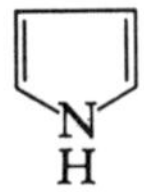

Pyrrole

The numbering of the carbon atoms and the designations for the pyrrole rings are indicated in the diagram. Individual porphyrins differ from each other with respect to the substituents or groups attached to the pyrrole units.

Porphyrin structures occur as prosthetic groups in a number of proteins. These include the cytochromes, hemoglobin, myoglobin, and the enzymes catalase and peroxidase.

The nitrogen atoms of animal porphyrins form coordination complexes with iron and other metal ions. In the cytochromes, the iron serves in the electron-transport sequences of the oxidative chain. The oxidized form of cytochromes contains iron in the ferric (Fe^{3+}) state. Upon reduction, the iron is changed to its ferrous (Fe^{2+}) form (page 135).

The iron in hemoglobin is in the ferrous state. When hemoglobin combines with oxygen to form oxyhemoglobin, the iron ion is coordinate-bonded with the oxygen, but it remains in the Fe^{2+} state.

Porphyrins also serve as the precursors for bile pigments, such as bilirubin and biliverdin.

10.2. Biosynthesis of Porphyrins

Porphyrins are synthesized in animal cells from comparatively simple precursors. Hence, they are not required in the diet.

The first step in the synthesis of porphyrins is the condensation of succinyl-coenzyme A with glycine to produce *δ-aminolevulinic acid*. The enzyme that catalyzes the process, *δ-aminolevulinate synthetase*, is associated with the mitochondrial portion of the cell. α-Amino-β-ketoadipic acid is presumed to be an intermediate in the process, but it has not been isolated:

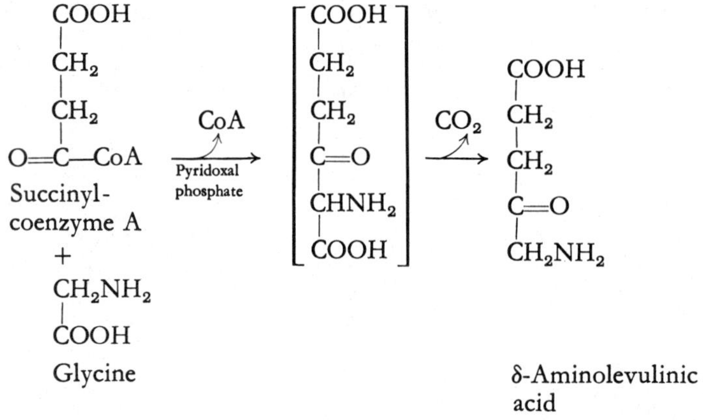

$$\begin{array}{c}\text{COOH}\\|\\\text{CH}_2\\|\\\text{CH}_2\\|\\\text{O=C-CoA}\end{array} \;\; \underset{\text{phosphate}}{\overset{\text{CoA}}{\underset{\text{Pyridoxal}}{\longrightarrow}}} \;\; \begin{bmatrix}\text{COOH}\\|\\\text{CH}_2\\|\\\text{CH}_2\\|\\\text{C=O}\\|\\\text{CHNH}_2\\|\\\text{COOH}\end{bmatrix} \;\; \overset{\text{CO}_2}{\longrightarrow} \;\; \begin{array}{c}\text{COOH}\\|\\\text{CH}_2\\|\\\text{CH}_2\\|\\\text{C=O}\\|\\\text{CH}_2\text{NH}_2\end{array}$$

Succinyl-coenzyme A + Glycine (CH$_2$NH$_2$–COOH) → δ-Aminolevulinic acid

The formation of δ-aminolevulinate is the committed step in porphyrin synthesis, and, as expected, it is subject to various controls. δ-Aminolevulinate synthetase is an allosteric enzyme and is inhibited by the final product in the pathway, heme or hemin (feedback inhibition). In the liver, heme can also inhibit the synthesis of δ-aminolevulinate through its action in decreasing the rate of synthesis of the enzyme δ-aminolevulinate synthetase.

The next reaction in porphyrin synthesis is catalyzed by *δ-aminolevulinate dehydratase* and involves the condensation of two molecules of δ-aminolevulinate to form *porphobilinogen*:

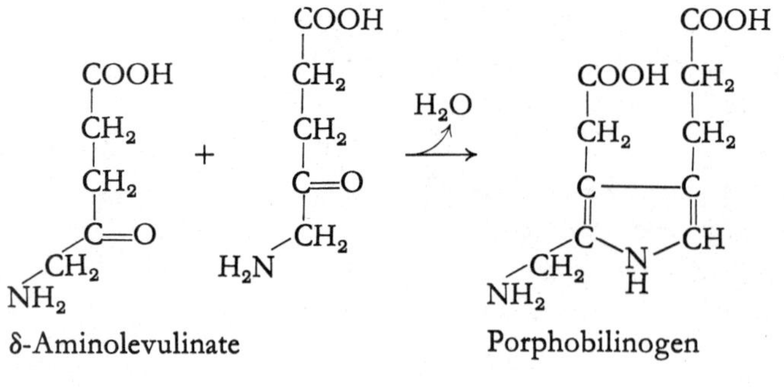

δ-Aminolevulinate + δ-Aminolevulinate $\xrightarrow{\text{H}_2\text{O}}$ Porphobilinogen

The enzyme that catalyzes this reaction is inhibited by lead. The enzyme is highly sensitive to this metal, which, in cases of lead poisoning, produces an inhibition of porphyrin synthesis and an accumulation of δ-aminolevulinate. Hence, although minimal amounts of the latter intermediate are normally excreted in urine (about 2 milligrams in 24 hours), the urinary concentration of δ-aminolevulinate increases considerably in lead poisoning and may be used as a diagnostic indicator for such cases.

In the next step toward the synthesis of porphyrins, four molecules of porphobilinogen react to produce *uroporphyrinogen III*. This requires the action of a dual enzyme system involving both porphobilinogen deaminase and porphobilinogen isomerase. The latter is required to bring about the inversion of substituents on ring IV. In the absence of the isomerase, *uroporphyrinogen I* is produced:

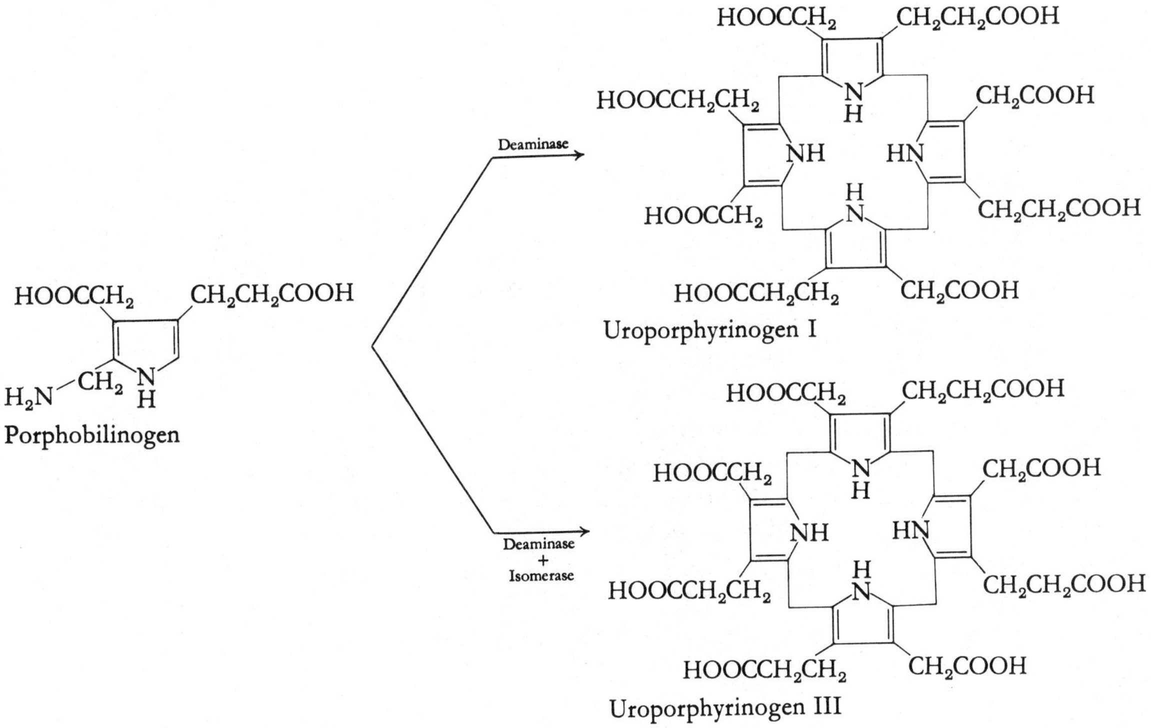

The action of uroporphyrinogen decarboxylase yields *coproporphyrinogen III* from uroporphyrinogen III. The porphyrin skeleton is desaturated, and two of the propionate side chains are converted to vinyl groups by coproporphyrinogen oxidase to yield *protoporphyrin IX*. Interaction of the latter with Fe^{2+} and the enzyme ferrochelatase results in the formation of *heme*:

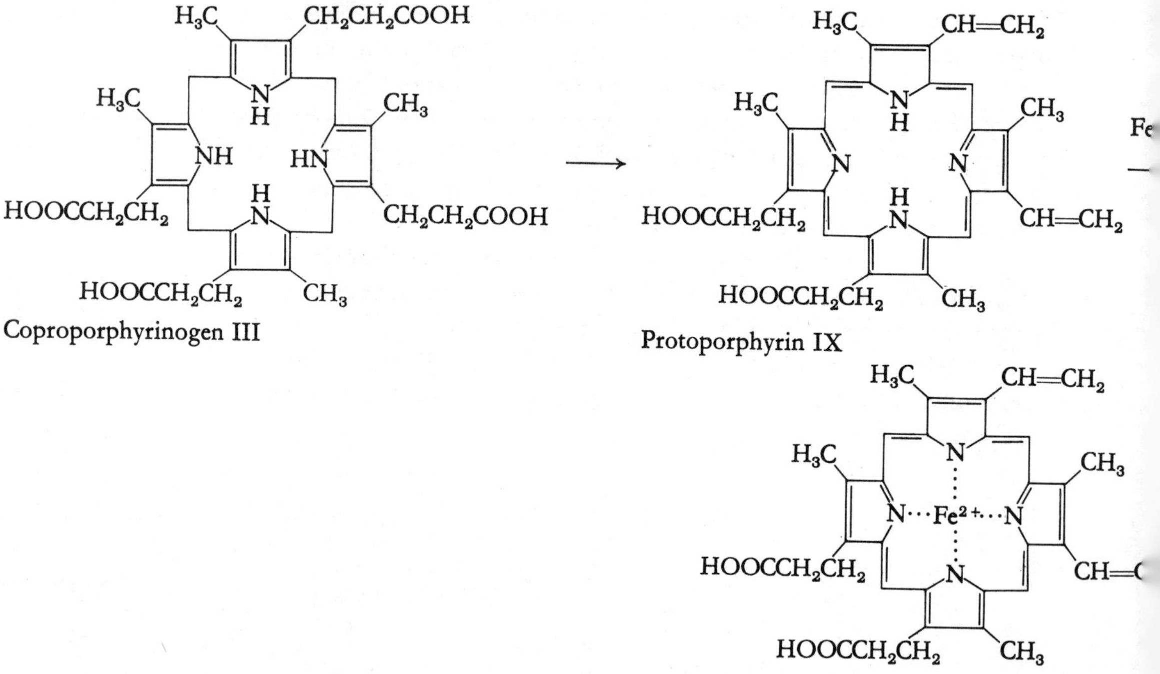

Coproporphyrinogen III

Protoporphyrin IX

Heme (ferroprotoporphyrin IX)

Ferroprotoporphyrin IX, or heme, is the prosthetic group of hemoglobin, myoglobin, cytochromes, catalase, and peroxidase. Over 85% of the ferroprotoporphyrin, which is continually synthesized, is incorporated into hemoglobin. About 10% is utilized in the production of myoglobin, and the remainder serves in the production of the other heme proteins.

Myoglobin is a component of muscle and is capable of binding oxygen. Since its affinity for oxygen is greater than that of hemoglobin but less than that of cytochrome oxidase, myoglobin can accept oxygen from hemoglobin and transfer it for utilization in the respiratory chain. This appears to be an important function of myoglobin in human heart muscle. The function of myoglobin in other muscle appears to be primarily in the intracellular movement of oxygen.

10.3. Methemoglobin and Other Derivatives of Hemoglobin

A small but significant amount of hemoglobin is oxidized by oxygen to the Fe^{3+} derivative, termed *methemoglobin*. Unlike the ferrous form of hemoglobin, methemoglobin does not combine with molecular oxygen and cannot function in the oxygen-transport process. Normally, methemoglobin is reduced to the ferrous state by the NADH-linked enzyme methemoglobin reductase. The amount of methemoglobin generally found in blood is about 300 milligrams per 100 milliliters (1.7% of the total hemoglobin), and this level is rarely exceeded. Concomitantly with the production of methemoglobin, the oxygen is transformed to the superoxide radical, O_2^-. This radical is converted to hydrogen peroxide by the copper-containing enzyme superoxide dismutase, which is present in erythrocytes:

$$2O_2^- + 2H^+ \longrightarrow H_2O_2 + O_2$$

The hydrogen peroxide is then decomposed by the catalase that is present in erythrocytes to oxygen and water:

$$2H_2O_2 \longrightarrow O_2 + 2H_2O$$

Thus, the action of these two enzymes serves to prevent damage to the erythrocyte by the toxic products O_2^- and H_2O_2.

Another reaction occurring in erythrocytes that also serves to destroy hydrogen peroxide is catalyzed by the enzyme glutathione peroxidase. If glutathione (γ-glutamylcysteinylglycine; section 8.3) is represented as G-SH, the reaction with H_2O_2 may be written as:

$$2\,G{-}SH + H_2O_2 \longrightarrow G{-}SS{-}G + 2H_2O$$

Oxidized glutathione (G-SS-G) is reduced by NADPH and glutathione reductase, to regenerate the original glutathione:

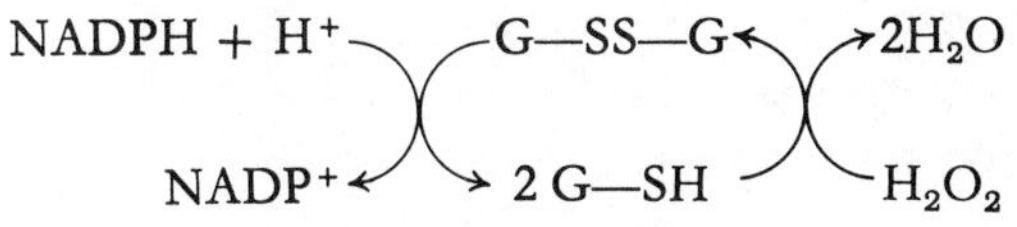

$$\begin{array}{c}
NADPH + H^+ \searrow \quad \nearrow G{-}SS{-}G \searrow \quad \nearrow 2H_2O \\
\qquad \qquad \times \qquad \qquad \times \\
NADP^+ \nearrow \quad \searrow 2\,G{-}SH \quad \nwarrow H_2O_2
\end{array}$$

NADPH is generated in erythrocytes as a result of the metabolism of glucose by the oxidative pentose pathway. It may be that the purpose of this pathway in these cells is to provide NADPH for glutathione regeneration.

Excessive amounts of methemoglobin (*methemoglobinemia*) can result if either the rate of methemoglobin formation is increased or its reduction mechanism is defective. Certain drugs—including salicylates, sulfonamides, acetanilid, and phenacetin—appear to accelerate the production of methemoglobin. A number of chemical agents, such as aniline, nitrites, and nitrates, are also known to have this effect.

A rare genetic disease, *familial methemoglobinemia,* is due to a deficiency of methemoglobin reductase in the erythrocytes. In individuals with this disorder, over 25% of the hemoglobin occurs in the ferric form. As would be expected in these patients there is a severe deficiency in the availability of oxygen to the cells.

Hemoglobin has a much higher affinity for carbon monoxide than for oxygen. The toxicity of CO is due to the formation of the highly stable carbon monoxide-hemoglobin complex and the consequent unavailability of the hemoglobin (and other heme porphyrins) for normal functions. Hemoglobin and cytochrome oxidase are also poisoned by sulfide and cyanide ions. The Fe^{3+} forms of these compounds form tight complexes with sulfide or cyanide.

10.4. Porphyrias

The synthesis of protoporphyrin IX requires the production of uroporphyrinogen III rather than uroporphyrinogen I, and the synthesis of the former requires the presence of both porphobilinogen deaminase and porphobilinogen isomerase (also called cosynthetase), as described in section 10.2. A deficiency in the isomerase, which occurs in *congenital erythropoietic porphyria*, results in the synthesis of abnormal porphyrin intermediates, such as uroporphyrinogen I and its product, coproporphyrinogen I. These substances, which cannot be converted to protoporphyrin IX, accumulate in the tissues and are excreted in the urine. The disease is transmitted by autosomal recessive inheritance. The teeth of patients with this disorder show a strong red fluorescence under ultraviolet light. These patients also excrete large amounts of uroporphyrinogen I, which gives the urine a red color. The disease is characterized by extreme sensitivity of the skin to light and enlargement of the spleen.

A relatively more common porphyria, which is transmitted as an autosomal dominant trait, is known as *erythropoietic protoporphyria*. Excessive amounts of protoporphyrin are produced in the erythrocytes of patients with this disorder.

Another autosomal dominant disorder involves excessive hepatic production of the intermediates of porphyrin synthesis. The disease, known as *acute intermittent porphyria,* is expressed by episodes of abdominal pain and mental derangement. Large amounts of δ-aminolevulinic acid and porphobilinogen are excreted in the urine; these substances give the urine a characteristic dark red color. The defect is attributed to the elevated and unregulated activity of δ-aminolevulinate synthetase in these patients.

10.5. Heme Catabolism

Erythrocytes in normal humans have a life span of about 120 days. When the cells degenerate, they are degraded in the spleen together with their hemoglobin component. The methylidyne bridge in the alpha position of the tetrapyrrole ring (section 10.1) is cleaved, and the iron ion is removed to yield *biliverdin*. This reaction is catalyzed by an NADPH-linked, mixed-function oxygenase, heme oxygenase. Biliverdin is reduced to *bilirubin* by an NADPH-linked reductase:

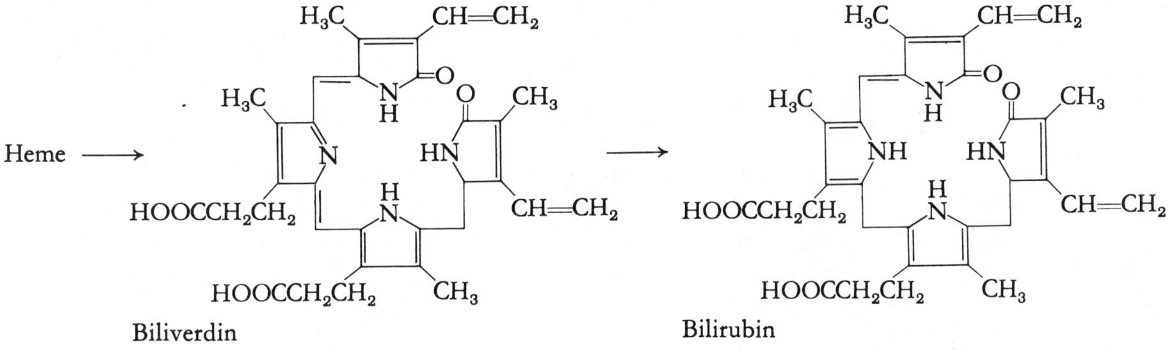

Heme ⟶ [Biliverdin structure] ⟶ [Bilirubin structure]

The orange-yellow, water-insoluble bilirubin is transported in the form of an albumin complex from the reticuloendothelial cells to the liver. Interaction of the bilirubin with UDP-glucuronic acid yields *bilirubin diglucuronide*. This conjugated derivative, which is water-soluble, is transported to the gallbladder and constitutes one of the components of bile. When bile is transferred to the intestine, the bilirubin diglucuronides are hydrolyzed by β-glucuronidase and are converted by the intestinal flora to a series of products, including *urobilinogen* and *stercobilinogen*. Normally, a large fraction of the bile products are absorbed by the portal system and taken up by the liver. Very small amounts of urobilinogen and its oxidation product, urobilin, are excreted in the urine. Stercobilinogen, on oxidation, yields stercobilin, which is removed with the feces.

About 250 milligrams of bile pigment is carried in the feces daily, and approximately 2 milligrams is excreted in the urine. The amount of bile pigment in the feces is an index of the quantity of bilirubin that appears in the gut. It is thus a measure of the relative amount of hemoglobin that has been degraded. Assuming there is no abnormality in hepatic or biliary function, an elevation in the amount of fecal stercobilinogen is suggestive of hemolytic disease. Increases in the amount of urinary urobilinogen indicate the presence of parenchymal hepatic disease that prevents the reabsorbed material from being removed from circulation.

Conjugated bilirubin—i.e., bilirubin diglucuronide—is differentiated from non-conjugated material by the *van den Bergh reaction,* which utilizes diazotized sulfanilic acid. The water-soluble, conjugated bilirubin reacts directly with the aqueous reagent. However, the water-insoluble, unconjugated pigment reacts only after the addition of ethanol (indirect van den Bergh reaction). Normally, serum shows primarily the indirect reaction, whereas bile gives the direct reaction.

In hemolytic diseases, where there is excessive destruction of erythrocytes and a high production of bile pigments, the capacity for bilirubin conjugation in the liver may be exceeded. There is then an increase in the amount of non-conjugated pigment in the blood. The occurrence of physiological jaundice in newborns appears to be due to the incomplete development of the enzyme that catalyzes the conjugation of bilirubin with glucuronate.

In biliary obstruction, the transport of bile pigments from the liver to the gallbladder is inhibited. Thus, the conjugated bilirubin that is formed in the liver is taken up into the circulation. If the biliary obstruction continues for an extended period, the liver cells are damaged and the conjugation reaction is diminished. This leads to the appearance of large amounts of free bilirubin in the blood.

Suggested Reading

Doso, M. (Ed.) *Porphyrins in Human Disease*. Basel: Karger, 1976.

Gidari, A., and Levere, R. D. Enzymatic Formation and Cell Regulation of Heme Synthesis. *Seminars in Hematology* 14:145, 1977.

Schmid, R. Bilirubin Metabolism in Man. *New England Journal of Medicine* 287:703, 1972.

11. Biosynthesis of Nucleic Acids and Proteins

11.1. Structure of DNA and Its Replication

The central role of deoxyribonucleic acid (DNA) in the transmission of hereditary traits is well established. Chromosomes are DNA-protein complexes or nucleoproteins. The information transmitted by the genes from one generation of cells to the next is stored in the structure of the cells' DNA or, more precisely, in the specific base sequences of the DNA macromolecules. When new DNA is synthesized and distributed during cell division, the DNA molecules provided to the daughter cells are copies or duplicates of those of the parent cell. The process by which each DNA molecule directs the synthesis and specifies the structure of newly formed DNA is called *replication*. The mechanism for the replicative process depends on the unique structure of DNA and the characteristics of the enzymes that catalyze DNA synthesis. Some fundamental aspects of this mechanism are discussed in this section and the one that follows.

DNA is composed of deoxyribonucleosides whose sugar units are linked to each other through phosphate bridges (Chap. 1, Section 2F). Specifically, the 5'-hydroxyl group of one deoxyribose unit is linked covalently to a phosphate group, which in turn is joined to the 3'-hydroxyl group of another nucleoside. This provides a backbone of alternating deoxyribose and phosphate groups. Each deoxyribose unit is also linked to one of four bases, either adenine, guanine, cytosine, or thymine. The abbreviated base-sequence notation for polyribonucleotide structures is shown on page 414. A, G, C, and T represent the nucleosides of adenine, guanine, cytosine, and thymine, respectively; P indicates a phosphate group and d denotes that the pentose is deoxyribose. In addition to the four principal bases, DNA also contains small amounts of methyl-substituted bases (e.g., 5-methylcytosine or 7-methyladenine), which are especially found in the DNA from viruses.

In the structural formulas of polynucleotides, the phosphodiester linkages are written in a specific direction, that is, from the 3'-hydroxyl group of one nucleoside to the 5'-hydroxyl group of the adjacent nucleoside.

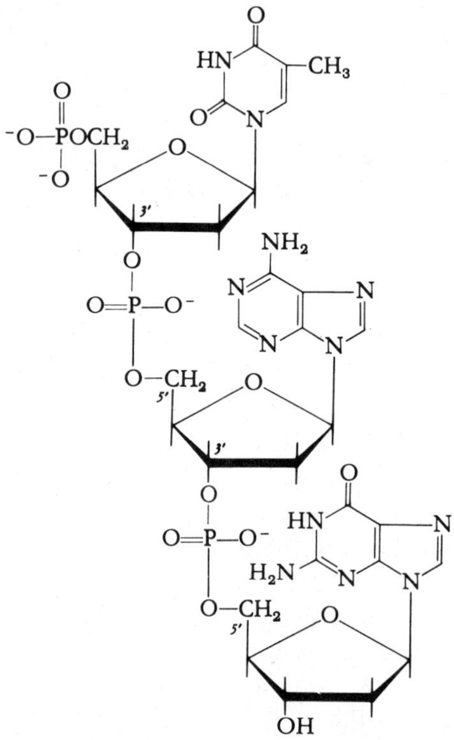

414   11. Biosynthesis of Nucleic Acids and Proteins

Thymine

pdTpdApdG
or
pdT—dA—dG
or
TAG

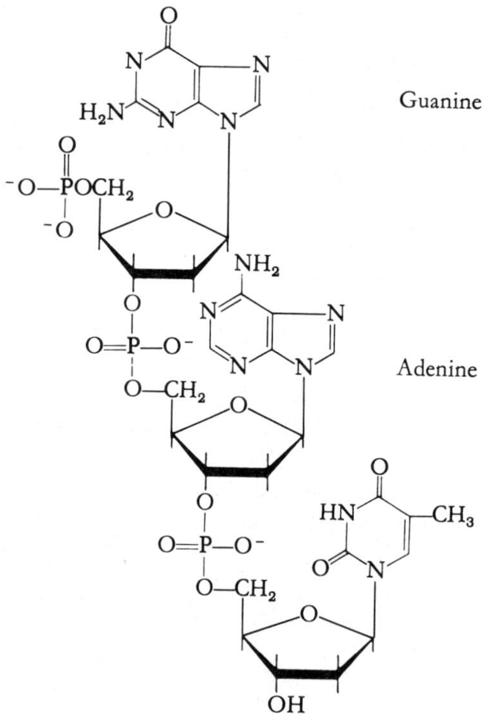

Adenine

Guanine

Guanine

pdGpdApdT
or
pdG—dA—dT
or
GAT

Adenine

Thymine

The structure on the bottom of page 414 differs from the top one in that the base sequence is reversed and the phosphodiester bridges have different sites of attachment on the respective nucleosides. In the first structure, the thymine nucleoside is linked to the adenine nucleoside via the 3'-hydroxyl group of the thymidine, whereas in the second structure the linkage to the neighboring adenosine is through the 5'-hydroxyl of the thymidine (in this sense, the chain has polarity). By convention, the notation TAG means that the 5'-hydroxyl of thymidine is not linked to another nucleoside; i.e., that it is a terminal group. The direction of the phosphodiester bridge in the notation system is from a 3'-hydroxyl to a 5'-hydroxyl group. The guanosine in TAG is therefore linked to its neighboring nucleoside through its 5'-hydroxyl, and the 3'-hydroxyl group of the guanosine is free. The two ends of a polynucleotide chain are designated according to the non-bridged hydroxyl group in the pentose of the nucleoside. Thus, the terminus that contains the nucleoside unit in which the 5'-hydroxyl is not linked in a phosphodiester bridge is called the 5' end. Similarly, the 3' end contains the nucleoside in which the 3'-hydroxyl is not involved in a phosphodiester bridge.

DNA molecules consist of two strands of polynucleotides that are linked to each other by hydrogen bonding between the respective nitrogenous bases. Thus, an adenine group from one strand will interact with a thymine group in the complementary strand, and similarly, guanine will be bonded to cytosine:

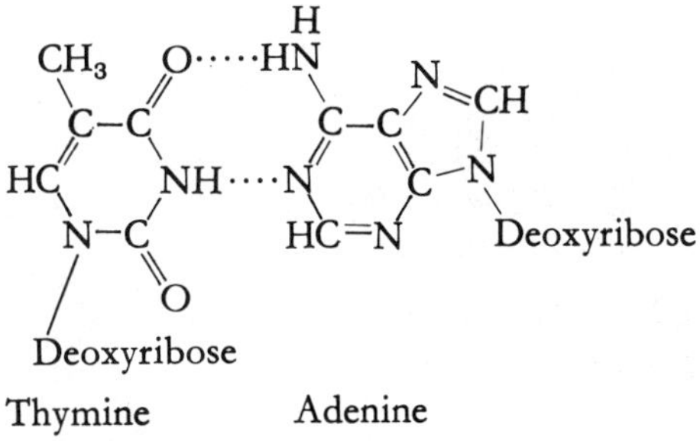

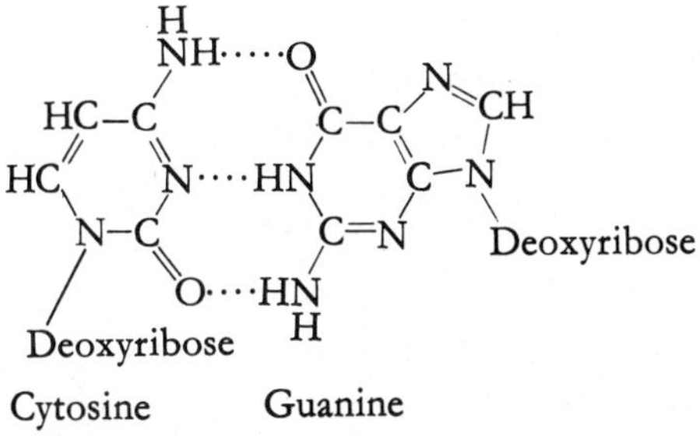

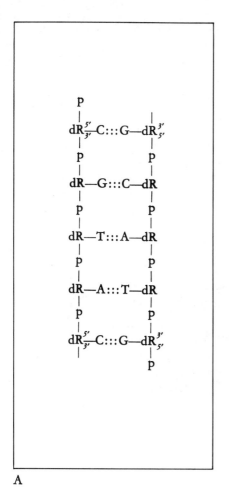 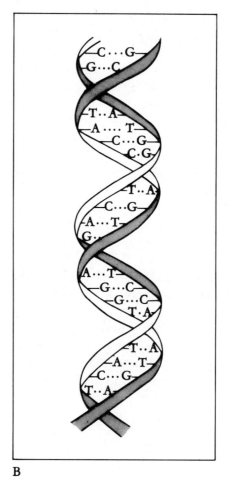

Fig. 11-1. A. Segment of DNA showing base pairing (hydrogen bonds) and antiparallel nature of complementary strands. B. Representation of a double helix.

A
B

A two-dimensional representation of two hydrogen-bonded strands is shown in Figure 11-1A. The three-dimensional characteristics of the DNA polynucleotide chain and the positions of the components were deduced by Watson and Crick from x-ray crystallographic studies. The two chains of alternating deoxyribose and phosphate units have helical configurations, and both are coiled around a common axis to form a double helix (Fig. 11-1B). The nitrogenous bases of each chain are oriented toward the axis of the helixes, and the hydrogen bonds between the bases effectively hold the two chains together. Optimal accommodation requires that a purine be adjacent to a pyrimidine; more specifically, adenine is almost always hydrogen-bonded to thymine and guanine to cytosine. The two strands have opposite polarities; i.e., they are *antiparallel* to one another (see Fig. 11-1A).

A necessary consequence of the specificity of hydrogen bonding in DNA is that there must be equimolar amounts of adenine and thymine and

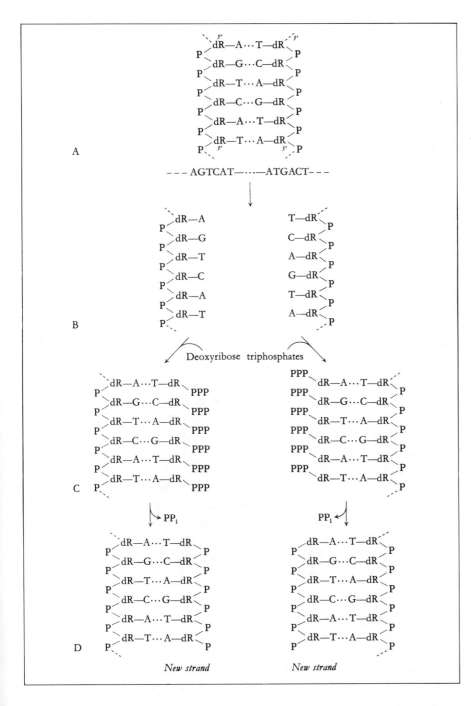

Fig. 11-2. Replication of DNA. A. Original double strand. B. Single-stranded section. C. Pairing of deoxynucleoside triphosphates with each single strand. D. Displacement of pyrophosphate and formation of new strands. (This representation is an overall view and does not show the stepwise mechanism.)

equimolar amounts of guanine and cytosine. This characteristic of DNA was actually known before its double-helical structure and hydrogen-bonding characteristics were elucidated.

As the carrier of genetic traits, DNA macromolecules must possess a self-contained system that enables them to provide each daughter cell with exact replicas of themselves in terms of base sequences. Replication involves the utilization of each single strand of DNA as a template in the synthesis of new complementary strands (Fig. 11-2). Sections of the double-stranded DNA may separate to produce single-stranded sections (Fig. 11-2B). The bases on each single strand can now form hydrogen bonds with free, complementary bases. The actual substrates in the biosynthesis of DNA are the deoxynucleoside 5'-triphosphates (Fig. 11-2C). Through the action of DNA polymerase, covalent bonding is produced between the 5'-phosphate groups and the 3'-hydroxyl groups of neighboring nucleotides, which provides a new polynucleotide strand and liberates inorganic pyrophosphate (Fig. 11-2D). It should be noted that the final polynucleotide products are identical with the original double strand of DNA.

The mechanism of DNA replication is extremely complex and numerous aspects remain to be clarified. Only certain details will be discussed in the subsequent paragraphs. Studies of the overall process revealed that after cell division, the double-stranded DNA in each new cell contains one strand from the parent cell and one newly synthesized strand (Fig. 11-3). The parental DNA in the diagram is shown as dark strands and the newly synthesized sections as light strands. This was verified experimentally by studies on the incorporation and distribution of the heavy isotope of nitrogen, ^{15}N, in the DNA of microorganisms. Cells with ^{15}N-labeled DNA were obtained by growing the microorganism in a medium containing ^{15}N-labeled precursors. The organisms were then transferred to a medium with the common isotope, ^{14}N, and the composition of the "heavy" DNA in the cells of the first and second generation was analyzed. The mechanism for the replication and transmission of DNA, confirmed by the findings in these studies (Fig. 11-3), is termed *semiconservative replication*.

11.2. Mechanisms of DNA Replication

A. REACTIONS INVOLVED IN REPLICATION AND THEIR PROBABLE SEQUENCE

Most of the original data on the details of DNA replication were obtained from studies of a prokaryotic organism, *Escherichia coli*. This is a comparatively simple cell that contains neither a nucleus nor mitochondria. Its only membranous structure is a cell membrane. It has only one chromosome, which is composed of a single molecule of DNA. The latter has a double-

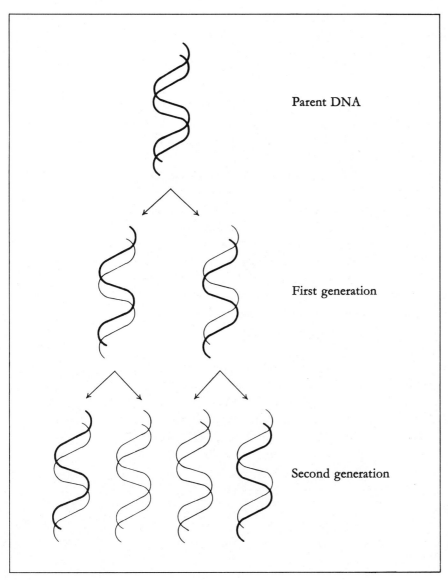

Fig. 11-3. *Semiconservative replication.*

helical structure, about four million nucleotide pairs, and a molecular weight of approximately 2.8×10^9. The double helix forms a closed loop or circle that is coiled and twisted around itself to form a dense nuclear area. The structure of this closely packed system is kept intact by the presence of a certain amount of RNA.

Some general aspects of the replication processes in *E. coli* have been found to apply to cells of eukaryotic or higher organisms; however, the greater complexity, as well as the differences in intracellular structures, of these organisms compared to prokaryotes clearly preclude complete extrapolation of the findings in *E. coli* to eukaryotic organisms. The information obtained from *E. coli* can only serve to provide hypotheses about what might occur in animal organisms.

Although all the details of DNA replication are not known even for prokaryotic cells, several generalizations can be made. First, both strands of the DNA double helix are utilized as templates in the formation of new DNA, so the original double-helical structure must unwind at the reaction sites. Second, an RNA polynucleotide primer must be synthesized prior to the formation of new DNA polynucleotide strands. Third, several types of DNA polymerases are involved in the replication process. Fourth, the synthesis of a complete strand is accomplished by forming smaller fragments that are subsequently linked to each other. Fifth, condensation of the nucleotides always occurs from the 3' side toward the 5' terminus of the template DNA; hence, the polynucleotide that is being synthesized starts from the 5' terminus and grows toward the 3' direction. On the basis of such information, it is possible to establish a plausible sequence of events for the replicative process. After this is described, some of the specific aspects of the reactions will be discussed.

At the site where the polynucleotide synthesis is initiated, a portion of the DNA chain interacts with DNA-directed RNA polymerase and an "unwinding protein" (Fig. 11-4). The latter uncoils a portion of the chain and produces a loop in the double helix (Fig. 11-4B). When the four nucleoside triphosphates (ATP, GTP, CTP, and UTP) are present, *RNA polymerase* catalyzes the synthesis of an RNA primer containing about 50 to 100 bases:

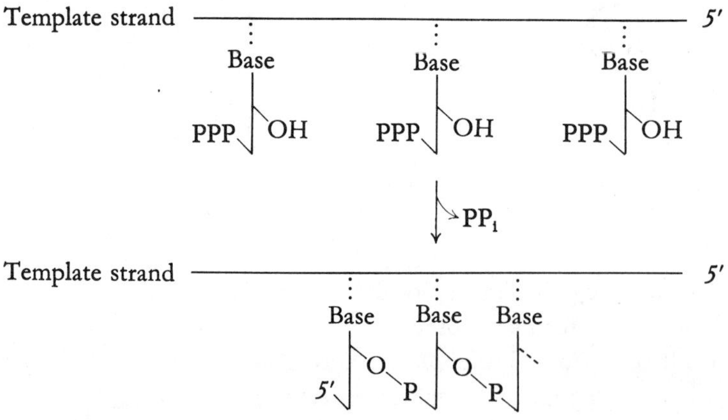

The sequence of bases in the polyribonucleotide is directed by that of the DNA; for example, for the sequence on the template 3'-ATGCAT-5', the

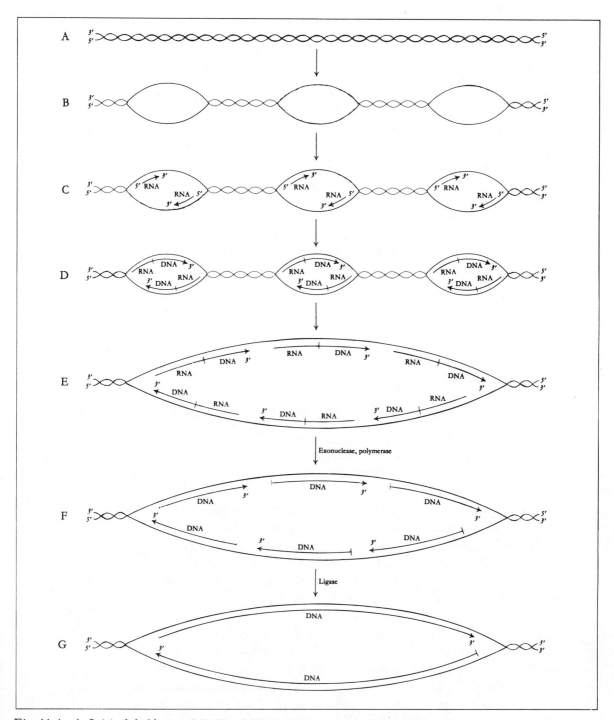

Fig. 11-4. A. Original double strand. B. Unwinding in various sections. C. Formation of RNA primer. D. Addition of DNA units to the primer and formation of Okazaki fragments. E. Large unwound section. F. Removal of primer sections and extension of DNA chains. G. Closure of gaps in DNA units.

order of the RNA primer will be 5′-UACGUA-3′-OH, since the base pairing between DNA and RNA is the same as that which occurs via hydrogen bonding between DNA strands (section 11.1), except that deoxyadenosine in DNA is paired with uridine in RNA (instead of with deoxythymidine, as within DNA). Since both strands of the DNA template are to be replicated and because this occurs in the 3′-to-5′ direction of the template, one RNA chain must grow in the direction away from the initiation site, while another grows toward that site (Fig. 11-4C).

At this stage, the *DNA-polymerase system* produces an elongation of the polynucleotide chain. The presence of the four deoxyribonucleoside triphosphates is required for the process. In *E. coli*, the particular enzyme for this reaction is termed *DNA polymerase III**. The initial step catalyzed by this enzyme requires ATP and a protein coenzyme termed *copolymerase III**. The deoxyribonucleoside triphosphates are arranged on the template DNA so they form complementary hydrogen-bonded base pairs:

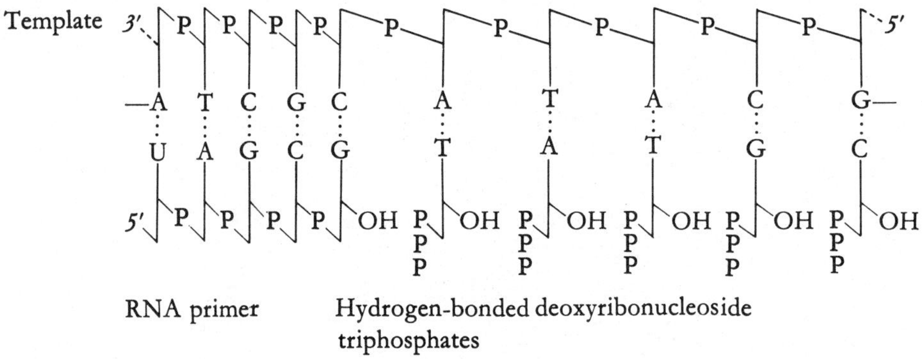

RNA primer   Hydrogen-bonded deoxyribonucleoside triphosphates

As a result of the action of DNA polymerase III*, about 500 to 1000 deoxyribonucleotide residues are condensed and linked to the RNA primer. Copolymerase III* is not required after the process is initiated, and it is dissociated from the enzyme-substrate system. A number of such RNA-DNA polynucleotides can form along the DNA template; these are called *Okazaki fragments*, after the discoverer who identified the fragments in *E. coli* (Fig. 11-4D,E).

The next step involves the removal of the RNA primer. In *E. coli*, this can be accomplished by *DNA polymerase I*. The latter has 5′-to-3′ *exonuclease activity*; i.e., it catalyzes the sequential hydrolysis of nucleotides from the 5′ terminus (an *endonuclease*, in contrast, can act on linkages anywhere in the chain, not only at a terminus). The products of this reaction are nucleoside 5′-phosphates. As the name implies, DNA polymerase I also brings about the polymerization of deoxynucleoside triphosphates, directing the synthesis from the 3' position of the previously synthesized DNA toward the 5' position of the DNA that lost the RNA primer (Fig. 11-4F).

The final reaction results in closing the gaps between the 3' terminus of one DNA fragment and the 5' terminus of an adjacent fragment. This reaction is catalyzed by *DNA ligase* (Fig. 11-4G). In *E. coli,* NAD^+ is utilized in this process, whereas in higher organisms, the energy-yielding cofactor is ATP.

B. ENZYMES INVOLVED IN DNA REPLICATION
1. *Initiation Point and Separation of Strands*

Initiation of the replication process in prokaryotic cells occurs at a specific site on the DNA macromolecule. So-called unwinding proteins cause the strands to separate so they can bind nucleoside triphosphates. Analogous processes are known to occur in eukaryotic cells. In the latter, initiation reactions occur simultaneously at many points, and the elongation steps proceed until newly formed polynucleotides approach each other and fuse.

2. *Polymerases*

DNA-directed RNA polymerase catalyzes the formation of the RNA primer from ribonucleoside 5'-triphosphates. Nucleoside triphosphates of the four bases—uracil, adenine, guanine, and cytosine—are required for RNA formation, and the base sequence in the resultant RNA is complementary with that on the DNA template. The direction of synthesis is from the 3' side toward the 5' terminus of the DNA template, while the newly synthesized RNA primer is built up from the 5' terminus toward the 3' terminus, or antiparallel to the DNA template. Subsequent growth of the polynucleotide involves the addition of deoxyribonucleoside 5'-phosphates, commencing with the formation of a $3' \rightarrow 5'$-phosphodiester through the 3'-OH terminus of the RNA primer.

Three DNA polymerases have been isolated from *E. coli,* which have been termed *DNA polymerase I, II, and III*. A more active form of DNA polymerase III has been designated as *polymerase III**. As mentioned already (section 11.2.A), this enzyme requires the cooperative action of a protein cofactor, *copolymerase III*,* for initiating the synthesis of the new DNA chain from the RNA primer.

DNA polymerase I has both synthetic activity and 5'-to-3' exonuclease activity. Its 5'-to-3' exonuclease activity allows it to catalyze the hydrolysis of the RNA section from Okazaki fragments. Polymerase I then effects the elongation of the new DNA from the 3' terminus of the Okazaki fragments following excision of the RNA portion. This polymerase has 3'-to-5' exonuclease activity as well, and it therefore can hydrolyze deoxynucleoside phosphates in the direction opposite to that of the replicative sequence. This

3'-to-5' exonuclease activity may serve to remove bases that were added erroneously, i.e., bases that are not complementary to the template DNA. In addition, this enzyme, through its 5'-to-3' exonuclease activity, can remove defective bases and clear the way for repair of damaged DNA.

DNA polymerase II also has 3'-to-5' exonuclease activity but not 5'-to-3' activity, and it may act in concert with polymerase I. However, the specific function of DNA polymerase II is not known.

DNA polymerase III* is the primary enzyme involved in the synthesis of new DNA on the RNA primer and the formation of Okazaki fragments. Copolymerase III* is required only for the initial linking of the first deoxyribonucleotide to the 3' terminus of the RNA primer.

In eukaryotic cells, there are three DNA polymerases, termed α, β, and γ. These differ from each other in their substrate specificities and probably in their function. A DNA polymerase is also involved in the synthesis of mitochondrial DNA. The α polymerase is the most abundant of the three enzymes. Although, as in prokaryotic cells, replication appears to involve an RNA primer, the actual requirement for this material in DNA synthesis in eukaryotes is not fully resolved.

3. DNA Ligase

This enzyme catalyzes the fusion of the 5'-phosphate from one DNA segment to the 3'-OH group of an adjacent segment (Fig. 11-4G), thus closing the gaps in the newly synthesized strands. The reaction requires the addition of an AMP unit to the terminal 5'-phosphate. In *E. coli*, NAD^+ functions as the donor of this unit (leaving nicotinamide mononucleotide, or nicotinamide ribose phosphate), whereas in higher animals ATP serves this purpose:

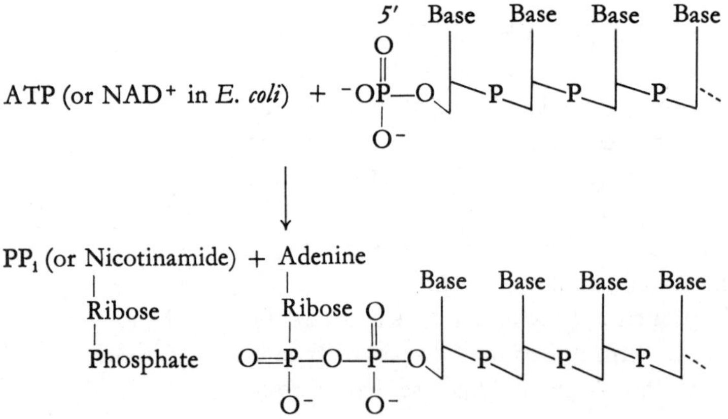

The ADP-deoxyribonucleotide fragment then condenses with another newly synthesized fragment, forming a phosphodiester bond and liberating the AMP unit:

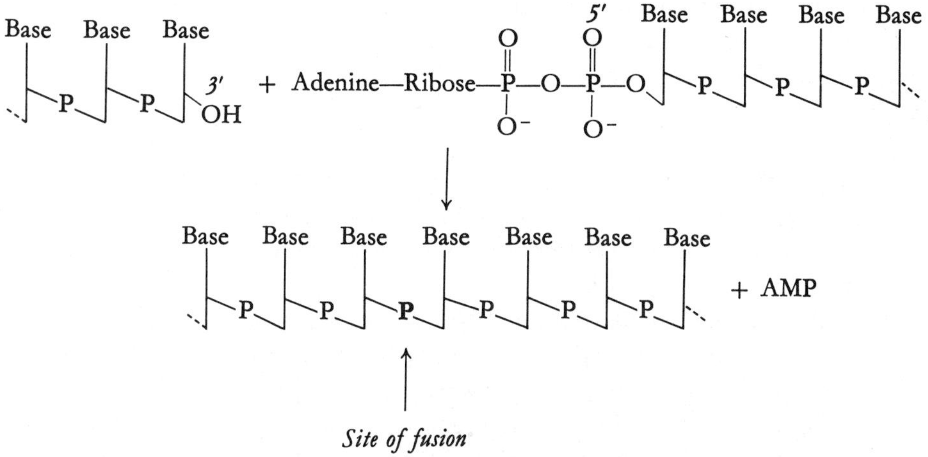

$$\text{Base-Base-Base-}\underset{\text{OH}}{\overset{3'}{|}} + \text{Adenine-Ribose-}\overset{O}{\underset{O^-}{\overset{\parallel}{P}}}-O-\overset{O}{\underset{O^-}{\overset{\parallel}{P}}}-O-\overset{5'}{|}\text{Base-Base-Base-Base}$$

$$\downarrow$$

$$\text{Base-Base-Base-Base-Base-Base-Base} + \text{AMP}$$

↑
Site of fusion

The ligase can also serve to repair nicks in DNA or to close the ring in circular DNA. In addition, this enzyme functions in joining DNA strands during genetic transformation and other processes in which "foreign" DNA is introduced into the chromosomes. It may also be involved in meiosis in eukaryotic cells.

11.3. Transcription

The sequence of reactions that results in the production of copies of preexisting DNA molecules is the underlying mechanism for the transmission of genetic data from parent cells to their progeny. Information for hereditary traits is carried in the base sequences of DNA, and the duplication of these nucleic acids serves to perpetuate genetic characteristics. The data stored in the DNA are utilized for programming the synthesis of proteins with specific amino-acid sequences. The inherent characteristics of organisms are determined by the nature and specificity of their proteins. Numerous proteins function as enzymes that control metabolism and biosynthesis of fundamental macromolecular components. Many proteins also serve as the structural units of cellular architecture. Thus, by controlling the primary structure of the proteins, the nucleic acids function to specify the essential characteristics of the organism.

The information carried by DNA is not transmitted to the proteins directly. The data stored in DNA are first incorporated into the structure of RNA. The RNA then functions in the synthesis and specification of the amino-acid sequence of proteins. DNA is utilized as a template in the biosynthesis of RNA in a manner analogous to that described for DNA replication. In the case of RNA synthesis, base pairing via hydrogen bonding occurs between adenine and uracil, and between guanine and cytosine. Thus, during RNA biosynthesis, the adenine, guanine, thymine, and cytosine of DNA will pair with the uracil, cytosine, adenine, and guanine, respectively, of the newly forming RNA. This specification of the sequence of RNA by that of DNA is called *transcription*.

The fundamental principles of transcription were learned from experiments with prokaryotic cells (*E. coli*), and it should be realized that the mechanism in higher organisms is much more complex. Transcription in *E. coli*, however, can serve to provide a basic picture of the process.

The substrates for RNA synthesis are the ribonucleoside triphosphates: ATP, GTP, CTP, and UTP. In the presence of these four triphosphates and a DNA template, the enzyme *DNA-dependent RNA polymerase* catalyzes the formation of $3' \rightarrow 5'$-phosphodiester linkages between adjacent ribonucleotides with the concurrent release of inorganic pyrophosphate (Fig. 11-5). New RNA chains are synthesized from their 5' end in the direction of the 3' terminus. In many RNA molecules, the 5' end is either ATP or GTP. As in the case of DNA replication, the growth of the macromolecule proceeds from the 3' to the 5' direction of the DNA template. The newly formed RNA is thus antiparallel to its template.

In contrast with DNA biosynthesis, the formation of RNA does not require a primer. Another difference is that only one of the complementary DNA strands is involved in directing the base sequence of RNA in vivo. Thus, although double-stranded DNA is required for the production of RNA, one of these strands does not function directly to transmit information. Unlike DNA, RNA is single-stranded and therefore constitutes a complete entity. Any RNA transcribed from the non-functional DNA strand would have a different structure and no apparent function. The synthesis of RNA must therefore be regulated with respect to the site of its initiation on the DNA template. Moreover, since only portions of the DNA are transcribed, the chain length must be determined by a control system that recognizes a termination site. Certain aspects of these control mechanisms are provided by the structure and activity of the enzyme, *RNA polymerase*.

The RNA polymerase of *E. coli* is a complex protein composed of so-called α, β, β', and σ subunits. The molecular weight of the total complex is about 500,000. A protein that binds to RNA polymerase is designated σ (Greek letter *sigma*). This protein is composed of four subunits and has a molecular weight of about 200,000. The concerted, regulated activities of the subunits of RNA polymerase and protein σ are essential for the synthesis of RNA.

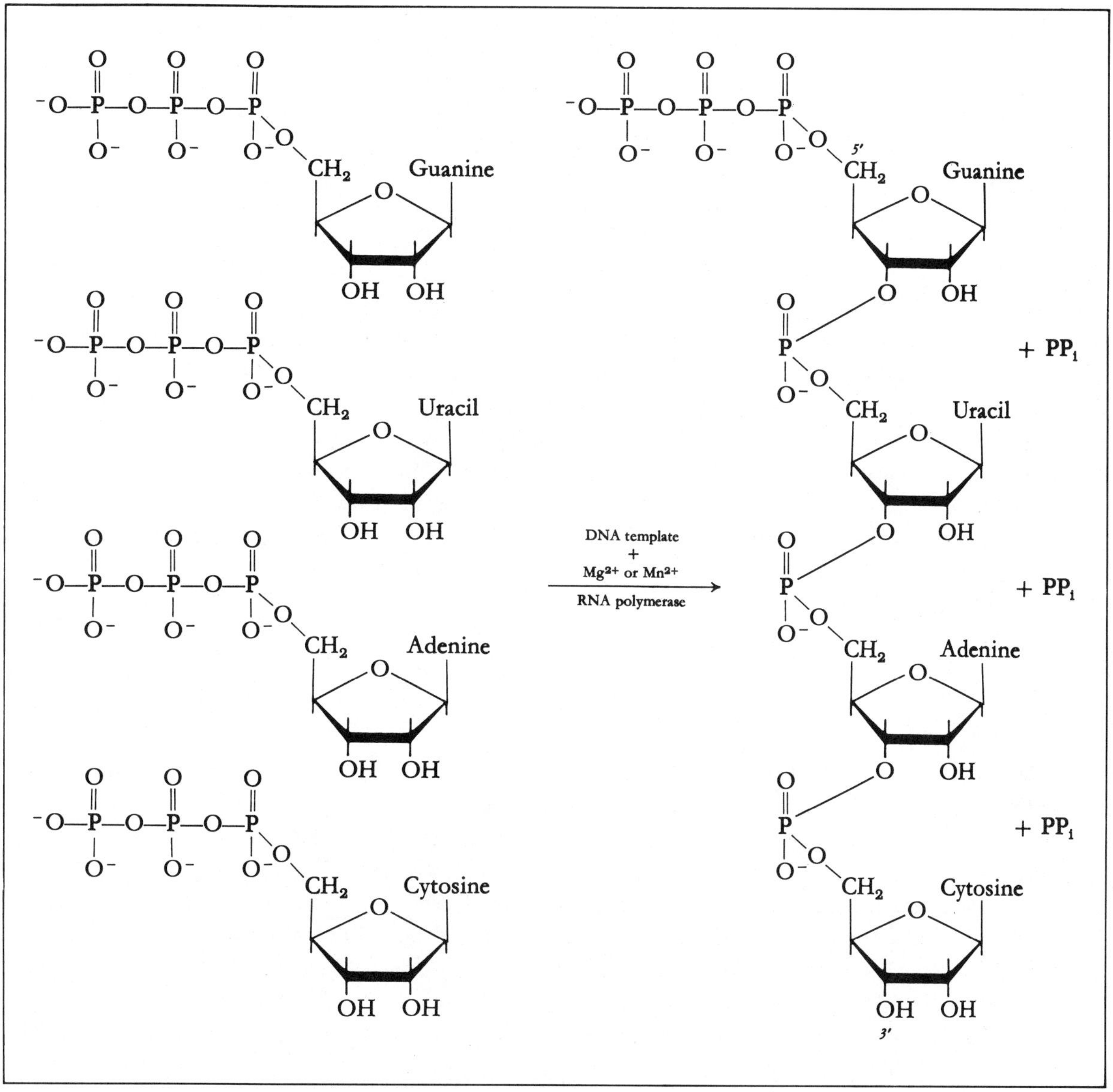

Fig. 11-5. *Formation of polyribonucleotide (RNA) from ribonucleoside triphosphates.*

Transcription is initiated on the DNA duplex in specific locations termed *promoter sites*. The β and β' subunits of RNA polymerase are apparently involved in the binding process. The affinity between enzyme and DNA is greatest at the promoter sites. When RNA polymerase binds to the DNA, one consequence is a partial unwinding of the double helix. The σ subunit in RNA polymerase functions in the initiation of transcription by specifying the primary polymerization at the promoter site and thus enhancing the rate of the reaction. A purine nucleoside triphosphate (ATP or GTP) binds to the correct strand of DNA and serves as the starting base in the formation of RNA. The other nucleoside triphosphates interact with the bases of DNA by hydrogen bonding, and RNA polymerase catalyzes the sequential formation of 3′,5′-phosphodiester bonds. The 5′ terminus of the newly formed RNA therefore contains a 5′ triphosphate, either ATP or GTP (see Fig. 11-5).

Following the initial polymerization steps, the σ subunit dissociates from the rest of the RNA polymerase as RNA polymerase moves along from the 3′-to-5′ direction of the DNA molecule. The portion of DNA that has been transcribed re-acquires its helical structure, while additional DNA in the direction of elongation is unwound.

The termination of the growth of the RNA molecule is signaled by specific sites or sequences of the DNA template. Additional termination sites are sensed by the ρ protein, which binds to RNA polymerase during the growth of the RNA and causes termination of synthesis when certain points on the template are reached. Rho protein is apparently important in producing RNA with characteristic molecular weights.

The probable sequence of events in the synthesis of RNA is shown in Figure 11-6.

11.4. Differentiation of RNA

DNA serves as a template for the synthesis of the three types of RNA: *messenger RNA* (mRNA), *transfer RNA* (tRNA), and *ribosomal RNA* (rRNA). Most of the specific nucleotide sequences and structural characteristics of the RNAs are established during the transcription from DNA. However, production of the completely functional RNA macromolecule requires additional modification of the released RNA. These alterations, called *posttranscriptional processing*, include (1) hydrolytic cleavage of nucleotide units or "trimming" of the chains, (2) addition of nucleotide sequences to the terminal groups, and (3) methylation and other modification of the nucleoside units.

A. MESSENGER RNA

The nuclear precursor of mRNA is called *heterogeneous nuclear RNA* (hnRNA). Although this material does not consist of a single component, it is known to contain base sequences similar to those of mRNA. Some of the terminal units of the precursor are cleaved, and about 200 adenylate units are added to the 3′ end. The 5′ terminus of mRNA is also modified by the addition of a GTP unit to its 5′-hydroxyl group. This site is thus characterized by a 5′-ppp (5′)G segment. In addition, mRNA is altered by methylation of the

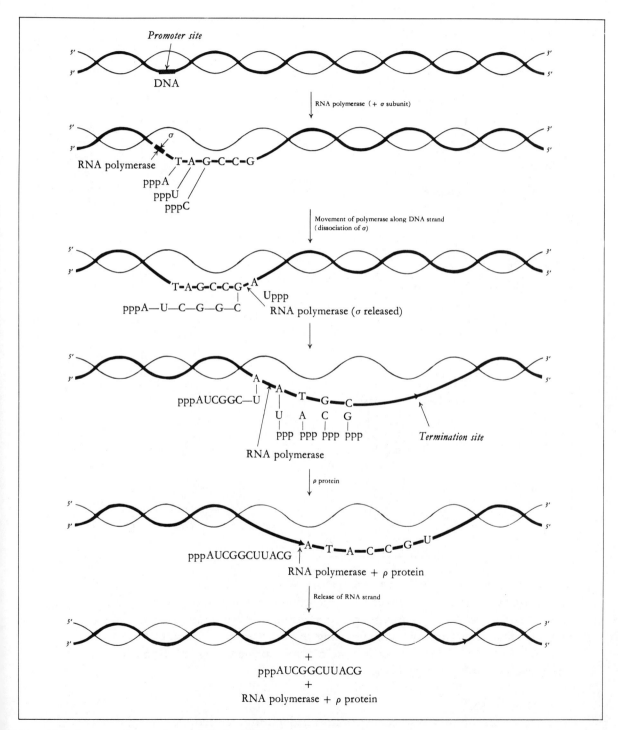

Fig. 11-6. *Sequence of events in transcription (RNA synthesis)*.

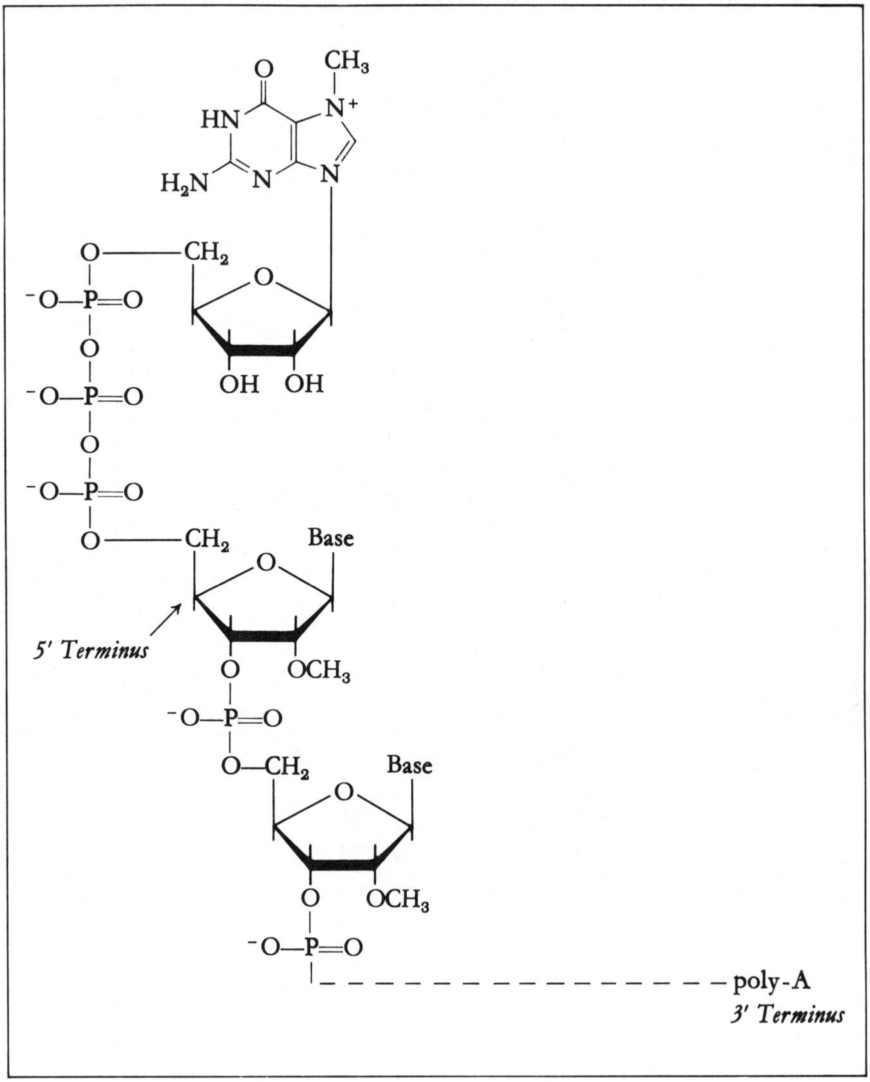

Fig. 11-7. Characteristic terminal residues on messenger RNA. A 7-methylguanosine group is attached via a 5'-triphosphate bridge to the 5' position of the ribose unit of the 5'-terminal nucleoside. The latter is methylated on the 2'-hydroxyl groups. The 3' terminus consists of about 200 adenylate residues (*poly-A*).

guanylate as well as several other terminal nucleotides. The methyl group in the former is introduced at position 7. The complete modified segment is referred to as the "cap." Through posttranscriptional modifications, the resultant functional mRNAs contain long poly-A segments at the 3' end and 7-methylguanylate units in 5'-to-5' triphosphate linkage at the 5' terminus of the mRNA (Fig. 11-7).

B. TRANSFER RNA

The precursors of tRNAs that are synthesized during transcription are larger than the final functional products. As a result of the posttranscriptional

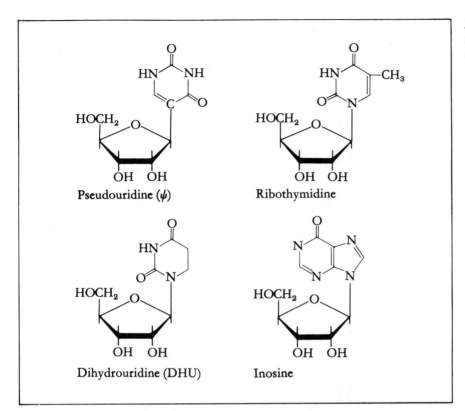

Fig. 11-8. Structures of uncommon nucleosides found in tRNA.

activity of various nucleases, about 20 to 30 nucleotides are removed. In certain instances, a CCA sequence is added to the 3′ terminus as part of post-transcriptional processing. All known tRNAs terminate with this sequence; however, in a number of cases, this trinucleotide unit may be introduced during transcription. Other modification reactions include methylation and alterations that produce nucleoside units such as dihydrouridine (DHU) and pseudouridine (ψ) (Fig. 11-8).

Cells contain over 20 tRNAs that differ in terms of base sequence; these correspond to each naturally occurring amino acid, since each amino acid requires at least one characteristic tRNA to transport it during protein synthesis. Actually, certain amino acids have several corresponding tRNAs, but a given tRNA is specific for only one amino acid. Although the tRNA for each amino acid has a characteristic base sequence and fine structure, all of them have certain common features. The tRNAs consist of less than 100 nucleotides. Hence, their molecular weight is about 25,000, which is low compared with other nucleic acids. Another common characteristic is the presence of a number of nucleosides other than adenosine, guanosine, cytidine, and uridine, which include the methyl derivatives of the usual

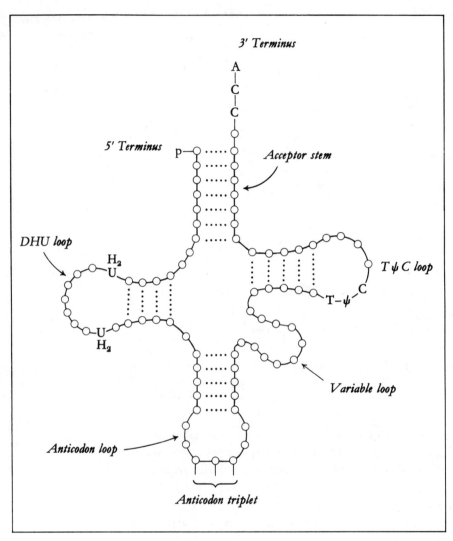

Fig. 11-9. Representation of tRNA showing some of the common structural features.

purines and pyrimidines as well as the rare nucleosides inosine, dihydrouridine (DHU), pseudouridine (ψ), and ribothymidine. The 3' terminus of tRNA molecules always consists of the sequence purine nucleoside-pCpCpA (Fig. 11-9). The nucleoside at the 5' end of tRNA is linked to a terminal phosphate group.

In spite of the fact that tRNAs are single-stranded polynucleotides, a large portion of each molecule has a helical, base-paired structure. Thus, sections of the tRNA are coiled in a specific manner to allow for intrachain bonding of complementary bases. All tRNAs have a common conformation that can be represented in a two-dimensional projection as a cloverleaf structure (Fig. 11-9). The sections that do not contain hydrogen-bonded

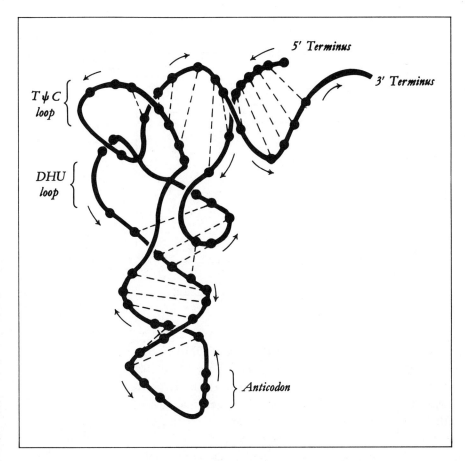

Fig. 11-10. Three-dimensional structure of tRNA. (Arrows show the 5'-to-3' direction.) Adapted from "The Three-dimensional Structure of Transfer RNA" by Alexander Rich and Sung Hou Kim. Copyright © January 1978 by Scientific American Inc. All rights reserved.

bases appear as loops in the pattern. Some of these loops are designated according to the characteristic components that they contain. Thus, the DHU loop contains dihydrouridine and the TψC loop has a ribothymidine-pseudouridine-cytosine sequence. The *anticodon* loop, which also contains a specific three-base sequence, is required for coding protein synthesis (section 11.11). Also, tRNAs may have an additional small loop with a variable number of residues that is therefore called the *variable* loop. The 3'-CCA terminus is the site of linkage between the tRNA and its specific amino acid (section 11.11). At the 5' end, the nucleoside (usually guanosine) is phosphorylated. The entire structure consisting of the base-paired 3' and 5' terminals is called the *acceptor stem* (Fig. 11-9).

Studies of the three-dimensional structure of tRNA employing x-ray crystallography have revealed that it has an L-shaped structure in which one leg of the L contains the anticodon and the other, the 3'-CCA terminus (Fig. 11-10).

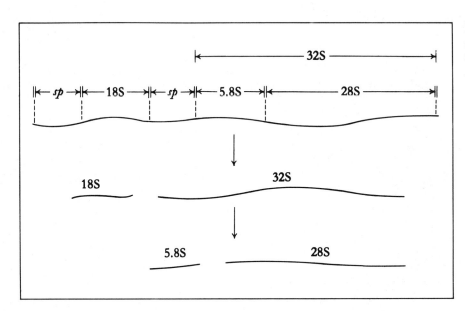

Fig. 11-11. Relationship of rRNA units to precursor (45S) rRNA (sp = spacer units).

C. RIBOSOMAL RNA

In animal cells, rRNA is transcribed from a section, or *cistron*, of DNA in the nucleolus. The precursor of rRNA has a molecular weight of about 4.1×10^6 and a sedimentation rate of 45S. Like the other RNAs, this material undergoes posttranscriptional processing that includes hydrolytic cleavages, nucleoside modification, and methylation. The nucleolus also appears to be the site where rRNA combines with proteins to produce the ribosomes.

The nascent or precursor rRNA consists of about 13000 nucleotides. This 45S material contains sections that ultimately give rise to functional rRNA as well as sequences that are not incorporated into the ribosomes. The latter sections are called *spacer* units, since they serve to separate particular portions in the precursor. The 45S precursor is sequentially cleaved to yield 18S, 5.8S, and 28S rRNAs in addition to spacer fragments (Fig. 11-11). Another RNA unit, termed *5S rRNA*, is not a component of the 45S precursor, but is present in the ribosomes.

A considerable number of nucleoside units on rRNA are methylated as a result of posttranscriptional processing reactions. Almost all the methyl groups that are inserted are introduced on 2′-hydroxyl groups (Fig. 11-12). Another posttranscriptional reaction is the conversion of certain uridine units to pseudouridine (ψ). The interaction of the RNA with protein and the excision of spacer units (see Fig. 11-12) also occur in the nucleolus. The functional ribosomal components of animal cells—the ribosomes—are 80S units that consist of both protein and RNA. The 80S eukaryotic ribosome is composed of two subunits with sedimentation rates of 60S and 40S. Each

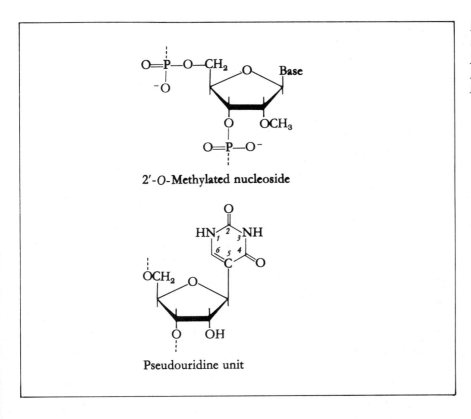

Fig. 11-12. Modified nucleoside units in rRNA produced in posttranscriptional processing.

of these subunits contains characteristic rRNA components (Table 11-1). The properties of ribosomes differ among species; thus, the values indicated in Table 11-1 are averages or approximations.

11.5. Translation and the Genetic Code

The data encoded in DNA are utilized in directing protein structure by means of the DNA functioning as a template for transcribing mRNA with specific base sequences. The mechanism for transcription involves complementary base-pairing between a strand of DNA and the mRNA that is being produced. The mRNA then moves out of the nucleus and combines with the ribosomes. The information that is stored as the sequence of bases in mRNA provides the data for the synthesis of a protein with a specific amino-acid sequence. Unlike transcription, this involves the transfer of information from a system with four components (two purine and two pyrimidine bases) to one of twenty different components (amino acids). Thus, more information than what could be transmitted by a single base is needed to serve as a signal or code for a specific amino acid. In other words, the information carried by the four-unit nucleotide system must be translated

Table 11-1. Properties of Eukaryotic and Prokaryotic Ribosomes

Eukaryotes

Sedimentation coefficient of ribosome		80S		
Sedimentation coefficient of subunits		60S		40S
Approximate number of protein units		40		30
Sedimentation coefficient of RNA subunits	28S	5.8S	5S	18S
Nucleotides per subunit	5000	158	120	2000
Approximate molecular weight of subunits	1.4×10^6	5×10^4	4×10^4	7×10
Methyl groups per subunit	68	2	0	45

Prokaryotes

Sedimentation coefficient of ribosome	70S	
Sedimentation coefficient of subunits	50S	30S
Approximate number of protein units	34	21
Sedimentation coefficient of RNA subunits	23S	5S	16S
Nucleotides per subunit	2700	120	1600
Approximate molecular weight of subunits	1.1×10^6	4×10^4	5.5×10^5
Methyl groups per subunit	16		14

in some way to the twenty-unit amino-acid system. The process by which the base sequence in mRNA codes for the amino-acid sequence in proteins is therefore called *translation*.

The signal that determines the position of each amino acid in a protein is a specific sequence of three nucleoside bases. For example, the sequence AGU (written in the 5'-to-3' direction) codes for the introduction of serine; similarly, GGU signals for glycine. Calculation of the number of *base triplets*, or *codons*, that can be generated by permutations of these four bases reveals that there are 64 possible combinations. Since only 20 are required for determining the sequences of the amino acids, this allows for what is termed *degeneracy*. This means that each amino acid is coded by more than one base sequence. For example, both UUU and UUC code for phenylalanine. Similarly, glycine is coded by the base triplets GGC, GGA, and GGG. Only 61 of the 64 possible triplets code for amino-acid incorporation (Table 11-2).

The triplet sequences UAA, UAG, and UGA function as signals for *termination*; that is, they indicate a stopping point in the synthesis of a given protein. In addition, since the code is not read from the end of the mRNA, another codon is required to identify the *initiation point* for the synthesis of the protein. The sequence AUG serves as the codon for the initiation of

Table 11-2. The Genetic Code—Base Triplet Sequences in Codons for Amino Acids*

5' Base	Center Base				3' Base
	U	C	A	G	
U	Phe	Ser	Tyr	Cys	U
U	Phe	Ser	Tyr	Cys	C
U	Leu	Ser	*Term*	*Term*	A
U	Leu	Ser	*Term*	Trp	G
C	Leu	Pro	His	Arg	U
C	Leu	Pro	His	Arg	C
C	Leu	Pro	Gln	Arg	A
C	Leu	Pro	Gln	Arg	G
A	Ile	Thr	Asn	Ser	U
A	Ile	Thr	Asn	Ser	C
A	Ile	Thr	Lys	Arg	A
A	Met†	Thr	Lys	Arg	G
G	Val	Ala	Asp	Gly	U
G	Val	Ala	Asp	Gly	C
G	Val	Ala	Glu	Gly	A
G	Val	Ala	Glu	Gly	G

* Each base triplet (read from left to right) constitutes the codon that is specific for the amino acid listed (e.g., the triplet UUU codes for Phe, or phenylalanine). *Term* = termination codon (see text).
† The base triplet AUG also serves as initiation codon (see text).

protein synthesis with methionine (the same codon also determines the position of a methionine residue in a peptide).

The genetic code for protein synthesis is universal; i.e., it is the same in all organisms (Table 11-2). In indicating the sequence of a codon, it is understood that it is read from left to right in the 5'-to-3' direction of mRNA. In Table 11-2, for example, the sequences UUU and UUC code for phenylalanine. Leucine is coded by six codons: UUA, UUG, CUU, CUC, CUA, and CUG.

When several base triplets code for the same amino acid (degeneracy), they usually differ only with respect to the third base. In the codons for valine (GUU, GUC, GUA, and GUG) or alanine (GCU, GCC, GCA, and GCG), for example, GU and GC are specific for valine and alanine, respectively; that is, the determination of the position of these amino acids in protein synthesis is not affected by the nature of the third (3') base in the codon.

The codons are read without interruption; i.e., there are no punctuation marks except for the termination signals UAA, UAG, and UGA. Thus, the sequence UCACUACACGAUUAUUAA codes for a serylleucylhistidyl-aspartyltyrosine unit.

11.6. Protein Synthesis

The sequence of reactions in which amino acids condense to produce a protein macromolecule involves (1) activation of each amino acid and combination with its specific transfer RNA (tRNA), (2) attachment of the tRNA-amino acid to messenger RNA (mRNA) on the ribosomes, (3) formation of peptide bonds between the amino acids, and (4) release of the formed protein from the ribosomes. Each of these steps is considered separately.

A. SYNTHESIS AND PROPERTIES OF AMINOACYL-tRNA

Each amino acid that is utilized as a component of proteins has an activating enzyme system, termed *aminoacyl-tRNA synthetase*. The reaction catalyzed by this enzyme is a two-step process. The first step—the activation reaction—is the interaction between the amino acid and ATP to yield an enzyme-bound aminoacyl adenylate and inorganic pyrophosphate:

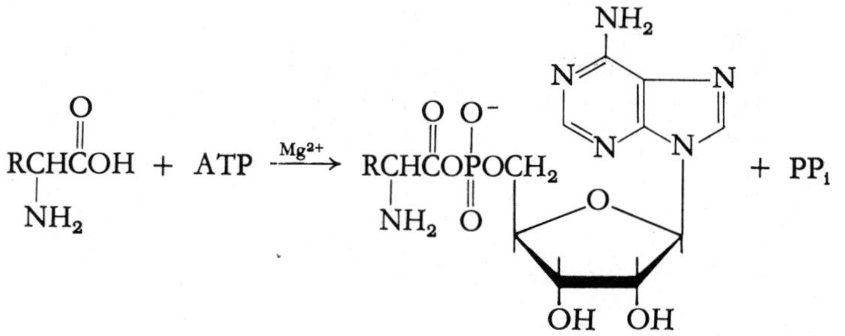

$$\text{RCHCOH} + \text{ATP} \xrightarrow{\text{Mg}^{2+}} \text{RCHCOPOCH}_2\text{-Adenosine} + \text{PP}_i$$

Aminoacyl adenylate

In the second step, the aminoacyl unit is transferred from adenylate to tRNA:

$$\text{RCHC—AMP} + \text{tRNA} \longrightarrow \text{RCHC—tRNA} + \text{AMP}$$

Aminoacyl adenylate   Aminoacyl-tRNA

This process is rendered irreversible because of the hydrolysis of the pyrophosphate to inorganic phosphate by the enzyme pyrophosphatase.

Aminoacyl-tRNA synthetase is specific for both the particular amino acid and its tRNA. When aminoacyl-tRNA is formed, the amino acid is esterified with the 3′-hydroxyl group of the 3′-terminal adenosine unit (Fig. 11-13).

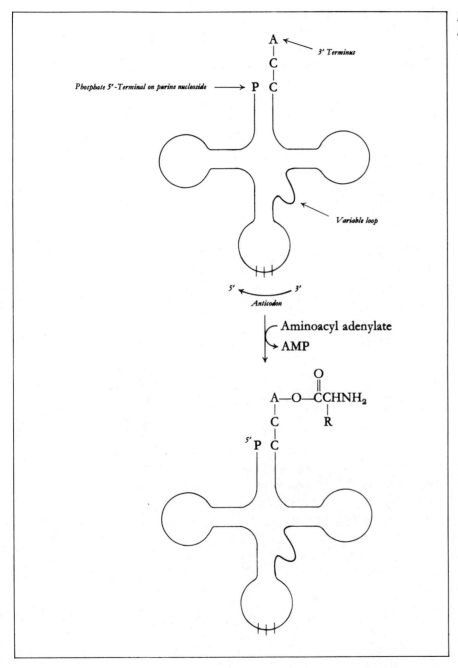

Fig. 11-13. *Formation of aminoacyl-tRNA.*

B. BASE-PAIRING BETWEEN tRNA AND mRNA

Aminoacyl-tRNA binds to mRNA by base pairing between the anticodon site of the former and the codon site of the latter. If, for example, the codon on the mRNA is UAC (reading in the 5'-to-3' direction), then the tRNA for the corresponding amino acid (tyrosine) will contain GUA as a possible anticodon:

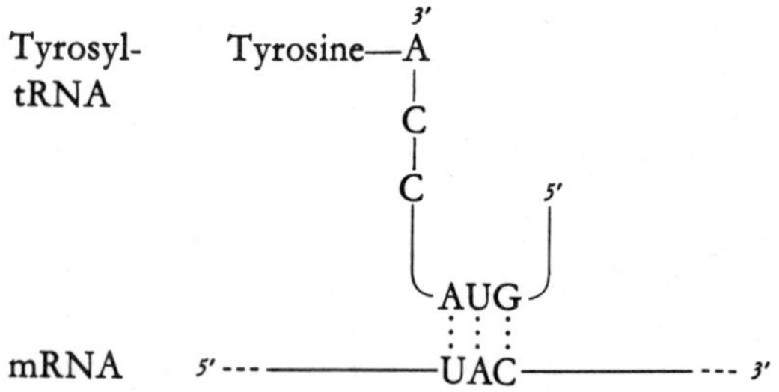

```
Tyrosyl-      Tyrosine—A  3'
tRNA                    |
                        C
                        |
                        C
                         \        5'
                          \      /
                           \    /
                            AUG
                            : : :
mRNA      5' ---————————UAC————————--- 3'
```

Similarly, one possible codon on mRNA for alanine is 5'-GCA-3'; it will pair with the anticodon 5'-UGC-3' on the tRNA:

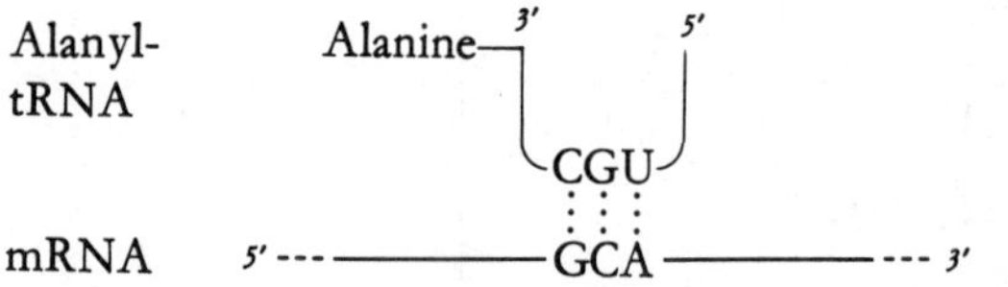

```
Alanyl-     Alanine—  3'     5'
tRNA               \        /
                    \      /
                     CGU
                     : : :
mRNA      5' ---————————GCA————————--- 3'
```

To some extent, the requirements for pairing on the third base of the mRNA codon are not as rigid as those for the first two bases. This is consistent with the variability of the third base with respect to the genetic code because of degeneracy. For example, the *first* base of the anticodon (reading, as usual, in the 5'-to-3' direction), which pairs with the *third* base on the mRNA codon, may not be the usual complementary base. Thus, G may function as the anticodon for either C or U on the 3' end of a codon. In some instances, inosine (I) appears as the first anticodon base, which can pair with A, C, or U when any of these appears as the third base of the messenger triplet. This relative lack of specificity is sometimes referred to as *wobble* in base pairing:

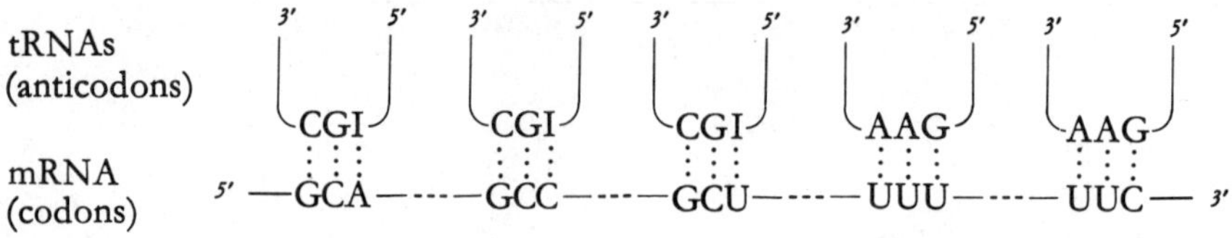

```
tRNAs          3'  5'   3'  5'   3'  5'   3'  5'   3'  5'
(anticodons)   \__/    \__/    \__/    \__/    \__/
                CGI     CGI     CGI     AAG     AAG
                : : :   : : :   : : :   : : :   : : :
mRNA      5' —GCA—----—GCC—----—GCU—----—UUU—----—UUC— 3'
(codons)
```

It should be noted that in no case does wobble in base pairing result in the introduction of a wrong amino acid into the protein whose synthesis is being determined.

An important aspect of the transport system for amino acids in protein synthesis is that the amino acid of the aminoacyl-tRNA is not directly involved in the binding with mRNA. Rather, the tRNA itself serves to interpret the code on the mRNA and correlates this with the corresponding amino acid. Each tRNA has three *recognition sites:* one for its specific amino acid, one for the aminoacyl-tRNA synthetase, and one (the anticodon) for the code site on the mRNA. The linking of the tRNA with its specific amino acid is therefore a critical factor in the incorporation of the amino acid into a particular protein.

C. INITIATION OF PROTEIN SYNTHESIS

Because the original studies on protein synthesis were performed with bacterial cells, details of the process are better understood in the case of prokaryotic organisms than in mammals. However, extensive investigations on the mechanisms of translation in mammalian cells have shown that the overall process is similar in all organisms. For example, protein synthesis always occurs on the ribosomes and is coded by the same sequences of bases on mRNA (as mentioned previously, the genetic code is identical for all types of cells). Other common features include the activation of the amino acids, their combination with tRNA, and the function of tRNA in positioning the amino acid at the correct site.

Prokaryotes differ from eukaryotes with respect to specific structural aspects of their mRNA as well as in the physical characteristics of the respective ribosomes. A number of differences also exist in certain details of the actual process in protein synthesis, for example, in specific steps in the initiation and elongation of polypeptide chains. The mechanisms and reaction sequences, described in the following discussion, are those that have been found to apply to mammalian cells. Some of the steps in which prokaryotes differ markedly from higher animals will be pointed out in subsequent paragraphs.

During protein synthesis, mRNA is translated from the 5'-to-3' direction. The first amino acids introduced are the ones that will ultimately be near the amino terminus of the protein. Thus, growth of the polypeptide chain progresses toward the carboxyl-terminal amino acid. All the reactions occur on the mRNA, which is bound in a complex system with the ribosome.

The first process in protein synthesis, termed *initiation,* involves the placement of the initial aminoacyl-tRNA on its specific site on the mRNA; the mRNA in turn is associated with a ribosomal unit which, in eukaryotes, has

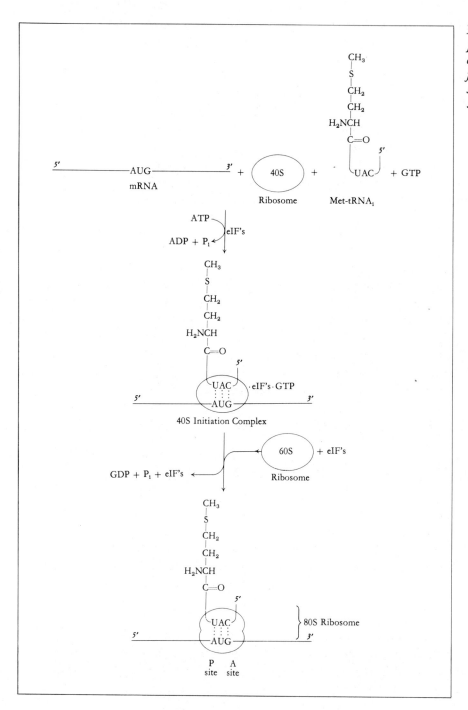

Fig. 11-14. Initiation of protein synthesis in eukaryot cells (eIF's = initiation factors; P site = peptidyl site; A site = aminoacyl site).

a sedimentation value of 40S. Since protein synthesis does not begin at the terminus of the mRNA polynucleotide chain, there must be a definite signal on the mRNA for the position of the initial amino acid. This is provided by an AUG triplet at a specific site on the messenger. This codon signals for the initiation of synthesis to begin with an initiator *methionyl-tRNA*. It should be noted that AUG when present at other points on the mRNA also codes for the incorporation of non-initiating methionyl-tRNA. However, the tRNA for initiation with methionine (Met-tRNA$_i$) differs from that of the non-initiating methionine-related tRNA (Met-tRNA$_m$). Although the anticodons for both are identical (CAU), other features in the structure (to be discussed) serve to identify the initiating methionyl-tRNA.

Methionine is activated and converted to aminoacyl-tRNA$_i$ as described in section 11.6.A. Aminoacyl-tRNA reacts with GTP and a protein called eukaryotic *initiation factor 2* (*eIF-2*) to form a ternary complex that combines with a 40S ribosome. The 40S ribosome that functions in this process is bound to an initiation factor designated as *eIF-3*. A specific initiation factor (*eIF-4C*) is also required for the combination of the 40S ribosome with methionyl-tRNA$_i$. The complex, composed of methionyl-tRNA$_i$, 40S ribosome, GTP, and initiation factors, then combines with mRNA to form a product termed the *40S initiation complex* (Fig. 11-14). This reaction utilizes additional initiation factors and consumes 1 mole of ATP. The following step involves the binding of a 60S ribosome with the 40S ribosome on the initiation complex, release of the initiation factors, and hydrolysis of GTP to GDP and inorganic phosphate. A specific initiation factor is also required for this step. The total ribosome on the mRNA now has a sedimentation coefficient of 80S (Fig. 11-14). The initiation factors (Table 11-3) function to select the tRNA for binding to specific sites on the ribosomes, to regulate the combination of the ribosomes with mRNA, and probably to control the sequence of the individual steps.

At this stage, the mRNA may be visualized as partially sandwiched within the 80S ribosome. The latter can accommodate about eight nucleoside units of mRNA or at least two nucleoside triplets. Each codon triplet is held within a specific site of the ribosome. One of these sites is called the *P site* (peptidyl site) and the other is designated as the *A site* (aminoacyl site). The reason for this terminology will be pointed out in the discussion of polypeptide elongation (section 11.6.D). For the present, it should be noted that the initiating codon (AUG) and the complementary anticodon (CAU) of the methionyl-tRNA are situated at the P site of the ribosome.

In *prokaryotes*, the initiating amino acid is *N*-formylmethionine, which is produced by the transfer of a single-carbon unit from tetrahydrofolate to methionine:

Table 11-3. Protein Factors in Polypeptide Synthesis

	Prokaryotic Cells	Eukaryotic Cells
Initiation factors*	IF-1 IF-2 IF-3	eIF-1 eIF-2 eIF-3 eIF-4A, 4B, 4C, 4D eIF-5
Elongation factors†	EF-T, EF-T_u EF-G EF-T_s	EF-1 EF-2
Releasing factors‡	RF-1 RF-2 RF-3	RF

* Described in section 11.6.C. (The functions of the eukaryotic initiation factors do not always correspond with those of the prokaryotic factors.)
† Section 11.6.D.
‡ Section 11.6.E.

$$CH_3SCH_2CH_2\overset{\overset{\displaystyle NHCHO}{|}}{C}HCOOH$$

N-Formylmethionine

This amino-acid derivative is activated and reacts to form its aminoacyl-tRNA. The anticodon on the N-formylmethionyl-tRNA is the same as that for methionyl-tRNA, that is, 5'-CAU-3'. Another feature unique to prokaryotes is that the ribosomal units involved in protein synthesis have sedimentation coefficients of 30S and 50S; the combined ribosome is a 70S unit. The 30S ribosome first combines with *prokaryotic initiation factor 3* (IF-3) and the mRNA at the initiation site. Then, the mRNA-30S ribosome-IF-3 complex interacts with N-formylmethionyl-tRNA, prokaryotic IF-2, and GTP. The next steps are (1) the release of IF-3, (2) the binding of a 50S ribosome to the 30S ribosome, and finally (3) the release of GDP, P_i, and IF-2. Prokaryotes have an initiation factor, *IF-1*, which appears to function in recycling the other initiation factors.

The initiation factors, as well as various other soluble protein factors, appear to have analogous functions in both eukaryotes and prokaryotes. Eukaryotic initiation factors are distinguished from prokaryotic factors by prefixing the letter *e* to the acronyms (e.g., initiation factor 2 for prokaryotes is symbolized as IF-2, whereas the factor in eukaryotes is eIF-2). The acronyms for some of the protein factors that have analogous functions (initiation or elongation) in prokaryotes and eukaryotes are listed in Table 11-3.

D. ELONGATION OF THE POLYPEPTIDE CHAIN

The sequence of reactions after initiation is called *elongation*. The first step in elongation is the binding of a second aminoacyl-tRNA to the next codon on the mRNA. This requires a protein called *elongation factor 1* (EF-1), and it involves the hydrolysis of GTP to GDP and inorganic phosphate. The new aminoacyl-tRNA is placed on the A site of the ribosome (Fig. 11-15).

The next step involves the transfer of methionine from its corresponding tRNA, which is hydrogen-bonded to the P site, to the amino group of the newly added, adjacent amino acid. This reaction forms a peptide bond between the carboxyl group of methionine and the amino group of the new amino acid; it is catalyzed by *peptidyl transferase*. Formation of the peptide bond thus results in a dipeptide that is linked to a single tRNA on the A site of the ribosome (Fig. 11-15).

The reaction that follows utilizes another protein, termed *elongation factor 2* (EF-2). The protein combines with GTP and causes a shift of the dipeptidyl-tRNA and the respective mRNA codon from the A site to the P site of the ribosome. Energy for this translocation is provided by the hydrolysis of the GTP to GDP and inorganic phosphate. The amino acid that is coded by the next three bases of the mRNA can then be inserted, via its specific tRNA, into the vacated A site on the ribosome. Sequential repetition of the reactions catalyzed by EF-1 and EF-2 results in the production of a polypeptide linked to ribosome-bound mRNA.

Although certain aspects of the elongation mechanism in prokaryotes differ from those in higher organisms, the overall sequence of events is similar. The elongation factors for prokaryotic protein synthesis are listed in Table 11-3. In spite of the fact that N-formylmethionine is the initial N-terminal amino acid of the growing polypeptide in prokaryotes, this unit does not appear in the final protein. Specific enzymes catalyze the removal of the N-formyl group as well as the methionine, so the ultimate protein molecule does not necessarily contain an N-terminal methionine.

In eukaryotes, the methionine that functions in initiation is similarly released from the polypeptide at an early stage in the elongation process.

E. TERMINATION IN PROTEIN SYNTHESIS

Elongation proceeds until a termination codon (UAA, UAG, or UGA) is encountered. When one of these sequences in mRNA appears in the A site of the ribosome, additional aminoacyl-tRNA cannot be incorporated. A protein termed *releasing factor* (RF) binds to the A site together with GTP. This factor causes the removal of the polypeptide from the P site of the

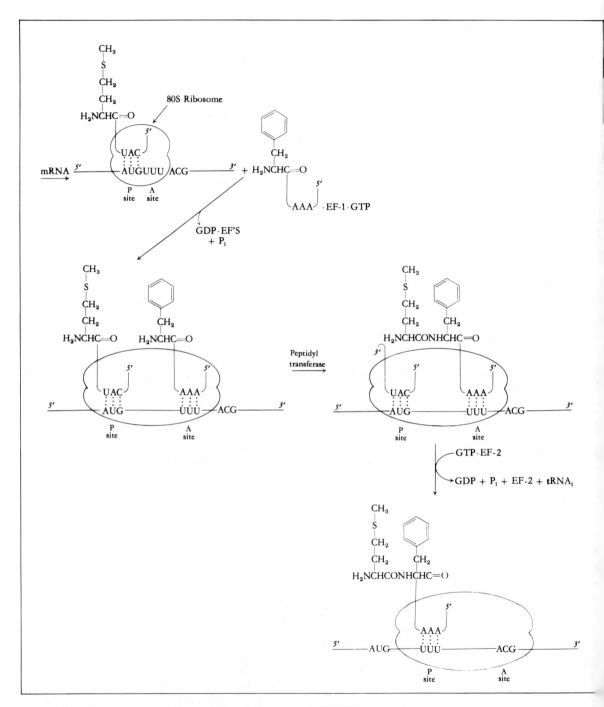

Fig. 11-15. Elongation sequence. Translation of the sequence AUGUUU... and formation of methionyl phenylalanine dipeptide (EF-1 and EF-2 = elongation factors).

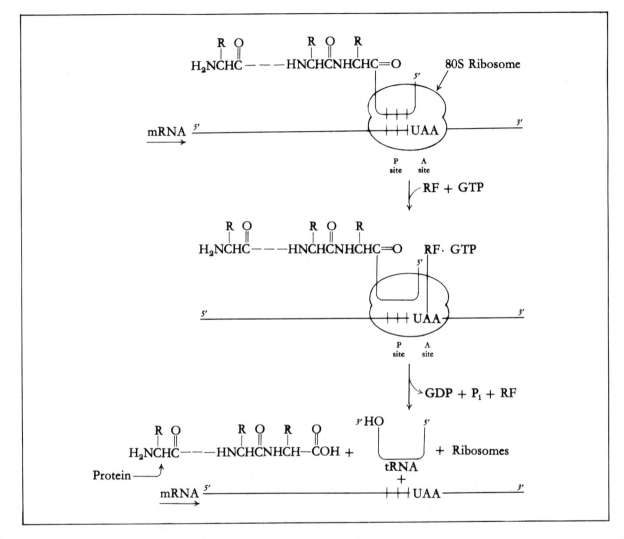

Fig. 11-16. *Termination of protein synthesis and release of the protein* (RF = *releasing factor*).

ribosome. Dissociation of the releasing factor from the ribosome proceeds with the hydrolysis of GTP to GDP and inorganic phosphate (Fig. 11-16).

Termination in prokaryotes involves at least two releasing factors, RF-1 and RF-2 (Table 11-3). An additional protein (RF-3) also appears to promote the release reaction. The releasing process in these organisms does not require GTP.

In this sequence of events in protein synthesis, only one 80S ribosome was considered to be on the mRNA. However, the total synthesis of a particular

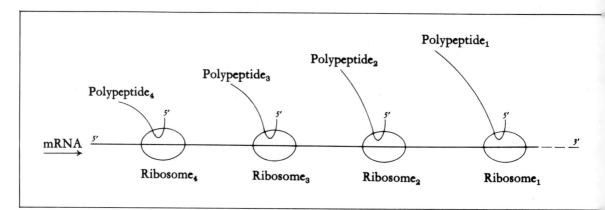

Fig. 11-17. *Polyribosome structure.*

protein need not be completed before the production of another protein can be initiated on the same mRNA. Actually, after the ribosome moves a certain distance from the initiation site, other ribosomes can be bound and the synthesis of additional polypeptides can be initiated. Thus, numerous ribosomes can move along a single mRNA in an assembly-line fashion. Each polypeptide contained in different ribosomes on the mRNA has a different chain length (Fig. 11-17). The structure composed of mRNA bound to many ribosomes is called a *polyribosome*.

F. POSTTRANSLATIONAL MODIFICATION OF PROTEINS

The structures of many proteins are modified after they are released from the ribosomes. Such posttranslational modifications include addition of various non-peptide components, alteration of certain amino-acid residues, and scission of specific bonds. Thus, the enzyme-catalyzed addition of sugar units to specific amino-acid components of a polypeptide gives rise to glycoproteins (sections 12.3.C and 14.3.E). Similarly, the carboxylation of specific glutamic acid in certain precursor proteins yields several of the blood coagulant proteins (section 12.3.B). The hydroxyproline residues in collagen are produced on the collagen precursor after it is released from the ribosome (section 14.3.B). Various functional proteins are generated as a result of the hydrolysis of specific peptide bonds in an inactive precursor. An example of this type of modification is the conversion of proinsulin to insulin. It should also be noted that disulfide bonds between cysteine residues are formed after the translational sequence.

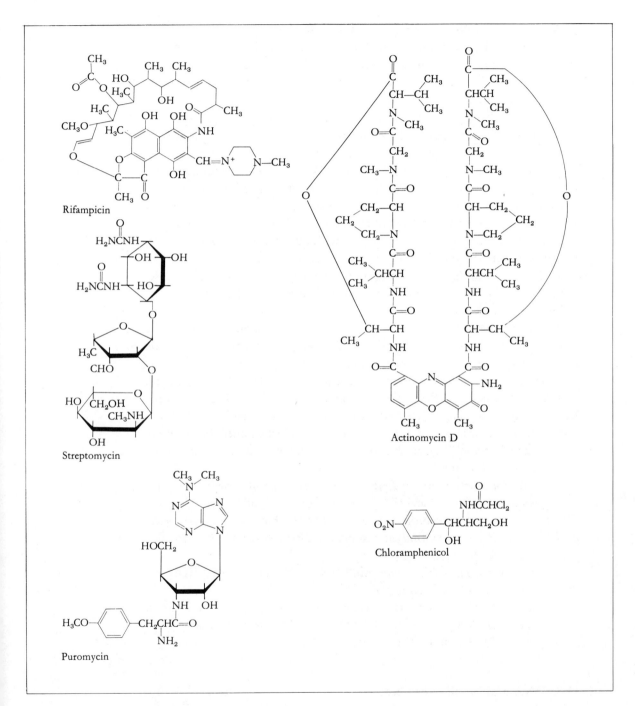

Fig. 11-18. *Inhibitors of protein synthesis.*

11.7. Inhibitors of Transcription and Translation

Several antibiotics, amino-acid analogs, and certain bacterial toxins are potent inhibitors of protein synthesis. In addition to their utility in the treatment of various diseases, they also serve as probes in elucidating many aspects of the mechanisms of protein synthesis. These materials interfere specifically with certain steps in transcription or translation (Fig. 11-18).

The antibiotics *rifampicin* and *actinomycin D* interfere with the transcription of DNA to RNA. The former inhibits the initial reactions in transcription, while the latter binds tightly to DNA and thus interferes with its function as a template. Actinomycin D inserts (*intercalates*) between guanine and cytosine pairs of the double-helical DNA and consequently inhibits propagation of RNA chains. Controlled amounts of rifampicin will not affect the elongation of RNA, but actinomycin D will interfere with the process. Neither of these, however, affects the translation steps in protein synthesis. It may be noted that rifampicin inhibits initiation by binding to bacterial RNA polymerase, but it has a minimal effect on the eukaryotic polymerase, thereby providing it with therapeutic utility.

Various antibiotics function by interfering with different aspects of the translation sequence of protein synthesis in bacteria. For example, *streptomycin* interacts with the 30S ribosomes of prokaryotes. This weakens ribosomal binding of aminoacyl-tRNA, which consequently produces "misreadings" of the mRNA codons. *Tetracycline* also binds to the 30S subunits of prokaryotic ribosomes and prevents their combination with aminoacyl-tRNA. *Chloramphenicol* blocks peptidyl transfer steps on 70S ribosomes. *Erythromycin* inhibits translocation (i.e., the shift of peptidyl-tRNA from the A site to the P site) on 70S ribosomes.

The antibiotic action of *puromycin* results from its having a structure analogous to that of aminoacyl-tRNA. It can therefore bind to the A site of a ribosome during the elongation sequence. It then combines with the peptidyl-tRNA at the P site, resulting in premature dissociation of the growing peptide from the ribosome.

Another antibiotic, *cycloheximide*, inhibits peptidyl transferase activity on the 80S ribosomes of eukaryotic cells.

Diphtheria toxin inhibits translocation in eukaryotes. It promotes a reaction between elongation factor 2 (EF-2) and NAD^+ to yield a covalently linked, inert ADP-ribose-EF-2 complex.

11.8. DNA and Nuclear Proteins

The DNA in the chromatin of eukaryotic cells is associated with basic proteins called *histones*. These proteins contain about 25% lysine and arginine; these amino acids are concentrated or clustered in the amino-terminal area

Table 11-4. Properties of Histones

Histone Type	Approximate Molecular Weight	Number of Amino Acids	Molar Ratio of Lysine-to-Arginine
I	2.1×10^4	215	22.0
IIb1	1.5×10^4	130	2.5
IIb2	1.4×10^4	125	2.5
III	1.5×10^4	130	0.8
IV	1.2×10^4	100	0.7

of the polypeptide chain. Some characteristics of different types of histones are given in Table 11-4. The histones are classified as lysine-rich (types I and II) or arginine-rich (types III and IV) according to the predominant basic amino acid. In chromosomes, the histones are apparently packed within specific grooves of the DNA helixes. In addition to the histones, chromatin also contains non-histone acidic proteins.

The exact function of the nuclear proteins is not known. Histones have been shown to inhibit some activities of DNA, and the possibility has been suggested that they function as gene regulators. In view of the limited number of different types of histones and their widespread occurrence in all types of cells, it is questionable to what extent they can serve as control agents in cellular differentiation. It seems that their primary action is in the maintenance of the structural integrity of chromatin.

11.9. Control of Gene Expression

Although the genes with their characteristic DNAs can provide the information for the cellular synthesis of numerous proteins, there are considerable differences in the amount of each protein that is normally produced. Moreover, certain proteins are synthesized only under specific conditions. Additionally, in higher animals, cells of different tissues do not synthesize the same repertoire of proteins, even though they contain the identical genetic components. Thus, certain cells synthesize a given protein only in the presence of specific stimulating factors, and others never realize their full genetic potential. This indicates that various regulatory systems are present that control or completely suppress the expression of certain genetic information.

Many aspects of the control mechanism for prokaryotic cells (e.g., *E. coli*) have been elucidated. The overall system in these organisms differs from that in higher animals. Nonetheless, certain principles have been established from studies with prokaryotes that are useful in understanding such regulatory processes in the specialized, differentiated cells of complex organisms.

Intracellular enzymes may be divided into two categories: *constitutive* enzymes, or those that are present at all times, and *induced* enzymes, or those that appear only when they are required. When lactose, for example, is introduced in a medium with *E. coli*, enzymes are generated within the organism that promote the transport of the substrate into the cell (permease) and catalyze its hydrolysis (β-galactosidase). When other carbohydrate nutrients are utilized, the cells do not produce the enzymes for the metabolism of lactose. Thus, lactose *induces* the synthesis of the enzymes required for its metabolic disposition, such as lactose permease, β-galactosidase, and transacetylase.

The phenomenon of enzyme induction should be distinguished from enzyme activation. The former is associated with the synthesis of the enzyme protein, whereas the latter involves the enhancement of activity of a preformed enzyme.

The genetic aspects of enzyme induction have been explained by the operon model. An *operon* is a region of DNA from which the mRNAs for several metabolically related enzymes are transcribed (Fig. 11-20). The operon also includes a section of DNA that controls the transcription process. This is called the *operator gene*. The portions of the DNA that provide the mRNAs for the enzymes are called *structural genes*. Contiguous with the operator gene are the *promoter locus* and the *inhibitory locus*. The inhibitory locus functions as a template to provide mRNA that codes for a *repressor protein*, which when bound to the operator gene, blocks transcription of the structural genes for the enzymes (see Fig. 11-19).

The *lac* operon that controls lactose metabolism may be taken as an example. In the absence of lactose, the repressor protein is bound to the operator gene. This interaction causes repression of the structural genes which are regulated by the operator. The mechanism for enzyme induction by lactose involves the neutralization of the repressor by the inducer molecule, i.e., lactose. (Methyl galactoside may have the same effect as lactose.) As a result of this action, the repressor is deactivated and is unable to affect the operator. Thus, transcription of the structural genes is de-repressed, and the lactose-metabolizing enzymes can be synthesized (see Fig. 11-19).

The *lac* operon is one example of a system that controls biosynthesis of a group of related proteins. A number of operons utilizing similar mechanisms have been demonstrated in bacteria. For example, the *pyr* operon involves the production of five enzymes that catalyze the formation of UMP. Another operon, which involves the generation of ten enzymes that are required for histidine synthesis (*his* operon), has been studied in *Salmonella typhimurium*. The operon and its associated control system thus appear to provide a common mode of regulation of gene expression in prokaryotes. It should be

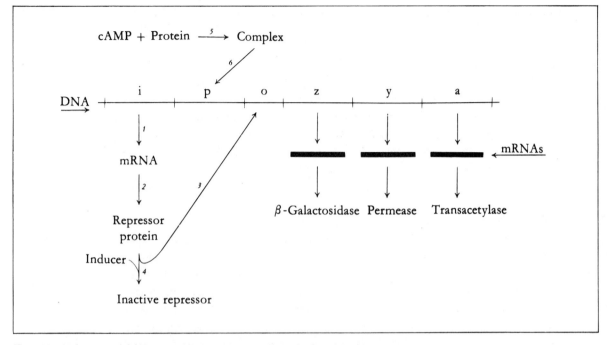

Fig. 11-19. Operon model (lac operon). *Sequence 1→2→3 leads to repression; sequence 1→2→4 (presence of lactose or methyl galactoside) results in enzyme induction; sequence 5→6 results in promotion of transcription (i = inhibitory locus; p = promoter locus; o = operator gene; z, y, and a = structural genes for synthesis of lactose-metabolizing enzymes).*

noted that the operon control site is at the *transcriptional* stage rather than translational.

Another site of transcriptional control of genetic expression is at the promoter locus of the DNA operon. In the *lac* operon of *E. coli*, the transcription of the structural genes is repressed by glucose. This action of glucose is overcome by cyclic AMP (cAMP); that is, cAMP enhances the transcription of operons repressed by glucose. The enhancement process first requires that cAMP be bound to a specific receptor protein. Interaction of the cAMP-protein complex with a site in the promoter locus of the DNA then stimulates transcription of the operon. Hence, in contrast to substrate induction (e.g., by lactose), which induces transcription by inactivating the repressor protein, cAMP enhances protein synthesis by its action on the promoter site.

This type of promotion of transcription depends upon two factors: (1) the inverse relationship between the concentrations of glucose and cAMP and (2) the need for the enzymes of lactose metabolism (e.g., β-galactosidase).

When glucose is abundant, the level of cAMP is minimal. Also, glucose causes a decrease in the synthesis of β-galactosidase, which catalyzes the metabolism at lactose to glucose. This repression by a lactose metabolite (i.e., glucose) is clearly advantageous, since the enzyme is not required when glucose levels are high, and the energy expended in the synthesis of the enzyme would be wasted. The mechanism by which glucose decreases the amounts of cAMP is not known; however, the level of cAMP, which antagonizes the repressive effect of glucose, is increased when glucose is absent or its level is low. When there are low glucose and elevated cAMP levels, the action of the cAMP on the promoter locus stimulates transcription of the structural gene for β-galactosidase, thereby enhancing the metabolism of lactose to glucose.

The operon mechanism appears to be limited to prokaryotic organisms. Messenger RNA generated from an operon unit in prokaryotic DNA is *polycistronic*; that is, it contains the required base sequences to code for several proteins. Hence, when the DNA from a specific operon is transcribed, a group of metabolically related enzymes are generated. In the cells of mammalian tissues, however, the mRNAs are *monocistronic*, so each mRNA codes the synthesis of a single polypeptide or protein. Moreover, the structural genes of related enzymes may be distributed in several different chromosomes. An additional complication that has come to light in recent studies is that in eukaryotic systems, following transcription, portions of the primary transcript are removed and the remainder is "spliced" together in a subsequent reaction.

One mode of control of protein synthesis in higher organisms is through the activity of certain hormones. The synthesis of various enzymes in animal tissues is inducible by glucocorticoids. For example, cortisol stimulates the production of tryptophan pyrrolase, tyrosine aminotransferase, and phosphoenolpyruvate carboxykinase. A number of steroids (estrogens, androgens, progesterone, and vitamin D) and the thyroid hormones are effective in promoting the synthesis of certain proteins in their target organs. Unlike insulin or epinephrine (which interact with cell membranes and relay their message via cAMP), steroid hormones enter the cell and bind with specific cytoplasmic receptor proteins. The resulting hormone-receptor complexes are transported into the nuclear compartment, where they bind to chromatin and promote transcription and synthesis of mRNAs. The latter are transferred back to the cytoplasm, where they generate their respective proteins. Although there are differences in specific details, the general pathways for steroid action that have been investigated appear to be similar (Fig. 11-20).

As a result of the differentiation and specialization of the cells of animal tissues, many cellular genetic capabilities are completely suppressed. Thus, the enzyme compositions of liver, muscle, and nerve cells differ both

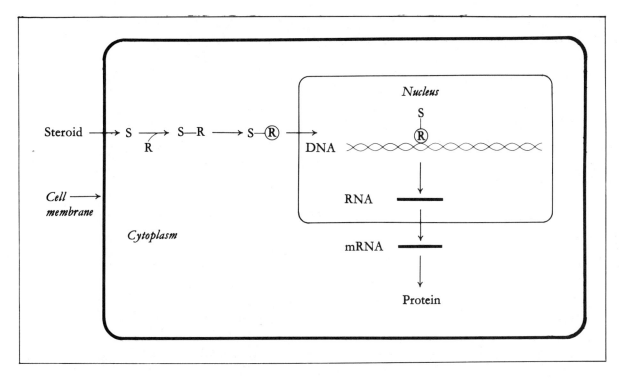

Fig. 11-20. General mechanism for effects of steroid hormones on protein synthesis in eukaryotic cells (R = receptor protein; S = steroid hormone).

quantitatively and qualitatively. Although the total genetic potentials of each of these types of cells are probably identical, the expression of certain genomes in each kind of cell is inhibited to permit differentiation of tissues. The major controls probably operate at the level of transcription; however, certain aspects may also involve translation and protein synthesis. Some of the processes of differentiation and development are regulated by hormones, but most of the fundamental mechanisms remain to be elucidated.

11.10. Viruses

Viruses are particles of nucleic acid enveloped by a protein coat. Some viruses also contain small amounts of carbohydrates and lipids. The nucleic acid component is either DNA or RNA, and each of these may be either single or double stranded. Viruses range in molecular weight from about 2×10^6 (tobacco mosaic virus) to 2×10^9 (vaccinia virus). By themselves, viruses cannot undergo replication or direct protein synthesis. When they enter a cell, they can replicate and synthesize their characteristic proteins by utilizing the substances and enzymes of the host.

Viruses are specific with respect to the host cells that they infect. Those that attack bacteria are called *bacteriophages,* or *phages.* Infection by viruses involves the introduction of their nucleic acids into the cells. The ultimate action of the viral nucleic acid is to direct the formation of mRNA and the production of viral proteins. These activities monopolize the synthetic machinery of the host cell and cause irreparable damage. When viruses undergo extensive intracellular replication, the cell is disrupted (lysis) and the viral progeny are released. As may be expected, there are differences among the intracellular biochemical reactions produced by viruses having different types of nucleic acid. Some aspects of these variations are described in the ensuing paragraphs.

An example of a double-stranded *DNA virus* is T_4 bacteriophage, which infects *E. coli.* It is composed of three sections: a head, a tail, and tail fibers. Its DNA, which contains over 150 genes, is enveloped by protein in the head section. When it interacts with the bacterium, the virus binds to the cells by its tail fibers. Contraction of the virus then results in the injection of its DNA into the bacterial cell. The normal synthesis of DNA, RNA, and proteins in the host cell is halted, while synthesis of viral materials takes over. The activities of the viruses include replication, transcription to produce mRNA, and subsequent synthesis of their characteristic proteins in a manner analogous to the activities of cellular chromosomal DNA. In order to accomplish these activities, the viruses utilize various host nucleotide polymerases and ribosomal synthetic mechanisms. Thus, the overall processes that are carried out by double-stranded DNA viruses, though complex, involve the same general sequences and type of information transfer that were described for cellular systems.

A virus that contains single-stranded DNA forms a complementary strand before replication. Thus, ϕX174 bacteriophage, which contains a single strand of circular DNA, can utilize DNA polymerase I (section 11.2.A) of *E. coli* to produce a double-stranded molecule for replication purposes. The process involves a number of complex steps that utilize the enzymes of the *E. coli* host cells.

RNA viruses undergo duplication and promote protein synthesis by utilizing base-pairing mechanisms. However, since the host cells do not contain the polymerase required for the duplication of RNA (i.e., RNA-directed RNA polymerase, or *RNA replicase*), this enzymatic protein must first be synthesized. An initial reaction therefore involves the synthesis of this enzyme, which in the case of polio virus, for example, is bound to the original viral RNA. In this type of RNA virus, the RNA functions as a messenger RNA for the synthesis of the replicase. The replicase then catalyzes the RNA-directed synthesis of a new RNA strand $(-)$ complementary to the original RNA. The new complementary strand $(-)$ functions as a template

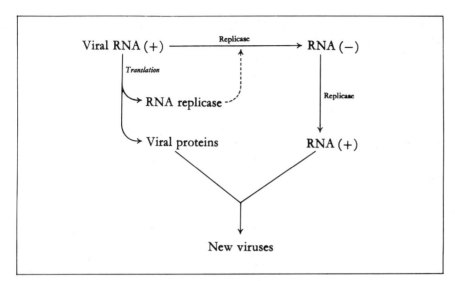

Fig. 11-21. *Sequence of intracellular activities of a single-stranded RNA virus. The original viral RNA (+) directs the synthesis of its complementary strand (−), which serves as the intermediate in the synthesis of new + strands.*

for the production of numerous copies of the original RNA (+). (The original viral RNA is indicated by + to distinguish it from its complementary strand, indicated by −, which serves as the intermediate in the synthesis of new + strands.) In addition to coding for RNA replicase, the viral RNA directs the synthesis of its coat proteins (Fig. 11-21).

Other types of viruses (e.g., rabies virus) carry a replicase together with the RNA. Within the host cell, the viral RNA directs the synthesis of a complementary strand that serves as an mRNA to code for the production of viral proteins and new RNA. Certain RNA viruses are also double stranded (e.g., reovirus). These double-stranded RNA viruses also carry their specific replicase.

In addition to undergoing replication and promoting the synthesis of their characteristic proteins (which are *lytic* processes), some viruses incorporate their DNA into the chromosomes of the host cells. The latter phenomenon, termed *lysogenicity*, occurs with certain types of bacteriophages and with a number of cancer-producing viruses. Since the viral DNA is integrated with that of the host genes, the characteristics that are produced are heritable; i.e., they are transmitted to the progeny of the infected cells. In such instances, the host cells may continue to thrive, but various structural features, such as their membrane components, may be altered.

An important mechanism whereby bacteria can counteract the effect of DNA viruses is through the action of intracellular enzymes, termed *restriction endonucleases* and *modification methylases*. The former recognize specific sequences in "foreign" viral DNA and catalyze hydrolysis at those sites. The host cells are protected from the effects of their restriction endonuclease because they have modified their own DNA at the sites that are susceptible to attack from the endonuclease. Such modifications usually involve the addition of methyl groups to certain critical adenine nucleosides of the host DNA. The modification reactions are catalyzed by methylases that utilize S-adenosylmethionine as the methyl donor (page 317). This restriction system thus involves the concerted action of matched modification and endonuclease enzymes; the modification methylase system recognizes and protectively alters the same sequence in the cellular DNA that its endonuclease cleaves in the viral DNA.

Restriction endonucleases are believed to play an important role in degrading invasive organisms. In addition to their natural function, however, these enzymes have been found valuable in studies of DNA structure and activity; for example, they are used in investigations involving the determination of nucleotide sequences, the mapping of chromosomes, the isolation of DNA fragments (genes), and the construction or recombination of DNA molecules.

11.11. Tumor Viruses

Several cancer viruses are known to operate by mechanisms analogous to lysogenicity (section 11.10). Such viruses are known as *oncogenic viruses*. These include both DNA viruses (e.g., polyoma virus, papilloma virus, and SV40 simian virus) and RNA viruses (e.g., Rous sarcoma virus, avian myeloblastosis virus, and various animal leukemia viruses). These viruses, upon alteration of the cellular chromosomal DNA, induce abnormal and uncontrolled growth of the infected cells. When grown in tissue culture, they retain their tumorigenic activity, and, when injected into susceptible animals, they produce characteristic tumors.

RNA tumor viruses first direct the synthesis of complementary DNA via reactions catalyzed by an RNA-directed DNA polymerase termed *reverse transcriptase*. The newly synthesized DNA is then incorporated into the genome of the affected cell. This type of sequence—i.e., one involving the flow of information from RNA to DNA, instead of the reverse (which is the usual direction)—may be a distinguishing feature of viral tumorigenesis and cells that carry cancerous genomes. The reverse transcriptase enzyme has been shown to be present in apparently normal cells. Its function under physiologic conditions has not yet been determined, and its possible relation to cancerous states is under active study.

11.12. Mutations

Mutations are produced either by alterations in the gross structure of chromosomes or by changes in single bases of DNA; the latter are called *point mutations*. Point mutations may be lethal, or they may manifest themselves in

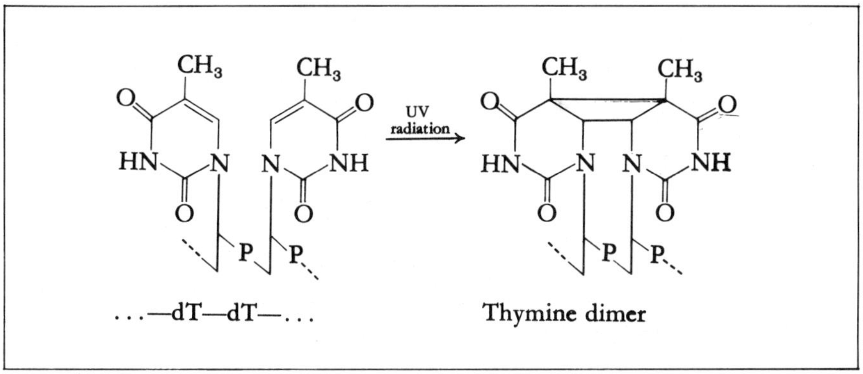

Fig. 11-22. Formation of a thymine dimer.

the synthesis of proteins with defects in their amino-acid sequence. Some point mutations may occur spontaneously; others result from irradiation or treatment with mutagenic reagents.

A common type of DNA alteration produced by irradiation with ultraviolet light is the formation of thymine dimers. This involves a reaction between adjacent thymine units that results in linkages between the bases (Fig. 11-22). The presence of such units blocks replication of the DNA. Other forms of radiation, such as x-ray and γ-ray, also produce mutations and may cause various types of tumors.

One class of point mutations involves substitution of one base for another. A *transitional* mutation is one in which a purine nucleoside is replaced by another purine nucleoside or a pyrimidine nucleoside is replaced by a different pyrimidine nucleoside; e.g., a transitional mutation results when a guanine group takes the place of an adenine unit. In this case, the respective substitution with regard to the adjacent base in the complementary strand would then require a cytosine unit to replace a thymine. Expressed in another way, this transitional mutation involves the substitution of a G-C base pair for A-T. An alternative transition is the substitution of an A-T base pair at the site of a G-C base pair.

The mutagenic activity of nitrous acid (HNO_2) is due to its deamination reactions. For example, it can convert the adenine in DNA to hypoxanthine:

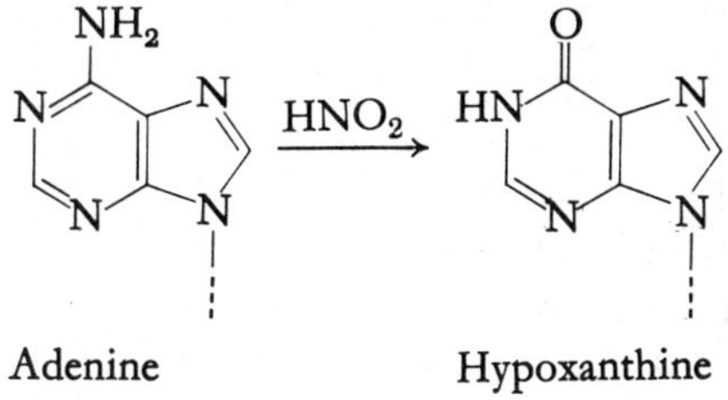

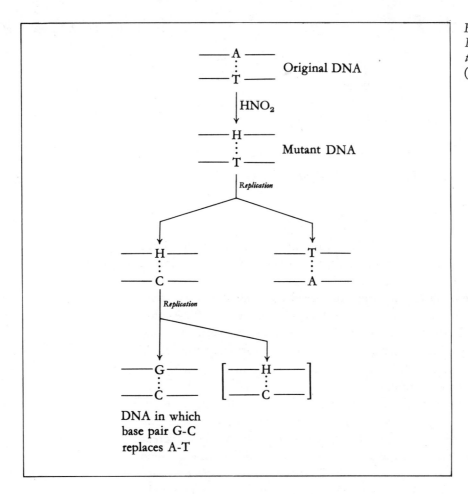

Fig. 11-23. Effect of HNO_2 in producing a transitional mutation (H = hypoxanthine).

During replication, hypoxanthine will pair with cytosine, which, in subsequent DNA synthesis, will direct the incorporation of guanine into the complementary strand (Fig. 11-23). The end result is a transitional mutation in which the base pair G-C is substituted for A-T.

A similar transitional mutation is produced by 5-bromouridine. This pyrimidine analog is incorporated in place of thymine during DNA synthesis. During the formation of the complementary DNA strand, 5-bromouridine will direct the incorporation of guanine. Subsequent replication results in the placement of cytosine in the adjacent site of the new strand. Hence, the original T-A pair is replaced by C-G.

Spontaneous transitional mutations may occur as the result of the pairing of an adenine tautomer with cytosine instead of the usual thymine. During replication, guanine will then be inserted in the adjacent strand at the site originally occupied by adenine. It may be noted that the sequence of events

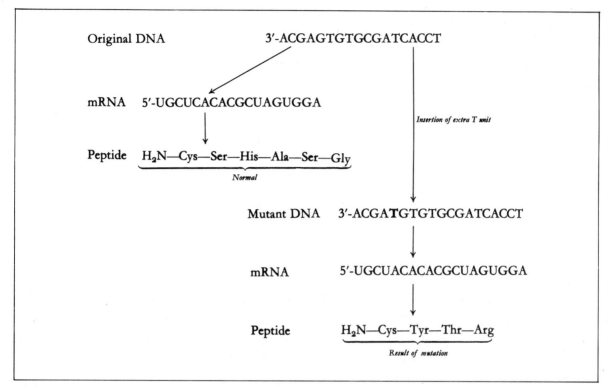

Fig. 11-24. *Effect of a frame-shift mutation on protein synthesis.* (*Mutant peptide ends at arginine residue, since UAG codon on mRNA signals termination of synthesis.*)

for the production of all transitional mutations involves DNA replication. Thus, synthesis of new DNA is necessary for these mutations to arise.

A second class of point mutation is one in which a pyrimidine-purine base pair (e.g., G-C) is replaced by a purine-pyrimidine pair (e.g., C-G); this change is called *transversion*. There are no specific reagents that produce transversions; however, this type appears to be common in spontaneous mutations.

Insertion constitutes a third class of point mutations. This involves the addition of one or several extra nucleotides between those already present in the DNA chain. Mutations of this type are produced by the dyes acridine and proflavine. These materials intercalate (see section 11.7) between adjacent bases in DNA strands and produce a gap that causes errors during replication.

A fourth class of point mutation, called *deletion*, involves the deletion of one or more nucleotides in DNA. Such mutations may be produced by the modification and removal of a base pair, followed by ligation of the chains.

Deletions can be caused by deaminating reagents or by proflavine. Conditions that enhance the hydrolysis of nucleosides may also lead to deletions.

Insertions and deletions are manifested by changes in the amino-acid sequence of the newly synthesized polypeptide during the translation of mRNA to protein. Since the mRNA is read as continuous base triplets, the readout is shifted after the deletion or insertion (Fig. 11-24). Such mutations are therefore referred to as *frame-shift mutations*. Thus, whereas transitional or transversional mutations result in the replacement of only one amino acid during protein synthesis, frame-shift mutations may yield a large number of changes in the amino-acid residues. As a consequence, frame-shift mutations are generally lethal; those involving substitutions, however, often do not affect the function of the protein.

Suggested Reading

Campbell, A. M. How Viruses Insert Their DNA into the DNA of Host Cells. *Scientific American* 235: No. 7, 103–113, 1976.

Chan, L., and O'Malley, B. W. Mechanism of Action of the Sex Steroid Hormones. *New England Journal of Medicine* 294:1322, 1372, 1430, 1976.

Crick, F. H. C. The Genetic Code. *Scientific American* 207:66, 1962.

Crick, F. H. C. The Genetic Code III. *Scientific American* 215:55, 1966.

Garen, A. Sense and Nonsense in the Genetic Code. *Science* 160:149, 1968.

Kornberg, A. *DNA Synthesis*. San Francisco: W. H. Freeman, 1974.

Maniatis, T., and Ptashne, M. A DNA Operator-Repressor System. *Scientific American* 234:64, 1976.

Mazia, D. The Cell Cycle. *Scientific American* 2130:54, 1974.

Miller, O. L., Jr. The Visualization of Genes in Action. *Scientific American* 228:34, 1973.

Nienhuis, A. W., and Benz, E. J., Jr. Regulation of Hemoglobin Synthesis. *New England Journal of Medicine* 297:1318, 1371, and 1430, 1977.

Nirenberg, M. W. The Genetic Code II. *Scientific American* 208:80, 1963.

Perry, R. P. Processing of RNA. *Annual Review of Biochemistry* 45:605, 1976.

Rich, A., and Kim, S. H. The Three-dimensional Structure of Transfer RNA. *Scientific American* 238 (1): 52, 1978.

Rich, A., and Rajbhandary, U. L. Transfer RNA: Molecular Structure, Sequence, and Properties. *Annual Review of Biochemistry* 45:805, 1976.

Schekman, R., Weiner, A., and Kornberg, A. Multienzyme Systems of DNA Replication. *Science* 186:987, 1974.

Stein, G. S., Stein, J. S., and Kleinsmith, L. J. Chromosomal Proteins and Gene Regulation. *Scientific American* 232:46, 1975.

Temin, H. M. RNA-Directed DNA Synthesis. *Scientific American* 226:24, 1972.

Uy, R., and Wold, F. Posttranslational Modification of Proteins. *Science* 198:890, 1977.

Watson, J. D. *Molecular Biology of the Gene* (3rd ed.). Menlo Park, Calif.: Benjamin, 1975.

Weissbach, H., and Ochoa, S. Soluble Factors Required for Eukaryotic Protein Synthesis. *Annual Review of Biochemistry* 45:191, 1976.

12. Blood

12.1. Composition and Function

Blood is a homogeneous suspension of cells in an aqueous solution of proteins, low-molecular-weight organic substances, and inorganic material. The cellular components consist of *erythrocytes* (red blood cells), *leukocytes* (white blood cells), and *thrombocytes* (platelets); these are designated collectively as the *formed elements*. Each cubic millimeter of blood contains about 5×10^6 erythrocytes, 5 to 10×10^3 leukocytes, and 1 to 3×10^5 platelets. The red color of blood is due to the hemoglobin in the erythrocytes. *Plasma*, the blood fraction obtained after removal of the cellular components, constitutes about 55% of the total blood volume. Separation of the protein fibrinogen from plasma yields the blood fraction called *serum*. Serum is generally obtained by allowing the blood to clot. In this process, fibrinogen is converted to an insoluble protein, fibrin, which is easily removed.

Plasma contains about 6 to 8 grams of protein per 100 milliliters. It also contains a wide array of lower molecular-weight organic compounds, of which the most prominent are cholesterol, phospholipids, glucose, urea, various amino acids, and lactic acid (Table 12-1). In addition to organic constituents, plasma contains a number of characteristic cations and anions, such as sodium, potassium, calcium, chloride, bicarbonate, and phosphate (Table 12-2). The specific gravity of whole blood is 1.055 to 1.065, whereas the specific gravities of plasma and serum are 1.025 to 1.029 and 1.024 to 1.028, respectively.

Blood accounts for about 8% of the body weight; an average human adult has about 5 or 6 liters of blood. The primary function of blood is transport, i.e., it transports nutrients and oxygen to the cells and carries away cellular waste products. Its characteristic electrolyte composition and pH serve to provide a constant, optimal milieu for all the tissues. Various mechanisms are available for maintaining the pH and concentration of most of the blood components within narrow limits. Blood and its constituents are also involved in various regulatory activities and protective mechanisms that are vital to the survival of the organism.

Plasma is separated from the interstitial fluid by the walls of the arteries and veins and the endothelium of the capillaries. Although some lymphatic fluid drains into the venous circulation via various ducts, the principal interaction between interstitial fluid and plasma occurs through the capillary endothelium. Protein does not move readily from the plasma to interstitial fluid; however, various ions do pass through the endothelial membranes.

Table 12-1. Major Organic Constituents of Plasma (Average Ranges)

Constituent	Concentration (mg per 100 ml)
Protein	$6-8 \times 10^3$
Amino acids	35–70
Urea	20–30
Uric acid	2–6
Creatinine	1–2
Creatine	0.5–1
Bilirubin (total)	0.2–1.4
Glucose	70–100
Cholesterol (total)	130–250
Phospholipids	150–250
Fatty acids (esterified and free)	200–500
Lactic acid	6–17
Citric acid	1.5–3.0
Ketone bodies	0.5–2.5

In the liver, where the capillaries have a sinusoidal structure, significant amounts of plasma protein are transferred to interstitial fluid.

Water is transferred from one body-fluid compartment to another by two processes. As a result of hydrostatic pressure in the arteries, water and low molecular-weight plasma components are "filtered" from the circulatory system to the interstitial compartment. Water also moves in the opposite direction as a result of osmotic pressure (section 1.3.C), which is due to the higher concentration of salts and proteins in plasma as compared to interstitial fluid.

As a result of the movement of water and electrolytes and the restriction of proteins, the concentrations of most inorganic ions in plasma are similar to those of interstitial fluid. The latter contains only minimal amounts of protein, whereas the protein concentration in plasma is about 7 grams per 100 milliliters.

12.2. Cellular Components of Blood

When blood is centrifuged, the cellular components (or formed elements) separate as a distinct layer. The percentage by volume of packed erythrocytes in a given volume of blood is called the *hematocrit*. In normal adults, the hematocrit value is about 45% to 47%.

The blood cells arise from undifferentiated stem cells (hemocytoblasts).

Table 12-2. Major Inorganic Electrolytes of Plasma

	Concentration	
Component	mg per 100 ml	mEq per Liter
Sodium	310–355	135–155
Potassium	14–21	3.5–5.5
Calcium	9–11	4.5–5.5
Magnesium	1.2–3.0	1–2.5
Chloride	355–390	100–110
Bicarbonate	152–183	25–30
Phosphates	3.0–4.5	1.5–2.5

In the fetus, the production of blood cells—or *hematopoiesis*—occurs in the liver, spleen, and bones, whereas in the adult, hematopoietic processes are confined to the marrow of flat bones, ribs, and sternum and the ends of specific long bones. Some leukocytes are also produced in the spleen and lymph nodes in adults.

1. *Erythrocytes*

Erythrocytes, as they appear in blood, have neither a nucleus nor mitochondria. Their shape is that of a biconcave disk with a thickness of about 1 micron in the center and 2 to 2.5 microns along the periphery. Their diameter is about 6 to 9 microns. The principal function of the erythrocytes is the transport of oxygen and carbon dioxide, which is carried out by the hemoglobin of the erythrocyte (Chap. 13).

The precursor cells of erythrocytes, termed *pronormoblasts*, contain large nuclei and mitochondria. In later stages of development (normoblastic), these cells synthesize protoporphyrin (section 10.2), globin, and hemoglobin. The nucleus and mitochondria are lost in the final stages of differentiation, so the erythrocytes released into the bloodstream do not contain these organelles and already have their full allotment of hemoglobin (approximately 3×10^{-8} milligrams per cell). In terms of whole blood, the concentration of hemoglobin is about 15 grams per 100 milliliters, since there are approximately 5×10^9 erythrocytes per milliliter of blood. It may be noted that this is more than twice the concentration of the plasma proteins.

The principal modes of glucose metabolism in mature erythrocytes are anaerobic glycolysis and oxidation via the pentose-phosphate pathway (section 4.6). Erythrocytes do not contain significant amounts of nucleic acids, and their synthetic activity is minimal. Their life span subsequent to their release into the bloodstream is about 126 days. Thus, erythrocytes undergo constant turnover, and their concentration in blood represents an equilibrium between formation and destruction. The degenerated cells are phagocytized by macrophages, and hemoglobin is degraded as outlined previously (section 10.5).

The role of glucose 6-phosphate dehydrogenase in erythrocyte metabolism and the action of glutathione reductase have been discussed already (section 6.7.F). The role of erythrocytic hemoglobin in the regulation of blood pH is considered in the next chapter (sections 13.3 and 13.5).

The cytoplasmic membrane of erythrocytes is composed of about 49% protein, 44% lipid, and 7% carbohydrate. Among the lipid components are cholesterol, phosphatidylcholine, phosphatidylethanolamine, and gangliosides. Integrated within the structure of the membrane are various glycoproteins and glycosphingolipids. Certain oligosaccharide units on these substances are responsible for the immunologic reactivity of the membranes and determine the characteristic blood-group types. The essential antigenic determinants of the blood-group substances are the units at the terminus of the oligosaccharide chain. The carbohydrate structures associated with blood-group substances A, B, and H are shown in Figure 12-1.

2. *Leukocytes*

Leukocytes, unlike erythrocytes, are true cells with nuclei and mitochondria. They contain the necessary enzyme systems for respiratory processes, glycolysis, the oxidative pentose-phosphate pathway, and glycogen metabolism. Most of the cell types of leukocytes have a comparatively high rate of cellular proliferation and RNA turnover.

On the basis of their intracellular appearance, leukocytes may be classified as either agranulocytes or granulocytes. Of the former type, most are lymphocytes; these represent about 25% to 35% of the leukocyte population. Lymphocytes play a central role in immune responses inasmuch as they contain and release immunoglobulins (section 12.5). They are also involved in cell-mediated immunity (e.g., as expressed by rejection of tissue grafts).

Most of the granulocytes show ameboid movement and perform a protective function by engulfing and destroying bacteria and foreign substances. The lysosomes of these phagocytizing cells are rich in proteinases and other hydrolases that degrade the captured materials.

3. *Platelets*

Platelets are non-nucleated cells that arise from megakaryocytes in the bone marrow. They are rich in glycolytic and mitochondrial enzymes and produce ATP via both of these systems. Platelets also synthesize lipids, proteins, and glycogen. The potent vasoconstrictor serotonin (page 370) is present in platelets and accounts for practically all of this material found in blood. The

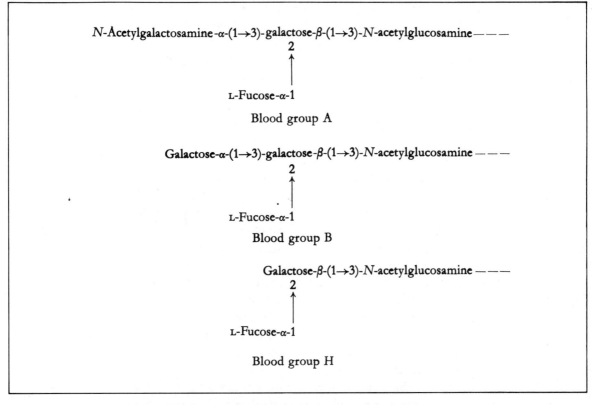

Fig. 12-1. *Oligosaccharide composition of antigenic determinants of blood-group substances. These structures apply for type 1 systems; in type 2 systems, the 1→3 linkages between galactose and N-acetylglucosamine are replaced by 1→4 linkages.*

most important function of platelets appears to be their action in hemostatic processes (section 12.4).

12.3. Plasma Proteins

A. PROTEIN FRACTIONS

The concentration of protein in plasma ranges from 5.7 to 8.0 grams per 100 milliliters. On the basis of their electrophoretic mobility, the plasma proteins are classified as *albumin* and *α-, β-, and γ-globulins*. Thus, at pH 8.6, the albumin fraction moves most rapidly toward the anode, while γ-globulin has the lowest mobility. Fibrinogen moves between the β- and γ-globulin peaks. Each of these fractions represents a mixture that consists of numerous proteins, as demonstrated by further electrophoretic subfractionation and other separation methods. Electrophoresis on starch or polyacrylamide gels, for example, separates the serum proteins into over 20 components. These alternative methods reveal a fraction with a greater mobility than albumin, designated as the *pre-albumin fraction*, as well as other substances.

As noted in the discussion of lipoproteins (section 6.4.A), plasma or serum proteins can also be fractionated by centrifugation procedures. The isolation of pure proteins generally involves the sequential application of several methods (e.g., chromatography, gel filtration, and so on). Special procedures are also available for obtaining concentrates or groups of specific proteins, for example, certain glycoproteins or clotting components.

B. ALBUMIN

The most abundant of the plasma proteins is albumin, which constitutes about 60% of the serum proteins. It has a molecular weight of 68,000 and is composed of a single polypeptide chain of about 580 amino acids. The comparatively low isoelectric point (pH 4.7) of albumin confers a considerable negative charge on the protein at the physiologic pH of blood. This high negative charge, the relatively low molecular weight, and the high concentration of albumin contribute to making albumin the principal component in the maintenance of the osmotic pressure of blood. Albumin also serves in the transport of free fatty acids and in the uptake and transport of bilirubin. In addition, albumin transports a significant fraction of circulating thyroxine, although the more specific thyroxine-binding globulin in plasma appears to play an important role in this process.

The synthesis of albumin occurs in the liver. It represents about 25% of the protein produced in this organ. Following its synthesis on the ribosomes, the protein moves via the smooth endoplasmic reticulum to the Golgi complex and finally through the cell membrane to the sinusoidal capillary. Essentially all the albumin produced is destined for excretion. In a human weighing 70 kg, the liver manufactures about 14 grams of albumin each day. A fraction of the plasma albumin constantly passes out of the circulatory system and finds its way into the lymph and extracellular fluids of tissues. This is eventually returned to the blood, primarily via the thoracic duct. Hence, only about 40% of the total albumin is found in circulatory blood at any given time.

Albumin is taken up into the cells by pinocytotic mechanisms and is then digested to yield amino acids. A small fraction escapes into the gut. The plasma albumin level is therefore a reflection of a steady state that is the resultant of synthesis, removal, return, and degradation. The average half-life of circulating albumin is about 19 days.

The amount of albumin synthesis in the liver is highly dependent on the nutritional state of the organism, especially on the amount of protein ingested. Thus, protein starvation results in a sharp decrease in albumin production and consequently in lowered plasma albumin levels. As indicated previously, the proteins in plasma, especially albumin, maintain the osmotic pressure across the capillary walls, which counteracts the hydrostatic pressure of the blood. Thus, any situation that results in low levels of albumin leads to a loss of water from the blood vessels to the tissues; this condition is called *edema*. In addition to its occurrence in protein starvation, edema is frequently present in nephrosis, when considerable amounts of albumin pass through the damaged kidneys into the urine.

C. GLOBULINS

On the basis of their electrophoretic mobilities, globulins can be separated into α_1, α_2, β_1, β_2 and γ fractions. Various types of proteins with common structural units occur in more than one of the globulin fractions. Some lipoproteins, e.g., are found in the α_1-globulin group, while others are present in α_2- and β-globulin fractions (Table 12-3). Similarly, various carbohydrate-containing conjugated proteins occur in both the α and β fractions. Thus, the classification of proteins according to their electrophoretic mobility under specific conditions is insufficient to define their structure or their functional characteristics. Nonetheless, the separation and quantitation of globulins by means of electrophoretic techniques are useful in research procedures and in the diagnosis of certain diseases that exhibit abnormalities in distribution patterns.

Some of the components of globulins are listed in Table 12-3. The functions of many of these proteins are known, although the specific contributions of others have not been defined. The lipoproteins serve in the transport of cholesterol, phospholipids, and esterified fatty acids. Various globular proteins are involved in the transport of metal ions; for example, ceruloplasmin and transferrin function in the binding and transport of Cu^{2+} and Fe^{2+}, respectively. Haptoglobins bind the hemoglobin that is released from erythrocytes, and hemopexin combines with heme.

A number of the globulin components are involved in blood coagulation; these include prothrombin and the various clotting factors. In addition, several globulins are protease inhibitors and coagulation inhibitors (e.g., α_2-macroglobulin, α_1-antitrypsin, and antithrombin III). Fibrinogen, which is the substrate in clot production (section 12.4), is usually not included among the globulins, although its electrophoretic mobility is in the same range as that of these materials.

The globulins also contain carrier proteins for various hormones, such as thyroxine-binding globulin (TBG) and the transcortin that binds cortisol. Other components of the plasma globulin fractions are the immunoglobulins and complement components, which will be discussed in later sections (sections 12.5 and 12.6).

The normal concentrations and relative amounts of albumin and globulins

Table 12-3. Components of Plasma Globulins

Component	Electrophoretic Fraction	Concentration (g/100 ml)
Lipoproteins	α_1	0.3–0.8
	α_2	0.18–0.22
	β_1	0.4–1
Glycoproteins	α_1	0.05–0.15
	α_2	0.8–1.4
	β_2	0.03
Ceruloplasmin	α_2	0.02–0.06
Transferrin	β_1	0.2–0.4
Hemopexin	β_1	0.05–0.11
Haptoglobin	α_2	0.1–0.2
Prothrombin	α_2	0.01
α_2-Macroglobulin	α_2	0.15–0.42
α_1-Antitrypsin	α_1	0.2–0.4
Antithrombin III	α_2	0.02–0.03
Fibrinogen	—	0.3
Thyroxine-binding globulin	α_1	0.0001
Transcortin	α_1	0.0007
Immunoglobulins	β, γ	0.95–2.5
Complement components	α, β, γ	0.25

in plasma are shown in Table 12-4. Specific alterations in these values sometimes serve as indicators in the diagnosis of certain diseases.

12.4. Blood Coagulation and Its Controls

A. THE COAGULATION PROCESS

When tissues are injured, the loss of extensive amounts of blood is prevented by three primary defense mechanisms. The blood vessels at the affected site constrict, platelets accumulate and plug up the opening, and the wound is sealed by a clot. Normally, blood will clot when it is collected in a container unless appropriate measures are taken to prevent the process (page 477). The steps leading to blood coagulation are highly complex and involve a number of components that accelerate the reactions of the coagulation system when bleeding occurs. Since it is critical that clots do not form intravascularly, various mechanisms function to inhibit or antagonize reactions that might result in blood coagulation. Thus, coagulation is normally under rigid control; i.e., it is rapid and efficient when blood is shed, but it is inhibited in circulating blood. Although certain aspects of the system have not been completely elucidated, some reactions involved in the regulatory process are well defined.

Table 12-4. Normal Concentrations of Serum Protein Classes

Class	Concentration (g/100 ml)	Percentage of Total Protein
Albumin	4–5	50–60
α_1-Globulin	0.09–0.22	4.2–7.2
α_2-Globulin	0.14–0.35	6.8–12
β-Globulin	0.20–0.45	9.4–15
γ-Globulin	0.25–0.70	13–23
Fibrinogen	0.3	7

The insoluble component of the blood clot is the protein, *fibrin*. During the coagulation process, the latter is produced from *fibrinogen*. The normal concentration of fibrinogen in plasma is about 300 milligrams per 100 milliliters. The molecular weight of fibrinogen is approximately 330,000. It is composed of three types of polypeptide chains, termed α, β, and γ, that have molecular weights of 64,000, 56,000, and 47,000, respectively. The chains are linked to each other by disulfide bridges. Each fibrinogen macromolecule consists of a pair of each of the three chains.

Formation of insoluble fibrin from fibrinogen involves three types of reactions. The first is the proteolytic cleavage, in the presence of *thrombin,* of specific arginylglycine bonds in the amino-terminal regions of the α and β chains of fibrinogen. These cleavages yield *fibrinopeptides A* and *B*, respectively, as well as *fibrin monomer.* The fibrinopeptides are small units, containing about 18 to 20 amino-acid residues. All four newly exposed amino-terminal units on the α and β chains are glycine residues.

The second step in fibrin formation is the association and polymerization of fibrin monomer units to yield a partially soluble, or *soft*, clot. The third process involves the action of the enzyme, *active factor XIII*, on the soft clot. This enzyme functions as a transglutaminase to produce cross links between the γ-amide units of glutamine residues of one fibrin monomer and the ϵ-amino groups of the lysine residues of another unit (Fig. 12-2). As shown in this diagram, the cross-linkage reaction requires the presence of calcium ions. The cross-linked fibrin that is produced is an insoluble, high-molecular weight compound.

Activated factor XIII is formed from a proenzyme, *factor XIII,** by the action of thrombin. The activation reaction involves the specific proteolysis

---

* In naming the enzymes involved in blood coagulation, the activated form is designated by the suffix "a." Thus, factor XIII is the inactive proenzyme and factor XIIIa is the active enzyme. The coagulation factors are designated by Roman numerals. These factors are plasma proteins or glycoproteins that have been isolated and characterized.

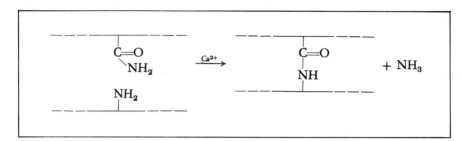

Fig. 12-2. Formation of cross-links between strands of fibrin monomers by transglutaminase action of factor XIIIa.

1. Fibrinogen $\xrightarrow{\text{Thrombin}}$ Fibrin monomer + 2 Fibrinopeptide A + 2 Fibrinopeptide B
2. Fibrin monomer ⟶ Fibrin polymer (soft clot)
3. Fibrin polymer $\xrightarrow{Ca^{2+}}$ Insoluble fibrin
   Factor XIIIa
   ↑
   | Thrombin
   Factor XIII

Fig. 12-3. Formation of insoluble fibrin from fibrinogen.

of factor XIII. The reactions in the formation of insoluble fibrin from fibrinogen are summarized in Figure 12-3.

Thrombin does not circulate in the blood; rather, it is produced during the coagulation process from a protein constituent of plasma, *prothrombin*. The latter is converted to thrombin by the action of *factor Xa* in conjunction with *factor V*, calcium ions, and phospholipid. Although factor V normally has some activity, its potency is enhanced considerably by thrombin. The precursor of factor Xa is *factor X*, which is a normal component of plasma and has no activity unless activated.

There are two mechanisms for the conversion of factor X to factor Xa; these are referred to as the *extrinsic* and *intrinsic* systems. The extrinsic system (Fig. 12-4) involves the action of a plasma protein termed *factor VII*, a lipoprotein isolated from tissues called *tissue thromboplastin*, or *tissue factor*, and calcium ions. The complex formed from these three components functions as a highly specific, proteolytic enzyme that produces factor Xa from factor X. This mode of activation of factor X is called the extrinsic process because it requires a non-plasma component, i.e., tissue factor. Tissue factor is found in most tissues; however, lung and brain are especially rich in this material.

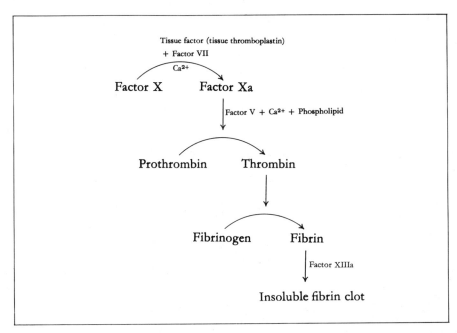

Fig. 12-4. *Extrinsic pathway for blood coagulation.*

In this pathway, coagulation is initiated by the release of tissue factor from the severed tissue.

The intrinsic system (Fig. 12-5) for the activation of factor X is initiated by the activation of a plasma protein termed *factor XII*. This clotting factor is converted to the active form, *factor XIIa*, by contact with various types of surfaces other than that of intact vascular lumen. Subendothelial components, such as collagen or basement membrane, that are exposed in traumatized blood vessels, could trigger the activation of factor XII. Factor XII is also activated by contact with glass. The activated product, factor XIIa, then catalyzes the conversion of another protein, *factor XI*, to its enzymatically active form, *factor XIa*. For optimum efficiency, this action of factor XIIa requires the presence of a plasma component termed *high-molecular-weight kininogen*. Factor XIa also functions as an enzyme, and, in conjunction with calcium ions, it converts *factor IX* to *factor IXa*. The latter interacts with another plasma protein (*factor VIII*), calcium ions, and a phospholipid to produce a complex that promotes the conversion of *factor X* to *factor Xa*. Since all the components for this pathway are normally present in the bloodstream, it is called the intrinsic system.

Factor XIIa, in addition to activating factor XI, also catalyzes the conversion of blood component called *prekallikrein* (page 491) to an active form, termed *kallikrein*. The latter is effective in transforming factor XII to factor XIIa. The overall effect of these reactions is thus a rapid, autocatalytic generation of factor XIIa and an amplification of the coagulation process.

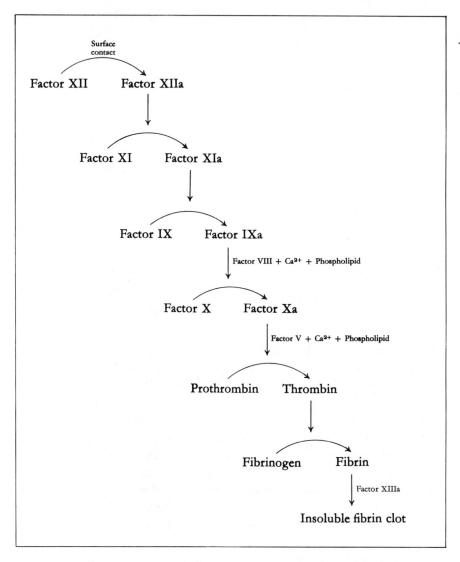

Fig. 12-5. Intrinsic pathway for blood coagulation.

The underlying scheme of both coagulation pathways is the sequential conversion of specific circulating proteins (proenzymes) to proteolytic enzymes with defined characteristics. The activation of the first proenzyme (e.g., factor XII in the intrinsic pathway) yields a product that activates the next proenzyme in the process. Thus, in this pathway, factors XI, IX, X, and prothrombin are substrates for the previous enzyme in the sequence, whereas the activated forms (factors XIa, IXa, Xa, and thrombin, respectively) function as enzymes to generate the subsequent activated components. This type of "cascade" phenomenon can act as an amplification system in which minute quantities of clotting factors early in the sequence generate considerable amounts of the end products, thrombin and fibrin. Since each of the activated factors are enzymes, the action of the clotting system is rapid and normally produces the critical response required to prevent excessive loss of blood.

Although relatively slight quantities of material trigger the sequence of steps that results in the formation of fibrin, intravascular clotting will not occur under normal conditions. Several elements enter into the control of such undesirable clotting. Plasma contains proteins that inhibit various clotting factors; these include antithrombin III, α_2-macroglobulin, and α_1-antitrypsin (see Table 12-3). Antithrombin III is an inhibitor of thrombin, factor Xa, and several other clotting factors. α_2-Macroglobulin interferes with the activity of thrombin, and α_1-antitrypsin blocks the action of a number of proteases, including some of the activated clotting components. In addition to their neutralization by inhibitors, the activated clotting components are removed by hepatocytes. Another critical factor in preventing the generation of fibrin is the continuous, rapid movement of blood and the consequent minimalization of interaction between clotting enzymes and their substrates. This effect is evidenced by the comparatively higher incidence of fibrin clots in constricted areas of the veins.

It should be noted that several steps in the clotting pathway require calcium ions. Substances that form complexes with calcium—such as citrate, oxalate, or ethylenediaminotetraacetate (EDTA)—therefore prevent clot formation. Advantage is taken of these reactions in clinical and research operations for storage of blood or plasma. The most common reagent employed for this purpose is citrate.

Two reactions of the coagulation sequence—the activation of factor X and the conversion of prothrombin to thrombin—also require intermediate complexes with phospholipids. Phospholipids that function in these steps are lecithins or cephalins. In vivo, platelets serve as the source of these materials.

It would appear that the extrinsic and intrinsic coagulation systems could function as independent, alternative pathways. That this is not the case physiologically is evidenced by the bleeding problems that arise when there is a deficiency in a factor for only one pathway. Notable examples are the recessive sex-linked genetic disorders, *hemophilia A* and *hemophilia B*. Although these diseases are due to defects in factors VIII and IX (respectively) of the intrinsic pathway, the affected individuals are prone to excessive hemorrhaging and have critical problems in arresting any loss of blood. Thus, both coagulation mechanisms are required for efficient blood clotting. The extrinsic system seems to act more rapidly, but it apparently does not have the capacity to generate the required amount of thrombin and fibrin to seal the damaged vessel.

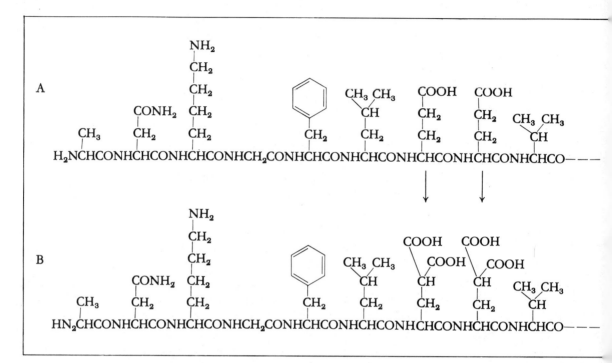

Fig. 12-6. Carboxylation of glutamate residues to form γ-carboxyl derivatives in the biosynthesis of prothrombin. A. Inactive prothrombin precursor. B. Prothrombin.

B. VITAMIN K AND DICUMAROL

The biosynthesis of prothrombin and factors X, IX, and VII is dependent on *vitamin K*. Therefore, any deficiency or inhibition of this vitamin leads to a decrease in the levels of the associated coagulation factors and a delay in blood clotting. The synthesis of these proteins on the ribosomes provides precursors that must be carboxylated to yield the functional procoagulants. This posttranslational modification requires vitamin K. The process may be illustrated with prothrombin. Figure 12-6 shows the amino-terminal octapeptide of this protein as it is released from the ribosomes. By the action of a specific carboxylase, CO_2 is introduced on the first ten glutamyl units (only two are shown in Figure 12-6) to provide γ-carboxyglutamate residues. The protein thus produced has the capacity to function as prothrombin. When there is a deficiency in vitamin K, this necessary modification of the prothrombin precursor does not occur. A common feature of the vitamin K-dependent coagulation proteins is that in order to be active, they must bind with calcium and phospholipid. The "defective" non-carboxylated materials produced in vitamin K deficiency do not bind to calcium and consequently cannot form active coagulant-calcium-phospholipid complexes.

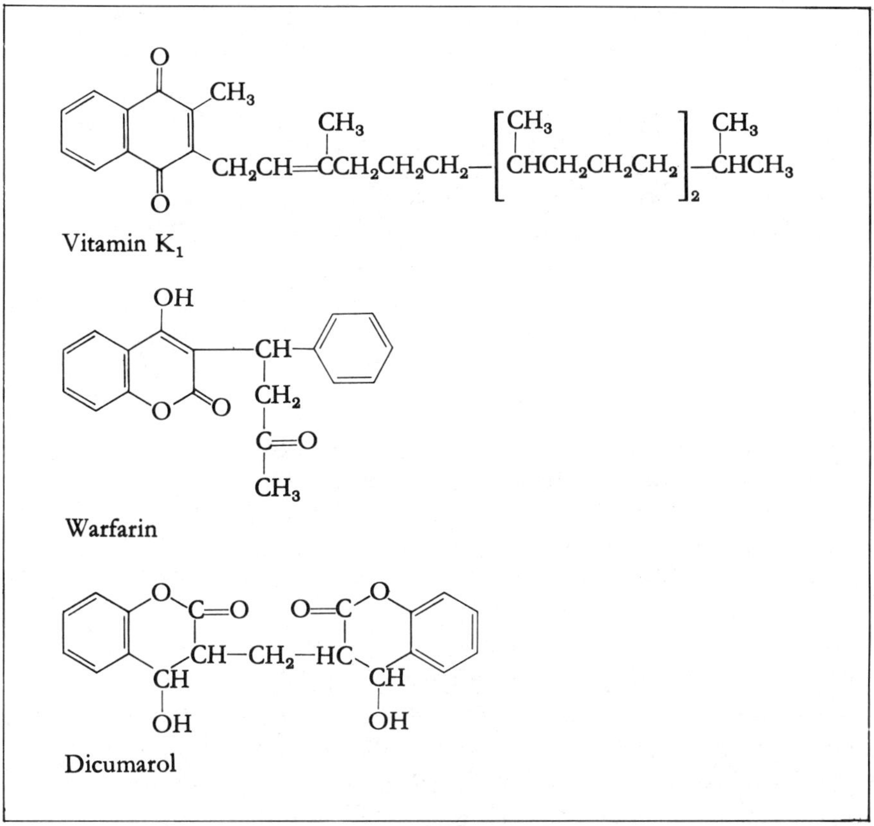

Fig. 12-7. Structures of vitamin K_1, warfarin, and dicumarol.

The action of vitamin K is antagonized by its structural analogs, *dicumarol* and *warfarin* (Fig. 12-7). These substances decrease the synthesis in the liver of prothrombin and other vitamin K-dependent proteins. The resulting deficiencies of coagulation components give rise to hemorrhagic conditions that may be lethal. By administering controlled amounts of dicumarol or warfarin, it is possible to lower the level of plasma prothrombin so that the rate of coagulation is decreased but not drastically inhibited. These drugs are employed clinically in the treatment of thrombosis and other diseases where there may be a tendency for intravascular clotting.

C. HEPARIN

Another drug employed in the treatment of thrombotic diseases is *heparin*, a glycosaminoglycan (section 14.3.D) or mucopolysaccharide that has been isolated from various tissues such as the lung, intestinal mucosa, and liver. The structure of heparin is not defined completely; however, its general aspects are shown in Figure 12-8. Heparin enhances the activity of antithrombin III, which is a circulating inhibitor of thrombin and factors Xa,

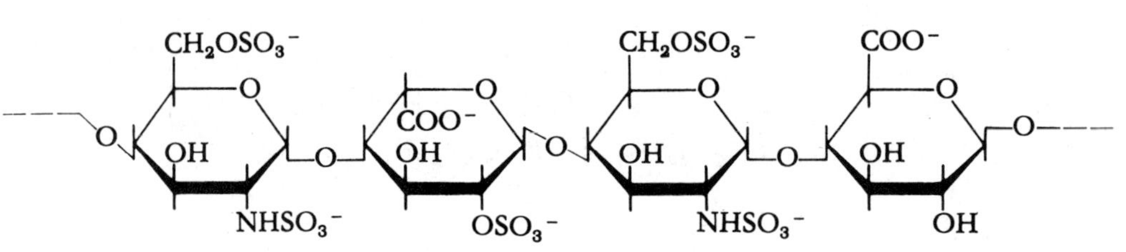

Fig. 12-8. General structure of heparin.

IXa, and XIa. Unlike dicumarol, which can be taken orally, heparin must be injected intravenously. As may be expected from the differences in their mechanisms of action, dicumarol does not have an anticoagulant effect until the available amounts of plasma prothrombin are turned over, whereas anticoagulation by heparin is almost immediate.

D. REACTIONS OF THE BLOOD TO TISSUE INJURY

The immediate physiologic reaction to injuries in which blood vessels are damaged is the adhesion of platelets to exposed collagen at the affected site. This is followed by agglutination of additional platelets to the adhering platelets. If the injury to the blood vessel is minimal, these responses are sufficient to plug up the opening and arrest the bleeding. Otherwise, the effect of platelet agglutination is reinforced, and the opening is sealed by deposition of fibrin. Thus, platelets function in concert with the coagulation cascade to prevent excessive loss of blood.

In addition to serving as mechanical barriers at the site of vessel injury, platelets release serotonin and catecholamines that cause constriction of the blood vessel (vasoconstriction). The platelets also provide phospholipids that are utilized in the activation of factor X and prothrombin. Furthermore, platelets contain factor XIII and a component that activates factor XI, which also contribute to production of the fibrin clot.

Plasma also contains a potential fibrin-digesting system. The circulating precursor, or zymogen, for the fibrinolytic enzyme is *plasminogen*. Specific proteolysis of the latter by a plasminogen activator generates the enzyme *plasmin*. The activator itself is produced through the action of factor XIIa. Plasmin catalyzes the hydrolysis of various peptide bonds in fibrinogen to yield soluble products.

The activation of many of the zymogens or proenzymes of the coagulation cascade involves the proteolysis of specific peptide bonds, a process referred to as *limited proteolysis*. The active clotting enzymes are *serine proteases*; that

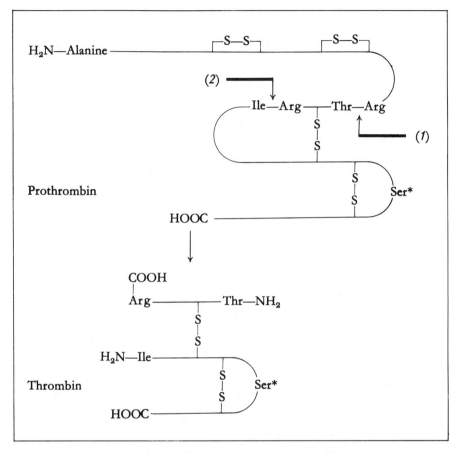

Fig. 12-9. One of the possible sequences for the conversion of prothrombin to thrombin as catalyzed by the factor-Xa complex. The first step (1) is the hydrolysis of an oligopeptide from the N-terminal portion of the chain, which exposes a threonine residue. This is followed by another cleavage (2) that generates a double chain linked by a disulfide bond. Asterisks indicate the serine residue at the active site of thrombin.

is, they contain a serine residue at their active site (section 2.12). Examples of serine proteases in the clotting sequence are factors XIa, IXa, Xa, and thrombin. Activation generally involves hydrolyses that remove peptide units as well as cleavages that yield double or multiple chains. One example of this type of reaction is the conversion of prothrombin to thrombin (Fig. 12-9).

12.5. Immunoglobulins

A. ANTIBODIES

Antibodies are proteins that react with substances, termed *antigens*, that are foreign to the organism. Antibodies are synthesized by the animal in response to exposure to the antigen. Hence, each particular antigen is associated with a unique antibody. Generally, antigens are macromolecules, such as proteins, polysaccharides, or nucleic acids. Certain low-molecular-weight compounds can also be effective antigens when they are bound to macromolecules; small molecules of this type are called *haptens*. The structural unit or section in an antigen to which a specific antibody is directed (and with which it interacts) is termed the *antigenic determinant*. Antibodies and structurally related materials are classified as *immunoglobulins*. Upon electrophoretic separation of serum proteins, the immunoglobulins are found in the γ-globulin and β-globulin fractions.

The cell types involved in the synthesis of antibodies are the lymphocytes and plasma cells. Actually, two varieties of lymphocytes are operative in immune responses: *B lymphocytes*, which arise primarily in bone marrow, and *T lymphocytes* from the thymus. The former cells function in the generation of circulating or humoral antibodies, while the latter are involved in cell-mediated immunity and the rejection of tissue grafts. There is, however, a certain degree of interplay and cooperation between the activities of the two systems. Unlike the B lymphocytes, the T cells do not secrete antibodies, yet they may act to enhance the activity of the B lymphocytes.

B lymphocytes give rise to plasma cells upon interaction of the lymphocytes with antigenic macromolecules, bacteria, or viruses. The plasma cells synthesize and secrete immunoglobulins with binding sites for the antigen. Serum containing a specific antibody is called *antiserum*. Synthesis of the antibody continues for some time, even indefinitely, so subsequent introduction of the antigen into the animal that carries the antibody (or the addition of the antigen to the antiserum) results in the combination of the two materials to form an *antigen–antibody complex* (Ag-Ab complex). In the circulation, the Ag-Ab complex is engulfed by macrophages or analogous cell types.

The affinity of an antibody for the antigen that elicited its synthesis is highly specific. If a given antibody is produced against a particular protein, it will be specific for that protein. Moreover, antibodies are sensitive to minute differences in the structure of the antigen and can generally discriminate between analogous proteins from different species; e.g., antibodies produced by rabbits toward bovine albumin react weakly, or not at all, with albumin from other species. This high specificity of antibodies makes them extremely useful in the identification of minute amounts of various materials and for determining the homogeneity of preparations of macromolecules.

B. IMMUNOGLOBULIN STRUCTURE

Although each immunoglobulin has a unique structural unit that allows it to combine with its specific antigen, all immunoglobulins also have certain common features in their gross structure. Immunoglobulins are composed of two identical high-molecular-weight chains, termed *heavy* or *H* chains, and two smaller polypeptide chains, designated as *light* or *L* chains. The four chains are linked to each other through disulfide bonds, and the total protein

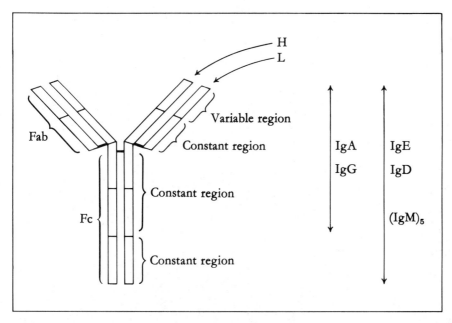

Fig. 12-10. *Structural aspects of an immunoglobulin molecule. The heavy chains (H) span the entire distance of the Y, while the light chains (L) constitute part of the arms. Disulfide bonds link the two H chains as well as each L chain to its respective H chain. In addition, intrachain disulfide bonds occur along the polypeptide chains. The variable regions are composed of approximately the first 110 residues starting from the amino terminals; the remainder of the molecule is composed of constant regions. Some of the details in the diagram are true only of IgG. The relative lengths of the molecules of each class of Ig are indicated by the scale on the right. Fab and Fc denote the two types of fragments obtained from IgG upon papain digestion (see text).*

molecule assumes a Y-shaped structure (Fig. 12-10). Five different types of heavy chains are found in human immunoglobulins, termed γ, α, μ, δ, and ϵ heavy chains. Similarly, there are two different types of light chains, termed κ and λ. In addition to the basic polypeptide sequences, the heavy chains also contain oligosaccharide units. The four-chain structure of these proteins thus allows for ten possible combinations: $\kappa_2\gamma_2$, $\lambda_2\gamma_2$, $\kappa_2\alpha_2$, $\lambda_2\alpha_2$, $\kappa_2\mu_2$, $\lambda_2\mu_2$, $\kappa_2\delta_2$, $\lambda_2\delta_2$, $\kappa_2\epsilon_2$, and $\lambda_2\epsilon_2$ (Fig. 12-11). In some cases, these exist in the form of polymers of the four-chain units, e.g., $(\kappa_2\alpha_2)_n$.

Immunoglobulins (Ig) are classified according to their characteristic heavy chains: *IgG, IgA, IgM, IgD,* and *IgE* correspond to immunoglobulins with H chains γ, α, μ, δ, and ϵ, respectively. On the basis of the wide differences in their concentration in plasma (Table 12-5), the immunoglobulins are classed as either *major* or *minor*. The major immunoglobulins include IgG, IgA, and IgM; the minor ones are IgD and IgE.

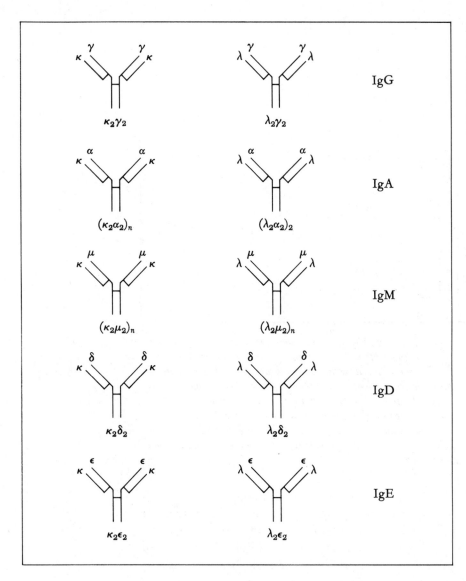

Fig. 12-11. The five classes of immunoglobulins corresponding to heavy-chain types and the ten different types based on light-chain composition. Heavy chains: γ, α, μ, δ, and ϵ. Light chains: κ and λ.

As shown in Figure 12-10, the majority of the protein chains of a molecule belonging to a particular immunoglobulin class (e.g., IgG) contain the same amino-acid sequences as all other molecules of that class. These sequences constitute the *constant region* of the molecule and characterize the immunoglobulin class. The *variable region* of the H or L chain consists of the sequence of about 110 amino-acid residues beginning at the amino-terminal ends of the arms; the amino-acid compositions of these regions differ, even within the same class of immunoglobulins. The variability in the amino-acid

Table 12-5. Immunoglobulins in Plasma

Class	Plasma Concentration (g/100 ml)	Molecular Weight
IgG	0.8–1.8	150,000
IgA	0.09–0.45	180,000*
IgM	0.06–0.25	190,000*
IgD	~0.01	175,000
IgE	~0.0001	200,000

* IgA also occurs as dimers or trimers of units with this molecular weight; IgM occurs as a pentamer, i.e., with molecular weight $(190,000)_5$.

sequences of these regions permits molecules of the same immunoglobulin class to be specific for a variety of antigens.

Some of the characteristics of immunoglobulins may be elucidated by considering the IgG class as an example. Each normal person will demonstrate numerous serum globulins of the IgG class that differ from each other with respect to their light-chain composition and the amino-acid sequence of their variable regions. Each heavy chain is of the γ type and has a molecular weight of about 50,000. The L-chain units may be either κ or λ in type; in either case, the molecular weight of each L chain is about 25,000. The H chains also contain oligosaccharide units. The amino terminals of all four chains are located at the tips of the arms of the Y (see Fig. 12-10). The carboxyl terminals of the H chains are at the base of the trunk of the Y, and the carboxyl terminals of the L chains are near the angles of the Y. The carboxyl-terminal amino acids of the L chains are cysteine residues, through which disulfide linkages are formed with adjacent cysteine residues on the H chains. Near where the L chains are bound, the two H chains are bound to each other by similar disulfide bridges. All IgG molecules appear to have the same amino-acid sequence in about 75% of the H chain, starting from the carboxyl-terminal end. Similarly, about half of each L chain (again from the carboxyl terminus) has essentially the same sequence in all molecules. These amino-acid sequences constitute the constant regions, and the remaining sections near the amino terminals are the variable regions.

Limited digestion of IgG with the proteolytic enzyme papain yields three fragments that retain some of the activities of the original immunoglobulin. Two fragments are similar and are designated as *Fab*; a third fragment, which differs from the others, is termed *Fc* (Fig. 12-12). It may be noted that the Fab fragment contains an intact L chain and the amino-terminal segment from an H chain. The Fc fragment consists of the carboxyl-terminal portions of two H chains that are still bonded together via disulfide bridges. Fragment

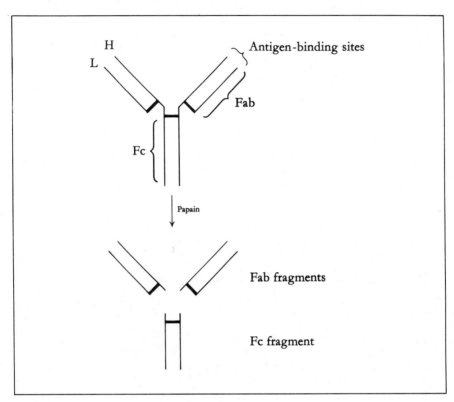

Fig. 12-12. Fragmentation of IgG by papain.

Fab binds to antigen, but the complex does not precipitate from solution, as occurs in the case of the complex of antigen and intact IgG. Fc can bind complement (section 12.6) and can cross the placental membrane. These findings, together with other evidence, demonstrated that the antigen-binding sites are in the amino-terminal areas—i.e., the variable regions—of the immunoglobulin. The reaction of antigen with intact immunoglobulin molecules results in the binding of the antigen with the two amino-terminal regions of the immunoglobulins and the cross-linking of a number of macromolecules. This reaction yields an Ag-IgG complex that is insoluble; however, such precipitation does not usually occur with the smaller Fab fragments and antigen.

Each class of immunoglobulins fulfills a specific function. IgM is the initial antibody induced upon the injection of an antigen. The level of this antibody then decreases, while that of IgG rises. IgG is quantitatively the most important class of immunoglobulins. Most of the antibody action against antigenic materials that reach the bloodstream is provided by molecules of the IgG class. IgM molecules associate to form pentamers, or *macroglobulins*, with molecular weights of about 950,000 (see Table 12-5). IgG, however, is a monomer consisting of two H chains and two L chains linked by disulfide bonds (see Fig. 12-10). IgA immunoglobulins are protein polymers (e.g., dimers or trimers) and are found in secretions, e.g., in bronchial and intestinal mucus, saliva, and tears. They probably serve to defend the animal against the entry of organisms via membranous structures. The functions of IgD and IgE are not clear. It is known, however, that IgE is involved in producing the allergic reactions expressed in asthma or various types of dermatitis.

C. MYELOMA PROTEINS

Considering the great number of possible antigenic substances, it is not surprising that there must be a vast variety of immunoglobulin structures in order for a specific structure to be associated with each antigen. The differences among molecules of immunoglobulin of the same class (e.g., IgG) are due to the presence of unique amino-acid residues in the variable regions of their heavy and light chains. Thus, each immunoglobulin macromolecule has a structural unit that may be directed to a specific antigen. Isolation from plasma of a certain class or subclass of immunoglobulins therefore yields material that is still heterogeneous with respect to the amino-acid sequences in the variable regions of the molecules.

In patients with aberrations of plasma cells, e.g., those that occur in multiple myeloma or macroglobulinemia, immunoglobulins with particular structures are synthesized in comparatively large quantities. Individuals with myeloma tumors in tissues of the antibody-forming system produce abnormal amounts of a unique immunoglobulin—commonly called *myeloma protein*—and thus contain high levels of this material in the bloodstream. The plasma levels of particular immunoglobulins in these patients may rise to as high as 5 or 10 grams per 100 milliliters. Large amounts of proteins—termed *Bence Jones proteins*—are also found in the urine of these patients. Bence Jones proteins are composed of the light chains (κ or λ) of circulating immunoglobulin; they may consist of either free or dimeric light chains. It is significant that myeloma proteins differ in certain aspects of amino-acid sequence from one patient to another. Thus, in each case, one particular immunoglobulin is selected for biosynthesis by the tumor cell.

Myeloma proteins actually served as prototypes for elucidating the structure of immunoglobulins. Studies of their amino-acid sequences and physical properties have revealed the common characteristics of immunoglobulins and the nature of their individual differences.

D. IMMUNOGLOBULIN SYNTHESIS

The fact that a vast variety of immunoglobulins can be produced, each in response to different foreign materials, presents some fundamental problems with regard to the genetic mechanism of their production. Since immunoglobulin molecules consist of constant and variable regions, it appears that each of these regions is coded by different genes. Thus, the formation of the complete protein may involve posttranslational covalent combination of individual polypeptide regions. Alternatively, messenger RNAs for each region may be linked prior to the initiation of protein synthesis. A third possibility is that the genes for constant regions and those for variable regions are integrated before transcription of the DNA. It remains to be determined which of these three possible mechanisms actually occurs in the immunoglobulin-synthesizing cell, or whether a completely different process is involved.

Another fundamental problem is how there can be a sufficient variety of DNA in cells to code for all possible immunoglobulin antibodies. Even if it is accepted that the amount of DNA is not limiting, the question remains as to how there can be DNA units already present to code for antibodies to all possible antigenic substances, including synthetic macromolecules. One proposal is that point mutations arise on the gene that codes for the variable region of immunoglobulin. These mutations might be induced, directly or indirectly, by foreign antigens. Another suggestion is that there are groups of genes for the variable regions that have a high incidence of crossover or recombination. This would generate a large variety of different genes from a comparatively small number of primary genetic components. Both of the proposals discussed assume that these numerous specific DNAs are produced in the somatic cells (lymphocytes), rather than in the germ cells. Whichever mechanism operates, it is generally accepted that each plasma cell produces a specific type of antibody. When it comes in contact with its respective antigen, the cell is stimulated to divide and produce increased quantities of the antibody. The overall process results in the growth of clones of antibody-generating cells that produce the respective antibodies and are sensitive to subsequent stimulation.

12.6. Complement

The complexes produced by the interaction of antibodies with antigens may give rise to a variety of changes in cellular structures. These include lethal alterations in cell membranes, increased capillary permeability, enhanced affinity of platelets and leukocytes, and increased phagocytosis by polymorphonuclear leukocytes. The transformations are mediated by the group of globulins that form the *complement system*. This system comprises a number of components that operate in concert via cascade processes somewhat similar to those of the coagulation pathways.

One route in the complement sequence, called the *classical pathway*, is

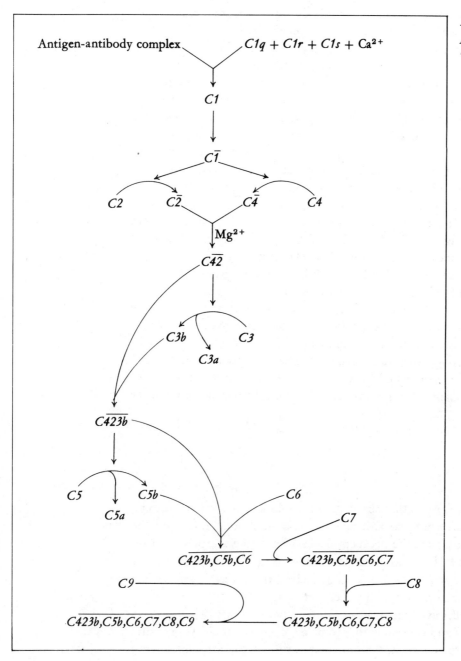

Fig. 12-13. Classical pathway for complement activation.

initiated by the reaction of the antigen-antibody (IgG or IgM) complex with $C1$ (termed the *first component of complement**) and calcium ions (Fig. 12-13). The $C1$ component consists of three distinct proteins designated as $C1q$, $C1r$, and $C1s$. Internal rearrangements and specific proteolytic cleavages of the complex yield an activated product called $\overline{C1}$, the activated form of the first component of complement. The latter catalyzes the conversion of another complement protein, $C4$, to its active form, $\overline{C4}$. Additionally, $\overline{C1}$ activates component $C2$ to provide $\overline{C2}$. $\overline{C4}$ binds to antigen-antibody complexes and, together with Mg^{2+}, combines with $\overline{C2}$ to form the product $\overline{C42}$, which functions as an enzyme that cleaves $C3$ to $C3a$ and $C3b$. Combination of $C3b$ with $\overline{C42}$ yields $\overline{C423b}$. This protein, which also contains the immune complex, cleaves $C5$ to $C5a$ and $C5b$. When $C5b$ combines with $C6$ and $\overline{C423b}$, the product is $\overline{C423b,C5b,C6}$. Subsequent sequential interactions with $C7$, $C8$, and $C9$ yield the active cytolytic agent $\overline{C423b,C5b,C6,C7,C8,C9}$. Several of the intermediate complexes also appear to enhance phagocytic actions. An important regulator of the sequence is a circulating inhibitor of $\overline{C1}$ called $\overline{C1}$-*inactivator* or $\overline{C1}$-INH, which binds to $C1$ and destroys its esterase activity.

Another process for the activation of the complement system, termed the *properdin pathway* or *alternate pathway*, bypasses the steps involving $C1$, $C4$, and $C2$. This pathway may be triggered by any of the immunoglobulins, as well as by certain polysaccharides, snake venoms, and lipopolysaccharides. In this pathway, $C3$ is activated by a group of serum components, one of which is properdin, and the reactions that follow are the same as those of the classical pathway.

12.7. Kinins

The *kinins* are polypeptides that circulate in the plasma and are known to be highly potent homeostatic control agents. Notable among these are the nonapeptide *bradykinin* and the decapeptide *kallidin*. These are produced by proteolytic cleavage of a protein called *kininogen* (Fig. 12-14). The enzyme that catalyzes the reaction, *kallikrein*, is generated from a zymogen termed *prekallikrein* or *Fletcher factor* (see also section 12.4.A).

The kinins are effective in increasing capillary permeability. They are also potent vasodilators and consequently produce a decrease in blood pressure. In addition to these actions, the kinins play a role in smooth-muscle contraction. High-molecular-weight kininogen is involved in the activation of factor XI of the blood coagulation cascade (section 12.4).

---

* Each component of complement is designated by a number, and the activated forms are written with a bar over the numeral.

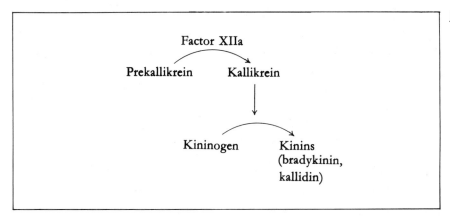

Fig. 12-14. Kinin production.

12.8. Diagnostic Enzymes

Plasma contains a wide variety of enzymes, of which most are intracellular in origin and appear in the blood as a result of normal turnover or leakage of cells. Some of the enzymes known to occur in plasma are listed in Table 12-6.

Since certain enzymes or definite isozymes are particularly abundant in specific tissues, an elevated serum level of such enzymes is indicative of degeneration of these tissues. Thus, the determination of the serum concentration of certain enzymes serves diagnostic purposes. Several examples of abnormalities in serum enzymes that are utilized in the diagnosis of disease will illustrate this method.

A sharp increase of *amylase* (page 164) in the serum indicates inflammation or degenerative changes in the pancreas. The cells of the pancreas are active in synthesizing this digestive enzyme, and the appearance in the serum of

Table 12-6. Representative Plasma Enzymes

Acid phosphatase
Alkaline phosphatase
Aldolase
Amylase
Carbonic anhydrase
Creatine phosphokinase
β-Glucuronidase
Glutamate-oxaloacetate transaminase (glutamate-aspartate transaminase)
Glutamate-pyruvate transaminase (glutamate-alanine transaminase)
Hydroxybutyrate dehydrogenase
Lactate dehydrogenase
Lipase
Peptidase
Ribonuclease

inordinate amounts of amylase must therefore be due to its leakage from the pancreas. Such leakage would occur when cells of the pancreas undergo extensive breakdown, e.g., during inflammation. (In addition to the pancreas, amylase is also synthesized and secreted by salivary glands; in most cases, however, analysis for serum amylase is performed for differential diagnosis of an abnormality in an internal organ.)

Alkaline phosphatase is a phosphate-hydrolyzing enzyme with a pH optimum at about 9.5. It is involved in the metabolism of bone, and an elevation of its level in serum suggests an abnormality in bone tissue. The serum level of alkaline phosphatase is elevated in patients with osteomalacia, bone cancers, hyperparathyroidism, Paget's disease, as well as in patients with hepatitis and liver tumors. For diagnostic purposes, then, the possibility of liver disease must be eliminated by other methods.

The analysis of serum for enzymes that are particularly abundant in heart muscle is useful in confirming diagnoses of myocardial infarction and in following the progress of the patient. These enzymes include *glutamate-oxaloacetate transaminase, lactate dehydrogenase,* and *creatine phosphokinase*. Since the level of glutamate-oxaloacetate transaminase in serum is normally low, an increase in the concentration of this enzyme is evident even in cases where cardiac damage is comparatively small. An increase in the serum concentration of this enzyme may also occur as a result of liver damage; however, in the latter case, there is also a rise in *glutamate-pyruvate transaminase* levels. Hence, simultaneous analysis for the two enzymes is required for accurate diagnosis.

Creatine phosphokinase is found in both heart muscle and skeletal muscle. Similarly, lactate dehydrogenase is abundant in both of these tissues, in addition to kidney and liver. Although serum analyses for both of these enzymes in conjunction with determinations of the transaminases are utilized in diagnosis, more specific data can be obtained from assays for the isozyme ratios. Each tissue has a characteristic distribution pattern of lactate-dehydrogenase isozymes (section 6.2.B) and creatine-phosphokinase isozymes, and an elevation in the serum level of the particular isozyme that is normally high in a tissue is an indicator of damage at that site.

In many instances, there are overlaps in the tissue distribution of isozymes. For example, the serum lactate-dehydrogenase isozyme pattern is similar in myocardial infarction, hemolytic anemia, and renal tubular necrosis. However, assays for a combination of several enzymes may serve to differentiate the signs of one disease from those of another. Moreover, enzyme determinations are usually performed in addition to other diagnostic procedures; often, such enzyme assays are utilized primarily to confirm a diagnosis or to substantiate a doubtful finding.

Suggested Reading

Capra, J. D., and Edmundson, A. B.   The Antibody Combining Site. *Scientific American* 236 (1): 50–59, 1977.

Edelman, G. M.   Antibody Structure and Molecular Immunology. *Science* 180:830–840, 1973.

Jerne, N. K.   The Immune System. *Scientific American* 229 (7): 52–60, 1973.

Meyer, M. M.   Complement, Past and Present. *The Harvey Lectures* (1976–1977), Series 72, pp. 139–193. New York: Academic Press, 1978.

Poljak, J. R.   Correlations Between Three-Dimensional Structure and Function of Immunoglobulins. *CRC Critical Reviews in Biochemistry* 5:45–84, 1978.

Putnam, F. W.   *The Plasma Proteins,* Vol. 1. New York: Academic Press, 1975.

Ratnoff, O. D., and Bennett, B.   The Genetics of Hereditary Disorders of Blood Coagulation. *Science* 179:1291–1298, 1973.

Wilkinson, J. H.   *The Principles and Practice of Diagnostic Enzymology.* Chicago: Year Book Medical Publishers, 1976.

13. Respiration and Electrolyte Balance: Functions of the Lungs and Kidneys

In order to thrive and maintain normal metabolism, the organism requires a system for delivering a constant supply of oxygen as well as a mechanism for the efficient removal of carbon dioxide. In higher animals, both functions are performed by the integrated actions of the pulmonary and circulatory systems. Oxygen in the air inhaled through the lungs passes from the alveoli into the bloodstream and is then transported through the arteries while bound to hemoglobin in erythrocytes. As the erythrocytes traverse the capillaries, oxygen is discharged and made available to the cells. Concurrently, carbon dioxide is released from the cells and is carried to the lungs via the venous circulation.

On the molecular level, respiration involves (1) diffusion and transport of the gases, (2) combination of oxygen with hemoglobin, (3) binding of carbon dioxide with hemoglobin, and (4) interactions between carbon dioxide, water, hydrogen ions, and other electrolytes.

13.1. Transport of Oxygen and Carbon Dioxide

Air is a mixture of gases composed of approximately 79% nitrogen, 20% oxygen, and 0.04% carbon dioxide. In the lungs, a portion of the oxygen diffuses into the blood via the alveolar membrane. Concurrently, the gaseous mixture in the alveoli is enriched in carbon dioxide, which diffuses from the capillaries of the pulmonary vascular system. Superimposed on these processes is the constant inhalation of atmospheric air and exhalation of the alveolar mixture. The average composition of alveolar gas reflects a relatively balanced state that in turn represents the resultant of all these events. The alveolar gas contains approximately 18% oxygen and 5.3% carbon dioxide.

Table 13-1. Average Partial Pressures of Oxygen and Carbon Dioxide in Different Body Compartments

	P_{O_2} (torr)	P_{CO_2} (torr)
Alveolar gas mixture	105	40
Arterial circulation	100	40
Capillaries	20	60
Cells	< 20	> 60
Venous circulation	40	46
Inspired air	159	0.3
Expired air	115	32

The concentrations of individual gases in a mixture are expressed most conveniently in terms of their *partial pressures*, which are defined as the pressure each gaseous component would exert if it were alone in the same volume. Since each gas in a mixture may be assumed for present purposes to contribute to the total pressure in proportion to its concentration, the partial pressure of a gaseous constituent can be obtained by multiplying its percentage composition times the total pressure. The unit for pressure is the *torr*, which is the pressure exerted by a 1-millimeter column of mercury at 0°C. (Gas pressures are commonly specified in terms of millimeters of mercury or *mmHg*. This unit is essentially equivalent to the torr.) The partial pressures of oxygen (P_{O_2}) and carbon dioxide (P_{CO_2}) at different anatomic sites are shown in Table 13-1.

The average partial pressures of gases in each body compartment are the resultant of the steady infusion of oxygen into the system and the concurrent removal of carbon dioxide. Although the compositions of inspired and expired air differ considerably, the steady-state concentrations of oxygen and carbon dioxide at each anatomic site fluctuate to a relatively small extent in the normal individual. As noted in the preceding discussion, this balance is produced by the regulated uptake of oxygen from the lungs and the release of carbon dioxide to the lungs.

Gases diffuse between the alveoli and their blood supply via the alveolar membrane. The direction of movement is governed by the differences between the partial pressures of each gaseous component in the blood and those in the alveolar gas mixture. For example, if the P_{CO_2} in the blood entering the lungs (venous blood) is 46 torr and the P_{CO_2} in the alveoli is 38 to 41 torr, the difference in pressure produces a movement of carbon dioxide from the blood to the alveoli. Similarly, the alveolar P_{O_2} is normally about 105 torr and the venous P_{O_2} about 40 torr. Consequently, oxygen will diffuse from the lungs into the blood. As a result of these transfers, the P_{O_2}

and P_{CO_2} in the blood leaving the lungs are 100 and 40 torr, respectively. Since nitrogen is not metabolized, its partial pressure in the lungs does not change significantly.

On the average, an individual *at rest* consumes about 250 ml of oxygen and produces 200 ml of carbon dioxide per minute. In a day, an active person who consumes 3000 calories of a balanced diet requires about 600 liters (about 27 moles) of oxygen and evolves 480 liters (about 22 moles) of carbon dioxide. In order to provide the necessary amounts of oxygen for the cells and, concomitantly, to remove sufficient carbon dioxide, approximately 5 liters of blood circulates through the pulmonary vascular system, and 8 to 10 liters of air is inhaled every minute. About 500 milliliters of air is taken into the lungs with each inhalation. Although this quantity of air contains approximately 100 milliliters oxygen, only 15 or 16 milliliters of O_2 diffuses into the blood. It is to be noted that these are average values that may vary somewhat under different conditions.

Almost all the oxygen that diffuses from the alveoli into the arterial circulation combines with hemoglobin within the erythrocytes. The solubility of oxygen in plasma is less than 0.3%, but as a result of the chemical combination of oxygen with hemoglobin, the oxygen-carrying capacity of blood is increased 50-fold. The average concentration of hemoglobin in blood is approximately 15 grams per 100 milliliters. Since each gram of hemoglobin can combine with as much as 1.34 milliliters of oxygen, each 100 milliliters of blood can carry about 20 milliliters of oxygen. The mechanism and regulatory factors involved in the uptake and delivery of oxygen by hemoglobin depend on the unique structure and chemical properties of the hemoglobin molecule. It should be noted that hemoglobin also functions in the transport of carbon dioxide and the maintenance of blood pH. The properties of this vital protein and the chemistry of its physiologic reactions are discussed in the next section.

13.2. Hemoglobin and Oxyhemoglobin

Hemoglobin is a globular, conjugated protein that contains four ferrous porphyrin units, termed *heme* (see top of page 498). The concentration of hemoglobin in erythrocytes is 31.3 to 39.6 grams per 100 milliliters of cells. This corresponds to 12 to 17 grams of hemoglobin per 100 milliters of whole blood. The apoprotein component, *globin,* is composed of four polypeptide chains that are held together by non-covalent forces. In normal adults, these polypeptide chains (except for a small number of variants) consist of two identical units called α *chains* and two identical units called β *chains*. The total hemoglobin macromolecule with four polypeptide-heme units has a molecular weight of 68,000; this is called *hemoglobin* A_1 and its polypeptide

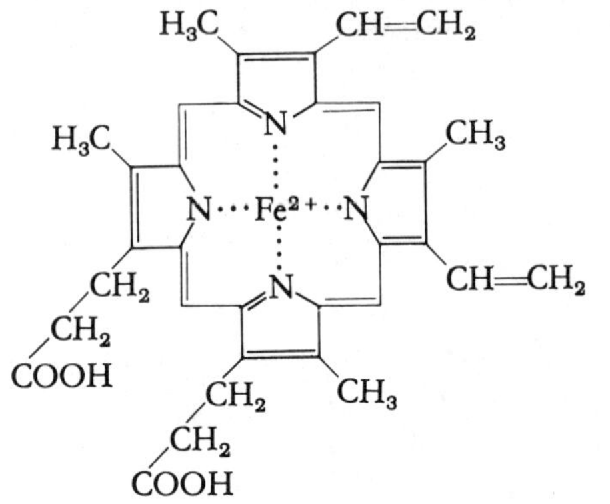

Heme

composition is designated as $\alpha_2\beta_2$ (see also section 1.2.C.6). Other variants and fetal hemoglobin will be discussed in section 13.4.

The hemoglobin macromolecule may be visualized as a spherical structure in which the four polypeptide chains are positioned in a tetrahedral relationship (Fig. 1-19, page 41). The heme groups of each subunit occur in niches close to the exterior of the macromolecule. The ferrous iron atom is located somewhat out of the plane of the nitrogen atoms of the tetrapyrrole groups. In addition to being bound to the four nitrogens of the porphyrin ring, each iron atom is coordinately bonded with the nitrogen atoms of two histidine residues of the protein. The folding of the polypeptide chain is such that these histidine units are perpendicular to the plane of the porphyrin ring. The crevices containing the heme and the interior, or core, of the hemoglobin macromolecule are populated with non-polar amino-acid residues.

Hemoglobin combines with oxygen to form *oxyhemoglobin*. When completely saturated, each molecule of hemoglobin (Hb) binds four molecules of oxygen:

$$Hb + 4O_2 \rightleftharpoons Hb(O_2)_4$$

Although this is an oxidation reaction, it does not involve a change in the charge on the iron atom. The iron in hemoglobin remains in the ferrous (Fe^{2+}) state, unlike the case when cytochromes are oxidized and the ferrous ion is oxidized to the ferric (Fe^{3+}) state.

When hemoglobin reacts with oxygen, its quaternary structure is altered, i.e., the hemoglobin molecule undergoes a conformational change. One effect of this change is that the iron atom, which was initially outside the plane of the porphyrin ring, is drawn into the same plane as the four pyrrole groups. One of the histidine residues that is coordinately bonded with the iron atom is displaced, causing a rearrangement in the entire chain. As a result of this movement, the spatial relationship between the subunits is reordered to yield a new conformation with a more compact structure than the original.

Another consequence of this conformational transformation is that certain histidine residues and the carboxyl-terminal amino acid (aspartate) of the β chains are shifted to new environments with lower negative charge. Consequently, there is an increase in the tendency for these residues to release hydrogen ions. Expressed in another way, oxygenation of hemoglobin produces a decrease in the pK_a of certain amino-acid residues of the protein. The overall effect is that the degree of dissociation or acidity of oxyhemoglobin is greater than that of hemoglobin (pK_a 6.17 versus 7.71). Specifically, for each molecule of oxygen bound to hemoglobin, 0.7 equivalent of H^+ is released. The reaction between oxygen and a hemoglobin subunit may thus be described by the equation:

$$\text{Subunit} + O_2 \rightleftharpoons \text{Subunit-}O_2 + 0.7H^+$$

A decrease in blood pH, then, will shift this reaction from right to left and thus enhance the dissociation of oxyhemoglobin. Conversely, an increase in pH should favor the binding of oxygen to hemoglobin.

The increase in acidity of hemoglobin upon oxygenation is important physiologically because it provides one of the mechanisms for the release of oxygen to the tissues. As blood moves through the tissue capillaries, it takes up acidic metabolites and carbon dioxide, which increase the hydrogen-ion concentration in the blood. This decrease in pH then enhances the dissociation of oxyhemoglobin, and the oxygen that is released diffuses into the cells. Since deoxygenation decreases the acidity (increases the pK_a) of hemoglobin, the latter also serves to neutralize some of the acidic metabolites and thus functions as one of the buffering systems in the blood.

The degree to which hemoglobin combines with oxygen is dependent on the concentration of oxygen, as may be anticipated from the equation:

$$Hb + 4O_2 \rightleftharpoons Hb(O_2)_4$$

However, if the oxyhemoglobin concentration is plotted against the concentration of oxygen (expressed as P_{O_2}), a sigmoid curve is obtained rather than the expected hyperbolic curve (Fig. 13-1). The relationship expressed by the curve in Figure 13-1—termed the *dissociation curve* for oxyhemoglobin— is similar to that found for the effect of substrate concentration on the initial reaction velocity of allosteric enzymes. Such a sigmoid curve indicates

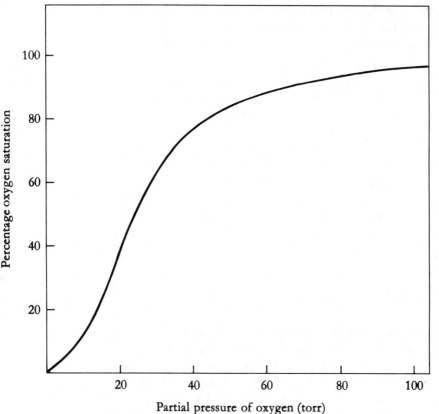

Fig. 13-1. Effect of oxygen concentration (expressed as the partial pressure of oxygen, P_{O_2}) on the binding of hemoglobin with oxygen (P_{CO_2} for this curve is 20 torr, pH 7.5). This graph is also called the dissociation curve for oxyhemoglobin. Note that significant unsaturation does not occur until P_{O_2} falls below 60 torr.

a cooperative effect on the binding process, i.e., the binding of the first oxygen to the hemoglobin produces a conformational change that increases the affinity of the hemoglobin molecule for additional oxygen. Although the individual oxygen-binding sites occur on different polypeptide subunits, they are not independent of each other. As explained in the preceding discussion, oxygen binding causes the β chains to move toward each other and this effect produces a profound conformational change in the total macromolecule. The transformation induced by the binding of each oxygen molecule produces an intermediate structure that has an even greater affinity for additional oxygen. As may be expected, the converse is also true; i.e., the dissociation of one of the oxygens from a fully saturated hemoglobin molecule increases the tendency for the remaining oxygen molecules to dissociate. The oxygen dissociation curve is also influenced by a number of other factors, e.g., pH and P_{CO_2}.

The partial pressure of oxygen in the alveoli is somewhat higher than 100 torr, and that in the blood leaving the lungs is about the same (see Table 13-1).

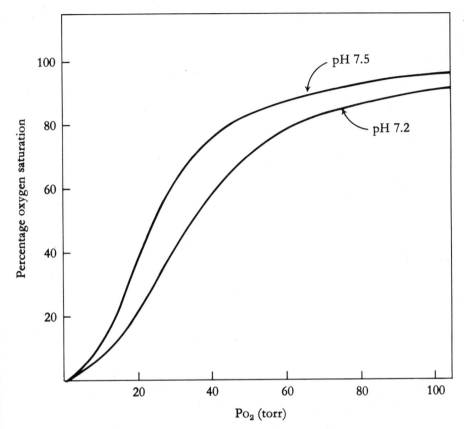

Fig. 13-2. Relationship between pH and the dissociation curve for oxyhemoglobin.

At this pressure, essentially all the hemoglobin forms oxyhemoglobin (Fig. 13-1). As the oxyhemoglobin passes through the capillary circulation of tissues (e.g., muscle) where the P_{O_2} is 20 torr or less, over 75% of the oxygen (Fig. 13-1) is released from the oxyhemoglobin and taken up by the cells. Thus, the characteristic dissociation properties of hemoglobin provide an important mechanism for the transfer of oxygen from the lungs to the cells of all other tissues.

As noted earlier, a decrease in pH enhances the dissociation of oxyhemoglobin. As shown in Figure 13-2, a decrease in pH brings about the displacement of the oxyhemoglobin dissociation curve to the right. Thus, at any given P_{O_2}, hemoglobin will bind less oxygen if the pH of the medium is decreased. For example, in the muscle capillaries where the P_{O_2} is 20 torr, the increase in acidity produced by the release of CO_2 or lactate enhances the release of oxygen from hemoglobin to the adjacent cells.

The binding of oxygen to hemoglobin is also sensitive to the concentration of carbon dioxide, or P_{CO_2} (Fig. 13-3). This phenomenon is known as the *Bohr effect*. As the partial pressure of carbon dioxide is increased (e.g., to 80

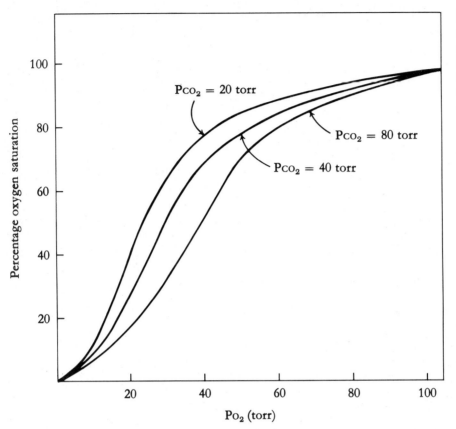

Fig. 13-3. Effect of carbon dioxide concentration on the dissociation of oxyhemoglobin (Bohr effect). As the P_{CO_2} increases, oxygen will tend to dissociate more readily from hemoglobin.

torr), the P_{O_2} required for saturating the hemoglobin is increased. Expressed in another way, for a given P_{O_2}, the degree of dissociation of oxyhemoglobin is increased as the P_{CO_2} is increased. To a certain extent, this phenomenon may be attributed to the action of CO_2 in acidifying the medium:

$$CO_2 + H_2O \rightleftharpoons H_2CO_3 \rightleftharpoons H^+ + HCO_3^-$$

Since an increase in hydrogen-ion concentration effectively increases the dissociation of oxyhemoglobin, it follows that an elevation in P_{CO_2} will produce the same result, though indirectly. However, even when the concentration of CO_2 is increased at constant pH, there is still a significant effect on the dissociation of oxyhemoglobin.

The direct effect of carbon dioxide (i.e., exclusive of its influence on pH) on the oxyhemoglobin dissociation curve arises from its combination with the amino terminals of the hemoglobin subunits to form *carbaminohemoglobin*:

$$Hb-NH_2 + CO_2 \rightleftharpoons Hb-NHCOO^- + H^+$$

This type of reaction is not specific for hemoglobin; it occurs with plasma proteins in general. The point, however, is that carbaminohemoglobin has a decreased affinity for oxygen, and thus its formation is reflected by a shift in the oxyhemoglobin dissociation curve to the right.

The Bohr effect is the result of the influence of hydrogen ions and carbon dioxide on the binding of oxygen to hemoglobin: both H^+ and CO_2 reduce the affinity of hemoglobin for oxygen, and conversely, they will enhance the dissociation of oxyhemoglobin. This interplay between H^+, CO_2, and O_2 is critical in the transport of oxygen from the alveoli to the cells of tissues. For example, in arterial blood leaving the lungs, the P_{CO_2} is approximately 40 torr and the P_{O_2} is 100 torr (see Table 13-1), which will favor the combination of oxygen with hemoglobin. In tissue capillaries, however, the P_{CO_2} is more than 60 torr and the P_{O_2} is about 20 torr (Table 13-1). These conditions promote the dissociation of oxyhemoglobin and the uptake of oxygen by the tissues. This transfer of oxygen is also enhanced by the release of metabolic acids from the tissues. Such acids decrease the pH in the microcirculation and consequently increase the dissociation of oxyhemoglobin.

The oxygen-binding properties of hemoglobin are also influenced by a component present in erythrocytes, *2,3-diphosphoglycerate* (DPG):

$$\begin{array}{l} COOH \\ | \\ CHOPO_3H_2 \\ | \\ CH_2OPO_3H_2 \end{array}$$

2,3-Diphosphoglyceric acid

The effect of DPG on the dissociation of oxyhemoglobin is shown in Figure 13-4. Diphosphoglycerate combines with hemoglobin within the central cavity of the macromolecule via ionic interactions with lysine, histidine, and terminal amino groups. The cross-linking produced by these reactions stabilizes the deoxygenated hemoglobin and thus decreases its affinity for oxygen. When oxygen combines with hemoglobin and the resultant conformational change produces a more compact form of hemoglobin, DPG is extruded and released:

$$Hb\text{—}DPG + 4O_2 \rightleftharpoons Hb(O_2)_4 + DPG$$

The importance of the DPG-hemoglobin complex in the release of oxygen to the tissues is shown by the dissociation curves in Figure 13-4. In the presence of DPG, considerable amounts of oxygen can be provided to the cells even when the P_{O_2} at the site is quite low (e.g., 20 torr).

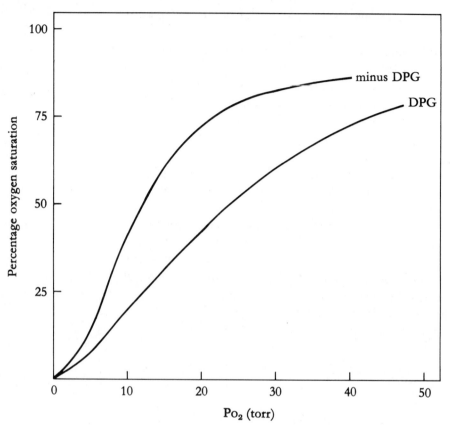

Fig. 13-4. Effect of 2,3-diphosphoglycerate on the dissociation of oxyhemoglobin.

13.3. Transport of Carbon Dioxide

Carbon dioxide from tissue cells diffuses through the interstitial fluid into the capillaries. In the erythrocytes, part of the CO_2 combines with hemoglobin to form the carbamino derivative. Erythrocytes also contain the enzyme *carbonic anhydrase*, which catalyzes the reaction:

$$CO_2 + H_2O \rightleftharpoons H_2CO_3$$

Dissociation of H_2CO_3 provides HCO_3^- and H^+. Some of the hydrogen ions are taken up by deoxygenated hemoglobin. Thus, the increase in the pK_a of hemoglobin upon oxygenation serves to counteract the potential decrease in pH that can be produced by CO_2 in the blood. This neutralizing action of hemoglobin is known as the *isohydric shift*. Since the negative charge of the hemoglobin molecule is originally counterbalanced by positively charged potassium ions, the effect of the isohydric shift is to release K^+ in addition to HCO_3^-. As mentioned in section 13.2, the dissociation of 1 mole of oxygen from oxyhemoglobin results in the uptake of 0.7 equivalent of hydrogen ion. Hence, as a consequence of the isohydric shift, 0.7 milliequivalent of HCO_3^- will be produced for each millimole of oxygen released.

The processes described in the previous paragraph lead to an increase in the concentration of HCO_3^- within the erythrocytes. This creates two imbalances that must be readjusted: (1) the required intracellular ratio of bicarbonate to chloride ions is shifted, and (2) the increase in HCO_3^- produces an alteration in the ratio of the concentration of diffusible ions within the erythrocyte to that in the plasma, and it thus causes an osmotic imbalance in the system (see section 1.3.C). These imbalances are rectified by (1) diffusion of HCO_3^- from the erythrocytes to the plasma and movement of chloride in the opposite direction (*chloride shift*) and (2) uptake of water by the erythrocytes.

Upon reaching the lungs where the P_{O_2} is comparatively high, hemoglobin combines with oxygen. The change in the pK_a of hemoglobin caused by this reaction results in the release of hydrogen ions, which interact with the bicarbonate ions (taken up from plasma) to form carbonic acid. This H_2CO_3 in turn decomposes to carbon dioxide, which diffuses into the alveoli:

$$H^+ + HCO_3^- \rightleftharpoons H_2CO_3 \longrightarrow CO_2 + H_2O$$

As would be expected, the movement of HCO_3^- from the plasma to the erythrocytes must be compensated by a corresponding efflux of chloride from these cells. The change in osmotic pressure brought about by these movements is adjusted by diffusion of water from the erythrocytes into the plasma. A general effect of these changes is that the erythrocytes of venous blood contain more Cl^- than those in arterial blood. Also, as a result of the influx of water, erythrocytes are larger when they are in the veins than they are in the arteries.

Although the solubility of carbon dioxide in water is less than 3 milliliters per 100 milliliters when the P_{CO_2} is 40 torr, blood generally carries from 50 to 60 milliliters CO_2 per 100 milliliters (venous blood 55% to 60%; arterial blood 50% to 52%). This is possible because practically all the carbon dioxide taken up by the blood is converted to soluble derivatives. As indicated previously, a considerable fraction (about 20% to 25% of the CO_2 content) is bound to hemoglobin and other proteins as carbamino derivatives. Almost all the unbound portion is in the form of bicarbonate ion, while minimal amounts are present as carbonic acid and dissolved carbon dioxide. The mixture of carbonic acid and bicarbonate constitutes a critical buffering system in the blood, and they will be discussed in detail in a later section (section 13.5).

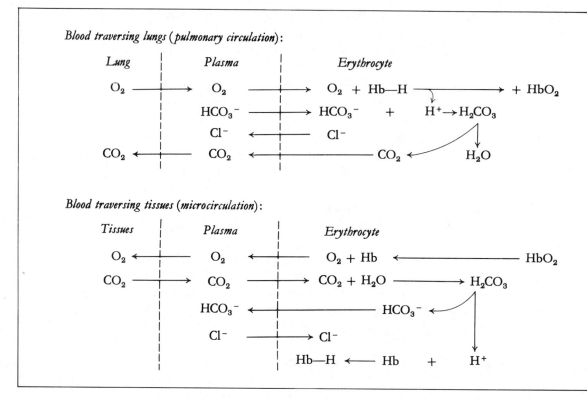

Fig. 13-5. Transport of oxygen from the lungs to the tissues and of carbon dioxide from the tissues to the lungs (carbamino derivatives are not shown).

The function of hemoglobin and the erythrocytes in the transport of oxygen and carbon dioxide will be reviewed at this point (Fig. 13-5). Oxygen diffuses from the alveoli of the lung through the plasma and into the erythrocytes. The partial pressure of oxygen in this vicinity is approximately 100 torr and that of carbon dioxide is about 40 torr. The hemoglobin therefore becomes over 96% saturated with oxygen. The hydrogen ions released as a result of the change in pK_a of hemoglobin interact with HCO_3^-, which is drawn into the erythrocytes to form H_2CO_3. The latter is converted to carbon dioxide and water by the action of carbonic anhydrase. The carbon dioxide thus formed then diffuses into the plasma and out through the lungs.

Oxyhemoglobin is carried through the arterial circulation and into the tissue capillaries. Since the P_{O_2} in interstitial fluid is considerably lower than that in arterial blood, oxyhemoglobin dissociates and the released oxygen is taken up by the cells. Also, CO_2 from tissue cells diffuses into the interstitial fluid and thence the capillaries. The increase in CO_2 concentration further enhances the dissociation of oxyhemoglobin (Bohr effect). By the time this blood reaches the venous circulation, the P_{CO_2} and P_{O_2} are 46 and 40 torr, respectively.

Each gram of hemoglobin when saturated carries 1.34 milliliters of oxygen. Since 100 milliliters of blood contains about 15 grams of hemoglobin, more than 20 milliliters of oxygen can be carried in this volume. Approximately 32% of the oxyhemoglobin is dissociated as the blood passes through the capillaries from the arterial to the venous circulation. Thus, 6.4 milliliters of oxygen is supplied to the cells as 100 milliliters of blood circulates through the capillaries.

Carbon dioxide released from the tissues diffuses through the plasma and into the erythrocytes. About 20% to 25% of the CO_2 is bound to hemoglobin and plasma proteins as carbamino derivatives. Most of the remaining CO_2 is converted in the erythrocytes to H_2CO_3 by the action of carbonic anhydrase. Dissociation of H_2CO_3 yields HCO_3^- and H^+. A considerable portion of this H^+ is neutralized by hemoglobin that has undergone a change in pK_a as a result of the loss of oxygen; the remainder is buffered by other systems to be discussed in the next section. The HCO_3^- diffuses out of the erythrocytes, and chloride ions move in (chloride shift). When the blood reaches the lungs, CO_2 diffuses into the alveoli, since the P_{CO_2} in the venous circulation is 46 torr while that in the lungs is 40 torr. The HCO_3^- from plasma moves into the erythrocytes (while Cl^- moves out), where it combines with H^+ released from oxygenated hemoglobin. Carbonic anhydrase drives the decomposition of H_2CO_3 to water and CO_2, and the latter diffuses into the lungs.

13.4. Variations in Hemoglobin Structure

The principal form of hemoglobin in erythrocytes of normal human adults, designated as *HbA*, consists of two α and two β polypeptide chains. Each type of chain is coded by a separate gene. In contrast to the polypeptide components of adult hemoglobin, fetal hemoglobin consists of α chains and γ chains. The latter differ from β chains in several amino-acid residues and in the sequences at specific sites in the polypeptides. Thus, fetal hemoglobin, *HbF*, is denoted as $\alpha_2\gamma_2$. At about two months before birth, the biosynthesis of γ chains decreases gradually and β chains are produced instead. Shortly after birth, therefore, most of the hemoglobin consists of the HbA variety and in the adult HbF comprises less than 0.5% of the total hemoglobin.

Another variant in hemoglobin structure, termed *HbA₂* is due to the replacement of β chains by δ chains, i.e., the polypeptide composition of HbA_2 is $\alpha_2\delta_2$. The composition of the δ chains differs from that of the β chains in about 10 amino-acid residues. HbA_2 comprises about 2.5% of the total hemoglobin content of normal adults.

In addition to the hemoglobin variants that arise from the biosynthesis of different polypeptide chains, erythrocytes also contain hemoglobins in which the β chains undergo posttranslational modification. For example, hemoglobins, A_{1b} and A_{1c}, are formed from HbA by a non-enzymatic combination of glucose 6-phosphate with amino-terminal valine residues in the β chains. A rearrangement of the adduct yields a fructosyl derivative of HbA:

$$
\begin{array}{c}
\text{HC=O} \\
\text{HCOH} \\
\text{HOCH} \\
\text{HCOH} \\
\text{HCOH} \\
\text{CH}_2\text{OPO}_3\text{H}_2
\end{array}
\quad + \quad
\begin{array}{c}
\text{O} \\
\parallel \\
\text{H}_2\text{NCHC}\cdots\beta \\
\mid \\
\text{CH(CH}_3)_2
\end{array}
\quad \rightleftharpoons \quad
\begin{array}{c}
\text{O} \\
\text{HC=NHCHC}\cdots\beta \parallel \\
\mid \quad\quad \mid \\
\text{HCOH} \quad \text{CH(CH}_3)_2 \\
\text{HOCH} \\
\text{HCOH} \\
\text{HCOH} \\
\text{CH}_2\text{OPO}_3\text{H}_2
\end{array}
\quad \longrightarrow
$$

Glucose 6-phosphate • N-Terminal valine of β chain • Sugar adduct

$$
\begin{array}{c}
\text{O} \\
\text{CH}_2\text{NHCHC}\cdots\beta \parallel \\
\mid \quad\quad\quad \mid \\
\text{C=O} \quad \text{CH(CH}_3)_2 \\
\text{HOCH} \\
\text{HCOH} \\
\text{HCOH} \\
\text{CH}_2\text{OPO}_3\text{H}_2
\end{array}
\quad \longrightarrow \quad
\begin{array}{c}
\text{O} \\
\text{CH}_2\text{NHCHC}\cdots\beta \parallel \\
\mid \quad\quad\quad \mid \\
\text{C=O} \quad \text{CH(CH}_3)_2 \\
\text{HOCH} \\
\text{HCOH} \\
\text{HCOH} \\
\text{CH}_2\text{OH}
\end{array}
$$

Sugar on HbA_{1b} (1-Deoxy-1-N-valylfructose-6-phosphate) • Sugar on HbA_{1c} (1-Deoxy-1-N-valylfructose)

HbA_{1b} and HbA_{1c} represent about 3% to 6% of the hemoglobin in normal individuals.

    The glycosylated hemoglobins are of special interest in view of the finding that their levels in blood are increased, as much as two- to three-fold in patients with diabetes mellitus. This is ascribed to the elevated concentration of glucose in the blood and the consequent increase in the rate of interaction between the sugar, or its phosphate derivative, with hemoglobin. Analysis for glycosylated hemoglobins is a valuable aid in assessing diabetic control and in monitoring the diabetic patient.

In addition to the minor hemoglobin components that are present in normal individuals, a wide variety of structurally abnormal hemoglobins occur in patients with inherited disorders. These aberrations can arise from point mutations or frame shifts in the DNA (section 11.12) and are transmitted genetically. For example, the hemoglobin of individuals with sickle cell anemia contains a variant designated as HbS, in which a specific glutamate residue of the β chain is replaced by a valine residue. This results from a point mutation that gives rise to a change from GAA (or GAG) base sequence to GUA (or GUG) in the codon of the mRNA. HbS can be distinguished from HbA by electrophoretic analysis, since the two hemoglobins differ from each other in their net charge (Figure 1-20, page 41). The solubility of non-oxygenated HbS is considerably lower than that of deoxy-HbA. It, therefore, precipitates as a semisolid gel and causes the erythrocytes to assume a crescent, or sickled, form. The deformed cells cannot pass through the capillaries and consequently occlude the vessels, causing thromboses and infarctions. In addition, sickled erythrocytes are more fragile than normal erythrocytes. The comparatively high degree of hemolysis of these cells is the reason for the occurrence of severe anemia in individuals with sickle cell disease.

Another example of an abnormal hemoglobin is HbI. This variant differs from HbA in that one of the lysine residues in the β chain is replaced by a glutamate unit. The resulting change in the net charge of the hemoglobin affects the electrophoretic mobility of the protein (Figure 1-20, page 41). This can also arise from a point mutation that changes the mRNA codon from AAA to GAA (see Table 11-2, page 436). Numerous other variations in hemoglobin structure have been reported; the two types that were defined in this section serve only to describe some of the effects of the structural changes on the chemistry and function of the hemoglobin.

13.5. Blood Buffers

The normal hydrogen-ion concentration of blood is 3.7 to 4.2×10^{-8} equivalents per liter; this corresponds to pH 7.38 to 7.45. Since a decrease in pH beyond 7.0 or an increase above 7.8 is not compatible with life, the organism must maintain the pH of blood within narrow limits. This is accomplished by various mechanisms that are capable of buffering possible fluctuations in hydrogen-ion concentration.

Normal intracellular metabolic processes generate acidic products. The oxidation of the carbon skeleton of most nutrients—especially carbohydrates and fatty acids—generates carbon dioxide. In this way, about 15 to 20 moles of CO_2 are produced daily by the average human adult. When released to the bloodstream, some of the CO_2 forms H_2CO_3. Part of the H_2CO_3 ionizes to bicarbonate and hydrogen ions. In addition to CO_2, a significant amount of sulfuric acid is produced in metabolism, mainly as a result of the oxidation of cysteine (approximately 1 millimole per kilogram of body weight). Organic acids are also produced during normal metabolism; these include lactic, acetoacetic, and β-hydroxybutyric acids.

The continual challenge to physiologic pH that is presented by metabolic products is met by several chemical buffering systems that operate in conjunction with the actions of the lungs and kidneys. The buffer systems of the blood are provided by bicarbonate-carbonic acid, ionized and non-ionized plasma proteins, hemoglobin-oxyhemoglobin, and dihydrogen phosphate-monohydrogen phosphate systems. Each of these will first be considered individually, and then the interactions among them will be described.

The components of the bicarbonate buffer system are carbon dioxide, carbonic acid, and bicarbonate:

(1) CO_2 (gaseous phase) $\rightleftharpoons$ CO_2 (dissolved in plasma)

(2) CO_2 (dissolved) + H_2O $\rightleftharpoons$ H_2CO_3

(3) H_2CO_3 $\rightleftharpoons$ H^+ + HCO_3^-

The first reaction indicates the dependence of the amount of carbon dioxide dissolved in plasma on its concentration in the gaseous phase interacting with the plasma (e.g., the alveolar air mixture). According to *Henry's law*, the concentration of dissolved CO_2 is proportional to the partial pressure of gaseous CO_2 (P_{CO_2}) in equilibrium with that solution:

$$[CO_2] = kP_{CO_2} \qquad (13\text{-}1)$$

The proportionality constant, k, is dependent on the temperature and the nature of the other substances in solution.

The second reaction describes the interconvertibility between carbon dioxide and carbonic acid. As indicated above, this reaction attains equilibrium rapidly as a result of the action of carbonic anhydrase in the erythrocytes. The effective concentration of carbon dioxide in the blood may thus be considered to be the sum of the CO_2 and H_2CO_3 concentrations. The proportionality constant in equation 13-1 is 3.01×10^{-2} at 38°C when the concentration of CO_2 *plus* H_2CO_3 is in terms of millimoles per liter of plasma and when the P_{CO_2} is expressed in torr. Hence, for CO_2 in plasma, equation 13-1 can be written as:

$$[CO_2 + H_2CO_3] = 3.01 \times 10^{-2} P_{CO_2}$$

In other words, both the CO_2 and the H_2CO_3 concentrations in plasma are regulated by the P_{CO_2} in the alveoli. Since the partial pressure of carbon dioxide in alveolar air is about 40 torr, the concentration of CO_2 plus H_2CO_3 in plasma is 1.2 millimoles per liter.

The rapid establishment of the equilibrium between carbon dioxide and carbonic acid that is brought about by the action of carbonic anhydrase also influences reaction 3. Thus, in considering the ionization of carbonic acid, the effective concentration is not simply that of H_2CO_3, but the sum of the CO_2 and H_2CO_3 concentrations. Moreover, although the pK_a for H_2CO_3 in aqueous solution is 3.8, that for H_2CO_3 in the presence of CO_2 in plasma is 6.1. Hence, the expression describing the relationship between the pH of plasma and the concentrations of H_2CO_3 and HCO_3^- (Henderson-Hasselbalch equation; see Chap. 1, section 1.2.A) is:

$$pH = 6.1 + \log \frac{[HCO_3^-]}{[H_2CO_3 + CO_2]} \qquad (13\text{-}2)$$

Substituting the expression derived from Henry's law, the relationship becomes:

$$pH = 6.1 + \log \frac{[HCO_3^-] \text{ millimole/liter}}{3.01 \times 10^{-2} P_{CO_2}} \qquad (13\text{-}3)$$

From equation 13-2, it can be seen that in arterial blood where the pH is 7.4, the ratio $[HCO_3^-]/[H_2CO_3 + CO_2]$ is 20 to 1:

$$7.4 = 6.1 + \log \frac{[HCO_3^-]}{[H_2CO_3 + CO_2]} \qquad (13\text{-}4)$$

$$\log \frac{[HCO_3^-]}{[H_2CO_3 + CO_2]} = 1.3$$

$$\frac{[HCO_3^-]}{[H_2CO_3 + CO_2]} = 20$$

Also, since the P_{CO_2} in arterial blood is about 40 torr (Table 13-1), the value of $[CO_2 + H_2CO_3]$ is 1.2 millimoles per liter. Substitution of this value in equation 13-4 shows that the concentration of bicarbonate in plasma is about 24 millimoles per liter.

Although a buffer system in which the acid has a pK_a of 6.1 would not be expected to function efficiently in maintaining a pH of 7.4 (see section 1.2.A), the bicarbonate-carbonic acid system does provide buffering at this pH under physiologic conditions. This is due to the fact that the partial pressure of CO_2 is kept relatively constant in alveolar air and arterial blood by the respiratory system. Any metabolic factor that causes a decrease in concentration of CO_2 plus H_2CO_3 in the blood—and consequently an increase in pH—is compensated by a rapid readjustment of the CO_2 concentration in alveolar air. Alternatively, when the level of organic acids (e.g., lactic or acetoacetic acids) is increased in the bloodstream and HCO_3^- is converted to H_2CO_3 plus CO_2, the decrease in pH is rectified by the removal of excess CO_2 via the lungs. As a result of such compensation, the ratio $[HCO_3^-]/[H_2CO_3 + CO_2]$ is kept within the required limits to maintain the pH at 7.4; i.e., the value of this ratio is held at about 20:1.

The vital factor in the effectiveness of the bicarbonate buffer system is that it operates in conjunction with the respiratory system, which in turn regulates the P_{CO_2}. Therefore, when a change in pH is due to some malfunction of the respiratory system that results in an abnormally high or low P_{CO_2}, the bicarbonate-carbonic acid system cannot provide the necessary buffering. Another way of expressing this idea is to say that physiologic buffering by the bicarbonate system is restricted to pH changes produced by acids other than H_2CO_3 or CO_2. Conditions involving pH changes due to breathing abnormalities—which are termed *respiratory acidosis* or *alkalosis*—require compensation by other systems, such as the renal compensatory mechanisms to be described subsequently and in section 13.6.

Regulation of blood pH by the lung operates via the nervous system; the respiratory center of the medulla is sensitive to changes in pH and P_{CO_2}. When the blood pH decreases, respiration is stimulated and the rate of removal of CO_2 is increased. The decrease in the ratio $[HCO_3^-]/[H_2CO_3 + CO_2]$ that was brought about by acid conditions is then corrected: the ratio is restored to the required value of 20:1 and the pH returns to 7.4. Alternatively, when there is an increase in pH, the rate of respiration is decreased, CO_2 and H_2CO_3 levels are increased, and the pH is readjusted to its normal value. The efficiency and utility of respiratory compensation will be discussed in a later section of this chapter (section 13.7).

The buffering action of the bicarbonate-carbonic acid system may be illustrated by considering a specific example. If 10 millimoles of a strong acid is added to a liter of blood containing 24 millimoles HCO_3^- and 1.2 millimoles H_2CO_3, the concentration of HCO_3^- will be decreased and that of H_2CO_3 increased as a result of the reaction:

$$HCO_3^- + H^+ \longrightarrow H_2CO_3$$

The concentrations of these two components will then be 14 millimoles HCO_3^- and 11.2 millimoles H_2CO_3 per liter, and the resulting pH will be:

$$pH = 6.1 + \log \frac{14}{11.2} = 6.2$$

However, in a physiologic system, the combined H_2CO_3 concentration—i.e., $[CO_2 + H_2CO_3]$— is maintained at approximately 1.2 millimoles per liter by the action of carbonic anhydrase and the removal of CO_2 through the lungs:

$$H_2CO_3 \longrightarrow H_2O + CO_2 \uparrow$$

As a result, the actual effect of the added acid on the pH of the blood is to reduce it by only 0.2 unit:

$$pH = 6.1 + \log \frac{14}{1.2} = 7.2$$

It is thus the maintenance of a constant concentration of H_2CO_3 through respiratory control that makes the bicarbonate system an efficient buffer against non-carbonic acids.

The bicarbonate buffer system also has a certain effect in counteracting pH changes brought about by the introduction of alkali. The hydroxyl ions react with H_2CO_3 according to the equation:

$$H_2CO_3 + OH^- \longrightarrow HCO_3^- + H_2O$$

If 1 millimole of OH^- is added to a liter of blood and there is no respiratory compensation, the resultant pH would be 8.2:

$$pH = 6.1 + \log \frac{25}{0.2} = 8.2$$

However, since the combined H_2CO_3 concentration is maintained constant at 1.2 millimoles per liter in the physiologic situation, the actual pH will be:

$$pH = 6.1 + \log \frac{25}{1.2} = 7.4$$

Other systems for the regulation of physiologic pH are provided by the blood proteins and phosphate ions. The blood proteins contain a number of ionizable side-chain groups (e.g., carboxyl, amino, imidazole, and guanidino); however, only those with a pK_a in the vicinity of 7.4 can exert a significant effect in regulating the pH of blood. The two that provide buffering at pH 7.4 are the imidazole group of histidine ($pK_a \approx 6.5$) and the amino group of the N-terminal amino acid of the proteins ($pK_a \approx 8.0$). The acid (proton-donor) and conjugate base (proton-acceptor) forms of histidine are depicted as follows:

$$\text{R}-\underset{\underset{\text{Imidazolium}}{}}{\text{imidazolium ring}-\text{NH}^+} \rightleftharpoons \text{R}-\underset{\underset{\text{Imidazole}}{}}{\text{imidazole ring}-\text{N}} + \text{H}^+$$

$$\text{R}-\text{NH}_3^+ \rightleftharpoons \text{R}-\text{NH}_2 + \text{H}^+$$
Ammonium   Amino

Since proteins contain only one N-terminal amino acid, this group makes a relatively small contribution as a buffer. However, histidine units are quite prevalent in a number of proteins, and their component imidazole groups are responsible for most of the buffering capacity of plasma proteins. For each protein, the pK_a value of particular imidazole groups is affected by the environment of the histidine residues within the macromolecule; however, an average of 7.4 may be assumed. Thus, in terms of the Henderson-Hasselbalch equation, the relationship between the pH and the ratio of ionized to non-ionized imidazole groups in the protein may be expressed as follows:

$$\text{pH} = 7.4 + \log \frac{[\text{Pr}]}{[\text{Pr}-\text{H}^+]}$$

where Pr—H$^+$ is the protein with proton-donating imidazole groups and Pr is the form with proton-accepting groups.

Of all the blood proteins, hemoglobin contributes the major buffering effect, primarily because of its high concentration. It should also be remembered that in addition to simple buffering, hemoglobin also functions to regulate the blood pH via the isohydric shift. This action is not that of a buffer in the strict sense of the term; rather, it is due to the difference in degree of ionization between hemoglobin and oxyhemoglobin. Specifically, at pH 7.4 an average of approximately 1.8 H$^+$ are released from each hemoglobin subunit, whereas 2.4 H$^+$ are dissociated from oxyhemoglobin. Consequently, as oxygen is released from hemoglobin in the bloodstream, its capacity to bind hydrogen ions is increased. This process (in addition to actual buffering) serves to counteract a large part of the acidifying effect produced by carbon dioxide from the cells.

The other significant buffer in the circulation is the dihydrogen phosphate anion, $H_2PO_4^-$. The pK_a for the ionization of $H_2PO_4^-$ to HPO_4^{2-} plus H$^+$

is 6.8. The dihydrogen phosphate-monohydrogen phosphate system is therefore an efficient buffer at pH 7.4. It is less important quantitatively than the other systems, because the phosphate concentration in plasma is comparatively low. The relationship between pH and the concentration of phosphate ions can also be expressed by the Henderson-Hasselbalch equation:

$$\text{pH} = 6.8 + \log \frac{[\text{HPO}_4^{2-}]}{[\text{H}_2\text{PO}_4^-]}$$

The metabolic activities of the cells result in the continuous release of carbon dioxide. As indicated in the previous discussion, the acidifying effect of CO_2 and H_2CO_3 cannot be buffered by the bicarbonate system. The principal buffers for this purpose are hemoglobin and the plasma proteins. The phosphate buffer system makes a small but significant contribution. Phosphate ions are also important in pH control by the kidneys (section 13.6).

Another mechanism to control blood pH, which may play a role after prolonged acidosis, is the interaction of hydrogen ions with bone. In acidic medium, calcium phosphate in bone binds hydrogen ions:

$$\text{Ca}_3(\text{PO}_4)_2 + 2\text{H}^+ \rightleftharpoons 3\text{Ca}^{2+} + 2\text{HPO}_4^{2-}$$

Also, concurrently with renal acidification of urine (section 13.6), each equivalent of HPO_4^{2-} can accept a proton from another acid, e.g., H_2CO_3:

$$\text{HPO}_4^{2-} + \text{H}_2\text{CO}_3 \longrightarrow \text{H}_2\text{PO}_4^- + \text{HCO}_3^-$$

Thus, each mole of $Ca_3(PO_4)_2$ can effectively neutralize four equivalents of acid. Although this process can serve as an efficient neutralization mechanism, it should not be considered as a normal functional process, because it leads to a serious loss of minerals from bone tissue. Actually, this reaction becomes significant only when acidosis is extended for a long period.

13.6. Renal Regulation of Body pH

In addition to their function in eliminating various metabolic waste products from the circulatory system, the kidneys serve to regulate blood pH and to maintain optimal concentrations of various vital components in the blood. Each kidney contains about a million functional units called *nephrons*. Several aspects of the anatomy and activities of the nephron should be noted (Fig. 13-6). The nephron contains a network of capillaries, the *glomerulus*, which is surrounded by a funnel-like epithelial sac, termed *Bowman's capsule*; the latter

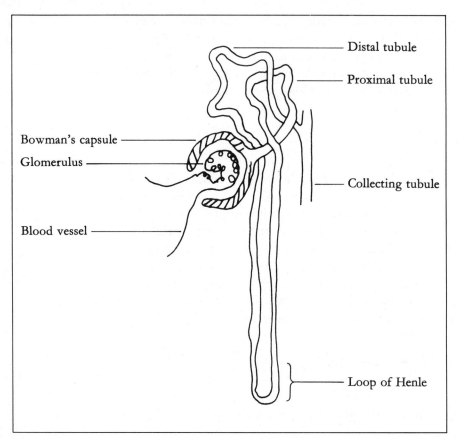

Fig. 13-6. Anatomy of the nephron.

leads into a series of complex tubules that empty into the urinary bladder. The successive sections of the renal tubule, beginning with the section nearest to Bowman's capsule, are called the *proximal tubule, loop of Henle, distal tubule,* and *collecting tubule.*

Blood that passes through the glomerulus is filtered, and the filtrate passes into Bowman's capsule. When the glomerular filtrate enters the proximal tubule, it contains most of the non-protein components of plasma, i.e., water, salts, glucose, amino acids, creatinine, urea, and uric acid. Acetoacetate, β-hydroxybutyrate, lactate, phosphate, and sulfate, together with equivalent amounts of cations (mainly sodium), are also filtered through the glomeruli. Normally, the glucose and most of the water are reabsorbed in the tubules. The latter also function in the selective reabsorption of sodium, chloride, and bicarbonate ions as well as in the secretion of hydrogen ions.

When anions of acids are removed from the bloodstream by glomerular filtration, they must be transferred with equivalent amounts of cations (principally sodium) in order to maintain charge neutrality. Most of the bicarbonate and sodium ions are then reabsorbed as they pass through the tubules. Reabsorption of Na$^+$ occurs with the concurrent release of H$^+$. These H$^+$ ions are eventually neutralized, so the pH of urine is maintained between 5 and 8. Over 4 moles of NaHCO$_3$ are filtered through the glomeruli each day in a normal adult. Since HCO$_3^-$ is the major anion in the extracellular fluid of the body, significant loss of this material would lead to a contraction of fluid volume. In addition, a decrease in HCO$_3^-$ would result in a diminution of buffering activity and lead to acidosis. It is therefore vital that HCO$_3^-$ be taken back by the system; this is accomplished in the renal tubules.

Sodium ions that pass through the renal tubules are reabsorbed by the epithelial cells (Fig. 13-7). Sodium reabsorption operates in conjunction with the release of H$^+$ ions from the cells into the lumen of the tubules. Interaction of the H$^+$ ions with HCO$_3^-$ from the glomerular filtrate yields carbonic acid. Carbon dioxide is evolved from the latter and diffuses back into the epithelial cells, where the catalytic action of carbonic anhydrase reconverts it into carbonic acid which again dissociates into HCO$_3^-$ and H$^+$ ions. The net process provides for the effective conservation of bicarbonate in plasma. Additionally, the H$^+$ ions produced are available for secretion into the tubular lumen in exchange for Na$^+$.

Hydrogen ions in the renal tubules are also neutralized by various anions other than bicarbonate in the glomerular filtrate. These anions include phosphate, creatinine, urate, and lactate. At pH 7.4, phosphate, which constitutes the majority of these anions, is primarily in the form of HPO$_4^{2-}$. This can interact with available H$^+$ ions to form H$_2$PO$_4^-$:

$$HPO_4^{2-} + H^+ \rightleftharpoons H_2PO_4^-$$

The decrease in pH that can result from this neutralization depends on the amount of H$^+$ neutralized, or the ratio of HPO$_4^{2-}$ to H$_2$PO$_4^-$. Even when this ratio is 1:6, the pH is not less than 6. Since urine can tolerate an even lower pH, neutralization of excess H$^+$ by an anion such as phosphate provides another mechanism for the removal of hydrogen ions (see Fig. 13-7).

A third mechanism for neutralizing H$^+$ ions in the renal tubules is provided by the secretion of ammonia from the epithelial cells (see Fig. 13-7). Production of ammonia from glutamine in these cells is catalyzed by the enzyme *glutaminase* (see section 7.4):

$$\underset{\text{H}_2\text{NCCH}_2\text{CH}_2\text{CHCOOH}}{\overset{\overset{\text{O}}{\|}\quad\quad\overset{\text{NH}_2}{|}}{}} + H_2O \longrightarrow NH_3 + \underset{\text{HOCCH}_2\text{CH}_2\text{CHCOOH}}{\overset{\overset{\text{O}}{\|}\quad\quad\overset{\text{NH}_2}{|}}{}}$$

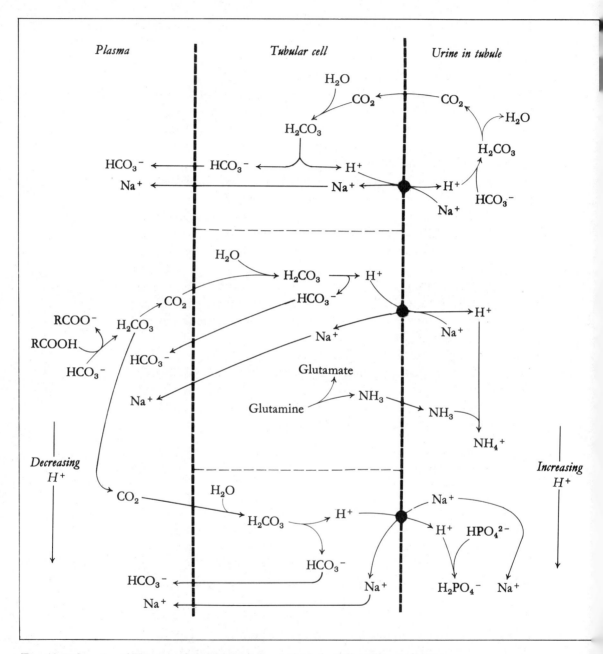

Fig. 13-7. Secretion of H^+ in exchange for Na^+ by renal tubular cells and neutralization of the H^+ ions. The three processes shown here—reabsorption of Na^+, ammonia production, and phosphate neutralization—serve to remove acidic substances from plasma and replace them with HCO_3^-.

The reaction of H^+ with ammonia then yields ammonium ions:

$$NH_3 + H^+ \rightleftharpoons NH_4^+$$

Since the kidneys react to a rise in the plasma hydrogen-ion concentration by an increase in the tubular secretion of H^+, the overall effect of the process provides a mechanism for regulating the pH of plasma. The neutralization mechanism of HPO_4^{2-} (which results in acidification of the urine) is more rapid than the neutralization mechanism involving the generation of NH_3 and the formation of NH_4^+. Nonetheless, the latter process is more important quantitatively. In prolonged acidosis, the capacity of the ammonia-production mechanism is considerably greater than the direct acidification process.

By decreasing the excretion of H^+ into the renal tubules, the kidneys can also counteract *increases* in the pH of blood and extracellular fluids. Under these conditions, the reabsorption of Na^+ and HCO_3^- is reduced, while HCO_3^- and HPO_4^{2-} together with Na^+ are transferred into the urine of the renal tubule. Dihydrogen phosphate ions ($H_2PO_4^-$) are transferred from the cells to the extracellular fluid, and they react with HCO_3^- to form HPO_4^{2-}:

$$HCO_3^- + H_2PO_4^- \rightleftharpoons H_2CO_3 + HPO_4^{2-}$$

Excretion of the monohydrogen phosphate brings about a decrease in HCO_3^- levels in the blood and a concomitant decrease in pH. Since HPO_4^{2-} is excreted with two equivalents of Na^+, the process entails a loss of water, which may be sufficient to preclude the maintenance of normal osmotic pressure. Thus, renal compensation for alkalosis has the auxiliary effect of decreasing the volume of the extracellular fluid and causing possible dehydration.

13.7. Compensatory Control of Blood pH

Pulmonary and renal control of blood pH involves compensatory reactions as well as normal maintenance processes. This section will deal with the specific reactions of these organ systems for readjusting momentary changes of blood pH to the normal value.

The concentration of H_2CO_3 in the circulatory system is a function of the P_{CO_2} in the alveolar gas mixture, while the latter represents the resultant of the rate at which CO_2 is expelled from the lungs and the rate of its dilution by air inhaled from the atmosphere. Thus, the partial pressure of CO_2 in the lungs depends on the rate and depth of respiration: rapid and deep respiration results in a decrease of P_{CO_2}, whereas slow or inhibited respiration has the opposite effect. Since the respiratory center of the medulla is stimulated when the blood pH is decreased, neural control of respiration serves to counterbalance sharp fluctuations in the plasma H^+ concentration.

The critical factor in blood pH regulation is the molar *ratio* of bicarbonate to carbonic acid, rather than their actual concentrations. The actual amounts of the two components are important only in determining the capacity of the buffer system with respect to the amount of acid or base it can counteract before becoming exhausted. Respiratory control of blood pH operates by adjusting the bicarbonate-to-carbonic acid ratio to provide a pH of 7.2 to 7.4. For example, when acid is released into the bloodstream and the level of bicarbonate is reduced ($HCO_3^- + H^+ \rightarrow H_2CO_3 \rightarrow H_2O + CO_2$), the ratio of bicarbonate to carbonic acid is decreased and the pH goes down. This condition stimulates the rate of respiration, which in turn lowers the P_{CO_2} in the alveolar air. As a consequence, the blood concentration of H_2CO_3 is decreased and the ratio of bicarbonate to carbonic acid approaches its normal value. However, although the pulmonary response is rapid, it does not produce complete restoration to normal pH, because the rate of respiration decreases rapidly with the decrease in P_{CO_2}.

Respiratory compensation also operates to counteract an increase in blood pH (i.e., alkalosis). Since this type of compensation is brought about by a decrease in the rate of respiration, the partial pressure of CO_2 in the alveolar air is increased. The increased concentration of HCO_3^- produced by the pH elevation is thus counterbalanced by an increased amount of H_2CO_3. This can restore the pH toward normal but not completely, because the rate of respiration returns to its normal value sooner than does the blood pH.

The process of respiratory compensation thus serves to increase the effectiveness of the bicarbonate-carbonic acid buffer system. It is estimated that the latter alone can maintain a pH of 6.8 to 7.8, even on addition of 16 milliequivalents of acid or 29 milliequivalents of alkali per liter of plasma. As a result of respiratory compensation, these amounts can be increased to 23 and 80 milliequivalents per liter, respectively. This control is accomplished by adjusting the molar concentration of H_2CO_3 so that it is approximately one-twentieth that of HCO_3^-.

In contrast to the respiratory system, the kidneys control the blood pH by regulating the concentration of bicarbonate, rather than carbonic acid. The renal reaction to an increase in H^+ concentration is to cause an increase in the concentration of HCO_3^- in the plasma. The two principal mechanisms by which this is accomplished are acidification of the urine and excretion of ammonium ions. The essential feature of these processes is the promotion of dissociation of H_2CO_3 by excretion of H^+ into the glomerular filtrate in exchange for Na^+. This dissociation effectively increases the level of HCO_3^-

ions, which are returned to the blood (see Fig. 13-7). The renal reaction to an increase in pH, on the other hand, is a decrease in the secretion of H^+, which effectively decreases the level of HCO_3^- in the blood.

Abnormalities involving malfunctions of the regulatory systems or the excessive release of acid or alkali into the bloodstream may result in acidosis or alkalosis. When such conditions are due to defective pulmonary control, they are termed *respiratory acidosis* and *respiratory alkalosis*. If the abnormal blood pH arises from increased levels of metabolically produced non-carbonic acids or bases, the condition is termed *metabolic acidosis* or *metabolic alkalosis*. Each of these conditions has distinctive features by which they can be differentiated from each other.

Respiratory acidosis is caused by states that result in hypoventilation, e.g., pneumonia, pulmonary edema, morphine poisoning, or respiratory paralysis. The increase in P_{CO_2} and the concomitant rise in the plasma H_2CO_3 level bring about a decrease in pH. As a result of this pH change, the oxygen-binding capacity of hemoglobin decreases and its pK_a is decreased. This in turn increases the acid-buffering capacity of the hemoglobin and gives rise to increased levels of HCO_3^-. Thus, respiratory acidosis is characterized by increased blood concentration of both CO_2 and HCO_3^-. Furthermore, the pH of the urine is decreased due to the action of renal compensation mechanisms.

The exact opposite phenomena occur in respiratory alkalosis; i.e., there is an increase in the pH of blood and urine and a decrease in the P_{CO_2} and levels of HCO_3^-. Respiratory alkalosis is brought about during hyperventilation and occurs commonly in hysteria or in children with meningitis.

Metabolic acidosis results from an increase in the cellular production and release of acids that are stronger (i.e., have a lower pK_a) than carbonic acid. Such acids include acetoacetic, β-hydroxybutyric, and lactic acids. These acids react with bicarbonate to form H_2CO_3 and CO_2. They also combine with the anions (i.e., conjugate bases) of other blood buffers. The effect of these reactions is a decrease in the ratio $[HCO_3^-]/[H_2CO_3 + CO_2]$; the pH of the blood decreases correspondingly. The respiratory response to the increase in H^+ concentration is to lower the P_{CO_2} through hyperventilation. In the untreated diabetic with considerable ketosis, this compensatory reaction does not provide complete relief, and the buffering capacity of the blood may be exceeded. Renal compensation—i.e., through tubular excretion of H^+ and ammonia—occurs, but it might not be rapid enough to prevent acute acidosis. The characteristic result in metabolic acidosis is thus a decrease in the level of intracellular bicarbonate and a decrease in the pH of the urine. Plasma H_2CO_3 levels decrease as a result of compensatory reactions.

Metabolic alkalosis is characterized by increased levels of bicarbonate in plasma. Alkaline urine is excreted as a result of renal compensation, and

Table 13-2. Characteristic Features in Acid-Base Disturbances

Condition	Blood pH	Plasma H_2CO_3 Level (mmol/liter)	Plasma HCO_3^- Level (mEq/liter)	Urinary pH
Normal	7.4	1.25	25	6–7
Respiratory acidosis	↓	↑	↑	↓
Respiratory alkalosis	↑	↓	↓	↑
Metabolic acidosis	↓	↓	↓	↓
Metabolic alkalosis	↑	↑	↑	↑

respiratory compensation (hypoventilation) gives rise to increases in the P_{CO_2} and the plasma H_2CO_3 levels.

The evaluation of a particular type of abnormality in acid-base balance can generally be made on the basis of the several factors given in Table 13-2 (the increases or decreases indicated in the table include the effects of compensatory reactions).

13.8. Compartmentalization of Fluids and Electrolytes

A. WATER

Water constitutes 45% to 60% of the total body weight. This variation is due to the difference in water content between adipose tissue and most of the other tissues. The former contains about 10% water, while the average for the rest of the organism is about 75%. Thus, the proportion of total body water is inversely related to the relative amount of adipose tissue: in an obese individual, water accounts for a smaller fraction of the total body weight than it does in a lean person.

The total water can be visualized as distributed between two general compartments: intracellular and extracellular. The former contains about 55% of the total water, whereas the latter contains approximately 45%. The extracellular fluid compartment comprises the blood (4.5% of body weight); the interstitial fluid, including lymph (16% of body weight); and the transcellular fluids, including various secretions and the cerebrospinal and synovial fluids (1% to 3% of body weight).

B. ELECTROLYTES

Although the electrolytic solutes in different tissue compartments are in a dynamic state, they tend to produce solutions in which the osmotic pressures across membranes are in equilibrium. This state is maintained by the movement of ions or small molecules to which the membranes are permeable.

Table 13-3. Electrolyte Concentrations in Body Fluid Compartments*

Electrolyte	Serum (mEq/liter)	Interstitial Fluids (mEq/liter)	Intracellular Fluid (mEq/liter)
Cations:			
Na^+	141	141	20
K^+	5	5	140
Ca^{2+}	5	5	Minimal
Mg^{2+}	2	2	~40
Anions:			
Cl^-	103	110	10
HCO_3^-	26	30	20
HPO_4^{2-}	2	2	80
SO_4^{2-}	1	1	10
Organic acids	5	5	—
Proteins	16	5	80

* Figures for serum and interstitial fluids are average values. The concentrations indicated for intracellular fluid are not completely defined and may differ in various tissues.

In terms of concentration, the major cation in extracellular fluids is sodium, whereas potassium ions predominate within the cells (Table 13-3). The intracellular concentrations of the other ions vary with different tissues, and the values given in Table 13-3 are only general averages. It may be noted, however, that the major anions within the cell are phosphate, sulfate, and protein.

Cell membranes are generally permeable to water. The direction of water movement is dependent on the difference in osmotic pressure between the intracellular and the interstitial solutions; that is, water will move in the direction that will lead to balanced osmotic pressures.

Exchange between vascular and interstitial compartments is a function of the hydrostatic pressure on the capillary bed and the osmotic pressure of the plasma protein solution. The high concentration of protein in plasma tends to draw water into the blood vessels. Normally, this is balanced by hydrostatic filtering forces in the capillaries. In effect, water is filtered at the arteriolar end of the capillaries and returned at the venular end.

C. EXCRETION OF URINE

Normal water and electrolyte levels are the result of a balance between the intake of these components and their disposition. The average daily loss of water in an adult human is between 3 and 3.5 liters. Approximately 1000

Table 13-4. Average Concentrations of Urinary Components*

Component	Concentration (mg/100 ml)	Ratio of Urinary Concentration to Plasma Concentration
Sodium	350	1
Potassium	150	7.5
Magnesium	15	—
Calcium	20	1.5
Chloride	500	1.4
Phosphate	150	50
Sulfate	150	50
Ammonia	60	—
Urea	2000	70
Uric acid	50	12
Creatinine	100	80

* Values for 24-hour urine sample from normal adult. The concentrations vary with the diet.

milliliters is removed through the lungs and via the skin, 100 to 200 milliliters in the stool, and about 1500 milliliters as urine.

Water, electrolytes, glucose, and low-molecular-weight organic compounds are filtered through the renal glomeruli. However, about 75% of the water, sodium ions, and chloride ions are reabsorbed in the renal tubules, and a portion of the sodium ions are exchanged for potassium ions in the distal tubules. Essentially all the glucose in the glomerular filtrate is reabsorbed as it passes through the tubules. The major components of a 24-hour urine sample in a normal adult are listed in Table 13-4.

Pronounced changes in specific urinary constituents or the appearance of abnormal materials reflect certain diseased states and may therefore serve as diagnostic indicators. In certain cases of renal insufficiency, for example, the urinary levels of urea, uric acid, and creatinine, as well as of potassium, sulfate, and phosphate ions, are increased. The amount of uric acid excreted is elevated in patients with gout. Glucose appears in the urine of diabetics, and, in addition, the concentration of ketone bodies is increased. The occurrence of significant amounts of albumin in urine is common in individuals with acute glomerulonephritis and nephrotic syndrome.

Abnormal materials also appear in the urine of individuals with various inborn errors of metabolism or genetic diseases. Examples of such cases include the presence of galactose in galactosemia, fructose in fructosemia, and homogentisic acid in alkaptonuria. The metabolic aberrations in these and other genetic diseases were discussed in conjunction with specific metabolic pathways.

Suggested Reading

Bunn, H. F., Gabbay, K. H., and Gallop, P. M. The Glycosylation of Hemoglobin: Relevance to Diabetes Mellitus. *Science* 200:21–27, 1978.

Earley, L. E., and Daughary, T. M. Sodium Metabolism. *New England Journal of Medicine* 281:72–92, 1969.

Finch, C. A., and Lenfant, C. Oxygen Transport in Man. *New England Journal of Medicine* 286:407–432, 1972.

Lang, A., and Lorkin, P. A. Genetics of Human Haemoglobins. *British Medical Bulletin* 32:239–245, 1976.

Rouiller, C., and Muller, A. *The Kidney: Morphology, Biochemistry and Physiology*, Vols. 1–4. New York: Academic Press, 1971.

Welt, L. G. *Clinical Disorders of Hydration and Acid-Base Equilibrium* (3rd ed.). Boston: Little, Brown, 1970.

14. Specialized Tissues: Their Structures and Functions

14.1. Muscle Contraction

A. THE MUSCLE CELL AND CONTRACTILE PROTEINS

A typical skeletal muscle cell, referred to as a *muscle fiber*, is a cylindrical, multinucleated entity enclosed in a membrane termed the *sarcolemma*. The cell is filled with a structured, repeating sequence of filaments (*myofibrils*) suspended in its cytoplasm (*sarcoplasm*). Nuclei and mitochondria generally occur at the periphery, adjacent to the sarcolemma.

Muscle contraction is effected by the activities of a complex system of filamentous proteins. Two types of filaments can be distinguished on the basis of their general appearance: *thick filaments* and *thin filaments*. The arrangement of these filaments in skeletal muscle is shown in Figure 14-1. Contraction is the result of a lateral movement of the thin filaments with respect to the thick filaments in which the ends of the two groups of thin filaments come close to each other (see Fig. 14-1). The length of the filaments themselves does not change; rather, the *sarcomere*, which is composed of interdigitating thick and thin filaments, shortens.

The thick filaments in muscle fibers are composed of the protein *myosin*. Myosin molecules are rod-shaped and have two globular sections or "heads" at one end (Fig. 14-2). A number of myosin units associate with each other to form the thick-filament system. The thick filaments contain parallel strands of myosin with the globular portions protruding out of the filament. Most of these globular portions occur near the ends of the filament, so the midsection of the filament appears as a bundle of smooth cylinders (see Fig. 14-2).

In contrast with the thick filaments, which are composed primarily of one type of protein, the thin filaments consist of assemblies of three proteins: *actin, tropomyosin,* and *troponin*. Isolated actin units are spherical; however, in the filament, they form a double-helical structure (Fig. 14-2). Superimposed on the actin double helix are strands of tropomyosin and globular units of

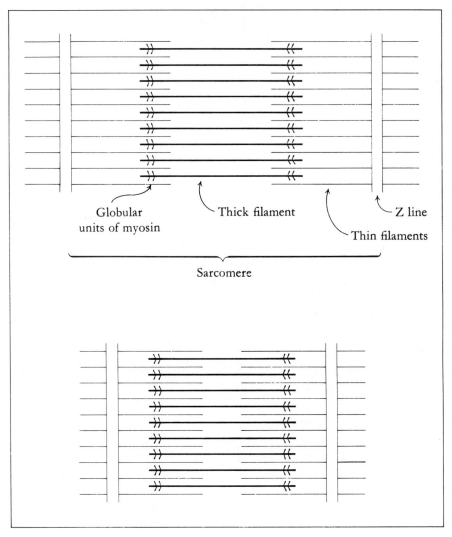

Fig. 14-1. Filament system in skeletal muscle. The upper diagram shows the myofibril in the relaxed state. Thin filaments are attached to the Z line, which forms the boundary of the sarcomere. These thin filaments partially overlap the thick filaments. This regular structure gives skeletal muscle fibers a striated appearance. The lower diagram shows the muscle in the contracted state. Note that the length of the sarcomere (the distance between Z lines) decreases, rather than the length of the filaments.

troponin to form the complete actin-tropomyosin-troponin complex (Fig. 14-2).

The globular units of the myosin in the thick filaments serve as bridges to the thin filaments. In contraction, these globular units undergo specific movements to bring the two sets of thin filaments closer to each other (see Fig. 14-1) and cause the sarcomere to contract. The distance between the Z lines, which delineate the sarcomere, thus becomes considerably shorter than in the relaxed state of the system.

The energy for muscle contraction is derived from the hydrolysis of ATP to ADP and inorganic phosphate. In addition to its contribution to the structure of muscle fibers, myosin has enzymatic activity that catalyzes

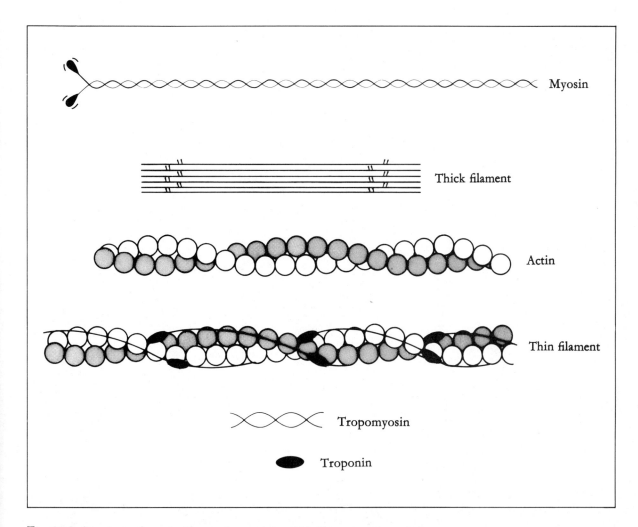

Fig. 14-2. Structures of myosin, the myosin-containing thick filament, actin, and the actin-tropomyosin-troponin complex (thin filament).

the breakdown of ATP; i.e., it acts as an ATPase. The energy released in the hydrolysis of ATP is translated into the appropriate movements of the globular heads of the myosin molecules. Since these units serve as bridges to the thin filaments, their directed motion produces the sliding of the filaments and the consequent contraction of the sarcomere.

B. STRUCTURAL ASPECTS OF CONTRACTILE PROTEINS

The molecular weight of myosin is about 500,000. It consists of six polypeptide chains: two identical units of molecular weight 200,000 and four

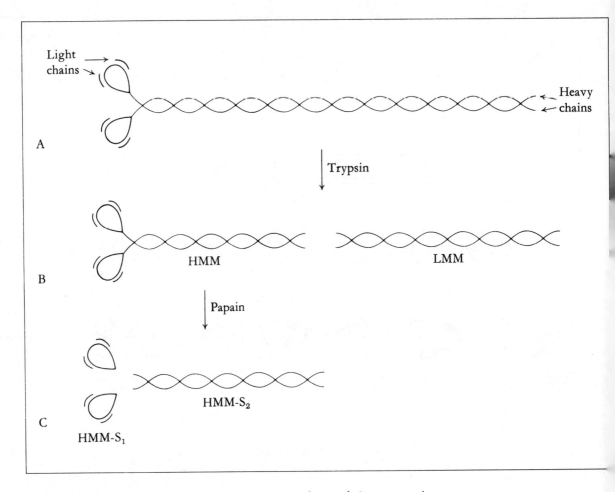

Fig. 14-3. Fragments of myosin obtained upon digestion with proteolytic enzymes. A. Intact myosin molecule (note the two light chains associated with each globular unit). B. Heavy (HMM) and light (LMM) meromyosin units. C. HMM subunits (two S_1 subunits and one S_2 subunit are formed).

smaller chains of molecular weight 20,000. The high-molecular-weight chains associate with each other to produce a double-stranded α-helical structure that constitutes the rod-shaped section of the protein. The remaining portion, which is essentially a continuation of the heavy-chain strands, consists of two separate globular units. A pair of light chains is associated with each of the globular regions (Fig. 14-3A).

When myosin is subjected to proteolytic digestion with trypsin, two products are obtained: *light meromyosin* (LMM) from the helical portion and *heavy meromyosin* (HMM), which consists of the globular sections and a portion of the helix (Fig. 14-3B). Further degradation of HMM with the proteolytic enzyme papain yields three subunits: two subunits (HMM-S_1) consist of the globular sections plus the light chains, and the third subunit (HMM-S_2) consists of a helical section (Fig. 14-3C). Studies of the fragments obtained by proteolysis have revealed that HMM-S_1 has the ATPase activity. Additionally, this portion of myosin contains the site that binds with actin.

The simplest monomeric unit in actin is a globular protein with a molecular weight of 42,000, termed *G-actin* (globular actin). In physiologic medium, the G-actin units associate to form polymeric strands of *F-actin* (fibrous actin), which interact to produce a double-stranded helix (see Fig. 14-2).

Tropomyosin is a protein with a molecular weight of about 70,000. It also has the structure of double-stranded α helix (see Fig. 14-2). It is a component of the thin filaments, and its strands run along the grooves between the strands of actin (Fig. 14-2).

Troponin is a complex of three polypeptide chains with molecular weights of 18,000, 24,000, and 37,000. This globular protein is bound to tropomyosin as shown in Figure 14-2.

Troponin and tropomyosin serve as modulators that inhibit or stimulate the combination of actin with myosin. The sequence that leads to the activation of actin is dependent on calcium ions, which function as the overall regulators of muscle action. When calcium ions bind to troponin, the latter undergoes a change that in turn affects the structure of tropomyosin. These structural alterations produce an alteration in the conformation of actin so it can form a cross-bridge with the globular unit of myosin. In the absence of calcium ions, troponin and tropomyosin prevent the binding of actin with myosin, whereas in the presence of these ions, they function to stimulate the combination.

C. REACTION SEQUENCE IN MUSCLE CONTRACTION

The energy for muscle contraction is provided by the hydrolysis of ATP to ADP and inorganic phosphate. Contraction results from the binding of myosin to actin via the globular unit of myosin and the *directed* movement of this globular component to effect the sliding of the thin filaments toward each other. Myosin is known to combine with actin to yield a complex protein termed *actomyosin*. Actomyosin has a thread-like structure and has been shown to undergo contraction upon the addition of ATP and magnesium and potassium ions. In addition, both myosin and actomyosin have ATPase activity. Thus, muscle contraction apparently occurs as a result of the interaction of ATP with cross-linked actin and myosin (actomyosin) and the concomitant release of energy from ATP hydrolysis, although the exact mechanism for channeling this energy into the specific, directed movement of the myosin globular unit is not yet known.

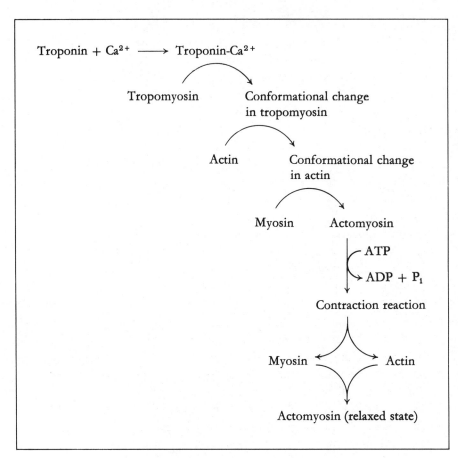

Fig. 14-4. *Transformations during muscle contraction. The sequence is initiated by the release of calcium ions from the sarcoplasmic reticulum, which combine with troponin. Subsequently, the energy released by the hydrolysis of ATP is utilized to move the globular units of myosin (see Fig. 14-1) in such a way that the actin strands of the thin filaments slide toward each other (contraction reaction). When the calcium is removed (i.e., is pumped back into the sarcoplasmic reticulum) following contraction, the actomyosin dissociates and then reaggregates to form the complex in the relaxed state.*

In the relaxed or resting state, the calcium ions in muscle cells are confined within a network of channels termed the *sarcoplasmic reticulum*. The accumulation of calcium ions in these organelles is accomplished by an ATP-requiring active-transport system that drives the ions from the cytosol against a concentration gradient across the membrane of the sarcoplasmic reticulum. When a nerve impulse is transmitted to muscle, the calcium ions are released from the sarcoplasmic reticulum in order for them to bind with troponin. As outlined in section 14.1.B, this ultimately leads to the appropriate conformational changes in actin, and muscle contraction occurs (Fig. 14-4).

D. MYOGLOBIN

As indicated in section 14.1.C, the immediate energy source for muscle contraction is provided by the hydrolysis of ATP. The latter is generated from the oxidation of fatty acids, acetoacetic acid, glycolysis (aerobic and anaerobic), and oxidation of acetyl-coenzyme A via the citric acid cycle. Except for anaerobic glycolysis, which constitutes a relatively small fraction of metabolic activity, these transformations and the oxidative phosphorylations that occur in conjunction with them consume considerable amounts of oxygen. An important muscle component that functions in the uptake and storage of oxygen is the protein *myoglobin*.

Myoglobin can accept oxygen from circulating oxygenated hemoglobin and thus increases the availability of oxygen in muscle cells. Myoglobin consists of a single polypeptide chain with a molecular weight of about 17,000. It is conjugated with a ferrous porphyrin group, which, like that of hemoglobin, can accommodate a molecule of oxygen. Its three-dimensional configuration resembles that of a single unit of hemoglobin (Figure 1-13, page 30), although it is quite different with respect to its primary structure. Myoglobin is a compact molecule in which the non-polar side chains are located in the interior. The interactions between the non-polar residues effectively stabilize the structure of the molecule. The single heme unit is located within the non-polar interior environment. In addition to its linkages with the four tetrapyrrole nitrogen atoms of the heme unit, the divalent iron atom has a coordinate bond with an imidazole nitrogen atom of a histidine residue in the polypeptide. An oxygen molecule can occupy the sixth coordination site on the iron atom.

Unlike hemoglobin, myoglobin is not an allosteric protein. Thus, instead of having a sigmoid-shaped oxygen saturation curve, myoglobin shows a simple hyperbolic relationship between the degree of saturation and oxygen tension (Fig. 14-5). Moreover, the oxygen saturation curve of myoglobin is displaced to the left of that of hemoglobin; that is, the affinity of myoglobin for oxygen is greater than that of hemoglobin. At any given partial pressure of oxygen, myoglobin is more efficient in binding oxygen and will therefore tend to draw it from hemoglobin. If the P_{O_2} in the capillaries of muscle is about 20 torr, for example, the hemoglobin reaching the tissues will be less than 25% saturated with oxygen, whereas myoglobin will be almost completely saturated (see Fig. 14-5). Myoglobin therefore has the capacity to store oxygen and increase its level within the muscle cells.

Myoglobin occurs in the sarcoplasm of muscle fibers. The color of the red muscle tissues is due to the presence of high concentrations of myoglobin. Light-colored muscle, or white muscle, contains minimal amounts of myoglobin. Generally, muscles involved in continuous or prolonged activity are red, whereas those required for short bursts of activity have a white coloration (this is seen in the difference between the leg muscle and breast muscle of chickens). The muscles of humans usually consist of a mixture of both types of fibers.

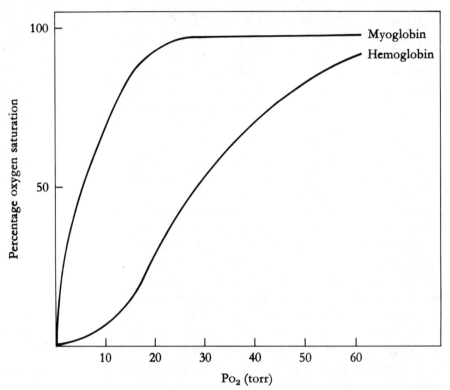

Fig. 14-5. Relationship between the partial pressure of oxygen (P_{O_2}) and the degree of oxygen binding to hemoglobin and myoglobin. Note the hyperbolic shape of the oxygen saturation curve for myoglobin versus the sigmoid shape of the curve for hemoglobin.

E. ROLE OF PHOSPHOCREATINE

The high utilization of energy by muscle gives rise to the need for auxiliary storage forms of energy. This requirement is met by the presence in muscle of the ATP-generation mechanisms provided by creatine, creatine kinase, and myokinase.

The synthesis of creatine from arginine, glycine, and methionine was described in Chapter 7 (section 7.9). The enzyme *creatine kinase* catalyzes the reversible transfer of the terminal phosphate from ATP to creatine:

$$
\begin{array}{c}
\text{NH}_2 \\
| \\
\text{C}=\text{NH} \\
| \\
\text{N}-\text{CH}_3 \\
| \\
\text{CH}_2 \\
| \\
\text{COOH}
\end{array}
+ \text{ATP} \xrightarrow{\text{Mg}^{2+}}
\begin{array}{c}
\text{NHPO}_3\text{H}_2 \\
| \\
\text{C}=\text{NH} \\
| \\
\text{N}-\text{CH}_3 \\
| \\
\text{CH}_2 \\
| \\
\text{COOH}
\end{array}
+ \text{ADP}
$$

Creatine   Phosphocreatine

Since the reaction can be reversed—i.e., ATP can be formed from ADP and phosphocreatine—the latter can function as reservoir for "energized" phosphate groups. Hence, as ATP is depleted by its conversion to ADP plus P_i during muscle contraction, it can be regenerated through the reaction between ADP and phosphocreatine. The capacity of the system is limited, and phosphocreatine is rather unstable. This reaction, however, adds to the immediate energy sources of muscle cells, although ultimately, energy must be provided by the oxidation of metabolites.

Another reaction that can produce ATP is catalyzed by the enzyme *myokinase*:

$$2ADP \rightleftharpoons ATP + AMP$$

This mechanism is probably important when there is an acute shortage of ATP, but it does not appear to function significantly under normal conditions.

14.2. Neural Transmission

A. THE NERVE CELL

The nerve cell, or *neuron*, contains a large nucleus, mitochondria, Golgi apparatus, and the common intracellular organelles. The distinguishing structural features of neurons are the thread-like extensions from the cell body termed *neuronal processes*; these include the *axons* and *dendrites*. The axon is the longer of the two processes, ranging in length from microscopic distances to as much as 1 meter. Neurons that have more than two axons are called *multipolar*. Neurons also contain short, branched processes emerging from the cell body, termed *dendrites*.

Nerve impulses generally travel in one direction along the neuron. In multipolar neurons, the dendrites send the impulses to the cell, while the axons carry them away for transmission to other neurons or muscle fibers. Impulses usually travel through a series of neurons, each of which transmits the stimulus to its adjacent neuron. The site of communication of impulses between neurons is called the *synapse*. Since the synapse constitutes a space or gap between adjacent neurons, the transfer of neural information must occur by means of the directed movement of substances termed *neurotransmitters* across the synapse.

B. NEURONAL STIMULATION

Neurons, like most other cells, have considerably higher concentrations of potassium ions than the surrounding extracellular fluid, whereas the opposite holds for sodium-ion concentration. The intracellular concentrations of K^+

and Na^+ are about 150 and 8 millimolar, respectively. The concentration of K^+ in the extracellular medium, however, is less than 4 millimolar, whereas that of Na^+ is about 150 millimolar. These differences are maintained by energy-requiring (ATP-requiring) active-transport mechanisms that pump the ions across the cell membranes. These imbalances, in addition to those produced by electrochemical gradients of other charged substances, make the inner side of the membrane more negative than the outer side. Such electrical and ionic concentration gradients give rise to a measurable potential difference of about -70 millivolts, termed the *membrane potential* or *resting potential*.

When a nerve impulse is transmitted through the cell body of the neuron, there is an abrupt alteration in membrane permeability so that Na^+ can move into the cell and K^+ can diffuse out. The transient change in potential created by these events is called *depolarization*; this gives rise to the *action potential*, which may be visualized as a spike-shaped voltage pulse. Impulses are propagated rapidly along the axon with resultant changes in membrane potential that are followed by rapid readjustment to the original resting state.

C. SYNAPTIC TRANSMISSION

The terminal sections of axons have numerous vesicles that contain the neurotransmitter substances. Most prominent among these are *acetylcholine* and *norepinephrine*. In addition to their involvement in transmitting impulses across synapses, these compounds function in stimulating muscle changes at the neuromuscular junctions. Acetylcholine serves as a transmitter to striated muscle, whereas norepinephrine operates at junctions between sympathetic nerves and smooth muscles.

The arrival of an action potential at the terminus of an axon containing *cholinergic* vesicles triggers the release of acetylcholine from the neuron into the synaptic gap or cleft. The acetylcholine then diffuses through the membrane of the neuron on the other side of the synapse (the *postsynaptic membrane*) and combines with specific receptor substances. This causes a dramatic alteration in membrane permeability that allows for the influx of Na^+ into neuron and movement of K^+ to the extracellular medium. The membrane depolarization that is responsible for the nerve impulse is propagated along the neuron as described previously. The resting potential is restored by the rapid hydrolysis of acetylcholine, which is catalyzed by the enzyme *acetylcholinesterase*.

In order to carry out these processes, cholinergic neurons must actively synthesize acetylcholine and have the capacity to effect its hydrolysis. The biosynthetic reaction is catalyzed by the enzyme, *choline acetyltransferase*, and acetyl-coenzyme A serves as the donor of the acetyl group:

$$(CH_3)_3N^+CH_2CH_2OH + CH_3\overset{O}{\overset{\|}{C}}-S-CoA \longrightarrow (CH_3)_3N^+CH_2CH_2O\overset{O}{\overset{\|}{C}}CH_3 + CoA-SH$$

Choline   Acetylcholine

Hydrolysis of acetylcholine to choline and acetate results in its deactivation.

The other neurotransmitters, which are especially common in the sympathetic nervous system, are norepinephrine and the related catecholamines:

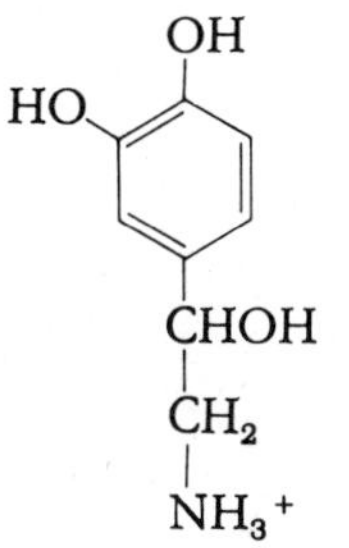

Norepinephrine 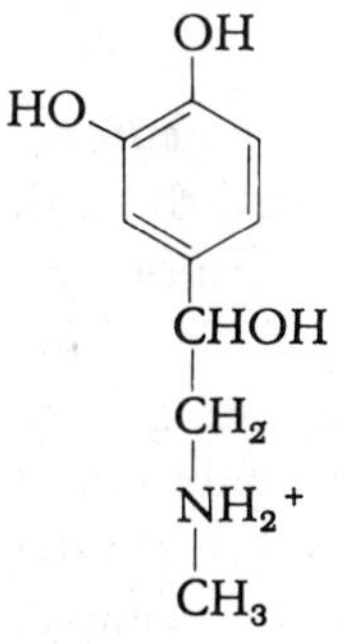   Epinephrine 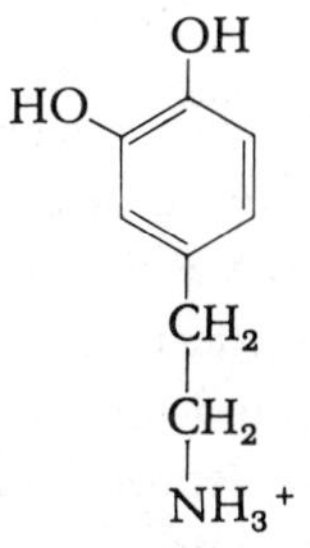   Dopamine (3,4-dihydroxyphenylethylamine)

These are synthesized from tyrosine by reaction sequences described previously (section 8.16).

There are two known mechanisms for inactivation of the catecholamines. One of these is the methylation of the hydroxyl group on carbon-3. This reaction is catalyzed by *catechol O-methyltransferase*, which utilizes S-adenosylmethionine as the methyl donor:

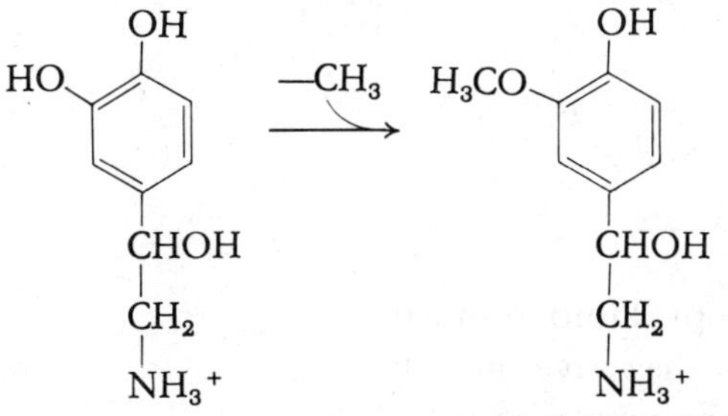

3-O-Methylnorepinephrine

Another deactivation reaction is oxidative deamination, which yields the respective aldehyde. This is catalyzed by *monoamine oxidase*:

$$\underset{\substack{\text{HO}\\\text{CHOH}\\\text{CH}_2\\\text{NH}_3^+}}{\overset{\text{OH}}{\bigcirc}} \xrightarrow{\text{NH}_3} \underset{\substack{\text{HO}\\\text{CHOH}\\\text{CHO}\\\text{3,4-Dihydroxyphenyl-}\\\text{glycolaldehyde}}}{\overset{\text{OH}}{\bigcirc}}$$

The binding of a neurotransmitter such as acetylcholine to the neuronal receptor is critical for propagating the impulse or action potential through the involved neuron or muscle and for the subsequent depolarization and repolarization processes. The poisonous effect of curare results from the activity of one of its components, *d*-tubocurarine, in competing with acetylcholine for the binding site on the receptor. This causes inhibition of the depolarization of the end-plate at the neuromuscular junction. A similar action is also produced by toxins from various snake venoms, for example, cobra toxin or bungarotoxin.

D. ACETYLCHOLINESTERASE INHIBITORS

Acetylcholinesterase, the enzyme that catalyzes the hydrolysis of acetylcholine, contains a serine residue at its active site. The mechanism of its action involves an acyl-enzyme intermediate that undergoes hydrolysis to regenerate the active enzyme:

$$\underset{\text{Active enzyme}}{\text{Enz—Ser—OH}} + \underset{\text{Acetylcholine}}{\text{CH}_3\overset{\overset{\text{O}}{\|}}{\text{C}}\text{OCH}_2\text{CH}_2\text{N}^+(\text{CH}_3)_3} \longrightarrow \underset{\text{Acyl-enzyme}}{\text{Enz—Ser—O}\overset{\overset{\text{O}}{\|}}{\text{C}}\text{CH}_3} + \underset{\text{Choline}}{\text{HOCH}_2\text{CH}_2\text{N}^+(\text{CH}_3)_3}$$

$$\underset{\text{Acyl-enzyme}}{\text{Enz—Ser—O}\overset{\overset{\text{O}}{\|}}{\text{C}}\text{CH}_3} + \text{H}_2\text{O} \longrightarrow \underset{\text{Active enzyme}}{\text{Enz—Ser—OH}} + \underset{\text{Acetate}}{\text{HO}\overset{\overset{\text{O}}{\|}}{\text{C}}\text{CH}_3}$$

Acetylcholinesterase is completely inhibited by diisopropylfluorophosphate or DFP (section 2.12), which blocks the serine residue at the active site. The effect of DFP and analogous compounds is to inhibit the repolarization of nerve membranes or end-plates at neuromuscular junctions. Its effect on diaphragm muscle produces respiratory paralysis and may be lethal. Compounds of this type are utilized as insecticides, and they have been prepared for use as nerve gases in chemical warfare.

Certain compounds bind to acetylcholinesterase to form esters, but their acyl group is released from the enzyme less rapidly than acetate; these compounds are utilized as muscle relaxants. Such materials decrease the rate of hydrolysis of acetylcholine, but only to a moderate extent. One of these, *succinylcholine*, is employed to produce muscle relaxation prior to surgery:

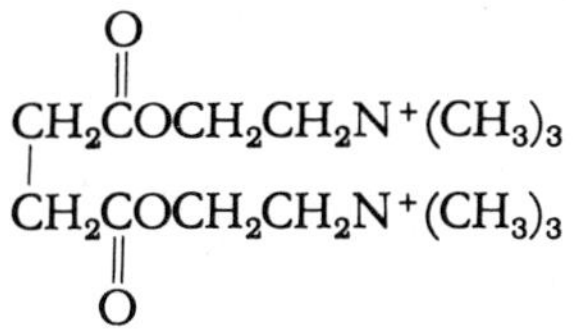

$$\begin{array}{l} CH_2COCH_2CH_2N^+(CH_3)_3 \\ | \\ CH_2COCH_2CH_2N^+(CH_3)_3 \end{array}$$
(with C=O carbonyls on each CH₂CO group)

Succinylcholine

The effect of succinylcholine on acetylcholinesterase and end-plate depolarization is transient since it is hydrolyzed by the acylcholinesterases of liver and plasma.

Another acetylcholinesterase inhibitor, *physostigmine*, is used in the treatment of glaucoma:

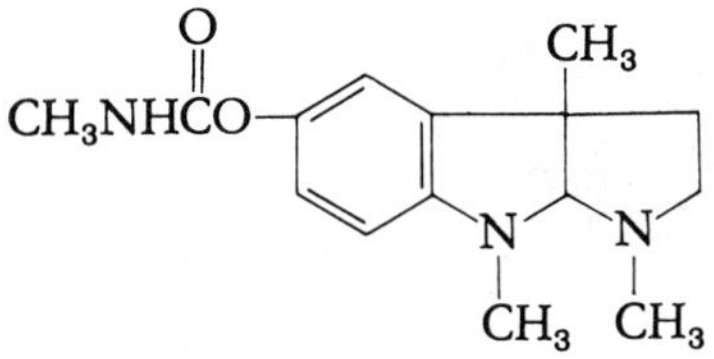

Physostigmine

14.3. Connective Tissue

A. GENERAL COMPOSITION

Connective tissue consists of cellular components surrounded by proteins, polysaccharides, and inorganic salts. It is found in numerous organs, and it functions to support or bind together various anatomic units of living organisms. Connective tissue is found in bone, cartilage, tendons, ligaments, dermis, and blood vessels. The principal cells of fibrous connective tissue are the *fibroblasts*. These produce *collagen fibrils* and particular *glycosaminoglycans* (*mucopolysaccharides*), which occur in the extracellular fluid. In the elastic connective tissue of arterial walls, the primary fibrous protein is *elastin*. Bone tissue also contains crystalline calcium phosphate in addition to fibrous proteins and glycosaminoglycans.

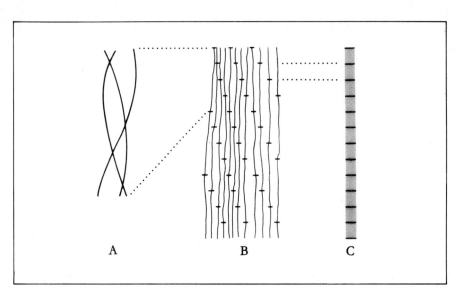

Fig. 14-6. Structure of collagen: A. Tropocollagen triple helix. B. Association of tropocollagen units in collagen. C. Electron-micrographic appearance of collagen.

B. COLLAGEN

Collagen is the principal protein of connective tissue. It is the most abundant protein of animal organisms, constituting about 30% of the total body protein. Tissues that contain especially high amounts of collagen are cartilage, bone, teeth, tendons, and skin.

Collagen has a peculiar amino-acid composition that is quite different from that of the common proteins. It contains about 35% glycine, 12% proline, and 11% alanine. It also contains residues of 4-hydroxyproline and 5-hydroxylysine:

$$\underset{\text{4-Hydroxyproline}}{\begin{array}{c} HOCH-CH_2 \\ | \quad\quad | \\ H_2C\underset{H}{\diagdown N \diagup} CHCOOH \end{array}} \quad\quad \underset{\text{5-Hydroxylysine}}{H_2NCH_2\underset{OH}{CH}CH_2CH_2\underset{NH_2}{CH}COOH}$$

Structurally, collagen is a fibrous protein that consists of units of *tropocollagen* linked to each other in a head-to-tail manner. Tropocollagen itself is composed of three polypeptide chains wound around each other in the form of a triple helix (Fig. 14-6A). In collagen, sequences of these tropocollagen units are arranged in parallel bundles. The sites of the head-to-tail tropocollagen linkages are staggered in a uniform manner, as shown in Figure 14-6B. This gives rise to the periodicity and striated appearance of the collagen fibers seen in electron micrographs (Fig. 14-6C).

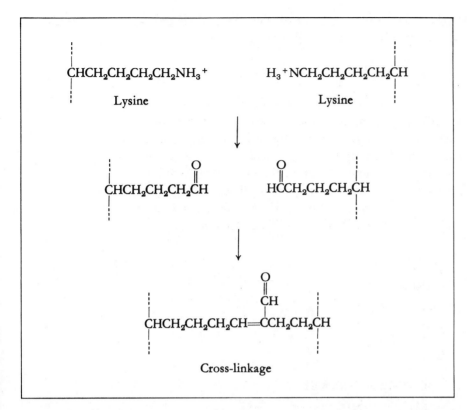

Fig. 14-7. *Formation of collagen interchain cross-linkage via an aldol condensation reaction.*

The structure of collagen is maintained by a unique type of cross-linkage between the polypeptide strands of the tropocollagen units. The carbon atoms adjacent to specific ε-amino groups of lysine residues are oxidized to form aldehyde groups, which in turn interact with each other in an aldol condensation reaction. This reaction results in the formation of a carbon-to-carbon double bond that cross-links the polypeptide chains (Fig. 14-7). Additional linkages with these covalently cross-linked bridges occur as a result of interactions with histidine and hydroxylysine residues of adjacent tropocollagen chains. These interactions provide insoluble fibers with the high tensile and mechanical strength characteristic of collagen.

Several aspects of the biosynthesis of collagen are of special significance. A precursor of collagen, *procollagen*, is synthesized by fibroblasts. Hydroxylation of proline and lysine occurs after the protein is formed on the ribosomes. This posttranslational reaction (page 352), which is catalyzed by *procollagen hydroxylase*, requires molecular oxygen, ascorbic acid, Fe^{2+}, and α-ketoglutarate. The enzyme system does not hydroxylate free proline or lysine.

Procollagen is secreted from the fibroblasts into the extracellular fluid, where tropocollagen is produced by limited proteolysis of the precursor.

Combination of the tropocollagen units and the production of their cross-linkages occur in the extracellular medium and give rise to the insoluble collagen fibers.

C. ELASTIN

Those connective tissues that can stretch and return to their original form when the tension is released contain high concentrations of the protein *elastin*. Examples of such tissues are ligaments and the walls of blood vessels. Like collagen, elastin is rich in glycine and proline residues; however, it does not appear to contain significant amounts of hydroxylysine or hydroxyproline. It also contains cross-linkages analogous to those of collagen. *Desmosine*, which has been isolated from elastin, may be involved in a unique type of cross-linkage involving the condensation of four lysine residues:

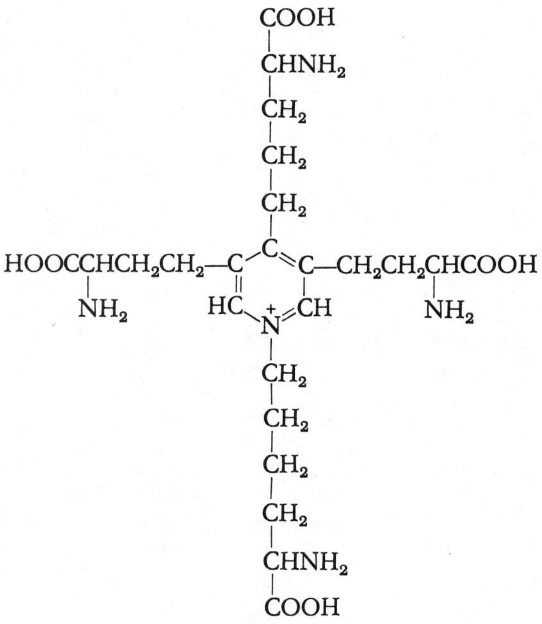

$$\begin{array}{c}
COOH \\
| \\
CHNH_2 \\
| \\
CH_2 \\
| \\
CH_2 \\
| \\
CH_2 \\
| \\
HOOCCHCH_2CH_2-C\diagup^C\diagdown_{C-CH_2CH_2CHCOOH} \\
\quad | \qquad\qquad\quad \| \quad\;\; \| \qquad\qquad\qquad | \\
\;NH_2 \qquad\qquad HC\diagdown_{\overset{+}{N}}\diagup CH \qquad\quad NH_2 \\
| \\
CH_2 \\
| \\
CH_2 \\
| \\
CH_2 \\
| \\
CHNH_2 \\
| \\
COOH
\end{array}$$

Desmosine

D. GLYCOSAMINOGLYCAN STRUCTURE

The medium in which the connective tissue cells and fibrous proteins are suspended is a viscous solution of inorganic salts and protein-heteropolysaccharide complexes termed *glycosaminoglycans* or *mucopolysaccharides*. As the name implies, the common feature of this class of polysaccharides is the presence of hexosamine moieties (Fig. 14-8). Glycosaminoglycans also contain either galactose or a uronic acid (glucuronic acid or iduronic acid,

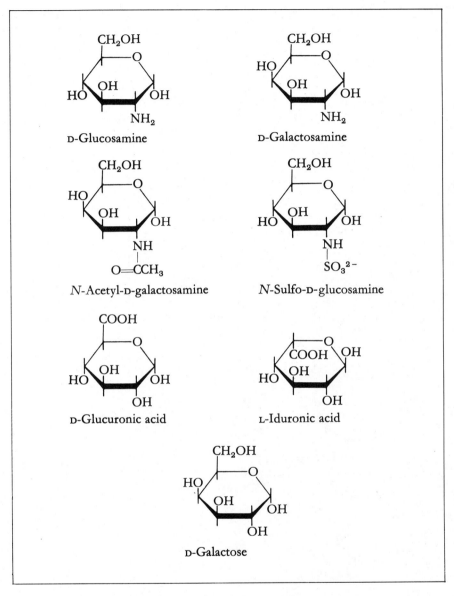

Fig. 14-8. Sugar constituents of glycosaminoglycans. In most glycosaminoglycans, the hexosamine is N-acetylated, as represented by the structure of N-acetylgalactosamine. Heparin and heparan sulfate also contain N-sulfated glucosamine. Uronic acid components are D-glucuronic or L-iduronic acid; these differ in their configuration at carbon-5. Keratan sulfate does not contain a uronic acid but has a galactose unit instead. (The sugars are shown in the α configuration.)

or both). The polysaccharide components generally consist of repeating disaccharide units in which the hexosamine and the uronic acid or galactose are present in equimolar proportions (Fig. 14-9).

Particular glycosaminoglycans differ from each other with respect to the hexosamine unit (e.g., glucosamine versus galactosamine), the uronic acid or hexose component, and substituent groups on the sugar backbone. The nitrogen atoms of the hexosamine residues may be either acetylated or

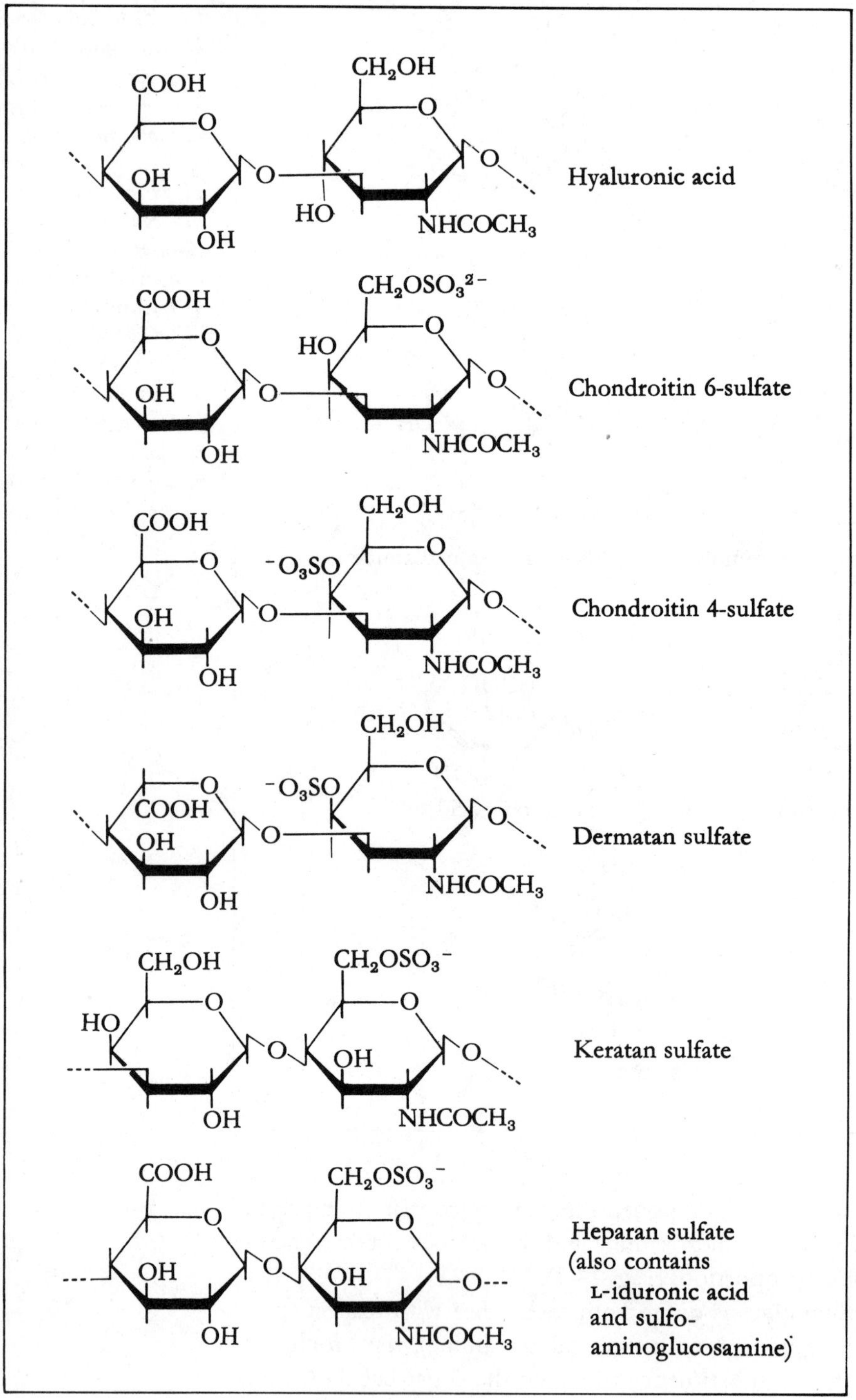

Fig. 14-9. Repeating units (disaccharide components) of common glycosaminoglycans.

Table 14-1. Glycosaminoglycans

	Monosaccharide Components	Mol. Wt.	Occurrence
Hyaluronic acid	N-Acetylglucosamine, glucuronic acid	$1–3 \times 10^6$	Synovial fluid, skin, vitreous humor, umbilical cord
Chondroitin 4-sulfate	N-Acetylgalactosamine 4-sulfate, glucuronic acid	$2–5 \times 10^4$	Cartilage, aorta
Chondroitin 6-sulfate	N-Acetylgalactosamine 6-sulfate, glucuronic acid	$2–5 \times 10^4$	Heart valve, umbilical cord
Dermatan sulfate	N-Acetylgalactosamine 4-sulfate, iduronic acid	$2–5 \times 10^4$	Skin, blood vessels, heart valve
Heparan sulfate	N-Acetylglucosamine 6-sulfate, sulfoaminoglucosamine, glucuronic acid, iduronic acid	$\sim 10^4$	Blood vessels, cell surfaces
Keratan sulfate	N-Acetylglucosamine 6-sulfate, galactose	$1–2 \times 10^4$	Cornea, cartilage
Heparin	Sulfoaminoglucosamine 6-sulfate, glucuronic acid, 2-sulfate, iduronic acid	$1–3 \times 10^4$	Mast cells, lung, intestine, skin

sulfated. In addition, various glycosaminoglycans contain sulfate ester units attached through carbon atoms of the polysaccharide backbone (see Table 14-1).

Hyaluronic acid is the only glycosaminoglycan that does not contain sulfate groups. Aqueous solutions of hyaluronic acid are very viscous and may therefore function within the tissue as molecular sieves that control the movement of ions and proteins around the cells. Animal tissues and certain bacteria contain *hyaluronidases*, which degrade these polysaccharides to smaller units and thus enhance penetration of various substances through the extracellular matrix. Hyaluronic acid also serves as a lubricant in joints. It is of interest that the concentration of hyaluronic acid is elevated in the synovial fluid of the joints of many patients with arthritis. The molecular weight of the hyaluronic acid found in arthritis appears to be lower than that in normal individuals. It has been suggested that this may be a factor in producing the symptoms of the disease.

Except for hyaluronic acid, all the glycosaminoglycans are linked to proteins to form complexes termed *proteoglycans*. The linkages between most sulfated glycosaminoglycans and their associated proteins occur via a bridge composed of galactosylgalactosylxylose, in which the xylose moiety is linked to a serine residue in the protein (Fig. 14-10).

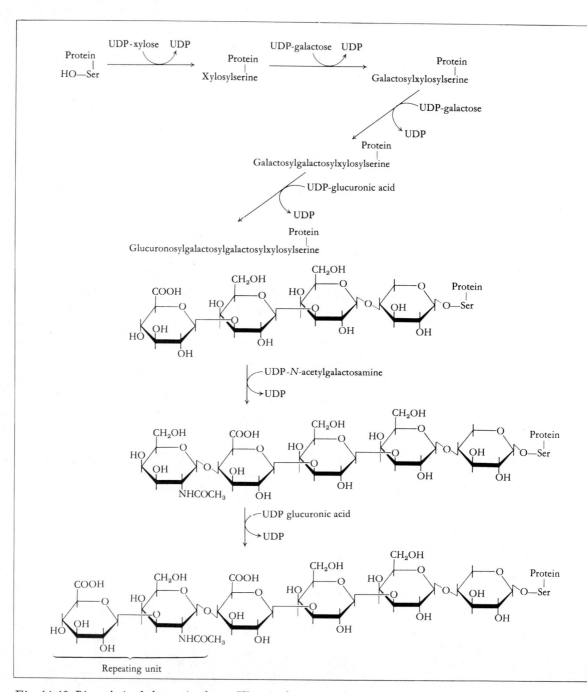

Fig. 14-10. Biosynthesis of glycosaminoglycans. The initial steps are the same for all compounds of this class. Following the formation of glucuronosylgalactosylgalactosylxylosylserine, the synthesis of a chondroitin sulfate is shown; the sulfate groups are introduced later in a separate process (see text).

E. METABOLISM OF GLYCOSAMINOGLYCANS

The biosynthesis of the heteropolysaccharide chains involves the transfer of monosaccharide units from uridine-diphosphate (UDP) sugars to the non-reducing terminus of a growing chain. Each step is catalyzed by a specific *glycosyltransferase*. The initial reaction is the transfer of a xylose unit from UDP-xylose to a serine residue of the protein (Fig. 14-10). This is followed by the sequential addition of two galactose units. The subsequent steps are the additions of the uronic acid and hexosamine units of the particular glycosaminoglycan, e.g., the chondroitin sulfates, as illustrated in Figure 14-10 (separate processes are required for introduction of the sulfate groups in either position 4 or position 6; see the following discussion). In the chondroitin sulfates, alternate attachment of glucuronic acid and N-acetylgalactosamine units results in the formation of the long-chain polymer with repeating units.

In the syntheses of sulfated glycosaminoglycans, the sulfate groups are introduced in specific positions of the polysaccharide chain by an activated sulfate donor, *3'-phosphoadenosylphosphosulfate* (PAPS). The latter is formed as a result of a two-step process (Fig. 14-11). The first reaction is the formation of adenosylphosphosulfate (APS) from ATP and inorganic sulfate; this is followed by the interaction of APS with a second mole of ATP to yield PAPS. The transfer of sulfate groups from PAPS to glycosaminoglycans is catalyzed by specific sulfotransferases. The sulfotransferase that catalyzes the formation of chondroitin 6-sulfate, for example, effects the esterification of sulfate groups at carbon-6 of N-acetylgalactosamines in the polysaccharide.

Chondroitin 4-sulfate and dermatan sulfate are sulfated on carbon-4 of N-acetylgalactosamine units (see Fig. 14-9). Heparin and heparan sulfate contain additional sulfate-ester groups on carbon-2 of iduronic-acid moieties. The nitrogen atoms of the glucosamine units in heparin are sulfated (sulfo-amino groups); in the chondroitin sulfates, dermatan sulfate and hyaluronic acid, the nitrogens of the hexosamine units are acetylated (acetamido groups). Heparan sulfate contains both N-sulfated and N-acetylated hexosamine units. It is significant that variations with respect to the position of sulfate groups occur within the polysaccharide chain. Thus, dermatan sulfates may contain some 6-sulfo-N-acetylgalactosamine units as well as units with sulfate groups at position 4. These variations within the polysaccharide chains are termed *microheterogeneities*.

Heparin, strictly speaking, is not classified as a connective-tissue glycosaminoglycan, since it does not occur in the extracellular matrix of these tissues. It is found in mast-cell granules and does not appear to be secreted. The effects of heparin on blood coagulation and lipoprotein lipase were discussed in previous chapters (sections 12.4.B and 6.4, respectively).

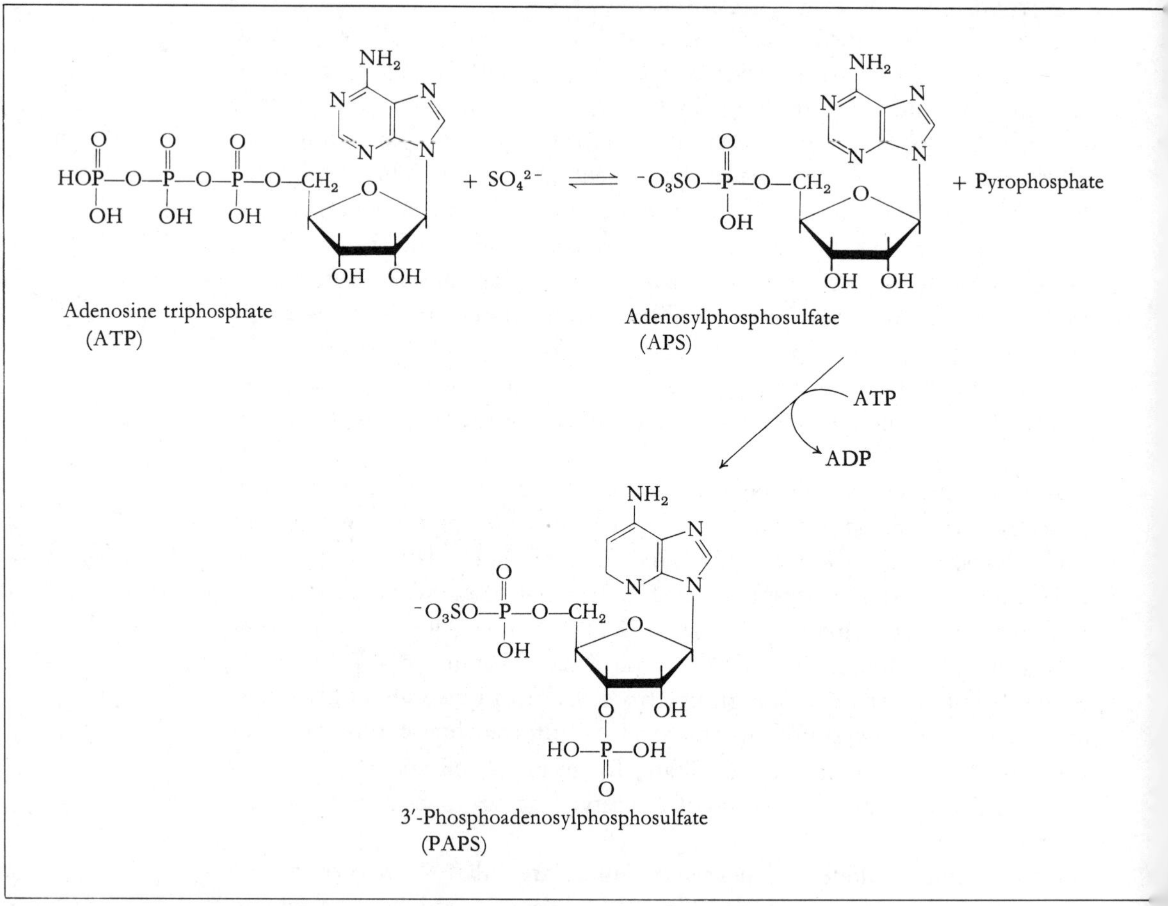

Fig. 14-11. Synthesis of 3'-phosphoadenosylphosphosulfate (PAPS).

A number of genetic diseases are due to defects in the metabolism of glycosaminoglycans (Table 14-2). Most notable are Hurler's and Hunter's diseases.

Hurler's disease is transmitted by an autosomal recessive mode of inheritance. Individuals with this disease have skeletal abnormalities and are mentally retarded. Generally, they have a short life expectancy. The enzyme that is deficient in this disease is α-L-iduronidase. Both dermatan sulfate and heparan sulfate accumulate in the tissues. As much as 100 milligrams of low-molecular-weight forms of the combined glycosaminoglycans may appear in the urine each day.

Hunter's disease produces symptoms similar to those of Hurler's syndrome. Although it does not involve the same enzyme deficiency (Table 14-2), the aberration affects the same glycosaminoglycans. This disease is transmitted

Table 14-2. Genetic Diseases Involving Glycosaminoglycan Metabolism

Disease	Symptoms	Deficient Enzyme	Affected Substances
Hurler's syndrome	Corneal clouding, skeletal abnormalities, mental deficiency, early death	α-L-Iduronidase	Dermatan sulfate and heparan sulfate
Scheie's syndrome	Corneal clouding, joint stiffness, aortic regurgitation, but normal life span and intelligence	α-L-Iduronidase	Dermatan sulfate and heparan sulfate
Maroteaux-Lamy syndrome	Growth retardation, corneal opacity, bone deformities	N-Acetylgalactosamine sulfatase	Dermatan sulfate
Hunter's syndrome	Similar to Hurler's but without corneal effects; some patients with variant defects survive to age 50; normal intelligence develops	Iduronate sulfatase	Dermatan sulfate and heparan sulfate
Sanfilippo's syndrome	Severe effects on central nervous system		
Type A		N-Sulfatase	Heparan sulfate
Type B		Acetylglucosaminidase	
Morquio's syndrome	Corneal clouding, defects in bone structure	(?)	Keratan sulfate

by a sex-linked gene; it is carried by females but appears only in male offspring.

The other disorders listed in Table 14-2 show analogous effects, although different glycosaminoglycans may be involved. They also differ in their mode of inheritance and in their effects on brain development.

F. BONE AND TEETH

Bone serves as the primary supporting material of the body. Although the different types of bone vary in structure and composition, the tissue contains certain basic common constituents, including common cellular components, fibrous proteins, and extracellular matrix materials.

The characteristic cells of bone are *osteoblasts* and *osteoclasts*. The former are responsible for the synthesis of collagen and proteoglycans and their deposition in the extracellular fluid. Osteoblasts are also involved in creating the high local concentration of calcium and phosphate ions that gives rise to crystallization of a specific form of calcium phosphate, $Ca_{10}(PO_4)_6(OH)_2$, termed *hydroxyapatite*. Osteoclasts function in degradative reactions, known as *bone resorption*, through which some bone is lost. Bone, like other body constituents, is thus in a dynamic state.

The maintenance of a normal steady state in bone is dependent on the actions of certain vitamins (e.g., vitamins A, D, and C) and hormones (e.g., parathyroid hormone, calcitonin, growth hormone, and thyroid hormone). A critical enzyme in bone metabolism is *alkaline phosphatase*. This enzyme catalyzes the hydrolysis of organic phosphate intermediates and yields the inorganic phosphate required for the production of hydroxyapatite.

In addition to the mineral component of bone, which constitutes about 25% of its volume, bone contains high amounts of collagen fibers. These are embedded in a viscous matrix of proteoglycans and glycosaminoglycans (section 14.3.D).

Disturbances in bone structure arise from imbalances of bone metabolism or turnover. For example, *hypertrophic osteoarthritis*, which may occur in middle-aged individuals, is due to an overgrowth of bones. Aberrations in the production of the mineral component of bone result in *osteomalacia* and fragility of the bones. Abnormalities in the synthesis of matrix constituents (including collagen) produces the condition known as *osteoporosis*, which is observed during ascorbic-acid (vitamin C) deficiency, estrogen deficiency, or other conditions.

Teeth are composed of three layers of calcified tissue: dentin, enamel, and cementum. *Dentin*, the innermost mineral layer and major constituent, encloses the pulp cavity, which contains blood vessels and nerves. Mineral components (mainly hydroxyapatite) make up most of the dentin. Like bone, dentin also contains collagen and glycosaminoglycans. *Enamel*, which covers the dentin in the exposed portion of the tooth, contains even higher concentrations of hydroxyapatite (about 75%) and only about 5% water. The organic matrix of enamel consists of a keratin-like protein and proteoglycans. The compact formation of these materials gives enamel the hardness required for the efficient functioning of teeth. The section of the dentin submerged within the gum is covered with the *cementum*, a bone-like tissue.

In addition to calcium phosphate, the minerals present in teeth include magnesium, sodium, potassium, carbonate, and chloride, which are found in small but significant amounts. Among the relatively minor components are iron and fluoride ions, of which the latter is known to be required for the normal development of teeth. The prevalence of dental caries is decreased considerably in areas where the water contains about 0.9 milligram of fluoride per liter; however, concentrations greater than 1.5 milligrams of fluoride per liter can produce irregular chalky patches in the teeth called *mottled enamel*. Both effects of fluoride are exhibited only during the early years when the teeth are developing.

14.4. Endocrine System

The actions of certain hormones on the metabolism of carbohydrates, lipids, and proteins were discussed in conjunction with the description of individual pathways. This section will deal with the characteristics of some hormones not discussed previously, and the tissue sources and activities of most of the prominent endocrine secretions will be reviewed.

A. EPINEPHRINE

The adrenal medulla is the major site of synthesis of the catecholamines *epinephrine* and *norepinephrine* (sections 8.16 and 14.2.C). These hormones are released from tissues as a result of specific stimuli generated by sympathetic nerves.

By binding to specific receptors on the membranes of adipose tissue cells, epinephrine (or norepinephrine) activates adenyl cyclase and thus produces an increase in intracellular cyclic AMP (sections 4.7.C and 6.6.B). This promotes the action of tissue lipase, which catalyzes the hydrolysis of triglycerides. Hence, epinephrine is effective in mobilizing lipids and elevating the plasma levels of free fatty acids (section 6.4).

Another effect of epinephrine is that it promotes an increase in the concentration of blood glucose. This, too, is produced via the action of cyclic AMP, which enhances liver phosphorylase (section 4.7.C) and depresses glycogen synthetase activity. Epinephrine also inhibits the release of insulin from the pancreas. As a result of increased glycogenolysis in muscle and consequent glycolysis, the levels of lactate in plasma are also increased.

The administration of epinephrine causes an elevation in blood pressure and an increase in cardiac output. The action of epinephrine on blood pressure (as well as on lipid mobilization) is inhibited by various prostaglandins (section 14.4.I).

B. INSULIN

Insulin is produced in the beta cells of the pancreas. The active hormone is a protein that consists of two polypeptide chains linked to each other by disulfide bridges (Fig. 14-12). It is formed from a single-chain precursor, *proinsulin,* by limited proteolysis that removes a specific polypeptide unit (see Fig. 14-12). The sequence of steps in insulin production and secretion is: (1) the synthesis and release of proinsulin by the ribosome, (2) the uptake of proinsulin by storage granules and conversion to insulin, and (3) the secretion of insulin into the bloodstream by a process involving the fusion of storage granules with the plasma membrane.

The metabolic effects of insulin are triggered by its binding to receptor sites on the membranes of sensitive cells. In addition, the binding of insulin to the cell membrane counteracts the effects of epinephrine and glucagon on

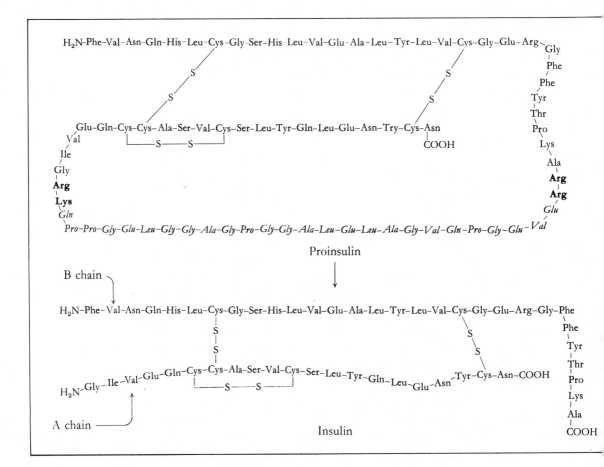

Fig. 14-12. Conversion of proinsulin to insulin. Residues shown in boldface type are excised, and the section shown in italics is removed. (This structure is for beef insulin. Some species differences are known to exist in the A chain; for example, residues 8, 9, and 10 in the A chain of sheep insulin are Ala–Gly–Val.)

adenyl cyclase. The major target cells affected by insulin are those of skeletal muscle and adipose tissue. Although certain aspects of liver metabolism are modulated by insulin, the mechanism of action on this organ may differ from that on the other tissues. Insulin is also known to affect certain enzymes, most notable of which is liver glucokinase (section 6.3).

The overall effects of insulin are to increase the rate of uptake of glucose and amino acids by cells of certain tissues (e.g., adipose tissue and skeletal muscle), to enhance glycogen synthesis and glycolytic processes, and to decrease the rate of degradation of glycogen and stored triglycerides. Hence, insulin is generally characterized as an *anabolic hormone*.

C. GLUCAGON

Glucagon is a polypeptide hormone produced and secreted by the alpha cells of the pancreas. It binds to specific receptor sites on the membranes of target cells and promotes the action of adenyl cyclase. In the liver, the resulting increase of cyclic AMP leads to accelerated degradation of glycogen to glucose 1-phosphate. The conversion of the latter to glucose 6-phosphate and the hydrolysis of this product by the action of glucose 6-phosphate phosphatase cause an elevation in the levels of circulating glucose (section 6.3). Glucagon also increases the rate of gluconeogenesis and activates hepatic lipase. Its effect on adipose tissue is similar to that of epinephrine; that is, it enhances the hydrolysis of triglycerides to fatty acids and glycerol.

The overall effect of glucagon is to increase the concentrations of glucose and free fatty acids in the blood. In contrast with insulin, glucagon is a *catabolic hormone*. Moreover, it inhibits the incorporation of amino acids into proteins and may cause a loss in the mass of liver and muscle tissue.

D. THYROID HORMONES AND THYROID-STIMULATING HORMONE

The follicular cells of the thyroid gland have the capacity to concentrate iodide ions which are incorporated into the tyrosyl units of a protein termed *thyroglobulin*. The thyroid hormones *thyroxine* (3,3′,5,5′-tetraiodothyronine, or T_4) and, to a lesser extent, *3,3′,5-triiodothyronine* (T_3) are released from the protein upon proteolysis (Fig. 14-13). Thus, thyroglobulin serves to store thyroxine. Upon its release into the bloodstream, thyroxine is bound to several plasma proteins. It appears, however, that the primary carrier is a globulin called *thyroxine-binding globulin* (TBG).

The release of thyroxine from the thyroid gland is triggered by the action of another hormone, *thyroid-stimulating hormone* (TSH), which is elaborated by the anterior pituitary gland. Since TSH also increases the level of cyclic AMP, it is conceivable that the cyclic AMP functions as a mediator in the thyroxine-releasing process. When the blood level of thyroxine is decreased, the rate of release of TSH from the anterior pituitary increases. Conversely, elevated levels of thyroxine inhibit the release of TSH. This interrelationship provides a negative feedback control and serves as a mechanism for maintaining normal amounts of thyroxine in the blood. The release of TSH into the circulation is also under neural control. This is probably the mechanism involved in the thyroid stimulation that occurs due to psychogenic factors or as a result of decreases in body temperature. In addition to inhibiting the release of thyroxine, TSH decreases the rate of transfer of iodide from the blood to the thyroid gland and inhibits the incorporation of iodide into thyroglobulin.

One physiologic effect of thyroxine is to increase the *basal metabolic rate*

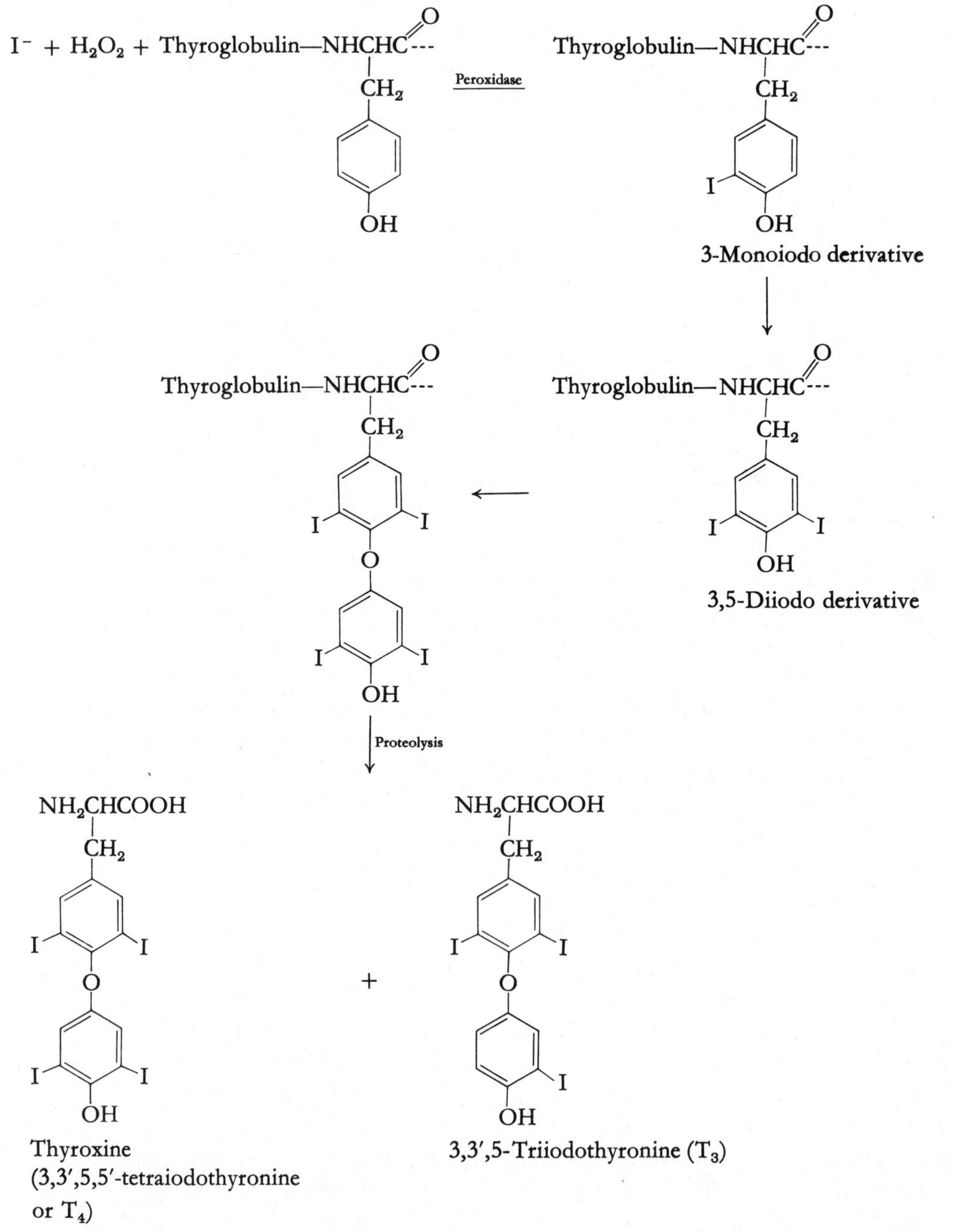

Fig. 14-13. *Biosynthesis of thyroxine* (T_4) *and triiodothyronine* (T_3).

(section 15.1.C). With respect to intermediary metabolism, thyroxine increases the rate of glucose oxidation. It also enhances mitochondrial ATPase activity, the production of adenyl cyclase, and the synthesis of RNA via DNA-dependent RNA polymerase (section 11.3). The increase in protein synthesis upon thyroxine stimulation results in elevated levels of numerous enzymes. The overall effect of thyroxine is to promote and regulate the growth of developing organisms.

Advantage is taken of the iodide-concentrating function of the thyroid gland in the treatment of hyperactivity of this gland. The administration of radioactive iodide (^{131}I) results in high levels of the isotope in the thyroid gland, and the γ rays and β particles emitted by the isotope are effective in destroying a portion of the gland's cells. The measurement of the rate of turnover of the radioactive isotope in the thyroid gland is also employed in the diagnosis of hormonal abnormalities.

The iodination step in the production of thyroid hormone is inhibited by *thiouracil*:

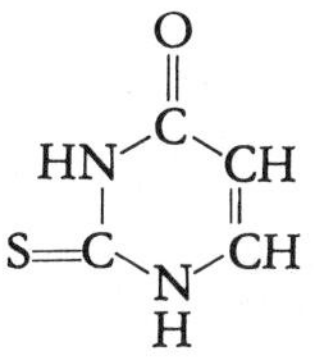

Thiouracil

This agent and analogous products are useful drugs in the treatment of hyperthyroid states.

E. PARATHYROID HORMONE AND CALCITONIN

The parathyroid gland regulates calcium metabolism by means of two secretions: *parathyroid hormone* and *calcitonin*. The former is a polypeptide that enhances the release of calcium ions from bone and thus raises the Ca^{2+} level in the blood. Another effect of parathyroid hormone is to decrease the rate of reabsorption of phosphate in the renal tubules, which in turn produces a decrease in the plasma concentration of phosphate.

The other hormone secreted by the parathyroid gland is the polypeptide, calcitonin. This has the opposite effect of parathyroid hormone; i.e., it decreases the blood level of calcium ions.

Abnormalities of the parathyroid gland result in changes in the normal blood level of calcium ions, which is approximately 10 milligrams per 100 milliliters of plasma. Since calcium ions are required in the functioning of muscle, a sharp decrease in their level in the circulation results in tetany and may be fatal. On the other hand, excessive or extended increases in circulating calcium may result in abnormal deposition of the mineral in the bones.

F. GROWTH HORMONE

Growth hormone (GH), or somatotropin, is a polypeptide that is secreted by the adenohypophysis (anterior pituitary). By stimulating the synthesis of RNA and proteins, this hormone promotes the growth of bone and other tissues during the early stages of life. Thus, dysfunction of the pituitary gland results in incomplete growth or *dwarfism*. Conversely, hyperfunctioning of the gland in young animals leads to *giantism*. When abnormalities occur in adults (e.g., due to tumors of the pituitary), there is excessive growth of the bones; this condition is called *acromegaly*.

Administration of growth hormone to animals produces a gain in weight and retention of nitrogen (positive nitrogen balance). Analysis of the plasma shows a considerable decrease in concentration of amino acids. These effects are due to the increased incorporation of amino acids into proteins.

Growth hormone is also effective in producing hyperglycemia and an elevation in the levels of free fatty acids in plasma.

G. SEX HORMONES

The pathways for the metabolism of steroid hormones and some of their activities were described previously (sections 5.6 and 6.6.D). This section will summarize some of the physiologic aspects of the hormones and their tissue relationships.

The principal male sex hormones are *testosterone* and *dihydrotestosterone*. These, as well as other androgenic steroids, are synthesized in the testes. Small but significant amounts are also produced in the adrenal gland, placenta, and ovary. The release of testosterone from the gonads is stimulated by pituitary hormones, which in turn are regulated by secretions from the hypothalamus. Since the secretion of pituitary gonadotropic hormones is affected by the levels of circulating testosterone, the system can be considered self-regulating.

Testosterone is required for the development of secondary sex characteristics as well as for the normal development of male gonads and accessory glands and the production of sperm. It binds to specific receptor sites on the membranes of target cells and subsequently promotes RNA and protein synthesis.

Female sex hormones are produced from testosterone. The primary functioning steroid, *17-β-estradiol*, is produced primarily in the ovaries (section 5.6.E), and small amounts are also formed in the placenta, adrenal cortex, and testes. The release of female sex steroids is under the control of hormones from the adenohypophysis, and their blood levels are regulated by negative feedback mechanisms analogous to those that operate for male hormones. Estrogens promote the growth of the ovaries and uterus, and they are required for the development of secondary sex characteristics. The binding of estrogens to their target cells and their effect on protein synthesis was described in a previous chapter (section 11.9).

Another type of female hormone is the steroid, *progesterone* (section 5.6.B), which is secreted from the corpus luteum. It functions to produce the necessary conditions in the uterus for implantation of the ovum and the maintenance of normal pregnancy. Secretion of progesterone occurs during the second half of the menstrual cycle. When pregnancy occurs, secretion of the hormone continues to the end of the gestation period.

The administration of progesterone during specific periods of the menstrual cycle inhibits normal ovulation. The employment of progesterone and analogous substances for contraceptive purposes is based on this effect.

H. ADRENOCORTICAL HORMONES

The cells of the adrenal cortex synthesize cholesterol from acetate (section 5.5.A). The conversion of cholesterol to pregnenolone is followed by transformation of the latter to progesterone, which serves as the precursor for the synthesis of *cortisol* and *aldosterone* (section 5.6.C).

Cortisol and its 11-keto derivative, *cortisone*, are classified as *glucocorticoids*. These steroids enhance the catabolism of proteins to amino acids, gluconeogenesis, and the deposition of glycogen in liver. They also decrease the peripheral utilization of glucose. As a result of their actions, cortisol and cortisone bring about an increase in the levels of blood glucose. Some of the mechanisms for these processes were discussed in conjunction with the control of glucose metabolism (section 6.6.D).

The functioning of the adrenal cortex in elaborating glucocorticoids is modulated by *adrenocorticotropic hormone* (ACTH). This polypeptide hormone, which is also produced and secreted by the anterior pituitary or adenohypophysis, binds to a receptor on the membrane of adrenocortical cells. The interaction leads to the activation of adenyl cyclase and the concomitant formation of cyclic AMP. The latter compound promotes a series of reactions that induce an increase in the synthesis of the enzymes required for the conversion of cholesterol to pregnenolone and the cortical steroids.

Aldosterone, 11-deoxycorticosterone, and related steroids are termed *mineralocorticoids* (section 5.6.C). These hormones regulate the levels of sodium and potassium ions in the circulatory system by causing the retention of sodium or the depletion of potassium in serum (section 13.7). It should be noted that most of the adrenocortical steroids affect blood glucose levels and sodium retention. They differ, however, in their relative potencies in producing such specific effects. Thus, cortisol has the greatest action on gluconeogenesis and

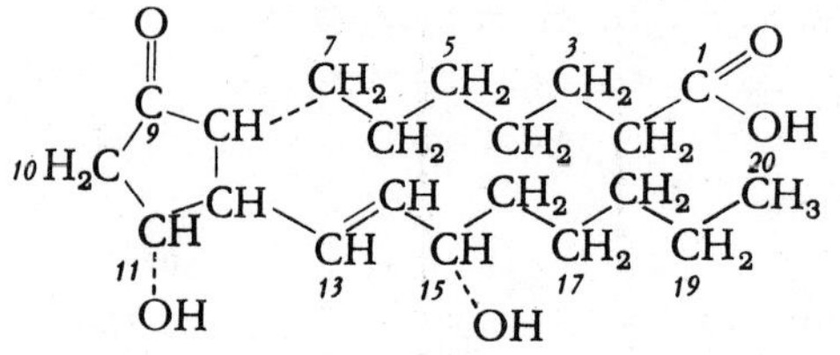

PGE_1

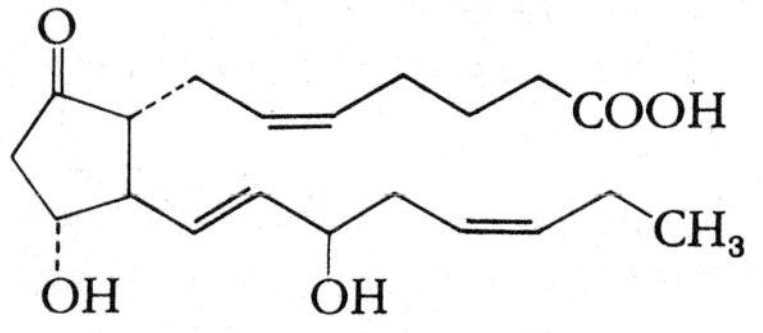

PGE_3

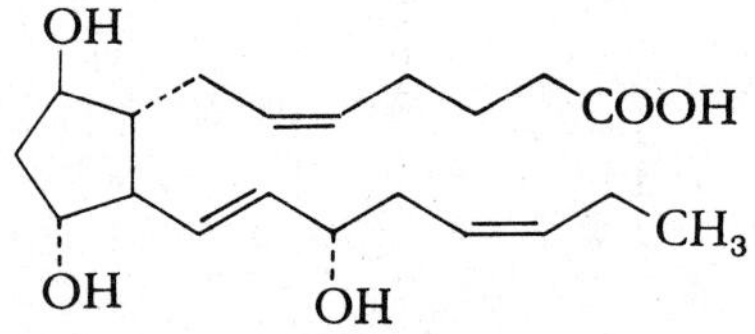

$PGF_{3\alpha}$

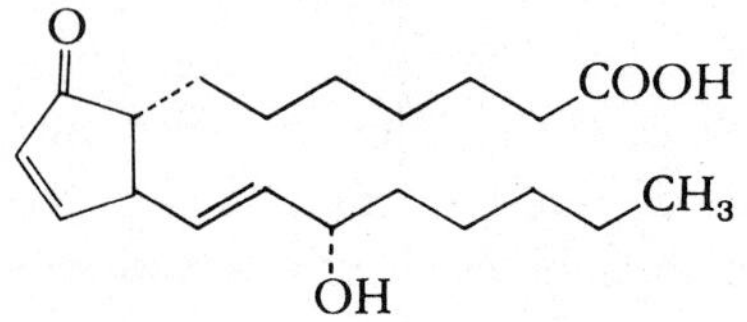

PGA_1

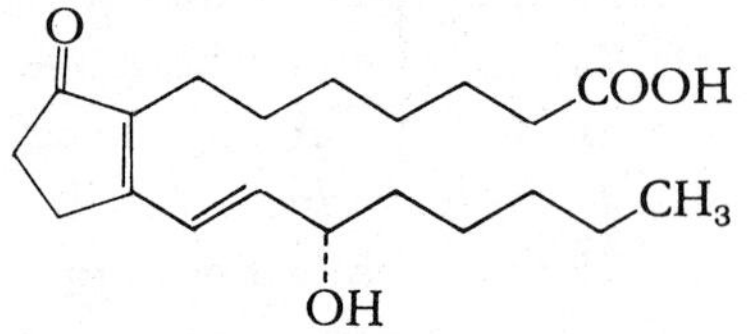

PGB_1

Fig. 14-14. Structures of representative prostaglandins.

blood glucose levels, whereas aldosterone is the most potent promoter of sodium retention.

I. PROSTAGLANDINS

The *prostaglandins* are a group of fatty acids that contain 20 carbon atoms and a five-membered ring (Fig. 14-14). In some prostaglandins, a keto group is attached to the ring; others have hydroxyl groups. The different classes of prostaglandins are denoted as *PGA*, *PGB*, *PGE*, and *PGF*; subscript numerals are added to indicate the number of carbon-carbon double bonds in the straight chains attached to the ring.

The biologic effects of the prostaglandins are quite diverse. Individual types differ in their action, and the same prostaglandin may have varying effects on different tissues. PGE_1, for example, inhibits the formation of cyclic AMP in adipose tissue and thus counteracts the effects of epinephrine or glucagon. It has the opposite effect, however, on platelets, thyroid, corpus luteum, and lung; that is, it increases the level of cyclic AMP in these tissues.

Other effects of prostaglandins include stimulating smooth-muscle contraction, increasing vascular permeability, and enhancing ion transport across certain membranes. The fact that prostaglandins are effective even at extremely low concentrations (e.g., 1 nanogram per milliliter) suggests that they have hormone-like activity. Numerous studies have been made on their pharmacologic effects, and they are being investigated for such diverse applications as the treatment of asthma, contraceptive effect, the control of blood pressure, and termination of pregnancy.

The prostaglandins are synthesized from arachidonic acid (section 3.1). A key step in their biosynthesis is the oxidation of the fatty acid by molecular oxygen to form a peroxide intermediate; this reaction is catalyzed by *cyclooxygenase*. In addition to yielding prostaglandins, the peroxide intermediate serves as a precursor for the synthesis of related compounds called *thromboxanes* and *prostacyclins*. It is of interest that cyclooxygenase is inhibited by aspirin and that many actions of this drug are probably due to the decrease in levels of prostaglandins, thromboxanes, and prostacyclins in various cells or tissues.

Suggested Reading
MUSCLE
Murray, J. M., and Weber, A. The Cooperative Action of Muscle Proteins. *Scientific American* 230 (2): 58–71, 1974.

Nobbs, C. L., Watson, H. C., and Kendrew, J. C. Structure of Deoxymyoglobin: A Crystallographic Study. *Nature* 209:339–341, 1966.

Pollard, T. D., and Weihing, R. R.   Actin and Myosin and Cell Movement. *CRC Critical Reviews in Biochemistry* 2:1–33, 1974.

Weber, A., and Murray, J. M.   Molecular Control Mechanisms in Muscle Contraction. *Physiological Reviews* 53:612–673, 1973.

Wittenberg, J. B.   Myoglobin-Facilitated Oxygen Diffusion. *Physiological Reviews* 50:559–636, 1970.

NERVE

Albers, R. W., Agranoff, B. W., Katzman, R., and Siegel, G. J. (Eds.).   *Basic Neurochemistry*. Boston: Little, Brown, 1972.

Axelrod, J.   Neurotransmitters. *Scientific American* 230 (6):59–71, 1974.

Nathanson, J. A., and Greengard, P.   Second Messengers in the Brain. *Scientific American* 237 (2): 108–119, 1977.

Snyder, S. H.   Receptors, Neurotransmitters and Drug Responses. *New England Journal of Medicine* 300:465–472, 1979.

CONNECTIVE TISSUE

Horowitz, M. I., and Pigman, W. (Eds.).   *The Glycoconjugates*, Vols. I and II. New York: Academic Press, 1977, 1978.

Neufeld, E. F., Lim, T. W., and Shapiro, L. J.   Inherited Disorders of Lysosomal Metabolism. *Annual Review of Biochemistry* 44:357–376, 1975.

Schubert, M., and Hamerman, D.   *A Primer on Connective Tissue Biochemistry*. Philadelphia: Lea & Febiger, 1968.

HORMONES*

Axelrod, J., and Weinshilbaum, R.   Catecholamines. *New England Journal of Medicine* 287:237–252, 1972.

Chan, L., and O'Malley, B. W.   Mechanism of Action of the Sex Steroid Hormones. *New England Journal of Medicine* 294:1322–1328, 1372–1381, 1430–1437, 1976.

Flier, J. S., Kahn, C. R., and Roth, J.   Receptors, Antireceptor Antibodies and Mechanisms of Insulin Resistance. *New England Journal of Medicine* 300:413–419, 1979.

Oppenheimer, J. H.   Thyroid Hormone Action at the Cellular Level. *Science* 203:971–979, 1979.

* Other references on the metabolic effects of hormones are found at the end of relevant chapters.

15. Nutrition

The subject of nutrition deals with the dietary requirements for the maintenance of optimal physiologic activities. Foods serve a variety of functions. Some nutrients are utilized primarily as fuels to provide energy; triglycerides, for example, yield fatty acids, which are the most concentrated source of biologic energy. Carbohydrates furnish the energy (as well as numerous intermediates) required for the synthesis of vital tissue constituents. Although the metabolism of amino acids also provides energy, the central function of these substances is to furnish building blocks for the synthesis of proteins. Various metal cations ingested with the diet serve as integral components of certain proteins. They are also critical for the action of certain enzymes. A number of electrolytes (both anions and cations) are essential components of body fluids and tissues. In addition, various vitamins are needed to maintain metabolic processes. Nutrition involves both qualitative and quantitative dietary requirements and the effects of decreased or excessive food intake.

15.1. Food Energy and Calories
A. CALORIC CONTRIBUTION OF NUTRIENTS

The energy required for the performance of physiologic functions is derived from the cellular oxidation of organic materials. The substrates for these reactions are the carbon compounds ingested in the diet. Individual foodstuffs vary with regard to the amount of energy they can yield. The specific energy value for each nutrient can be quantitated by calorimetric measurements and expressed as the change in enthalpy, or ΔH, of combustion (section 2.1). On the basis of thermodynamic principles, the amount of energy released upon the complete conversion of organic materials to carbon dioxide and water is independent of the specific mechanisms of the intermediary steps. Thus, the value of ΔH for the stepwise biologic oxidation of glucose is the same as that obtained when the same quantity of glucose is combusted in a calorimeter.

The energy value for each foodstuff is generally expressed in terms of heat units, i.e., kilocalories. These are sometimes called "large calories" or "kilogram calories" and are also denoted as *Calories* or *Cal* (with a capital "C"). One *kilocalorie* is the amount of heat required to raise the temperature of a kilogram of water from 15°C to 16°C at one atmosphere pressure. Experimental measurements of ΔH are performed by calorimetric procedures, and the values yielded represent the heat evolved in the complete oxidation of

Table 15-1. Average Caloric Values of Each Class of Nutrient

	Protein (kcal/gram)	Carbohydrate (kcal/gram)	Fat (kcal/gram)
Values obtained by calorimetry	5.6	4.1	9.4
Correction for production of urea	1.3	—	—
Net values	4.3	4.1	9.4
Values (rounded off) applied in nutritional calculations	4	4	9

all the elements in a compound to their oxides. Hence, the enthalpy data for nutrients that contain only carbon, hydrogen, and oxygen (e.g., carbohydrates and triglycerides) can be utilized directly, since in the organism these compounds are oxidized completely. In the case of amino acids and proteins, however, most of the nitrogen atoms are metabolized to urea, rather than to oxides of nitrogen. As a consequence of the incomplete oxidation of proteins, the values obtained by calorimetric methods must be corrected and reduced by approximately 1.3 kilocalories per gram.

It should also be noted that the caloric values of compounds within each class of nutrients (i.e., carbohydrates, proteins, and lipids) vary to some extent. One gram of polysaccharide, for example, provides more calories than the same amount of monosaccharide (4.1 and 3.8 kilocalories per gram, respectively). There are similar variations among individual proteins or triglycerides. Another important factor in determining the energy yield of various foodstuffs is the degree to which they are absorbed or digested. Generally, about 98% of dietary carbohydrates are utilized, whereas 95% of lipids and 92% of proteins are digested.

As a result of some of these variations and uncertainties, the figures used for the caloric value of various nutrients are rounded off, as shown in Table 15-1. These values may be utilized in the calculation of the caloric value of particular foods. A sample of bread, for example, contains 52% carbohydrate, 8% protein, and 3% fat. The number of kilocalories (Calories) in 25 grams of bread is obtained as follows:

Carbohydrate: 25 grams × 0.52 × 4 kilocalories/gram = 52 kilocalories
Protein: 25 grams × 0.08 × 4 kilocalories/gram = 8 kilocalories
Fat: 25 grams × 0.03 × 9 kilocalories/gram = 6.75 kilocalories

Total: 66.75, or 67 kilocalories (Cal)

Tables of the caloric values for different foods (e.g., Table 15-2) are constructed using calculations according to the method shown above.

Table 15-2. Composition and Caloric Values of Representative Foods*

Food	Weight (grams)	Protein (grams)	Carbohydrate (grams)	Fat (grams)	Kilo-calories
Bread (rye), 1 slice	23	2	12	Trace	56
Butter, 1 pat	7	Trace	Trace	6	54
Beef, 4 oz.	113	18	0	21	261
Egg, 1 medium	50	6	Trace	6	78
Milk, 1 cup	244	9	12	9	165
Sugar, 1 tablespoon	12	0	12	0	48
Orange, 1 medium	180	2	25	Trace	108
Apple, 1 medium	150	Trace	16	1	73
Pear, 1 small	75	1	12	Trace	52
Carrots, 1 cup	160	1	11	Trace	48
Potato, 1 medium	100	3	21	Trace	96
Rice, 1 cup	193	4	47	Trace	204

* Abstracted from Latham, M. C., McGandy, R. B., McCann, M. B., and Stare, F. J. *Scope Manual of Nutrition.* Kalamazoo, Michigan: The Upjohn Co., 1975.

B. PHYSIOLOGIC ENERGY EXPENDITURE

In order to maintain a steady energy state, the caloric content of the diet should be similar to the daily energy requirements. Intake of food in excess of the expended number of calories results in the storage of unused material in the form of triglycerides; i.e., a gain in weight occurs. Conversely, ingestion of low amounts of energy-providing nutrients relative to normal expenditure leads to the utilization of stored lipids and carbohydrates and a concomitant decrease in body weight. A considerable amount of energy is consumed in maintaining the integrity of the cells and the vital functions of the organism. Among these energy-requiring activities are the intracellular endergonic processes, the active transport of ions and metabolites, pumping of blood, the contraction of involuntary muscles, the operation of the central nervous system, and the osmotic work of the kidneys. Over and above this basal amount, energy is expended in numerous daily activities involving muscle contraction and various types of movements (see Table 15-3). The latter type of activity introduces considerable variations in the caloric requirements among different individuals.

Since most of the energy derived and utilized by animal organisms is ultimately evolved in the form of heat, the amount of energy expended in a specific time interval can be assessed by determining the amount of heat that is released. Actual measurements of this type—termed *direct calorimetry*—involve complex equipment and procedures. A more common approach is to utilize the respiratory quotient of nutrients together with data on the

Table 15-3. Energy Expended per Hour during Various Activities*

Activity	Kilocalories/Kilogram Body Weight	Kilocalories/Pound Body Weight
Sleeping	0.93	0.42
Sitting at rest	1.43	0.65
Standing relaxed	1.50	0.68
Walking slowly (2.6 mph)	2.86	1.30
Walking downstairs	5.20	2.36
Walking upstairs	15.8	7.18
Running (5.3 mph)	8.14	3.70
Swimming	6.86	3.12

* Adapted from *Recommended Dietary Allowances* (8th ed.). Washington, D.C.: National Academy of Sciences, 1974.

amount of excreted nitrogen. The *respiratory quotient* (RQ) of a nutrient is the molar ratio of the amount of carbon dioxide produced to the amount of oxygen consumed in the metabolic oxidation of the nutrient. Since equal volumes of different gases contain the same number of molecules, the respiratory quotient can be obtained from measurements of the volumes of CO_2 and O_2. As will be seen, the RQ can, at certain times and under specific conditions, serve as an indicator of which nutrient was oxidized, since the value of the RQ will reflect the composition of the material metabolized. Conversely, tabulated data on the RQs of individual nutrients and mixtures of different nutrients can be utilized for calculating and defining caloric expenditures.

On the basis of the stoichiometry for the complete oxidation of glucose, it can be seen that it has an RQ of one:

$$C_6H_{12}O_6 + 6O_2 \longrightarrow 6CO_2 + 6H_2O$$

$$RQ = \frac{6 \text{ moles } CO_2}{6 \text{ moles } O_2} = 1$$

Similar analysis of triglyceride oxidation (e.g., glyceryl tristearate) shows that the RQ is 0.7:

$$C_{57}H_{110}O_6 + 81\tfrac{1}{2}O_2 \longrightarrow 57CO_2 + 55H_2O$$

$$RQ = \frac{57 \text{ moles } CO_2}{81.5 \text{ moles } O_2} = 0.7$$

Although there are some small variations among different carbohydrates (e.g., among monosaccharides, disaccharides, and polysaccharides) or fats (e.g., between saturated and unsaturated fats), the average values of 1.0 and 0.7, respectively, are generally employed as their RQs.

The RQ for proteins cannot be calculated from the stoichiometry for their complete oxidation. As noted in the previous discussion, the physiologic oxidation of proteins is incomplete. About 96% of the ingested nitrogen of proteins is excreted in the urine (mainly as urea), and the other 4% is removed through the feces. When corrected for these factors, the RQ for proteins is calculated to have an average value of 0.8.

The RQ at any given time is thus dependent on the composition of the nutrients ingested and metabolized at that time. For a diet consisting only of fats, the expected RQ is 0.7. Similarly, for carbohydrate or protein diets, the expected RQ values are 1.0 and 0.8, respectively. Common diets, however, consist of mixtures of nutrients, and the RQ value will reflect their composition. Under basal conditions (see the following discussion), a value of 0.82 is generally assumed. The RQ value during the rapid production of fatty acids from glucose can also rise significantly above unity. This can be explained by the stoichiometry of lipogenesis; in the following reaction, for example, the RQ = 8:

$$4C_6H_{12}O_6 + O_2 \longrightarrow C_{16}H_{32}O_2 + 8CO_2 + 8H_2O$$

This effect is seen when animals are given a high carbohydrate diet in order to fatten them. On the other hand, in conditions of impaired carbohydrate metabolism (e.g., diabetes), the increased utilization of fat leads to a significant decrease in the RQ from the normal value.

A relationship may be derived with respect to the amount of oxygen consumed, the amount of nutrient metabolized, and the number of calories released. The calculations involved may be exemplified by considering the oxidation of the fat, tristearin or glyceryl tristearate. During its oxidation, 2 moles of the triglyceride and 163 moles of oxygen are consumed:

$$2C_{57}H_{110}O_6 + 163O_2 \longrightarrow 114CO_2 + 110H_2O$$

Since each mole of O_2 occupies 22.4 liters (at standard temperature and pressure), 163 moles of O_2 are equivalent to $163 \times 22.4 = 3651$ liters. Similarly, the amount of CO_2 produced is $114 \times 22.4 = 2554$ liters. These volumes of the gases are for the oxidation of 2 moles of tristearin (molecular weight 877), or 1754 grams. Hence, oxidation of each gram of the fat consumes 2.0 liters of O_2 and releases 1.46 liters of CO_2. Since the oxidation of a gram of fat also releases about 9.4 kilocalories (see Table 15-1), each liter of oxygen consumed is equivalent to 4.7 kilocalories of heat produced.

Table 15-4. Gas Consumption and Production, Respiratory Quotient, and Heat Production in Oxidation for Each Class of Nutrient

	Carbohydrate	Fat	Protein
Liters of O_2 consumed per gram	0.83	2.0	0.94
Liters of CO_2 produced per gram	0.83	1.42	0.77
Respiratory quotient (RQ)	1.00	0.71	0.81
Kilocalories produced per liter O_2 consumed	4.82	4.50	4.26

Similar calculations for the oxidation of proteins and carbohydrates provide the respective data for the kilocalories produced per liter of oxygen consumed (Table 15-4).

In the organism, the amounts of carbohydrate, fat, and protein utilized for energy in a given time can be estimated from the oxygen consumption, carbon dioxide production, and nitrogen excretion. The approach involves the application of the general principles already discussed, and it is best illustrated by a specific example. Suppose an individual consumes 440 liters of O_2 (corrected to standard temperature and pressure) and releases 370 liters of CO_2 in a day. His urinary nitrogen excretion during this period is 13.6 grams. How many grams of each nutrient are consumed and how many kilocalories of heat are released?

To answer these questions, the total quantity of protein metabolized must first be assessed from the value of the urinary nitrogen excretion. (The amount of nitrogen from proteins that is lost through the feces is about 4% of the total and does not affect the following calculations significantly.) Since nitrogen constitutes 16% of the average protein, multiplication of the urinary nitrogen value by 6.25 yields the value for the amount of protein metabolized. In the present example, the result is:

$13.6 \times 6.25 = 85$ grams of protein

Metabolic oxidation of 1 gram of protein consumes 0.94 liter of O_2 and releases 0.77 liter of CO_2. For the oxidation of 85 grams of protein, the respective volumes are:

$85 \times 0.94 = 79.9$ liters O_2

$85 \times 0.77 = 65.5$ liters CO_2

(Generally, the volumes of the respective gases are obtained directly from the urinary nitrogen value by means of specific conversion factors. Thus, as may be seen from the calculation, each gram of urinary nitrogen corresponds to 5.88 liters of O_2 consumed and 4.82 liters of CO_2 produced in protein metabolism.)

Table 15-5. Relationship of Non-protein RQ to Quantity of Fat and Carbohydrate Metabolized and Heat Evolved per Liter of Oxygen Consumed*

Non-protein RQ (Fats plus Carbohydrates)	Fat (grams)	Carbohydrate (grams)	Heat Evolved (kcal)
0.72	0.482	0.055	4.702
0.74	0.450	0.134	4.727
0.77	0.400	0.254	4.764
0.80	0.350	0.375	4.081
0.82	0.317	0.456	4.825
0.85	0.267	0.580	4.862
0.88	0.215	0.708	4.899
0.90	0.180	0.793	4.924
0.93	0.127	0.922	4.961
0.96	0.073	1.053	4.998
0.98	0.036	1.142	5.022

* From Bodansky, M. *Introduction to Physiological Chemistry* (4th ed.). New York: Wiley, 1938 (Abbreviated).

The volumes of O_2 and CO_2 that are exchanged in the metabolism of carbohydrates and fats are then:

440 − 80 = 360 liters O_2 consumed

370 − 65 = 305 liters CO_2 produced

The *non-protein RQ* is therefore 305/360 = 0.847 ≈ 0.85. Inspection of Table 15-5 shows that for a non-protein RQ of 0.85, each liter of oxygen consumed represents the metabolism of 0.580 gram of carbohydrate and 0.267 gram of fat. Hence, for 360 liters of oxygen, the amounts metabolized are:

360 × 0.580 gram/liter O_2 = 209 grams of carbohydrate

360 × 0.267 gram/liter O_2 = 96 grams of fat

The total heat released is therefore:

Protein: 85 grams × 4 kilocalories/gram = 340 kilocalories
Carbohydrate: 209 grams × 4 kilocalories/gram = 836 kilocalories
Fat: 96 grams × 9 kilocalories/gram = 864 kilocalories
Total: 2040 kilocalories

The calculation of the heat released according to the procedures described in this example is called *indirect calorimetry*. (It should be noted that there may be minor inconsistencies in results when data or tables from different laboratories are employed in calculations of this type. These are usually due to the number of significant figures that are employed and the "rounding off" of the data.)

C. BASAL METABOLISM

The *basal metabolism* is defined as the minimal heat produced by an individual at rest and in the postabsorptive state. It is a reflection of the energy consumed in the maintenance of the basic physiologic functions, i.e., the so-called vegetative functions. The *basal metabolic rate* (BMR) is expressed as the heat evolved (kilocalories) per square meter of body surface per hour. Measurements of the BMR are conducted after the subject has fasted a minimum of 12 hours; the subject is at rest (lying down) but awake. The amount of heat produced is derived from determinations of the amount of O_2 consumed and CO_2 released in a given time interval. This method allows for the calculation of the RQ and the amount of calories contributed by each nutrient, as described above.

The common clinical procedure is merely to determine the amount of O_2 consumed within a given time interval (e.g., 6 minutes) and to assume an average value of 0.82 for the non-protein RQ. From Table 15-5, it is seen that this RQ is equivalent to 4.825 kilocalories per liter of O_2, which is the rate of heat evolution from the metabolism of carbohydrate and fat. (The relative effect of protein metabolism on oxygen consumption is considered to be minimal under basal conditions, and the error introduced by this factor is negligible.) Nomograms are available that yield values for the body surface area when the height and weight are known. These nomograms are based on the following equation:

$$A = H^{0.725} \times W^{0.425} \times 71.84 \times 10^{-4}$$

where A is the area in square meters, H is the height in centimeters, and W is the weight in kilograms.

The method for determining the BMR can be illustrated by the following example. Suppose a 20-year-old male (weight 68 kg; height 177.8 cm) consumes 1.3 liters (corrected to 0°C and 760 torr) of O_2 in 6 minutes. Using the formula, we find that the surface area of this individual is:

$$177.8^{0.725} \times 68^{0.425} \times 71.84 \times 10^{-4} = 1.8 \text{ square meters}$$

The amount of O_2 consumed per hour is 13 liters. Multiplication of the latter value by 4.825 kilocalories per liter of O_2 yields 62.7 kilocalories evolved per hour. Hence, the BMR is:

$$62.7/1.8 = 34.8 \text{ kilocalories/square meter/hour}$$

The BMR may also be reported as a percentage of the average value for an individual of a given age. Since the average value for a 20-year-old male is

38.6 kilocalories per square meter per hour, the lower figure obtained in this case is subtracted from the average value in calculating the BMR as a percentage:

$$\frac{(38.6 - 34.8) \times 100}{38.6} = -9.8\%$$

Generally, findings within 15% of the tabulated averages are considered to be normal. The BMR is increased in patients with hyperthyroidism and in those with hyperfunction of the adrenal or pituitary glands. Conversely, the values may fall to -30% to -50% in cases of hypofunction of these endocrine organs. The BMR also decreases during periods of fasting.

D. SPECIFIC DYNAMIC ACTION

When food is ingested by an individual at rest, his heat production and oxygen consumption are increased. This increase is proportional to the caloric intake, but it varies with different nutrients. For proteins, the increase is about 30% of the caloric value, but for carbohydrates and fats, it is only 6% and 4%, respectively. A subject given 1000 kilocalories of protein, for example, will expend 300 kilocalories during the following 12 to 18 hours in addition to the basal amount. The ingestion of the same number of calories of fat will lead to the expenditure of 40 kilocalories over the basal amount. The corresponding value for carbohydrate is 60 kilocalories. This increase in heat release or energy consumption above the basal rate that follows the ingestion of nutrients is known as *specific dynamic action* (SDA).

The specific dynamic action of combinations of foodstuffs is not the sum of their individual contributions; generally, it is less than the calculated value. For a mixed American diet, an average value of 6% of the caloric intake is employed as the SDA figure. Although it is accepted that the SDA effect is due to the energy involved in the metabolism of food, the quantitative details are not clear. It appears that at least one factor is the energy utilized in the conversion of nutrients to fats and in the subsequent disposition of these components. Whatever the reason for the energy expenditure may be, it is necessary to compensate for the SDA when calculating dietary caloric requirements.

E. ENERGY REQUIREMENTS

The principal consideration in describing nutritional energy requirements is the balance between caloric intake and the amount expended. The specific caloric needs depend on the age, sex, occupation, and life habits of the individual. A male student in his early twenties, for example, consumes about

2800 kilocalories per day, whereas a female of the same age consumes an average of about 2400 kilocalories. Similarly, a clerical worker at age 40 expends approximately 2400 kilocalories, but a coal miner at the same age may utilize 4000 to 5000 kilocalories. During pregnancy or lactation, the BMR is increased, and additional calories are required to maintain a steady state.

The major compounds for providing energy reserves are the triglycerides that are stored in adipose tissue. Thus, an average 70-kilogram man has about 15 kilograms of triglycerides but only 70 grams of liver glycogen as stored reserves. These are equivalent to energy reserves of over 140,000 kilocalories for the lipid and less than 500 kilocalories for glycogen. The ingestion of excess food calories results in a weight gain, which is due essentially to the deposition of lipids in the adipose tissue. It is estimated that intake of excessive triglycerides equivalent to 3500 kilocalories in energy value produces a gain in body weight of 1 pound (0.45 kilogram). Conversely, a deficit in the same number of calories is required for the loss of 1 pound of body weight. Generally, dietary calories derive from a mixture of nutrients. Since components other than triglycerides are less efficient in providing energy, a weight gain of 1 pound requires an intake of 4000 to 5000 kilocalories of non-triglycerides.

From the standpoint of energy needs, the required calories could be derived from any one of the basic nutrients. Other factors, however, make it necessary that the diet should consist of a mixture of proteins, fats, and carbohydrates. Thus, in order to maintain metabolic protein balance, about 0.8 gram of protein should be ingested each day per kilogram of body weight. Similarly, approximately 0.5 to 1 gram of fat per kilogram of body weight per day serves to meet the requirements for essential fatty acids and for the utilization of fat-soluble vitamins. Carbohydrates play a critical role in the production of NADPH, nucleotides, and various amino acids. Although glucose can be synthesized from amino acids and glycerol (via gluconeogenesis), the dietary intake of carbohydrates is beneficial for conserving proteins and minimizing their rate of degradation. Additionally, the availability of a certain amount of glucose serves to forestall ketosis, its attendant acidosis, and the loss of inorganic ions from the blood. It is estimated that at least 5 to 7 grams of glucose per kilogram of body weight be included in the daily diet. On the basis of these considerations, it is recommended that the caloric intake consist of at least 50% carbohydrate, 20% fat, and 10% protein.

In addition to carbohydrates, lipids, and proteins, a significant fraction of the calories in certain diets is derived from alcohol (ethanol). Each gram of ethanol provides 7 kilocalories. For example, an average bottle of beer (3.6% alcohol) yields about 340 kilocalories. When calculating the total caloric intake, the contribution from alcoholic beverages should be included in many instances. Although important for the normal diet, minerals, water, vitamins, and cellulose do not contribute to the caloric intake.

15.2. Protein Requirements

Proteins are required in the diet to provide amino acids and nitrogen for the synthesis of tissue proteins. Although the catabolism of the carbon chains of proteins proceeds with a release of energy, the caloric contribution is not the primary function of these nutrients. Like other body constituents, tissue proteins are constantly degraded and must be replenished by cellular biosynthesis. The turnover rate varies with individual proteins. Most of the liver proteins, for example, have short half-lives, while those of connective tissue (especially the collagen in bone) have comparatively long turnover times. The average half-life of muscle proteins is between the half-life of collagen and that of the liver proteins. A deficiency in dietary protein will thus affect these tissues to different degrees.

The degree to which metabolized tissue is being replaced by ingested protein may be calculated from the *nitrogen balance* of the individual. This involves a determination of the total nitrogen ingested within a specific period, followed by measurement of the nitrogen excreted in the urine and feces. When the two values are similar, the individual is said to be in nitrogen balance. *Negative* nitrogen balance—i.e., the condition in which the output of nitrogen exceeds its intake—occurs during illness, wasting diseases, or starvation. Conversely, *positive* nitrogen balance—which occurs when intake exceeds output—is a normal finding in growing children and in individuals convalescing after a debilitating disease.

Amino acids are not stored to any significant extent and must be made available through regular protein intake. Since the cellular biosynthesis of proteins requires the presence of all the amino acids, the unavailability or low levels of a single amino acid will prevent the formation of a given protein. Hence, the relative nutritive value of ingested proteins depends on the degree to which they provide the complete array of amino acids. Of special interest are the essential amino acids that cannot be synthesized by the animal organism and must be derived from dietary proteins (section 7.11). These include isoleucine, leucine, lysine, methionine, phenylalanine, threonine, tryptophan, and valine. During early growth periods, exogenous histidine is also essential. A critical factor in fulfilling the protein requirement is to supply the necessary amounts of the essential amino acids, as is the provision of sufficient nitrogen for the synthesis of the non-essential amino acids (section 7.11.B).

Assuming that the ingested proteins contain the necessary amounts of essential amino acids, the minimal protein requirement is 300 to 400 milligrams per kilogram of body weight per day. In view of the differences in the dietary value of common proteins and the variations in individual physical activities, it is recommended that for health, at least 800 milligrams of protein per kilogram of body weight be included in the daily diet. Generally, proteins from animal sources are more apt to fulfill the requirement of providing essential amino acids. Those from cereals and vegetables may be deficient in this respect to a greater or lesser degree.

Fulfillment of the protein requirement is especially critical during infancy and early development. In these stages, greater amounts of proteins must be ingested in order to provide for tissue growth. For example, about 2.2 grams of protein per kilogram of body weight should be furnished daily during the first 6 months of infancy. In the period from the age of 6 months to 1 year, approximately 1.8 grams of protein are required per kilogram of body weight per day. Protein deficiency (even when carbohydrate intake is adequate) during childhood results in *kwashiorkor*, a form of malnutrition that is common in a number of developing countries and poverty-stricken areas. Characteristic manifestations of this disease are anemia, growth retardation, and various consequences of hypoproteinemia. The latter may include edema, fatty infiltration of the liver, and fibrosis. Other lesions (e.g., atrophy of the pancreatic tissue and renal defects) may develop if the protein deficiency is not corrected.

Extended deprivation of both carbohydrate and protein in early childhood produces *nutritional marasmus*. This disorder is expressed in arrested growth, and it cannot be corrected by subsequent nutritional adjustments. Adults can adjust to limited starvation conditions by various metabolic mechanisms. Poor nutrition during early childhood, however, can produce deformities and diseases that result in permanent conditions.

15.3. Carbohydrates

Since carbohydrates are metabolically interconvertible, there is no specific hexose that must be included in the diet. Although they are not essential, certain carbohydrates enhance various processes in the gastrointestinal tract. Non-digestible celluloses (section 4.1) and the pentosans of vegetables or fruits serve as roughage for maintaining normal movement of materials through the intestines. Also, a small fraction of these polysaccharides are degraded by intestinal flora to carbon dioxide and organic acids, which stimulate peristalsis. The disaccharide lactose (milk sugar) has been shown to enhance calcium absorption from the ileum. Similar effects on absorption are attributed to other carbohydrates. Nonetheless, such functions are not so critical as to render any of the sugars indispensable.

The actual source of dietary carbohydrate varies among specific populations and geographic areas. Prior to the present century, the principal carbohydrate in most European and North American diets was starch, mainly from cereals. This is still the case in underdeveloped countries. In more affluent areas, a considerable fraction of dietary carbohydrate is now provided by the disaccharide, sucrose. The utilization of this sugar has increased significantly during the last 50 years; at present, it constitutes a considerable fraction of the carbohydrate consumed with the diet. It may be noted that various types of data have been presented that correlate the ingestion of sucrose with coronary heart disease. However, the evidence is far from compelling, and most of the findings can be attributed to overconsumption of food and resulting obesity.

15.4. Lipids

It was indicated in Chapter 5 (section 5.2.E) that certain unsaturated fatty acids cannot be synthesized by mammalian organisms; these are termed the *essential fatty acids*. They include linoleic (C_{18}, $\Delta^{9,12}$), linolenic (C_{18}, $\Delta^{9,12,15}$), and arachidonic (C_{20}, $\Delta^{5,8,11,14}$) acids. The last can be synthesized from linolenic acid and is sometimes not included among the essential fatty acids. These acids are utilized for the synthesis of prostaglandins (section 14.4.I), and they may also have a role in the formation of cell membranes. The recommended dietary allowances for essential fatty acids is about 1% to 2% of the total caloric intake.

The nature of the dietary triglycerides has a significant influence on the level of cholesterol in blood. Saturated fatty acids (e.g., palmitic or stearic acids) produce an increase in blood cholesterol levels, whereas polyunsaturated acids (linoleic, linolenic, and so on) have the opposite action. Since high levels of blood cholesterol have been implicated in atherosclerosis and coronary heart disease, the intake of fat (especially of triglycerides containing saturated fatty acids) should be kept within certain limits. The present recommendations are that dietary fat should not exceed 30% of the total caloric intake. The fat should be composed of at least 30% polyunsaturated fatty acids and no more than 30% saturated fatty acids. The remainder may include monounsaturated acids.

Those fats that are solids at room temperature consist primarily of saturated fatty acids. Liquid fats (oils) are composed mainly of unsaturated acids. Butter, for example, has about 65% saturated fatty acids and 35% unsaturated acids. Soybean oil consists of 60% polyunsaturated, 25% monounsaturated, and 15% saturated fatty acids. It should be noted that all natural fats consist of mixed triglycerides (section 3.1), and the physical properties (e.g., melting point) can only serve as a general guide of their composition.

Tables with data on the proportions of saturated, monounsaturated, and polyunsaturated fatty acids in various substances are available and provide more specific information.

15.5. Water and Minerals

Various aspects of the role of inorganic ions in metabolic transformations and the maintenance of homeostasis have been discussed previously. The primary emphasis in this chapter will be on the nutritional requirements for these materials.

The amounts of each inorganic ion, vitamin, or other specific nutrient that should be included in the diet are expressed in terms of the minimal daily requirement or the recommended daily allowance. The *minimal daily requirement* (MDR) is the amount required to prevent development of deficiency symptoms. However, this amount may not be enough to meet the needs for optimal physiologic function. The *recommended daily allowance* (RDA) is the amount of nutrient that may be expected to fulfill the total body requirements, including a certain excess to provide for average individual variations. Ideally, the RDA values should be used in dietary calculations.

1. *Water*

Water functions as the solvent or vehicle for the absorption, transport, and metabolic disposition of nutrients. Except for the storage of triglycerides and certain transformations that occur on membranes, all the cellular and tissue reactions involve water as the medium. Additionally, water has a vital function in preventing undue increases in body temperature by means of its evaporation from the skin and lungs.

Although water does not add to the dietary calories, it comprises about 60% of the body weight. About 30 milliliters (7 ounces) of water per day provide for the metabolic needs of the normal individual. This amount is over and above that present in solid foods or produced during the oxidation of nutrients. It is a well-known fact that humans can survive much longer without food than when deprived of water.

2. *Calcium*

The daily calcium requirement for an average adult is about 400 to 500 milligrams. Additional amounts of calcium ion are required during pregnancy and lactation. Infants and children during their growth period also have a relatively higher demand for calcium. The RDAs are 360 to 540 milligrams for infants, 800 milligrams for children before puberty, and 1200 milligrams for children during postpubertal growth periods and for women during pregnancy and lactation.

The primary structural function of calcium is to provide a component of bones and teeth. It is deposited in these tissues as calcium phosphate or hydroxyapatite (section 14.3.F). About 99% of the 1250 grams of calcium present in the normal human adult is found in this form, and most of the rest of the calcium circulates in the blood. Serum contains approximately 10 milligrams of calcium per 100 milliliters. Sixty percent of the serum calcium occurs in the ionized form; the remainder is bound to serum proteins. Small amounts of calcium are present in extracellular fluid and soft tissues. As indicated previously (section 14.1.C), calcium ions have a critical function in muscle action. Abnormally low levels of calcium therefore lead to tetany. High levels also interfere with muscle function and may produce respiratory or cardiac failure.

Milk and cheese are rich food sources of calcium. Cow's milk contains approximately 120 milligrams of calcium per milliliter. It may be noted that this concentration is considerably higher than that of human milk, which has about 30 milligrams of calcium per milliliter. Other sources of calcium are whole grain cereals, leafy vegetables, and nuts.

The absorption of calcium from the intestines is dependent on the action of vitamin D (section 15.6.B). Other substances that enhance calcium absorption are lactose, citrate, and various amino acids. High amounts of dietary phosphate or oxalate have an inhibitory effect on the intestinal absorption of calcium.

3. Phosphorus

Since phosphate is a common constituent of all biologic materials and foods, the required amounts are normally ingested when the protein and caloric needs are met. The human adult contains approximately 650 grams of phosphorus; about 80% of the total is combined with calcium in the form of calcium phosphate in the bones and teeth. A portion of the body phosphate circulates in the blood (about 4 milligrams of phosphorus per 100 milliliters of plasma). The ratio of HPO_4^{2-} to $H_2PO_4^-$ at the pH of plasma is approximately 4:1. The rest of the phosphate found in tissues is in the form of organic phosphate, i.e., that phosphate which is combined in nucleic acids, nucleotides, monosaccharide derivatives, and phospholipids.

Phosphate is found in substantial concentrations in meat, fish, eggs, and cereal products. Absorption of inorganic phosphate occurs in the small intestine.

4. Sodium and Potassium

Both sodium and potassium ions function in regulating the fluid volume of the organism and in maintaining the blood pressure. The concentration of sodium ions is highest in the extracellular fluid, where it is balanced by chloride and bicarbonate ions. Potassium is the principal cation within the cell. The relative amounts distributed between the extracellular and intracellular compartments and the balanced movement of ions between these two compartments are critical for normal homeostasis. Both cations are involved in the maintenance of blood pH, since they can be exchanged in the kidneys for hydrogen ions or reabsorbed within the renal tubules (section 13.5). Various other physiologic activities are also dependent on these cations; for example, potassium is required for the secretion of insulin from the pancreas, and sodium functions in the movement of glucose across the intestinal membrane.

The dietary requirement for sodium and potassium is not established. It is estimated that less than 1 gram of each of these ions per day can meet the normal body needs. Actually, much more than this amount is generally consumed.

Most foods of plant or animal origin contain the required amounts of potassium and sodium. Furthermore, most of the sodium ingested in the average American diet comes from the salt added to enhance the taste of foods.

Chronic overindulgence in sodium results in an expansion of the extracellular fluid volume. High intake of sodium has also been implicated in the causation of hypertension. Conversely, limiting the amount of salt in the diet has generally (though not universally) been found to be beneficial in lowering the blood pressure of individuals with hypertension.

Sodium depletion occurs during abnormal and prolonged sweating, during excessive intake of diuretics, and in certain renal diseases. This may produce nausea, exhaustion, vomiting, and respiratory failure. When the loss of sodium is not due to an organic abnormality, the symptoms can be prevented by adding salt to the drinking water.

Excessive loss of potassium can induce muscle weakness, cardiac abnormalities, and respiratory failure.

5. *Magnesium*

The average human adult contains about 25 grams of magnesium. More than 60% of the body magnesium is found in bone, and almost 30% is present in muscle. The liver and pancreas also contain about 200 milligrams of magnesium per kilogram. Magnesium is also present to some degree in all cells, and it plays a key role in numerous cellular reactions.

Although magnesium is present in foods from animal sources, vegetables, fruits, and cereals provide a better source. The RDA of magnesium for adults is 300 to 350 milligrams per day. Normal serum levels are 1 to 2 milligrams per 100 milliliters. Low serum magnesium levels have been seen in alcoholics.

The symptoms in magnesium deficiency include irritability, emotional lability, and tetany.

6. *Chloride*

Chloride ions occur in body fluids together with sodium, potassium, and bicarbonate ions. The actions of chloride are related to those of the associated cations. Among these functions are the maintenance of fluid balance and osmotic pressure. In addition, chloride is important in the production of the hydrochloric acid of gastric juice.

Since the chloride in foods generally occurs as sodium chloride, normal salt intake provides the necessary amount of this anion.

7. *Sulfur*

Sulfur is sometimes included among the nutrient minerals. Although inorganic sulfate is absorbed from the intestine, it is questionable whether it is utilized to a significant extent. Most of the sulfate required for metabolic reactions arises from the oxidation of cysteine (page 342). The average requirement for this element is fulfilled by normal protein intake.

8. *Iron*

The total iron content of an average adult is 4 to 4.5 grams. About 65% of the iron in the body occurs in circulating hemoglobin, and 3% is found in myoglobin. Most of the remainder is stored in the liver, spleen, bone marrow, and muscle in the form of iron-containing proteins, *ferritin* (15%) and *hemosiderin*, or as the transport protein, *transferrin* (Chap. 10, section 5). Less than 1% of the total iron is found in various enzymes, e.g., catalase, cytochromes, or peroxidases.

Most of the iron in the body is conserved and reutilized during normal metabolism. The average loss (about 1 milligram per day) is due mainly to shedding of the epithelial cells that line the alimentary and urinary tract. As a result of menstrual bleeding, women during child-bearing years lose greater amounts of iron. Similarly, during pregnancy and lactation, more than the average amounts of iron are removed from the body.

Iron absorption occurs primarily in the duodenum. The amount absorbed is regulated by the mucosal cells of the intestinal tract and is dependent on the levels of circulating transferrin. Thus, the degree of iron uptake is relatively greater in anemia, when the levels of stored ferritin and plasma transferrin decrease. Control of absorption is the principal mechanism for preventing overloading of iron, as well as for maintaining the desired amounts of the storage forms.

Generally, only about 10% of the iron consumed with foods is utilized. An important factor in controlling the degree of absorption of dietary iron, in addition to the limits imposed by the levels of iron stored in the body, is the difference in the uptake efficiency among the different forms of iron found in foods. Divalent iron (Fe^{2+}) appears to be more absorbable than the trivalent ion (Fe^{3+}). Heme complexes, as found in liver and other meats, are more readily taken up than the salts found in wheat and other cereals. Phytic acid and phosphates in foods impair iron absorption, whereas ascorbic acid, which reduces ferric iron to the divalent state, renders the iron more absorbable.

The RDA of iron for an average adult is about 10 to 15 milligrams. Adjustments in this value must be made to provide additional iron during growth periods, bleeding, pregnancy, and lactation.

Inadequate iron intake leads to iron-deficiency anemia. This condition—which may range from relatively mild to severe—is the most common nutritional deficiency in this country. It occurs mainly in individuals who have a temporary or extended increased demand for iron. Infants, especially those born to anemic mothers, may either develop a deficiency or be born with the condition. It is noteworthy that milk is a poor source of iron. Growing children, adolescent girls, and women during the reproductive period are also prone to anemia. Individuals with bleeding ulcers or who undergo an operation are also subject to iron deficiency. In severe cases, iron deficiency is manifested as hypochromic, microcytic anemia, a condition in which the red blood cells show a low hemoglobin level and reduced size.

Since iron is not excreted to a significant extent, its excessive consumption results in an overload of the storage forms, or *hemosiderosis*. This condition may also occur after frequent blood transfusions. The iron-storage disease *hemochromatosis* appears to arise from an inborn error of metabolism involving poor control in the absorption of iron. The unregulated transport of the ion from the intestine, coupled with the absence of a disposal mechanism, produces excessive concentrations of the storage form of iron in the tissues.

9. *Copper*

The average 70-kilogram adult contains approximately 150 milligrams of copper. The normal plasma level of copper is about 1 microgram per milliliter. Approximately 90% of plasma copper circulates as the metalloprotein, *ceruloplasmin*. Ceruloplasmin contains 3 milligrams of copper per gram of protein, has ferroxidase activity (i.e., oxidizes Fe^{2+} to Fe^{3+}), and catalyzes the oxidation of aromatic amines. It also functions to transport copper to the tissues for incorporation into a number of metabolic enzymes, e.g., cytochrome oxidase, δ-aminolevulinic acid dehydratase, tyrosinase, and lysyl oxidase.

Copper is thus an essential nutrient. In addition to its contributions to the structure of the various metalloenzymes, its presence is critical for the incorporation of iron into hemoglobin and in the production of the connective-tissue proteins, collagen and elastin. Dietary copper salts, following their absorption from the stomach and duodenum, are carried by albumin to the liver, where their copper is incorporated into ceruloplasmin. The suggested daily allowance of copper is 2.5 milligrams for normal adults. Less than half of this amount is absorbed; the remainder is eliminated through the intestinal tract. The nutritional requirement for copper is generally met by the average intake of meats and legumes.

Although comparatively rare, copper deficiencies have been reported in humans, e.g., in infants on a milk diet. The deficiency results in microcytic, normochromic anemia and may be manifested by unusual pallor, edema, and retarded growth. Animal studies have revealed that copper deficiency may lead to incomplete cross-linking of collagen and elastin (section 14.3). The consequences of such abnormalities include bone fragility and other connective-tissue dysfunctions. Poorly cross-linked collagen and elastin cannot maintain the integrity of blood vessels. The condition is therefore reflected in subcutaneous and internal hemorrhages and consequent cardiovascular disorders.

A dysfunction in the production of ceruloplasmin may occur in the abnormality known as *Wilson's disease*. This disorder is characterized by the accumulation of excessive amounts of copper in the tissues and results in neurologic degeneration and cirrhotic liver changes.

10. *Iodine*

The requirement for iodine is due to its utilization in the production of thyroid hormones (section 14.4.D). About 1 microgram of iodine per kilogram of body weight is an adequate intake for a normal adult. Prolonged deprivation of iodine produces the condition known as *goiter*, or hypertrophy or enlargement of the thyroid gland, which results from excessive secretory activity of the tissue.

11. *Fluoride*

The contribution of fluoride to the structure of teeth was discussed previously (section 14.3.F). The major need for this anion occurs during infancy and early childhood. An intake of 1 to 2 milligrams per day appears to be adequate.

12. *Zinc, Manganese, and Molybdenum*

A number of enzymes are known to contain zinc; these include carbonic anhydrase, alkaline phosphatase, alcohol dehydrogenase, carboxypeptidase, and various enzymes involved in nucleic-acid function. Zinc is also a component of insulin, and it appears to be involved in the function of vitamin A. About 10 to 15 milligrams of zinc is a recommended daily allowance. As a rule, the needed amounts occur in usual diets. Zinc deficiency appears to result in hypogonadism and mild anemia.

Manganese is also a component of a number of enzymes, such as carboxylase, cholinesterase, and various phosphatases. Manganese is needed for the normal functioning of the central nervous system, reproduction, and the maintenance of bone structure. About 5 milligrams daily of dietary manganese appears to fulfill the requirement.

Molybdenum is regarded as essential because of its presence in certain enzymes, of which the most prominent is xanthine oxidase. Although no deficiency disease is known, a daily intake of 100 micrograms is suggested. This amount is generally present in the average diet.

13. *Other Trace Constituents of Tissue*

A number of metals are present in tissues as trace amounts. Examples are strontium, silicon, chromium, and nickel. These are consumed with ingested food, but there is no evidence that they have a physiologic function. Cobalt is an integral component of vitamin B_{12}. However, since humans cannot synthesize this vitamin, cobalt cannot be required for this purpose.

15.6. Vitamins

Vitamins are organic compounds (other than the common nutrients) that are required for the normal functioning of the organism, but which cannot be produced by internal biosynthetic mechanisms. Therefore, they must be included with the diet, albeit in comparatively small amounts. The vitamins are classified into two groups, *fat-soluble* and *water-soluble*. The former group includes vitamins A, D, E, and K. These substances are insoluble in aqueous medium and are transported by and stored with lipids or lipoproteins. Although the overall functions of these vitamins are known, the detailed mechanisms of their action have not yet been elucidated. The water-soluble vitamins comprise vitamin C, the B-complex group, and related substances. Most of the water-soluble vitamins serve as coenzymes, and their sites of action are therefore better defined. The ensuing discussion will deal with some of the biochemical characteristics and nutritional requirements of each of the vitamins.

A. VITAMIN A

Vitamin A is a 20-carbon alcohol termed *retinol* (Fig. 15-1). It is found in foods from animal sources, e.g., liver, eggs, and cream. In addition, various yellow and green vegetables contain a group of pigments, called *carotenes* (α-, β-, and γ-carotene), that can be transformed into vitamin A in animal

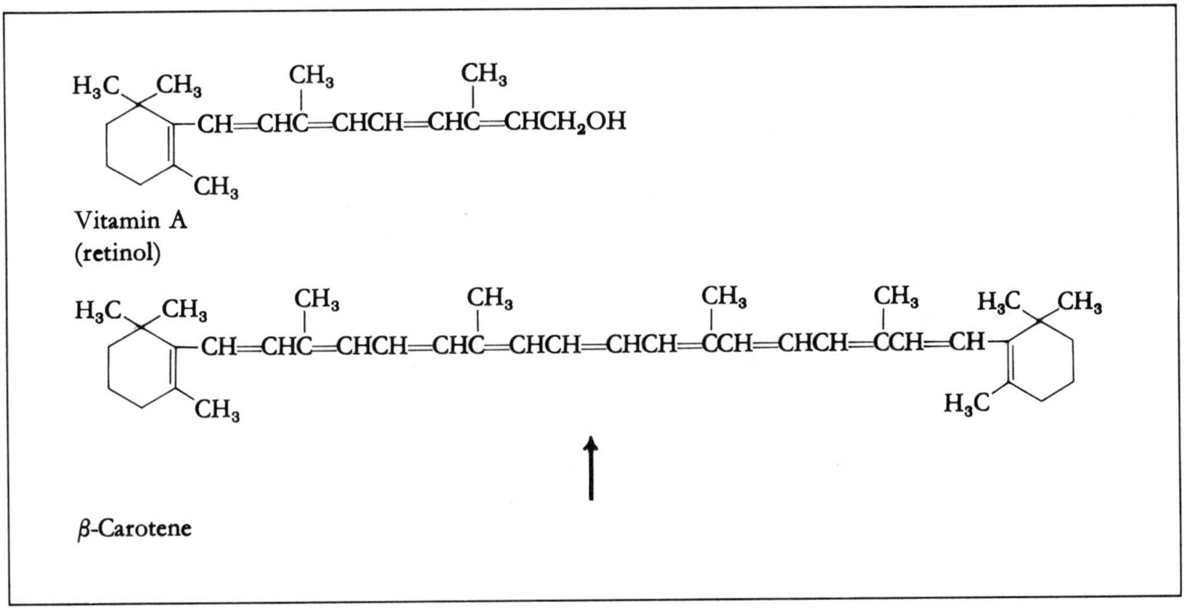

[Structure of Vitamin A (retinol) and β-Carotene shown; arrow indicates site of cleavage]

Fig. 15-1. *Structures of vitamin A and β-carotene. The arrow shows the site of cleavage for the transformation of carotene to retinol.*

organisms. Thus, carotene functions as a provitamin that may effectively provide the required retinol (see Fig. 15-1).

Vitamin A appears to have a number of metabolic and physiologic actions. It is critical for normal growth and reproduction. Its functions in maintaining the stability of membranes and is necessary for the maintenance of epithelial cells. Retinol and its derivatives are also important components in vision processes. The contribution of vitamin A to vision is the function that has been defined most clearly on a molecular level.

The retina of the eye contains two types of structures for sensing of light, the *rods* and *cones*. Each of these serves a different purpose. The rods are involved in vision at low light intensities or dim light, while the cones function in color vision or in processing high-intensity light. The rod cells contain a light-sensing conjugated protein termed *rhodopsin*. The prosthetic group in this protein is an aldehyde derivative of retinol called *cis*-retinal (Fig. 15-2). The absorption of light by rhodopsin-containing vesicles of the rod cells produces a change in the conformation of the protein and dissociation of the retinal as all-*trans*-retinal (see Fig. 15-2). The unconjugated protein is designated as *opsin*. For optimal eye adjustment, rhodopsin must be regenerated rapidly. This involves the transformation of all-*trans*-retinal to the *cis* isomer (catalyzed by retinal isomerase) and recombination of the

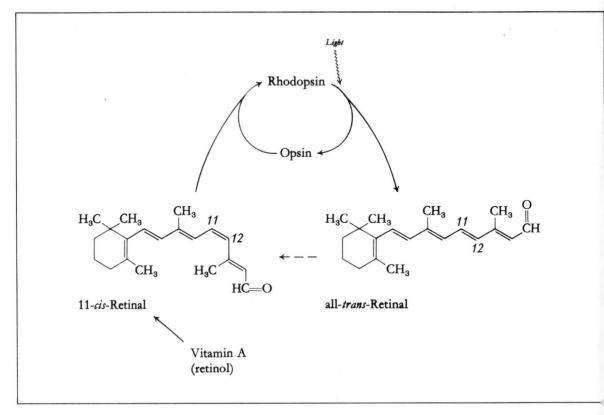

Fig. 15-2. The visual cycle and the mechanism whereby vitamin A participates in the process. Carbon-11 and carbon-12, located at the site where the two isomers differ, are indicated by italics. The point of entry of vitamin A into the cycle is also shown. The transformation of all-trans-retinal to 11-cis-retinal is indicated by a dashed arrow because several steps may be involved.

product with opsin. The complete process is called the *visual cycle*. Vitamin A (retinol) serves as the substrate for providing the required aldehyde. An NAD^+-linked dehydrogenase enzyme catalyzes the production of *cis*-retinal from retinol. Thus, vitamin A is utilized to replenish the supply of retinal and to provide the levels necessary for the optimal synthesis of rhodopsin.

Ingested vitamin A and carotene are absorbed from the intestine and esterified with a long-chain fatty acid, usually palmitic acid. The ester is taken up into the lymphatic system and carried to the liver with the chylomicrons. When transported from the liver to the tissues, the vitamin is complexed with an α_2-globulin termed *retinol-binding protein*. The rate at which rhodopsin can be regenerated is dependent on the blood level of the vitamin. Normal concentrations are 30 to 50 micrograms per 100 milliliters of plasma.

Carotenes are not absorbed as readily as pre-formed retinol, so their vitamin activity cannot be calculated on the basis of stoichiometric considerations. Studies on animals indicate that on a weight basis, carotenes are half as effective as retinol. The unit activity of the vitamin can be expressed as retinol equivalents or as international units. One *retinol equivalent* is equal in activity to 1 microgram of retinol, and an *international unit* (IU) *of vitamin A* is equal to 0.3 retinol equivalents. The recommended daily allowances (RDA) for average adults are 5000 IU for males and 4000 IU for females. Higher amounts (relative to the body weight) are recommended for infants and growing children.

In addition to resulting from low dietary intake, a deficiency of vitamin A may arise either from poor absorption caused by malfunction of the mucosal cells or from liver disease (poor bile flow). One of the early deficiency symptoms is night blindness, which is due to a decrease in availability of retinol for the regeneration of rhodopsin. A deficiency of the vitamin in children results in defective growth and in arrested development of the bones and nervous system. Degenerative changes occur in epithelial cells with concomitant keratinization of the affected tissues. Such changes result in disorders of the mucous membranes lining various tracts and ducts and degenerative transformations in the epithelial cells of the eye. The condition known as *xerophthalmia*, a consequence of these effects on the lacrimal glands and cornea, may lead to blindness of the affected eye.

Excessive amounts of vitamin A produce toxic effects. The symptoms include headache, nausea, weight loss, and dermatitis. Generally, the afflicted individual recovers when the vitamin is withdrawn. Acute overdose may result in death.

B. VITAMIN D

The term *vitamin D* is generally applied to two steroid alcohols. One of these —called *cholecalciferol* or *vitamin* D_3—is formed in mammalian skin as a result of the action of ultraviolet light. The other form of vitamin D is produced by irradiation of ergosterol, a sterol from yeast. The product, *ergocalciferol*, is also called *vitamin* D_2 (Fig. 15-3). Commercial vitamin D is generally prepared from the yeast precursor. One *international unit of vitamin D* is defined as 0.025 microgram of the vitamin.

The provitamin of cholecalciferol—i.e., 7-dehydrocholesterol—is a metabolite of cholesterol that is deposited in skin. Vitamin D_3 is formed internally from 7-dehydrocholesterol upon exposure of the skin to sunlight. The dietary requirement for the vitamin therefore depends upon the individual's environment and life-style. Persons working outdoors in sunny regions require minimal amounts of the vitamin, whereas the opposite is true for individuals who have little exposure to the sun. Another factor is the degree

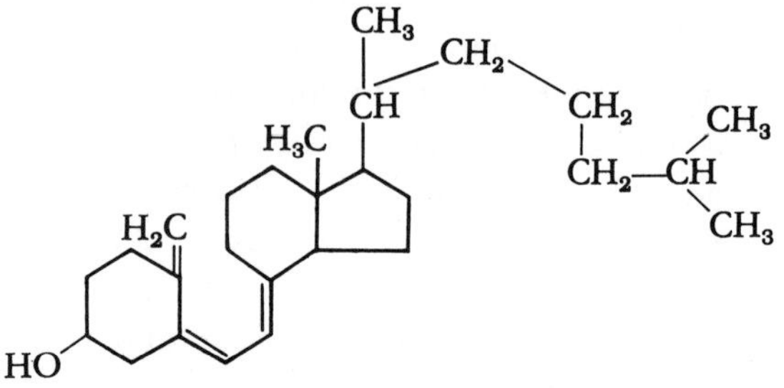

Cholecalciferol
(vitamin D_3)

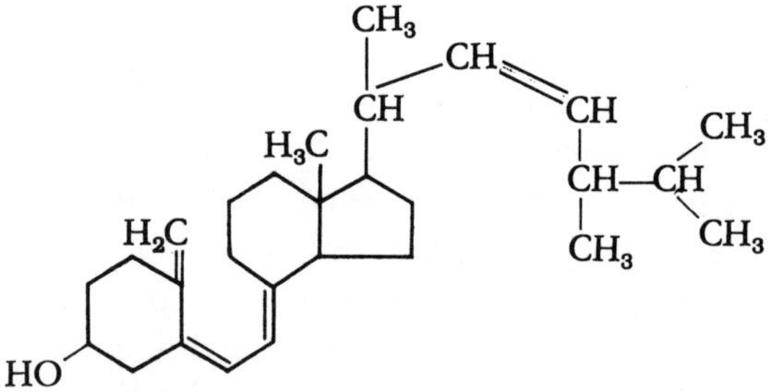

Ergocalciferol
(vitamin D_2)

Fig. 15-3. Structures of the major forms of vitamin D.

of pigmentation and color of the skin since ultraviolet radiation is filtered by skin components. Deficiency diseases can be prevented by an intake of vitamin D of about 100 IU per day; however, 400 IU is generally recommended, especially for infants.

The biologic effects of cholecalciferol or ergocalciferol are actually produced by metabolites of the vitamins. In the liver, cholecalciferol is transformed to 25-hydroxycholecalciferol, which is the principal circulating form of the vitamin. 25-Hydroxycholecalciferol is converted in the kidneys to 1,25-dihydroxycholecalciferol (Fig. 15-4); this metabolite is the primary active form of vitamin D. Analogous transformations appear to occur with ergocalciferol.

1,25-Dihydroxycholecalciferol is extremely effective in stimulating the absorption of calcium ions from the small intestine. The calcium circulating in the bloodstream is thus made available for the production of bone tissue. The mechanism of action of 1,25-dihydroxycholecalciferol seems to involve stimulation of the synthesis of a specific calcium-binding protein in the intestinal epithelial cells, which apparently occurs as a result of enhancement of

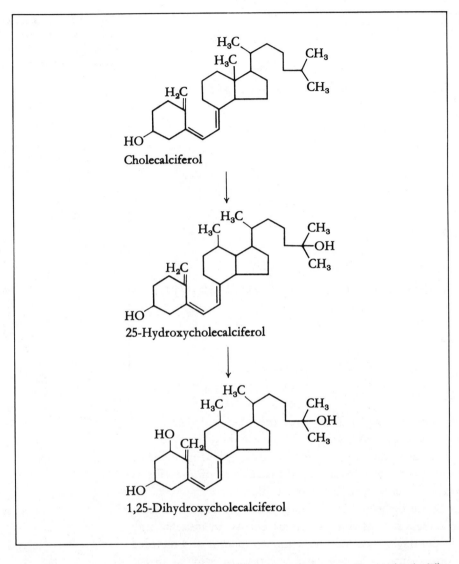

Fig. 15-4. Metabolic transformations of vitamin D_3.

the DNA activity that generates the mRNA required for protein synthesis. The calcium-binding protein that is formed as a result of this activity functions to transport calcium ions to the bloodstream.

The production of 1,25-dihydroxycholecalciferol in the kidneys seems to be regulated by parathyroid hormone (section 14.4.E). Secretion of this hormone is elevated when the blood levels of calcium are decreased. The hormone effectively enhances the action of the vitamin, which in turn functions to increase the blood calcium levels.

The overall physiologic action of vitamin D is thus to promote the normal formation of bone. In children, a deficiency of the vitamin causes rickets, a

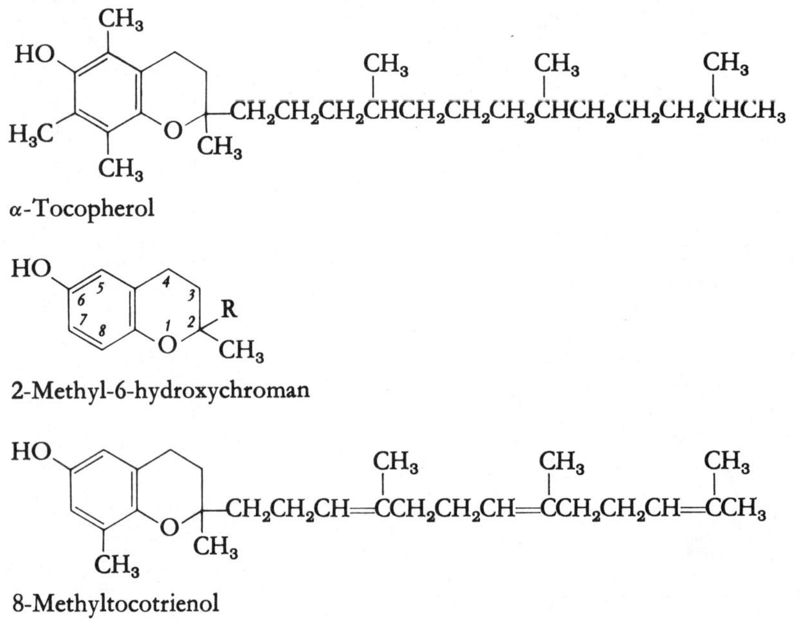

Fig. 15-5. Structures of α-tocopherol (vitamin E), 2-methyl-6-hydroxychroman, and 8-methyltocotrienol.

disease that involves abnormalities in bone structure. Poor calcification of the leg bones results in curvature or bowing of the legs, and the development of other bone structures also tends to be defective. Thus, the fontanelle of the skull may not close. In some instances, the structure of the ribs is irregular, and the teeth are malformed and prone to decay.

Although vitamin D can be stored in the body for considerable periods, the ingestion of excessive amounts produces toxic effects. In infants, high levels (e.g., 3000 units per day) may result in the loss of appetite and retarded growth. Higher amounts produce abnormal deposition of calcium in the arteries, kidneys, and various other organs (calcinosis).

C. VITAMIN E

The biologic activities of vitamin E are exhibited by a number of compounds with analogous structures, of which the most potent is α-*tocopherol* (Fig. 15-5). All forms of vitamin E have a 2-methyl-6-hydroxychroman ring (see Fig. 15-5) but differ from each other with respect to the number of methyl groups on the ring or the unsaturated bonds in the side chain. An example of a compound of this class with an unsaturated structure, 8-methyltocotrienol, is also shown in Figure 15-5.

Vitamin E is found in vegetable oils as well as in wheat germ oil. Relatively significant amounts also occur in beef liver, eggs, milk, and butter. The units of vitamin E are expressed in terms of α-tocopherol acetate; that is, one *international unit of vitamin E* is equal to 1 milligram of the ester. The RDA of vitamin E is 10 to 15 IU for adults and about 5 IU for infants.

The requirement for vitamin E and the effects of its deficiencies were demonstrated in different animal species. Although numerous activities have been ascribed to its effect in humans, many of these claims remain to be verified. Moreover, even animals show differences among species with regard to the deficiency symptoms. The absence of definitive data for humans is probably due to the impossibility of direct experimentation. The possibility that this vitamin is required, however, cannot be excluded.

Deficiency of vitamin E in the male rat results in sterility arising from degeneration of the testicular germinal epithelium. The vitamin is not required for conception in female rats, but in its absence the fetus dies during gestation. The effect of vitamin E deficiency in guinea pigs, rabbits, and monkeys is degeneration of skeletal muscle accompanied by replacement with connective tissue and fat. The disorder is similar in this respect to human muscular dystrophy. There is no established relationship, however, between the action of vitamin E and muscular dystrophy in man. Chicks are affected by a deficiency of the vitamin in still another manner: they develop encephalomalacia and accumulation of fluid under the skin.

Several manifestations of vitamin E deficiency are especially prominent when animals are fed a diet rich in unsaturated fatty acids. The symptoms include hepatic necrosis and yellow fat disease. The latter appears to be due to oxidation and polymerization of the fatty acids. That vitamin E functions as an important nutritional antioxidant is also evidenced by the finding that erythrocytes of deficient animals are hemolyzed more readily by peroxides than are those of non-deficient animals. The different effects of vitamin E as well as the variety of diseases resulting from its deficiency may all involve its antioxidant action.

Numerous pharmacologic actions have been ascribed to high doses of vitamin E in humans. These include increased fertility, increased male potency, strengthening of female breasts, improvement of various skin defects, and relief from ischemic heart disease. The scientific basis for all these claims, however, is still open to wide criticism.

As are other fat-soluble substances, vitamin E is absorbed and transported with lipids and lipoproteins. Consequently, malfunctions of the pancreas or liver or disorders in bile secretion will lead to a decrease in the absorption of the vitamin. The average plasma levels of vitamin E are about 0.8 milligram per milliliter; however, there are no apparent disease symptoms even when the levels are very low.

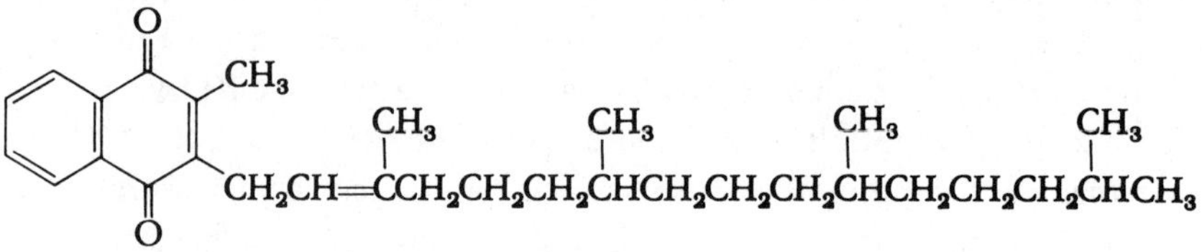

Vitamin K_1

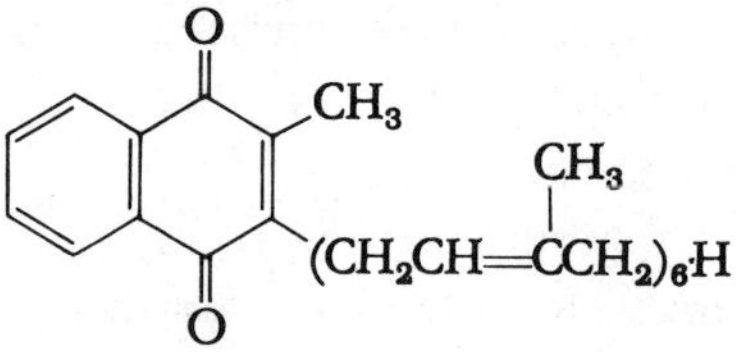

Vitamin K_2

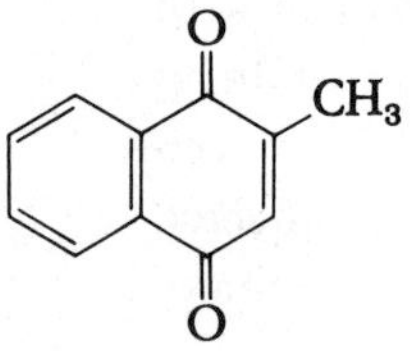

Vitamin K_3

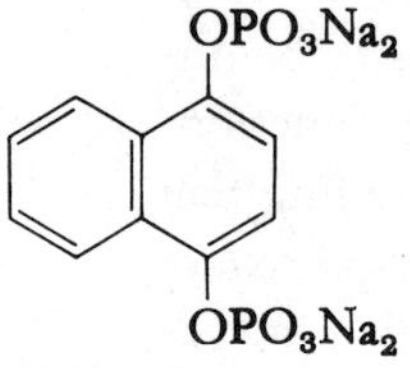

Synkayvite
(menadiol sodium diphosphate)

Fig. 15-6. Structures of the K vitamins and synthetic analogs.

D. VITAMIN K

Vitamin K occurs naturally in two forms. One compound, which is found in green leaves and other plant parts, is 2-methyl-3-phytyl-1,4-naphthoquinone or *vitamin* K_1 (Fig. 15-6). Another analog, *vitamin* K_2, is synthesized by the intestinal flora. As a result of the latter source, deficiencies of the vitamin do not occur when the dietary intake is minimal. A synthetic fat-soluble preparation that is twice as active as vitamin K_1—termed *menadione* or *vitamin K_3*—is also available.

Vitamin K is required for the production of normal prothrombin as well as of blood-clotting factors VII, IX, and X. The participation of the vitamin in the posttranscriptional γ-carboxylation of prothrombin was discussed in conjunction with the blood coagulation process (section 12.4.B).

Although vitamin K deficiency does not result from insufficient dietary intake, it may occur in newborns or in individuals with malabsorption disorders. Various water-soluble preparations are available for the treatment of these diseases, such as the sodium salt of menadiol diphosphate or Synkayvite (see Fig. 15-6).

Vitamins K_1 and K_2 are not toxic, even in high doses. However, overdoses of the synthetic preparations (e.g., vitamin K_3 or Synkayvite) produce toxic reactions, which are manifested as hyperbilirubinemia, kernicterus, or hemolytic anemia.

E. B-COMPLEX VITAMINS

The B-complex vitamins constitute a group of water-soluble vitamins that generally occur in the same food sources, though not necessarily in similar proportions. The class comprises thiamine, niacin, riboflavin, pyridoxine, biotin, pantothenic acid, folic acid, and vitamin B_{12} (cobalamin). Other substances that are sometimes included as B-type vitamins are choline, inositol, *p*-aminobenzoic acid, lipoic acid, and coenzyme Q. The B vitamins are absorbed from the small intestine and distributed to the tissues via the systemic circulation; the highest amounts are generally found in the liver. The degree of storage of the water-soluble vitamins is relatively insignificant, so their intake should be regular and frequent. Excess amounts are excreted in the urine.

The functions of many of the B vitamins were indicated in the previous discussions of intermediary metabolism. Their characteristics and requirements will be described in the following discussion.

1. Thiamine

Thiamine or *vitamin B_1* (Fig. 15-7) is utilized for the intracellular synthesis of thiamine pyrophosphate (TPP). The latter is formed by the transfer of a pyrophosphate group from ATP to thiamine. The role of thiamine pyrophosphate in the decarboxylation of pyruvate (section 4.4) and α-ketoglutarate (section 3.4) and its function in the transketolase reaction (section 4.6) were described previously.

The food source with the highest concentration of thiamine is wheat germ (2.05 milligrams per 100 grams). Other sources of the vitamin are pork, beef

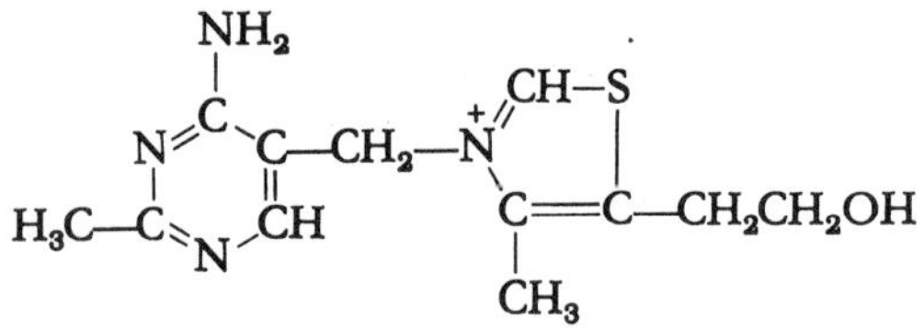

Fig. 15-7. Structures of thiamine and thiamine pyrophosphate.

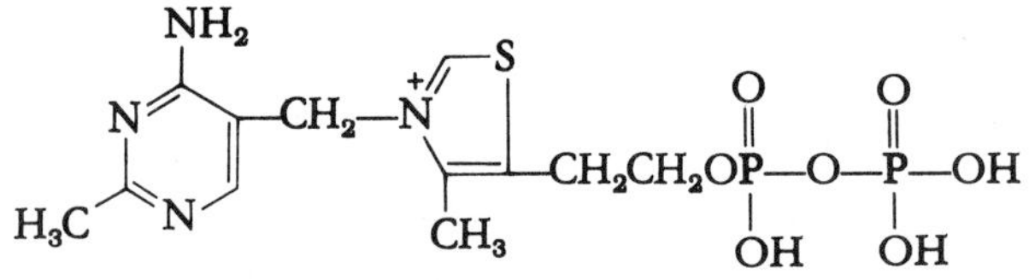

liver, eggs, peas, and potatoes. The RDA is 0.5 milligram of thiamine for each 1000 kilocalories taken in the diet. This corresponds to an average of 1.2 to 1.5 milligrams per day for a normal adult.

A deficiency in thiamine results in the disease generally termed *beriberi*. As may be expected from biochemical considerations, the levels of pyruvate and lactate increase in the blood during thiamine deficiency; the accumulation of these acids is especially notable in the brain. Impairments also occur in the nervous, cardiovascular, and gastrointestinal systems. The affected individual tires easily, weakness of the hands and legs is common, and swelling may occur around the ankles.

In addition to these general features, the disease may be expressed as "wet" or "dry" beriberi. The former is characterized by accumulation of fluid (edema) in the legs, chest pains, and cardiac palpitations. Wet beriberi may terminate in circulatory failure and death. The dry form of beriberi does not involve the accumulation of edematous fluid. The common features of both disorders are peripheral neuritis and muscular weakness. Other possible developments are mental confusion, anxiety, and the loss of deep reflexes. In infants, thiamine deficiency may produce diarrhea, convulsions, and dyspnea. If the disease is not treated adequately, the infant may develop cyanosis and die of cardiac failure.

Alcoholics are prone to develop deficiencies in a number of nutrients, including thiamine. The Wernicke-Korsakoff syndrome is attributed to thiamine deficiency, since the symptoms are alleviated (and may even disappear) upon treatment with large doses of thiamine. Among the clinical

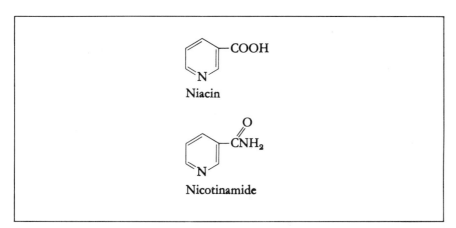

Fig. 15-8. Structures of niacin (nicotinic acid) and nicotinamide.

features of the disease are weakness in the eye muscles, ataxia of gait, loss of memory, and confusion. It may be noted that the action of thiamine depends on its prior conversion to thiamine pyrophosphate. As a consequence, alcoholics with cirrhosis may exhibit a delay in response to thiamine therapy that is due to defective activation in the diseased liver.

2. Niacin (Nicotinic Acid)

Niacin (Fig. 15-8) provides a structural component of the pyridine nucleotide coenzymes, NAD^+ and $NADP^+$. Both nicotinic acid and its amide (nicotinamide) can be incorporated into these coenzymes by the animal organism. The roles of NAD^+ and $NADP^+$ in numerous dehydrogenase reactions are well recognized, so the function of niacin is understood in biochemical terms. Although humans have the ability to produce niacin from tryptophan (section 8.18), the total requirement for the vitamin cannot be fulfilled by the average tryptophan intake. Thus, the RDA of niacin for an adult is 14 to 18 milligrams; only 5 to 10 milligrams, however, could be derived from the average American protein diet.

Pre-formed niacin occurs in liver, lean meats, and whole-grain cereals. Refined flours and cereals contain significantly lesser amounts of the vitamin.

Deficient intake of niacin results in *pellagra*. The characteristics of the disease are dermatitis, especially in areas of the body exposed to sunlight, stomatitis, soreness, dark coloration of the tongue ("black tongue"), diarrhea, and difficulty in digesting food. Affected individuals are prone to mental disturbance and early death.

3. Riboflavin

Dietary intake of riboflavin or *vitamin B_2* (Fig. 15-9) is required for the intracellular production of the flavin nucleotides FMN and FAD. Good food

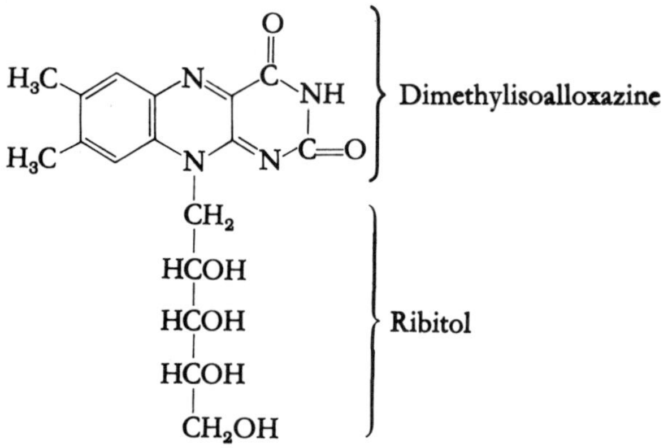

Fig. 15-9. Structure of riboflavin.

sources of riboflavin include beef liver, eggs, milk, and green leafy vegetables. The RDA of the vitamin is 1.5 milligrams.

In spite of the numerous vital functions of flavoprotein enzymes (e.g., mitochondrial electron transport, fatty-acid oxidation, and purine and amino-acid catabolism), no specific disease can be attributed to riboflavin deficiency. The general effects of its deficiency are dermatitis, dark coloration of the tongue, and fissures in the lips and the corners of the mouth. It is known that there are considerable differences in the turnover rates of the actual enzymes. The rate of depletion of intracellular flavin units of vital enzymes may be comparatively slow, so acute reactions are not detected.

4. Pyridoxine

Pyridoxine (Fig. 15-10) is a member of the *vitamin B_6* group. Vitamin activity is also exhibited by its two derivatives, *pyridoxal* and *pyridoxamine* (see Fig. 15-10). These compounds are converted to pyridoxal phosphate and, to a small extent, pyridoxamine phosphate, which have a number of diverse functions. Most notable among the roles of pyridoxal phosphate is its action as a cofactor in transamination reactions (section 7.2). The coenzyme is also associated with the actions of kynureninase (section 8.18), δ-aminolevulinic acid synthase (section 10.2), and glycogen phosphorylase (section 4.7.B). Pyridoxal phosphate seems to be involved in stabilizing the structure of the last-mentioned enzyme.

Among the food sources of pyridoxine are liver, muscle meats, wheat, corn, and eggs. The RDA of the vitamin for adults is 2 milligrams. In view of its central role in amino-acid metabolism, however, this requirement probably varies, depending on the amount of dietary protein.

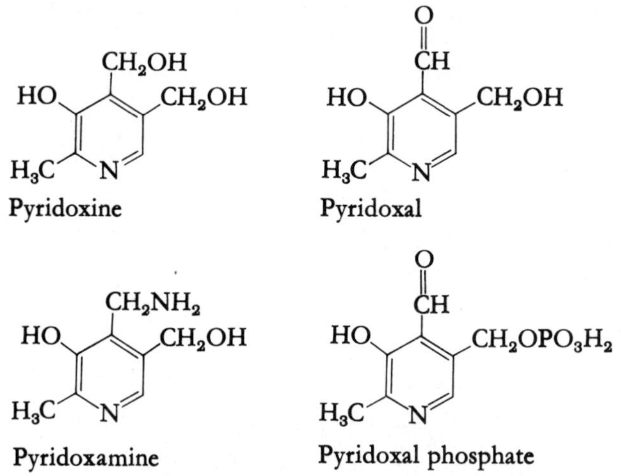

Fig. 15-10. Structures of pyridoxine and its derivatives.

Deficiency of the vitamin may result in poor growth, anemia, decreased antibody production, and convulsions, as well as renal, hepatic, and skin lesions. Infants on diets with inadequate amounts of the vitamin develop convulsions, which may be relieved by the administration of pyridoxine. These symptoms are also seen in babies with an inborn error of metabolism characterized by an abnormally high requirement for pyridoxine. Several types of measurements have been employed to detect a deficiency in the vitamin, including determinations of the levels of excreted tryptophan metabolites (e.g., xanthurenic acid) and serum transaminase activity. In adults, there is no characteristic syndrome that is specific for pyridoxine deficiency. The general symptoms are nausea, dermatitis, and polyneuritis. Tryptophan metabolites (e.g., xanthurenic acid) are commonly excreted after ingestion of high amounts of protein.

Isonicotinoylhydrazine (isoniazid), which is used in the treatment of tuberculosis, produces a vitamin B_6 deficiency by combining with pyridoxal phosphate to form a hydrazone. Another compound, deoxypyridoxine, competes with pyridoxal for binding to the active site on pyridoxal phosphate-specific enzymes. The symptoms produced by these materials can be reversed by treatment with the vitamin.

5. Biotin

Biotin (Fig. 15-11) is involved as a cofactor in carboxylase-catalyzed reactions. These include the syntheses of malonyl-coenzyme A from acetyl-coenzyme A (section 5.2.A), oxaloacetate from pyruvate (section 4.8), and

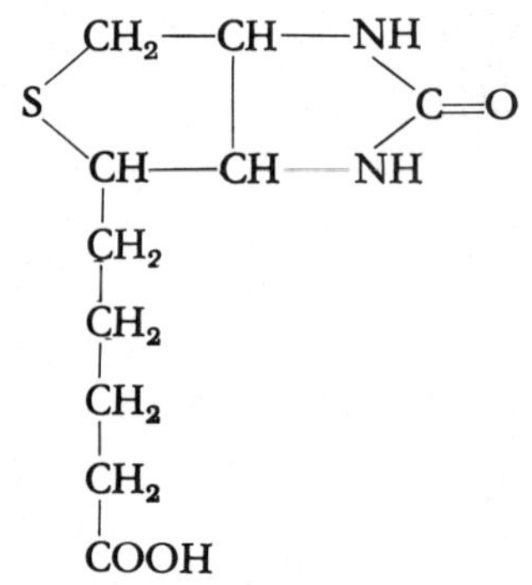

$$\begin{array}{c}\text{CH}_2-\text{CH}-\text{NH}\\ \text{S}\diagdown\quad\diagup\quad\diagdown\\ \quad\text{CH}-\text{CH}\quad\text{C}=\text{O}\\ \quad\quad|\quad\quad\diagup\\ \quad\text{CH}_2\quad\text{NH}\\ \quad|\\ \text{CH}_2\\ |\\ \text{CH}_2\\ |\\ \text{CH}_2\\ |\\ \text{COOH}\end{array}$$

Fig. 15-11. *Structure of biotin.*

methylmalonyl-coenzyme A from propionyl-coenzyme A (section 8.4). The vitamin thus plays important roles in carbohydrate, protein, and lipid metabolism. In humans, about 150 micrograms per day seem to be required.

Biotin occurs in almost all foods; however, the richest sources are liver, milk, egg yolk, and yeast. It is estimated that the average daily diet contains 150 to 300 micrograms of the vitamin. In addition, biotin is produced by the intestinal flora which may provide most or all of the required amounts.

Since biotin is a highly available compound and comparatively small amounts are required, deficiencies rarely occur unless induced experimentally. Raw egg white contains a protein called *avidin,* that binds biotin so that it is not absorbed. However, to produce a biotin deficiency, an inordinately high proportion of the protein intake would have to be derived from raw egg whites.

6. *Pantothenic Acid*

Pantothenic acid (Fig. 15-12) is required for the structure of coenzyme A and acyl-carrier protein. In spite of the critical requirement for these materials in catabolic and anabolic processes, a dietary deficiency of the vitamin does not produce any specific symptoms. Deficiency effects, however, include generalized fatigue, sleep disturbances, and abdominal distress. Although precise data on the daily requirement of this vitamin are not available, it appears that about 5 to 10 milligrams fulfills the daily need.

7. *Folic Acid*

The structures of folic acid, tetrahydrofolic acid (THFA), and their derivatives were discussed previously (section 7.10). As indicated, these substances serve as carriers of single-carbon units and thus play vital functions in the metabolism of amino acids and nucleotides. The interconversions of the

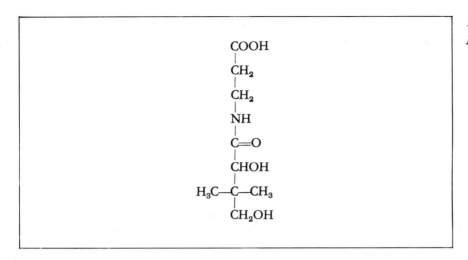

Fig. 15-12. Structure of pantothenic acid.

different derivatives of folic acid and the reactions in which they play a role are summarized in Figure 15-13.

Folic acid and its polyglutamate derivatives occur in liver, green leafy vegetables, and whole-grain cereals. Polyglutamates are hydrolyzed to the monoglutamate (folic acid) during absorption in the epithelial cells of the intestine. It is estimated that the minimum daily requirement (MDR) is approximately 50 micrograms; the RDA for adults is 400 micrograms. Higher amounts should be ingested during growth, pregnancy, and lactation.

The metabolic effects of a folic-acid deficit are blocks in the synthesis of purine nucleotides and in the conversion of deoxyuridine monophosphate (dUMP) to thymidine phosphates. As a consequence, the synthesis of DNA cannot proceed normally. Tissues with a high degree of cell multiplication are therefore the first to be affected. Morphologic changes occur in nucleated red blood cells and bone marrow (megaloblastic anemia). Other cell types that are prone to derangement are leukocytes and the epithelial cells lining the stomach and small intestine.

In addition to the deficiencies that arise from inadequate dietary intake, folate may become effectively unavailable because of poor absorption or utilization. Alcohol when consumed in excessive amounts interferes with the absorption and metabolism of folic acid. Various drugs employed in the treatment of malignancies (e.g., methotrexate) produce the symptoms of folic-acid deficiency. The abnormalities in red blood cells and bone marrow that result from folate deficiency also appear when there is a deficit in vitamin B_{12}; this is attributed to the requirement for the latter in the normal functioning of tetrahydrofolate. The metabolic link between folic acid and vitamin B_{12} will be discussed in the next section.

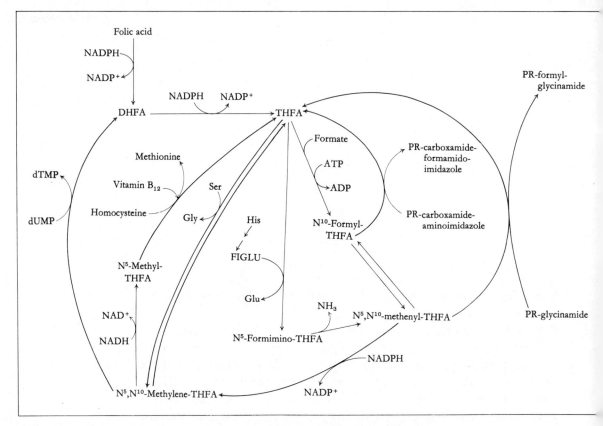

Fig. 15-13. Roles of folic acid, its derivatives, and vitamin B_{12} (DHFA = dihydrofolate; THFA = tetrahydrofolate; dUMP = deoxyuridine monophosphate; dTMP = deoxythymidine monophosphate; FIGLU = N-formiminoglutamate; PR = phosphoribosyl; Ser = serine; Glu = glutamate; Gly = glycine; His = histidine.) Additional details of interconversions of the folic acid derivatives and the reversibility of the reactions are discussed in sections 7.10, 8.9, 9.1, and 9.5.

8. Vitamin B_{12}

Structurally, vitamin B_{12} is a cobalt-corrin complex designated as *cobalamin*. It is isolated as a cyanide complex, *cyanocobalamin* (Fig. 15-14); however, the natural form is probably a hydroxide. The metabolically functional forms are the coenzymes *5'-deoxyadenosylcobalamin* and *methylcobalamin* (see Fig. 15-14).

The vitamin is synthesized exclusively by microorganisms and does not occur in plant foods. Animal foods contain the vitamin as a result of their contact and interaction with microorganisms. The best sources of vitamin B_{12} are beef liver, whole milk, eggs, and chicken. Although bacteria in the human colon produce the vitamin, it cannot be utilized, because its absorption

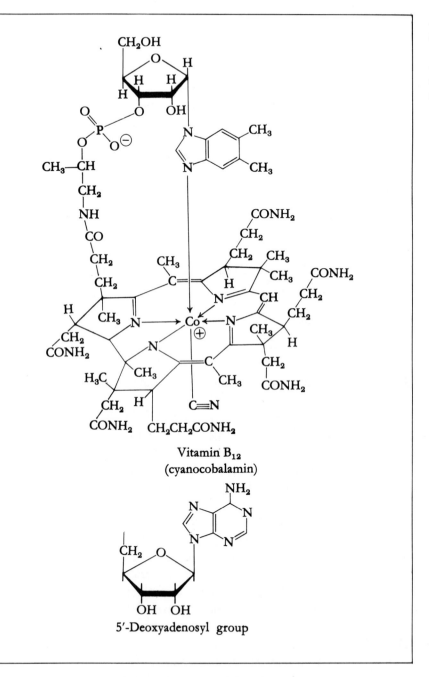

Fig. 15-14. Structure of vitamin B_{12}. In the biologically active forms, the cyanide group is replaced by either a methyl group ($-CH_3$) or a 5′-deoxyadenosyl group.

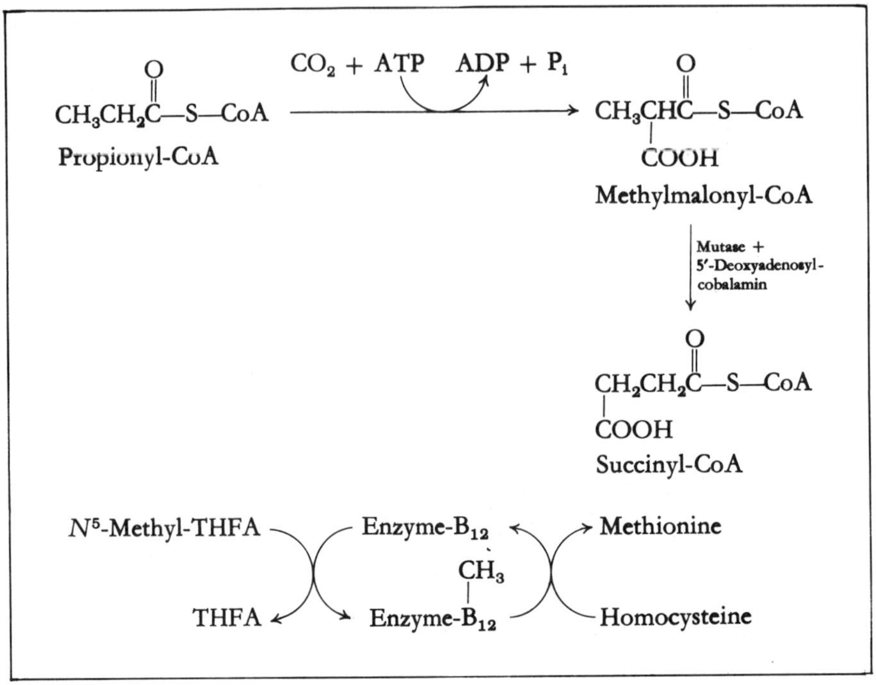

Fig. 15-15. Metabolic activities of cobalamin coenzymes. Top: Conversion of propionyl-CoA to succinyl-CoA via methylmalonyl-CoA (an intermediate step involving a racemase is not shown). Bottom: Formation of methionine from homocysteine, utilizing methylcobalamin as a coenzyme. This reaction also serves to regenerate tetrahydrofolate (THFA).

must occur in the small intestine. The RDA of vitamin B_{12} for adults is 3 micrograms.

Gastric parietal cells secrete a glycoprotein, termed *intrinsic factor*, that is essential for the absorption of vitamin B_{12} from the ileum. Cobalamin is carried to the liver by a specific transport protein and is then delivered to hematopoietic tissues by another transport protein called *transcobalamin II*. The normal plasma concentration of the vitamin is 200 to 900 micrograms per 100 milliliters.

At least two metabolic functions are ascribed to vitamin B_{12}. The coenzyme *5′-deoxyadenosylcobalamin* serves in conjunction with a specific mutase to produce succinyl-coenzyme A from methylmalonyl-coenzyme A. As shown in Figure 15-15, this reaction is required for the metabolic utilization of propionyl-coenzyme A. A deficiency in vitamin B_{12} results in a block at the mutase reaction. Hence, most individuals with this condition excrete considerable amounts of methylmalonate in the urine (up to 300 milligrams per day); normally, only 1.5 to 2 milligrams are excreted daily.

Another coenzyme form of the vitamin, *methylcobalamin*, acts with N^5-methyltetrahydrofolate in converting homocysteine to methionine (see Fig. 15-15). This process links the activity of vitamin B_{12} with that of folic acid, and it is believed to be critical for providing tetrahydrofolate (THFA) from its N-methyl derivative (section 7.10).

Since the primary acceptor and carrier for single-carbon units is THFA, it is essential that the mechanisms for its regeneration from all its derivatives are operative. The effect of vitamin B_{12} in returning N^5-methyl-THFA to the THFA pool serves to make the latter available for cellular functions. The reaction involving the formation of methylcobalamin (Fig. 15-15) is of special importance in view of the finding that N^5-methyl-THFA is the predominant form of folic acid in human serum and liver.

A deficiency in vitamin B_{12} produces, in effect, a deficiency in folic acid. This phenomenon—referred to as the "folate trap" hypothesis—explains why the hematologic effects of a vitamin B_{12} deficiency are indistinguishable from those of a folate deficiency. In both deficiencies, there is an associated deficit in N^5,N^{10}-methenyl-THFA that results in defective DNA synthesis.

The metabolic dependence of folic-acid activity on vitamin B_{12} is also evidenced by the fact that the excretion of formiminoglutamic acid (FIGLU) is increased in vitamin B_{12} deficiency, as it is in folic-acid deficiency. Similarly, 5'-phosphoribosyl-4-carboxamide-5-aminoimidazole—a substrate in a folic acid-dependent reaction in purine biosynthesis (section 9.1)—accumulates in the organism when there is a deficit in either of the two vitamins. In these cases, a partially degraded form of the metabolic intermediate, i.e., 4-carboxamide-5-aminoimidazole, is excreted in increased levels.

A dietary deficiency of vitamin B_{12} may occur in individuals who do not eat meat products, eggs, or milk and live exclusively on vegetables. A more common cause for the deficiency is an impairment in the secretion of intrinsic factor. The disease *pernicious anemia* is essentially a manifestation of impaired vitamin B_{12} absorption due to the unavailability of intrinsic factor in the gastric secretions. When it is orally administered to patients with the disease, vitamin B_{12} can be recovered in the stool. In this disorder, the formation of megaloblasts occurs as a result of the deficient synthesis of DNA. Other cells that divide rapidly, such as those of the lining of the gastrointestinal tract, are also vulnerable. In addition, the deficiency leads to degeneration of nerve fibers and abnormalities of the central nervous system. The symptoms of pernicious anemia are alleviated by the intravenous administration of vitamin B_{12}.

F. VITAMIN C

Vitamin C or *ascorbic acid* (Fig. 15-16) is an enediol derivative of L-gulonolactone. Small amounts of metal cations (e.g., copper or iron) cause the oxidation of ascorbic acid to dehydroascorbic acid (see Fig. 15-16). Such oxidation does not affect the vitamin activity to an appreciable extent.

Ascorbic acid is especially abundant in citrus fruits, tomatoes, and fresh

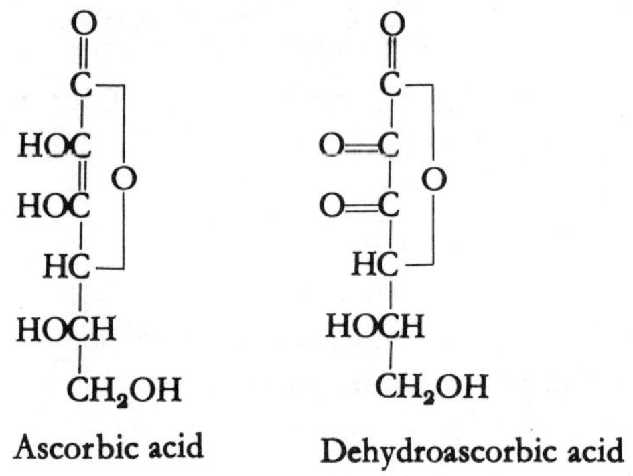

Fig. 15-16. *Structures of vitamin C (ascorbic acid) and its oxidized form (dehydroascorbic acid).*

vegetables such as broccoli. The vitamin is thermolabile and is destroyed in cooking.

Most animal species have the biosynthetic capability to produce ascorbic acid. It is essential, however, for man, guinea pigs, and various primates. The minimal daily requirement (MDR) for the vitamin is 30 to 50 milligrams. Ascorbic acid is not stored in the tissues to any significant extent, and ingestion of excessive amounts results in its rapid excretion in the urine.

Vitamin C deficiency disease is known as *scurvy*. The symptoms of this disorder include bleeding due to capillary fragility, swollen gums, defective bone structure, and impaired wound healing. Ascorbic acid plays a critical role in the production of connective-tissue components, especially collagen. It is necessary for the hydroxylation of proline and lysine (section 14.3.B), as well as for the normal growth of fibroblasts. Deficiency in the vitamin results in the production of collagen that cannot form normal fibers. The lesions of the skin, capillaries, and bones can be ascribed to physical weakness of the supporting collagen at these sites.

G. OTHER VITAMIN-LIKE NUTRIENTS

A number of substances perform important metabolic functions and, when added to the diet, correct certain disorders. These compounds do not appear to be essential under all conditions. In addition, some of them can be synthesized, provided that other precursors are available in the diet. Generally included in this group are lipoic acid, choline, ubiquinones, and *myo-inositol* (Fig. 15-17).

Lipoic acid is a component of the multienzyme systems that convert pyruvate to acetyl-coenzyme A (section 4.4) and α-ketoglutarate to succinyl-coenzyme A (section 3.4). Lipoic acid is synthesized in adequate amounts in mammals and is probably not required as a dietary component.

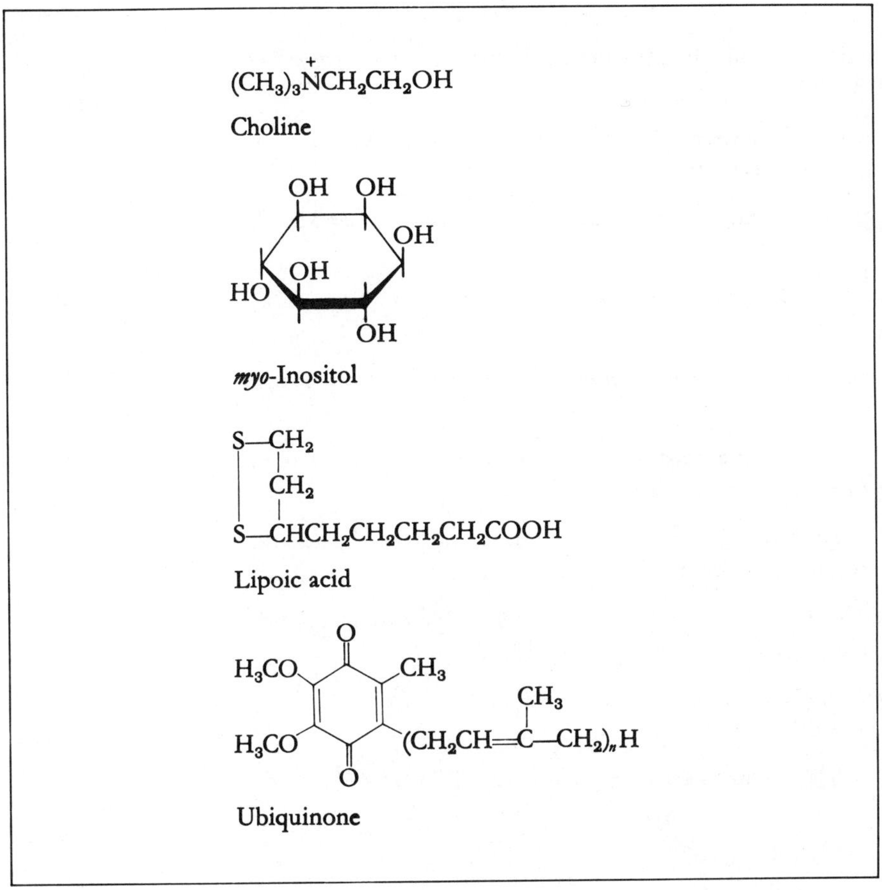

Fig. 15-17. Structures of choline, myo-inositol, lipoic acid, and ubiquinone.

Choline is a structural component of phosphatidylcholine and sphingomyelin. Although it can be produced in the cell, its synthesis is dependent on adequate amounts of methionine (section 7.8). Growing animals who are on a low-protein diet that is deficient in choline develop fatty liver, which can be alleviated by the administration of choline. Choline is thus classified as a *lipotropic* substance (section 6.4.B).

myo-Inositol has a similar effect in reversing the development of fatty liver. Additionally, a dietary deficiency in *myo*-inositol in mice results in alopecia and retardation in growth. In chicks, the deficiency produces encephalomalacia. However, *myo*-inositol does not seem to be a nutritional requirement for humans.

Ubiquinones include various forms of coenzyme Q (section 3.6). These have pivotal roles as electron carriers in the mitochondrial oxidative chain. Although coenzyme Q can be synthesized in the cell, an external source of its aromatic ring may be required.

Suggested Reading

DeLuca, H. F. Metabolism of Vitamin D: Current Status. *American Journal of Clinical Nutrition* 29:1258, 1976.

Food and Nutrition Board. *Recommended Dietary Allowances* (8th ed.). Washington, D.C.: National Academy of Sciences, 1974.

Hegsted, D. M. Energy Needs and Energy Utilization. *Nutrition Reviews* 32: 33, 1974.

Jacobs, A., and Worwood, M. (Eds.). *Iron in Biochemistry and Medicine*. New York: Academic Press, 1974.

Munro, H. N. (Ed.). *Mammalian Protein Metabolism*. New York: Academic Press, 1964.

Nixon, P. F., and Bertino, J. R. Interrelationships of Vitamin B_{12} and Folate in Man. *American Journal of Medicine* 48:555, 1970.

Nutrition Reviews. *Present Knowledge of Nutrition* (4th ed.). New York and Washington, D.C.: Nutrition Foundation, Inc., 1976.

Schneider, H. A., Anderson, C. E., and Coursin, D. B. (Eds.). *Nutritional Support of Medical Practice*. Hagerstown, Md.: Harper & Row, 1977.

Underwood, E. J. *Trace Elements in Human and Animal Nutrition* (3rd ed.). New York: Academic Press, 1971.

Yudkin, J. Diet and Coronary Thrombosis. Hypothesis and Fact. *Lancet* 11:155, 1957.

Index

Index

Absorption, intestinal
  of calcium, 575
  of carbohydrates, 163–166
  of glycerides, 114–116
  of iron, 557
  of nucleotides, 393
  of proteins, 297–301
  of vitamin B_{12}, 598
Acetaldehyde, 338, 340
Acetic acid
  buffers with acetate salts, 9, 10
  degradation to carbon dioxide, 122, 128, 150
  ionization of, 2
  pH of solutions of, 5
  titration curve of, 9
Acetoacetate, 254–256, 284, 358, 362, 521
Acetoacetyl-ACP, 222
Acetoacetyl-coenzyme A, 121, 254
  from amino acids, 314, 358, 360, 367, 369
  and cholesterol synthesis, 234
  from fatty acids, 121, 234
Acetone, 255, 256
Acetylcholine, 536–537
Acetylcholinesterase, 536
  inhibitors of, 538–539
Acetyl-coenzyme A, 122, 214, 219, 537
  from amino acids, 314, 357, 358
  and cholesterol synthesis, 234
  citrate from, 122
  energy from, 145
  from fatty acids, 121, 254
  and fatty acid synthesis, 219–222
  and gluconeogenesis, 207, 275
  intracellular transport of, 219
  pyruvate conversion to, 175–179
Acetyl-coenzyme A carboxylase, 220
N-Acetylgalactosamine, 543
N-Acetylglucosamine, 545
N-Acetylglucosamine 1-phosphate, 215
N-Acetylglucosamine 6-phosphate, 215
N-Acetylglutamate, 309
N-Acetylneuraminic acid, 49
  in gangliosides, 249
  in glycoproteins, 48
N-Acetylserotonin, 371

Acetyl transacylase, 221
Acid(s), 1
  conjugate bases of, 6–7
  ionization of, 2–6
  organic, in fluid compartments, 523
  polyprotic, 5
  titration curves for, 8
Acid-base balance, disorders in, 521–522
Acidosis
  in diabetics, 256–257, 279
  metabolic, 256, 521, 522
  respiratory, 512, 521, 522
Aconitase, 123, 129, 147
Aconitic acid, 123, 128
ACP. *See* Acyl-carrier protein
Acromegaly, 556
ACTH. *See* Adrenocorticotropic hormone
Actin, 527, 531
Actinomycin D, 449, 450
Action potential, 536
Activation, energy of, 82
Active transport, 61
Actomyosin, 531
Acyl-carnitine transferase, 118
Acyl-carrier protein (ACP), 221
Acyl-coenzyme A, 117
  desaturation of, 225, 226
  oxidation of, 135
  and synthesis of glycerides, 227
  and synthesis of phosphatidic acids, 227
Acyl-coenzyme A dehydrogenase, 119
Acyl-coenzyme A synthetase, 118–119
Acyl-enzyme, 106, 538
Addison's disease, gluconeogenesis in, 277
Adenase, 396
Adenine, 49
  catabolism of, 396
  hydrogen bonding in DNA, 415
Adenine phosphoribosyl transferase, 388
Adenosine, 50, 53–54
Adenosine diphosphate (ADP), 51, 79–81
  and isocitrate dehydrogenase activity, 153
  phosphorylation to ATP, 143
  ratio to AMP and ATP, 275
    and glucose metabolism, 207, 275–277
    and regulation of gluconeogenesis and glycolysis, 275

Adenosine monophosphate (AMP), 51, 81, 377, 389
  cyclic. See Cyclic adenosine monophosphate
  deamination of, 395
  ratio to ADP and ATP, 275
  and regulation of gluconeogensis and glycolysis, 275–277
  synthesis of, 377, 389
Adenosine triphosphatase, 155, 157
Adenosine triphosphate (ATP), 51, 78–82
  cyclic AMP from, 198
  formation of, 143–144
    electron transport and, 155–156, 159–161
    glycolysis and, 172, 173
  hydrolysis of, 79–80
    and free energy change, 80–81
  and isocitrate dehydrogenase inhibition, 150, 153
  and muscle contraction, 528–529, 531–532
  and phosphofructokinase inhibition, 183, 184, 207
  ratio to AMP and ADP, 275
    and glucose metabolism, 207
  and regulation of glycolysis and gluconeogenesis, 275–277
  yield in aerobic oxidation of glucose, 181
  yield in citric acid cycle, 146
  yield in fatty acid degradation to acetyl-CoA, 145
  yield in glycolysis, 181
S-Adenosylhomocysteine, 317, 318, 327
S-Adenosylmethionine, 317, 318
Adenyl cyclase, 198
  and ACTH, 557
  and epinephrine, 198–199, 551
  and glucagon, 199, 553
  and prostaglandins, 558
  and thyroxine, 555
Adenylic acid. See Adenosine monophosphate
Adenylosuccinate lyase, 376, 378
Adenylosuccinate synthetase, 377, 378
Adipose tissue, metabolism of carbohydrates and lipids in, 284–286
ADP. See Adenosine diphosphate
Adrenocortical hormones, 557
Adrenocorticotropic hormone (ACTH), 557
  and pregnenolone formation, 241

Alanine, 15, 17, 316, 325, 368
  in collagen, 540
  formation from pyruvate, 304
  metabolism of, 331–332
  sources of, 328
  synthesis of, 324
β-Alanine, 394
L-Alanine:2-oxoglutarate aminotransferase, 304
Alanine transaminases, 304
Alanylserine, 22
Albinism, 364
Albumin, in plasma, 470–471
Alcohol
  calories in, 570
  and hyperuricemia, 399
  and thiamine deficiency, 590–591
Aldohexoses, 42
Aldolase, 170, 187
  and gluconeogenesis, 202
  isozyme, and absence of, 213
Aldoses, 42
Aldosterone, 242, 557
Alimentary lipemia, 265
Alkaline phosphatase
  and bone metabolism, 550
  plasma levels of, 492
Alkalosis
  metabolic, 521, 522
  respiratory, 512, 521, 522
Alkaptonuria, 363
Allopurinol
  in gout, 401
  with 6-mercaptopurine, in chemotherapy, 402
Allosteric enzymes. See under Enzyme(s)
Amethopterin, action of, 390
Amidotransferases, 374
  and glucosamine 6-phosphate formation, 214
  inhibition of, 379, 389
  and nucleotide synthesis, 374, 375, 383
Aminoacetone, formation of, 339, 340
Amino acid(s), 13–21
  acidic, 16
  basic, 17
  catabolism of carbon skeleton, 313–314
  codons for, 436–438
  in collagen, 29, 540
  deamination of, 306–308
  dietary requirements, 571
  essential, 323–324
  glucogenic, 314–316

formation of aminoacyl adenylate, 438
formation of aminoacyl-tRNA, 438
helix-stabilizing, 31
in hemoglobin chains, 40
intestinal absorption of, 300
isoelectric points of, 17
ketogenic, 314–316
metabolism of, 297–328, 331–371
    relation to metabolism of lipids and carbohydrates, 314
neutral, 15
non-essential, 322–324
    synthesis of, 324–328
plasma levels of, 466
properties of, 17
structures of, 15–17
transamination of, 301–306
transmethylation of, 316–318
L-Amino-acid oxidases, 307
Aminoacyl adenylate, 438
Aminoacyl site, of ribosomes, 443
Aminoacyl-tRNA synthetase, 438
α-Aminoadipic acid, 360
α-Aminoadipic semialdehyde, 360
p-Aminobenzoic acid, 390
γ-Aminobutyric acid, 346–347
2-Amino-3-carboxymuconic acid semialdehyde, 368, 369
α-Amino groups, pK value of, 17, 37
ϵ-Amino groups, pK value of, 17, 37
β-Aminoisobutyrate (BAIB), 394–395
2-Amino-3-ketobutyric acid, 339, 340
δ-Aminolevulinate dehydratase, 406
δ-Aminolevulinate synthetase, 406
δ-Aminolevulinic acid, 406
    as donor to tetrahydrofolic acid, 321
    elevated activity of, 410
    in porphyrin synthesis, 406
2-Aminomuconic acid, 368, 369
2-Aminomuconic acid semialdehyde, 368, 371
Aminopeptidases, 297, 299, 301
α-Aminophenol, 371
Aminopterin, action of, 390
Aminotransferase. *See* Transaminases
Ammonia
    from amino acids, 307
    from asparagine, 306, 343
    and carbamoyl-phosphate synthesis, 309
    from cysteine, 341, 353
    from glutamate, 306, 307, 345
    in glutamate synthesis, 325, 345
    from glutamine, 305, 308

    in glutamine synthesis, 325, 345
    from histidine, 308, 343
    metabolism of, 308–310
    secretion of, 517–519
    from serine, 308, 332
    from threonine, 339
    in urine, 524
Ammonium ions, and urinary pH, 517
AMP. *See* Adenosine monophosphate
Amphoteric materials, 6
Amylase
    pancreatic, 164
    plasma levels of, 491–492
    salivary, 164
Amylo-1,6-glucosidase, 196
Amylopectin, 47
Amylose, 47
Amylo-(1,4→1,6)-transglycosylase, 193
    deficiency of, 291
Amytal, effect on oxidative chain, 156
Anabolism, 64
Anaerobic glycolysis, 181
Anaplerotic reactions, 210
Andersen's disease, 291
Androgens, 237, 243–244
Androstane, 237
Androstenedione, 243, 244
Anemia
    iron-deficiency, 578
    pernicious, 599–600
Anomeric carbon, 44
Anserine, 394
Antibodies, 481–482
Anticodon loop. *See* Codons
Antigen(s), 481
Antigen-antibody complex, 482
Antilipotropic substances, 274
Antimetabolites, action of, 389
Antimycin A, 156
Antiserum, 482
Antithrombin III, 472, 477
α_1-Antitrypsin, 472, 477
Apoenzyme, 95
Apolipoproteins, 270
Arabinose, intestinal absorption of, 164
Arabinosylcytosine, 391, 392
Arachidic acid, 112
Arachidonic acid, 112, 558, 573
Arginase, 310, 311
Arginine, 17, 310, 311, 316, 324, 343
    and creatine formation, 319
    metabolism of, 347–348
    synthesis of, 328
Argininosuccinate, 343, 347

Argininosuccinate lyase, 311
Argininosuccinate synthetase, 311
Ascorbic acid, 362, 600
    mitochondrial oxidation of, 156
Asparagine, 16, 17, 316, 324
    ammonia from, 306, 343
    metabolism of, 344
    sources of, 328
    synthesis of, 325
    transamination with, 306
Asparagine synthetase, 325
Aspartate carbamoyltransferase, 381, 383–384
Aspartate-glutamate transaminases, 204, 302, 304, 343
Aspartate transcarbamoylase, 381, 383–384
Aspartic acid, 16, 316, 324
    dissociation of, 18–19
    function in intracellular shuttles, 204
    metabolism of, 342–344
    sources of, 328
    synthesis of, 325
Atoms percent, and excess of, 68
ATP. *See* Adenosine triphosphate
ATPase, 155
    activity in mitochondria, 157
Avidin, 594
Axons, 535
Azaserine, 374, 389

Bacteriophage, 456
Basal metabolic rate (BMR), 552, 568–569
Basal metabolism, 565, 568–569
Base(s), 1
    conjugate bases of acids, 6–7
Base triplets, for amino acid sequences, 436–437
Bence Jones proteins, 487
Benzoic acid, conversion to hippurate, 338
Beriberi, 590
Betaine, and transmethylation, 317
Bicarbonate
    buffer system, 510–513
    and control of blood pH, 510–513, 517–519, 520–522
    and evaluation of acid-base balance, 522
    in fluid compartments, 523
    and mineralocorticoids, 243
    plasma levels of, 467
Bile salts, 114, 115, 239–240

Bilirubin, 441
    plasma levels of, 466
Bilirubin diglucuronide, 411
Biliverdin, 411
Biotin, 95, 593–594
    in acetyl-coenzyme A carboxylase, 220
    in pyruvate carboxylase, 203
Blood, 465–492
    buffers in, 509–515
    cellular components of, 465, 466–469
    coagulation of, 472–481
    complement in, 488–490
    composition of, 465–466
    enzymes in, 262–264, 491–492
    immunoglobulins in, 481–488
    kinins in, 490
    and lipid disorders, 291–294
    lipoproteins in, 267–273
    pH of, 256, 509
        compensatory control of, 519–522
    protein in, 469–472
    regulation of glucose in, 165, 262–264
    transport of lipids in, 264–267
Blood groups, 468, 469
BMR. *See* Basal metabolic rate
Bohr effect, 501, 503, 506
Bone, 549–550
    interaction with hydrogen ions, 515
    and parathyroid secretions, 556
    resorption of, 550
    and somatotropin, 556
    and vitamin A, 583
    and vitamin D, 585–586
Bowman's capsule, 515
Bradykinin, 490
Brain, metabolism of carbohydrates and lipids in, 286. *See also* Neurotransmitters
    and sphingolipidoses, 250–251
Branching enzymes, 193
    deficiency of (Andersen's disease), 291
Bromodeoxyuridine, action of, 391, 392
5-Bromouridine, mutagenic activity of, 460
Buffers, 8–9
    in blood, 509–515
Bungarotoxin (cobra toxin), 538
Butyric acid, 113
Butyryl-ACP, 222
Butyryl-CoA, 254

Calcinosis, 586
Calcitonin, 555
Calcium
    absorption of, 575
    and blood coagulation, 477

in bones, 549-550
and complement system, 489
dietary requirements, 574-575
and enzyme activity, 96
in fluid compartments, 523
and muscular contraction, 532
and parathyroid hormones, 555
plasma levels of, 467
in urine, 524
Calories, 75, 561-563
Calorimetry, 561
  direct, 563
  indirect, 567
Capric acid, 113
Caproic acid, 113
Carbaminohemoglobin, 502-503
Carbamoyl aspartate, 343
Carbamoyl phosphate, 311
  synthesis of, 309-310
Carbamoyl-phosphate synthetase, 309, 379-381
  glutamine-dependent, 384-385
Carbohydrates, 42-49, 163-216
  caloric value of, 561
  dietary requirements, 570, 572-573
  digestion and absorption of, 163-166
  disaccharides, 46-47
  fructose metabolism, 212-213
  galactose metabolism, 211-212
  gluconeogenesis, 201-210
  glucose oxidation, aerobic, 179-184. *See also* Glucose
  glycogen metabolism, 190-201
  glycolysis, 169-175
  glycosaminoglycans, 542-549
  hexosamines, 213-215
  metabolism of
    in adipose tissue, 284-286
    in brain, 286
    epinephrine affecting, 199, 277, 279-281, 551
    in erythrocytes, 286-288
    glucagon affecting, 199, 281
    glucocorticoids affecting, 281
    hormones affecting, 196-201, 277-281, 551-553, 557-558
    insulin affecting, 199-200, 277-279, 551-553
    in kidney, 284
    in liver, 282-284
    in muscle, 284-285
    relation to metabolism of lipids and amino acids, 314
    thyroid hormone affecting, 281, 553, 555
  monosaccharides, 42-46
  and nucleotide sugar interconversions, 210-211
  pentose-phosphate pathway, 184-190
  polysaccharides, 47-49
  pyruvate conversion to acetyl-coenzyme A, 175-179
  respiratory quotient of, 564-565
  tissue disposition and transport of, 253-294
Carbon-14, 68
Carbon dioxide
  and acetic acid degradation, 122, 128, 150
  in bicarbonate buffer system, 510-513
  and citric acid cycle, 129, 147-151
  and fatty acid synthesis, 220
  and gluconeogenesis, 203
  and oxygen binding to hemoglobin, 501
  partial pressure of, 496
  from pentose-phosphate pathway, 185-187
  in propionyl-CoA metabolism, 339
  in purine synthesis, 376
  from pyruvate, 176
  transport of, 495-497, 504-507
Carbon monoxide
  and hemoglobin, 410
  and oxidative chain, 156
Carbonic acid
  in bicarbonate buffer system, 510
  conjugate base of, 7
  ionization of, 5-6
Carbonic anhydrase, 504
γ-Carboxyglutamic acid, 478
Carboxyl groups, pK values of, 17, 37
Carboxypeptidases, 297, 299, 301
Cardiolipin, 231, 232
Carnitine, 117-118
Carnosine, 317, 318, 394
Carotenes, 580-581
Cartilage, 538
Catabolism, 64
Catalase, 397
Catecholamines, 551
  as neurotransmitters, 537
Catechol O-methyltransferase, 537
CDP-choline, 228-229, 246
CDP-diglyceride, 231-232
CDP-ethanolamine, 229
Cells
  as blood components, 465, 466-469
  chemistry of, 1-58
  internal structure of, 69-70
  membrane structure of, 58-60
  metabolic processes in, 64-69
  organelles of, 69
    fractionation of, 66

Cellulose, 48
Cementum, 550
Centrifugation, differential, 66
Cephalins, 56, 228
Ceramides, 57, 246
  derivatives of, 248
Cerebronic acid, 113
Cerebrosides, 57, 247
Cerebrospinal fluid, 522
Ceruloplasmin, 472, 578
Charge relay system, 106
Chemiosmotic coupling theory, 158–159
Chemotherapy, and inhibition of nucleotide synthesis, 387, 389–393
Chloramphenicol, 449, 450
Chloride
  dietary requirements, 577
  in fluid compartments, 523
  plasma levels of, 467
  in urine, 524
Chloride shift, 505, 507
Cholane, 237
Cholecalciferol, 583
Cholestane, 237
Cholestanol, 240
Cholesterol, 59
  biosynthesis of, 232–237
  esterification of, 238
  hypercholesterolemia, familial, 293
  in lipoproteins, 270
  plasma levels of, 265, 466, 573
    and atherosclerosis, 293
  precursor of steroid hormones, 232–240
  and stone formation, 240
  turnover and elimination of, 239–240
Choline, 56, 601
  from acetylcholine, 538
  deficiency of, 274
  and single-carbon transfer, 323
  and synthesis of phospholipids, 228–231
  and transmethylation, 316–318
Choline acetyltransferase, 536
Choline phosphate, 228
Cholinergic vesicles, in axons, 536
Choloyl-coenzyme A, 239
Chondroitin 4-sulfate, 544, 545, 547
Chondroitin 6-sulfate, 544, 545, 547
Chromatin, 451
Chromosomes, 413
Chylomicrons, 115, 265, 269
  genetic defects of, 292, 293
  and lipid transport, 270, 273
Chymotrypsin, 299, 301

Chymotrypsinogen, 299, 301
Citrate
  acetyl-CoA from, 219
  formation of, 122
  and glycolysis, 275
  phosphofructokinase inhibition by, 183, 184
  plasma levels of, 466
Citrate-cleaving enzyme, 205, 219
Citrate lyase, 205, 219
Citrate synthetase, 122
Citric acid cycle, 122–128
  and acetate degradation to carbon dioxide, 122, 128, 150
  and asymmetry of citrate reactions, 147–149
  ATP yield from, 146
  and carbohydrate biosynthesis, 208–210
  and fate of carbons from acetate, 150
  and randomization after succinate, 149–150
  and reduction of oxygen to water, 140–141
  special features of, 147–150
  and urea cycle, 314
Citrulline, 311, 312, 314
Clotting factors, 473–476
Coagulation, 472–481
  extrinsic pathway for, 474–475
  heparin affecting, 479–480
  intrinsic pathway for, 475–476
  as reaction to tissue injury, 480
  and vitamin K action, 478–479
Cobalamin, 596
Cobalt-60, 68
Cobamide, 95
Cobra toxin (bungarotoxin), 538
Codons, 436–438
  and anticodon loop, 433
Coenzyme(s), 95–96
Coenzyme A, interaction with fatty acids, 116–122
Coenzyme Q, 136–137, 601
Cofactors, enzyme, 95
Colchicine, 399
Collagen, 29, 540–542
  in bone, 550
Collecting tubule, renal, 516
Competitive inhibition of enzymes, 107–108
Complement, 472, 488–490
Cones and rods, ocular, 581
Connective tissue, 539–550
  collagen in, 540–542

composition of, 539
elastin in, 542
glycosaminoglycans in metabolism of, 547–549
structure of, 542–546
Contractile proteins, 527–531
Cooperative binding, of substrate to enzymes, 152
Copolymerase III*, and DNA synthesis, 422, 423
Copper
dietary requirements, 578–579
and enzyme activity, 96
Coproporphyrinogen III, 407
Coprostanol, 240
Cori cycle, 257–259
Corticosterone, 242
Cortisol, 242, 557
and gluconeogenesis, 277
Cortisone, 557
and gluconeogenesis, 277
Coupling and coupling factors, 157–159, 160–161
Crabtree effect, 182
Creatine, 534
metabolism of, 319–320
plasma levels of, 466
Creatine kinase, 319, 534
Creatine phosphate, 319–320
Creatine phosphokinase, isozymes in serum, 492
Creatinine, 320
plasma levels of, 466
in urine, 524
Cristae, 154
Crotonyl-ACP, 222
Crotonyl-CoA, 254
CTP synthetase, 382
Curie, 68
Cyanide, effect on oxidative chain, 156
Cyanocobalamin, 596
Cyclic adenosine monophosphate, 198
ACTH and, 557
in bacterial cells, 453–454
epinephrine and, 198
and glycogen phosphorylase, 198–199
and glycogen synthetase, 199
insulin and, 199–200
prostaglandins and, 558
and tissue lipase, 266
Cyclic AMP. See Cyclic adenosine monophosphate
Cycloheximide, 450
Cyclooxygenase, 558
Cystathionase, 327

Cystathionine, 327, 352–353
Cystathionine synthetase, 327
Cysteine, 16, 17, 324
metabolism of, 315, 316, 340–342
and cystine, 36
glutathione from, 335
pK of sulfhydryl in, 37
pyruvate from, 344
sources of, 328
synthesis of, 327–328
Cysteine desulfhydrase, 341
Cysteine sulfinic acid, 340
Cysteinylglutamylvalylmethionine, 23
Cysteinylglycine, 336
Cystine, 16, 17
from cysteine, 327
reduction of, 36
Cytidine, 50, 53–54
Cytidine monophosphate (CMP), 51
Cytidine triphosphate (CTP), 382–383
and phospholipid synthesis, 228–229, 231
Cytidine triphosphate synthetase, 382
Cytidylic acid. See Cytidine monophosphate
Cytochrome a-a_3, 139–141, 142–143, 155–156
Cytochrome b, 137–138, 140–141, 143, 155–156
Cytochrome b_5, 225–226
Cytochrome c, 138–141, 142–143, 155–156
Cytochrome c_1, 138, 139–141, 142–143, 155–156
Cytochrome oxidase, 139–141, 142–143, 155–156
Cytosine, 49, 54

Deamidases, 305
Deamination reactions, 306–308
non-oxidative, 307–308
oxidative, 306–307
Debranching enzymes, 195–196
deficiency of, 290
Dehydroascorbic acid, 599
7-Dehydrocholesterol, 583
Dehydroepiandrosterone, 243
Dehydrogenases, 96, 97–101
Deletions, mutations, 461–462
Denaturation, of protein, 35
Dendrites, 535
Dentin, 550
5′-Deoxyadenosylcobalamin, 596, 598
11-Deoxycorticosterone, 557
Deoxypyridoxine, 593

Deoxyribonucleic acid (DNA), 413–416
    antiparallel strands, 416
    bases in, 53, 413
    circular, 419
    complementary chains, 416
    guanine in, 415
    hydrogen-bonding in, 415, 416, 417
    methylated bases in, 413
    and mutations, 458–462
    notation for nucleotides in, 414–415
    Okazaki fragments, 421, 423
    and operator gene, 452, 453
    phosphodiester bridges in, 414
    polarity of stands in, 416
    promoter sites, 427
    radiation and, 458
    recombination of, 458
    repair of, 425
    replication of, 413–418
        enzymes involved in, 423–425
        mechanisms in, 418–425
        reactions involved in, 418–423
        unwinding protein, 420, 423
    and restriction endonucleases, 457–458
    and reverse transcriptase, 458
    semiconservative replication, 418, 419
    and structural genes, 452, 453
    transcription of, 425–428
    in viruses, 455–458
    Watson-Crick model of, 416
Deoxyribonucleosides, 53
Deoxyribonucleotides, biosynthesis of, 385–387
2-Deoxyribose, 45
Deoxythymidine 5′-phosphate (dTMP), 386
Deoxyuridine 5′-phosphate (dUMP), 386
1-Deoxy-N-valylfructose, 508
Depot lipids, 266
    mobilization of, 266–267
Dermatan sulfate, 544, 545, 547
Desmolase, 240
Desmosine, 542
Desmosterol, 236
Detergents, 115
Dextrin, limit, 196
    and limit dextrinosis, 290
DFP, 103, 106, 538
Diabetes
    acidosis in, 256, 521
    gluconeogenesis in, 277
    glucose levels in, 278, 279
    hyperuricemia in, 399
    ketonuria in, 256

levels of HbA_c in, 508
lipid mobilization in, 274, 279
6-Diazo-5-oxonorleucine (DON), 374, 389
Dicumarol, 157, 479
Diet. See Nutrition
Diffusion, of gases, 496–497
Digestion
    of carbohydrates, 163–166
    of nucleic acids, 393
    of proteins, 297–300
    of triglycerides, 114–116
Diglyceride, 55
    and triglyceride formation, 227
Diglycosylceramide, 248
Dihydrobiopterin, 365–366
Dihydrofolate (DHFA), 386
Dihydrofolate reductase, 386
Dihydrolipoamide, 125
Dihydrolipoyl dehydrogenase, 123, 177
Dihydrolipoyl transacetylase, 177
Dihydrolipoyl transsuccinylase, 123, 124, 125
Dihydroorotate dehydrogenase, 381, 385
Dihydroorotic acid, 381
Dihydrosphingosine, 246
Dihydrotestosterone, 556
Dihydrouridine (DHU), 431
Dihydroxyacetone phosphate, 170, 179, 209, 213
1,25-Dihydroxycholecalciferol, 584–585
20α,22-Dihydroxycholesterol, 240–241
Dihydroxyphenylalanine, 363
3,4-Dihydroxyphenylglycolaldehyde, 538
Diisopropylfluorophosphate. See DFP
2,4-Dinitrophenol, 157, 159
Dipeptides, 22
Diphosphates, nucleoside, 51
Diphosphatidylglycerol, 231
1,3-Diphosphoglyceric acid, 171
2,3-Diphosphoglyceric acid (DPG), 503
Diphtheria toxin, 450
Dipolar ion, 15
Disaccharides, 46–47, 164
DNA. See Deoxyribonucleic acid
DNA ligase, 423, 424–425
DNA polymerase, 422, 423–424
    in eukaryotic cells, 424
    RNA-directed, 458
Dopamine, 364
    as neurotransmitter, 537
Dulcitol, 212
Dwarfism, 556

Edema, 471
Effectors of allosteric enzymes, 150

Elastase, 299
Elastin, 542
Electrode potential, 130
Electrolytes, 1–13, 522–523
  plasma levels of, 467
Electrophoresis, 39
  of hemoglobin, 41–42
  of lipoproteins, 267–268
  of proteins, 40, 469
Elongation factors, 445
Elongation of fatty acids, 223
Enamel, of teeth, 550
  mottled, from fluoride, 550
Endergonic processes, 75
Endocrine system, 551–559. *See also* Hormones
Endonucleases, 422
  restriction, 457–458
Endopeptidases, 297
Endoplasmic reticulum, 70
Energy
  of activation, 82
  and dietary requirements, 569–571
  expenditures of, 563–567
  from fatty-acid catabolism, 144–147
  free-energy change, 74–78
    standard, 76–78
    and standard redox potentials, 131
  from glucose metabolism, 181
  from oxidations, 141–144
  requirements of muscle, 260
Enolase, 172
  and gluconeogenesis, 202
Enoyl-ACP hydratase, 222
Enoyl-ACP reductase, 222
Enoyl hydratase, 120
Enterokinase, 299
Enthalpy change, 74
Entropy change, 74
Enzyme(s), 73–110
  active site of, 101–106
  allosteric, 150–152, 383
    catalytic sites of, 152
    catalytic subunits of, 152
    negative effectors of, 383
    reaction velocity in, 152
  branching, 193
  classification of, 97
  coenzymes and cofactors, 95–96
  constitutive, 452
  cooperative binding, 152
  induced, 452
  induced-fit theory, 102
  inhibition of activity, 103–104, 106–110
    competitive, 106, 107–108
    feedback, 379, 383
    non-competitive, 106, 108–109
    uncompetitive, 106, 109–110
  and maximum velocity (V_{max}), 89–92
  metal ions affecting, 96
  and Michaelis-Menten equation, 89
  in mitochondrial matrix, 155
  nomenclature systems for, 96–97
  oxidoreductases, 97–101
  pH affecting, 93–94
  in plasma, 491–492
  prosthetic groups, 95
  reaction rate, 82–85
    in allosteric enzymes, 152
    enzyme concentration affecting, 92–93
    substrate concentration affecting, 85–92, 152
  regulatory sites of, 152
  regulatory subunits of, 152
  specificity of, 83, 102
  stereospecificity of, 148
  temperature affecting, 94–95
  zinc and activity of, 96
Enzyme-activity unit, 92–93
Enzyme-substrate complex, 86–88
Epinephrine, 364
  and adenyl cyclase, 198–199, 551
  and carbohydrate metabolism, 277, 279–281
  and gluconeogenesis, 277
  and glycogen metabolism, 198, 199
  and lipid metabolism, 279–281
  as neurotransmitter, 537
Equilibrium constant, 75–77
Ergocalciferol, 583
Erucic acid, 113
Erythrocytes, 465, 467–468
  metabolism of carbohydrates and lipids, 286–288
Erythromycin, 450
Erythrose 4-phosphate, 186, 187, 189
Escherichia coli, DNA replication in, 418–420
Essential amino acids, 323–324
Essential fatty acids, 226, 573
17β-Estradiol, 244, 556–557
Estrogens, 237, 244
Estrone, 244
Ethanol, 97, 570
Ethanolamine, 228
Ethanolamine phosphate, 228
Eukaryotic cells, 69
  compared to prokaryotes, 441
  DNA polymerase in, 424

Exergonic processes, 74
Exonucleases, 422
Exopeptidases, 297
Extracellular fluid, 522
Eyes, visual cycle in, 581–582

F_0 in mitochondrial membrane, 160–161
F_1 ATPase, 160–161
Fab fragment, 485–486
Fabry's disease, 251
Facilitated diffusion, 61
Factor V, 474
Factor VII, 474
Factor VIII, 475
  deficiency of, 477
Factor IX, deficiency of, 477
Factor X, 474
Factor XI, 475
Factor XII, 475
Factor XIII, 473
FAD. See Flavin adenine dinucleotide
Familial high density lipoprotein deficiency, 292
Familial hyperbetalipoproteinemia, 292
Familial hypercholesterolemia, 293
Familial methemoglobinemia, 409
Farber's disease, 251
Farnesyl pyrophosphate, 235
Fat
  caloric value of, 562
  dietary requirements, 570, 573–574
  respiratory quotient of, 565
Fatty acid(s), 112–113
  activation of, 116, 118–119
  biosynthesis of, 220–226
    NADPH requirements in, 223–225
  catabolism of
    energy from, 144–147
    oxidation steps in, 132–133
  degradation to acetyl-coenzyme A, 116–122
  elongation of, 223
  essential, 226, 573
  free, 55
  glycerol esters of, 55
  and glycolysis, 275
  plasma levels of, 265, 466
  saturated, 112, 113, 285, 573
    degradation of, 116–120
  unsaturated, 112, 113, 285, 573
    biosynthesis of, 225–226
    degradation of, 120–121
Fatty-acid synthetase system, 220–223
Fatty liver, 273–274
Favism, 287

Fc fragment, 485–486
Feedback inhibition, 182, 237, 379, 383, 406
Ferritin, 577
Fibrin, 473–474
  dissolution of, 480
Fibrinogen, 465, 471, 472, 473
Fibrinopeptides A and B, 473
Fibroblasts, 539
Filaments, in muscle, 527–529
Flavin adenine dinucleotide (FAD), 95, 119, 135–136
Flavin mononucleotide (FMN), 95, 133–134
Flavoproteins, 36, 135
Fletcher factor, 491
Fluid mosaic model, 60
Fluids of body, 522
Fluoride
  dietary requirements, 579
  enzyme inhibition by, 104
  mottled enamel from, 550
5-Fluorouracil, 391, 392
FMN. See Flavin mononucleotide
Folic acid, 320, 594–595
  action of antagonists, 390–391
  deficiency of, 599
  derivatives of, 321
  reduced, 320, 323
  and vitamin B_{12}, 595
Food intake. See Nutrition
Forbe's disease, 290
Formaldehyde, as donor to tetrahydrofolic acid, 321
Formate, as donor to tetrahydrofolic acid, 321
Formiminoglutamic acid (FIGLU), 349, 599
N^5-Formiminotetrahydrofolic acid, 321
N-Formylkynurenine, 367
N-Formylmethionine, 443–444
N^5-Formyltetrahydrofolic acid, 321
N^{10}-Formyltetrahydrofolic acid, 321, 596
Frame-shift mutations, 462
Fructokinase, 212
  deficiency of, 213
Fructose, 42
  metabolism of, 212–213
  structure of, 45
Fructose diphosphatase, 170, 187, 202, 207, 275
  and gluconeogenesis, 202, 208
  insulin affecting synthesis of, 278
Fructose 1,6-diphosphate, 169–170, 201–202

Fructose 1-phosphate, 212–213
Fructose 6-phosphate, 169, 186, 187, 275
   reaction with glutamine, 214
L-Fucose, 248
Fumarase, 127
Fumaric acid, 127, 136, 311, 343, 362
Fumarylacetoacetate, 366
Furanose ring, 44

Galactitol, 212
Galactocerebroside, 57
Galactokinase, 211
   deficiency of, 212
Galactosamine, 48, 213–214
Galactose, 43
   in glycosaminoglycans, 542–543
   metabolism of, 211–212
Galactosemia, 212
Galactose 1-phosphate, 211–212
Galactose 1-phosphate uridylyl transferase, 212
Galactosylceramide, 247
Gangliosides, 58, 249–250
Gangliosidoses, 250–251
Gases, diffusion of, 496–497
Gaucher's disease, 251
GDP. *See* Guanosine diphosphate
Genes
   control of expression of, 451–455
   operator, 452
   structural, 452
Genetic code, 435–438
Geranyl pyrophosphate, 234–235
Giantism, 556
Gibbs-Donnan equilibrium, 61–64
Globin, 497
Globular proteins, 29–35
Globulins, in plasma, 471–472
Glomerular filtration, 516
Glomerulus, 515
Glucagon, 553
   and adenyl cyclase stimulation, 199, 553
   and carbohydrate metabolism, 281
   and gluconeogenesis, 277
   and lipid metabolism, 281
Glucocorticoids, 242, 557
   and carbohydrate metabolism, 281
Glucogenic amino acids, 201, 316
Glucokinase, 166, 263–264
   insulin affecting synthesis of, 278
Gluconeogenesis, 174, 201–210
   in liver, 282
   regulation of, 274–277
Gluconic acid 6-phosphate, 184

Gluconic acid 6-phosphate dehydrogenase, 184, 185
Gluconolactone 6-phosphate, 184
Glucopyranose, 44
Glucopyranoside, 46
Glucosamine, 48, 213–214
Glucosamine 6-phosphate, 168, 214
Glucose, 42, 44. *See also* Carbohydrates
   glycolysis, 169–175
   hyperglycemia, 278
   hypoglycemia, 278
      in von Gierke's disease, 288
   metabolism of
      aerobic, 175, 179–184
      anaerobic, 166–175
      energy from, 181
      pentose-phosphate pathway, 184–190
      regulation of direction of, 207–208
   plasma levels of, 466
      regulation of, 165, 262–264
   transport of, 164, 165
Glucose 6-phosphatase, 168, 264, 278
   deficiency in von Gierke's disease, 288
Glucose 6-phosphate, 166–168, 169, 192, 282
   conversion to glucose 1-phosphate, 192
   degradation to pyruvate, 207
   oxidation by NADP, 184
Glucose 6-phosphate dehydrogenase, 98
   deficiency in erythrocytes, 286–288
Glucose-phosphate isomerase, 169
   and gluconeogenesis, 202
α-Glucosidase, 196
α-1,4-Glucosidase, deficiency in Pompe's disease, 290
Glucosylceramides, 248
Glucosyl phosphate uridylyl transferase, 192
Glucuronic acid, 48
   in glycosaminoglycans, 543
Glucuronides, 211, 411–412
Glutamate-alanine transaminase, 304
Glutamate-aspartate transaminase, 302
Glutamate dehydrogenase, 306–307
Glutamate-oxaloacetate transaminase (GOT), 302, 305
   plasma levels of, 492
Glutamate-pyruvate transaminase (GPT), 305
   plasma levels of, 492
Glutamic acid, 16, 17, 316, 324
   metabolism of, 344–347
   sources of, 328
   synthesis of, 325

Glutamic semialdehyde, 345
Glutaminase, 517
Glutamine, 16, 316, 324
  metabolism of, 347
  reaction with fructose 6-phosphate, 214
  sources of, 328
  synthesis of, 309, 325
  transamination from, 305
Glutamine amidotransferase, 214, 374, 375, 379, 383, 389
Glutamine synthetase, 309
γ-Glutamyl cycle, 336
Glutathione reductase, 335
Glyceraldehyde, 213
Glyceraldehyde 3-phosphate, 170, 171, 186, 187, 189–190
Glyceraldehyde 3-phosphate dehydrogenase, 171
  and gluconeogenesis, 202
Glycerol, 55
Glycerol kinase, 228
Glycerol phosphate, 282
Glycerol-phosphate dehydrogenase, 179–180, 227
Glycerol phosphate shuttle, 179–180
Glycerol phospholipids, 55–56
  biosynthesis of, 228–232
Glycine, 15, 17, 316, 324
  in collagen, 540
  and creatine formation, 319
  as donor to tetrahydrofolic acid, 321
  metabolism of, 334–338
  sources of, 328
  synthesis of, 327
  titration curve for, 14–15
Glycocholic acid, 115, 239, 336
Glycogen, 48, 163, 190–192, 192–196
  debranching enzyme system deficiency, 290
  metabolism of, 190–201
    hormones affecting, 196–201
  storage of, 282, 570
  storage diseases, 288–291
    type I, 288–290
    type II, 290
    type III, 290–291
    type IV, 291
    type V, 291
    type VI, 291
Glycogenolysis, 195–196
Glycogen phosphorylase, 195, 196–199
Glycogen synthetase, 193, 196, 199–201
  and gluconeogenesis, 202
Glycogen synthetase D, 199, 264
Glycogen synthetase D phosphatase, 199
Glycogen synthetase I, 199
  epinephrine inhibiting formation of, 279
Glycolipids, 163
Glycolysis, 169–175
  anaerobic, 181
  fatty acids and, 275
  feedback inhibition of, 183
  inhibition by oxygen, 181–184
  regulation of, 274–277
Glycoproteins, 37, 103
  N-acetylneuraminic acid in, 48
  biosynthesis of, 210
  plasma levels of, 472
Glycosaminoglycans
  biosynthesis of, 546, 547
  and carbohydrates, 542–549
  metabolism of, 547–549
  structure of, 542–546
Glycosides, 46
Glycosphingolipids, 247
Glycosyltransferases, 247, 250, 547
GMP. *See* Guanosine monophosphate
Goiter, 579
Gout, 398–399, 401
  tophi in, 398
Granulocytes, 468
Growth hormone, 556
Guanidinoacetic acid, 319, 348
  as acceptor for methyl groups, 317, 318
Guanine, 49, 54
  in DNA structure, 415
Guanosine, 50
Guanosine diphosphate (GDP), 378
Guanosine monophosphate (GMP), 51, 378
Guanosine monophosphate synthetase, 378
Guanylic acid. *See* Guanosine monophosphate

Haptaglobins, 471, 472
Haptens, 482
Hartnup's disease, 370
Hematocrit, 466
Hematopoiesis, 467
Heme, 406, 407, 497
  catabolism of, 410–412
Hemochromatosis, 578
Hemoglobin, 35, 36, 39–42, 405, 467, 497–498
  α chain, 39, 40, 41
  β chain, 40, 41–42

buffering effect of, 514
derivatives of, 408–410
disorders in, 509
electrophoresis of, 41–42
oxyhemoglobin, 498–503, 506–507
variations in, 507–509
Hemopexin, 471, 472
Hemophilia, 477
Hemosiderin, 577
Hemosiderosis, 578
Henderson-Hasselbalch equation, 8–9, 511
Henle, loop of, 516
Henry's law, 510
Heparan sulfate, 544, 545, 547
Heparin, 479–480, 545, 547
Hers' disease, 291
Hexokinase, 166, 169, 202
and glucose concentrations, 263
Hexosamines, 213–215
in glycosaminoglycans, 542–543
Hexosaminidase A, 251
Hexose(s), 42
Hexose-monophosphate shunt, 184, 188
Hexose-phosphate isomerase, 187
Hippuric acid, 338
Histamine, 350–351
Histidase, 307
Histidine, 17, 308, 316, 324
dissociation of, 19–21
as donor to tetrahydrofolic acid, 321
metabolism of, 348–351
titration curve for, 21
Histones, 450–451
Holoenzyme, 95
Homocysteine, 327, 333
Homogentisic acid, 361–362
Hormones, 551–559
adrenocortical, 242, 557
anabolic, 552
calcitonin, 555
and carbohydrate metabolism, 277–281
catabolic, 553
epinephrine, 551
glucagon, 553
and gluconeogenesis, 277
and glycogen metabolism, 196–201
growth, 556
insulin, 551–552
and lipid metabolism, 277–281
parathyroid, 555
prostaglandins, 558–559
and protein synthesis, 454
sex, 556–557

steroid, 58, 240–244, 556–557
thyroid, 553–555
Hunter's disease, 548–549
Hurler's disease, 548, 549
Hyaluronic acid, 544, 545
Hyaluronidases, 545
Hydantoin 5-propionate, 350
Hydrocortisone, 242, 557
and gluconeogenesis, 277
Hydrogen-bonding, 27, 29, 33, 415–416, 417
Hydrogen ions
concentration of, 2–5. *See also* pH
Hydrogen peroxide, 397
Hydrolase, 97
Hydrophobic interactions, 31–32
L-3-Hydroxyacyl-coenzyme A dehydrogenase, 120
3-Hydroxyanthranilic acid, 368
Hydroxyapatite, 549
β-Hydroxybutyrate, 255, 256
β-Hydroxybutyrate dehydrogenase, 255
D-β-Hydroxybutyryl-ACP, 222
β-Hydroxybutyryl-CoA, 254, 355, 360
α-Hydroxy-γ-carboxypropyl-thiamine pyrophosphate, 124
25-Hydroxycholecalciferol, 584
α-Hydroxyethyl thiamine pyrophosphate, 176
5-Hydroxyindoleacetic acid, 370
β-Hydroxyisobutyrate, 355
β-Hydroxyisobutyryl-CoA, 355
3-Hydroxykynurenine, 368
Hydroxylysine, 17
in collagen, 540
β-Hydroxy-β-methylglutaryl-coenzyme A, 234, 254, 358–359
3-Hydroxy-3-methylglutaryl-coenzyme A, 234, 254, 358–359
β-Hydroxy-β-methylglutaryl-coenzyme A reductase, 234, 237
α-Hydroxynervonic acid, 113
p-Hydroxyphenyllactate, 362
p-Hydroxyphenylpyruvate oxidase, 362–363
p-Hydroxyphenylpyruvic acid, 361
17α-Hydroxypregnenolone, 243
Hydroxyproline, 16, 17, 316, 324, 352
in collagen, 540
sources of, 328
synthesis of, 328
β-Hydroxypyruvic acid, 333
5-Hydroxytryptamine, 369–370
in platelets, 468

5-Hydroxytryptophan, 370
Hyperglycemia, 278
Hyperlipidemias, 292–294
Hyperuricemia, 398–402
Hypoglycemia, 278
Hypoxanthine, 396
Hypoxanthine-guanine phosphoribosyl transferase, 388
Hypoxanthine ribonucleotide, 376

Iduronic acid, 48
  in glycosaminoglycans, 543
Imidazole group, 20
  and buffer effect, 513–514
  in histidine, 20
  and hydrogen bonding, 33
  pK of, 37
Imidazolone 5-propionic acid, 348–349
Immune complex, 482
Immunoglobulins, 472, 481–488
  and antibody formation, 481–482
  characteristics of, 485
  constant region of, 484
  Fab fragments of, 485–486
  Fc fragment of, 485
  heavy chains of, 482, 483
  IgA, 485, 487
  IgD, 485, 487
  IgE, 485, 487
  IgG, 485–486
  IgM, 485, 486
  light chains of, 482, 483
  structure of, 482–487
  synthesis of, 487–488
  variable region of, 484
Imuran, action of, 392–393
Initiation factors, in protein synthesis, 436, 443–444
Inosinic acid (IMP), 376–377, 388, 395
Inosinic acid cyclohydrolase, 376
Inosinic acid dehydrogenase, 378
Inositol, 56. See also myo-Inositol
Insertions, mutagenic, 461, 462
Insulin, 165–166, 551–552
  bovine, structure of, 23
  and carbohydrate metabolism, 277–279
  and glycogen metabolism, 199–200
  and lipid metabolism, 257, 277–279
Interstitial fluid, 522
Intestinal absorption. See Absorption, intestinal
Intracellular fluid, 522
Intrinsic factor, 598
Iodine, dietary requirements, 579

Iodine-131, 68, 135
  uptake by thyroid, 555
Iodoacetate, 104, 106
Ionization constant, of acids, 2
Ion-product, of water, 2
Iron
  in cytochrome system, 138–141
  dietary requirements, 577–578
  and enzyme activity, 96
  in hemoglobin, 498
Iron-59, 68
Iron-porphyrin complex, 95
Iron-sulfur proteins, 136–137
Isobutyryl-coenzyme A, 355
Isocitrate, 123, 128
Isocitrate dehydrogenase, 123
  as allosteric enzyme, 150
  kinetic properties of, 153
  NAD-linked, 150
  NADP-dependent, 225
Isoelectric point, 15, 37
Isohydric shift, 504
Isoleucine, 15, 17, 316, 324
  metabolism of, 356–358
Isomaltase, 164, 167
Isomerase, 97
Isoniazid, 593
3-Isopentenyl pyrophosphate, 234
Isotopes, 66–69
  properties of, 68
  radioactive, and half-life of, 68
Isovaleryl-coenzyme A, 358
Isozymes
  of lactate dehydrogenase, 259–262, 492
    distribution in tissues, 262
  serum levels of, 491–492

Kallidin, 491
Kallikrein, 475, 491
Katal, 93
Keratan sulfate, 544, 545
Keratins, 29
β-Ketoacyl-ACP reductase, 222
β-Ketoacyl-ACP synthetase, 221
β-Ketoacyl-coenzyme A, 120
α-Ketoadipic acid, 360, 368–369
α-Ketobutyrate, 353
Ketogenic amino acids, 316
α-Ketoglutarate, 123
  in citric acid cycle, 123
  in glutamate dehydrogenase reaction, 344–345
  and transamination, 301–302
α-Ketoglutarate decarboxylase, 123, 124

α-Ketoglutarate-dehydrogenase complex, 123
Ketohexoses, 42
α-Ketoisocaproic acid, 358
Ketone bodies
  in diabetes, 256
  formation and disposition of, 234, 253–257
  plasma levels of, 466
Ketonemia, 256, 273
Ketonuria, 256
Ketoses, 42
Ketosis, 256
Kidney
  metabolism of carbohydrates and lipids, 284
  and pH regulation, 515–519
Kilocalories, 561
Kinin(s), 490
Kininogen, 490
  high-molecular-weight, 475
Krabbe's disease, 251
Krebs cycle. *See* Citric acid cycle
Kwashiorkor, 572
Kynureninase, 369
Kynurenine, 368

Lactase, 164, 167
Lactate dehydrogenase, 98, 173
  isozymes of, 259–262, 492
    distribution in tissues, 262
Lactic acid, 173–174
  and gluconeogenesis, 207
  metabolism of, 257–262
  plasma levels of, 466
Lactonase, 184
Lactose, 46, 47
  hydrolysis of, 211
Lanosterol, 235–236
Lauric acid, 113
Lead, and inhibition of porphyrin synthesis, 407
Lecithin, 56, 228
Lecithin-cholesterol acyltransferase (LCAT), 238, 270
Lesch-Nyhan syndrome, 401
Leucine, 15, 17, 316, 324
  metabolism of, 358–359
Leucine aminopeptidase, 301
Leukocytes, 465, 468
Ligase, 97
Lignoceric acid, 113
Limit dextrin, 196
Limit dextrinosis, 290
Lineweaver-Burk equation, 90

Linoleic acid, 112, 573
  and prostaglandin biosynthesis, 226
Linolenic acid, 112, 573
Lipase
  lipoprotein, 115, 265, 270
  pancreatic, 114, 240
  tissue, 266
Lipemia, alimentary 265
Lipid(s), 55–58, 111
  bilayer of membranes, 58–60
  biosynthesis of, 219–251
    and acetyl-coenzyme A transport, 219
    fatty acids, 220–226
    phospholipids, 228–232
    sphingolipids, 244–251
    steroid hormones, 240–244
    triglycerides, 226–228
  and cholesterol metabolism, 232–240
  circulation and mobilization of, 264–274
  dietary requirements, 570, 573–574
  hyperlipidemias, 265, 292–294
    serum characteristics in, 294
    type I, 292
    type II, 293
    type III, 293
    type IV, 293–294
    type V, 294
  metabolism of
    in adipose tissue, 284–286
    in brain, 296
    disorders of, 291–294
    epinephrine affecting, 279–281
    in erythrocytes, 286–288
    glucagon affecting, 281
    hormones affecting, 277–281
    insulin affecting, 277–279
    in kidney, 284
    in liver, 282–284
    in muscle, 284
    relation to metabolism of amino acids and carbohydrates, 314
  plasma levels of, 269–270
    physiologic significance of, 273–274
  tissue disposition and transport of, 253–294
Lipidemias, 265, 292–294
Lipoamide, 125
Lipoic acid, 95, 124–125, 600
Lipoprotein(s), 37, 267–273
  α-, 267
  β-, 267
    abetalipoproteinemia, 292
    broad beta disease, 293
    hypobetalipoproteinemia, familial, 292

Lipoprotein(s)—Continued
  composition and function of, 269–270
  density fractionation of, 269
  electrophoretic mobility of, 267–268
  high-density (HDL), 266, 269, 273
    familial deficiency of, 292
  low-density (LDL), 266, 269, 271–273
  metabolic transformations of, 270–273
  plasma levels of, 472
  pre-β, 266, 267
    hyperlipidemia, 293–294
  Svedberg flotation units for, 269
  very-low-density (VLDL), 269, 270
Lipoprotein lipase, 115, 265–266, 270
  deficiency of, 292–293
Lipotropic compounds, 274, 601
Liver
  fatty, 273–274
  glycogen storage in, 282, 570
  metabolism of carbohydrates and lipids, 282–284
Lungs
  and control of blood pH, 519–520
  and exchange of gases, 495–497, 506
  respiratory acidoses, 521–522
  respiratory alkalosis, 521–522
Lyase, 97
Lymph, 522
Lymphocytes, 468
  B and T cells, 482
Lysine, 17, 316, 324
  and collagen cross-links, 540–542
  dissociation of, 19
  metabolism of, 359–361
  titration curve for, 19
Lysogenicity, 457
Lysosomes, metabolism disorders of, 251

Macroglobulin(s), 486
α_2-Macroglobulin, 472, 477
Magnesium
  dietary requirements, 576
  and enzyme activity, 96
  in fluid compartments, 523
  plasma levels of, 467
  in urine, 524
Malate dehydrogenase, 98, 127, 205
Malate shuttle, 180, 205
Malic acid, 127
  conversion to pyruvate, 225
Malonic acid, 107–108
Malonyl-coenzyme A, 220
Malonyl transacylase, 221
Maltase, 164, 167
Maltose, 46, 47

Manganese
  dietary requirements, 580
  and enzyme activity, 96
Maple-syrup urine disease, 359
Marasmus, nutritional, 572
Maroteaux-Lamy syndrome, 549
Maximum velocity (V_{max}), 89–92
McArdle's disease, 291
Melanin, 363
Melatonin, 370–371
Membrane potential, 536
Membranes
  fluid-mosaic model of, 60
  gangliosides in, 249–250
  lipid bilayer of, 58–60
  transport through, 61–64
    and Gibbs-Donnan equilibrium, 61–64
    and osmotic pressure, 62
Menadione, 589
β-Mercaptoethanol, 35, 36
6-Mercaptopurine, 391, 392
  with allopurinol, in chemotherapy, 402
β-Mercaptopyruvic acid, 341
Meromyosin, 530–531
Metabolism, 64–69
  basal, 568–569
    thyroxine affecting, 553
  experimental approaches to, 65–66
  intermediary, 64
  isotope studies of, 66–69
Methemoglobin, 408–409
Methemoglobinemia, 409–410
N^5,N^{10}-Methenyltetrahydrofolic acid, 321, 375, 596
Methionine, 15, 17, 316, 324, 598
  and creatine formation, 319
  metabolism of, 352–355
  and transmethylation, 317
Methionyl-tRNA, 443
Methotrexate, action of, 390
Methyl-adenosyl transferase, 317
Methylases, modification, 457–458
α-Methylbutyryl-coenzyme A, 357
Methylcobalamin, 596, 598
3-Methylcrotonyl-coenzyme A, 358
N^5,N^{10}-Methylenetetrahydrofolic acid, 321, 322
3-Methylglutaconyl-coenzyme A, 358
Methyl groups, transmethylation of, 316–318
α-Methyl-β-hydroxybutyryl-coenzyme A, 357
2-Methyl-6-hydroxychroman, 586
Methylmalonyl-coenzyme A, 355

N^5-Methyltetrahydrofolic acid, 321
8-Methyltocotrienol, 586
Methyltransferases, 317
Mevalonic acid, 234, 237
Michaelis constant, 88
  for enzymatic reactions, 89–92
Michaelis-Menten equation, 89
Microcurie, 68
Microheterogeneities, polysaccharide, 547
Millicurie, 68
Mineralocorticoids, 243, 557
Mitochondria
  ATPase activity in, 157
  oxidation reactions in, 132, 156
  shuttle systems for transport through membranes, 180
  structure of, 154
  transfer of substances through membranes
    acetyl-coenzyme A, 219
    aspartate, 204
    fatty acids, 117–118
    α-ketoglutarate, 204
    oxaloacetate, 204–206
Molybdenum, dietary requirements for, 580
Monoamine oxidase, 537
Monoglyceride, 55
Monosaccharides, 42–46
  intestinal absorption of, 164
Morquio's syndrome, 549
Mucopolysaccharides, 163, 542
  synthesis of, 210, 211, 215
Muscle, 527–535
  contractile proteins of, 527–531
  creatinine in, 320
  metabolism of carbohydrates and lipids, 284
  myoglobin in, 532–533
  phosphocreatine affecting, 534–535
  reaction sequence in contraction of, 531–532
Mutations, 458–462
  deletion in, 461–462
  frame-shift, 462
  induction by antigens, 488
  insertion in, 461
  transitional, 459–461
  transversion in, 461
Myeloma proteins, 487
Myofibrils, 527
Myoglobin, 408, 532–533
myo-Inositol, 600, 601
Myokinase, 535

Myosin, 527–530
Myristic acid, 112

NAD. *See* Nicotinamide adenine dinucleotide
NADP. *See* Nicotinamide adenine dinucleotide phosphate
Nephrons, 515
Nephrosis, 471
Nerve impulses, 535–539
Nervonic acid, 113
Neurons, 535
Neurotransmitters, 535
  catecholamines as, 537
Niacin, 369, 591
Nicotinamide, 369, 591
  as acceptor for methyl groups, 317, 318
Nicotinamide adenine dinucleotide (NAD), 95, 99
  reduced (NADH), 99, 120
    NADH dehydrogenase, 133
    oxidation of, 144
    reaction with dihydroxyacetone phosphate, 179
    reaction with oxaloacetate, 180
Nicotinamide adenine dinucleotide phosphate (NADP), 95, 99
  reduced (NADPH), 99, 163, 386
    and fatty acid synthesis, 223–225
    and mixed function oxygenases, 225, 235
    production in pentose pathway, 184–190
    requirements and sources, 223, 225
Nicotinic acid, 369, 591
Niemann-Pick disease, 251
Nitrogen
  balance of, 571
  urinary, 566
Nitrous acid, mutagenic activity of, 459–460
Norepinephrine, 364, 551
  as acceptor for methyl groups, 317, 318
  as neurotransmitter, 537
Nucleic acids, 52–55
  biosynthesis of, 413–438
Nucleoside(s), 40
Nucleoside diphosphate, 51
Nucleoside-diphosphate kinase, 378, 382, 387
Nucleoside-monophosphate kinase, 378, 382, 387
Nucleoside triphosphates, 51
Nucleotides, 49, 51
  digestion and absorption of, 393

Nucleotides—*Continued*
  metabolism of, 373, 402
    deoxyribonucleotides, 385–387
    and inhibitors of synthesis, 387, 389–393
    purine catabolism, 395–397
    purine ribonucleotides, 373–379
    pyrimidine catabolism, 393–395
    pyrimidine ribonucleotides, 379–385
    salvage pathways in, 387–389
    and uric acid levels in blood, 397–402
Nutrition, 561–601
  and basal metabolism, 568–569
  calcium in, 574–575
  and calories, 561–563
  carbohydrates in, 570, 572–573
  chloride in, 577
  copper in, 578–579
  and energy expenditure, 563–567
  and energy requirements, 569–571
  fluoride in, 579
  iodine in, 579
  iron in, 577–578
  lipids in, 570, 573–574
  magnesium in, 576
  manganese in, 580
  minerals in, 574–580
  minimal daily requirements in (MDR), 574
  molybdenum in, 580
  potassium in, 575–576
  protein in, 570, 571–572
  recommended daily allowances in (RDA), 574
  and respiratory quotient of nutrients, 564–565
    non-protein, 567
  sodium in, 575–576
  and specific dynamic action of foods, 569
  sulfur in, 577
  vitamin-like nutrients in, 600–601
  vitamin requirements in, 580–600
  water in, 574
  zinc in, 579–580

Ochronosis, 363
Okazaki fragments, 421, 422, 423
Oleic acid, 112, 285
  degradation of, 120–121
Oligo-(1,4→1,4)-glucantransferase, 196
Operons, 452–454
Opsin, 581

Ornithine, 310, 319, 345, 348
Ornithine transcarbamoylase, 311
Orotic acid, 381–382
Oroticaciduria, 382
Orotidine 5′-phosphate decarboxylase, 382, 385
Orthophosphate cleavage, 82
Osteoarthritis, hypertrophic, 550
Osteoblasts, 549
Osteoclasts, 549
Osteomalacia, 550
Osteoporosis, 550
Oxaloacetic acid, 127
  and carbohydrate biosynthesis, 208–210
  conversion to phosphoenolpyruvate, 203–204, 276
  from pyruvate, 275
  reaction with NADH, 180
  shuttle for, 204–206
  and transamination, 302
Oxidation, 128–141
  aerobic, of glucose, 179–184
  and deamination reactions, 306–307
  energy from, 141–144
  and pentose-phosphate pathway, 184–190
  and redox potentials, 142
Oxidative chain, 132–141
Oxidative phosphorylation, 143, 154–161
  uncoupling of, 156–157
Oxidizing agents, 131
Oxidoreductases, 97–101
Oxygen
  binding to myoglobin, 533
  combining with hemoglobin, 497, 498
  and glycolysis inhibition, 182
  partial pressure of, 496
  ratio to phosphate, and oxidations, 144
  reduction to water, 140–141
  transport of, 495–497, 506–507
Oxygenases, mixed-function, 225
  desmolase, 240
Oxyhemoglobin, 498–503, 506–507
  dissociation curve for, 499

Palmitic acid, 112, 285
Palmitoleic acid, 112
Palmitoyl-ACP, 223
Palmitoyl-CoA, and sphingosine synthesis, 246
Pancreatic amylase, 164
Pancreatic lipase, 114
Pantothenic acid, 116, 594

Parathyroid hormone, 555
Pasteur effect, 181–184
Pellagra, 370, 591
Pentose(s), 42, 45
Pentose-phosphate epimerase, 185
Pentose-phosphate isomerase, 185
Pentose-phosphate pathway, 184–190
Pepsin, 297, 301
Pepsinogen, 297, 301
Peptides, 21–24
Peptidyl site (P site), of ribosomes, 443
Peptidyl transferase, 445
pH, 3
  of acetic acid solutions, 5
  of blood, 256, 509
    compensatory control of, 519–522
    and dissociation of oxyhemoglobin, 499, 501, 502
    and enzyme activity, 93–94
    and Henderson-Hasselbalch equation, 8
    physiologic, 21
    renal regulation of, 515–519
    and titration curve, 8
Phages, 456
Phenylalanine, 15, 17, 316, 324
  metabolism of, 365–367
Phenylalanine hydroxylase, 365
  deficiency of, 328
Phenylketonuria, 328, 367
Phosphatases, 96
Phosphate
  dietary requirements, 575
  in fluid compartments, 523
  inorganic, 79
    ratio to oxygen, and oxidations, 144
  plasma levels of, 467
  transformations in kidney, 517
  in urine, 524
Phosphatidic acids, 56, 227, 231
Phosphatidylcholine, 56
  biosynthesis of, 228–231
  and transmethylations, 317, 318
Phosphatidylethanolamine, 56
  as acceptor for methyl groups, 317, 318
  biosynthesis of, 228–230
Phosphatidylglycerol, 231, 232
Phosphatidylinositol, 56, 231–232
Phosphatidylserine, 56
  biosynthesis of, 230
3′-Phosphoadenosylphosphosulfate (PAPS), 547, 548
Phosphocreatine, 319–320
  and muscle contraction, 534–535

Phosphoenolpyruvate, 172–173, 275
  and gluconeogenesis, 202
  synthesis from pyruvate, 206
Phosphoenolpyruvate carboxykinase, 203, 207, 276
  and gluconeogenesis, 208
  insulin affecting synthesis of, 278
Phosphofructokinase, 169–170, 207, 275
  inhibition by citrate, 183, 184
  inhibition by ATP, 183, 184, 207
Phosphoglucomutase, 202
Phosphogluconate pathway, 184
6-Phosphogluconic acid, 184
Phosphoglycerate kinase, 172
  and gluconeogenesis, 202
2-Phosphoglyceric acid, 172
3-Phosphoglyceric acid, 172
Phosphoglycerides, 55–56
  biosynthesis of, 228–232
Phosphoglyceromutase, 172
  and gluconeogenesis, 202
Phospholipids
  biosynthesis of, 228–232
  glycerol, 55–56
  in lipoproteins, 270
  plasma levels of, 265, 466
3-Phospho-5-pyrophosphomevalonic acid, 234
Phosphoribosylaminoimidazole-carboxamide formyltransferase, 376
Phosphoribosylaminoimidazole carboxylase, 376
Phosphoribosylaminoimidazole-succinocarboxamide synthetase, 376
Phosphoribosylaminoimidazole synthetase, 375
Phosphoribosylformylglycinamide synthetase, 375
Phosphoribosylglycinamide formyltransferase, 375
Phosphoribosylglycinamide synthetase, 374
5-Phosphoribosyl-pyrophosphate (PRPP), 374, 382, 388
5-Phosphoribosyl-pyrophosphate amidotransferase, 374
Phosphoric acid, 11–12
  conjugate base of, 7
  ionization of, 6
Phosphorus-32, 68
Phosphorus, dietary requirements for, 575. *See also* Phosphate
Phosphorylase a, 197–198

Phosphorylase a phosphatase, 198
Phosphorylase b, 197–198
Phosphorylase b kinase, 197–198
Phosphorylase, deficiency of, 291
Phosphorylation
   oxidative, 143, 154–161
      coupling in, 157–159
      mechanisms of, 157–161
      uncouplers of, 156–157
   substrate-level, 144, 172
Physostigmine, and acetylcholinesterase inhibition, 539
Plasma, 465
Plasma cells, 482
Plasmin, 480
Plasminogen, 480
Platelets, 465, 468–469
   and clot formation, 480
Polypeptides, 21–24
   alpha helical structure of, 26
   beta(β) pleated sheet of, 27–28
   hydrogen bonding sites in
      interchain, 27, 28, 29
      intrachain, 27, 29
Polyribosome, 448
Polysaccharides, 47–49, 163
   digestion and absorption of, 164
   glycosaminoglycans, 542–549
Pompe's disease, 290
Porphobilinogen, 406
Porphobilinogen isomerase, 407
   deficiency of, 410
Porphyrias, 410
Porphyrins, 138, 405–412
   biosynthesis of, 406–408
   structure and functions of, 405
Portal circulation, monosaccharide transport to, 164
Postsynaptic membrane, 536
Potassium
   dietary requirements, 575–576
   and enzyme activity, 96
   in fluid compartments, 523
   plasma levels of, 467
   in urine, 524
Potentials, standard electrode, 130–132, 142
Pregnenolone, 240–241
Prekallikrein, 475, 490
Procollagen, 541
Procollagen hydroxylase, 541
Progesterone, 241–242, 557
Proinsulin, 551
Prokaryotic cells, 69
   compared to eukaryotes, 441

Proline, 15, 17, 316, 324, 345
   in collagen, 540
   metabolism of, 351–352
   sources of, 328
   synthesis of, 325–326
Pronormoblasts, 467
Properdin pathway, 490
Propionyl-coenzyme A, 122, 357
Prostacyclins, 558
Prostaglandins, 226, 558–559
Prosthetic groups, enzyme, 95
Proteases, 96
Protein, 21–42
   acidic, 38
   acyl-carrier, 221
   basic, 38
   caloric value of, 562
   conjugated, 36
   contractile, 527–531
   denaturation of, 35
   dietary requirements, 570, 571–572
   digestion and absorption of, 297–301
   electrolyte characteristics of, 37–38
   electrophoresis of, 39
   fibrous, 29
   in fluid compartments, 523
   globular, 29–31
      in lipid matrix of membranes, 60
      random coils in, 29
   hemoglobin, 39–42
   iron-sulfur, 136–137
   isoelectric point of, 37
   in lipoproteins, 270
   metabolism of, 297–328
   myeloma, 487
   peptides and polypeptides, 21–24
   in plasma, 466, 469–472, 473
   primary structure of, 24
   quaternary structure of, 25, 35
   renaturation of, 35
   respiratory quotient of, 565
   retinol-binding, 582
   secondary structure of, 24
   sedimentation coefficient for, 38–39
   separation and quantitative assay of, 38–39
   spatial characteristics of, 24–36
   synthesis of, 438–450
      and codons for amino acids, 436–438
      and control of gene expression, 451–455
      and elongation of polypeptide chain, 445
      inhibitors of, 449–450
      initiation of, 436, 441–445

mutations in, 458–462
and posttranslational modifications, 448
termination of, 436, 445–448
in viruses, 455–458
tertiary structure of, 24
stabilization of, 31–35
ultracentrifugation of, 38–39
Protein kinase, 198
Proteoglycans, 545
Proteolysis, limited, 480
Prothrombin, 472, 474
Protoheme IX, 138
Protoporphyria, erythropoietic, 410
Protoporphyrin IX, 138, 407
Pseudouridine, 431
Pteroic acid, 320
Pteroylglutamic acid, 320. See also Folic acid
Ptyalin, 164
Purine(s), 49
Purine nucleotides. See also Adenosine triphosphate
biosynthesis of, 373–379
catabolism of, 393–397
Puromycin, 449, 450
Pyranose ring, 44
Pyridoxal, 592
Pyridoxal phosphate, 95, 302
Pyridoxamine, 592
Pyridoxamine phosphate, 303
Pyridoxine, 592–593
deficiency of, 370
Pyrimidine nucleotides, 382–383
biosynthesis of, 379–385
catabolism of, 393–395
Pyrophosphatase, 374
inorganic, 117
and UDP-glucose pyrophosphorylase reaction, 193
Pyrophosphate
cleavage of, 82
inorganic, 81
Pyruvate, 173
and alanine formation, 304
conversion to acetyl-coenzyme A, 175–179
conversion to oxaloacetate, 275
conversion to phosphoenolpyruvate, 203
as non-oxidative deamination product, 307–308
reduction to lactate, 257
Pyruvate carboxylase, 203, 207, 275
and gluconeogenesis, 208
insulin affecting synthesis of, 278

Pyruvate dehydrogenase system, 176–179, 208
Pyruvate kinase, 173, 207, 275
inhibition by ATP, 207

Quaternary structure of proteins, 25, 35

Radiation, mutations from, 459
Radioactive isotopes, 68
Redox potentials, 130–132
and free-energy change, 131
for substances in biologic oxidations, 142
Redox reactions, 97, 130
Reducing agents, 131
Reduction reactions, 97, 130
Releasing factor, and termination of protein synthesis, 445–447
Renaturation, of protein, 35
Repression, 452–453
Respiratory chain, 132–141
Respiratory quotient
of fat, 565
non-protein, 567
of nutrients, 564–565
Respiratory system
and control of blood pH, 519–520
and transport of gases, 495–497, 504–507
Resting potential, 536
Retinol, 580
Retinol-binding protein, 582
Reverse transcriptase, 458
Reversible reactions, 75
Rhodopsin, 581
Riboflavin, 119, 591–592
Ribofuranose, 45
Ribonuclease, 31
Ribonucleic acid (RNA), 52
biosynthesis of, 426–428
and posttranscriptional reactions, 428, 430, 434
differentiation of, 428–435
messenger, 52, 428–430
base-pairing with tRNA, 440–441
monocistronic, 454
polycistronic, 454
ribosomal, 52, 434–435
transfer, 52, 430–433
base-pairing with mRNA, 440–441
recognition sites in, 441
translation of data from, 435–436
in viruses, 455, 456
Ribonucleoside(s), 53

Ribonucleoside-diphosphate reductase, 385
Ribonucleotides
　purine, biosynthesis of, 373–379
　　control mechanisms in, 378–379
　pyrimidine, biosynthesis of, 379–385
　　control mechanisms in, 383–385
Ribose 5-phosphate, 185, 190, 373
Ribose-phosphate pyrophosphokinase, 373
Ribosomes, 52–53, 70
　aminoacyl site of, 443
　peptidyl site of, 443
Ribothymidine, 431
D-Ribulose 5-phosphate, 185
Rickets, 585–586
Rifampicin, 449, 450
RNA. *See* Ribonucleic acid
RNA polymerase, 420, 426–428
　DNA-dependent, 423, 426
　RNA-directed, 456
RNA replicase, 456
Rods and cones, ocular, 581
Rotenone, and ATP formation, 156

Saccharopine, 359
Salt, dietary, 575–576
Salvage pathways, in nucleotide biosynthesis, 387–389
Sandhoff's disease, 251
Sanfilippo's syndrome, 549
Sarcolemma, 527
Sarcomere, 517
　Z lines of, 528
Sarcoplasm, 527
Sarcoplasmic reticulum, 532
Sarcosine, as donor to tetrahydrofolic acid, 321
Scheie's syndrome, 549
Scurvy, 600
Secondary structure of proteins, 24
Sedimentation coefficient, for proteins, 38–39
Sedoheptulose 7-phosphate, 185, 189–190
Serine, 16, 17, 316, 324
　ammonia from, 308, 332
　as donor to tetrahydrofolic acid, 321
　metabolism of, 332–334
　phosphatidylserine formation, 230
　sources of, 328
　and sphingosine synthesis, 246
　synthesis of, 326
Serine dehydratase, 307

Serine hydroxymethyltransferase, 327, 338
Serine proteases, 104–106, 480–481
Serotonin, 369–370
　in platelets, 468
Serum, 465
　antisera, 482
Serylalanine, 22
Sex hormones, 556–557
Shuttle systems, 179–181
　glycerol-phosphate, 179–180
　malate, 180
Sialic acid, 49
　in gangliosides, 249
Sigmoid relationships, 152, 384, 500–502
Soaps, 114
Sodium
　dietary requirements, 575–576
　and enzyme activity, 96
　in fluid compartments, 523
　plasma levels of, 467
　tubular reabsorption of, 517
　in urine, 524
Sodium-24, 68
Somatotropin, 556
Sphingolipid(s), 57–58
　biosynthesis of, 244–251
Sphingolipidoses, 250–251
Sphingomyelins, 57, 246–247
Sphingosine, 57, 246
Squalene, 235
Squalene 2,3-epoxide, 235
Squalene synthetase, 235
Starch, 47–48
　digestion of, 164
Stearic acid, 112
　degradation of, 120
Stercobilinogen, 411
Stereospecificity of aconitase, 147–149
Steroid hormones, 58, 240–244, 556–557
　biosynthesis of, 232
　nomenclature of, 237–238
Sterols, 58, 238
Streptomycin, 449, 450
Substrate
　concentrations affecting enzyme reaction rate, 85–92, 152
　cooperative binding to enzymes, 152
　induced-fit theory of, 102
Succinate dehydrogenase, 127
　malonic acid affecting activity of, 107–108
Succinate, oxidation of, 135, 144, 155–156

Succinate thiokinase, 126
Succinic semialdehyde, 347
Succinylcholine, and acetylcholinesterase inhibition, 539
Succinyl-coenzyme A, 255, 339, 353, 355, 598
Succinyl dihydrolipoamide, 125
Sucrase, 164, 167
Sucrose, 46, 47
   dietary, 573
Sugars, reducing, 46. See also Carbohydrates
Sulfanilamide, 390
Sulfate
   in fluid compartments, 523
   in urine, 524
Sulfhydryl, pK of in cysteine, 37
Sulfite oxidase, 341
Sulfotransferases, 547
Sulfur
   dietary requirements, 577
   iron-sulfur proteins, 136–137
Sulfur-35, 68
Svedberg flotation units (S_f), 269
Synapse, 535
Synaptic transmission, 536–538
Synkayvite, 588, 589
Synovial fluid, 522

Tangier disease, 273, 292
Taurocholic acid, 155, 341
Tay-Sachs disease, 251
Teeth, 550
Temperature
   and enzyme activity, 94–95
   and equilibrium constant, 75–77
Termination, in protein synthesis, 436, 445–448
Tertiary structure of proteins, 23, 31–35
Testosterone, 243, 244, 556
Tetany, 555
Tetracycline, 450
Tetraglycosylceramide, 248
Tetrahydrofolic acid (THFA), 321–322, 386, 594, 598–599
Tetrahydropteridine dehydrogenase, 366
Tetraiodothyronine, 553
Tetrapeptides, 23
Thermodynamics, first law of, 74
Thiamine, 589–591
Thiamine pyrophosphate (TPP), 95, 124–125, 589
   binding to pyruvate, 176
   and transketolase reaction, 186
Thiokinase, 116–117

Thiolase, 120
Thiophorase, 255
Thioredoxin, 385–386
Thiouracil, 555
Threonine, 16, 17, 316, 324
   metabolism of, 338–340
Threonine aldolase, 338
Threonine deaminase, 338
Thrombin, 473
Thrombocytes. See Platelets
Thromboplastin, tissue, 474
Thromboxanes, 558
Thymidine, 50, 53
Thymidylate synthetase, 386
Thymine, 49
   in DNA structure, 415
Thyroid hormones, 553–555
   and carbohydrate metabolism, 281
Thyroid-stimulating hormone (TSH), 553
Thyroxine, 364–365, 553–555
   and uncoupling of oxidative phosphorylation, 157
Thyroxine-binding globulin, 472, 553
Tissue factor, 474
Tissue lipase, 266
Titration curves, 8
   for acetic acid, 9
   for aspartic acid, 18
   for glycine, 14–15
   for histidine, 21
   for lysine, 20
   for phosphoric acid, 12
α-Tocopherol, 586
Tophi, in gout, 398
Torr, 496
Transaldolase, 186–187, 189
Transaminases, 96, 302–305
Transaminations, 301–306
Transcellular fluids, 522
Transcobalamin II, 598
Transcriptase, reverse, 458
Transcription, 425–428
   and control of genetic expression, 453
   inhibitors of, 450
   and posttranscriptional reactions, 428, 430, 434
Transferase, 97
Transferrin, 472, 577
Transketolase reactions, 186, 187, 189
Translation, 435–436
   inhibitors of, 450
   and posttranslational modifications, 448, 478, 540–541
Transmethylation, 316–318
Transport through membranes, 61–64

Transversions, 461
Tricarboxylic acid cycle. *See* Citric acid cycle
Triglycerides, 55
  biosynthesis of, 226–228
  catabolism of, 111–161
  dietary, 573
  digestion and absorption of, 114–116
  fatty acids in. *See* Fatty acid(s)
  plasma levels of, 265
  storage in adipose tissue, 570
  structural chemistry of, 111–114
Triglycosylceramide, 248
Triiodothyronine, 365, 553
Triose-phosphate isomerase, 170, 187
  and gluconeogenesis, 202
Tripeptides, 23
Triphosphates, nucleoside, 51
Triplets, base codons for amino acids, 436–437
Tropocollagen, 540
Tropomyosin, 527, 531
Troponin, 527, 531
Trypsin, 299, 301
  active site of, 104
Trypsinogen, 299, 301
Tryptophan, 16, 17, 316, 324
  as donor to tetrahydrofolic acid, 321
  excretion of metabolites, 593
  metabolism of, 367–371
Tryptophan pyrrolase, 367, 369
Tubules, renal, 516
  reabsorption in, 517, 524
Tumor viruses, 458
Tyrosinase, 363
Tyrosine, 316, 324
  metabolism of, 361–365
  sources of, 328
  synthesis of, 327–328
Tyrosine-glutamate transaminase, 362
Tyrosinemia, 363
Tyrosinosis, 363
Tyrosylvalylthreonine, 23

Ubiquinone, 137, 601
UDP. *See* Uridine diphosphate, and glycogen synthesis
UDP-N-acetylgalactosamine, 215
UDP-N-acetylglucosamine, 214–215
UDP-galactose, 210
  and cerebroside synthesis, 247
  conversion to UDP-glucose, 212
UDP-glucose, 282
  and cerebroside synthesis, 247
  conversion to UDP-galactose, 210

UDP-glucose dehydrogenase, 210
UDP-glucose 4-epimerase, 210, 212
UDP-glucose pyrophosphorylase, 192
  and gluconeogenesis, 202
UDP-glucuronic acid, 210–211, 282
Ultracentrifugation, 38–39
Ultraviolet light, mutations from, 459
UMP. *See* Uridine monophosphate
Uncoupling, of oxidative phosphorylation, 156–157
Uracil, 49, 54
Urea
  formation of, 310–313
  plasma levels of, 466
  in urine, 524
Urea cycle, 311, 314, 343, 348, 380
Uric acid, 397–402
  hyperuricemia, 398–402
    in von Gierke's disease, 290
  plasma levels of, 398, 466
  in urine, 524
Uricolysis, 398
Uridine, 50, 53
Uridine diphosphate, and glycogen synthesis, 201
Uridine monophosphate (UMP), 51, 382
Uridine triphosphate (UTP), 382–383
  reaction with glucose 1-phosphate, 192
Uridylic acid. *See* Uridine monophosphate
Urine, 523–525
Urobilinogen, 411
Urocanic acid, 308, 348
Uronic acid, in glycosaminoglycans, 542–543
Uroporphyrinogens, 407
UTP. *See* Uridine triphosphate

Valine, 15, 17, 316, 324
  metabolism of, 355–356
van den Bergh reaction, 411–412
Velocity
  initial reaction, 85
  maximum, 86
Viruses
  DNA and RNA in, 455–458
  tumor, 458
Visual cycle, 582
Vitamin(s), 580–600
  fat-soluble, 580
  water-soluble, 580
Vitamin A, 580–583
Vitamin B_{12}, 595, 596–600
Vitamin B complex, 589–596
  biotin, 95, 220, 593–594

folic acid, 320, 594–595
   deficiency of, 599
niacin, 369, 591
pantothenic acid, 594
pyridoxine, 592–593
   deficiency of, 370
riboflavin, 119, 591–592
thiamine, 589–591
Vitamin C, 600
Vitamin D, 583–586
Vitamin E, 586–587
Vitamin K, 588–589
   and coagulation, 478–479
Vitamin-like nutrients, 600–601
von Gierke's disease, 288–290
   hyperuricemia in, 399

Warfarin, as vitamin K antagonist, 479
Water
   cellular, 1
   compartmentalization in body, 522
   conjugate base for, 7
   ion-product of, 2
   movement of, 466
   and nutrition, 574
   oxygen reduction to, 140–141
Watson-Crick model of DNA, 416
Wernicke-Korsakoff syndrome, 590–591
Wilson's disease, 579

Xanthine, 396
Xanthine oxidase, 396–397
Xanthurenic acid, 369, 593
Xanthylic acid, 388
Xerophthalmia, 583
D-Xylulose 5-phosphate, 185, 189–190

Zinc
   dietary requirements, 579–580
   and enzyme activity, 96
Zwitterion, 15
Zymosterol, 236